Representations of Groups

The representation theory of finite groups has seen rapid growth in recent years with the development of efficient algorithms and computer algebra systems. This is the first book to provide an introduction to the ordinary and modular representation theory of finite groups with special emphasis on the computational aspects of the subject.

Evolving from courses taught at Aachen University, this well-paced text is ideal for graduate-level study. The authors provide over 200 exercises, both theoretical and computational, and include worked examples using the computer algebra system GAP. These make the abstract theory tangible and engage students in real hands-on work. GAP is freely available from www.gap-system.org and readers can download source code and solutions to selected exercises from the book's webpage.

Klaus Lux is Associate Professor in the Department of Mathematics at the University of Arizona, Tucson.

Herbert Pahlings is Professor Emeritus at RWTH Aachen University in Germany.

Representations of Groups

A Computational Approach

KLAUS LUX
University of Arizona

HERBERT PAHLINGS
RWTH Aachen University

CAMBRIDGE
UNIVERSITY PRESS

CAMBRIDGE
UNIVERSITY PRESS

University Printing House, Cambridge CB2 8BS, United Kingdom

One Liberty Plaza, 20th Floor, New York, NY 10006, USA

477 Williamstown Road, Port Melbourne, VIC 3207, Australia

314-321, 3rd Floor, Plot 3, Splendor Forum, Jasola District Centre, New Delhi - 110025, India

79 Anson Road, #06-04/06, Singapore 079906

Cambridge University Press is part of the University of Cambridge.

It furthers the University's mission by disseminating knowledge in the pursuit of education, learning and research at the highest international levels of excellence.

www.cambridge.org
Information on this title: www.cambridge.org/9780521768078

First published 2010

A catalogue record for this publication is available from the British Library

Library of Congress Cataloging in Publication data
Lux, Klaus.
Representations of groups : a computational approach / Klaus Lux, Herbert Pahlings.
p.   cm. – (Cambridge studies in advanced mathematics ; 124)
Includes bibliographical references and index.
ISBN 978-0-521-76807-8
1. Representations of groups–Data processing.   I. Pahlings, H. (Herbert)   II. Title.
QA176.L89 2010
512′.22–dc22   2010017822

ISBN 978-0-521-76807-8 Hardback

Additional resources for this publication at
www.math.rwth-aachen.de/~RepresentationsOfGroups

Contents

v

Preface

The representation theory of finite groups was developed around 1900 by Frobenius, Schur and Burnside. The theory was first concerned with representing groups by groups of matrices over the complex numbers or a field of characteristic zero. Representations over fields of prime characteristic, called "modular representations" (as opposed to "ordinary" ones), were considered somewhat later, and its theory began with fundamental papers by R. Brauer starting in 1935. Despite its age, the representation theory of finite groups is still developing vigorously and remains a very attractive area of research. In fact, the theory is notorious for its large number of longstanding open problems and challenging conjectures. The availability of computers, the development of algorithms and computer algebra systems within the last few decades have had some impact on representation theory, perhaps most noticeable by the appearance of the *ATLAS of Finite Groups* in 1985 (see [38]). Note that we refer to this as the ATLAS in the text.

The present book gives an introduction into representation theory of finite groups with some emphasis on the computational aspects of the subject. The book grew out of some sets of courses that the senior of the authors has given at Aachen University since the early 1990s. It was our experience that many students appreciated having many concrete examples illustrating the abstract theory.

The range of examples in the area is rather limited if one restricts oneself to paper and pencil work, but can be greatly enhanced by using a computer algebra system such as GAP or MAGMA. For the examples and exercises in this book we have chosen GAP, which can be freely obtained from http://www.gap-system.org and for which the source code is publicly available. We did not want to use these systems as mysterious black boxes, so we have explained along with the theory the most important algorithms in the field, leaving out technical details or complexity questions altogether. Instead we have included in some examples (commented and sometimes edited) GAP-code mainly to give the unexperienced reader an impression of how easily most of these calculations can be done. The complete (and unedited) GAP-code and all special GAP-programs used for the examples and exercises in this book appear on the homepage of this book: http://www.math.rwth-aachen.de/~RepresentationsOfGroups. Here one can also find solutions to some of the exercises in the book. It is also planned to include additional material and a list of errata.

We have treated ordinary and modular representation theory together, not only because this seems to be economic, but also since there are so many interactions. For that reason we also did not refrain from occasional forward references, in particular in the examples given.

The book presupposes some knowledge on basic topics in abstract algebra, such as the Sylow theorems and occasionally some Galois theory. Although modules over algebras are defined, some familiarity with these notions will be assumed. Also the reader should be familiar with linear algebra, including normal forms of matrices. Tensor products are introduced, but for most of their basic properties we refer to standard text books in algebra.

The first chapter introduces the basic notions of representation theory and describes as examples the representations of cyclic groups and algebras. Permutation modules are then discussed in some detail, because of their importance for practical examples. Simple modules are treated in Section 1.3, including Norton's criterion and algorithms for proving or disproving simplicity often referred to under the key-word "Meataxe." The chapter also includes the relevant material on projective modules and blocks.

Ordinary characters of finite groups are treated in the second chapter. We give several applications of characters in different areas of algebra. We also include several algorithms for computing character tables of groups, such as the Dixon–Schneider algorithm, which can be applied if one can compute within a group sufficiently well to find the conjugacy classes. Other methods apply when one knows just the centralizer orders and perhaps a few characters. The chapter finishes with an example in which the character table of a simple group is computed using only the order of the group.

The third chapter covers the interplay between representations of groups and subgroups, which is, of course, vital for the representation theory of groups. We include a section on tables of marks as introduced already by Burnside. Marks can be interpreted as extensions of permutation characters, and tables of marks may be extremely useful when dealing with particular groups. Of course, Clifford theory and projective representations are covered. We also describe B. Fischer's method of Clifford matrices to compute character tables of certain group extensions which often occur as local (or maximal) subgroups of simple groups. The method is somewhat technical and perhaps best explained by giving examples, which we do. The chapter closes with Brauer's characterization of characters including some applications.

The last and longest chapter is devoted to modular representation theory. Our use of p-modular systems differs slightly from the one in the literature and we introduce standard p-modular systems in order to arrive at uniquely defined Brauer character tables, which we introduce in Section 4.2 using Conway polynomials. This is important, especially when one is dealing with Brauer characters of different groups at the same time, which very often is the case in concrete problems. We give examples for computing Brauer character tables using basic sets and other methods, in particular condensation. The chapter includes an exposition on Brauer's main theorems on blocks and the Green correspondence. Here the book is not entirely self-contained. There are a few cases where we omit proofs and give instead proper references to the literature, for instance Green's indecomposability theorem in Section 4.8. Trivial source modules are treated including Conlon's Induction Theorem, which was already used in an example in Section 3.5. They also provide easy examples for the Green correspondence. We don't give a proof of Brauer's theorem on blocks of defect one, but instead include some applications. Modular representations of p-solvable groups are treated only to an extent to be able to prove the Fong–Swan theorem and to explain the connection between the $k(GV)$-problem and Brauer's $k(B)$-problem for solvable groups. Modular representation theory abounds in longstanding open problems and conjectures. The final section mentions some of the most famous ones and verifies them in some examples.

Finally we would like to point to the literature we have used and also alternative treatments which might be useful for the reader. The most comprehensive monographs on representation theory of groups are found in [41] and [42]. Students who might find our first section a bit daunting should perhaps consult some slower-paced introductory text such as [3], [73] or [97]. A standard text mainly on ordinary character theory is [92]. Concerning modular representation theory, [1] is an accessible introduction dealing only with modules, where [68] deals only with characters. All aspects of modular representation theory are covered in [57], which also contains a full proof of the Brauer–Dade theorem on blocks with cyclic defect groups. References [125] and [126], which we have used frequently, are not as comprehensive but are more easily accessible. For alternative treatments see [50], [51], [109] and [10]. Of course there are many topics that we have barely touched, or omitted altogether. We mention just two, the theory of exceptional characters, for which we refer to [33], and the representation theory of finite groups of Lie type covered in [25].

This book would not have been written without the existence and availability of the GAP system. So we wish to thank the whole GAP team for its work and in particular our colleague Joachim Neubüser, the "father" of GAP. Special thanks are due to Thomas Breuer, who frequently helped us when we had questions or problems with the system and who also carefully read an early version of the manuscript suggesting a large number of improvements. We also would like to thank the participants of the Representation Theory courses one of the authors taught at the University of Arizona for pointing out mistakes in preliminary versions of the manuscript.

Frequently used symbols

$\mathrm{Aut}(G)$	group of automorphisms of G
$\mathrm{cf}(G,K)$	K-vector space of class functions on G
$\mathbf{C}_G(g)$	centralizer of g in G
$\mathbb{F}_q$	finite field with q elements
$g \in_G H$	$g \in H^x$ for some $x \in G$
$G = N.H$	extension of N by H, thus $N \trianglelefteq G$ and $G/N \cong H$
$G' = [G,G]$	commutator subgroup of G
$\mathrm{GL}_n(K)$	group of invertible elements in $K^{n \times n}$, $\mathrm{GL}_n(q) := \mathrm{GL}_n(\mathbb{F}_q)$
$H \leq G$	H is a subgroup of G
$H_1 =_G H_2$	$H_1 = H_2^g$ for some $g \in G$
$H_1 \leq_G H_2$	$H_1 \leq H_2^g$ for some $g \in G$
H^g, ${}^gH$	$H^g := g^{-1}Hg$, ${}^gH := gHg^{-1}$ for $H \leq G$ and $g \in G$
id_V	identity map from V to V
$\mathbf{I}_n, \mathbf{0}_n$	$n \times n$-identity matrix ($\mathbf{I}_n := [\delta_{i,j}]_{1 \leq i,j \leq n}$),$n \times n$-zero matrix
$\mathrm{Irr}_K(G)$	irreducible characters of G over K, $\mathrm{Irr}(G) := \mathrm{Irr}_\mathbb{C}(G)$
$\mathrm{IBr}_p(G)$	irreducible p-Brauer characters of G
$K^{n \times n}$	ring of $n \times n$-matrices over a commutative ring K
KG	group algebra of the group G over a commutative ring K
$\mathbb{N}, \mathbb{N}_0, \mathbb{Z}$	natural numbers, natural numbers with 0, and integers
$N \rtimes H$	$(= N : H)$ split extension (semidirect product) of N by H
$N \cdot H$	non-split extension of N by H
$\mathbf{N}_G(H)$	normalizer of H in G
$\mathbf{1}_G$	trivial character of G
1_K or 1	one of the commutative ring K.
$\mathrm{Out}(G)$	group of outer automorphisms of a group G
$\mathbb{Q}, \mathbb{R}, \mathbb{C}$	rational, real and complex numbers
$R^\times$	multiplicative group of units (= invertible elements) of a ring R
$\mathrm{S}_n, \mathrm{A}_n$	symmetric and alternating group of degree n
$\mathrm{SL}_n(K) :=$	$\{g \in \mathrm{GL}_n(K) \mid \det g = 1\}$, $\mathrm{SL}_n(q) := \mathrm{SL}_n(\mathbb{F}_q)$
$\mathbf{Z}(G), \mathbf{Z}(A)$	the center of a group G or a ring A
$\delta_{i,j}$	Kronecker delta
$\zeta_m, \mathbb{Q}_m$	$\zeta_m := \exp(\frac{2\pi i}{m}) \in \mathbb{C}$, $\mathbb{Q}_m := \mathbb{Q}(\zeta_m)$ for $m \in \mathbb{N}$
$\varphi^\mathrm{T}, A^\mathrm{T}$	transposed linear map or matrix
$(\chi, \psi)_G :=$	$\frac{1}{|G|} \sum_{g \in G} \chi(g)\psi(g^{-1})$ for $\chi, \psi \in \mathrm{cf}(G,K)$

1

Representations and modules

1.1   Basic concepts

In this section we introduce the basic concepts of representation theory, fix most of the notation used in this book and give a large number of concrete examples for representations.

Definition 1.1.1 A **representation** of a group G over a field K is a homomorphism $\delta\colon G \to \mathrm{GL}(V)$ of G into the group of invertible K-endomorphisms of a finite dimensional vector space V over K. The dimension of this vector space is called the **degree** of δ. A **matrix representation** of G of degree n is a homomorphism $\boldsymbol{\delta}\colon G \to \mathrm{GL}_n(K)$ of G into the full linear group $\mathrm{GL}_n(K)$ over K of some degree n. If δ or $\boldsymbol{\delta}$ is injective, it is called **faithful**.

If V is a K-vector space of dimension n and $B := (v_1, \dots, v_n)$ is a K-basis of V, then by assigning to each endomorphism the matrix representing it with respect to the basis B one obtains a group isomorphism

$$\mathrm{GL}(V) \to \mathrm{GL}_n(K), \quad \varphi \mapsto [\varphi]_B,$$

where the matrix $[\varphi]_B = [a_{ij}] \in K^{n \times n}$ is defined by

$$\varphi(v_j) = \sum_{i=1}^{n} a_{ij} v_i \qquad (1 \le j \le n).$$

Thus for any representation $\delta\colon G \to \mathrm{GL}(V)$ and any choice of a basis B of V one obtains a matrix representation

$$\boldsymbol{\delta}_B\colon G \to \mathrm{GL}_n(K), \quad g \mapsto [\delta(g)]_B.$$

Observe that the multiplication in $\mathrm{GL}(V)$ and also in the ring $\mathrm{End}_K V$ of all K-endomorphisms of V is defined by $\varphi \circ \psi(v) = \varphi(\psi(v))$ for $v \in V$. We will usually omit the symbol "$\circ$."

Definition 1.1.2 Two representations $\delta\colon G \to \mathrm{GL}(V)$ and $\delta'\colon G \to \mathrm{GL}(W)$ are called **equivalent** if there is a K-vector space isomorphism $\varphi\colon V \to W$ such that

$$\delta'(g) = \varphi\,\delta(g)\,\varphi^{-1} \quad \text{for all} \quad g \in G.$$

Similarly two matrix representations $\boldsymbol{\delta}\colon G \to \mathrm{GL}_n(K)$ and $\boldsymbol{\delta}'\colon G \to \mathrm{GL}_n(K)$ are called **equivalent** if there is a matrix $T \in \mathrm{GL}_n(K)$ such that

$$\boldsymbol{\delta}'(g) = T\,\boldsymbol{\delta}(g)\,T^{-1} \quad \text{for all} \quad g \in G.$$

So obviously different matrix representations corresponding to the same representation are equivalent.

It is convenient to use the language of modules over rings or better yet of modules over algebras over a commutative ring K. In this book "ring" means *associative ring* having a unit element, which we denote by 1, or 1_A if the ring is called A. For convenience we recall some of the relevant definitions.

Definition 1.1.3 If K is a commutative ring, a K**-algebra** is a ring A together with a ring homomorphism $\lambda_A\colon K \to \mathbf{Z}(A)$ satisfying $\lambda_A(1_K) = 1_A$. Here $\mathbf{Z}(A)$ is the **center** of A, defined by

$$\mathbf{Z}(A) := \{z \in A \mid a\,z = z\,a \ \text{ for all } \ a \in A\}\,.$$

If (A, λ_A), $(A', \lambda_{A'})$ are K-algebras, then a ring homomorphism $\varphi\colon A \to A'$ is a K-algebra homomorphism if $\lambda_{A'} = \varphi \circ \lambda_A$.

Observe that any ring can be considered as a $\mathbb{Z}$-algebra in a unique way.

Definition 1.1.4 If A is a ring, an A**-module**, or, more precisely, a left A-module, is an abelian group V together with a map $A \times V \to V$, $(a,v) \mapsto a \cdot v$, satisfying

$$(a+b)\cdot v = a\cdot v + b\cdot v, \quad a\cdot(v+w) = a\cdot v + a\cdot w, \quad 1_A\cdot v = v \quad \text{for } a, b \in A,\ v, w \in V.$$

A right A-module is defined similarly with a map $V \times A \to V$, $(v,a) \mapsto v \cdot a$. It is equivalent to an A^{op}-module, where A^{op} stands for the **opposite ring** of A with multiplication changed to $a * a' := a' \cdot a$.

Remark 1.1.5 A together with the ring multiplication $A \times A \to A$ is an A-module, called the **left regular** A**-module**, often written as $_AA$. If (A, λ_A) is a K-algebra over a commutative ring K, then any A-module V becomes also a K-module by defining $\alpha \cdot v := \lambda_A(\alpha) \cdot v$ for $\alpha \in K$ and $v \in V$ with

$$\alpha \cdot (a \cdot v) = a \cdot (\alpha \cdot v) = (\lambda_A(\alpha)\,a) \cdot v \quad \text{for} \quad \alpha \in K,\, a \in A,\, v \in V\,.$$

In particular, for $V = {_AA}$ and $a = v = 1_A$ we get $\lambda_A(\alpha) = \alpha \cdot 1_A$. This is the reason why the notation λ_A will hardly ever be used and we will usually talk about an algebra A over K instead of (A, λ_A), regarding A as a K-module with $\alpha \cdot 1_A \in \mathbf{Z}(A)$ for all $\alpha \in K$. Occasionally we take the liberty to abbreviate $av := a \cdot v$ and $\alpha v := \alpha \cdot v$.

Important examples of algebras are group algebras. For simplicity we restrict ourselves to finite groups.

Definition 1.1.6 Let K be a commutative ring and $(G, \cdot)$ be a finite group. Put $KG := K^G$, the set of all maps from G to K, and define for $a, b \in KG$, $\alpha \in K$ and $g \in G$

$$(a+b)(g) := a(g)+b(g), \quad (ab)(g) := \sum_{h \in G} \sum a(h)b(h^{-1} \cdot g), \quad (\alpha a)(g) := \alpha a(g).$$

For $g \in G$ let $g^o \in KG$ be defined by

$$g^o(h) := \left\{ \begin{array}{ll} 0 & \text{for } h \in G \setminus \{g\}, \\ 1_K & \text{for } h = g \end{array} \right.$$

Then it is readily verified that KG is a K-algebra with unit 1°, where 1 is the unit in G. It is called the **group algebra** of G over K.

Remark 1.1.7 If $a \in KG$ then clearly $a = \sum_{g \in G} a(g)g^o$. Thus KG is a free K-module with basis $(g^o)_{g \in G}$, that is, every element $a \in KG$ can be uniquely written in the form $a = \sum_{g \in G} \alpha_g g^o$ with $\alpha_g \in K$. Also $g^o h^o = (gh)^o$, so that $g \mapsto g^o$ gives an embedding (= injective group homomorphism) from G to the group of units $(KG)^\times$ (= multiplicative group of invertible elements) of KG. It is common practice to identify $g \in G$ with $g^o \in KG$. Then

$$KG = \{ \sum_{g \in G} \alpha_g g \mid \alpha_g \in K \}.$$

Of course, KG is commutative if and only if G is abelian.

Example 1.1.8 We consider the symmetric group $\mathrm{S}_n = (\mathrm{S}_n, \circ)$ of degree n, the set of all permutations of $\{1, \ldots, n\}$ with multiplication $\circ$ defined by $(\sigma \circ \tau)(i) := \sigma(\tau(i))$ for $\sigma, \tau \in \mathrm{S}_n$ and $i \in \{1, \ldots, n\}$. We will use the familiar cycle notation for elements of S_n. So, for instance, $\sigma = (2, 3)(5, 7, 8) \in \mathrm{S}_9$ is the permutation with $\sigma(2) = 3$, $\sigma(3) = 2$, $\sigma(5) = 7$, $\sigma(7) = 8$, $\sigma(8) = 5$ and $\sigma(i) = i$ for $i \in \{1, 4, 6, 9\}$. If K is a commutative ring then

$$K\mathrm{S}_3 = \{\alpha_1() + \alpha_2(1,2,3) + \alpha_3(1,3,2) + \alpha_4(1,2) + \alpha_5(2,3) + \alpha_6(1,3) \mid \alpha_i \in K\},$$

and in $K\mathrm{S}_3$ we can compute, for instance with $a := (() + (1,2,3) + (1,3,2))$,

$$a \cdot (1,2) = (1,2) + (1,3) + (2,3), \qquad a \cdot (() - (1,3,2)) = 0, \qquad a^2 = 3\,a.$$

Assumption: For the rest of this section we will assume that G is a finite group and K a commutative ring. If we require that K is a field we will say so, unless it is clear from the context, e.g. if we are talking about K-vector spaces.

Notation: If M is a subset of G we put

$$M^+ := \sum_{g \in M} g \in KG. \tag{1.1}$$

As usual we write $g^h := h^{-1}gh$ for $g, h \in G$. Furthermore $g^G := \{g^h \mid h \in G\}$ and $\mathrm{cl}(G) := \{g^G \mid g \in G\}$, the set of conjugacy classes of G.

The center of a group algebra is easily described as follows.

Lemma 1.1.9 $(C^+)_{C \in \mathrm{cl}(G)}$ *is a K-basis of the center $\mathbf{Z}(KG)$ of KG.*

Proof. If $z = \sum_{g \in G} \alpha_g g \in KG$, then $z \in \mathbf{Z}(KG)$ if and only if $h^{-1}zh = z$ for all $h \in G$, so if and only if

$$\sum_{g \in G} \alpha_g h^{-1}gh = \sum_{g \in G} \alpha_{hgh^{-1}} g = \sum_{g \in G} \alpha_g g;$$

hence if and only if $\alpha_g = \alpha_{hgh^{-1}}$ for all $h, g \in G$. This condition is equivalent to saying that α_g must be constant on conjugacy classes, or in other words that $z = \sum_{C \in \mathrm{cl}(G)} \alpha_C C^+$, where $\alpha_C = \alpha_g$ if $g \in C$. $\square$

Any representation $\delta \colon G \to \mathrm{GL}(V)$ of a finite group G over a field K (and likewise any matrix representation $\boldsymbol{\delta} \colon G \to \mathrm{GL}_n(K)$) extends by K-linearity naturally to a K-algebra homomorphism $\delta \colon KG \to \mathrm{End}_K V$ (resp. $\boldsymbol{\delta} \colon KG \to K^{n \times n}$), which will be denoted by the same symbol and which is called a representation (resp. matrix representation) of the group algebra. Also V becomes a (left) KG-module via $a \cdot v := \delta(a)(v)$ for $a \in KG$ and $v \in V$. Conversely, if V is any KG-module, which has finite dimension as a K-vector space, then one obtains a representation $\delta \colon KG \to \mathrm{End}_K V$ by defining $\delta(a)(v) := a \cdot v$ and one obtains a representation of G by restricting δ to G. The KG-module V is often called the **representation module** of the representation $\delta \colon G \to \mathrm{GL}(V)$ and δ is called the representation "afforded by V." Obviously, equivalent representations have representation modules that are isomorphic as KG-modules and vice versa.

Although we are usually considering representations (and algebras) over a field, there are occasions where representations (and algebras) over a commutative ring K come up. In this case we will make provisions that the K-modules considered have a finite K-basis.

Definition 1.1.10 Let K be a commutative ring and A a K-algebra. A **matrix representation** of A over K is a K-algebra homomorphism $\boldsymbol{\delta} \colon A \to K^{n \times n}$ for some $n \in \mathbb{N}$. If V is a free K-module with a finite K-basis and $\mathrm{End}_K V$ is the K-algebra of all K-linear maps from V to V, a K-algebra homomorphism $\delta \colon A \to \mathrm{End}_K V$ is called a **representation** of A.

Remark 1.1.11 If $\boldsymbol{\delta} \colon A \to K^{n \times n}$ is a matrix representation of the K-algebra A then the free K-module $K^n := K^{n \times 1}$ becomes an A-module by

$$a \cdot v := \boldsymbol{\delta}(a) v \qquad \text{for} \qquad a \in A, \; v \in K^n.$$

Conversely, if V is an A-module which is free as a K-module with K-basis $B := (v_1, \ldots, v_n)$, then we obtain a matrix representation $\boldsymbol{\delta}_B \colon A \to K^{n \times n}$ by defining

$$\boldsymbol{\delta}_B(a) := [\alpha_{ij}] \in K^{n \times n} \qquad \text{if} \qquad a \cdot v_j = \sum_{i=1}^{n} \alpha_{ij} v_i \qquad (a \in A, \; 1 \leq j \leq n).$$

However, if V is an A-module which does not have a finite K-basis, V does not give rise to a matrix representation.

Notation: Let V and W be A-modules for a K-algebra A. Furthermore let $A' \subseteq A$, $a \in A$ and $U \subseteq V$.

(1) $U \leq_A V$ means "U is an A-submodule of V," that is, U is a subgroup of the additive group of V and $A \cdot U := \{a \cdot u \mid a \in A, u \in U\} \subseteq U$.

(2) $\ker_V A' := \{v \in V \mid A' \cdot v = 0\} \subseteq V$ and $\ker_V(a) := \ker_V\{a\}$.

(3) $\operatorname{ann}_A U := \{a \in A \mid a \cdot U = 0\} \leq_A {}_A A$, a left ideal in A.

(4)

$$
\begin{aligned}
\operatorname{Hom}_A(V, W) := \ & \{\varphi \colon V \to W \mid \varphi(a \cdot v_1 + v_2) = a \cdot \varphi(v_1) + \varphi(v_2), \\
& \text{for all } a \in A \text{ and } v_1, v_2 \in V\}
\end{aligned}
$$

is the **K-module of all A-homomorphisms of V to W.**

(5) $\operatorname{End}_A V := \operatorname{Hom}_A(V, V)$ is the A-endomorphism ring of V with identity id_V, the identity map of V, and multiplication $\varphi \psi := \varphi \circ \psi$ for $\varphi, \psi \in \operatorname{End}_A V$, the composition of φ and ψ.

(6) $\delta_V \colon A \to \operatorname{End}_K V$ is defined by $\delta_V(a)(v) := a \cdot v$ for $a \in A$, $v \in V$. This is obviously a K-algebra homomorphism with kernel $\operatorname{ann}_A V$. If V has a finite K-basis, δ_V is called **the representation of A afforded by V.**

Observe that by Remark 1.1.5 A-modules V, W are also K-modules and it is clear that $\operatorname{Hom}_A(V, W) \subseteq \operatorname{Hom}_K(V, W)$. Remark 1.1.5 also shows that $E := \operatorname{End}_A V$ is a K-algebra with $\lambda_E(\alpha) \colon v \mapsto \alpha v$. The following is an obvious consequence from the definitions.

Lemma 1.1.12 *Let A be a K-algebra and V an A-module with $E := \operatorname{End}_A V$. Then V can be considered as an E-module via $\varphi \cdot v := \varphi(v)$ for $\varphi \in E$, $v \in V$, and*

(a) $\ker_V A' \leq_E V$ *for* $A' \subseteq A$,

(b) *if* $U \leq_A V$, *then* $\operatorname{ann}_A U$ *is a two-sided ideal in A, in symbols:* $\operatorname{ann}_A U \trianglelefteq A$.

Definition 1.1.13 Let A be a K-algebra and V an A-module with submodules $V_1, \ldots, V_n$. Then V is called the **direct sum** of $V_1, \ldots, V_n$, written

$$
V = \bigoplus_{i=1}^n V_i = V_1 \oplus \cdots \oplus V_n \tag{1.2}
$$

if every $v \in V$ can be written uniquely in the form $v = v_1 + \cdots + v_n$ with $v_i \in V_i$. For $1 \leq i \leq n$ define $\pi_i \colon V \to V_i$, $v \mapsto v_i$ and $\iota_i \colon V_i \to V$, $v_i \mapsto v_i$. Then $\pi_i \in \operatorname{Hom}_A(V, V_i)$, $\iota_i \in \operatorname{Hom}_A(V_i, V)$ are called the projections and injections corresponding to the decomposition (1.2). They satisfy

$$
\operatorname{id}_V = \sum_{i=1}^n \iota_i \circ \pi_i \quad \text{and} \quad \pi_i \circ \iota_j = \begin{cases} 0 & \text{for } i \neq j, \\ \operatorname{id}_{V_i} & \text{for } i = j. \end{cases} \tag{1.3}
$$

Remark 1.1.14 V in (1.2) is sometimes called the "internal direct sum" of the submodules $V_1, \ldots, V_n$ in contrast to the "external direct sum," which can be defined for arbitrary (not necessarily distinct) A-modules $V_1, \ldots, V_n$ as follows. Put $\hat{V} := V_1 \hat{\oplus} \cdots \hat{\oplus} V_n := \{(v_1, \ldots, v_n) \mid v_i \in V_i\} = \{(v_i)_{i=1}^n \mid v_i \in V_i\}$ and

$$(v_i)_{i=1}^n + (v_i')_{i=1}^n := (v_i + v_i')_{i=1}^n \quad \text{and} \quad a \cdot (v_i)_{i=1}^n := (a \cdot v_i)_{i=1}^n.$$

Then $\hat{V}$ is an A-module and we have the obvious embeddings $\hat{\iota}_i \in \operatorname{Hom}_A(V_i, \hat{V})$ so that $V_1 \hat{\oplus} \cdots \hat{\oplus} V_n = \hat{\iota}_1(V_1) \oplus \cdots \oplus \hat{\iota}_n(V_n)$. We will identify V_i with $\hat{\iota}_i(V_i)$ and use only $\oplus$, leaving it to the reader to decide from the context whether the internal or external direct sum is meant.

Theorem 1.1.15 *Let A be a K-algebra and let $V := \bigoplus_{i=1}^m V_i$, $W := \bigoplus_{j=1}^n W_j$ be A-modules. Then*

$$\operatorname{Hom}_A(V, W) \cong_K \bigoplus_{i=1}^m \bigoplus_{j=1}^n \operatorname{Hom}_A(V_i, W_j),$$

where "$\cong_K$" means being isomorphic as K-modules.

Proof. Let $(\pi_i, \iota_i)_{i=1}^m$ and $(\pi_j', \iota_j')_{j=1}^n$ be the families of projections and injections corresponding to $V := \bigoplus_{j=1}^m V_i$ and $W := \bigoplus_{i=1}^n W_j$, respectively. Using (1.3) it is easily checked that

$$\operatorname{Hom}_A(V, W) \to \bigoplus_{i=1}^m \bigoplus_{j=1}^n \operatorname{Hom}_A(V_i, W_j) \, , \ \varphi \mapsto (\pi_j' \circ \varphi \circ \iota_i)_{1 \leq i \leq m, 1 \leq j \leq n}$$

gives the desired K-isomorphism. $\qquad\qquad\qquad\qquad\qquad\qquad\qquad\qquad\square$

Definition 1.1.16 Let V be a K-module with a finite K-basis. A representation $\delta \colon G \to \operatorname{GL}(V)$ of a group G over K or a representation $\delta_V \colon A \to \operatorname{End}_K V$ of a K-algebra A is called **irreducible** if the corresponding KG-module or A-module V is **simple**, i.e. has exactly two submodules, namely $\{0\}$ and V, and **reducible** otherwise. (So the zero module $\{0\}$ is by definition not simple.) Often simple modules are also called irreducible.

Likewise δ or δ_V as above is called **indecomposable** if the corresponding module is indecomposable (i.e. cannot be written as a direct sum of two nontrivial submodules) and is called **decomposable** otherwise.

If K is a field, we see that δ is reducible if and only if a basis $B = (v_1, \ldots, v_n)$ of V can be found such that

$$[\delta(g)]_B = \left[\begin{array}{c|c} \delta_1(g) & C(g) \\ \hline 0 & \delta_2(g) \end{array} \right] \quad \text{for all} \ \ g \in G \tag{1.4}$$

with square matrices $\delta_1(g) \in K^{m \times m}$, $\delta_2(g) \in K^{(n-m) \times (n-m)}$ and $C(g) \in K^{m \times (n-m)}$ for some $0 < m < n = \dim_K V$. Here

$$\delta_1 \colon g \mapsto \delta_1(g) \qquad \text{and} \qquad \delta_2 \colon g \mapsto \delta_2(g)$$

are matrix representations of G afforded by the submodule $W = \langle v_1, \ldots, v_m \rangle_K$, the K-span of $v_1, \ldots, v_n$, with respect to the basis $(v_1, \ldots, v_m)$ and the factor module V/W with respect to the basis $(v_{m+1}+W, \ldots, v_n+W)$. A representation $\delta\colon G \to \mathrm{GL}(V)$ over a field K is decomposable if and only if a basis $B = (v_1, \ldots, v_n)$ can be found such that

$$[\delta(g)]_B = \left[\begin{array}{c|c} \delta_1(g) & 0 \\ \hline 0 & \delta_2(g) \end{array}\right] \quad \text{for all } g \in G.$$

In this case W, as defined above, has a complement $W' = \langle v_{m+1}, \ldots, v_n \rangle_K$ which is also a KG-module and δ_2 is also a matrix representation of G afforded by W' (with respect to the basis $(v_{m+1}, \ldots, v_n)$).

Example 1.1.17 If V is any K-vector space we get for any group G a "trivial" representation $\delta\colon G \to \mathrm{GL}(V)$, $g \mapsto \mathrm{id}_V$. This representation is irreducible if and only if $\dim_K V = 1$. ◆

Definition 1.1.18 If V is a KG-module then

$$\mathrm{Inv}_G(V) := \{v \in V \mid g \cdot v = v \quad \text{for all } g \in G\}$$

is called the submodule of G-**invariants** of V. We also define

$$\mathrm{Inv}^G(V) := \langle (g-1)v \mid g \in G, \ v \in V \rangle_K.$$

The notations $\mathbf{C}_V(G) := \mathrm{Inv}_G(V)$ and $[G, V] := \mathrm{Inv}^G(V)$ are also in use.

Obviously $\mathrm{Inv}_G(V)$ and $\mathrm{Inv}^G(V)$ are KG-modules, the largest submodule with trivial action of G and the smallest submodule W of V such that G acts trivially on V/W.

Example 1.1.19 The (left) regular KG-module ${}_{KG}KG$ for a finite group G leads to the so-called (left) **regular representation** $\rho_G\colon G \to \mathrm{GL}(KG)$ of G. It is easily seen that

$$\mathrm{Inv}_G(KG) = K \cdot \sum_{g \in G} g \quad \text{and} \quad \mathrm{Inv}^G(KG) = \{\sum_{g \in G} \alpha_g g \mid \sum_{g \in G} \alpha_g = 0\}.$$

In particular, the regular representation ρ_G is always reducible if G is not the trivial group. ◆

Definition 1.1.20 If A is a ring and $\varphi\colon A' \to A$ is a ring homomorphism, then any A-module V can be turned into an A'-module by defining

$$a' \cdot v := \varphi(a')v \quad \text{for} \quad a' \in A', \ v \in V.$$

This A'-module will be denoted by $\mathrm{Inf}_\varphi V$ and is called the **inflated module**. If $\varphi\colon A' \to A$ is a homomorphism of K-algebras and V affords the representation δ of A then $\mathrm{Inf}_\varphi V$ affords the representation $\delta \circ \varphi$ of A', called the **inflated representation**. The same concept applies to group representations: if $\varphi\colon G_1 \to G$ is a group homomorphism and $\delta\colon G \to \mathrm{GL}(V)$ is a representation then $\delta \circ \varphi$ is the inflated group representation.

If A, A' are K-algebras, one should note that the K-module structures of V and $\mathrm{Inf}_\varphi V$ may be different, unless φ is a homomorphism of K-algebras; see Exercise 1.1.12 for an example.

Remark 1.1.21 Obviously, if $\mathrm{Inf}_\varphi V$ is simple (indecomposable) then V itself must be simple (resp. indecomposable); the converse holds too, provided that φ is surjective. Also, if V_1, V_2 are A-modules, then

$$\mathrm{Hom}_A(V_1, V_2) \subseteq \mathrm{Hom}_{A'}(\mathrm{Inf}_\varphi V_1 , \mathrm{Inf}_\varphi V_2)$$

with equality, if φ is surjective.

The special and simple case that φ is an embedding needs special attention, since it is used so frequently. If A' is a subalgebra of A with embedding $\varphi \colon A' \to A$ and V is an A-module as before, then we write $V_{A'} := \mathrm{Inf}_\varphi V$. It is just the module obtained by restricting the action. In particular, if H is a subgroup of a group G and V is a KG-module we have an embedding of K-algebras $\varphi \colon KH \to KG$, and we denote the KH-module $\mathrm{Inf}_\varphi V$ by V_H.

Remark 1.1.22 If A is a K-algebra and V is an A-module which affords a matrix representation $\delta \colon A \to K^{n \times n}$ with respect to some K-basis of V, then

$$V \cong_A \mathrm{Inf}_\delta K^n,$$

where K^n is considered as a $K^{n \times n}$-module in the natural way.

In the following two examples we show that the representation theory of cyclic groups and, more generally, cyclic algebras over a field K is quite simple. In some important cases we will write down explicitly all irreducible and indecomposable representations up to equivalence.

Example 1.1.23 (Cyclic groups) If $C_m = \langle g \rangle$ is a cyclic group of order m (with generator g), giving a matrix representation δ of C_m of degree n over a field K amounts to giving a matrix $a := \delta(g) \in K^{n \times n}$ with $a^m = \mathbf{I}_n$, where $\mathbf{I}_n$ denotes the identity matrix in degree n, or, to put it otherwise, to giving a matrix $a \in K^{n \times n}$ with minimal polynomial μ_a dividing $X^m - 1$ in the polynomial ring $K[X]$. Similar matrices define equivalent representations and vice versa. It is well known that a matrix a is diagonalizable in $K^{n \times n}$ (that is, similar to a diagonal matrix) if and only if the minimal polynomial μ_a decomposes into linear factors in $K[X]$ without multiple roots (see [153], theorem 8.11, p. 166). The polynomial $X^m - 1$ has no multiple roots if and only if char $K \nmid m$. Thus it follows that if K is a field of characteristic not dividing m containing a primitive mth root ζ_m of unity, then any matrix representation of C_m is equivalent to one that maps the generator g to a diagonal matrix with mth roots of unity on the diagonal. In particular every irreducible matrix representation of C_m over such a field K (as for instance $\mathbb{C}$) is of degree one and of the form

$$g \mapsto \zeta_m^i \qquad \text{for} \quad i \in \{0, \ldots, m-1\}.$$

Also, every indecomposable representation is irreducible in this case.

On the other hand, if char $K = p$, a prime, $m = p^s$ and $\delta\colon C_m \to GL_n(K)$ is a matrix representation, then $\delta(g)$ has a minimal polynomial dividing $(X-1)^m$, and hence δ is equivalent to the representation given by the Jordan normal form of $\delta(g)$ (see [153], theorem 8.6, p. 159):

$$\delta\colon g \mapsto \begin{bmatrix} 1 & 0 & \cdots & \cdots & 0 \\ \epsilon_1 & 1 & \ddots & \ddots & 0 \\ 0 & \epsilon_2 & \ddots & \ddots & 0 \\ \vdots & \ddots & \ddots & \ddots & 0 \\ 0 & \cdots & 0 & \epsilon_{n-1} & 1 \end{bmatrix}$$

with $\epsilon_i \in \{0,1\}$, which is indecomposable if and only if all ϵ_i equal unity. Hence C_m has for $j = 1, \ldots, m$ up to equivalence exactly one indecomposable representation of degree j and the trivial representation is the only irreducible one.

The structure of the group algebra of a cyclic group C_m of order m can also be easily described: keeping the notation as above we have an algebra epimorphism $K[X] \to K\, C_m$, $X \mapsto g$, with kernel the principal ideal $(X^m - 1) \trianglelefteq K[X]$. Thus

$$K\, C_m \cong K[X]/(X^m - 1).$$

By the isomorphism theorem for rings the ideals of $K\, C_m$ correspond bijectively to those ideals of $K[X]$ which contain $(X^m - 1)$, thus to the monic divisors of $X^m - 1 \in K[X]$. Hence the poset of ideals of $K\, C_m$ (with inclusions) is anti-isomorphic to the poset of monic divisors of $X^m - 1$ (with divisibility as order relation). If char $K \nmid m$ then $X^m - 1 = \prod_{i=1}^r f_i$ with pairwise distinct monic irreducible polynomials $f_i \in K[X]$, and

$$K\, C_m \cong K[X]/(f_1) \oplus \cdots \oplus K[X]/(f_r)$$

is a direct sum of fields with the ith projection being given by $g \mapsto X + (f_i)$. In particular, if in addition K contains a primitive mth root of unity ζ_m then we may choose $f_i := X - \zeta_m^i$ ($1 \le i \le r = m$), and it follows that

$$K\, C_m \cong \underbrace{K \oplus \cdots \oplus K}_{m}.$$

The m distinct projections are exactly the m irreducible representations of $K\, C_m$ over K, the ith one yielding the representation defined by $g \mapsto \zeta_m^i$.

On the other hand, if char $K = p > 0$ and $m = p^s$, the poset of ideals of $K\, C_m$ is isomorphic to the poset of monic divisors of $X^m - 1 = (X - 1)^m$ and $K\, C_m$ has exactly $m+1$ ideals I_i, all appearing in the unique composition series

$$K\, C_m = I_m \triangleright I_{m-1} \triangleright \ldots \triangleright I_0 = \{0\}$$

with $\dim_K I_i = i$. ◆

Example 1.1.24 (Cyclic algebras) We generalize Example 1.1.23 and let A be a cyclic K-algebra over a field K, i.e. an algebra having one algebra generator, say a. So $A = K[a]$ is a homomorphic image of the polynomial ring $K[X]$ over K in one indeterminate X. Hence $A \cong K[X]/(f)$ for some $f \in K[X]$. If V is an A-module (affording the representation δ_V) then V can also be considered as a $K[X]$-module (see Remark 1.1.21) with $f \cdot v = 0$ for all $v \in V$ and, if $\dim_K V < \infty$, we can invoke the theorem on finitely generated modules over principal ideal domains (see [110], theorem III.7.5, p. 149) to conclude that

$$V \cong_{K[X]} K[X]/(q_1) \oplus \cdots \oplus K[X]/(q_m),$$

where $q_1, \ldots, q_m$ are powers of monic irreducible polynomials. The q_i are uniquely determined up to the ordering and are divisors of f. They are usually called the **elementary divisors** of $\delta_V(a)$ (see [153], p. 135). The modules $V_{q_i} := K[X]/(q_i)$ are annihilated by f and can thus be considered as A-modules, and the generating element a acts on V_{q_i} as X does; this means that we can find a K-basis B of V such that

$$[\delta_V(a)]_B = \begin{bmatrix} M_{q_1} & & 0 \\ & \ddots & \\ 0 & & M_{q_m} \end{bmatrix},$$

where M_q is the companion matrix (see [153], p. 146) of the monic polynomial q; i.e. if $q = \sum_{i=0}^{d} \alpha_i X^i$ with $\alpha_d = 1$ then

$$M_q = \begin{bmatrix} 0 & 0 & \cdots & 0 & -\alpha_0 \\ 1 & 0 & & & -\alpha_1 \\ 0 & \ddots & \ddots & & \vdots \\ & & \ddots & 0 & \\ 0 & 0 & & 1 & -\alpha_{d-1} \end{bmatrix} \in K^{d \times d}.$$

Observe that the characteristic polynomial of $\delta_V(a)$ is exactly $\prod_{i=1}^{m} q_i$. The poset of submodules of V_{q_i} is isomorphic to the poset of divisors of q_i in $K[X]$. In particular V_{q_i} is always an indecomposable A-module and it is simple if and only if q_i is irreducible in $K[X]$. If $p \in K[X]$ irreducible of degree s and $q = p^e$ then by Exercise 1.1.11 M_q is similar to

$$M_q' := \begin{bmatrix} M_p & 0 & \cdots & 0 \\ E & M_p & \ddots & \vdots \\ \vdots & \ddots & \ddots & 0 \\ 0 & \cdots & E & M_p \end{bmatrix}, \text{ with } E = \begin{bmatrix} 0 & \cdots & 0 & 1 \\ 0 & \ddots & & 0 \\ \vdots & & & \vdots \\ 0 & \cdots & & 0 \end{bmatrix} \in K^{s \times s}. \quad (1.5)$$

For later use (in Lemma 1.3.9) we note that

$$\dim_K \ker_V(p_i(a)) \geq \deg p \quad (1.6)$$

because M_p has minimal polynomial p. If equality holds in (1.6) then there is exactly one elementary divisor q_i which is a power of p.   ◆

Remark 1.1.25 Let $a_1, \ldots, a_r$ be algebra generators of a K-algebra A. Then any matrix representation $\delta\colon A \to K^{n\times n}$ of A is completely determined by the r matrices $\delta(a_1), \ldots, \delta(a_r)$. If a K-algebra presentation of A is given with generators $a_1, \ldots, a_r$, then for any r matrices $\mathbf{a}_1, \ldots, \mathbf{a}_r \in K^{n\times n}$ there is a representation $\delta\colon A \to K^{n\times n}$ with $\delta(a_i) = \mathbf{a}_i$ for $i = 1, \ldots, r$ if and only if the matrices $\mathbf{a}_i$ satisfy the defining relations of the presentation. Note that for group algebras a presentation of the underlying group very conveniently also yields a presentation of the group algebra – one just has to add relations assuring that the group generators are invertible if need be.

Example 1.1.26 The alternating group $G := \mathrm{A}_4$ of degree four has the following presentation: $\langle a, b \mid a^2 = b^3 = (a \cdot b)^3 = 1 \rangle$ (see also Remark 2.5.11). We want to find all non-trivial representations $\delta\colon G \to \mathrm{GL}_2(\mathbb{Q})$ up to equivalence. If δ is non-trivial, then $\delta(a) \neq \mathbf{I}_2$, because $\delta(a) = \mathbf{I}_2$ implies $\delta(b)^2 = \delta(b)^3 = \mathbf{I}_2$. So we may choose $\delta(b)$ in rational canonical form. So we may assume that

$$\delta(a) = \begin{bmatrix} x & y \\ z & t \end{bmatrix}, \quad \delta(b) = \begin{bmatrix} 0 & -1 \\ 1 & -1 \end{bmatrix} \quad (\text{so that} \quad \delta(ab) = \begin{bmatrix} y & -x-y \\ t & -z-t \end{bmatrix})$$

with unknowns x, y, z, t. The equations

$$\delta(a)^2 = \mathbf{I}_2, \qquad (\delta(a) \cdot \delta(b))^3 = \mathbf{I}_2 \qquad\qquad (1.7)$$

yield polynomial equations for x, y, z, t. The first one gives $\delta(a) = \mathbf{I}_2$ or $t = -x$ and $x^2 + yz = 1$. Since $\delta(ab)$ has order three, its trace must be $-1 = y - z - t$, so that we can eliminate z. If $\delta(a) \neq \mathbf{I}_2$ we obtain from (1.7)

$$\det \delta(a) = -x^2 - xy - y^2 - y = -1 \quad \text{and} \quad \det(\delta(a)\delta(b)) = -x^2 - xy - y^2 - y = 1,$$

which shows that up to equivalence the only non-trivial rational representation of G of degree two is given by $a \mapsto \mathbf{I}_2$, $b \mapsto \begin{bmatrix} 0 & -1 \\ 1 & -1 \end{bmatrix}$.

On the other hand, it is easy to verify that the matrices

$$A := \begin{bmatrix} 0 & 1 & -1 \\ 1 & 0 & -1 \\ 0 & 0 & -1 \end{bmatrix}, \qquad B := \begin{bmatrix} 0 & 0 & 1 \\ 1 & 0 & 0 \\ 0 & 1 & 0 \end{bmatrix}$$

satisfy $A^2 = B^3 = (A \cdot B)^3 = \mathbf{I}_3$, so that $a \mapsto A$, $b \mapsto B$ defines a (faithful) representation of G (over any field).

In [148] the method is used to construct representations of the (infinite) group $\langle a, b \mid a^2 = b^3 = (a \cdot b)^7 = 1 \rangle$ of degree up to seven over fields of characteristic zero. ◆

If $\delta\colon G \to \mathrm{GL}_n(K)$ is a representation of G over the field K and $L \supseteq K$ is a field extension, it is clear that δ can also be considered as a representation over L using the embedding $\mathrm{GL}_n(K) \subseteq \mathrm{GL}_n(L)$. To extend this simple idea (and for later purposes as well) we have to recall the concept of tensor products.

Definition 1.1.27 Let K be a commutative ring and let V, W be K-modules. We denote the free K-module with basis $V \times W$ by $F(V, W)$ and consider the following subsets of $F(V, W)$:

$$M_1 := \{(v + v', w) - (v, w) - (v', w) \mid v, v' \in V,\ w \in W\},$$
$$M_2 := \{(v, w + w') - (v, w) - (v, w') \mid v \in V,\ w, w' \in W\},$$
$$M_3 := \{(\alpha v, w) - \alpha(v, w),\ (v, \alpha w) - \alpha(v, w) \mid \alpha \in K,\ v \in V,\ w \in W\}$$

and $U_{V,W} := \langle M_1 \cup M_2 \cup M_3 \rangle_K \leq_K F(V, W)$. Then $V \otimes_K W := F(V, W)/U_{V,W}$ is called the **tensor product** of V and W. For $v \in V$ and $w \in W$ one defines $v \otimes w := (v, w) + U_{V,W}$.

Remark 1.1.28 The definition implies $V \otimes_K W = \langle v \otimes w \mid v \in V,\ w \in W \rangle_K$ and the following basic relations:

- $(v + v') \otimes w = v \otimes w + v' \otimes w,$
- $v \otimes (w + w') = v \otimes w + v \otimes w',$
- $\alpha(v \otimes w) = (\alpha v) \otimes w = v \otimes (\alpha w),$

for $\alpha \in K$, $v, v' \in V$, $w, w' \in W$.

Observe that in general $V \otimes_K W \neq \{v \otimes w \mid v \in V,\ w \in W\}$. If V and W are free K-modules with K-bases $B = (v_1, \dots, v_n)$ and $B' = (w_1, \dots, w_m)$, respectively, then one can check that $V \otimes_K W$ is a free K-module with K-basis

$$B \otimes B' := (v_1 \otimes w_1, \dots, v_1 \otimes w_m,\ \dots\ ,v_n \otimes w_1, \dots, v_n \otimes w_m).$$

Assume that $\Phi\colon V \times W \to K$ is a K-bilinear map. Then Φ extends uniquely to a K-linear map $F(V, W) \to K$ and this obviously has $U_{V,W}$ in its kernel, so factors over a K-linear map

$$\varphi\colon V \otimes_K W \to K \qquad \text{with} \qquad \varphi(v \otimes w) = \Phi(v, w) \quad \text{for} \quad v \in V,\ w \in W.$$

This is usually called the *universal property* of the tensor product.

If L is a commutative ring with $L \supseteq K$ then L may also be considered as a K-module and $L \otimes_K W$ may be turned into an L-module in the following way. For $\gamma \in L$ let $\lambda_\gamma\colon F(L, W) \to L \otimes_K W$ be the K-linear map with $\lambda_\gamma(\beta, w) = \gamma\beta \otimes w$. Then $U_{L,W} \leq \ker \lambda_\gamma$ and we obtain a K-linear map

$$\bar{\lambda}_\gamma\colon L \otimes_K W \to L \otimes_K W \quad \text{with} \quad \beta \otimes w \mapsto \gamma\beta \otimes w.$$

For $x \in L \otimes_K W$ we define $\gamma \cdot x := \bar{\lambda}_\gamma(x)$. If W is free with K-basis $B := (w_1, \dots, w_m)$, then it is easily checked that $L \otimes_K W$ is free with L-basis $1 \otimes B := 1 \otimes w_1, \dots, 1 \otimes w_m$.

If A, A' are K-algebras then the K-module $A \otimes_K A'$ may be turned into a K-algebra satisfying

$$(a_1 \otimes a_1') \cdot (a_2 \otimes a_2') = a_1 a_2 \otimes a_1' a_2' \qquad \text{for} \qquad a_1, a_2 \in A,\ a_1', a_2' \in A'.$$

This is particularly easily seen provided that A, A' are free and finitely generated as K-modules, the case we are usually dealing with, but holds in general as well. If A is a K-algebra and L is as above, then $LA := L \otimes_K A$ is an L-algebra. If $(a_1, \ldots, a_n)$ is a K-basis of A, then $(1 \otimes a_1, \ldots, 1 \otimes a_n)$ is an L-basis of LA. Furthermore, for an A-module V we can turn the L-module $LV := L \otimes_K V$ into an LA-module LV with

$$(\beta \otimes a) \cdot (\gamma \otimes v) = \beta\gamma \otimes a\,v \qquad \text{for} \qquad \beta, \gamma \in L,\ a \in A,\ v \in V.$$

Definition 1.1.29 Let A be a K-algebra for some field K and let $L \supseteq K$ be an extension field. If V is an A-module (with $\dim_K V < \infty$) affording the representation $\delta \colon A \to \operatorname{End}_K V$ then the representation of LA afforded by LV will be denoted by $^L\delta$. The A-module V is called **absolutely simple** or **absolutely irreducible** if LV is simple for any field extension $L \supseteq K$. A group representation $\delta \colon G \to \operatorname{GL}(V)$ over the field K is called **absolutely irreducible** if the corresponding representation module V is absolutely simple.

Observe that if $\delta \colon G \to \operatorname{GL}(V)$ is a representation of a group G over a field K and B is a K-basis of V then for an extension field $L \supseteq K$ we have $[\delta(g)]_B = [^L\delta(g)]_{1 \otimes B}$ for all $g \in G$, where $1 \otimes B$ is as in Remark 1.1.28.

Example 1.1.30 Obviously every representation of a group of degree one is absolutely irreducible. $\blacklozenge$

Example 1.1.31 Let $G = \langle g \rangle$ be a cyclic group of prime order $p > 2$. It follows from Example 1.1.24 that we have an irreducible rational matrix representation

$$\delta \colon G \to \operatorname{GL}_{p-1}(\mathbb{Q}) ,\ g \mapsto \begin{bmatrix} 0 & 0 & \cdots & & -1 \\ 1 & 0 & 0 & & -1 \\ 0 & 1 & 0 & & -1 \\ \vdots & & \ddots & & \vdots \\ 0 & & 0 & 1 & -1 \end{bmatrix},$$

because $X^{p-1} + \cdots + X + 1 \in \mathbb{Q}[X]$ is irreducible. But δ is not absolutely irreducible because over the complex numbers $^{\mathbb{C}}\delta$ is equivalent to

$$\delta' \colon G \to \operatorname{GL}_{p-1}(\mathbb{C}) ,\ g \mapsto \operatorname{diag}(\zeta, \zeta^2, \ldots, \zeta^{p-1})$$

for a primitive pth root ζ in $\mathbb{C}$, where $\operatorname{diag}(\alpha_1, \ldots, \alpha_n)$ denotes a diagonal matrix with $\alpha_1, \ldots, \alpha_n$ on the diagonal. $\blacklozenge$

Definition 1.1.32 If A is a K-algebra and V is a (left) A-module, then $V^* := \operatorname{Hom}_K(V, K)$ becomes a right A-module with

$$(x \star a)(v) := x\,(a\,v) \quad \text{for}\ a \in A,\ x \in V^*\ \text{and}\ v \in V,$$

which is called the **dual module** to V.

If V is an A-module which is free as a K-module this means, in terms of representations, the following. If $\delta\colon A \to \operatorname{End}_K V$ is the K-representation corresponding to V (i.e. $\delta(a)(v) = a \cdot v$) then corresponding to $V^\star$ we have an "anti-representation"

$$\delta^\star\colon A \to \operatorname{End}_K V^\star \quad \text{defined by} \quad \delta^\star(a) = \delta(a)^{\mathrm{T}} \quad \text{(the transposed map)},$$

that is $\delta^\star(a)(x) := x \circ \delta(a)$ for $x \in V^\star$. This is an anti-representation because, since $V^\star$ is a right A-module, we have $\delta^\star(a \cdot a') = \delta^\star(a') \circ \delta^\star(a)$ for $a, a' \in A$.

For practical purposes, it is convenient to consider instead of the action of a K-algebra A on an A-module V of K-rank n the corresponding matrix representation of A with respect to a chosen K-basis $B := (v_1, \ldots, v_n)$ of V. Moreover, it suffices to give the matrix representation evaluated on a given generating set of A only. These matrices act from the left by multiplication on the column-vectors in $K^{n \times 1}$, which we interpret as the coordinate-vectors with respect to the basis B of V. The multiplication of these matrices with the row-vectors in $K^{1 \times n}$ describes the action of A (from the right) on $V^\star$ with respect to the dual basis $B^\star$. Recall that $B^\star = (x_1, \ldots, x_n)$ is defined by $x_i(\sum_{j=1}^n \alpha_j v_j) = \alpha_i$, where $\alpha_1, \ldots, \alpha_n \in K$.

It is clear that one can likewise define the dual of a right A-module, which is, of course, a left A-module. Moreover the dual of an (A, A)-bimodule is an (A, A)-bimodule as defined below.

Definition 1.1.33 If A and A' are K-algebras an (A, A')-bimodule V is a left A-module which is at the same time a right A'-module satisfying

$$a \cdot (v \cdot a') = (a \cdot v) \cdot a' \quad \text{for all} \quad a \in A,\ a' \in A',\ v \in V$$

and

$$(\alpha\, 1_A) \cdot v = v \cdot (\alpha\, 1_{A'}) \quad \text{for all} \quad \alpha \in K,\ v \in V.$$

A homomorphism of (A, A')-bimodules is a K-linear map which is a homomorphism of left and right modules.

A K-algebra A can be viewed as an (A, A)-bimodule in a natural way, and the observation preceding Definition 1.1.33 shows that $A^\star$ is also an (A, A)-bimodule.

If K is a field and $\dim_K V < \infty$ one obtains the following as a slight generalization of the duality theorem of linear algebra.

Theorem 1.1.34 (Duality theorem) *Let A be a K-algebra for a field K and V be a left A-module with $\dim_K V = n < \infty$; then $(V^\star)^\star \cong_A V$ and the map*

$$W \mapsto W^\circ := \{x \in V^\star \mid x(w) = 0 \text{ for all } w \in W\}$$

defines an inclusion reversing bijection, in fact an anti-isomorphism of posets, of the poset of A-submodules of V onto the poset of right A-submodules of $V^\star$. For A-submodules $U \leq W$ of V one has $U^\circ / W^\circ \cong_A (W/U)^\star$. In particular $\dim_K W^\circ = n - \dim_K W$.

Proof. For $v \in V$ define $v^{\bullet} \colon V^{\star} \to K$, $\lambda \mapsto \lambda(v)$. Then $v^{\bullet} \in (V^{\star})^{\star}$ and $\Psi \colon V \to (V^{\star})^{\star}$, $v \mapsto v^{\bullet}$ is an injective A-linear map, which is an isomorphism, because $\dim_K V = \dim_K (V^{\star})^{\star} = n$. If $W \leq_A V$ then the restriction map $\Phi \colon V^{\star} \to W^{\star}$, $\lambda \mapsto \lambda|_W$ is an A-linear map with kernel W°, so $W^{\circ} \leq_A V^{\star}$. It is surjective, since any $\mu \in W^{\star}$ can be extended to an element of $V^{\star}$. If $\Lambda \leq_A V^{\star}$ then $W := \{v \in V \mid \lambda(v) = 0 \text{ for all } \lambda \in \Lambda\} \leq_A V$ and $\Lambda = W^{\circ}$. Hence $W \mapsto W^{\circ}$ indeed defines a bijection between the A-submodules of V and $V^{\star}$. The remaining assertions now follow readily. $\qquad\square$

In the case of a group algebra $A = KG$ of a finite group G one may turn the dual $V^{\star}$ of a KG-module V again into a (left) KG-module as follows.

Definition 1.1.35 If V is a KG-module we define

$$\Big(\sum_{g \in G} \alpha_g g\Big) \cdot x := x \star \Big(\sum_{g \in G} \alpha_g g^{-1}\Big) \quad \text{for} \quad x \in V^{\star},\ \alpha_g \in K \ .$$

Then $(V^{\star}, \cdot)$ is a (left) KG-module, called the **contragredient** module to V.

In essence this means that we define for $x \in V^{\star}$

$$(g \cdot x)(v) := x(g^{-1}v) \qquad \text{for} \qquad v \in V,\ g \in G$$

and extend the action K-linearly. We will often write gx instead of $g \cdot x$. In terms of representations, this means that if V affords the representation $\delta \colon G \to \mathrm{GL}(V)$ then the contragredient module $V^{\star}$ affords $\delta^{\star} \colon G \to \mathrm{GL}(V^{\star})$, $g \mapsto (\delta(g^{-1}))^{\mathrm{T}}$. If B is a K-basis of V and $B^{\star}$ is the dual basis of $V^{\star}$ then $[\delta^{\star}(g)]_{B^{\star}} = [\delta(g^{-1})]_B^{\mathrm{T}}$, the transposed matrix of $[\delta(g^{-1})]_B$. Thus, if $\boldsymbol{\delta} \colon G \to \mathrm{GL}_n(K)$ is a matrix representation, the contragredient matrix representation is defined to be

$$\boldsymbol{\delta}^{\star} \colon G \to \mathrm{GL}_n(K) \ , \ g \mapsto \boldsymbol{\delta}(g^{-1})^{\mathrm{T}}.$$

Corollary 1.1.36 *If V is a KG-module then* $(\mathrm{Inv}^G(V))^{\circ} = \mathrm{Inv}_G(V^{\star})$.

Making $V^{\star}$ into a (left) KG-module is a special case of the following.

Definition 1.1.37 If V and W are KG-modules the K-module $\mathrm{Hom}_K(V, W)$ of all K-linear maps from V and W becomes a KG-module by defining

$$(g \cdot \varphi)(v) := g(\varphi(g^{-1}v)) \qquad \text{for} \quad g \in G, \quad \varphi \in \mathrm{Hom}_K(V, W), \quad v \in V,$$

and by K-linear extension to KG.

Obviously, $\mathrm{Hom}_{KG}(V, W) = \mathrm{Inv}_G(\mathrm{Hom}_K(V, W))$.

Another important construction is the following lemma.

Lemma 1.1.38 *Let V, W be KG-modules. Then $V \otimes_K W$ may be turned into a KG-module with $g \cdot (v \otimes w) = gv \otimes gw$. If V, W are free and finitely generated* *as K-modules and $\delta := \delta_V$, $\delta' := \delta_W$ are the corresponding representations we get a new representation*

$$\delta \otimes \delta' : G \to \mathrm{GL}(V \otimes_K W) \ , \ g \mapsto \delta(g) \otimes \delta'(g).$$

If $B = (v_1, \ldots, v_n)$, $B' = (w_1, \ldots, w_m)$ are K-bases of V and W respectively, and if

$$[\delta(g)]_B = D(g) = [d_{ij}(g)] \ , \quad [\delta'(g)]_{B'} = D'(g) \qquad for \quad g \in G,$$

then

$$[(\delta \otimes \delta')(g)]_{B \otimes B'} = \begin{bmatrix} d_{11}(g)D'(g) & \cdots & d_{1n}(g)D'(g) \\ \vdots & & \vdots \\ d_{n1}(g)D'(g) & \cdots & d_{nn}(g)D'(g) \end{bmatrix},$$

where $B \otimes B' := (v_1 \otimes w_1, \ldots, v_1 \otimes w_m, \ \ldots \ , v_n \otimes w_1, \ldots, v_n \otimes w_m)$.

Proof. This is a simple exercise in multilinear algebra. One shows that the map $V \times W \to V \otimes_K W$, $(v, w) \mapsto gv \otimes gw$, is K-bilinear for any $g \in G$, and hence that it factors over $V \otimes_K W$, and that this, in fact, defines a proper action of G on $V \otimes_K W$.   □

This action of KG on $V \otimes_K W$ is usually called the **diagonal action**. The matrix $[(\delta \otimes \delta')(g)]_{B \otimes B'}$ defined above is usually called the **Kronecker product** of $D(g)$ and $D'(g)$ and is often denoted by $D(g) \otimes D'(g)$.

A basic and simple property of the tensor product is given by

Lemma 1.1.39 *If V_1, V_2, W are KG-modules then*

$$(V_1 \oplus V_2) \otimes_K W \cong_{KG} (V_1 \otimes_K W) \oplus (V_2 \otimes_K W).$$

Proof. See [110], corollary XVI.2.2, p. 608.   □

The tensor product modules and the Hom-modules are related by

Lemma 1.1.40 *If V and W are KG-modules with finite K-bases, then*

$$\mathrm{Hom}_K(V, W) \cong_{KG} V^\star \otimes_K W.$$

Proof. We get a KG-linear map $\Psi : V^\star \otimes_K W \to \mathrm{Hom}_K(V, W)$ with $\Psi(\lambda \otimes w) : v \mapsto \lambda(v)w$. Let $(v_1, \ldots, v_m)$ and $(w_1, \ldots, w_n)$ be K-bases of V and W, respectively, with dual bases $(v_1^\star, \ldots, v_m^\star)$ and $(w_1^\star, \ldots, w_n^\star)$. Then

$$\Psi(v_i^\star \otimes w_j)(v_k) = \delta_{i,k} \, w_j \qquad (1 \leq i, k \leq m, \ 1 \leq j \leq n),$$

where $\delta_{i,k}$ is the usual Kronecker delta, which is unity for $i = k$ and zero if $i \neq k$. From this it readily follows that $(\Psi(v_i^\star \otimes w_j))_{1 \leq i \leq m, \, 1 \leq j \leq n}$ is a K-basis of $\mathrm{Hom}_K(V, W)$. Hence Ψ is bijective.   □

It is not hard to see that the assertion of Lemma 1.1.40 also holds if only V or W has a finite K-basis, but we will not need this.

Definition 1.1.41 If V is a KG-module then the K-module of bilinear forms on V

$$\text{Bifo}_K V := \{\Phi \colon V \times V \to K \mid \Phi \text{ is } K\text{-bilinear }\}$$

is a KG-module with $(g\Phi)(v,w) := \Phi(g^{-1}v, g^{-1}w)$ for all $g \in G$ and $v, w \in V$.

If $B := (v_1, \ldots, v_n)$ is a K-basis of V, and $\Phi \in \text{Bifo}_K V$, then

$$[\Phi]_B := [\Phi(v_i, v_j)]_{i,j=1,\ldots,n}$$

is called the "Gram-matrix" of Φ with respect to B. If $\delta \colon G \to \text{GL}_n(K)$ is the matrix representation afforded by V with respect to the basis B, then clearly

$$[g \cdot \Phi]_B = \delta(g^{-1})^{\mathrm{T}} [\Phi]_B \, \delta(g^{-1}) \qquad \text{for} \quad g \in G.$$

The next simple lemma will be useful when studying "selfdual" modules, i.e. KG-modules V with $V \cong_{KG} V^\star$.

Lemma 1.1.42 *Let V be a KG-module with a finite K-basis. Then*

$$(V \otimes_K V)^\star \cong_{KG} \text{Bifo}_K V \cong_{KG} \text{Hom}_K(V, V^\star).$$

Proof. The universal property of the tensor product (see Remark 1.1.28) gives a K-isomorphism $\text{Bifo}_K V \cong \text{Hom}_K(V \otimes_K V, K) = (V \otimes_K V)^\star$, and it is readily checked that this is, in fact, a KG-isomorphism. As for the second isomorphism, observe that we have a K-linear isomorphism

$$\Psi \colon \text{Hom}_K(V, V^\star) \to \text{Bifo}_K V \quad \text{with} \quad \Psi(\varphi) \colon (v, v') \mapsto \varphi(v)(v') \in K$$

for $v, v' \in V$ and $\varphi \in \text{Hom}_K(V, V^\star)$. Since

$$\Psi(g\varphi)(v, v') = (g\varphi)(v)(v') = (g\varphi(g^{-1}v))(v') = \varphi(g^{-1}v)(g^{-1}v') = (g\Psi(\varphi))(v, v')$$

this is actually a KG-isomorphism $\qquad\qquad\qquad\qquad\qquad\qquad\qquad\qquad\square$

There is an important class of algebras A, including group algebras, for which $A^\star$ is strongly related to A.

Definition 1.1.43 Let A be a K-algebra which is finitely generated and free over K. A **trace function** on A is a K-linear map $\tau \colon A \to K$ satisfying $\tau(aa') = \tau(a'a)$ for all $a, a' \in A$. The algebra A is called a **symmetric algebra** if a trace function τ on A exists such that the map $\lambda \colon A \to A^\star$ with $\lambda(a)(x) := \tau(ax)$ for $a, x \in A$ is bijective. In this case τ is called a **symmetrizing trace** for the symmetric algebra A.

Remark 1.1.44 Let A be a symmetric K-algebra with symmetrizing trace τ.
(a) The map $\lambda \colon A \to A^*$ in Definition 1.1.43 is an isomorphism of right A-modules, in fact even of (A, A)-bimodules.
(b) For any K-basis $B = (b_1, \ldots, b_n)$ of A there is a uniquely defined "contragredient basis" $B^\tau = (b_1^\tau, \ldots, b_n^\tau)$ of A such that $\tau(b_i\, b_j^\tau) = \delta_{i,j}$.

Example 1.1.45 For $A = K^{n \times n}$ the usual trace function

$$\tau = \text{trace} \colon A \to K \quad , \quad [a_{ij}] \mapsto \sum_{i=1}^{n} a_{ii},$$

qualifies as a symmetrizing trace in the sense of Definition 1.1.43. In fact, let $e_{i,j}$ be the matrix in A with (i, j)-entry 1 and all other entries 0; then for $a = [a_{ij}] \in A$ we have $\text{trace}(e_{i,j} \cdot a) = a_{ji}$. In particular

$$\text{trace}(e_{i,j}e_{k,l}) = \delta_{i,l}\delta_{j,k} = \begin{cases} 1 & \text{for } (i,j) = (l,k), \\ 0 & \text{for } (i,j) \neq (l,k). \end{cases}$$

Thus in the notation of the remark, $e_{i,j}^\tau = e_{j,i}$. ◆

Example 1.1.46 For a group algebra KG of a finite group G the map

$$\tau : KG \to K \quad , \quad \sum_{g \in G} \alpha_g\, g \mapsto \alpha_1,$$

is also easily seen to be a trace function. For $g, h \in G$ we have

$$\tau(g\, h) = \begin{cases} 1 & \text{for } h = g^{-1}, \\ 0 & \text{for } h \neq g^{-1}. \end{cases}$$

Thus τ is a symmetrizing trace and KG is a symmetric algebra. ◆

Up to now we have considered tensor products only over commutative rings. In later sections we will need a somewhat more general construction.

Definition 1.1.47 Assume that A, A' are K-algebras, V is an (A', A)-bimodule and W is an A-module. We use the notation of Definition 1.1.27 and define in addition

$$M_4 := \{(va, w) - (v, aw) \mid v \in V,\ w \in W,\ a \in A\}$$

and $U_{V,W,A} := \langle M_1 \cup M_2 \cup M_3 \cup M_4 \rangle_K \leq_K F(V, W)$. Then $V \otimes_A W := F(V, W)/U_{V,W,A}$. For $(v, w) \in V \times W$ we put $v \otimes w := (v, w) + U_{V,W,A} \in V \otimes_A W$.

Remark 1.1.48 The definition yields $V \otimes_A W = \langle v \otimes w \mid v \in V,\ w \in W \rangle_K$ and an additional basic relation

$$va \otimes w = v \otimes a\, w \qquad \text{for} \qquad a \in A,\ v \in V,\ w \in W.$$

Also $V \otimes_A W$ can be turned into an A'-module in the following way. For $a' \in A'$ let $\lambda_{a'} : F(V, W) \to V \otimes_A W$ be the K-linear map with $\lambda_{a'}(v, w) = a'v \otimes w$. Then $U_{V,W,A} \leq \ker \lambda_{a'}$ and hence factors over a K-linear map

$$\bar{\lambda}_{a'} : V \otimes_A W \to V \otimes_A W \qquad \text{with} \qquad v \otimes w \mapsto a'v \otimes w.$$

For $x \in V \otimes_A W$ we define $a' \cdot x := \bar{\lambda}_{a'}(x)$, and $V \otimes_A W$ becomes an A'-module.

Example 1.1.49 GAP provides the possibility to perform some computations in group algebras of (small) groups. We give an example computing some representations of $G := S_4$ over $K := \mathbb{Q}$:

```
gap> G  := Group( (1,2), (1,2,3,4) );;   K := Rationals;;
gap> KG := GroupRing( K, G );;           o := Embedding( G, KG );;
```

This defines the group algebra of the permutation group generated by the cycles $(1, 2)$ and $(1, 2, 3, 4)$ (which is obviously the symmetric group of degree four) over the field $\mathbb{Q}$ of rational numbers. In GAP groups act from the right, $i^\sigma := \sigma(i)$ for $\sigma \in S_n$ and $i \in \{1, \ldots, n\}$, and the product for $\sigma, \tau \in S_n$ is defined as $\sigma * \tau := \tau \circ \sigma$. The isomorphism $(S_n, \circ) \to (S_n, *)$, $\sigma \mapsto \sigma^{-1}$ provides a way to translate results.

GAP does not identify the group elements with the corresponding elements in the group algebra, although the way elements of a group algebra are displayed seems to indicate this. See Exercise 1.1.1 for an example where such an identification could lead to contradictions. So we use o to embed G into KG. Note that in GAP one may suppress the output of a command (which might be lengthy or uninteresting) by using double semicolons.

We first check the little computation done in Example 1.1.8:

```
gap> a := ()^o + (1,2,3)^o + (1,3,2)^o;;
gap> Print( a*(1,2)^o ,",     ", a*(()^o - (1,2,3)^o) ,",    ", a*a = 3*a );
(1)*(2,3)+(1)*(1,2)+(1)*(1,3),   <zero> of ...,    true
```

Next we want to compute non-trivial submodules of the regular module $_{KG}KG$, that is non-trivial left ideals of KG. Of course, $KG \cdot g = KG$ for all $g \in G$. In order to obtain proper left ideals we try $KG \cdot (1 - g)$ for all $g \in G$:

```
gap> Set( List( G, g -> Dimension( LeftIdeal (KG, [()^o - g^o]) ) ) );
[ 0, 12, 16, 18 ]
gap> Filtered(G, g -> Dimension( LeftIdeal (KG, [()^o - g^o] ) ) = 12 );
[ (1,4)(2,3),  (1,2)(3,4),  (1,3)(2,4),  (3,4),  (1,2),  (2,4),  (1,3),
(2,3),   (1,4) ]
gap> a := ()^o - (1,2)^o;;
```

We see that $\dim_K KG \cdot (1 - g) \in \{0, 12, 16, 18\}$ for all $g \in G$. We have chosen $a := () - (1, 2)$ so that $\dim_K KG \cdot a = 12$. We may look for submodules of smaller dimension and observe that $KG \cdot ba \leq_{KG} KG \cdot a$:

```
gap> Set(List(G, g -> Dimension( LeftIdeal (KG, [(() - g^o) * a] ) ) ) );
[ 0, 6, 8, 9, 11, 12 ]
gap> Filtered(G, g -> Dimension( LeftIdeal (KG, [(() - g^o) * a] ) )= 6);
```

```
[ (1,2)(3,4), (3,4) ]
gap> b := (()^o - (3,4)^o) * a;;
gap> V := LeftIdeal (KG, [b]);; B := Basis (V);;
```

V is a six-dimensional KG-module. We have chosen a K-basis B of V and we can obtain the matrix $[\delta_V(g)]_B$ for any $g \in G$ with the GAP-commands

```
gap> g := (1,2,3);;   # for example
gap> dg := TransposedMat(List( B, v -> Coefficients( B, ((g^-1)^o*v) )));;
```

We have to transpose the matrices because GAP uses the "row convention," that is, the coefficients of a vector with respect to a fixed basis are given as a row-vector. But we want to analyze further the KG-module V. Exactly as above we find KG-submodules V_1, V_2 of V of dimensions 5 and 3 with

$$V \geq_{KG} V_1 := KG \cdot c \geq_{KG} V_2 := KG \cdot d \geq_{KG} \{0\}, \qquad (1.8)$$

where $c := (() - (2,4,3)) \cdot b$ and $d := (() - (1,4)) \cdot c$:

```
gap> c:= (()^o - (2,4,3)^o) * b;;    d:= (()^o - (1,4)^o) * c;;
gap> B1 := Basis ( LeftIdeal (KG, [c]) );;
gap> B2 := Basis ( LeftIdeal (KG, [d] ));;
```

We now construct a K-basis of V adapted to (1.8). That is, we start with the basis vectors of B2, add suitable vectors of B1 to obtain a K-basis of V_1 and finally add a vector of B to get basis vectors for V:

```
gap> adbas := [];; Append( adbas, BasisVectors(B2) );
gap> for x in BasisVectors(B1) do
>         if not x in Subspace( KG, adbas )  then Add(adbas ,x); fi;
>     od;
gap> for x in BasisVectors(B) do
>         if not x in Subspace( KG, adbas )  then Add(adbas ,x); fi;
>     od;
gap> BB := Basis ( V , adbas );;
```

Finally we compute the matrices $[\delta_V(g)]_{BB}$ for $g \in \{(1,2),(1,2,3,4)\}$:

```
gap> m1 := List( BB, x -> Coefficients( BB, (1,2)^o*x) );;
gap> m2 := List( BB, x -> Coefficients( BB, ((1,2,3,4)^-1)^o*x) );;
gap> m1 := TransposedMat(m1);; m2 := TransposedMat(m2);;
```

The result is the following:

$$
m1 = \begin{bmatrix}
-1 & 0 & 0 & 0 & 0 & 0 \\
1 & 1 & -1 & 0 & 0 & 0 \\
0 & 0 & -1 & 0 & 0 & 0 \\
0 & 0 & 0 & -1 & 0 & 0 \\
0 & 0 & 0 & 1 & 1 & 0 \\
0 & 0 & 0 & 0 & 0 & -1
\end{bmatrix},
\quad
m2 = \begin{bmatrix}
0 & 1 & -1 & 0 & 1 & 0 \\
1 & 0 & 1 & 1 & 0 & 0 \\
2 & 0 & 1 & 1 & 0 & 0 \\
0 & 0 & 0 & 0 & -1 & 1 \\
0 & 0 & 0 & -1 & 0 & -1 \\
0 & 0 & 0 & 0 & 0 & -1
\end{bmatrix}.
$$

We thus have produced matrix representations δ_i $(i = 1, 2, 3)$ of G of degrees three, two and one with representation spaces V_2, V_1/V_2 and V/V_1. Only δ_1 is faithful, as can be seen by

```
gap> Size( Group( m1{[1..3]}{[1..3]}, m2{[1..3]}{[1..3]} ) );
24
gap> Size( Group( m1{[4,5]}{[4,5]}, m2{[4,5]}{[4,5]} ) );
6
```

Thus δ_2 is the inflation of a representation of $S_4/V_4 \cong S_3$. That δ_i is irreducible (for $i = 1, 2, 3$) can be shown by verifying that $\delta_i((1,2))$ and $\delta_i((1,2,3,4))$ do not have a common eigenvector.

Instead of taking $\mathbb{Q}$ as underlying field one can also take cyclotomic fields (e.g. K := CyclotomicField(60)) or finite fields (K := GF(16)). One should keep in mind that computing in the group algebra as above is feasible only for groups of relatively small order, maybe up to a few hundreds. There are far better methods to compute and analyze representations of groups, as we will see shortly. The group algebra is mainly of theoretical importance and the above example is given just for illustration. ◆

Exercises

Exercise 1.1.1 Let $g := \begin{bmatrix} 0 & 1 \\ 1 & 1 \end{bmatrix} \in \mathrm{GL}_2(2) := \mathrm{GL}_2(\mathbb{F}_2)$. Show that $G := \langle g \rangle \cong C_3$ and that the group algebra $\mathbb{F}_2 G$ is **not** isomorphic to

$$\{ \sum_{i=1}^{3} a_i g^i \mid a_i \in \mathbb{F}_2 \} \subseteq \mathbb{F}_2^{2 \times 2}.$$

Exercise 1.1.2 Show that any group homomorphism $\varphi \colon G \to H$ can be uniquely extended to an algebra homomorphism $\hat{\varphi} \colon KG \to KH$ of the corresponding group algebras and that $\hat{\varphi}$ is injective or surjective if and only if φ has the corresponding property. Show that

$$\ker(\hat{\varphi}) = \{ \sum_{i=1}^{t} \sum_{u \in \ker\varphi \setminus \{1\}} \alpha_{i,u}\, g_i\,(u-1) \mid \alpha_{i,u} \in K \}$$

provided that $G = \dot{\cup}_{i=1}^{t} g_i \cdot \ker\varphi$.

Exercise 1.1.3 Let K be a field with char $K \neq 2$ containing a primitive fourth root of unity i and let $\langle g \rangle = C_4$ a cyclic group of order four. Put

$$a = \frac{1+i}{2}g + \frac{1-i}{2}g^3 \in K\,C_4, \qquad b = \frac{1-i}{2}g + \frac{1+i}{2}g^3 \in K\,C_4.$$

(a) Show that $\{1, g^2, a, b\} \subseteq K\,C_4$ is a subgroup of the unit group of $K\,C_4$ isomorphic to the Klein 4-group $V_4 \cong C_2 \times C_2$.
(b) Show that $K\,C_4 \cong KV_4$ as algebras over K.

Note: It follows from the above exercises that the group algebras of isomorphic groups are isomorphic, but also that the converse does not hold in general, at least if the ring of coefficients K is a field. In 1971, Dade [44] constructed examples of non-isomorphic groups having isomorphic group algebras over any field.

It had been an open problem for about 50 years whether or not non-isomorphic groups G, H would exist such that even $\mathbb{Z}G \cong \mathbb{Z}H$. In 1997, Hertweck [79] found such examples, the smallest having order $2^{21} \, 97^{28}$.

Exercise 1.1.4 Let S_3 be the symmetric group on three letters and let K be a commutative ring. Verify that there is a representation $\delta \colon S_3 \to \mathrm{GL}_2(K)$ with

$$\delta \colon (1,2) \mapsto \begin{bmatrix} -1 & 1 \\ 0 & 1 \end{bmatrix}, \quad (2,3) \mapsto \begin{bmatrix} 1 & 0 \\ 1 & -1 \end{bmatrix}.$$

Show that δ is equivalent to the representation $\delta' \colon S_3 \to \mathrm{GL}_2(K)$ with

$$\delta' \colon (1,2) \mapsto \begin{bmatrix} -2 & -1 \\ 3 & 2 \end{bmatrix}, \quad (2,3) \mapsto \begin{bmatrix} 1 & 0 \\ -3 & -1 \end{bmatrix}.$$

Suppose that K is a field. Prove that δ is irreducible if and only if $\operatorname{char} K \neq 3$.

Exercise 1.1.5 Let $Q_8 := \langle a, b \mid a^2 = b^2 = [a,b], \ a^4 = 1 \rangle$ be the quaternion group of order eight and let K be any field.

(i) If $\delta \colon Q_8 \to \mathrm{GL}_2(K)$ is a faithful representation, show that δ is equivalent to a representation $\delta_{\alpha,\beta}$ with

$$a \mapsto \begin{bmatrix} 0 & -1 \\ 1 & 0 \end{bmatrix}, \quad b \mapsto \begin{bmatrix} \alpha & \beta \\ \beta & -\alpha \end{bmatrix}, \tag{1.9}$$

where $\alpha, \beta \in K$ satisfy $\alpha^2 + \beta^2 = -1$.

(ii) If K is a field and $\alpha, \beta \in K$ satisfy $\alpha^2 + \beta^2 = -1$, then (1.9) defines an irreducible representation $\delta_{\alpha,\beta}$ of Q_8. If $\operatorname{char} K > 2$ then Q_8 has an absolutely irreducible representation of degree two over K.

(iii) Let α be transcendental over $\mathbb{Q}$ and let β be a root of $X^2 + \alpha^2 + 1 \in \mathbb{Q}(\alpha)[X]$. Putting $K := \mathbb{Q}(\alpha, \beta)$, show that the representation $\delta_{\alpha,\beta} \colon Q_8 \to \mathrm{GL}_2(K)$ is not equivalent to a matrix representation having algebraic entries. On the other hand, show that on replacing K by $K := \mathbb{Q}(\alpha, \beta, i)$, where $i^2 = -1$, it is equivalent to $\delta_{i,0}$.

Exercise 1.1.6 Let $n \geq 1$ and let $D_{2n} := \langle a, b \mid a^n = b^2 = 1, \ bab = a^{-1} \rangle$ be the dihedral group of order $2n$. Let $\zeta \in \mathbb{C}$ be an nth root of unity in $\mathbb{C}$. Show that the map

$$a \mapsto \begin{bmatrix} \zeta & 0 \\ 0 & \zeta^{-1} \end{bmatrix}, \quad b \mapsto \begin{bmatrix} 0 & 1 \\ 1 & 0 \end{bmatrix},$$

defines a representation $\delta_\zeta \colon D_{2n} \to \mathrm{GL}_2(\mathbb{C})$, which is irreducible if and only if $\zeta \neq \zeta^{-1}$.

Exercise 1.1.7 Let G' be the commutator subgroup of G and let $\delta \colon G \to \mathrm{GL}_n(K)$ be a representation of degree n over a field K. Show the following.

(i) If $n = 1$, then $G' \leq \ker \delta$.

(ii) $\delta(G') \subseteq \mathrm{SL}_n(K) := \{a \in \mathrm{GL}_n(K) \mid \det a = 1\}.$

Exercise 1.1.8 Let p be a prime, let $G := \mathrm{SL}_2(p) := \mathrm{SL}_2(\mathbb{F}_p)$ and let $V := \mathbb{F}_p[X, Y]$ be the ring of polynomials over $\mathbb{F}_p$ in the indeterminates X, Y.

(i) For $g = \begin{bmatrix} a & b \\ c & d \end{bmatrix} \in G$ we define $g.X := aX + cY$ and $g.Y := bX + dY$. Show that this extends uniquely to an automorphism of the algebra $\mathbb{F}_p[X, Y]$. Prove that V becomes thereby an $\mathbb{F}_pG$-module.

(ii) For $m \in \mathbb{N}$ let $V_m \leq V$ be the subspace of homogeneous polynomials of degree $m - 1$. Show that V_m is a submodule of the $\mathbb{F}_pG$-module V defined in part (i). What is $\dim_{\mathbb{F}_p} V_m$?

Exercise 1.1.9 Let G be a cyclic group and let V be a KG-module for a field K. Show that $\dim_K \mathrm{Inv}_G(V) = \dim_K \mathrm{Inv}_G(V^*)$.

Exercise 1.1.10 Let K be a field. Show that the regular module $V := {}_{KG}KG$ is selfdual, i.e. that V is isomorphic to the contragredient module $V^\star$.

Exercise 1.1.11 Let K be a field, let $p \in K[X]$ be a monic irreducible polynomial of degree s and let $q = p^e$ as in Example 1.1.24. Let $V := K[X]/(q)$ and $\varphi \in \mathrm{End}_K V$ be defined by $\varphi(f + (q)) := X \cdot f + (q)$. Starting with $v_1 := 1 + (q)$, define recursively $v_i := X \cdot v_{i-1}$ if $s \nmid i$ and $v_{js} := p^j + (q)$ for $j \in \mathbb{N}$. Show that $B := (v_1, \ldots, v_{es})$ is a K-basis of V and that $[\varphi]_B = M_q'$, as defined in (1.5). Deduce that the companion matrix M_q is similar to M_q'.

Exercise 1.1.12 (a) Consider $A := \{[\alpha_{ij}] \in \mathbb{C}^{2\times 2} \mid \alpha_{12} = 0\} \leq_{\mathbb{C}} \mathbb{C}^{2\times 2}$ and the ring homomorphism $\gamma \colon A \to A$ with $[\alpha_{ij}] \mapsto [\overline{\alpha_{ij}}]$ induced by complex conjugation. Let $V := \mathbb{C}^{2\times 1}$ be the A-module affording (with respect to the standard basis B of V) the matrix representation δ with $\delta(a) = a$ for $a \in A$. Check that γ is not a $\mathbb{C}$-algebra homomorphism. Show that the $\mathbb{C}$-module structures of V and $\mathrm{Inf}_\gamma V$ are different and that $\mathrm{Inf}_\gamma V$ affords the same matrix representation δ with respect to the basis B.

(b) Let $G := \langle g \rangle$ be a cyclic group of order four and let $\gamma \colon \mathbb{C}G \to \mathbb{C}G$ be the ring homomorphism mapping $\sum_{j=1}^{4} \alpha_j g^j$ to $\sum_{j=1}^{4} \overline{\alpha_j} g^j$. Let V be a $\mathbb{C}G$-module affording the representation $\delta \colon G \to \mathbb{C}$ with $\delta(g) = i \in \mathbb{C}$. Again γ is not a $\mathbb{C}$-algebra homomorphism and the $\mathbb{C}$-module structures of V and $\mathrm{Inf}_\gamma V$ are different. Show that $\mathrm{Inf}_\gamma V$ affords the representation $\bar{\delta}$ with $\bar{\delta}(g) = -i$.
Note: See Definition 1.8.2 for a generalization.

1.2 Permutation representations and G-sets

Permutation representations to be introduced in this section form an important class of representations of groups. In practice a permutation representation of small degree often forms the starting point for analyzing the structure of a particular group or its representations in arbitrary characteristic. A big advantage of these representations is that they are defined and available uniformly over arbitrary fields or even commutative rings K.

Before defining permutation representations we recall a few basic facts about actions of groups on sets and introduce the corresponding notions.

Assumption: Throughout this section G is supposed to be a finite group and K a commutative ring.

Definition 1.2.1 Let Ω be a finite non-empty set. Note that Ω or more precisely $(\Omega, \cdot)$ is called a (left) G-**set**, and G is said to act on Ω (from the left) if

$$\cdot : G \times \Omega \to \Omega \qquad (g, \omega) \mapsto g \cdot \omega$$

is a map satisfying

$$g_1 \cdot (g_2 \cdot \omega) = (g_1 g_2) \cdot \omega \quad \text{for all} \quad g_1, g_2 \in G, \quad \omega \in \Omega, \qquad (1.10)$$

$$1_G \cdot \omega = \omega \qquad \text{for all} \quad \omega \in \Omega. \qquad (1.11)$$

If Ω_1, Ω_2 are G-sets then a map $\varphi \colon \Omega_1 \to \Omega_2$ is called G-**equivariant** or a G-**map** if

$$\varphi(g \cdot \omega) = g \cdot \varphi(\omega) \quad \text{for all} \quad g \in G, \ \omega \in \Omega_1.$$

The set of G-maps from Ω_1 to Ω_2 is usually denoted by $\mathrm{Hom}_G(\Omega_1, \Omega_2)$. If, in addition, φ is bijective then φ is called a G-isomorphism, and if such a φ exists Ω_1, Ω_2 are called **isomorphic** G-sets; in symbols, $\Omega_1 \cong_G \Omega_2$. If Ω is a G-set then for any $\omega \in \Omega$ the set $\omega^G := G \cdot \omega := \{g \cdot \omega \mid g \in G\}$ is called a G-**orbit** in Ω, and Ω is called **transitive** if Ω is itself a G-orbit. Also for any $\omega \in \Omega$

$$\mathrm{Stab}_G(\omega) := \{g \in G \mid g \cdot \omega = \omega\} \leq G$$

is called the **stabilizer** of ω in G. Furthermore for $g \in G$ and $H \subseteq G$ we put

$$\mathrm{Fix}_\Omega(g) := \{\omega \in \Omega \mid g \cdot \omega = \omega\} \quad \text{and} \quad \mathrm{Fix}_\Omega(H) := \bigcap_{h \in H} \mathrm{Fix}_\Omega(h).$$

Remark 1.2.2 Every G-set is a disjoint union of G-orbits (see [110], p. 29). Any group G acts transitively (by left multiplication) on the set $G/H = \{gH \mid g \in G\}$ of left cosets of any subgroup $H \leq G$. Furthermore if Ω is any transitive G-set then $\Omega \cong_G G/H$ for $H = \mathrm{Stab}_G(\omega)$ for any $\omega \in \Omega$ and

$$G/H_1 \cong_G G/H_2 \quad \text{if and only if} \quad H_2 = g^{-1} H_1 g = H_1^g \quad \text{for some } g \in G.$$

Definition 1.2.3 If Ω is a finite G-set and K a commutative ring we define $K\Omega$ to be the free K-module with basis Ω and consider it as a KG-module by extending the action of G on Ω to a K-linear action of KG on $K\Omega$. Thus

$$\sum_{g \in G} a_g g \cdot \sum_{\omega \in \Omega} b_\omega \omega = \sum_{g \in G} \sum_{\omega \in \Omega} a_g b_\omega (g \cdot \omega) \quad \text{for} \quad a_g, b_\omega \in K;$$

$K\Omega$ is called the **permutation module** corresponding to Ω (and K). The corresponding representation $\delta_\Omega \colon KG \to \mathrm{End}_K K\Omega$ or its restriction to G is called a **permutation representation** of KG or G.

Observe that for any finite G-set Ω and any group element $g \in G$ the matrix $[\delta_\Omega(g)]_\Omega$ is a permutation matrix, that is, it has exactly one non-zero entry in each row and column, which is, in fact unity. In this context it is often convenient to use the elements of Ω also as labels for the rows and columns of the corresponding permutation matrices. It follows that

$$[\delta_\Omega(g)]_\Omega = [d_{\omega,\omega'}]_{\omega,\omega' \in \Omega} \quad \text{with} \quad d_{\omega,\omega'} = \delta_{\omega,g\cdot\omega'} = \begin{cases} 1 & \text{if } \omega = g \cdot \omega', \\ 0 & \text{else,} \end{cases}$$

the last δ being the usual Kronecker delta. Conversely, if $\delta \colon G \to \mathrm{GL}(V)$ is a representation for some finitely generated free K-module V, such that for some K-basis B of V the matrices $[\delta(g)]_B$ are permutation matrices for all $g \in G$, then δ is a permutation representation (for the G-set B).

A permutation representation is never irreducible except in the case of the trivial representation which corresponds to a one-element G-set.

Lemma 1.2.4 *Let Ω be a transitive G-set and $V = K\Omega$. Then*

$$\mathrm{Inv}_G(V) = \langle \sum_{\omega \in \Omega} \omega \rangle_K \quad \text{and}$$

$$\mathrm{Inv}^G(V) = \{ \sum_{\omega \in \Omega} \alpha_\omega \omega \mid \sum_{\omega \in \Omega} \alpha_\omega = 0 \,, \; \alpha_\omega \in K \}$$

are KG-submodules of V (recall Definition 1.1.18). If K is an integral domain then $V = \mathrm{Inv}_G(V) \oplus \mathrm{Inv}^G(V)$ if and only if $|\Omega|$ is invertible in K.

Proof. The first assertion follows immediately, since

$$g \cdot \sum_{\omega \in \Omega} \alpha_\omega \omega = \sum_{\omega \in \Omega} \alpha_\omega \omega \quad \text{if and only if} \quad \alpha_{g^{-1}\omega} = \alpha_\omega \quad \text{for all } \omega \in \Omega.$$

Since G acts transitively on Ω it follows that $\mathrm{Inv}_G(V) = \{\alpha \sum_{\omega \in \Omega} \omega \mid \alpha \in K\}$. Obviously $\mathrm{Inv}^G(V) \leq_{KG} V$, and if $W \leq_{KG} V$ is any submodule such that G acts trivially on V/W, then $\omega - g\omega \in W$ for all $\omega \in \Omega$ and $g \in G$. Since $\Omega = G\omega_0$ for some $\omega_0 \in \Omega$,

$$\langle \{\omega - \omega_0 \mid \omega \in \Omega\} \rangle_K = \mathrm{Inv}^G(V) \leq W.$$

Clearly $\mathrm{Inv}_G(V) \cap \mathrm{Inv}^G(V) \neq \{0\}$ if and only if $|\Omega| \cdot 1_K = 0$. If $\gamma|\Omega| = 1_K$ then any element in V can be written in the form

$$v = \sum_{\omega \in \Omega} \alpha_\omega \omega = \gamma\alpha \sum_{\omega \in \Omega} \omega + \sum_{\omega \in \Omega} (\alpha_\omega - \gamma\alpha)\omega \in \mathrm{Inv}_G(V) + \mathrm{Inv}^G(V)$$

with $\alpha = \sum_{\omega \in \Omega} \alpha_\omega$. $\qquad\square$

It is clear how to extend this result to non-transitive G-sets as follows.

Corollary 1.2.5 *Let* $\Omega = \mathcal{O}_1 \dot{\cup} \ldots \dot{\cup} \mathcal{O}_r$ *be a* G-set with orbits $\mathcal{O}_i, i = 1, \ldots, r.$
Then

$$\mathrm{Inv}_G(K\Omega) = \langle \mathcal{O}_1^+, \ldots, \mathcal{O}_r^+ \rangle_K,$$
$$\mathrm{Inv}^G(K\Omega) = \{ \sum_{w \in \Omega} \alpha_w w \mid \sum_{w \in \mathcal{O}_i} \alpha_w = 0 \quad for \quad 1 \leq i \leq r \}.$$

Example 1.2.6 The symmetric group S_n acts on an n-element set $\Omega_n = \{\omega_1, \ldots, \omega_n\}$ by $\sigma \cdot \omega_i = \omega_{\sigma(i)}$ and hence has a natural permutation representation of degree n. For $n = 3$ the matrices of this representation with respect to the basis Ω_3 are given by

$$(1,2) \mapsto \begin{bmatrix} 0 & 1 & 0 \\ 1 & 0 & 0 \\ 0 & 0 & 1 \end{bmatrix}, \quad (2,3) \mapsto \begin{bmatrix} 1 & 0 & 0 \\ 0 & 0 & 1 \\ 0 & 1 & 0 \end{bmatrix}.$$

Observe that S_3 is generated by $(1,2)$ and $(2,3)$, so that the representation is completely determined by these two matrices. A K-basis of $\mathrm{Inv}^{S_3}(K\Omega_3)$ is $B := (\omega_1 - \omega_2, \; \omega_2 - \omega_3)$, and the matrix representation with respect to B is given by

$$(1,2) \mapsto \begin{bmatrix} -1 & 1 \\ 0 & 1 \end{bmatrix}, \quad (2,3) \mapsto \begin{bmatrix} 1 & 0 \\ 1 & -1 \end{bmatrix};$$

compare with Exercise 1.1.4.   ◆

Remark 1.2.7 If Ω_1, Ω_2 are G-sets and $\varphi \colon \Omega_1 \to \Omega_2$ is a G-map then it is obvious that φ can be extended uniquely to a K-linear map $\varphi_K \colon K\Omega_1 \to K\Omega_2$ and the map φ_K is a KG-homomorphism. If φ is bijective then φ_K is a KG-isomorphism. Thus isomorphic G-sets lead to isomorphic KG-modules and equivalent permutation representations. But the converse does not hold, not even for $K = \mathbb{Z}$ (see Remark 3.4.5).

Definition 1.2.8 Let A be a ring and let e be an element in A. We call e an **idempotent** if $e^2 = e \neq 0$.

Observe that the regular module $_{KG}KG$ is itself a permutation module corresponding to the "regular" action of G on itself by left multiplication. In fact every permutation module corresponding to a transitive G-set is isomorphic to a submodule of the regular module $_{KG}KG$, that is to a left ideal of the group algebra KG:

Lemma 1.2.9 *Let* Ω *be a transitive* G-set, $\Omega = G \cdot \omega$ *with* $\mathrm{Stab}_G(\omega) = H$ *and*

$$H^+ = \sum_{h \in H} h \in KH \subseteq KG.$$

Then $g \cdot \omega \mapsto g \cdot H^+$ *extended* K-linearly defines an isomorphism $K\Omega \cong KG \cdot H^+$
of KG-modules. If $|H|$ *is invertible in* K *then*

$$e_H := \frac{1}{|H|} H^+ \in KH \quad satisfies \quad e_H^2 = e_H$$

that is, e_H is an idempotent, and $KG = KGe_H \oplus KG(1 - e_H)$; thus $K\Omega \cong KG \cdot H^+ = KGe_H$ is isomorphic to a direct summand of the regular module KG in this case.

Proof. Obviously the mapping $\varphi \colon \Omega \to \{gH^+ \in KG \mid g \in G\}$, $g\omega \mapsto gH^+$ is well-defined and an isomorphism of G-sets. With the notation of Remark 1.2.7, $\varphi_K \colon K\Omega \to KGH^+$ is a KG-isomorphism. Since $h \cdot H^+ = H^+$ for all $h \in H$ we have $(H^+)^2 = |H| \cdot H^+$ and e_H is idempotent if $|H|1_K \in K$ is invertible. From this it follows easily that $KGe_H \cap KG(1 - e_H) = \{0\}$. $\qquad\square$

Definition 1.2.10 If X, Y are G-sets then the Cartesian product $X \times Y$ is also a G-set in a natural way: define $g \cdot (x, y) := (g \cdot x, g \cdot y)$ for $g \in G, x \in X, y \in Y$. If $\mathcal{O}$ is a G-orbit on $X \times Y$ then $\mathcal{O}' = \{(y, x) \in Y \times X \mid (x, y) \in \mathcal{O}\}$ is a G-orbit on $Y \times X$. Also, if $a \in X, b \in Y$, then we define

$$\mathcal{O}(a) = \{y \in Y \mid (a, y) \in \mathcal{O}\} \ , \ \mathcal{O}'(b) = \{x \in X \mid (x, b) \in \mathcal{O}\}.$$

Obviously $\mathcal{O}(a)$ is empty or a $\mathrm{Stab}_G(a)$-orbit on Y and $\mathcal{O}'(b)$ is empty or a $\mathrm{Stab}_G(b)$-orbit on X. If $X = Y$ then we have a pairing $\mathcal{O} \mapsto \mathcal{O}'$ of G-orbits on $X \times X$. If $\mathcal{O} = \mathcal{O}'$ then $\mathcal{O}$ is called **self-paired**.

Lemma 1.2.11 *If X, Y are transitive G-sets, $H = \mathrm{Stab}_G(x_1)$, $U = \mathrm{Stab}_G(y_1)$ for $x_1 \in X$, $y_1 \in Y$, then the orbits of G on $X \times Y$ are in one-to-one correspondence with the orbits of H on Y and also with the double cosets HgU of H and U in G.*

Proof. If $\mathcal{O}$ is a G-orbit on $X \times Y$ then $\mathcal{O}(x_1)$ is obviously an orbit of H on Y. Observe that $\mathcal{O}(x_1) \neq \emptyset$ because G is transitive on X. Since G is also transitive on Y, any $y \in Y$ can be written in the form $y = g \cdot y_1$ for some $g \in G$ which is in a uniquely determined coset gU. The map $G \cdot (x_1, g \cdot y_1) \mapsto HgU$ defines the second one-to-one correspondence. $\qquad\square$

Remark 1.2.12 If $\mathcal{O}$ is a G-orbit on $X \times Y$ and $(x, y) \in \mathcal{O}$ then $\mathcal{O}(x) = \mathrm{Stab}_G(x) \cdot y$ and $\mathcal{O}'(y) = \mathrm{Stab}_G(y) \cdot x$. Hence

$$|\mathcal{O}| = [G : \mathrm{Stab}_G((x, y))] = [G : \mathrm{Stab}_G(x)]|\mathcal{O}(x)|$$
$$= [G : \mathrm{Stab}_G(y)]|\mathcal{O}'(y)|.$$

Hence, if $\Omega = X = Y$ is a transitive G-set, then

$$|\mathcal{O}(x)| = |\mathcal{O}'(y)| = \frac{|\mathcal{O}|}{|\Omega|} \quad \text{for arbitrary} \quad (x, y) \in \mathcal{O}.$$

Definition 1.2.13 If Ω is a transitive G-set then the number of orbits of G on $\Omega \times \Omega$ (which by Lemma 1.2.11 is the same as the number of orbits of $H = \mathrm{Stab}_G(\omega)$ on Ω for any $\omega \in \Omega$) is called the **rank** of Ω and also of $K\Omega$. If the rank is 2 then one also says that G acts **doubly transitively** on Ω. If $\mathcal{O}_1, \ldots, \mathcal{O}_r$

are the orbits of G on $\Omega \times \Omega$ then the cardinalities $|\mathcal{O}_1(\omega)|, \ldots, |\mathcal{O}_r(\omega)|$ are called the **subdegrees** of Ω. As seen before these are independent of the choice of $\omega \in \Omega$.

Example 1.2.14 Let $G = \langle (1,3,4)(2,5,6), (3,5)(4,6) \rangle$ acting on the set $\Omega = \{1,2,3,4,5,6\}$. It is easily checked that G is isomorphic to the alternating group A_4; G has four orbits on $\Omega \times \Omega$:

$\mathcal{O}_1 = G \cdot (1,1) = \{(i,i) \mid 1 \le i \le 6\}$,
$\mathcal{O}_2 = G \cdot (1,2) = \{(1,2),(2,1),(3,5),\ldots\}$,
$\mathcal{O}_3 = G \cdot (1,3) = \{(1,3),(1,5),(2,3),\ldots\}$,
$\mathcal{O}_4 = G \cdot (1,4) = \{(1,4),(1,6),(2,4),\ldots\}$.

	1	2	3	4	5	6
1	$\mathcal{O}_1$	$\mathcal{O}_2$	$\mathcal{O}_3$	$\mathcal{O}_4$	$\mathcal{O}_3$	$\mathcal{O}_4$
2	$\mathcal{O}_2$	$\mathcal{O}_1$	$\mathcal{O}_3$	$\mathcal{O}_4$	$\mathcal{O}_3$	$\mathcal{O}_4$
3	$\mathcal{O}_4$	$\mathcal{O}_4$	$\mathcal{O}_1$	$\mathcal{O}_3$	$\mathcal{O}_2$	$\mathcal{O}_3$
4	$\mathcal{O}_3$	$\mathcal{O}_3$	$\mathcal{O}_4$	$\mathcal{O}_1$	$\mathcal{O}_4$	$\mathcal{O}_2$
5	$\mathcal{O}_4$	$\mathcal{O}_4$	$\mathcal{O}_2$	$\mathcal{O}_3$	$\mathcal{O}_1$	$\mathcal{O}_3$
6	$\mathcal{O}_3$	$\mathcal{O}_3$	$\mathcal{O}_4$	$\mathcal{O}_2$	$\mathcal{O}_4$	$\mathcal{O}_1$

$\Omega \times \Omega$

These can be visualized as in the table, in which at position (i,j) the orbit of (i,j) can be found.

♦

We will now study KG-homomorphisms between permutation modules. Let $X = \{x_1, \ldots, x_n\}$, $Y = \{y_1, \ldots, y_m\}$ be G-sets. It is convenient to use the elements of X and Y also for labeling rows and columns of matrices. A KG-homomorphism $\varphi \colon KX \to KY$ is a K-linear map commuting with the action of the elements of G on KX and KY. Thus its matrix $_Y[\varphi]_X = [a_{yx}]_{x \in X, y \in Y} \in K^{m \times n}$ with respect to the K-bases X and Y must satisfy

$$[\delta_Y(g)]_Y \cdot [a_{yx}] = [a_{yx}] \cdot [\delta_X(g)]_X \quad \text{for all} \quad g \in G,$$

which is equivalent to

$$a_{g^{-1}y,x} = a_{y,gx} \quad \text{for all} \quad g \in G, \ x \in X, \ y \in Y.$$

This just means that the matrix entries of $_Y[\varphi]_X$ must be constant on the orbits of G on $Y \times X$. Hence we get the following complete description of $\mathrm{Hom}_{KG}(KX, KY)$.

Lemma 1.2.15 *If X, Y are G-sets and K a commutative ring then*

$$\mathrm{Hom}_{KG}(KX, KY) = \{\varphi \in \mathrm{Hom}_K(KX, KY) \mid {}_Y[\varphi]_X = [a_{yx}]$$

$$\text{with} \quad a_{y,x} = a_{gy,gx} \quad \text{for all} \quad g \in G, x \in X, y \in Y\}.$$

For any G-orbit $\mathcal{O}$ on $Y \times X$ let $\theta_{\mathcal{O}} \in \mathrm{Hom}_{KG}(KX, KY)$ be defined by $\theta_{\mathcal{O}}(x) = \mathcal{O}'(x)^+ := \sum_{y \in \mathcal{O}'(x)} y$. Thus the matrix of $\theta_{\mathcal{O}}$ with respect to the bases X and Y is $_Y[\theta_{\mathcal{O}}]_X = [a_{yx}]$ with

$$a_{yx} = \begin{cases} 1 & \text{for } (y,x) \in \mathcal{O}, \\ 0 & \text{else}. \end{cases}$$

*Then $\{\theta_{\mathcal{O}} \mid \mathcal{O} \text{ a } G\text{-orbit on } Y \times X\}$ is a K-basis for $\mathrm{Hom}_{KG}(KX, KY)$ often referred to as the **standard basis** of $\mathrm{Hom}_{KG}(KX, KY)$.*

Corollary 1.2.16 *If Ω is a transitive G-set of rank r then $\operatorname{End}_{KG} K\Omega$ is a free K-module of rank r.*

In the case that $X = Y = G/H$ the above standard basis for $\operatorname{End}_{KG} K(G/H)$ is often referred to as the **Schur basis**. $\operatorname{End}_{KG} K(G/H)$ is also sometimes called the **Hecke algebra** corresponding to H. If

$$G = \bigcup_{i=1}^{r} {}^{\cdot} H g_i H$$

is a double coset decomposition then the G-orbits on $G/H \times G/H$ are

$$\mathcal{O}_i = G \cdot (g_i H, 1H) = \{(g g_i H, gH) \mid g \in G\}$$

and $\mathcal{O}'_i(1H) = \{h g_i H \mid h \in H\}$. Observe that $h g_i H = h' g_i H$ for $h, h' \in H$ if and only if $h'^{-1}h \in H \cap {}^{g_i}H$. Here we use the standard notation for conjugate subgroups ${}^g H = H^{g^{-1}} = gHg^{-1}$ for $g \in G$. Thus if T_i is a left transversal of $H \cap {}^{g_i}H$ in H, that is $H = \bigcup_{h \in T_i} h(H \cap {}^{g_i}H)$, then the standard basis elements $\theta_i = \theta_{\mathcal{O}_i}$ are given by

$$\theta_i(1H) = \sum_{h \in T_i} h g_i H \in K(G/H).$$

Example 1.2.17 Continuing with Example 1.2.14 we see that for the alternating group A_4 with its action on $\Omega = \{1, 2, 3, 4, 5, 6\}$ we have

$$\operatorname{End}_{KG} K\Omega \cong_K \langle \mathbf{I}, \mathbf{a}_2, \mathbf{a}_3, \mathbf{a}_3^{\mathrm{T}} \rangle_K$$

with

$$\mathbf{a}_2 = \begin{bmatrix} 0 & 1 & 0 & 0 & 0 & 0 \\ 1 & 0 & 0 & 0 & 0 & 0 \\ 0 & 0 & 0 & 0 & 1 & 0 \\ 0 & 0 & 0 & 0 & 0 & 1 \\ 0 & 0 & 1 & 0 & 0 & 0 \\ 0 & 0 & 0 & 1 & 0 & 0 \end{bmatrix}, \quad \mathbf{a}_3 = \begin{bmatrix} 0 & 0 & 1 & 0 & 1 & 0 \\ 0 & 0 & 1 & 0 & 1 & 0 \\ 0 & 0 & 0 & 1 & 0 & 1 \\ 1 & 1 & 0 & 0 & 0 & 0 \\ 0 & 0 & 0 & 1 & 0 & 1 \\ 1 & 1 & 0 & 0 & 0 & 0 \end{bmatrix}.$$

$\blacklozenge$

Example 1.2.18 The orbits of S_n on $\Omega \times \Omega$ with $\Omega = \{1, \ldots, n\}$ resulting from the natural action are $\Delta_\Omega := \{(\omega, \omega) \mid \omega \in \Omega\}$ and $(\Omega \times \Omega) \setminus \Delta_\Omega$. Thus

$$\operatorname{End}_{K S_n} K\Omega \cong_K K \cdot \mathbf{I}_n \oplus K \cdot (J_n - \mathbf{I}_n)$$

as K-algebras, where $\mathbf{I}_n$ is the $n \times n$ identity matrix and J_n is the $n \times n$ all-1-matrix. Of course, we get the same result if we replace S_n by any group G acting doubly transitively on Ω. $\blacklozenge$

Sometimes the following version of Lemma 1.2.15 is useful.

Corollary 1.2.19 *Let H_1, H_2 be subgroups of G. Thus $G/H_1, G/H_2$ are transitive G-sets (see Remark 1.2.2). Assume that $\varphi\colon K(G/H_1) \to K(G/H_2)$ is a KG-homomorphism. Let $\mathcal{O}_1, \ldots, \mathcal{O}_r$ be the orbits of H_1 on G/H_2. Then*

$$\varphi(1H_1) = \sum_{i=1}^{r} a_i \mathcal{O}_i^+ \quad with \quad a_i \in K,$$

where $\mathcal{O}_i^+ = \sum_{w \in \mathcal{O}_i} w \in K(G/H_2)$ and φ is completely determined by the coefficients a_i. The map

$$\varphi \mapsto (a_1, \ldots, a_r), \qquad \mathrm{Hom}_{KG}(K(G/H_1), K(G/H_2)) \to K^r,$$

is an isomorphism of K-modules.

Observe that with the notation of Corollary 1.2.19 the orbits $\mathcal{O}_i$ of H_1 on G/H_2 correspond bijectively to the double cosets

$$H_1 g_i H_2 = \bigcup_{y \in \mathcal{O}_i} y \quad with \quad g_i H_2 \in \mathcal{O}_i.$$

The product of the Schur basis elements of $\mathrm{End}_{KG} KX$ can easily be described as follows,

Theorem 1.2.20 *For a finite G-set Ω let $\mathcal{O}_i$ $(1 \le i \le r)$ be the orbits of G on $\Omega \times \Omega$ and the $\theta_i = \theta_{\mathcal{O}_i}$ be the Schur basis elements of $E := \mathrm{End}_{KG} K\Omega$ (see Lemma 1.2.15). Then*

(a)

$$\theta_i \theta_j = \sum_{k=1}^{r} a_{ijk} \theta_k \quad with \quad a_{ijk} = |\mathcal{O}_i(x) \cap \mathcal{O}'_j(y)| \quad for \quad (x, y) \in \mathcal{O}_k.$$

*In particular the right hand side is independent of the choice of $(x, y) \in \mathcal{O}_k$. The a_{ijk} are often called the **intersection numbers** of the G-set Ω.*

(b) *If Ω is transitive and $\theta_{i'} := \theta_{\mathcal{O}'_i}$ (see Definition 1.2.10) then*

$$a_{ij'1} = |\mathcal{O}_i(\omega)| \cdot \delta_{i,j}$$

for any $\omega \in \Omega$, where $\mathcal{O}_1 = \{(\omega, \omega) \mid \omega \in \Omega\}$, so that $\theta_1 = 1_E$.

(c) *Assume again for simplicity that Ω is transitive. Then the K-linear map*

$$\zeta_1\colon E \to K \qquad with \qquad \zeta_1\colon \theta_{\mathcal{O}} \mapsto |\mathcal{O}(\omega)| \cdot 1_K \qquad for \ any \ \omega \in \Omega$$

*is a representation of E, usually called the **principal representation** or principal character of E.*

Proof. (a) For $1 \leq k \leq r$ let $[a_{xy}^k]_{x,y \in \Omega} \in K^{n \times n}$ with $n = |\Omega|$ the matrix of θ_k with respect to the basis Ω of $K\Omega$. Thus by Lemma 1.2.15

$$a_{x,u}^i = \begin{cases} 0 & \text{if } (x,u) \notin \mathcal{O}_i, \\ 1 & \text{if } (x,u) \in \mathcal{O}_i. \end{cases}$$

Then the entry at position (x,y) of the matrix of $\theta_i \theta_j$ with respect to the basis Ω is

$$\sum_{u \in \Omega} a_{xu}^i a_{uy}^j = \sum_{u \in \mathcal{O}_i(x)} a_{uy}^j = \sum_{u \in \mathcal{O}_i(x) \cap \mathcal{O}_j'(y)} 1 = |\mathcal{O}_i(x) \cap \mathcal{O}_j'(y)|$$

and this is constant for $(x,y) \in \mathcal{O}_k$. On the other hand the entry at position (x,y) of the matrix of $\sum_{k=1}^r a_{ijk} \theta_k$ is a_{ijk} if $(x,y) \in \mathcal{O}_k$.

(b) This follows immediately from the above since $\mathcal{O}_{j'}'(y) = \mathcal{O}_j(y)$.

(c) Obviously $\text{Inv}_G(K\Omega)$ is an E-submodule (usually called the principal submodule). Since Ω is supposed to be a transitive G-set, $\Omega^+ = \sum_{w \in \Omega} w$ is a K-basis of $\text{Inv}_G(K\Omega)$ by Lemma 1.2.4. Using Remark 1.2.12 we get

$$\theta_\mathcal{O}(\Omega^+) = \sum_{w \in \Omega} \mathcal{O}'(w)^+ = \sum_{w \in \Omega} \sum_{(w',w) \in \mathcal{O}} w' = |\mathcal{O}(w)| \Omega^+ \qquad \text{for} \qquad w \in \Omega$$

since Ω is transitive. □

Remark 1.2.21 Observe that using the notation of Theorem 1.2.20

$$\text{End}_{KG} K\Omega \to K^{r \times r} , \qquad \theta_i \mapsto A_i = [a_{ijk}]_{j,k=1,\dots,r} \quad (1 \leq i \leq r),$$

gives the regular matrix representation of $\text{End}_{KG} K\Omega$ with respect to the Schur basis. The matrices A_i are usually called the **intersection matrices** of Ω. If Ω is transitive then we may take $\mathcal{O}_1 = \Delta_\Omega$ and θ_1 is the identity element of $\text{End}_{KG} K\Omega$. Furthermore it follows from Theorem 1.2.20 that

$$a_{ijk} = a_{j'i'k'} \qquad \text{and} \qquad a_{i1k} = \delta_{i,k} \qquad \text{if} \quad \mathcal{O}_1 = \Delta_\Omega.$$

Corollary 1.2.22 *Let Ω be a transitive G-set and assume that $|\Omega| 1_K$ is invertible in K. Then $E := \text{End}_{KG} K\Omega$ is a symmetric algebra with symmetrizing trace $\tau \colon E \to K$, $\varphi \mapsto \text{trace}(\varphi)$. If $\mathcal{O}_1 := \Delta_\Omega, \mathcal{O}_2, \dots, \mathcal{O}_r$ are the orbits of G on $\Omega \times \Omega$ and $\theta_i = \theta_{\mathcal{O}_i}$ as above, then the contragredient basis to the Schur basis $(\theta_1, \dots, \theta_r)$ is given by*

$$(\theta_1^\tau, \dots, \theta_r^\tau) \qquad \text{with} \qquad \theta_i^\tau = \frac{1}{|\mathcal{O}_i|} \theta_{i'} \qquad \text{where} \quad \theta_{i'} := \theta_{\mathcal{O}_i'}.$$

Proof. It is obvious that $\text{trace}(\varphi\psi) = \text{trace}(\psi\varphi)$ for $\varphi, \psi \in E$. Observe that $\text{trace}(\theta_i) = \delta_{i,1} |\Omega|$. Hence

$$\text{trace}(\theta_i \theta_{j'}) = a_{ij'1} |\Omega| = |\Omega| |\mathcal{O}_i(w)| \delta_{i,j} = |\mathcal{O}_i| \delta_{i,j}$$

by Theorem 1.2.20 and Remark 1.2.12. □

Example 1.2.23 We continue with Example 1.2.17 considering the alternating group $G = A_4$ acting on six points. The intersection numbers can be read off from Table 1.1.

Table 1.1. $[\mathcal{O}_i(j)]_{i,j=1,2,3,4}$

	j				
	1	2	3	4	
$\mathcal{O}_1(j)$	$\{1\}$	$\{2\}$	$\{3\}$	$\{4\}$	$(1,1) \in \mathcal{O}_1 = \mathcal{O}_1'$
$\mathcal{O}_2(j)$	$\{2\}$	$\{1\}$	$\{5\}$	$\{6\}$	$(1,2) \in \mathcal{O}_2 = \mathcal{O}_2'$
$\mathcal{O}_3(j)$	$\{3,5\}$	$\{3,5\}$	$\{4,6\}$	$\{1,2\}$	$(1,3) \in \mathcal{O}_3 = \mathcal{O}_4'$
$\mathcal{O}_4(j)$	$\{4,6\}$	$\{4,6\}$	$\{1,2\}$	$\{3,5\}$	$(1,4) \in \mathcal{O}_4 = \mathcal{O}_3'$

Let $\mathbf{a}_i = [a_{i,j,k}]_{j,k=1,\dots,4}$, so that the regular representation of E (with respect to the standard basis) is given by $\theta_i \mapsto \mathbf{a}_i$ $(1 \le i \le 4)$. Then we have $\mathbf{a}_1 = \mathbf{I}_4$ and

$$\mathbf{a}_2 = \begin{bmatrix} 0 & 1 & 0 & 0 \\ 1 & 0 & 0 & 0 \\ 0 & 0 & 1 & 0 \\ 0 & 0 & 0 & 1 \end{bmatrix}, \ \mathbf{a}_3 = \begin{bmatrix} 0 & 0 & 1 & 0 \\ 0 & 0 & 1 & 0 \\ 0 & 0 & 0 & 2 \\ 2 & 2 & 0 & 0 \end{bmatrix}, \ \mathbf{a}_4 = \begin{bmatrix} 0 & 1 & 0 & 0 \\ 0 & 0 & 0 & 1 \\ 2 & 2 & 0 & 0 \\ 0 & 0 & 2 & 0 \end{bmatrix}.$$

If K is a field with char $K \ne 2$ containing a primitive third root of unity ϵ, we find that the the K-linearly independent vectors

$$[1,1,2,2]^{\mathrm{T}}, \ [1,-1,0,0]^{\mathrm{T}}, \ [1,1,2\epsilon,2\epsilon^2]^{\mathrm{T}}, \ [1,1,2\epsilon^2,2\epsilon]^{\mathrm{T}}$$

are common eigenvectors of the matrices $\mathbf{a}_i$, which can therefore simultaneously be diagonalized. Thus in this case E has four irreducible representations ζ_i, all of degree one, which are displayed in the following matrix $[\zeta_i(\theta_j)]$. This table is also called the "character table" of E.

	θ_1	θ_2	θ_3	θ_4
ζ_1	1	1	2	2
ζ_2	1	-1	0	0
ζ_3	1	0	2ϵ	$2\epsilon^2$
ζ_4	1	1	$2\epsilon^2$	2ϵ

Observe that the rows of this table coincide with the common eigenvectors of the $\mathbf{a}_i$.   ◆

Example 1.2.24 If G acts regularly on Ω, i.e. $\Omega \cong_G G/\{1\}$, then the orbits of G on $\Omega \times \Omega$ are

$$\mathcal{O}_g = \{(x, gx) \mid x \in \Omega\} \quad \text{for} \quad g \in G$$

and we get as intersection numbers for $g_1, g_2, g_3 \in G$

$$a_{g_1,g_2,g_3} = \begin{cases} 1 & \text{if} \quad g_2 g_1 = g_3, \\ 0 & \text{else.} \end{cases}$$

Furthermore

$$KG \to \mathrm{End}_{KG} K\Omega \qquad \sum_{g \in G} a_g g \mapsto \sum_{g \in G} a_g \theta_{\mathcal{O}_g} \quad (a_g \in K)$$

is an anti-isomorphism of K-algebras.   ◆

Example 1.2.25 The symmetric group S_n acts for $k \leq n$ in a natural way transitively on the set $\binom{\Omega}{k}$ of k-element subsets of the power set of $\Omega = \{1, \ldots, n\}$. Thus one obtains a permutation representation of S_n of degree $\binom{n}{k}$. It is easily seen that the action on $\binom{\Omega}{k}$ is equivalent to the action on $\binom{\Omega}{n-k}$, so it is enough to consider the case $k \leq \frac{n}{2}$. Since S_n is n-fold transitive on Ω it is clear that for $1 \leq k \leq l \leq \frac{n}{2}$ we obtain the following orbits of S_n on $Y = \binom{\Omega}{k} \times \binom{\Omega}{l}$:

$$\mathcal{O}_t = \{(A, B) \mid A \in \binom{\Omega}{k}, B \in \binom{\Omega}{l}, \ |A \cap B| = k - t + 1\} \qquad \text{for} \quad 1 \leq t \leq k + 1.$$

Hence

$$\mathrm{Hom}_{K S_n}(K\binom{\Omega}{l}, K\binom{\Omega}{k}) \cong K^{k+1} \qquad (1 \leq k \leq l \leq \frac{n}{2}).$$

In particular the rank of the S_n-set $\binom{\Omega}{k}$ is $k + 1$. Also all orbits on $\binom{\Omega}{k} \times \binom{\Omega}{k}$ are self-paired.

We now consider the case $k = l = 2$ and $n \geq 4$. Thus we have three orbits, $\mathcal{O}_1 = \Delta$, the diagonal, $\mathcal{O}_2 = S_n \cdot (\{1,2\}, \{1,3\})$ and $\mathcal{O}_3 = S_n \cdot (\{1,2\}, \{3,4\})$, and an easy computation shows that the intersection matrices $\mathbf{a}_i = [a_{ijk}]_{j,k=0,1,2}$ are as follows: $\mathbf{a}_1 = \mathbf{I}_3$ and

$$\mathbf{a}_2 = \begin{bmatrix} 0 & 1 & 0 \\ 2n-4 & n-2 & 4 \\ 0 & n-3 & 2n-8 \end{bmatrix}, \quad \mathbf{a}_3 = \begin{bmatrix} 0 & 0 & 1 \\ 0 & n-3 & 2n-8 \\ \binom{n-2}{2} & \binom{n-3}{2} & \binom{n-4}{2} \end{bmatrix}.$$

If K is a field with char $K \nmid (n-2)\binom{n}{2}$ we find that the $\mathbf{a}_i$ have the following common eigenvectors: $[1, 2n-4, \binom{n-2}{2}]^T$, $[1, n-4, -n+3]^T$, $[1, -2, 1]^T$. So the matrices $\mathbf{a}_i$ can be simultaneously diagonalized in this case and $E := \mathrm{End}_{K S_n} K\binom{\Omega}{2}$ has three irreducible representations $\zeta_1, \zeta_2, \zeta_3$, all of degree one with values

	θ_1	θ_2	θ_3
ζ_1	1	$2n-4$	$\binom{n-2}{2}$
ζ_2	1	$n-4$	$-n+3$
ζ_3	1	-2	1

where $\theta_i = \theta_{\mathcal{O}_i}$. Is is easily seen that we would get the same result in the case $k = l = 2$ if we replace S_n by another group acting four-fold transitively on Ω. ◆

Example 1.2.26 In GAP it is very easy to calculate the intersection matrices of a given permutation group. We demonstrate this with the sporadic simple Higman–Sims group G, usually denoted by HS (see [38]) in its well known action on $\Omega := \{1, \ldots, 100\}$. We take this permutation representation from the library of primitive groups in GAP and compute the orbits of G on $\Omega \times \Omega$:

```
gap> G := PrimitiveGroup( 100, 3 );;
gap> orb := Orbits( G , Tuples([1..100],2) , OnPairs);;
gap> List( orb , Length );
```

The last command returns the lengths of the orbits $\mathcal{O}_i := \text{orb[i]}$ of G on $\Omega \times \Omega$; they are 100, 7700 and 2200. Since $\mathcal{O}_1 = \Delta$ we obtain the subdegrees by

```
gap> List([1,2,3], i -> Length( Filtered( orb[i], p -> p[1] = 1) ) );
[ 1, 77, 22 ]
```

The following loop calculates all the intersection matrices `a[i]` for $i = 1, 2, 3$:

```
gap> a := [];; x := 1;; y := 1;;
gap> for i in [1..Length(orb)] do
>       a[i] := [];        # a[i] will be the i-th intersection matrix
>       for j in [1..Length(orb)] do
>         a[i][j]:= [];       # a[i][j] will be the j-th row of a[i]
>         for k in [1..Length(orb)] do
>           x:=orb[k][1][1]; y:=orb[k][1][2];    # [x,y] in orb[k]
>           a[i][j][k] := Size( Intersection (
>              Filtered([1..100] , z -> [x,z] in orb[i]),
>              Filtered([1..100] , z -> [y,z] in orb[j]) ) );
>         od;
>       od;
>     od;
```

The result is:

$$a[1] = \mathbf{I}_3, \quad a[2] = \begin{bmatrix} 0 & 1 & 0 \\ 77 & 60 & 56 \\ 0 & 16 & 21 \end{bmatrix}, \quad a[3] = \begin{bmatrix} 0 & 0 & 1 \\ 0 & 16 & 21 \\ 22 & 6 & 0 \end{bmatrix}.$$

Finally we calculate the common eigenspaces of the intersection matrices. Since GAP calculates eigenrowspaces we transpose the matrices and use the GAP-commands

```
gap> Display( Eigenspaces( Rationals, TransposedMat(a[2]) ));;
gap> Display( Eigenspaces( Rationals, TransposedMat(a[3]) ));;
```

We find that the "character table" $[\zeta_i(\theta_j)]_{1 \leq i,j \leq 3}$ over $\mathbb{Q}$ is

$$[\zeta_i(\theta_j)] = \begin{bmatrix} 1 & 77 & 22 \\ 1 & 7 & -8 \\ 1 & -3 & 2 \end{bmatrix}.$$

◆

With the notation of Theorem 1.2.20, $(\Omega, \{\mathcal{O}_1, \ldots, \mathcal{O}_r\})$ is an example of a *coherent configuration* as introduced by D. Higman ([81]) or an *association scheme* (see [6] and [173]). Some of the results in this section have been generalized to this more general setting.

Exercises

Exercise 1.2.1 Let Ω be a transitive G-set with $H = \text{Stab}(\omega)$ for some $\omega \in \Omega$ and let $\delta \colon G \to \text{GL}(K\Omega)$ be the corresponding representation. Show that

$$\ker \delta = \bigcap_{g \in G} H^g,$$

the largest normal subgroup of G contained in H.

Exercise 1.2.2 Let Ω be a transitive G-set. Show that there is exactly one KG-module epimorphism

$$\lambda_\Omega \colon K\Omega \to K$$

where K is considered as a KG-module with trivial G-action. Show also that for a KG-module isomorphism $\varphi \colon K\Omega_1 \to K\Omega_2$ with transitive G-sets Ω_1, Ω_2 one has

$$\lambda_{\Omega_1} = \lambda_{\Omega_2} \circ \varphi.$$

Exercise 1.2.3 Using the assumptions and the notation of Corollary 1.2.19 show that

$$1_K = \sum_{i=1}^{m} a_i |\mathcal{O}_i|$$

is a necessary condition for φ to be an isomorphism of KG-modules.

Exercise 1.2.4 (a) Let V be the natural permutation module of $G := S_n$ over a field K for $n \geq 3$. Show that V, $\mathrm{Inv}^G(V)$, $\mathrm{Inv}_G(V)$ and $\{0\}$ are the only submodules of V.
Hint: For $v = \sum_{\omega \in \{1,\ldots,n\}} \alpha_\omega \omega \in V$ let $l(v) = |\{\omega \in \{1,\ldots,n\} \mid \alpha_\omega \neq 0\}|$. For a submodule $U \leq_{K S_n} V$ consider an element $0 \neq v \in U$ with $l(v)$ minimal.

(b) Let V be the natural permutation module of the alternating group $G := A_n$ over a field K for $n \geq 4$. Prove that

(i)   if $n \geq 5$ or

(ii)   if char $K \neq 2$ or

(iii)   if char $K = 2$ and $\alpha^2 + \alpha + 1 \neq 0$ for all $\alpha \in K$,

then V, $\mathrm{Inv}^G(V)$, $\mathrm{Inv}_G(V)$ and $\{0\}$ are the only submodules of V.

Exercise 1.2.5 Let V be the natural permutation module of S_n over K for $n > 1$. Show that as K-algebras $\mathrm{End}_{K S_n} V \cong K[X]/(X \cdot (X - n + 1))$, where X denotes an indeterminate.

The following exercise gives an interpretation of the center of a group algebra as a Hecke algebra.

Exercise 1.2.6 Let G be a finite group, $G \times G$ be the direct product, K be a field and $\Delta_G := \{(g,g) \mid g \in G\}$ be the (diagonal) subgroup of $G \times G$. Show that the Hecke algebra $\mathrm{End}_{K(G \times G)} K(G \times G)/\Delta_G$ of the $G \times G$-set $(G \times G)/\Delta_G$ is isomorphic to $\mathbf{Z}(KG)$.

Exercise 1.2.7 The sporadic simple Janko group $G := J_1$ (see [38]) has a maximal subgroup of index 266 and thus acts primitively on $\Omega := \{1, \ldots, 266\}$. Show that G has five orbits on $\Omega \times \Omega$ of lengths 1, 11, 12, 110, 132. Compute the intersection matrices and show that

$$E := \mathrm{End}_{\mathbb{Q}G}\, \mathbb{Q}\Omega \cong \mathbb{Q} \oplus \mathbb{Q} \oplus \mathbb{Q} \oplus \mathbb{Q}(\sqrt{5}).$$

Compute all irreducible representations of E over $\mathbb{C}$.

1.3   Simple modules, the "Meataxe"

In this section A is assumed to be a K-algebra for some commutative ring K (which is often a field). Standard examples we have in mind are the group algebra $A = KG$ of a finite group G or $A = \mathrm{End}_{KG} V$ for a KG-module V.

We recall the Theorem of Jordan and Hölder (see [94], pp. 108, 109), which says that if an A-module V has a composition series

$$\{0\} = V_0 <_A V_1 <_A \cdots <_A V_n = V,$$

then any two composition series of V have the same length $n = l(V)$, which is called the **length** of V, and the composition factors which are, by definition, the simple A-modules V_i/V_{i-1} are uniquely determined up to isomorphism and ordering.

Definition 1.3.1 An A-module which has a composition series is said to have **finite length**, and A has **finite length** if the regular module $_A A$ has finite length. The length of a composition series of $_A A$ is also called the length $l(A)$ of A.

Of course, if A is a K-algebra of finite dimension over a field K then A and any finitely generated A-module V have finite length. For matrix representations of a finite group G over a field K the Jordan–Hölder theorem means that any representation $\delta \colon G \to \mathrm{GL}_n(K)$ is equivalent to one of the form

$$g \mapsto \begin{bmatrix} \delta_1(g) & * & & * \\ 0 & \delta_2(g) & & \\ & & \ddots & * \\ 0 & & & \delta_m(g) \end{bmatrix}$$

with irreducible matrix representations $\delta_i \colon G \to \mathrm{GL}_{m_i}(K)$ which are uniquely determined up to equivalence and ordering and which are called the **irreducible constituents** of δ.

Thus the simple modules or irreducible representations can be viewed as the building blocks of all others. It is an important property, that any algebra A of finite length has only a finite number of simple modules up to isomorphism because any such module is a composition factor of A. This follows from the first part of the following lemma.

Lemma 1.3.2 *Let V be an A-module. Then the following hold.*

(a) *If V is simple then $V \cong A/M$, where M is a maximal left ideal of A.*

(b) *V is simple if and only if $V = A \cdot v$ for all $v \in V$ with $v \neq 0$.*

(c) **(Schur's lemma)** *If V is simple then $\mathrm{End}_A V$ is a division ring. If K is a field, $\dim_K V < \infty$ and V is absolutely simple then $\mathrm{End}_A V = K \cdot \mathrm{id}_V$.*

Proof. (a) and (b) If we choose $v \in V$ the map $A \to V$, $a \mapsto a \cdot v$ defines an A-module homomorphism. Since V is simple, the image is $\{0\}$ or V. The second alternative must hold for any $v \neq 0$, because $v = 1 \cdot v$; recall our convention that any ring contains a one and any module is "unital." So (b) holds and (a) follows from the homomorphism theorem.

(c) Since the kernel and the image of any A-module homomorphism is a submodule, any non-zero A-endomorphism of a simple module V must be an isomorphism hence invertible in $\operatorname{End}_A V$.

If V is any A-module then K embeds into $\operatorname{End}_A V$ via $\alpha \mapsto \alpha \operatorname{id}_V$. If K is algebraically closed then the characteristic polynomial of any $\varphi \in \operatorname{End}_A V$ has a root $\alpha \in K$. For such an α the endomorphism $\varphi - \alpha \operatorname{id}_V \in \operatorname{End}_A V$ is not a unit, hence it must be zero if V is simple. Thus we have in this case $\operatorname{End}_A V = K \operatorname{id}_V$. For an arbitrary field K one may use Exercise 1.3.2. Alternatively one may argue as follows: the K-dimension of $\operatorname{End}_A V$ is the K-dimension of the space of solutions of the following system of homogeneous linear equations over K:

$$\mathbf{a}_k [x_{i,j}]_{i,j=1}^n - [x_{i,j}]_{i,j=1}^n \mathbf{a}_k = 0 \qquad (k = 1, \ldots, r),$$

where $n = \dim_K V$, $A = \langle a_1, \ldots, a_r \rangle_{\text{alg}}$ and $\mathbf{a}_k = [\delta_V(a_k)]_B \in K^{n \times n}$ for some K-basis B of V. This dimension does not change if we extend the field K to an algebraic closure, so it must be one if V is absolutely simple. $\qquad \square$

Part (b) of the preceding lemma can be used in principle to test whether or not a finite module is simple, assuming that A is a K-algebra for which one has a finite set of algebra generators and their action on V. The cyclic modules $A \cdot v$ can be computed using the following.

Spinning algorithm
Assume that K is a field and A is a finite dimensional K-algebra A with algebra generators $(a_1, \ldots, a_r)$ and that V is an A-module (given by the action of these generators on V).

Input: $0 \neq v \in V$ and algebra generators $(a_1, \ldots, a_r)$ of A
Output: a K-basis B for $A \cdot v$
Initialize: $B = (v)$
for b in B **do**
    for i from 1 to r **do**
        $w := a_i \cdot b$
        if w is not contained in $\langle B \rangle_K$ **then**
            append w to the list B
        end if
    end for
end for
return B

The algorithm terminates, since V was assumed to be finite dimensional. By construction B is linearly independent over K and $B \subseteq A \cdot v$. Furthermore

$\langle B \rangle_K$ is invariant under the action of the generators of A and hence under A. Thus $A \cdot v = \langle B \rangle_K$.

The process of calculating $A \cdot v$ using this algorithm is usually referred to as "*spinning up the vector v.*" A simple-minded test for simplicity for a finite A-module V would be to spin up every non-zero vector v of V and to check whether or not $A \cdot v = V$. Of course it would be sufficient to spin up just one vector v in each one-dimensional subspace. But one can do much better.

Theorem 1.3.3 (Norton's irreducibility criterion) *Let A be a K-algebra, K a field, and V an A-module which is finite dimensional over K. If $\ker_V(c) \neq 0$ for some $c \in A$ then V is simple if and only if*

(a) $V = A \cdot v$ *for* **all** $v \in \ker_V(c) \setminus \{0\}$ *and*

(b) $V^\star = x \cdot A$ *for* **some** $x \in \ker_{V^\star}(c)$.

The result is a simple consequence of the following observation from linear algebra.

Lemma 1.3.4 *Let V be a finite dimensional K-vector space and let $W \leq V$ be a φ-invariant subspace for some $\varphi \in \mathrm{End}_K V$. Then*

$$W \cap \ker(\varphi) = \{0\} \quad \text{implies} \quad \ker(\varphi^{\mathrm{T}}) \leq W^\circ = \{x \in V^\star \mid x|_W = 0\}.$$

Proof of Lemma 1.3.4. By definition of φ^{T} we have

$$\ker(\varphi^{\mathrm{T}}) = \{x \in V^\star \mid 0 = \varphi^{\mathrm{T}}(x) = x \circ \varphi\} = (\mathrm{im}(\varphi))^\circ. \tag{1.12}$$

If $W \cap \ker(\varphi) = \{0\}$, then the restriction of φ to W is injective, hence surjective, since $\dim_K V < \infty$. So $W \leq \mathrm{Im}(\varphi)$ and by duality $(\mathrm{Im}(\varphi))^\circ \leq W^\circ$, so by (1.12) one has $\ker(\varphi^{\mathrm{T}}) \leq W^\circ$.   □

Proof of Theorem 1.3.3. The necessity of the conditions (a) and (b) follows from Lemma 1.3.2 and the fact that $V^\star$ is simple if and only if V is (see Theorem 1.1.34).

We now assume that (a) and (b) hold. Let $W <_A V$. We have to show that $W = \{0\}$. Because of (a) we have $\ker_V(c) \cap W = \{0\}$, hence by Lemma 1.3.4 $\ker_{V^\star}(c) \leq W^\circ \leq_A V^\star$. From (b) we conclude that $W^\circ = V^\star$, hence $W = \{0\}$ by duality.   □

We include a rather trivial example:

Example 1.3.5 As in Exercise 1.1.4 we consider the matrix representation δ of $S_3 = \langle a, b \rangle$ with $a = (1,2)$ and $b = (2,3)$ over an arbitrary field K given by

$$\delta : a \mapsto \begin{bmatrix} -1 & 1 \\ 0 & 1 \end{bmatrix}, \quad b \mapsto \begin{bmatrix} 1 & 0 \\ 1 & -1 \end{bmatrix}.$$

So $V = K^{2 \times 1}$ and V^* may be identified with $K^{1 \times 2}$ on which $A = \delta(K\,S_3)$ acts from the right. It is obvious that

$$\ker_V(1 + a) = \langle \begin{bmatrix} 1 \\ 0 \end{bmatrix} \rangle_K \quad \text{and} \quad A \cdot \begin{bmatrix} 1 \\ 0 \end{bmatrix} = \langle \begin{bmatrix} -1 \\ 0 \end{bmatrix}, \begin{bmatrix} 1 \\ 1 \end{bmatrix} \rangle_K = V,$$

$$\ker_{V^*}(1 + a) = \langle [\, 2, -1 \,] \rangle_K \quad \text{and} \quad [\, 2, -1 \,] \cdot A = V_1 := \langle [\, -2, 1 \,], [\, 1, 1 \,] \rangle_K,$$

and $V_1 = V^*$ if and only if char $K \neq 3$. Thus, δ is (absolutely) irreducible if char $K \neq 3$, whereas for char $K = 3$ we have found a proper right submodule $U^\circ := \langle [\, 1, 1 \,] \rangle_K \leq V^*$. Obviously the corresponding submodule $U \leq V$ is $\langle [\, 1, -1 \,]^T \rangle_K$, on which a and b and hence all of S_3 act trivially. It is easily checked that $\mathrm{End}_{K\,S_3} V \cong K$ in all cases. This shows that the converse of Schur's lemma is not true, namely that one cannot conclude from $\mathrm{End}_A V \cong K$ for a K-algebra A that V is simple. ◆

Of course, Theorem 1.3.3 is only useful in the case that $\ker_V(c)$ is a proper subspace and it works better the smaller the dimension of this non-zero subspace is, because we have to spin up one vector v of each of the one-dimensional subspaces of $\ker_V(c)$. In the ideal case that $\dim_K(\ker_V(c)) = 1$ we have to spin up just one vector of $\ker_V(c)$, and in this case the criterion can even be used for infinite fields. Of course the problem remains of how to find an element $c \in V$ with $\dim_K(\ker_V(c)) = 1$. In fact, we will now investigate in which cases such an element does exist in A provided that V is a simple A-module.

Lemma 1.3.6 *Let V be a simple A-module, $D := \mathrm{End}_A V$ and $W \leq_D V$. Assume that W is finitely generated as a D-module. Then*

(a) $W = \ker_V(\mathrm{ann}_A W)$;

(b) *for any $\psi \in \mathrm{Hom}_D(W, V)$ there is an $a \in A$ with $\psi(w) = a \cdot w$ for all $w \in W$;*

(c) $\delta_V(A) = \mathrm{End}_D V$ *if $\dim_D V < \infty$.*

Observe that by Schur's lemma D is a division ring, so that V is a D-vector space. Assertion (b) is often cited as "Jacobson's density lemma" and (c) is usually called the "double centralizer property." It is clear that the hypothesis on W is always fulfilled, if K is a field and V is finite dimensional as a K-vector space, because K is embedded in D via $\alpha \mapsto \alpha \, \mathrm{id}_V$ and $\dim_K V = \dim_D V \cdot [D : K]$.

Proof. (a) Obviously $W \subseteq \ker_V(\mathrm{ann}_A W)$ holds in general. To prove the converse we use induction on $\dim_D W$. If $W = \{0\}$ then $\mathrm{ann}_A W = A$ and $\ker_V(\mathrm{ann}_A W) = \ker_V A = \{0\}$. Let $W \neq \{0\}$ and write $W = W' + D\,w_1$, with $\dim_D W' < \dim_D W$. Assume that $v \in \ker_V(\mathrm{ann}_A W)$; we have to show that $v \in W$. The assumption means that for any $a \in A$ with $a \cdot W = 0$ we have $a \cdot v = 0$. Hence

$$a \cdot W' = \{0\} \quad \text{and} \quad a \cdot w_1 = 0 \quad \text{implies} \quad a \cdot v = 0.$$

This means that the map

$$\varphi\colon \operatorname{ann}_A W' \cdot w_1 \to V, \quad a \cdot w_1 \mapsto a \cdot v$$

is a well-defined A-linear map between A-modules. Since V is simple the submodule $\operatorname{ann}_A W' \cdot w_1$ can only be $\{0\}$ or V. In the first case we also have $\operatorname{ann}_A W' \cdot v = \{0\}$, so $v \in \ker_V(\operatorname{ann}_A W') = W'$ by the induction hypothesis, hence $v \in W$ as asserted. In the second case $\operatorname{ann}_A W' \cdot w_1 = V$ and so $\varphi \in D$ and consequently $a \cdot (\varphi(w_1) - v) = \varphi(aw_1) - a \cdot v = 0$ for all $a \in \operatorname{ann}_A W'$, hence $\varphi(w_1) - v \in \ker_V(\operatorname{ann}_A W') = W'$ and $v \in W' + D w_1 = W$.

(b) Again we use induction on $\dim_D W$. For $W = \{0\}$ the result is trivial. Let $\{0\} \neq W = W' + D w_1$ be as above with $\dim_D W' < \dim_D W$. Let $\psi \in \operatorname{Hom}_D(W, V)$. By induction there is $a' \in A$ with $\psi(w') = a' \cdot w'$ for all $w' \in W'$. Since $w_1 \notin W' = \ker_V \operatorname{ann}_A W'$ by (a) we have $\{0\} \neq \operatorname{ann}_A W' \cdot w_1 \leq_A V$, so

$$\operatorname{ann}_A W' \cdot w_1 = V$$

because V is simple. Hence there is an $a'' \in \operatorname{ann}_A W'$ with $\psi(w_1) - a' \cdot w_1 = a'' \cdot w_1$. We now can write an arbitrary $w \in W$ in the form $w = w' + \varphi w_1$ with $w' \in W'$ and $\varphi \in D$ to get

$$\psi(w) = \psi(w') + \varphi(\psi(w_1))$$
$$= (a' + a'') \cdot w' + \varphi((a' + a'') \cdot w_1) = (a' + a'') \cdot w.$$

Thus $a' + a''$ is the required element a in the assertion.

(c) This follows immediately from (b), taking $W := V$ and recalling the definition of δ_V. $\square$

Corollary 1.3.7 *Let A be a K-algebra of finite length and $\delta\colon A \to \operatorname{End}_K V$ be an irreducible representation. Then*

(a) *$\delta(A) \cong (D^{op})^{n \times n}$ with $D := \operatorname{End}_A V$, a division ring, and $n \leq \mathrm{l}(A)$;*

(b) *Assume that K is a field. Then δ is absolutely irreducible if and only if δ is surjective and this holds if and only if $\operatorname{End}_A V = K \cdot \operatorname{id}_V$.*

Proof.   (a) Because of Lemma 1.3.6 we have to show that $\dim_D V < \infty$. Assume that $v_1, \ldots, v_n$ in V are linearly independent over D. Then

$$\{0\} < D v_1 < \cdots < D v_1 + \cdots + D v_n \leq V.$$

From Lemma 1.3.6 (a) we get

$$A >_A \operatorname{ann}_A D v_1 >_A \cdots >_A \operatorname{ann}_A(D v_1 + \cdots + D v_n) \geq_A \{0\}.$$

Hence the Jordan–Hölder theorem implies $n \leq \mathrm{l}(A)$. Now let $n = \dim_D V$ and $B = (v_1, \ldots, v_n)$ be a D-basis of V. For any $\varphi \in \operatorname{End}_D V$ there are $d_{i,j}(\varphi) \in$

D , $1 \le i, j \le n$ uniquely determined with $\varphi(v_j) = \sum_{i=1}^n d_{i,j}(\varphi) v_i$. If also $\psi \in \mathrm{End}_D V$ we find

$$\psi \circ \varphi(v_j) = \sum_{k=1}^n (\sum_{i=1}^n d_{i,j}(\varphi) d_{k,i}(\psi)) v_k,$$

from which it is apparent that

$$\varphi \mapsto [d_{i,j}(\varphi)] \quad \text{gives an isomorphism} \quad \mathrm{End}_D V \to (D^{\mathrm{op}})^{n \times n}.$$

(b) If δ is absolutely irreducible, then $\mathrm{End}_A V = K \, \mathrm{id}_V$ by Schur's lemma (Lemma 1.3.2). Conversely, if $\mathrm{End}_A V = K \, \mathrm{id}_V$ and δ is irreducible, then by Lemma 1.3.6 δ is surjective. But then $^L\delta \colon LA \to \mathrm{End}_L LV \cong L \otimes_K \mathrm{End}_K V$ is also surjective for any field $L \supseteq K$. But $\mathrm{GL}(LV)$ acts transitively on the non-zero vectors of LV, so there is no non-trivial LA-submodule in LV and hence V is absolutely simple. □

Theorem 1.3.8 *Let A be a K-algebra for a field K, and let V be a simple A-module, with $\dim_K V < \infty$ and $D = \mathrm{End}_A V$. Then $\dim_K \ker_V(a)$ is divisible by $[D : K]$ for all elements $a \in A$. Moreover, there are elements $a \in A$ for which $\dim_K \ker_V(a) = [D : K]$.*

Proof. We have that $\ker_V(a)$ is a D-subspace of V and

$$\dim_K \ker_V(a) = \dim_D \ker_V(a) \cdot [D : K].$$

For the second assertion, consider an endomorphism $\psi \in \mathrm{End}_D V$ with D-rank $\dim_D V - 1$, i.e. with $\dim_D(\ker_V \psi) = 1$; then $\dim_K(\ker_V \psi) = [D : K]$. By Lemma 1.3.6(c) there is an element $a \in A$ with $\delta(a) = \psi$. □

The above results yield an algorithm for testing whether or not a given A-module V is simple if A is an algebra over a finite field K and $\dim_K V < \infty$. This algorithm was originally developed by R. A. Parker and J. Thackray at the end of the 1970s. One of its first major applications was the existence proof of the Janko group J_4 by constructing a representation of J_4 of degree 112 over the field of two elements. Nowadays it is a standard tool in representation theory. The algorithm was named "Meataxe" by Parker ([142]) and in its basic form it runs as follows.

The Meataxe algorithm

We assume that K is a field and $\delta \colon A \to K^{n \times n}$ is a matrix representation with representation module $V := K^n$.

Input: An n-dimensional matrix representation δ of A in terms of matrices $(\mathbf{a}_1 := \delta(a_1), \ldots, \mathbf{a}_k := \delta(a_k))$ for a generating system $(a_1, \ldots, a_k)$ of A.

Output: Either the information that δ is irreducible, or matrix representations of A on an A-submodule W and on the factor module V/W given by matrices for the generating system $(a_1, \ldots, a_k)$ of A.

(1) Choose an element a in A (uniformly at random).

(2) Determine a K-basis of the (column) null space $\ker(\boldsymbol{\delta}(a))$.

(3) If $\ker(\boldsymbol{\delta}(a))$ is zero go back to step (1). Otherwise, for all non-zero vectors $v \in \ker(\boldsymbol{\delta}(a))$, up to K-scalar multiples,

  (a) run the spinning algorithm with input v and $(\mathbf{a}_1, \ldots, \mathbf{a}_k)$ and let $B :=$ $(b_1, \ldots, b_m)$ be the output of the spinning algorithm.

  (b) If $m < n$, extend B to a K-basis $(b_1, \ldots, b_m, b_{m+1}, \ldots, b_n)$ of K^n. For $i = 1, \ldots, k$ compute the matrices $\mathbf{a}_i'$ and $\mathbf{a}_i''$ of the actions of $\mathbf{a}_i$ on $\langle B \rangle_K$ and $K^n / \langle B \rangle_K$ with respect to the bases B and $(b_{m+1} + \langle B \rangle_K, \ldots, b_n + \langle B \rangle_K)$, respectively. Return $((\mathbf{a}_i')_{i=1}^k, (\mathbf{a}_i'')_{i=1}^k)$.

(4) If $m = n$ for all non-zero $v \in \ker(\boldsymbol{\delta}(a))$ up to K-scalar multiples, then transpose $\boldsymbol{\delta}(a)$ and compute one non-zero vector v such that $\boldsymbol{\delta}(a)^{\mathrm{T}} v = 0$.

  (a) Run the spinning algorithm with transposed input, namely v and $(\mathbf{a}_1^{\mathrm{T}}, \ldots, \mathbf{a}_k^{\mathrm{T}})$ and let $B = (b_1, \ldots, b_m)$ be the output of the spinning algorithm.

  (b) If $m < n$, then extend B to a K-basis $(b_1, \ldots, b_m, b_{m+1}, \ldots, b_n)$ of K^n. For $i = 1, \ldots, k$ compute the matrices $\mathbf{a}_i'$ and $\mathbf{a}_i''$ of the actions of $\mathbf{a}_i^{\mathrm{T}}$ on $K^n / \langle B \rangle_K$ and $\langle B \rangle_K$ with respect to the bases $(b_{m+1} + \langle B \rangle_K, \ldots, b_n + \langle B \rangle_K)$ and B, respectively. Return $((\mathbf{a}_i'^{\mathrm{T}})_{i=1}^k, (\mathbf{a}_i''^{\mathrm{T}})_{i=1}^k)$.

(5) If $m = n$ return the answer "$\boldsymbol{\delta}$ is irreducible."

Proof. If $(b_1, \ldots, b_m, b_{m+1}, \ldots, b_n)$ is as in step (3)(b) of the algorithm and X is the matrix with column vectors $b_1, \ldots, b_n$, then $X \in \mathrm{GL}_n(K)$ and

$$X^{-1}\mathbf{a}_i X = \left[\begin{array}{c|c} \mathbf{a}_i' & \star \\ \hline 0 & \mathbf{a}_i'' \end{array} \right] \quad \text{with} \quad \mathbf{a}_i' \in K^{m \times m}, \ \mathbf{a}_i'' \in K^{(n-m) \times (n-m)}$$

for $i = 1, \ldots, k$. Similarly, if $(b_1, \ldots, b_m, b_{m+1}, \ldots, b_n)$ is as in step (4)(b) and X is the matrix with column vectors $b_1, \ldots, b_n$, then $X \in \mathrm{GL}_n(K)$ and

$$X^{-1}\mathbf{a}_i^{\mathrm{T}} X = \left[\begin{array}{c|c} \mathbf{a}_i'' & \star \\ \hline 0 & \mathbf{a}_i' \end{array} \right] \quad \text{with} \quad \mathbf{a}_i'' \in K^{m \times m}, \ \mathbf{a}_i' \in K^{(n-m) \times (n-m)},$$

so that

$$X^{\mathrm{T}}\mathbf{a}_i (X^{\mathrm{T}})^{-1} = \left[\begin{array}{c|c} (\mathbf{a}_i'')^{\mathrm{T}} & 0 \\ \hline \star & (\mathbf{a}_i')^{\mathrm{T}} \end{array} \right] \quad \text{for} \quad i = 1, \ldots, k.$$

If in step (3)(a) we always get $m = n$ and also in step (4)(a) we obtain $m = n$ then by Theorem 1.3.3 $\boldsymbol{\delta}$ is irreducible, so the answer in step (5) is justified. $\qquad \square$

Of course, step (3) can be very time-consuming if $\dim_K \ker(\delta(a))$ is large and the matrix representation δ of A is irreducible, so the performance of the algorithm in this case depends very much on the likelihood of finding an element $\delta(a) \in \delta(A)$ in step one with a small kernel, ideally of dimension one. Using Lemma 1.3.6 one can compute the ratio of singular matrices in the algebra $\delta(A)$ when δ is irreducible and K is finite; one observes that this ratio is proportional to $\frac{1}{|K|}$ if δ is absolutely irreducible, so the Meataxe seems to work best for K a small sized field. In [143] Parker extends the idea of the Meataxe to characteristic zero. He uses a modified approach to find elements of small corank, and it turns out that this modified approach produces an amazing number of elements with a kernel of dimension one (see also Exercise 1.3.6).

There is a variant of the Meataxe algorithm due to Holt and Rees [86], which has a performance quite independent of the size of the finite field K that we are going to present. It is based on the following lemma.

Lemma 1.3.9 *Let A be an algebra over a field K and V be an A-module of finite dimension over K. Let f be an irreducible factor of the characteristic polynomial c_a of $\delta_V(a)$ for some $a \in A$. Assume that $\deg f = \dim_K \ker_V(f(a))$. Then V is a simple A-module if (and only if)*

(a) *$A \cdot v = V$ for some $0 \neq v \in \ker_V(f(a))$ and*

(b) *$x \cdot A = V^\star$ for some $0 \neq x \in \ker_{V^\star}(f(a))$.*

Proof. Recall from Example 1.1.24 that the irreducible factors of the characteristic polynomial c_a of $\delta_V(a)$ are in one-to-one correspondence with the composition factors of the $K[a]$-module $V_{K[a]}$. It follows from the same example that the assumption $\deg f = \dim_K \ker_V(f(a))$ implies that there is just one elementary divisor q which is a power of f or, equivalently, that there is exactly one simple $K[a]$-submodule of $V_{K[a]}$ which is isomorphic to the module $V_f = K[X]/(f)$, and this submodule is just $\ker_V(f(a))$. Recall that $V_q \cong K[X]/(q)$ has exactly one composition series and $\ker_V(f(a))$ as its unique minimal submodule. It also follows that $V_{K[a]}$ has exactly one maximal submodule M with $V/M \cong_{K[a]} \ker_V(f(a)$.

Now assume that V has a proper submodule $0 < W <_A V$ and that for some $0 \neq v \in \ker_V(f(a))$ we have $A \cdot v = V$. Then certainly $v \notin W$ and moreover $W_{K[a]}$ cannot have any composition factor isomorphic to V_f, for otherwise W would have to intersect V_q non-trivially and hence would have to contain its unique minimal submodule $\ker_V(f(a))$ and hence v. In particular we must have $W \subseteq M$ and hence $M^\circ \subseteq W^\circ \leq_A V^\star$. Observe that $f(a)V = M$, so that $\ker_{V^\star}(f(a)) = M^\circ$. Hence $x \cdot A \leq W^\circ < V^\star$ for all $0 \neq x \in \ker_{V^\star}(f(a))$. $\qquad\square$

Before formulating the algorithm we pause to include a very simple application.

Corollary 1.3.10 *Let G be a finite group acting on Ω with $|\Omega| = p$, a prime. Assume that K is a field with $\operatorname{char} K \neq p$ and let $V := K\Omega/(K\sum_{\omega \in \Omega} \omega)$.*

If G contains an element g of order p acting non-trivially on Ω and which is conjugate in G to all its powers $\neq 1$, then V is simple.

Proof. We may, assume without loss of generality, that K contains a primitive pth root of unity ζ, extending K to $K(\zeta)$, if need be. We apply Lemma 1.3.9 with $a := g$. By assumption, $\langle g \rangle$ permutes the $\omega \in \Omega$ transitively. Then the characteristic polynomial of $\delta_V(g)$ is $c_g = \frac{X^p - 1}{X - 1} = \prod_{i=1}^{p-1}(X - \zeta^i)$. Let $\langle v_i \rangle_K = \ker_V(g - \zeta^i \operatorname{id}_V)$. Then $B := (v_1, \ldots, v_{p-1})$ is a K-basis of V. Our assumption says that for any $i \in \{1, \ldots, p-1\}$ there is an element $h_i \in G$ with $h_i^{-1} g h_i = g^i$. Then $h_i^{-1} g h_i v_1 = g^i v_1 = \zeta^i v_1$, thus $h_i v_1 \in \langle v_i \rangle$ for $1 \leq i \leq p - 1$, hence $KG \cdot v_1 = V$. Similarly, if $(x_1, \ldots, x_{p-1})$ is the dual basis to B then $x_1 h_i \in \langle x_i \rangle$ for $1 \leq i \leq p - 1$, and $x_1 \cdot KG = V^\star$. $\qquad\square$

We now formulate the above mentioned variant of the Meataxe algorithm.

Holt–Rees algorithm

Input: An n-dimensional matrix representation δ of A in terms of matrices $(\mathbf{a}_1 := \delta(a_1), \ldots, \mathbf{a}_k := \delta(a_k))$ for a generating system $(a_1, \ldots, a_k)$ of A.

Output: Either the information that δ is irreducible, or matrix representations of A on an A-submodule W and on the factor module V/W given by matrices for the generating system $(a_1, \ldots, a_k)$ of A.

(1) Choose an element a in A (uniformly at random).

(2) Calculate the characteristic polynomial c of $\delta(a)$.

(3) Factor c into irreducible factors over K. Order the factors by increasing multiplicity and degree. Then for each irreducible factor p, do the following.

- Compute $a' := p(\delta(a))$.

- Choose a non-zero vector v in $\ker(a')$ (uniformly at random) and run the spinning algorithm with input v and $(\mathbf{a}_1, \ldots, \mathbf{a}_k)$. Let $B := (b_1, \ldots, b_m)$ be the output of the spinning algorithm. If $m < n$, then extend B to a basis $(b_1, \ldots, b_m, b_{m+1}, \ldots, b_n)$ of K^n. For $i = 1, \ldots, k$ compute the matrices $\mathbf{a}_i'$ and $\mathbf{a}_i''$ of the actions of $\mathbf{a}_i$ on $\langle B \rangle_K$ and $K^n/\langle B \rangle_K$ with respect to the bases B and $(b_{m+1} + \langle B \rangle_K, \ldots, b_n + \langle B \rangle_K)$, respectively. Return $((\mathbf{a}_i')_{i=1}^k, (\mathbf{a}_i'')_{i=1}^k)$.

- Choose a non-zero vector v in $\ker((a')^{\mathrm{T}})$ and run the spinning algorithm with input v and $(\mathbf{a}_1^{\mathrm{T}}, \ldots, \mathbf{a}_k^{\mathrm{T}})$. Let $B := (b_1, \ldots, b_m)$ be the output of the spinning algorithm. If $m < n$, extend B to a basis $(b_1, \ldots, b_m, b_{m+1} \ldots, b_n)$ of K^n. For $i = 1, \ldots, k$ compute the matrices $\mathbf{a}_i'$ and $\mathbf{a}_i''$ of the actions of $\mathbf{a}_i^{\mathrm{T}}$ on $K^n/\langle B \rangle_K$ and $\langle B \rangle_K$ with respect to the bases $(b_{m+1} + \langle B \rangle_K, \ldots, b_n + \langle B \rangle_K)$ and B, respectively. Return $((\mathbf{a}_i'^{\mathrm{T}})_{i=1}^k, (\mathbf{a}_i''^{\mathrm{T}})_{i=1}^k)$.

- If $m = n$ and $\dim_K(\ker(a')) = \deg(p)$, return the answer "$\delta$ is irreducible."

(4) Go back to step (1).

Both algorithms, the Meataxe and the one by Holt and Rees are examples of what is usually called a **Las Vegas** algorithm, which means that there is no certainty that it will terminate, but if it does then it will always return the correct answer. Holt and Rees [86] show that if δ is irreducible, then a constant proportion of elements a in A will have the property that $\dim_K \ker(p(\delta(a))) = \deg(p)$, for some irreducible factor of the characteristic polynomial of $\delta(a)$, and can therefore be used successfully to prove the irreducibility of δ. For a justification in the case of an absolutely irreducible representation, see Exercise 1.3.3. By Theorem 1.3.8 an element $a \in A$ with the property that $\dim_K \ker(p(\delta(a))) = \deg(p)$ for some irreducible factor p of degree one can be used to show that δ is absolutely irreducible.

The case of a reducible representation δ is also analyzed in [86], and, together with a result by Ivanyos and Lux, see [93], one obtains the result that a constant proportion of elements a in A can be used successfully to show that δ is reducible.

Before closing this section with a larger example we describe the simple KG-modules for a finite p-group G over a field K of characteristic p.

Theorem 1.3.11 *An irreducible representation of a finite p-group over a field of characteristic p is trivial of degree one.*

Proof. Let G be a finite p-group, K a field with $\operatorname{char} K = p$ and V a simple KG-module. For $0 \neq v \in V$ define $M_v := \{\sum_{g \in G} \alpha_g \, g \cdot v \mid \alpha_g \in \mathbb{F}_p\}$, where $\mathbb{F}_p$ is the prime field of K. Then M_v is a finite additive p-group on which G acts by left multiplication. Since the orbit lengths are powers of p, and $\{0\}$ is one orbit, there must be a further orbit $\{w\}$ of length one. Then $\langle w \rangle_K$ is a proper submodule on which G acts trivially. Since V is simple $\langle w \rangle_K = V$. $\qquad\square$

Example 1.3.12 We want to analyze the 100-dimensional permutation representation of the sporadic simple Higman–Sims group HS considered in Section 1.2. We construct the corresponding permutation module of HS over $\mathbb{F}_2$ and then ask for bases of all submodules:

```
gap> G := PrimitiveGroup( 100, 3 );;
gap> module := PermutationGModule( G , GF(2) );;
gap> bsm := MTX.BasesSubmodules( module );;
```

In GAP, modules are always right modules and are usually realized as subspaces of row spaces, in this example of $\mathtt{GF(2)^{100}} := \mathbb{F}_2^{1 \times 100}$ in our notation. To deal with left modules and column spaces we simply use

$$\mathbf{a} \cdot v = (v^{\mathrm{T}} \mathbf{a}^{\mathrm{T}})^{\mathrm{T}} \qquad \text{for} \qquad \mathbf{a} \in \mathbb{F}_2^{n \times n}, \ v \in \mathbb{F}_2^{n \times 1}.$$

The list $\mathtt{bsm}$ is a list of bases for all submodules of $\mathtt{module}$ ordered w.r.t. increasing dimensions. We construct the list $\mathtt{sm}$ of submodules, print their dimensions and ask for mutual inclusions, that is we compute the incidence matrix `mat` of the poset of submodules:

```
gap>  sm := List( bsm , bas -> Submodule( GF(2)^100 , bas ) );;
gap>  mat := [];;
gap> for i in [1..Length(bsm)] do
>      mat[i] :=   [];
>        for j in [1..Length(bsm)] do
>            if IsSubspace( sm[i], sm[j]) then mat[i][j] := 1 ;
>             else  mat[i][j] := 0;
>            fi;
>        od;
>    od;
```

We display the lower triangular part of the incidence matrix `mat`, giving as row headings the dimensions of the submodules in Figure 1.1. We can see from the

0	1												
1	1	1											
21	1	1	1										
22	1	1	1	1									
22	1	1	1	.	1								
22	1	1	1	.	.	1							
77	1	1	1	.	.	.	1						
23	1	1	1	1	1	1	.	1					
78	1	1	1	1	.	.	1	.	1				
78	1	1	1	.	1	.	1	.	.	1			
78	1	1	1	.	.	1	1	.	.	.	1		
79	1	1	1	1	1	1	1	1	1	1	1	1	
99	1	1	1	1	1	1	1	1	1	1	1	1	1
100	1	1	1	1	1	1	1	1	1	1	1	1	1

Figure 1.1. Lower triangular part of the incidence matrix of the poset of sub-modules.

incidence matrix that the permutation module has a composition series with composition factors of dimension $1, 20, 1, 56, 1, 20, 1$. But if one is just interested in these factors, there is a much easier way. GAP conveniently provides a command computing composition factors of a module and giving for each factor the multiplicity with which it occurs in a composition series up to isomorphism:

```
gap> cf := MTX.CollectedFactors( module );;
gap> List( cf, x->x[1].dimension );
[ 1, 20, 56 ]
gap> List( cf, x->x[2] ); # the multiplicities:
[ 4, 2, 1 ]
gap> List( cf, x -> MTX.IsAbsolutelyIrreducible(x[1]) );
[ true, true, true ]
```

Thus the permutation module has just three composition factors up to isomorphism with multiplicities $4, 2, 1$. The last command verifies that these modules are all absolutely simple.

If $F = \mathbb{F}_q$ is a finite field and $V = F^m$ then the **weight** $\mathrm{wt}(v)$ of a vector $v \in V$ is defined to be the number of non-zero components of v and the (Hamming-) distance of two vectors $v, w \in V$ is $\mathrm{wt}(v - w)$. This is of relevance in **coding theory**, where one is often interested in finding a subspace $W \leq V$ (which is there called a **linear code**) where the minimal distance of two different vectors in W ("code words") is not too small. If this minimal distance is $2d + 1$ then it is easily seen that W is able to correct d errors in the following sense: if $v \in V$ is obtained from a "code word" $w \in W$ by changing at most d components, then there is a unique element $w_c \in W$ such that $\mathrm{wt}(w_c - v) \leq d$ and thus $w = w_c$. It is obvious that the minimal distance of two different vectors in W is just the minimal weight of a non-zero element of W. We want to analyze the 21-dimensional $\mathbb{F}_2 G$-subspace W of $V = \mathbb{F}_2^{100}$ constructed above:

```
gap> W := sm[3];;
gap> wd := DistancesDistributionMatFFEVecFFE( bsm[3], GF(2), Zero(W) );
[ 1, 0, 0, 0, 0, 0, 0, 0, 0, 0, 0, 0, 0, 0, 0, 0, 0, 0, 0, 0, 0, 0,
  0, 0, 0, 0, 0, 0, 0, 0, 3850, 0, 0, 0, 4125, 0, 0, 0, 92400, 0, 0, 0,
  347600, 0, 0, 0, 600600, 0, 0, 0, 600600, 0, 0, 0, 347600, 0, 0, 0,
  92400, 0, 0, 0, 4125, 0, 0, 0, 3850, 0, 0, 0, 0, 0, 0, 0, 0, 0, 0, 0, 0,
  0, 0, 0, 0, 0, 0, 0, 0, 0, 0, 0, 0, 0, 0, 0, 0, 0, 0, 0, 1 ]
gap> Position( wd, 3850 );
33
```

Here for i from 0 to 100 the $(i + 1)$st entry $\mathtt{wd}[i + 1]$ of $\mathtt{wd}$ is the number of vectors in $\mathtt{W}$ of weight i. Thus the coefficients of $\mathtt{wd}$ should add up to 2^{21}. We see that there are 3850 vectors in $\mathtt{W}$ with minimal non-zero weight 32. Thus $\mathtt{W}$ can correct 15 errors in the above sense. It is clear that G acts on

$$W_i = \{w \in W \mid \mathrm{wt}(w) = i\} \qquad \text{for} \qquad i = 0, \ldots, 100.$$

We thus obtain new permutation representations of G of degree 3850, 4125, 92400 etc. which we might want to look at:

```
gap> repeat   w := Random( W );   until WeightVecFFE( w ) = 32 ;
```

Here we have randomly chosen elements $\mathtt{w} \in \mathtt{W}$ until we have found one with weight 32. Observe that the proportion of vectors of this weight is approximately $\frac{1}{544}$, so one has a good chance of finding such a vector quickly. We now compute the orbit of this vector $\mathtt{w}$ under the right action of the matrix group corresponding to G:

```
gap> Gmat := Group ( MTX.Generators(module) );;
gap> orbit := Orbit( Gmat , w , OnRight );;
gap> Length( orbit );
3850
```

Thus G acts transitively on the vectors of W_{32} and thus has a subgroup U of index 3850. Constructing the corresponding permutation representation we check that it is primitive and hence U is a maximal subgroup of G. In order to identify this group, we first compute a permutation representation of smaller degree for this group and then calculate the poset of normal subgroups, because it is obviously much faster to compute with a permutation group of degree 32 than with one of degree 3850:

```
gap> Gper := Image( ActionHomomorphism( Gmat, orbit) );;
gap> IsPrimitive( Gper );
true
gap> U := Stabilizer( Gper, 1 );
<permutation group of size 11520 with 3 generators>
gap>  U := Image ( SmallerDegreePermutationRepresentation( U ) ) ;;
gap> DegreeOperation( U );
32
gap> nsU := NormalSubgroups( U );
[ Group(()), <permutation group of size 16 with 4 generators>,
   <permutation group of size 5760 with 6 generators>,
   <permutation group of size 11520 with 3 generators> ]
gap> StructureDescription( nsU[2] );
"C2 x C2 x C2 x C2"
gap> StructureDescription( FactorGroup( U , nsU[2] ) );
"S6"
```

Thus $U \cong C_2^4 . S_6$, an elementary abelian 2-group of order 2^4 extended by S_6, which is in accordance with the information we get from the ATLAS ([38], p. 80). One can use the permutation representation of degree 3850 of G just constructed to obtain further absolutely irreducible matrix representations of G over $\mathbb{F}_2$; see Exercise 1.3.7. But observe that it would not be wise to try to calculate the complete submodule poset of this module in one go as above, since the number of submodules is rather large.

Similarly one finds that the vectors of W_{36} form one orbit under G and we get a primitive permutation representation and hence a maximal subgroup of index 4125 which we could identify using GAP or the ATLAS as being isomorphic to $C_4^3 \rtimes L_3(2)$. Also, W_{40} breaks up into orbits of lengths 77000 and 15400 with G acting primitively on the latter one, so that we find a maximal subgroup of index 15400, which turns out to be isomorphic to $C_2 \times (A_6 \cdot V_4)$.   ◆

Exercises

Exercise 1.3.1 Let q be a prime power and

$$G := \{ [\alpha_{i,j}] \in \mathrm{SL}_n(\mathbb{F}_q) \mid \alpha_{ij} = \delta_{i,j} \text{ for } i \geq j \}.$$

G acts on $V := \mathbb{F}_q^n$ by left multiplication. Show that $\mathrm{End}_{\mathbb{F}_q G} V \cong \mathbb{F}_q$, but V is not simple if $n > 1$.

Exercise 1.3.2 Let A be a finite dimensional K-algebra over a field K and let $L \supseteq K$ be an extension field. If W_1, W_2 are A-modules show that

$$\mathrm{Hom}_{LA}(L \otimes_K W_1, L \otimes_K W_2) \cong_L L \otimes_K \mathrm{Hom}_A(W_1, W_2).$$

If W_1, W_2 are simple, conclude that $\mathrm{Hom}_{LA}(L \otimes_K W_1, L \otimes_K W_2) \neq \{0\}$ implies $W_1 \cong_A W_2$.

Exercise 1.3.3 Let $K = \mathbb{F}_q$ be a finite field with q elements and V a K-vector space of dimension d. For $\lambda \in K$ and $\varphi \in \mathrm{End}_K V$ let c_φ denote the

characteristic polynomial of φ and $m_\lambda(\varphi)$ the multiplicity of λ as a root of c_φ.
(a) Show that $\dim_K \ker p(\varphi) = \deg p$ if $p = X - \lambda$ with $m_\lambda(\varphi) = 1$.
(b) In this part we will give a lower bound for $m_{d,q}/|\operatorname{End}_K V|$, where

$$m_{d,q} := |\{\varphi \in \operatorname{End}_K V \mid m_\lambda(\varphi) = 1 \quad \text{for some} \quad \lambda \in K\}|.$$

For $\lambda_1, \lambda_2 \in K$ put $n(\lambda_1, \lambda_2) := |\{\varphi \in \operatorname{End}_K V \mid m_{\lambda_1}(\varphi) = m_{\lambda_2}(\varphi) = 1\}|$ and $n_{d,q} := n(0,0)$. Show that for $\lambda, \lambda_1, \lambda_2 \in K$

$$n(\lambda, \lambda) = n_{d,q} = \frac{q^d - 1}{q - 1} q^{d-1} |\operatorname{GL}_{d-1}(q)| = \frac{|\operatorname{GL}_d(q)|}{q - 1},$$

$$n(\lambda_1, \lambda_2) \leq \frac{q^d - 1}{q - 1} q^{d-1} n_{d-1,q} \quad \text{for} \quad \lambda_1 \neq \lambda_2,$$

$$m_{d,q} \geq q\, n_{d,q} - \sum_{|\{\lambda_1, \lambda_2\}| = 2} n(\lambda_1, \lambda_2) \geq \frac{1}{2} q\, n_{d,q}.$$

Conclude (using elementary calculus) that

$$\frac{m_{d,q}}{|\operatorname{End}_K V|} \geq \frac{1}{2}(1 - q^{-2})(1 - q^{-3}) \cdots (1 - q^{-d}) \geq 0.288.$$

Hint: To compute $n_{d,q}$ observe that $\varphi \in \operatorname{End}_K V$ with $\ker \varphi = \langle v_1 \rangle_K \neq \{0\}$ and $m_0(\varphi)$ induces a $\bar{\varphi} \in \operatorname{GL}_K(V/\langle v_1 \rangle_K)$.

Exercise 1.3.4 Let $K = \mathbb{F}_p$ and $G = \operatorname{SL}_2(p)$. As in Exercise 1.1.8 let V_m be the KG-module of homogeneous polynomials in X, Y of degree $m - 1$. Let $\delta_m : G \to \operatorname{GL}_m(K)$ be the corresponding matrix representation of G on V_m with respect to the K-basis

$$(X^{m-1}, X^{m-2}Y, X^{m-3}Y^2, \ldots, XY^{m-2}, Y^{m-1}).$$

The aim of this exercise is to show that $\delta_1, \ldots, \delta_p$ are irreducible representations of G using Norton's irreducibility criterion.
(i) G is generated as a group by the elements $a = \begin{bmatrix} 1 & 1 \\ 0 & 1 \end{bmatrix}$ and $b = \begin{bmatrix} 1 & 0 \\ 1 & 1 \end{bmatrix}$.
(ii) $\delta_m(a)$ is an upper triangular matrix with 1's on the diagonal and the $(i, i+1)$st entry equal to i, for $i = 1, \ldots, m - 1$. Prove an analogous assertion for $\delta_m(b)$.
(iii) Consider the element $a - 1 \in KG$, and compute $\ker \delta_m(a - 1)$ and also $\ker \delta_m(a - 1)^{\mathrm{T}}$ and apply the "spinning algorithm."
(iv) Try to prove that V_m is reducible for $m > p$.
Note: It follows from (iii) and Theorem 1.3.8 that $\delta_1, \ldots, \delta_p$ are absolutely irreducible.

Exercise 1.3.5 Let $\Omega := \{1, \ldots, 5\}$ and $G := A_5$ act on $\binom{\Omega}{2}$ in a natural way. For $K \in \{\mathbb{F}_2, \mathbb{F}_4\}$ compute the composition factors and the poset of submodules of the corresponding permutation module $K\binom{\Omega}{2}$. Show that $\mathbb{F}_2\binom{\Omega}{2}$ has a composition factor of dimension four which is not absolutely simple.

Exercise 1.3.6 Similarly as in Example 1.1.49 compute a $\mathbb{Q}A_5$-module W of dimension five as a factor module of $\mathbb{Q}A_5(1-h)(1-g) \leq_{\mathbb{Q}A_5} \mathbb{Q}A_5$ for suitable $g, h \in A_5$. Compute the matrices $\delta(g_1), \delta(g_2)$ of the representation $\delta = \delta_W$ with respect to some basis of W, where $A_5 = \langle g_1, g_2 \rangle$. Find a matrix of rank four in $\langle \mathbf{I}_5, \delta(g_1), \delta(g_2) \rangle_{\mathbb{Z}}$ and use it to prove that δ is absolutely irreducible.

Exercise 1.3.7 We use the notation of the Example 1.3.12.

(a) Explain why $|W_i| = |W_{n-i}|$ for $i = 1, \ldots, n$. Is it always true that $\mathbb{F}_2 W_i \cong_{\mathbb{F}_2 G} \mathbb{F}_2 W_{n-i}$?

(b) Find the orbits of $G = $ HS on W_{44} and W_{48} and find the structure of the point stabilizers. Show that G has a maximal subgroup isomorphic to $L_3(4) \rtimes C_2$, a split extension of $L_3(4)$ with a cyclic group of order two.

(c) Compute a composition series for the permutation module $\mathbb{F}_2 W_{32}$ and verify that HS has absolutely irreducible representations of degrees 132, 518 and 1000 over $\mathbb{F}_2$ (in addition to those of degrees 20 and 56 we have seen before).

(d) Show that the permutation module of HS of dimension 100 over $\mathbb{F}_3$ is the direct sum of three absolutely irreducible modules of dimensions 1, 22 and 77. Although it does not appear to be feasible to compute the weight distribution of all vectors of the 22-dimensional submodule, one may do some statistics by computing the weight of a few million randomly chosen vectors. Compute the stabilizers of some vectors of small weight (say of weight < 43). What happens if one tries to compute the stabilizer of an arbitrary vector?

Exercise 1.3.8 Compute a composition series for the primitive permutation representation of the sporadic simple Janko group J_2 of degree 100 over $\mathbb{F}_2$ and show that J_2 has an irreducible representation of degree 12 over $\mathbb{F}_2$, which is not absolutely irreducible. Show that J_2 has three orbits on the vectors of this 12-dimensional $\mathbb{F}_2 J_2$ module. (The permutation representation may be found in GAP using the command "AllPrimitiveGroups(DegreeOperation, 100)".)

1.4   Structure of algebras

In this section we study the structure of a K-algebra A, where K is assumed to be a commutative ring K. Often, but not always, we will assume that A is of finite length. We first introduce some useful general notions.

Definition 1.4.1 Let V be an A-module and S a simple A-module. We put

$$\operatorname{Rad} V := \bigcap \{U \leq_A V \mid V/U \text{ is simple }\},$$

$$\operatorname{Soc} V := \sum \{U \leq_A V \mid U \text{ is simple }\},$$

$$\operatorname{H}_S(V) := \sum \{U \leq_A V \mid U \cong_A S\}, \quad \operatorname{H}^S(V) := \bigcap \{U \leq_A V \mid A/U \cong_A S\}.$$

$\operatorname{Rad} V$ is called the **radical**, $\operatorname{Soc} V$ the **socle** and $\operatorname{H}_S(V)$ the **S-homogeneous component** of V. The **Jacobson radical** of A is $\operatorname{J}(A) := \operatorname{Rad}{}_A A$.

Lemma 1.4.2 *If S is a simple A-module then $\mathrm{H}_S(A) \trianglelefteq A$ and*

$$\mathrm{Soc}(_AA) = \sum\{\mathrm{H}_S(A) \mid S \text{ simple } A\text{-module}\} \trianglelefteq A, \qquad (1.13)$$

$$\mathrm{J}(A) = \bigcap\{\mathrm{H}^S(A) \mid S \text{ simple } A\text{-module}\} \trianglelefteq A. \qquad (1.14)$$

Proof. If S is not isomorphic to a submodule of $_AA$ then $\mathrm{H}_S(A) = \{0\}$. So let $S \leq_A A$. For any $a \in A$ the map $S \to Sa$, $s \mapsto sa$ is an epimorphism of A-modules. Since S is simple we have $Sa = \{0\}$ or $Sa \cong_A S$. Thus $\mathrm{H}_S(_AA)$ is invariant under right multiplication and hence an ideal in A and (1.13) follows.

If $S \cong_A A/M$ is simple and $a \in \mathrm{ann}_A S$, then $aA \subseteq M$. Hence $\mathrm{ann}_A S \subseteq \mathrm{H}^S(A)$. Also $\mathrm{J}(A)S \leq_A S$, thus $\mathrm{J}(A)S = \{0\}$ or $\mathrm{J}(A)S = S$. But the latter is impossible by Exercise 1.4.1. Hence

$$\mathrm{J}(A) \subseteq \bigcap\{\mathrm{ann}_A S \mid S \text{ simple } A\text{-module}\}$$
$$\subseteq \bigcap\{\mathrm{H}^S(A) \mid S \text{ simple } A\text{-module}\} = \mathrm{J}(A)$$

and (1.14) follows, since $\mathrm{ann}_A S \trianglelefteq A$ by Lemma 1.1.12. $\qquad\square$

Whereas we have seen (see Lemma 1.3.2) that every simple A-module is an epimorphic image of the regular module $_AA$, in general it is not true that every simple A-module is isomorphic to a submodule of $_AA$ (see for instance Exercise 1.4.3). But for group algebras the latter also holds.

Lemma 1.4.3 *If G is a finite group and K is a field then*

$$\mathrm{Soc}\, KG \cong_{KG} KG/\mathrm{J}(KG).$$

Proof. The result follows, since $_{KG}KG$ is selfdual (see Exercise 1.1.10). $\qquad\square$

We recall from Corollary 1.3.7 that the image $\delta(A)$ of any irreducible representation $\delta\colon A \to \mathrm{End}_K V$ for a K-algebra A of finite length is isomorphic to a full matrix ring over a division ring D:

$$A/\ker\delta \cong \delta(A) \cong D^{m\times m} \quad \text{for} \quad D = (\mathrm{End}_A V)^{\mathrm{op}}. \qquad (1.15)$$

The structure of a full matrix ring over a division ring is well known and easily described as follows.

Remark 1.4.4 If D is a division ring and $m \in \mathbb{N}$ then $D^{m\times m}$ is a simple ring. Let $L_k := \{[d_{ij}] \in D^{m\times m} \mid d_{ij} = 0 \;\; \text{for} \;\; j \neq k\}$ for $k = 1, \ldots, m$. Then the L_k are minimal left ideals of $D^{m\times m}$ and

$$D^{m\times m} = L_1 \oplus \cdots \oplus L_m \quad \text{with} \quad L_i \cong_{D^{m\times m}} D^{m\times 1}.$$

Corollary 1.4.5 (a) *The image of an algebra of finite length under an irreducible representation is a simple algebra.*

(b) *Any simple K-algebra A of finite length is isomorphic to a full matrix ring $D^{m \times m}$ over a division ring D containing K in its center. $\mathrm{J}(A) = \mathrm{H}^V(A) = \{0\}$ and $A = \mathrm{H}_V(A)$ for a simple A-module V. Furthermore $D = (\mathrm{End}_A V)^{op}$ and $m = \mathrm{l}(_A A)$. If K is a field then $\dim_K V = m[D : K]$.*

Theorem 1.4.6 *Let A be of finite length and let $V_1, \ldots, V_r$ be a complete set of simple A-modules up to isomorphism. Moreover, let $\delta_i \colon A \to \mathrm{End}_K V_i$ be the corresponding irreducible representations and $D_i = (\mathrm{End}_A V_i)^{op}$ for $i = 1, \ldots, r$. Then*

$$\mathrm{J}(A) = \bigcap_{i=1}^r \ker \delta_i \quad and \quad A/\mathrm{J}(A) \cong \bigoplus_{i=1}^r D_i^{m_i \times m_i}.$$

Furthermore, if K is a field we have $\dim_K V_i = [D_i : K] m_i$.

Proof. By Lemma 1.3.2(a) and Lemma 1.4.2 $\mathrm{J}(A) = \bigcap_{i=1}^r \mathrm{H}^{V_i}(A)$. As we have already seen in the proof of Lemma 1.4.2 $\ker \delta_i = \mathrm{ann}_A V_i \subseteq \mathrm{H}^{V_i}(A)$. By Lemma 1.4.5 $\mathrm{H}^{V_i}(A/\mathrm{ann}_A V_i) = \{0\}$, thus $\mathrm{H}^{V_i}(A) = \ker \delta_i$. We thus obtain a K-algebra homomorphism

$$\delta \colon A \to A' := \bigoplus_{i=1}^r \delta_i(A) \quad a \mapsto (\delta_i(a))_{i=1}^r$$

with $\ker \delta = \mathrm{J}(A)$, inducing an injection $\bar\delta \colon A/\mathrm{J}(A) \to A'$. By Lemma 1.4.5 $\delta_i(A) \cong D_i^{m_i \times m_i}$, with $m_i = \mathrm{l}(A/\ker \delta_i) \le l_i$, the multiplicity of V_i in a composition series of $A/\mathrm{J}(A)$. Hence $\mathrm{l}(A/\mathrm{J}(A)) = \sum_{i=1}^r l_i \ge \sum_{i=1}^r m_i = \mathrm{l}(A')$ and $\bar\delta$ is surjective, hence an isomorphism. Finally, $\dim_{D_i} V_i = m_i$ and hence $\dim_K V_i = [D_i : K] m_i$ in the case that K is a field. $\qquad\square$

Studying irreducible representations of A is equivalent to studying those of the algebra $A/\mathrm{J}(A)$. The Jacobson radical is mapped to zero under any irreducible representation. From this, some properties of $\mathrm{J}(A)$ can be concluded indirectly as shows the proof of Corollary 1.4.7.

We recall that an element a in a ring A is called **nilpotent** if $a^n = 0$ for some $n \in \mathbb{N}$. Similarly $I \trianglelefteq A$ is called nilpotent if there is an $n \in \mathbb{N}$ with $L^n = \{0\}$ and a **nil ideal** if all $a \in I$ are nilpotent.

Corollary 1.4.7 (a) *If $I \trianglelefteq A$ is a nil ideal, then $I \le \mathrm{J}(A)$.*

(b) *If K is a field and $n = \dim_K A$ then $\mathrm{J}(A)^n = 0$. Thus $\mathrm{J}(A)$ is the largest nilpotent ideal of A.*

Proof. (a) For any irreducible representation $\delta \colon A \to \mathrm{End}_K V$ the image $\delta(I) \trianglelefteq \delta(A)$ is an ideal, and, since $\delta(A)$ is a simple ring, $\delta(I) = \{0\}$ or $\delta(I) = \delta(A)$. The latter cannot hold if I consists of nilpotent elements. Thus a nil ideal is in the kernel of every irreducible representation, hence by Theorem 1.4.6 in $\mathrm{J}(A)$.

(b) Consider the regular representation $\rho \colon A \to \mathrm{End}_K {}_A A$ and choose a K-basis B of A which is adapted to a composition series. Then for any $a \in \mathrm{J}(A)$

the matrix $[\rho(a)]_B$ is an upper triangular matrix with zeros on the diagonal, since $\delta_i(a) = 0$ for all irreducible representations δ_i. Since ρ is faithful, and the product of n upper triangular $n \times n$-matrices with zeros on the diagonal is the zero matrix, the assertion follows. $\qquad\square$

Example 1.4.8 Let $G = S_3$ and let K be a field. Let δ_1 be the trivial representation and δ_2 the sign representation. Thus $\delta_1(g) = 1$ and $\delta_2(g) = \mathrm{sgn}(g) = \pm 1$ for all $g \in S_3$. Of course $\delta_1 = \delta_2$ if char $K = 2$. Finally let $\boldsymbol{\delta_3} : K\,S_3 \to K^{2\times 2}$ be the representation defined in Example 1.3.5. As was shown there, δ_3 is absolutely irreducible if and only if char $K \neq 3$. We denote the corresponding $K\,S_3$-modules by $V_1 = K_{S_3}, V_2, V_3$, and let $\ker \delta_i = I_i \trianglelefteq K\,S_3$ for $i = 1, 2, 3$. If char $K \neq 2, 3$ we have $K\,S_3 \,/\, \mathrm{J}(K\,S_3) \cong K \oplus K \oplus K^{2\times 2}$, thus $\mathrm{J}(K\,S_3) = \{0\}$.

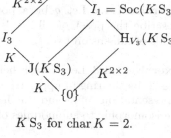

If char $K = 2$ the representations δ_1 and δ_3 are irreducible and $V_1 \cong \mathrm{Inv}_{S_3}(K\,S_3) = K(S_3)^+$ is a nilpotent ideal, because $((S_3)^+)^2 = |S_3|(S_3)^+ = 6\,(S_3)^+$. By Corollary 1.4.7 $K(S_3)^+ \leq \mathrm{J}(K\,S_3)$. It follows that

$$K\,S_3 \,/\, \mathrm{J}(K\,S_3) \cong K \oplus K^{2\times 2} \quad \text{and}$$

$$\mathrm{J}(K\,S_3) = K(S_3)^+ \cong K.$$

By Lemma 1.4.3 $\dim_K \mathrm{Soc}\, K\,S_3 = 5$ and so $\mathrm{Soc}\, K\,S_3 = I_1 = \mathrm{Inv}^{S_3}(K\,S_3)$ because

$K\,S_3$ for char $K = 2$.

this is, in our case, the only (left) ideal of codimension 1 in $K\,S_3$. Note that $K\,S_3 = \mathrm{H}_{V_3}(K\,S_3) \oplus I_3$.

Finally, if char $K = 3$ we get two maximal submodules

$$I_1 := \{\textstyle\sum_{g \in S_3} \alpha_g g \mid \sum_{g \in S_3} \alpha_g = 0\},$$
$$I_2 := \{\textstyle\sum_{g \in S_3} \alpha_g g \mid \sum_{g \in S_3} \mathrm{sgn}(g)\alpha_g = 0\},$$

and two simple submodules

$$V_1 := \langle\, \textstyle\sum_{g \in S_3} g \,\rangle_K,$$
$$V_2 := \langle\, \textstyle\sum_{g \in S_3} \mathrm{sgn}(g)g \,\rangle_K.$$

Obviously $V_1, V_2 \leq I_1 \cap I_2$. By Theorem 1.4.6 we cannot have an irreducible representation of G of degree two. Hence

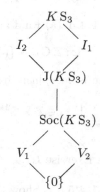

$K\,S_3$ for char $K = 3$.

we have found all irreducible representations (use Exercise 1.1.7). In particular, $\mathrm{J}(K\,S_3) = I_1 \cap I_2$ and $\mathrm{Soc}\, K\,S_3 = V_1 \oplus V_2$. It is also easily seen that $\mathrm{J}(K\,S_3)/\,\mathrm{Soc}(K\,S_3) \cong_{K\,S_3} V_1 \oplus V_2$. Of course, there are more $K\,S_3$-submodules (left ideals). In fact, $K\,S_3$ is the direct sum of two indecomposable submodules, as we will see later when dealing with idempotents. $\qquad\blacklozenge$

Exercises

Exercise 1.4.1 Let A be a ring. Prove

(a) If $a \in \mathrm{J}(A)$ then $1 - a$ has a left inverse in A.

(b) (Nakayama's lemma) If V is a finitely generated A-module and $\mathrm{J}(A)V = V$ then $V = \{0\}$.

(c) If V is a finitely generated A-module and $L \leq_A V$ with $L + \mathrm{J}(A)V = V$ then $L = V$.

Hint: (a) Show that $A = \mathrm{J}(A) + A(1-a)$ for $a \in \mathrm{J}(A)$. If $A(1-a) \neq A$ conclude using Zorn's lemma that $A(1 - a)$ is contained in a maximal left ideal.
(b) Let $V = \langle v_1, \ldots, v_n \rangle_A$ with n minimal. Assume that $\mathrm{J}(A)V = V$ and use part (a) to show that $v_n \in \langle v_1, \ldots, v_{n-1} \rangle_A$.

Exercise 1.4.2 Let q be a power of two and let V_4 be the Klein 4-group. Describe the poset of all ideals of $\mathbb{F}_q V_4$. Show that $\mathbb{F}_q V_4$ has exactly $q + 1$ ideals of dimension two.

Exercise 1.4.3 Let K be a field and $A := \{[a_{ij}] \in K^{n \times n} \mid a_{ij} = 0 \text{ for } i > j\}$, $n \in \mathbb{N}$, be the ring of upper triangular matrices over K. Find all the irreducible representations of A and the Jacobson radical $\mathrm{J}(A)$. Which simple A-modules are isomorphic to submodules of $_A A$? What is $\mathrm{Soc}\, A$? Show that

$$A \cong_A L_1 \oplus \cdots \oplus L_n$$

with $L_k = \{[a_{ij}] \in A \mid a_{ij} = 0 \text{ for } j \neq k\}$.

Exercise 1.4.4 (a) Find an isomorphism

$$\varphi \colon \mathbb{Z}\,C_2 \to \{(a, b) \in \mathbb{Z} \times \mathbb{Z} \mid a \equiv b \bmod 2\} \leq \mathbb{Z} \times \mathbb{Z}.$$

(b) Find an isomorphism from $\mathbb{Z}\,S_3$ to the subring

$$\{(a, \begin{bmatrix} b & c \\ d & e \end{bmatrix}, f) \in \mathbb{Z} \times \mathbb{Z}^{2 \times 2} \times \mathbb{Z} \mid a \equiv f \bmod 2,\ d \equiv 0,\ a \equiv b,\ e \equiv f \bmod 3\}$$

of $\mathbb{Z} \times \mathbb{Z}^{2 \times 2} \times \mathbb{Z}$.
Hint: Use Exercise 1.1.4.

Exercise 1.4.5 (a) Show that

$$\mathbb{F}_2\,A_5\,/\mathrm{J}(\mathbb{F}_2\,A_5) \cong \mathbb{F}_2 \oplus \mathbb{F}_4^{2 \times 2} \oplus \mathbb{F}_2^{4 \times 4}, \quad \mathbb{F}_4\,A_5\,/\mathrm{J}(\mathbb{F}_4\,A_5) \cong \mathbb{F}_4 \oplus \mathbb{F}_4^{2 \times 2} \oplus \mathbb{F}_4^{2 \times 2} \oplus \mathbb{F}_4^{4 \times 4}$$

as algebras over $\mathbb{F}_2$ and $\mathbb{F}_4$, respectively.
(b) For $K \in \{\mathbb{F}_2, \mathbb{F}_4\}$ and $n \in \mathbb{N}$ compute $\dim_K(\mathrm{J}(KA_5))^n$.
Hint: Use the GAP-command `RadicalOfAlgebra` to compute $\dim_K \mathrm{J}(K\,A_5)$ and Exercise 1.3.5. The GAP-command `ProductSpace` might also be helpful.

1.5 Semisimple rings and modules

Throughout this section A is assumed to be a K-algebra for some commutative ring K unless otherwise stated.

Definition 1.5.1 An A-module is called **semisimple** if it is a sum of simple modules. The corresponding representation is usually called **completely reducible**; A is called **semisimple** if the regular A-module ${}_A A$ is semisimple.

Lemma 1.5.2 *The following conditions for an A-module V are equivalent:*

(a) *V is semisimple;*

(b) *V is a direct sum of simple modules;*

(c) *every submodule of V has a complement, and hence is also a direct summand.*

Proof. Assuming (a), let $V = \sum_{i \in I} V_i$ with simple submodules $V_i \leq_A V$ and $U <_A V$. By Zorn's lemma there is $J \subseteq I$ maximal with the property that $\sum_{j \in J} V_j = \bigoplus_{j \in J} V_j$ and $U \cap \sum_{j \in J} V_j = \{0\}$. If $V_i \not\subseteq U \oplus \bigoplus_{j \in J} V_j$ for some $i \in I$ then $V_i \cap (U \oplus \bigoplus_{j \in J} V_j) = \{0\}$, which contradicts the maximality of J. Hence $V = U \oplus \bigoplus_{j \in J} V_j$, and we have proved (c). For $U = \{0\}$ we obtain (b).

Conversely, assuming (c) let $0 \neq v \in V$. By Zorn's lemma $Av \cong_A A/\operatorname{ann}_A v$ contains a maximal submodule U. By assumption there is $U' \leq_A V$ with $V = U \oplus U'$. Then $Av = U \oplus (Av \cap U')$ and $Av \cap U'$ is simple. Let $V' = \sum\{S \leq_A V \mid S \text{ simple}\}$. If $V \neq V'$ then $V = V' \oplus V''$ with $V'' \neq \{0\}$. Then V'' contains a simple submodule, which is a contradiction. Thus (a) follows. □

Corollary 1.5.3 (a) *Sums, submodules and factor modules of semisimple modules are semisimple.*

(b) *A is semisimple if and only if every A-module is semisimple.*

(c) *A is semisimple if and only if A has finite length and $J(A) = \{0\}$.*

Proof. (a) is obvious and (b) follows from (a) since every A-module is isomorphic to a factor module of a free A-module.

(c) If $A = \bigoplus_{i \in I} L_i$ with minimal left ideals L_i then $1 = \sum_{j=1}^m e_j$ with $e_j \in L_{i_j}$ and hence $A = \sum_{j=1}^m A e_j$. Hence I is finite and thus A is of finite length. Also it is clear that $J(A) = \{0\}$. Conversely, if $J(A) = \{0\}$ and $l(A) = m < \infty$, then by Exercise 1.5.1 A has m maximal left ideals $M_i \leq_A A$ such that ${}_A A \cong_A A/M_1 \oplus \cdots \oplus V/M_m$. □

Theorem 1.5.4 *Assume that V is a semisimple A-module, $V = \bigoplus_{i=1}^m \mathrm{H}_{L_i}(V)$, with simple A-modules L_i, where $L_i \not\cong_A L_j$ for $i \neq j$. Let*

$$\mathrm{H}_{L_i}(V) \cong_A \underbrace{L_i \oplus \cdots \oplus L_i}_{n_i} \qquad and \qquad D_i := \operatorname{End}_A L_i. \qquad (1.16)$$

Then $E := \operatorname{End}_A V$ is semisimple, $E \cong \bigoplus_{i=1}^m D_i^{n_i \times n_i}$ as K-algebras. Also

$$V \cong_E \bigoplus_{i=1}^m \underbrace{X_i \oplus \cdots \oplus X_i}_{z_i} \qquad \text{with} \qquad z_i := \dim_{D_i} L_i,$$

where X_i is a simple E-module with $X_i \not\cong_E X_j$ for $i \neq j$.

Proof. We put $H_i := \operatorname{H}_{L_i}(V)$. By Theorem 1.1.15 $\operatorname{Hom}_A(H_i, H_j) = \{0\}$ for $i \neq j$ and

$$\operatorname{End}_A V \cong_K \bigoplus_{i=1}^m \operatorname{End}_A H_i; \quad \text{in fact} \quad \varphi \mapsto (\varphi|_{H_i})_{i=1}^m$$

is a K-algebra isomorphism. In particular, $H_i \leq_E V$. For $j = 1, \ldots, n_i$ let $\pi_j \colon H_i \to L_i$ and $\iota_j \colon L_i \to H_i$ be the projections and injections corresponding to (1.16). Then we obtain an isomorphism of K-algebras (cf. Theorem 1.1.15)

$$E_i := \operatorname{End}_A H_i \to D_i^{n_i \times n_i}, \qquad \varphi \mapsto [(\pi_j \circ \varphi \circ \iota_k]_{1 \leq j, k \leq n_i}.$$

E_i has a unique simple module X_i up to isomorphism (see Remark 1.4.4) with $\dim_{D_i} X_i = n_i$. Since $\dim_{D_i} H_i = n_i z_i$ it follows (considering X_i as an E-module by inflation) that $H_i \cong_E \underbrace{X_i \oplus \cdots \oplus X_i}_{z_i}$. $\qquad\qquad\qquad\square$

Theorem 1.5.5 (Wedderburn) *Every semisimple algebra A is isomorphic to a direct sum of full matrix rings over division rings. If $\{L_1, \ldots, L_r\}$ is a set of representatives of the isomorphism classes of simple A-modules and $D_i = (\operatorname{End}_A L_i)^{\mathrm{op}}$ then*

$$A = \bigoplus_{i=1}^r A_i \quad \text{and} \quad A_i \cong_A \underbrace{L_i \oplus \cdots \oplus L_i}_{n_i} \quad \text{with} \quad n_i = \dim_{D_i} L_i,$$

and the two-sided ideal $A_i = \operatorname{H}_{L_i}(_A A)$ is isomorphic to $D_i^{n_i \times n_i}$ as an algebra. If A is a K-algebra over a field K then $\dim_K L_i = [D_i : K] n_i$.

Proof. Since $A \cong (\operatorname{End}_A {}_A A)^{\mathrm{op}}$ (see Exercise 1.5.3) this follows from the preceding theorem using an obvious algebra isomorphism $(D^{n \times n})^{\mathrm{op}} \cong (D^{\mathrm{op}})^{n \times n}$. Alternatively it is a consequence of Corollary 1.5.3(c) and Theorem 1.4.6. $\qquad\square$

For the remainder of this section we stick to the case of K-algebras over a field K.

Theorem 1.5.6 (Maschke) *If K is a field and G a finite group, then the group algebra KG is semisimple if and only if the characteristic of K is not a divisor of the group order $|G|$.*

Proof. (a) Assume char $K \nmid |G|$ and let $U \leq_{KG} KG$. We show that U has a complement as KG-module. We start with a K-vector space complement U' of U in KG and let $\pi \in \mathrm{End}_K KG$ be the corresponding projection with $\mathrm{im}(\pi) = U$ and $\ker(\pi) = U'$. We define $\tilde{\pi} \colon KG \to KG$ by

$$\tilde{\pi} \colon KG \to KG \ , \ a \mapsto \frac{1}{|G|} \sum_{g \in G} g\pi(g^{-1}a).$$

Observe that our assumption says that $\frac{1}{|G|} \in K$. A straightforward computation shows that $\tilde{\pi}$ is a KG-module homomorphism. Also $\tilde{\pi}^2 = \tilde{\pi}$, because U is G-invariant and so, clearly, $\mathrm{im}(\tilde{\pi}) = \mathrm{im}(\pi)$. It follows that $\ker(\tilde{\pi}) \leq_{KG} KG$ is a complement of U in KG.

(b) Conversely assume char $K \mid |G|$. Then by Example 1.1.19

$$\mathrm{Inv}_G(KG) = K \cdot \sum_{g \in G} g \subseteq \{\sum_{g \in G} \alpha_g g \mid \sum_{g \in G} \alpha_g = 0\} = \mathrm{Inv}^G(KG).$$

So $\mathrm{Inv}^G(KG)$ cannot have a KG-module as complement, since G would have to act trivially on it. $\qquad\square$

Example 1.5.7 Let $3 < n \in \mathbb{N}$ and let K be a field of characteristic zero or $p > n$. We consider the permutation modules for the symmetric group S_n corresponding to the natural actions on $\Omega = \{1, \ldots, n\}$ and $\binom{\Omega}{k} = \{M \subseteq \Omega \mid |M| = k\}$ for some $k \leq n$. By Maschke's theorem $K\,\mathrm{S}_n$ is semisimple, and by Example 1.2.18 $\mathrm{End}_{K\,\mathrm{S}_n} K\Omega \cong K \oplus K$. Thus

$$K\Omega \cong_{K\,\mathrm{S}_n} K \oplus L, \qquad L \text{ simple with } \dim_K(L) = n - 1,$$

with K denoting also the trivial $K\,\mathrm{S}_n$-module. Likewise by Example 1.2.25 $\mathrm{End}_{K\,\mathrm{S}_n} K\binom{\Omega}{2} \cong K \oplus K \oplus K$, thus

$$K\binom{\Omega}{2} \cong_{K\,\mathrm{S}_n} K \oplus L_1 \oplus L_2, \qquad \text{with simple } K\,\mathrm{S}_n\text{-modules } L_1, L_2.$$

Furthermore by Example 1.2.25 $\mathrm{Hom}_{K\,\mathrm{S}_n}(K\Omega, K\binom{\Omega}{k}) \cong_K K \oplus K$, so that $K\Omega$ must be isomorphic to a submodule of any $K\binom{\Omega}{k}$. In particular we can choose our notation so that $L \cong_{K\,\mathrm{S}_n} L_1$, and we conclude that

$$\dim_K L_2 = \binom{n}{2} - n = \frac{1}{2}n(n-3).$$

Thus S_n has irreducible representations of degrees $n-1$ and $\frac{1}{2}n(n-3)$. In fact, it is easy to see that these are absolutely irreducible representations. $\qquad\blacklozenge$

Definition 1.5.8 If A is an algebra over a field K, then an extension field $L \supseteq K$ is called a **splitting field** for A if every simple LA-module is absolutely simple. If G is a finite group and K is a splitting field for KG, then K is also called a splitting field for G.

Observe that an algebraically closed field is always a splitting field by Lemma 1.3.2. But for finite dimensional algebras we usually have smaller splitting fields.

Lemma 1.5.9 *If A is a K-algebra for a field K and $\dim_K A < \infty$ then there is a splitting field $L \supseteq K$ for A with $[L : K] < \infty$.*

Proof. Let $\bar{K}$ be the algebraic closure of K and let $(a_1, \ldots, a_m)$ be a K-basis of A. Assume that $V_1, \ldots, V_r$ are representatives of the isomorphism classes of simple $\bar{K}A$-modules and let $B_i := (v_1^i, \ldots, v_{n_i}^i)$ be a $\bar{K}$-basis of V_i for $i = 1, \ldots, r$. If V_i affords the matrix representation $\boldsymbol{\delta}_i$ with respect to the basis B_i, let L be the field obtained from K by adjoining all the entries of $\boldsymbol{\delta}_i(a_j)$ for $1 \le i \le r$ and $1 \le j \le m$. Then $V_i' = \langle v_1^i, \ldots, v_{n_i}^i \rangle_L$ is a simple LA-module with $\bar{K}V_i' = V_i$, hence absolutely simple. If V is any simple LA-module, then for some $i \in \{1, \ldots, r\}$

$$\{0\} \ne \operatorname{Hom}_{\bar{K}A}(\bar{K}V, \bar{K}V_i') \cong \bar{K} \otimes_L \operatorname{Hom}_{LA}(V, V_i')$$

(by Exercise 1.3.2), and hence $V \cong_{LA} V_i'$. So L is indeed a splitting field for A. $\qquad\square$

Theorem 1.5.10 *Let A be a semisimple K-algebra and let $\{L_1, \ldots, L_r\}$ be a complete set of representatives of the isomorphism classes of simple A-modules. Then the following are equivalent :*

(a) *K is a splitting field for A;*

(b) *$A \cong \bigoplus_{i=1}^r K^{n_i \times n_i}$ as a K-algebra;*

(c) *$\dim_K A = \sum_{i=1}^r (\dim_K L_i)^2$;*

(d) *$\operatorname{End}_A L_i = K \operatorname{id}_{L_i}$ for all $i \in \{1, \ldots, r\}$.*

Proof. The result follows immediately from Theorem 1.5.5, because Wedderburn's theorem gives in our case

$$\dim_K A = \sum_{i=1}^r n_i \dim_K L_i = \sum_{i=1}^r [D_i : K](\dim_{D_i} L_i)^2.$$

$\qquad\square$

Theorem 1.5.11 *Assume that K is a field.*

(a) *If A is semisimple and n is the number of simple A-modules (up to isomorphism), then $n \le \dim_K \mathbf{Z}(A)$ with equality if K is a splitting field for A.*

(b) *If $\operatorname{char} K \nmid |G|$ then the number of irreducible K-representations of G (up to equivalence) is less than or equal to the number of conjugacy classes of G, and equality holds if K is a splitting field for G.*

Proof. We assume the notation of Theorem 1.5.10. Observe that $\mathbf{Z}(D^{n \times n}) \cong \mathbf{Z}(D)$ for any division ring D, so

$$\mathbf{Z}(A) \cong \bigoplus_{i=1}^{r} \mathbf{Z}(D_i) \quad \text{and} \quad r \leq \dim_K \mathbf{Z}(A).$$

By Theorem 1.5.10 K is a splitting field for A if and only if $D_i = \mathbf{Z}(D_i) \cong K$ for all i, hence (a) follows.

(b) This follows from (a) and Lemma 1.1.9. $\qquad\square$

Corollary 1.5.12 *If G is a finite group and K is a splitting field for G of characteristic not dividing $|G|$ then $KG \cong \bigoplus_{i=1}^{r} K^{n_i \times n_i}$ with*

$$r = |\operatorname{cl}(G)| \quad \text{and} \quad |\{i \in \{1, \dots, r\} \mid n_i = 1\}| = [G : G'].$$

If G is abelian then $KG \cong K \oplus \cdots \oplus K$ with $|G|$ summands. In particular any two abelian groups of the same order have isomorphic group algebras over $\mathbb{C}$.

Proof. The corollary follows immediately for abelian groups G from Theorem 1.5.10, since KG must then be commutative. In general the number l of i's with $n_i = 1$ equals the number of irreducible representations of G over K of degree one. Since such a representation is a group homomorphism of G into the abelian group $K^{\times}$ the commutator subgroup G' is in the kernel of any such representation, hence the representations of G over K of degree one are inflations of irreducible representations of G/G'. So $l \leq [G : G']$. Conversely, $K(G/G')$ is a homomorphic image of KG and is commutative of dimension $[G : G']$, so $l = [G : G']$. $\qquad\square$

Example 1.5.13 Let G be a non-abelian group of order eight and let K be a splitting field for G of characteristic $\neq 2$, e.g. $K = \mathbb{C}$. Then

$$KG \cong K \oplus K \oplus K \oplus K \oplus K^{2 \times 2}$$

because $8 = 1 + 1 + 1 + 1 + 2^2$ is the only non-trivial way to write $|G|$ as a sum of squares with at least one summand being unity. Note that for any group we have the trivial representation, so that we may always assume in Theorem 1.5.10(b) that $n_1 = 1$ if A is a group algebra. As a consequence we may conclude in our example that any non-abelian group of order eight has exactly five conjugacy classes. It is well known that there are two such groups, namely the dihedral group D_8 and the quaternion group Q_8 of order eight. $\qquad\blacklozenge$

Remark 1.5.14 Let K be an algebraically closed field and $A \cong \bigoplus_{i=1}^{r} K^{n_i \times n_i}$. It is a natural but open problem to ask whether or not there is a finite group G such that $A \cong KG$ as a K-algebra. If so, we have of course $|G| = \sum_{i=1}^{r} n_i^2$, and r is the number of conjugacy classes of G and $|G|$ is not divisible by the characteristic of K by Maschke's theorem. There are a few necessary conditions, the second of which we will derive in chapter 2:

(a) $|\{i \in \{1, \ldots, r\} \mid n_i = 1\}|$ must be a divisor of $|G|$, because it is equal to $[G : G']$ by Corollary 1.5.12;

(b) all n_i must be divisors of $|G|$ (cf. Theorem 2.3.4).

But these conditions are far from being sufficient. For small n it is easy to list all integral solutions of

$$n = \sum x_i \cdot n_i^2 \qquad \text{with} \qquad n_1 = 1 \ , \ n_i | n \ , \ x_1 | n \ , \ 1 \le x_1 < n$$

(which describe the isomorphism types of non-commutative semisimple algebras, which might be group algebras). We do this for $n < 30$, where we have added the condition that $x_1 > 1$, since groups with order in this range are not perfect:

$6 = 2 \cdot 1^2 + 1 \cdot 2^2$

$8 = 4 \cdot 1^2 + 1 \cdot 2^2$

$10 = 2 \cdot 1^2 + 2 \cdot 2^2$

$12 = 3 \cdot 1^2 + 1 \cdot 3^2 \ = \ 4 \cdot 1^2 + 2 \cdot 2^2$

$14 = 2 \cdot 1^2 + 3 \cdot 2^2$

$16 = 4 \cdot 1^2 + 3 \cdot 2^2 \ = \ 8 \cdot 1^2 + 2 \cdot 2^2$

$18 = 2 \cdot 1^2 + 4 \cdot 2^2 \ = \ 6 \cdot 1^2 + 3 \cdot 2^2 \ = \ 9 \cdot 1^2 + 1 \cdot 3^2$

$20 = 4 \cdot 1^2 + 1 \cdot 4^2 \ = \ 4 \cdot 1^2 + 4 \cdot 2^2$

$21 = 3 \cdot 1^2 + 2 \cdot 3^2$

$22 = 2 \cdot 1^2 + 5 \cdot 2^2$

$24 = 2 \cdot 1^2 + 1 \cdot 2^2 + 2 \cdot 3^2 \ = \ 3 \cdot 1^2 + 3 \cdot 2^2 + 1 \cdot 3^2 \ = \ (4 \cdot 1^2 + 1 \cdot 2^2 + 1 \cdot 4^2)$
$\ = 4 \cdot 1^2 + 5 \cdot 2^2 \ = \ 6 \cdot 1^2 + 2 \cdot 3^2 \ = \ (8 \cdot 1^2 + 1 \cdot 4^2) \ = \ 8 \cdot 1^2 + 4 \cdot 2^2 \ = \ 12 \cdot 1^2 + 3 \cdot 2^2$

$26 = 2 \cdot 1^2 + 6 \cdot 2^2$

$27 = 9 \cdot 1^2 + 2 \cdot 3^2$

$28 = (4 \cdot 1^2 + 2 \cdot 2^2 + 1 \cdot 4^2) \ = \ 4 \cdot 1^2 + 6 \cdot 2^2$

The values in parentheses are not realized by any group, as can be seen for instance from the library of "small groups" in GAP from which the structure of all group algebras $\mathbb{C}G$ of groups G of order up to 1000 can be obtained via the character tables (see Chapter 2). The above cases can also easily be excluded using elementary group theoretical arguments (see Exercise 1.5.9).

We conclude with a practical application of the structure theorem for semisimple group algebras over a splitting field.

If for a finite group G and a field K an isomorphism

$$\varphi \colon KG \to \bigoplus_{i=1}^{r} K^{n_i \times n_i}$$

is explicitly given together with its inverse φ^{-1}, then this might be used for carrying out multiplications in the group algebra KG via

$$a \cdot b = \varphi^{-1}(\varphi(a)\varphi(b)) \quad \text{for} \quad a, b \in KG.$$

This can be efficient in particular if the degrees n_i are small or ideally one, as is the case for abelian groups. This is because a multiplication of elements of a group algebra KG of a group of order n costs approximately n^2 multiplications and additions of coefficients in addition to carrying out the multiplication of the group elements. On the other hand, for an abelian group G multiplication of elements in $\varphi(KG) = K^n$ requires just n multiplications.

The classical and most important case is that of a cyclic group C_n of order n. As shown in Example 1.1.23 the group algebra $\mathbb{C}\,C_n$ is naturally isomorphic to $R = \mathbb{C}[X]/(X^n - 1)$. Any multiplication of complex polynomials $f, g \in \mathbb{C}[X]$ can be reduced to the multiplication of the elements $f + (X^n - 1)$ and $g + (X^n - 1)$ in R, if one chooses $n > \deg f + \deg g$. The isomorphism mentioned in Corollary 1.5.12 can be given explicitly and is usually called the **discrete Fourier transform**. For this it is convenient to identify $\mathbb{C}^n$ with the set of polynomials over $\mathbb{C}$ of degree less than n – after all a polynomial is by its very definition simply the sequence of its coefficients – and write

$$a(z) = \sum_{i=0}^{n-1} a_i z^i \quad \text{for} \quad a = (a_0, \ldots, a_{n-1}) \in \mathbb{C}^n \quad \text{and} \quad z \in \mathbb{C}.$$

We then define for a primitive nth root of unity $\zeta \in \mathbb{C}$

$$\mathrm{DF}_\zeta \colon \mathbb{C}^n \to \mathbb{C}^n, \quad a \mapsto (a(\zeta^0), a(\zeta), \ldots, a(\zeta^{n-1})),$$

to obtain the following lemma.

Lemma 1.5.15 *With the above notation the map*

$$DFT \colon \mathbb{C}[X]/(X^n - 1) \to \mathbb{C} \oplus \cdots \oplus \mathbb{C}, \quad a + (X^n - 1) \mapsto DF_\zeta(a),$$

is an algebra isomorphism with inverse

$$IDFT \colon \mathbb{C}^n \to \mathbb{C}[X]/(X^n - 1), \quad a \mapsto \frac{1}{n} DF_{\zeta^{-1}}(a) + (X^n - 1).$$

Proof. If $a = (a_0, \ldots, a_{n-1})$ we find

$$\mathrm{DF}_{\zeta^{-1}}(\mathrm{DF}_\zeta(a)) = (b_0, \ldots, b_{n-1})$$

with

$$
\begin{aligned}
b_i &= \sum_{j=0}^{n-1} \left(\sum_{k=0}^{n-1} a_k \zeta^{jk} \right) \zeta^{-ij} = \sum_{k=0}^{n-1} a_k \left(\sum_{j=0}^{n-1} \zeta^{j(k-i)} \right) \\
&= \sum_{k=0}^{n-1} a_k \cdot n \delta_{i,k} = n \cdot a_i,
\end{aligned}
$$

where we used the well known identity

$$\sum_{i=0}^{n-1} \zeta^{ij} = \begin{cases} n & \text{if } j = 0, \\ 0 & \text{if } j \neq 0. \end{cases}$$

(See also Exercise 2.1.8.) $\qquad\square$

The discrete Fourier transform has many practical applications and can for instance be used to perform efficient multiplications of polynomials whose degrees add up to less than n requiring only n multiplications in addition to the operations DFT and IDFT, which can be carried out fast and recursively if n is a power of two. In fact, it can be shown that the multiplication of polynomials of degree $m < \frac{n}{2}$ can thus be carried out at cost $O(m \log m)$. For details and extensions of the method to other than cyclic groups we refer to [27]. We would also like to mention that there is an intriguing new application of group algebras to the complexity theory of matrix multiplication; see [30] and [31].

Exercises

Exercise 1.5.1 Let $V \neq \{0\}$ be an A-module of finite length m with $\operatorname{Rad} V = \{0\}$. Show (by induction on m) that V has m maximal submodules M_i such that $V \cong_A V/M_1 \oplus \cdots \oplus V/M_m$.

Exercise 1.5.2 (a) Let V be a semisimple A-module. Show that $\operatorname{End}_A V$ is commutative if and only if all composition factors of V have multiplicity one. Such a module (or the corresponding representation) is usually called **multiplicity-free**.
(b) Let G be a finite group and let Ω be a G-set of rank $r \leq 3$. Show that $K\Omega$ is multiplicity-free provided that $\operatorname{char} K \nmid |G|$.

Exercise 1.5.3 Show that $(\operatorname{End}_A {}_A A)^{\mathrm{op}} \to A$, $\varphi \mapsto \varphi(1)$ is an algebra isomorphism.
Hint: For $a \in A$ consider $\varphi_a \colon A \to A$, $x \mapsto xa$. In Theorem 1.6.8 we will generalize this result.

Exercise 1.5.4 For any $n \in \mathbb{N}$ we always use $\zeta_n := \exp(2\pi i/n) \in \mathbb{C}$.
(a) For $n = 2k$, $k \in \mathbb{N}$, let
$$f = (X^k - 1)q_1 + r_1 = (X^k + 1)q_2 + r_2 \in \mathbb{C}[X] \quad \text{with} \quad \deg r_1, \deg r_2 < k$$
and $q_1, q_2, r_1, r_2 \in \mathbb{C}[X]$. Show that for $0 \leq i < k$ one has
$$f(\zeta_n^{2i}) = r_1(\zeta_n^{2i}) \qquad \text{and} \qquad f(\zeta_n^{2i+1}) = r_2(\zeta_n^{2i+1}).$$
(b) Let $f := X^3 + 3X^2 + X + 1 \in \mathbb{C}[X]$ and $g := X^4 + 3X^2 + X + 1 \in \mathbb{C}[X]$. Compute the product $f \cdot g \in \mathbb{C}[X]$ via the discrete Fourier transform.

Exercise 1.5.5 For $n = 2k$, $k \in \mathbb{N}$, let $\mathbf{DF}_\zeta$ for $\zeta := \zeta_n$ be the matrix of the discrete Fourier transform $\mathrm{DF}_\zeta \colon \mathbb{C}^n \to \mathbb{C}^n$. Moreover, define $\mu := \zeta^2$ and denote by D the diagonal matrix with entries $\zeta^i, i = 0, \ldots, k-1$. Finally let P be the permutation matrix corresponding to the permutation π on $\{1, \ldots, n\}$ with $\pi(i) = k + i/2$ if i is even and $\pi(i) = (i+1)/2$ if i is odd. Show that $\mathbf{DF}_\zeta$ can be factorized as follows:
$$\mathbf{DF}_\zeta = \begin{bmatrix} \mathbf{I}_k & D \\ \mathbf{I}_k & -D \end{bmatrix} \cdot \begin{bmatrix} \mathbf{DF}_\mu & 0 \\ 0 & \mathbf{DF}_\mu \end{bmatrix} \cdot P.$$
Show that if n is of the form 2^m, the Fourier transform can be computed in $O(n \ln(n))$ steps.

Exercise 1.5.6 Show that $\mathbb{Q}\,\mathrm{C}_n \cong \bigoplus_{d|n} \mathbb{Q}(\zeta_d)$.

Exercise 1.5.7 Let V and W be A-modules and $E := \mathrm{End}_A V$. Show that $\mathrm{Hom}_A(W, V)$ is an E-module in a natural way. If A is semisimple, show that E is a semisimple K-algebra. Let $V_1, \ldots, V_r$ be a system of representatives for the isomorphism classes of simple A-submodules of V. Show that $\{\mathrm{Hom}_A(V_i, V) \mid 1 \leq i \leq r\}$ is the set of all simple E-modules up to isomorphism.

Exercise 1.5.8 Let A be a finite dimensional K-algebra for a field K and let $L \supseteq K$ be an extension field.

(a) If K is a splitting field for A, show that L is also a splitting field for A and that for every simple LA-module W there is a simple A-module V with $W \cong LV$.

(b) Conversely, assume that L is a splitting field for A and that for every simple LA-module W there is a simple A-module V with $W \cong LV$. Show that K is a splitting field for A.

Exercise 1.5.9 Show that

(a) a group G of order 24 with $[G : G'] = 4$ cannot have exactly six conjugacy classes;

(b) a group G of order 24 with $[G : G'] = 8$ cannot have exactly nine conjugacy classes;

(c) a group G of order 28 cannot have exactly seven conjugacy classes.

1.6 Direct sums and idempotents

In this section A is assumed to be a K-algebra for some commutative ring K and V is a finitely generated A-module.

We recall that a submodule $U \leq_A V$ is called a **direct summand** if there is a submodule $U' \leq_A V$ such that $V = U \oplus U'$ (see Definition 1.1.13). We will use the notation

$$U \mid_A V \quad \text{or just} \quad U \mid V$$

to mean that U is isomorphic as A-module to a direct summand of V. Recall that $e \in A$ is an idempotent if $e^2 = e \neq 0$ (see Definition 1.2.8).

Definition 1.6.1 Two idempotents $e', e'' \in A$ are called **orthogonal** if $e'e'' = e''e' = 0$. An idempotent $e \in A$ is called **primitive** if e cannot be written as a sum of two orthogonal idempotents in A.

Lemma 1.6.2 (a) $U \mid V$ if and only if $U = \mathrm{im}\,\varphi$ for some $\varphi \in \mathrm{End}_A V$ with $\varphi^2 = \varphi$. Then $V = \mathrm{im}\,\varphi \oplus \ker\varphi$ and U is indecomposable if and only if id_U is the only idempotent in $\mathrm{End}_A U$.

(b) $\{0\} \neq U \mid {}_AA$ if and only if $U = Ae$ for some idempotent $e \in A$. Then ${}_AA = Ae \oplus A(1 - e)$ and Ae is indecomposable if and only if e is primitive.

Proof.   (a) If $U = U_1, U_2 \leq_A V$ and $V = U_1 \oplus U_2$ with projections π_1, π_2 and injections ι_1, ι_2 as in Definition 1.6.2 then $\varphi := \iota_1 \circ \pi_1$ satisfies $\varphi^2 = \varphi$ and $U = \operatorname{im} \varphi$, $U_2 = \ker \varphi$. Conversely, let $\varphi^2 = \varphi \in \operatorname{End}_A V$ any $v \in V$. Then

$$v = \varphi(v) + (v - \varphi(v)) \in \operatorname{im} \varphi + \ker \varphi.$$

If $v \in \operatorname{im} \varphi \cap \ker \varphi$ then $v = \varphi(w)$ for some $w \in V$ and $0 = \varphi(v) = \varphi^2(w) = \varphi(w) = v$. Thus $V = \operatorname{im} \varphi \oplus \ker \varphi$. Applying the argument to U, we see that any idempotent in $\operatorname{End}_A U$ produces a direct summand $\neq \{o\}$ of U, and conversely.

   (b) If $_A A = U \oplus U'$ there are uniquely determined elements $e \in U$, $e' \in U'$ with $1_A = e + e'$. Then for any $a \in U$ we have $a + 0 = a = a 1_A = ae + ae'$ implies $ae = a$, $ae' = 0$. Thus $U = Ae$ and for $a := e$ we get $e + 0 = e = e 1_A = e^2 + ee'$ implies $e^2 = e$, $ee' = 0$. Similarly $U' = Ae'$ and $e'^2 = e'$, $e'e = 0$.

   Conversely, if $e^2 = e \in A$ then $_A A = Ae \oplus A(1 - e)$ is readily verified. $\square$

Remark 1.6.3 $\varphi^2 = \varphi \in \operatorname{End}_A V$ is equivalent to $\varphi(\varphi - \operatorname{id}_V) = 0$. Thus, if $\operatorname{End}_A V$ is a division ring, then V is indecomposable. Consequently, if V is semisimple and $\operatorname{End}_A V$ is a division ring then V is simple. This is a kind of converse to Schur's lemma.

We recall that an A-module V is called **artinian** (resp. **noetherian**) if every sequence of submodules of the form $V = V_1 >_A V_2 >_A \cdots >_A V_n >_A \cdots$ (resp. $\{0\} = V_1 <_A V_2 <_A \cdots <_A V_n <_A \cdots$) is finite. It is not hard to see (see [4], Proposition 11.1), that a module has finite length if and only if it is artinian and noetherian. A ring A is called **artinian** (resp. **noetherian**) if the regular module $_A A$ has this property. There is a well known theorem due to Hopkins (see [4], Theorem 15.20) saying that an artinian ring (with unit) is necessarily also noetherian. We will not prove this fact.

Lemma 1.6.4 (**Fitting's lemma**) *For $\varphi \in \operatorname{End}_A V$ the following hold.*

(a) $\ker \varphi^i \subseteq \ker \varphi^{i+1}$ *and* $\operatorname{im} \varphi^i \supseteq \operatorname{im} \varphi^{i+1}$.

(b) *If V is artinian, then there is an $r \in \mathbb{N}$ with $\operatorname{im} \varphi^r = \operatorname{im} \varphi^{r+i}$ for all $i \in \mathbb{N}$ and*

$$V = \operatorname{im} \varphi^r + \ker \varphi^r;$$

φ is an automorphism if and only if it is injective.

(c) *If V is noetherian, then there is an $s \in \mathbb{N}$ with $\ker \varphi^s = \ker \varphi^{s+i}$ for all $i \in \mathbb{N}$ and*

$$\operatorname{im} \varphi^s \cap \ker \varphi^s = \{0\};$$

φ is an automorphism if and only if it is surjective.

(d) *If V is artinian and noetherian then*

$$V = \operatorname{im} \varphi^m \oplus \ker \varphi^m$$

for sufficiently large m.

Proof. (a) is obvious.

(b) If $\operatorname{im}\varphi^r = \operatorname{im}\varphi^{r+1}$ then for any $v \in V$ there is a $w \in V$ with $\varphi^r(v) = \varphi^{2r}(w)$. Then

$$v = \varphi^r(w) + (v - \varphi^r(w)) \in \operatorname{im}\varphi^r + \ker\varphi^r$$

because $\varphi^r(v - \varphi^r(w)) = 0$.

(c) If $\ker\varphi^s = \ker\varphi^{s+1}$ then for any $v = \varphi^s(w) \in \operatorname{im}\varphi^s \cap \ker\varphi^s$ one finds $\varphi^{2s}(w) = \varphi^s(v) = 0$, so $w \in \ker\varphi^{2s} = \ker\varphi^s$, hence $v = \varphi^s(w) = 0$.

(d) is an immediate consequence of (b) and (c). $\qquad\square$

We recall that a ring is called **local** if its non-units form an ideal (which is then, of course, the only maximal ideal). Since a local ring obviously cannot contain any idempotent $\neq 1$, it follows that if $\operatorname{End}_A V$ is local then V must be indecomposable.

Corollary 1.6.5 *Assume that V has finite length. Then V is indecomposable if and only if $\operatorname{End}_A V$ is local.*

Proof. If $\varphi \in \operatorname{End}_A V$ is not a unit, then by Fitting's lemma part (c) it cannot be surjective and by (d) it must be nilpotent. For any $\psi \in \operatorname{End}_A V$ it follows that $\psi\varphi$ is not injective and $\varphi\psi$ is not surjective, hence these products are non-units. If ψ is also a non-unit in $\operatorname{End}_A V$ then $\varphi + \psi$ cannot be invertible, because otherwise $\varphi' = \varphi(\varphi + \psi)^{-1}$ would be a non-unit, and hence nilpotent, and $1 - \varphi'$ would be invertible (with inverse $\sum_{i \geq 0} \varphi'^i$). But $1 - \varphi' = \psi(\varphi + \psi)^{-1}$ is not surjective. $\qquad\square$

Theorem 1.6.6 (Krull–Schmidt) *If $V \neq \{0\}$ is an A-module and*

$$V = V_1 \oplus \cdots \oplus V_r = W_1 \oplus \cdots \oplus W_s$$

with indecomposable submodules V_i, W_j, such that $\operatorname{End}_A V_i$ is local for $i = 1, \ldots, r$, then $r = s$ and there is a permutation $\sigma \in S_r$ with $V_i \cong_A W_{\sigma(i)}$. If $V \neq \{0\}$ is an A-module of finite length, then V is a finite direct sum of indecomposable A-modules which are uniquely determined up to isomorphism and ordering.

Proof. We use induction with respect to $n := \min(r, s)$, the result being evidently true for $n = 1$.

Let $\varphi_i = \varphi_i^2 \in \operatorname{End}_A V$ and $\psi_j = \psi_j^2 \in \operatorname{End}_A V$ be the natural projections with

$$\operatorname{im}\varphi_i = V_i, \quad \ker\varphi_i = \bigoplus_{k \neq i} V_k, \quad \operatorname{im}\psi_j = W_j, \quad \ker\psi_j = \bigoplus_{k \neq j} W_k,$$

for $i = 1, \ldots, r$ and $j = 1, \ldots, s$. We have $\operatorname{id}_V = \sum_{i=1}^r \varphi_i = \sum_{j=1}^s \psi_j$ and hence

$$\operatorname{id}_{V_1} = \varphi_1|_{V_1} = \sum_{j=1}^s \varphi_1 \psi_j|_{V_1} \in \operatorname{End}_A V_1.$$

Since $\operatorname{End}_A V_1$ is a local ring by assumption, some summand $\varphi_1 \psi_j|_{V_1}$ must be a unit. We may choose the notation so that this is the case for $j = 1$. Hence $\varphi_1 \psi_1|_{V_1} : V_1 \to V_1$ is an isomorphism. We put

$$\varphi := \psi_1 \varphi_1 + \varphi_2 + \cdots + \varphi_r \in \operatorname{End}_A V.$$

Then $\varphi(V_i) = V_i$ for $i \geq 2$. If $v_1 \in V_1$ then $v_1 = \varphi_1(v_1) = \varphi_1 \psi_1(v_1')$ for some $v_1' \in V_1$. Hence

$$v_1 - \varphi(v_1') = v_1 - \psi_1(v_1') \in \ker \varphi_1 = V_2 \oplus \cdots \oplus V_r \subseteq \operatorname{im} \varphi.$$

So $v_1 \in \operatorname{im} \varphi$ and φ is surjective. If $v \in \ker \varphi$ then $0 = \varphi_1 \varphi(v) = \varphi_1 \psi_1 \varphi_1(v)$, so $\varphi_1(v) = 0$ and, since $\varphi = \operatorname{id}_V - \varphi_1 + \psi_1 \varphi_1$, we get $v = 0$. Thus φ is an isomorphism, and

$$V = \varphi(V_1) \oplus \cdots \oplus \varphi(V_r) = W_1 \oplus V_2 \oplus \cdots \oplus V_r$$

and

$$V_2 \oplus \cdots \oplus V_r \cong V/V_1 \cong \varphi(V)/\varphi(V_1) \cong V/W_1 \cong W_2 \oplus \cdots \oplus W_s.$$

The result now follows by induction and Corollary 1.6.5. $\square$

Remark 1.6.7 If V, V' are A-modules then $H := \operatorname{Hom}_A(V, V')$ is an (E', E)-bimodule with $E := \operatorname{End}_A V$, $E_A' := \operatorname{End}_A V'$ and

$$E' \times H \to H, \ (\epsilon', \varphi) \mapsto \epsilon' \circ \varphi, \qquad H \times E \to H, \ (\varphi, \epsilon) \mapsto \varphi \circ \epsilon.$$

Theorem 1.6.8 *If $e \in A$ is an idempotent and V is an A-module, then*

(a) $\operatorname{Hom}_A(Ae, V) \cong eV$ *as $(\operatorname{End}_A V)$-modules;*

(b) $\operatorname{End}_A Ae \cong (eAe)^{\operatorname{op}}$ *as K-algebras. In particular, $\operatorname{End}_A A \cong A^{\operatorname{op}}$;*

Proof. (a) A natural isomorphism from $\operatorname{Hom}_A(Ae, V)$ to eV is given by

$$\varphi \mapsto \varphi(e) = \varphi(e^2) = e\varphi(e) \in eV \quad \text{for} \ \varphi \in \operatorname{Hom}_A(Ae, V),$$

with inverse

$$eV \to \operatorname{Hom}_A(Ae, V), \ ev \mapsto \varphi_{ev} \ \text{with} \ \varphi_{ev}(ae) := aev \ \text{for} \ a \in A, v \in V.$$

   (b) The same map as defined in (a) gives for $V = Ae$ a ring isomorphism $\operatorname{End}_A Ae \to (eAe)^{\operatorname{op}}$, since

$$\varphi \circ \psi(e) = \varphi(\psi(e)e) = \psi(e)\varphi(e) \quad \text{for} \quad \varphi, \psi \in \operatorname{End}_A Ae.$$

$\square$

Remark 1.6.9 By Remark 1.6.7, $\text{Hom}_A(Ae, V)$ is a right $\text{End}_A\, Ae$-module, and hence by Theorem 1.6.8(b) it is a left eAe-module with action given by

$$(a \cdot \varphi)(x) := \varphi(xa) \quad \text{for} \quad a \in eAe \,, \ \varphi \in \text{Hom}_A(Ae, V) \,, \ x \in Ae.$$

The isomorphism given in Theorem 1.6.8(a) is an eAe-module isomorphism.

The results of the rest of this section will only be used in Chapter 4.

Lemma 1.6.10 (Rosenberg's lemma) *Let $e = e^2 \in A$ with $\text{End}\, Ae$ local. If $e \in I_1 + \cdots + I_m$ with $I_i \trianglelefteq A$ $(1 \le i \le m)$ then $e \in I_j$ for some $j \in \{1, \ldots, m\}$.*

Proof. By Theorem 1.6.8 eAe is local. Obviously $eI_1e, \ldots, eI_me$ are ideals in eAe and $e \in eI_1e + \cdots + eI_me$. Since $e \notin J(eAe)$ there must be some j with $eI_je \not\subseteq J(eAe)$. Hence $e \in eAe = eI_je \subseteq I_j$. □

Theorem 1.6.11 *Let $e \in A$ be an idempotent. Then $eAe \subseteq A$ is a K-algebra with identity e, and for any A-module V we get an eAe-module eV. Also, $\varphi \mapsto \varphi|_{eV}$ defines a map $\text{Hom}_A(V, W) \to \text{Hom}_{eAe}(eV, eW)$. Let V be an A-module.*

(a) *If $W \le_A V$ then $eW \le_{eAe} eV$ and $e(V/W) \cong_{eAe} eV/eW$.*

(b) *Every eAe-submodule of eV is of the form eW for some A-submodule W of V. In particular, if V is a simple A-module then $eV = \{0\}$ or eV is a simple eAe-module.*

(c) *Let $\{L_i \mid i \in I\}$ be a complete set of representatives of simple A-modules and $I' := \{i \in I \mid eL_i \ne \{0\}\}$. Then $\{eL_i \mid i \in I'\}$ is a complete set of representatives of simple eAe-modules.*

(d) *If A is artinian and $eL \ne \{0\}$ for all composition factors L of V then $W \mapsto eW$ defines a poset isomorphism from the poset of A-submodules of V to the poset of eAe-submodules of eV.*

Proof. (a) The first assertion is obvious and

$$e(V/W) \to eV/eW \quad, \quad e(v + W) \mapsto ev + eW$$

is clearly an isomorphism of eAe-modules.

(b) Let $\tilde{W} \le_{eAe} eV$. Then $e\tilde{W} = \tilde{W}$ and we obtain an A-submodule of V by defining $W := A\tilde{W}$. This satisfies $eW = eA\tilde{W} = e(Ae\tilde{W}) = (eAe)\tilde{W} = \tilde{W}$.

(c) Let $\tilde{V}$ be an eAe-module. Then $V := Ae \otimes_{eAe} \tilde{V}$ is an A-module because Ae is an (A, eAe)-bimodule and $eV = e \otimes_{eAe} \tilde{V} \cong_{eAe} \tilde{V}$. Let

$$V_{(e)} := \sum \{W \le_A V \mid eW = \{0\}\},$$

the largest A-submodule of V contained in $(1 - e)V$. Thus $eV_{(e)} = \{0\}$, and by (a) we have $e(V/V_{(e)}) \cong_{eAe} eV \cong_{eAe} \tilde{V}$. Now assume that $\tilde{V}$ is simple. Then

$V/V_{(e)} \neq \{0\}$ and we want to show that this is a simple A-module, or, in other words, that $V_{(e)}$ is a maximal submodule of V. Let W be a proper submodule of V. By (a) eW is an eAe-submodule of $eV \cong_{eAe} \tilde{V}$. If $eW \neq \{0\}$ then $eW = eV$ because this module is simple by assumption. Then $W \geq AeW = AeV = V$, which is a contradiction. Thus $V_{(e)}$ is a maximal submodule of V and $L := V/V_{(e)}$ is a simple A-module with $eL \cong_{eAe} \tilde{V}$.

(d) This obviously follows from (a) and (b). $\qquad\qquad\qquad\qquad\square$

Corollary 1.6.12 *Let $n \in \mathbb{N}$. Then A is semisimple if and only if the K-algebra $A^{n \times n}$ is semisimple.*

Proof. Let $A' := A^{n \times n}$ and let $\epsilon_{ij} \in A'$ be the matrix with 1_A at position (i,j) and zeros elsewhere. Then $e_i := \epsilon_{ii}$ is an idempotent and

$$A' = A'e_1 \oplus \cdots \oplus A'e_n.$$

Also $\eta_i \colon A \to e_i A' e_i$, $a \mapsto ae_i$ defines a ring isomorphism and $A'e_i \cong_{A'} A'e_j$ for $1 \leq i, j \leq n$. For $i \neq j$ put $\pi_{ij} := \epsilon_{ij} + \epsilon_{ji} + \sum_{k \neq i,j} e_k$, a permutation matrix with $\pi_{ij}^2 = 1_{A'}$. Then $\pi_{ij} e_j = e_i \pi_{ij}$. If $W \leq_{A'} A'e_i$ then

$$e_j W = \pi_{ij} \pi_{ij} e_j W \subseteq \pi_{ij} e_i W \subseteq e_j W.$$

Hence $W = \sum_{j=1}^n e_j W = \sum_{j=1}^n \pi_{ij} e_i W$, and $W \mapsto e_i W$ defines (by parts (a) and (b) of the theorem) an isomorphism of the poset of submodules of $A'e_i$ to the poset of left ideals of $e_i A' e_i \cong A$. From this the assertion follows. $\quad\square$

The eAe-module eV is often called the **condensed** module for V and eAe the **Hecke algebra**. Observe that eV will usually have a dimension which is much smaller than that of V (assuming that A is K-algebra over a field), so that sometimes the structure of eV can be investigated on a computer in cases when V would be intractable. Then part (d) of Theorem 1.6.11 may be used to obtain information on V itself. There is one practical problem though. If $a_1, \ldots, a_m$ are algebra generators for A (in most applications $m = 2$) there is in general no reason why $ea_1 e, \ldots, ea_m e$ should be algebra generators for eAe (see Example 1.6.13 below).

In the case of a group algebra over a field K, subgroups H of order prime to the characteristic of K can be used to get idempotents $e_H := \frac{1}{|H|} H^+$; see Lemma 1.2.9. In this context the idempotent e_H is called the fix idempotent of H, and the process described above is called **fix point condensation**. The reason for this is that for a KG-module V and H as above we obviously have

$$e_H V = \{v \in V \mid hv = v \text{ for all } h \in H\} = \mathrm{Inv}_H(V).$$

Recall that by Lemma 1.2.9 we have isomorphisms

$$\Psi \colon K[G/H] \to KGe_H, \qquad gH \mapsto g \cdot H^+$$

and

$$\Phi \colon (\operatorname{End}_{KG} K[G/H])^{\mathrm{op}} \to e_H KG e_H, \qquad \varphi \mapsto \Psi \circ \varphi \circ \Psi^{-1}(e_H) = \frac{1}{|H|} \Psi(\varphi(1H)).$$

Let $G = \bigcup_{i=1}^{r} H g_i H$. Then the basis corresponding to the Schur basis (see p. 29) of $\operatorname{End}_{KG} K[G/H]$ is given by

$$\{\Phi(\theta_1), \dots, \Phi(\theta_r)\} = \Big\{ \sum_{H = \bigcup h(H \cap g_i H)} h g_i e_H \mid 1 \le i \le r \Big\}.$$

Note that $\Phi(\theta_i) = \frac{|H|}{|H \cap g_i H|} e_H g_i e_H$.

Example 1.6.13 Let $G := S_4$ and $H := \langle (2,3,4) \rangle$. It is obvious that $G = \langle a_1, a_2 \rangle$ with $a_1 := (1,2,3,4)$ and $a_2 := (1,2)$. If K is a field with $\operatorname{char} K \ne 3$ and $e := e_H$ as above, it is easily checked that $e a_1 e = e a_2 e$. Moreover, if $\operatorname{char} K = 2$ then $(e a_1 e)^2 = e$ and hence $\dim_K \langle e a_1 e, e a_2 e \rangle_{\mathrm{alg}} = 2$. On the other hand it is easy to see that the number of (H,H)-double cosets in G is four and so by Lemma 1.2.11 and Corollary 1.2.16 $\dim_K (eKGe) = 4$ for any field K with $\operatorname{char} K \ne 3$. Thus, although a_1, a_2 are algebra generators for KG the elements $e a_1 e$, $e a_2 e$ may generate a proper subalgebra of $eKGe$. However, if we take $a_3 := (3,4)$, then $G = \langle a_1, a_3 \rangle$ and $e \mathbb{F}_2 G e = \langle e a_1 e, e a_3 e \rangle_{\mathrm{alg}} \cong \mathbb{F}_2 V_4$ (see Exercise 1.6.1). ◆

Example 1.6.14 We use GAP to compute matrices for condensed elements using the formula given in Exercise 1.6.3: we take the sporadic simple Mathieu group $G := M_{11}$ acting on $\Omega = \{1, \dots, 11\}$. Let M be the natural permutation $\mathbb{F}_2 M_{11}$-module $M := \mathbb{F}_2 \binom{\Omega}{2}$ on the subsets of size two of Ω. We choose the condensation subgroup H to be a Sylow 3-subgroup (of order nine). The following GAP-code computes the action on $e_H M$ for the elements $e_H g_1 e_H, e_H g_2 e_H$, where $g_1 := (1,4)(2,10,3)(5,11,8,6,7,9)$, $g_2 := (2,3,7,4,10,8,5,11)(6,9)$. Finally we determine the composition factors for the condensed module $e_H M$ as a module for $\langle e_H g_1 e_H, e_H g_2 e_H \rangle_{\mathrm{alg.}}$.

We take M_{11} from the GAP-library, verify that the chosen elements g_1, g_2 generate G, compute a Sylow 3-subgroup and proceed by determining the action of M_{11} on the subsets of size two. Applying the resulting homomorphism we construct permutations for the generators g_1, g_2 and the Sylow 3-subgroup H:

```
gap> G := MathieuGroup(11);
Group([ (1,2,3,4,5,6,7,8,9,10,11), (3,7,11,8)(4,10,5,6) ])
gap> g1 := (1,4)(2,10,3)(5,11,8,6,7,9);; g2 := (2,3,7,4,10,8,5,11)(6,9);;
gap> G = Group( g1, g2 );
true
gap> H := SylowSubgroup( G, 3 );;
gap> orb :=  Orbit( G , [1,2], OnSets );;
gap> hom := ActionHomomorphism( G, orb, OnSets );;
gap> g1 := Image( hom, g1 );; g2 := Image( hom, g2 );;
gap> H := Image( hom, H );;
```

The following GAP-code constructs the condensations of g_1, g_2 using the program **cond** given in Exercise 1.6.4. Finally, we determine the composition factors of the condensed module and their dimensions.

```
gap> mats := List( [g1, g2] , g ->  cond( H, 55, g, 2) );;
gap> Display(mats[1]);
 . . 1 . . . .
 . 1 1 . . . 1
 1 1 . . . . 1
 . . . 1 1 1 .
 . . . 1 1 1 .
 . . . 1 1 1 .
 . 1 1 . . . 1
gap> Display(mats[2]);
 . . . . . . 1
 . 1 . . . . .
 . . . . 1 . .
 . . . . . 1 .
 . . . 1 . . .
 . . 1 . . . .
 1 . . . . . .
gap> M := GModuleByMats( List( mats, TransposedMat ), GF(2) );;
gap> compfactors  := MTX.CompositionFactors( M );;
gap> List( compfactors, x -> x.dimension );
[ 2, 4, 1 ]
```

This shows that the condensed module has exactly three composition factors of dimensions four, two, and one. If we had chosen $g_2' := (1, 2, 3, 8, 7, 5, 4, 6)(10, 11)$ instead of g_2 then still $G = \langle g_1, g_2' \rangle$, but we would have obtained composition factors of dimensions two, two, one, one, and one, which clearly shows that $\langle e_H g_1 e_H, e_H g_2' e_H \rangle_{\mathrm{alg}} < e_H F G e_H$. ♦

If A is a semisimple algebra then every A-module is a direct summand of a free A-module. In general, modules with this property play an important role and deserve a proper name.

Definition 1.6.15 An A-module V is **projective** if V is a direct summand of a free A-module. A module V is called **projective indecomposable**, often abbreviated as *PIM*, if V is projective and indecomposable.

Occasionally we will also use other well known characterizations of projective modules.

Lemma 1.6.16 *V is projective if and only if one of the following holds.*

(a) *For each submodule U of an A-module W any $\bar{\varphi} \in \mathrm{Hom}_A(V, W/U)$ can be "lifted" to $\varphi \in \mathrm{Hom}_A(V, W)$, that is $\varphi(v) + U = \bar{\varphi}(v)$ for all $v \in V$.*

(b) *Any short exact sequence of A-homomorphisms*

$$\{0\} \longrightarrow V_1 \xrightarrow{\alpha} V_2 \xrightarrow{\beta} V \longrightarrow \{0\}$$

splits. That is, if α and β are injective and surjective, respectively, and if $\ker \beta = \operatorname{im} \alpha$ *then there is a* $\gamma \in \operatorname{Hom}_A(V, V_2)$ *with* $\beta \circ \gamma = \operatorname{id}_V$ *(and consequently* $V_2 = \operatorname{im} \alpha \oplus \operatorname{im} \gamma$*).*

Proof. This is a simple exercise. □

Lemma 1.6.17 *If G is a finite group, $H \leq G$ and V is a projective KG-module, then the restriction V_H is a projective KH-module.*

Proof. Let $G = \dot{\bigcup}_{i=1}^{t} H g_i$. Then $(KG)_H = \bigoplus_{i=1}^{t} KH\, g_i$ and $KH\, g_i \cong_{KH} KH$. So $(KG)_H$ is a free KH-module. If $V \mid KG \oplus \cdots \oplus KG$ then $V_H \mid (KG)_H \oplus \cdots \oplus (KG)_H$, and hence is projective. □

The following result should be compared with Theorem 1.1.15.

Theorem 1.6.18 (Fitting) *Let $E = (\operatorname{End}_A V)^{\mathrm{op}}$ and assume that*

$$V = V_1 \oplus \cdots \oplus V_n, \qquad V_i \leq_A V,$$

with projections $\pi_i \colon V \to V_i$ and embeddings $\iota_i \colon V_i \to V$ (see Definition 1.1.13). Put $E_i' = \operatorname{Hom}_A(V, V_i)$ and $E_i := \{\iota_i \circ \varphi \mid \varphi \in E_i'\}$. Then

(a) $E_i \leq_E E$ *for $1 \leq i \leq n$ and $E = E_1 \oplus \cdots \oplus E_n$,*

(b) $E_i \cong_E E_j$ *if and only if $V_i \cong_A V_j$,*

(c) E_i *is indecomposable (as an E-module) if and only if V_i is indecomposable as an A-module.*

Proof. (a) Observe that E acts on E_i' by $\psi \cdot \varphi := \varphi \circ \psi$ for $\psi \in E$ and $\varphi \in E_i'$ and similarly on E_i. Thus $E_i \leq_E E$.

Put $\epsilon_i := \iota_i \circ \pi_i$ for $i = 1, \ldots n$. Then $\operatorname{id}_V = \sum_{i=1}^{n} \epsilon_i$ and $\epsilon_i \cdot \epsilon_j = \delta_{i,j}\epsilon_i$. Thus the ϵ_i are orthogonal idempotents in E and $E = \sum_{i=1}^{n} E \cdot \epsilon_i$. Also, since $\operatorname{id}_{V_i} = \pi_i \circ \iota_i$ it is clear that $E_i = E \cdot \epsilon_i$.

(b) If $\varphi_{ij} \colon V_i \to V_j$ is an isomorphism of A-modules then

$$\boldsymbol{f}(\varphi_{ij}) \colon E_i \to E_j, \qquad \varphi \mapsto \iota_j \circ \varphi_{ij} \circ \pi_i \circ \varphi$$

is an isomorphism of E-modules (with inverse $\boldsymbol{f}(\varphi_{ij}^{-1})$).

Conversely, if $f_{ij} \colon E_i \to E_j$ is an isomorphism of E-modules, then we get an isomorphism of A-modules by

$$\boldsymbol{g}(f_{ij}) \colon V_i \to V_j, \qquad v_i \mapsto f_{ij}(\epsilon_i)(v_i)$$

with inverse $\boldsymbol{g}(f_{ij}^{-1})$. In fact, since f_{ij}^{-1} is E-linear, we get

$$\boldsymbol{g}(f_{ij}^{-1})(f_{ij}(\epsilon_i)(v_i)) = (f_{ij}^{-1}(\epsilon_j) \circ (f_{ij}(\epsilon_i)))(v_i) = (f_{ij}(\epsilon_i) \cdot f_{ij}^{-1}(\epsilon_j))(v_i)$$

$$= (f_{ij}^{-1}(f_{ij}(\epsilon_i) \cdot \epsilon_j))(v_i) \quad = \quad f_{ij}^{-1}(f_{ij}(\epsilon_i))(v_i) \ = \ v_i$$

(c) E_i is indecomposable if and only if $\iota_i \circ \pi_i$ is a primitive idempotent in E. Likewise, V_i is indecomposable if and only if $\mathrm{id}_{V_i} = \pi_i \circ \iota_i$ cannot be written as a sum of two idempotents (projections) in $\mathrm{End}_A V_i$, which is equivalent to $\iota_i \circ \pi_i$ being primitive in E. $\square$

Definition 1.6.19 A projective A-module P is called a **projective cover** of V if $P/\mathrm{Rad}(P) \cong_A V/\mathrm{Rad}(V)$.

Lemma 1.6.20 (a) *Let P be a projective cover of V. Then there is an epimorphism $\pi_P \colon P \to V$ lifting the epimorphism $P \to P/\mathrm{Rad}(P) \to V/\mathrm{Rad}(V)$. If $\psi \colon P' \to V$ is an epimorphism with P' projective then there is an epimorphism $\theta \colon P' \to P$ with $\psi = \pi_P \circ \theta$ and $P' \cong_A P \oplus \ker \theta$.*

(b) *If P, P' are projective covers of V then $P \cong_A P'$ and $\ker \pi_P \cong_A \ker \pi_{P'}$ with the notation of (a). Thus a projective cover (if it exists) is uniquely determined up to isomorphism; it is usually denoted by $P(V)$. Also $\Omega(V) := \ker \pi_{P(V)}$ is uniquely determined up to isomorphism and called the **Heller module** of V.*

(c) *If V is projective then $P(V) \cong_A V$ and $\Omega(V) = \{0\}$.*

Proof.   (a) By Lemma 1.6.16(a) there is a homomorphism $\pi_P \colon P \to V$ lifting the epimorphism $P \to V/\mathrm{Rad}(V)$. Then $\mathrm{im}\,\pi_P + \mathrm{Rad}(V) = V$, so $\mathrm{im}\,\pi_P$ cannot be contained in a maximal submodule of V and hence $\mathrm{im}\,\pi_P = V$. If ψ is as indicated, then by Lemma 1.6.16(a) there is $\theta \in \mathrm{Hom}_A(P', P)$ with $\psi = \pi_P \circ \theta$. Since ψ is surjective and π_P maps maximal submodules to maximal submodules, θ must be surjective; $P' \cong_A P \oplus \ker \theta$ follows from Lemma 1.6.16(b).

(b) By part (a) there is an epimorphism $\theta \colon P' \to P$ with $\pi_{P'} = \pi_P \circ \theta$. By Lemma 1.6.16(b) there is a monomorphism $\mu \colon P \to P'$ with $\theta \circ \mu = \mathrm{id}_P$. Then $\pi_{P'} = \pi_{P'} \circ \mu \circ \theta$, so $\mu \circ \theta$ and, consequently μ, is surjective. Thus θ is an isomorphism and the result follows.

(c) is obvious from the definition. $\square$

Definition 1.6.21 A ring A is called **semi-perfect** if 1_A is a sum of local orthogonal idempotents that means

$$1_A = e_1 + e_2 + \cdots + e_n,$$

where the $e_1, e_2, \dots, e_n$ are mutually orthogonal and $e_i A e_i$ is a local ring for $i = 1, \dots, n$. For a detailed study of semi-perfect rings, see, for example, chapter 23 of [108].

Remark 1.6.22 (a) By Corollary 1.6.5, if A is of finite length (for instance, if K is a field) then A is semi-perfect. But in Section 4.1 we will show that R-algebras over so-called complete discrete valuation rings R also share this property; see Theorem 4.1.21.

(b) If A is semi-perfect, then

$$A = Ae_1 \oplus Ae_2 \oplus \cdots Ae_n$$

and since $\operatorname{End}_A Ae_i \cong (e_i Ae_i)^{\operatorname{op}}$ is local, the indecomposable direct summands Ae_i are uniquely determined up to isomorphism and ordering by Theorem 1.6.6. Moreover, if $e \in A$ is a primitive idempotent, then Ae is isomorphic to Ae_i for some $i = 1, \ldots, n$ and the same holds for any projective indecomposable A-module.

Theorem 1.6.23 *Let A be a semi-perfect ring and let $e \in A$ be a primitive idempotent.*

(a) *Ae has a unique maximal submodule $\operatorname{Rad}(Ae) = \operatorname{J}(A)e$. The module Ae is a projective cover of the simple module $Ae/\operatorname{J}(A)e$.*

(b) *Every A-module V of finite length has a projective cover $P(V)$. If $V = V_1 \oplus \cdots \oplus V_m$ then*

$$P(V) \cong P(V_1) \oplus \cdots \oplus P(V_m), \qquad \Omega(V) \cong \Omega(V_1) \oplus \cdots \oplus \Omega(V_m). \qquad (1.17)$$

Proof. (a) Observe that by Zorn's lemma Ae contains a maximal submodule. Assume that M_1, M_2 are different maximal submodules of Ae. Then $Ae = M_1 + M_2$ and hence there are $v_1 \in M_1$ and $v_2 \in M_2$ with $e = v_1 + v_2$. Define $\varphi_i \in \operatorname{End}_A Ae$ by $\varphi_i \colon ae \mapsto aev_i$ for $i = 1, 2$. Then $\varphi_1 + \varphi_2 = \operatorname{id}_{Ae}$. Since $\operatorname{End}_A Ae$ is local by assumption, φ_1 or φ_2 must be invertible, which is absurd because $\operatorname{im} \varphi_i \subseteq M_i$. Thus $\operatorname{Rad}(Ae)$ is the unique maximal submodule of Ae. By Exercise 1.6.2 $\operatorname{J}(A) = \operatorname{Rad}(Ae) \oplus \operatorname{Rad}(A(1-e))$, hence $\operatorname{J}(A)e = \operatorname{Rad}(Ae)$.

(b) If V is a simple A-module there is a primitive idempotent e with $eV \neq \{0\}$ by Remark 1.6.22. Because of Theorem 1.6.8(a) $\operatorname{Hom}_A(Ae, V) \neq \{0\}$. Thus Ae is a projective cover of V. Since P is a projective cover of an A-module V if and only if it is a projective cover of $V/\operatorname{Rad}(V)$ it suffices (by Exercise 1.5.1) to consider semisimple modules. Equation (1.17) follows, since by Exercise 1.6.2

$$\bigoplus_{i=1}^m P(V_i)/\operatorname{Rad}(\bigoplus_{i=1}^m P(V_i)) \cong_A \bigoplus_{i=1}^m P(V_i)/\operatorname{Rad}(P(V_i)) \cong_A \bigoplus_{i=1}^m V_i = V.$$

$\square$

Theorem 1.6.24 *Let A be a semi-perfect ring and let $V_1, \ldots, V_r$ be the simple A-modules up to isomorphism and $D_i := \operatorname{End}_A V_i$. Then there are r projective indecomposable modules $P_1, \ldots, P_r$ with $P_i/\operatorname{Rad} P_i \cong_A V_i$ and*

$$_A A = \bigoplus_{i=1}^r (P_{i,1} \oplus \cdots \oplus P_{i,f_i}) \quad \text{with} \quad P_{i,j} \cong_A P_i,$$

where $f_i = \dim_{D_i} V_i$. If K is a splitting field for A then the f_i are just the degrees of the irreducible representations of A.

Proof. Let $1_A = e_1 + \cdots + e_n$ be a decomposition of 1_A into a sum of mutually orthogonal primitive idempotents. Then for $i \in \{1, \ldots, r\}$

$$V_i = 1_A \cdot V_i = (e_1 + \cdots + e_n) \cdot V_i = e_1 V_i \oplus \cdots \oplus e_n V_i$$

as vector spaces over D_i. By Theorem 1.6.8 we have $e_j V_i \cong \mathrm{Hom}_A(Ae_j, V_i)$ and by Theorem 1.6.23 we get

$$\mathrm{Hom}_A(Ae_j, V_i) \cong \begin{cases} \{0\} & \text{if } Ae_j/\mathrm{Rad}(Ae_j) \not\cong V_i, \\ D_i & \text{if } Ae_j/\mathrm{Rad}(Ae_j) \cong V_i. \end{cases} \qquad (1.18)$$

Observe that any A-module homomorphism $Ae_j \to V_i$ has $\mathrm{Rad}(Ae_j)$ in its kernel and thus induces an A-homomorphism $Ae_j/\mathrm{Rad}(Ae_j) \to V_i$, which by Schur's lemma is either zero or an isomorphism. Thus $\dim_{D_i} V_i < \infty$ and counting dimensions we find that $\dim_{D_i} V_i = |\{ j \mid 1 \le j \le n \,,\, e_j V_i \ne \{0\} \}|$. □

Corollary 1.6.25 *Let K be a field of prime characteristic p.*

(a) *If G is a finite p-group the regular module KG is indecomposable. Every projective KG-module is free.*

(b) *If G is a finite group with order $|G| = p^k a$ with $p \nmid a$ then the dimension of any projective KG-module is divisible by p^k.*

Proof. (a) follows, since by Theorem 1.3.11 the trivial one-dimensional module is the only simple KG-module in this case.

(b) follows, because by Lemma 1.6.17 the restriction of any projective KG-module to KP is projective for any subgroup P and hence (by (a)) free if we choose P to be a Sylow p-subgroup of G. □

Example 1.6.26 Let $G := S_3$ and let K be a field with char $K = 3$. We saw in Exercise 1.4.8 that there are exactly two simple $K S_3$-modules $V_1 = K_G, V_2$, both of dimension one, with V_2 being the module affording the sign representation $g \mapsto \mathrm{sign}(g)$. By Theorem 1.6.24 and Corollary 1.6.25(b) there must be primitive idempotents $e_1, e_2 \in K S_3$ with

$$K S_3 = K S_3\, e_1 \oplus K S_3\, e_2 \quad \text{and} \quad \dim_K(K S_3\, e_i) = 3 \quad \text{for} \quad i = 1, 2.$$

It is not difficult to find such idempotents, which are, of course, not unique, and to investigate the precise structure of the projective indecomposable modules $K S_3\, e_1, K S_3\, e_2$. In fact, it follows from our analysis that any idempotent $e \ne 1$ must generate a three-dimensional projective indecomposable module KGe. So we choose $e_1 = \frac{1}{2}(1 + (1,2))$ and find

$$P_1 = KGe_1 = \langle e_1 \,,\, (1,2,3)e_1 \,,\, (1,3,2)e_1 \rangle_K,$$
$$\mathrm{Rad}(P_1) = \langle e_1 - (1,3,2)e_1 \,,\, \mathbf{s}_G \rangle_K,$$
$$\mathrm{Soc}(P_1) = \langle \mathbf{s}_G \rangle_K,$$

with $\mathbf{s}_G = \sum_{g \in G} g = 2(e_1 + (1,2,3)e_1 + (1,3,2)e_1)$. Similarly with $e_2 = 1 - e_1 = \frac{1}{2}(1 - (1,2))$ we find

$$P_2 = KGe_2 = \langle e_2 \,,\, (1,2,3)e_2 \,,\, (1,3,2)e_2 \rangle_K,$$
$$\mathrm{Rad}(P_2) = \langle 1 + (1,2,3) + (1,3,2) \,,\, \mathbf{a}_G \rangle_K,$$
$$\mathrm{Soc}(P_2) = \langle \mathbf{a}_G \rangle_K,$$

with $\mathbf{a}_G = \sum_{g \in G} \operatorname{sgn}(g)g$. It is easy to see that the composition factors of P_1 and P_2 are V_1, V_2, V_1 and V_2, V_1, V_2, respectively.

If char $K = 2$ then as we have seen in Example 1.4.8 KG has simple modules $V_1 = K_G, V_2$ with $\dim_K V_2 = 2$. Thus

$$KG \cong P(V_1) \oplus P(V_2) \oplus P(V_2) \quad \text{with} \quad \dim_K P(V_i) = 2 \ (i = 1, 2).$$

♦

For simplicity we restrict ourselves to algebras over fields for the following result.

Theorem 1.6.27 *Let A be a symmetric algebra over a field K and let P be a projective and V an arbitrary A-module.*

(a) $P/\operatorname{Rad}(P) \cong_A \operatorname{Soc}(P)$.

(b) $P^\star$ *is a projective right A-module.*

(c) *If V is semisimple then $P(V)^\star \cong_A P(V^\star)$ as right A-modules where $P(V^\star)$ is the projective cover of the right A-module $V^\star$.*

(d) *If $\varphi: V \to W$ is injective and $\psi: V \to P$ then there is $\hat{\psi}: W \to P$ with $\psi = \varphi \circ \hat{\psi}$. If $P \leq_A V$ then $P \mid V$ ("projective modules are injective").*

(e) *There is a monomorphism $\psi: V \to P(\operatorname{Soc}(V))$. We define $\Omega^{-1}(V) := P(\operatorname{Soc}(V))/\psi(V)$. Then $\Omega^{-1}(V) \cong_A \Omega(V^\star)^\star$; in particular it is independent of the choice of ψ, up to isomorphism.*

(f) *The **projective-free part** $\Omega^0(V)$ of V is an A-module having no projective direct summand $\neq \{0\}$ such that $V = \Omega^0(V) \oplus P$ for some projective A-module P. Then*

$$\Omega^0(V) \cong_A \Omega(\Omega^{-1}(V)) \cong_A \Omega^{-1}(\Omega(V)).$$

Proof. (a), (b), (c) Because of (1.17) on p. 73 and

$$\operatorname{Rad}\left(\bigoplus_{i=1}^n V_i\right) \cong \bigoplus_{i=1}^n \operatorname{Rad}(V_i), \quad \operatorname{Soc}\left(\bigoplus_{i=1}^n V_i\right) \cong \bigoplus_{i=1}^n \operatorname{Soc}(V_i), \quad \left(\bigoplus_{i=1}^n V_i\right)^\star \cong \bigoplus_{i=1}^n V_i^\star$$

for A-modules V_i, it is sufficient to consider the case of a projective indecomposable module $P = Ae$ with $e^2 = e \in A$. Let $L := P/\operatorname{Rad}(P)$ and let

$$A = P \oplus U, \quad \text{hence} \quad A^\star = U^\circ \oplus P^\circ,$$

using Theorem 1.1.34. This theorem also yields $U^\circ \cong (A/U)^\star \cong P^\star$. By Remark 1.1.44(a) we have $A^\star \cong A$ as right A-modules. Thus U° is a projective indecomposable right A-module since indecomposability is preserved under duality. By Theorem 1.6.23 (applied to right modules) U° has a unique maximal submodule W° with $U <_A W$. Thus W/U is the unique minimal submodule of $A/U \cong_A P$ and $S := W \cap P$ is the unique minimal submodule of P. It remains to prove that $L \cong_A S$. By Theorem 1.6.23 it is sufficient to show that $eS \cong \operatorname{Hom}_A(Ae, S) \neq 0$. For this we use a symmetrizing trace $\tau: A \to K$ and choose an arbitrary non-zero element $y = ye \in S = Se \subseteq Ae$. Then there exists an $a \in A$ with

$$0 \neq \tau(aye) = \tau(eay) \quad \text{and thus} \quad 0 \neq eay \in eS.$$

(d) If φ is injective then $\varphi^{\mathrm{T}} \colon W^\star \to V^\star$ is surjective. Since $P^\star$ is projective, $\psi^{\mathrm{T}} \colon P^\star \to V^\star$ can be lifted to $\psi' \colon P^\star \to W^\star$ with $\psi^{\mathrm{T}} = \psi^{\mathrm{T}} \circ \psi'$. Putting $\hat{\psi} := \psi'^{\mathrm{T}}$ the first assertion follows. The second one follows from this by choosing $W := P$, $\psi := \mathrm{id}_P$ and $\varphi \colon V \to P$ the embedding.

(e) Let $S := \mathrm{Soc}(V)$ with embedding $\iota \colon S \to V$. Then $\iota^{\mathrm{T}} \colon V^\star \to S^\star$ is an epimorphism. By Lemma 1.6.20(a) there is an epimorphism $\varphi \colon P(S^\star) \to V^\star$. Then $\psi := \varphi^{\mathrm{T}} \colon V \cong V^{\star\star} \to P(S^\star)^\star \cong P(S)$ is a monomorphism.

Now, let $\psi \colon V \to P(S)$ be any monomorphism. Then $\psi^{\mathrm{T}} \colon P(S)^\star \to V^\star$ is an epimorphism. Observe that $V^\star / \mathrm{Rad}(V^\star) \cong S^\star$, so $P(V^\star) \cong P(S^\star) \cong P(S)^\star$ and $\ker \psi^{\mathrm{T}} \cong \Omega(V^\star)$ by Lemma 1.6.20. On the other hand, $\ker \psi^{\mathrm{T}} = (\psi(V))^\circ$, so $P(S)/\psi(V) \cong (\Omega(V^\star))^\star$ by duality.

(f) Since $\Omega(P) \cong \Omega^{-1}(P) = \{0\}$ by Lemma 1.6.20 and part (e), we may assume $V = \Omega^0(V)$. We have an embedding $\psi \colon V \to P(\mathrm{Soc}(V))$ and, by definition, $P(\mathrm{Soc}(V))/\psi(V) \cong \Omega^{-1}(V)$. By Lemma 1.6.20

$$P(\Omega^{-1}(V)) \mid P(\mathrm{Soc}(V))/\psi(V).$$

Since V has no projective direct summand we have equality and we get $\psi(V) = \Omega(\Omega^{-1}(V))$.

Since $\Omega(V) \subseteq \mathrm{Rad}(P(V))$ it is easily seen that $\mathrm{Soc}(P(V)) = \mathrm{Soc}(\Omega(V))$, hence

$$V \cong P(V)/\Omega(V) \cong P(\mathrm{Soc}(\Omega(V))) \cong \Omega^{-1}(\Omega(V)).$$

$\square$

Corollary 1.6.28 *Under the assumption of Theorem 1.6.27 V is a non-projective indecomposable A-module if and only if $\Omega(V)$ and $\Omega^{-1}(V)$ are non-projective indecomposable A-modules.*

Proof. This follows immediately from Theorems 1.6.27 and 1.6.23. $\square$

Definition 1.6.29 Let V, W be A-modules. Then $\varphi \in \mathrm{Hom}_A(V, W)$ is called **projective** if it factors through a projective A-module P, that is, if there are $\psi' \in \mathrm{Hom}_A(V, P)$, $\psi \in \mathrm{Hom}_A(P, W)$ such that $\varphi = \psi \circ \psi'$. If φ factors through P, then also any scalar multiple $\alpha \varphi$ for $\alpha \in K$. If in addition $\varphi' \in \mathrm{Hom}_A(V, W)$ factors through a projective A-module P', then $\varphi + \varphi'$ factors through $P \oplus P'$. Thus $\mathrm{pHom}_A(V, W) := \{\varphi \in \mathrm{Hom}_A(V, W) \mid \varphi \text{ projective}\} \leq_K \mathrm{Hom}_A(V, W)$ and we define

$$\overline{\mathrm{Hom}}_A(V, W) := \mathrm{Hom}_A(V, W)/\mathrm{pHom}_A(V, W).$$

Lemma 1.6.30 *Let A be a symmetric algebra over a field K and let $\varphi \colon V \to W$ be a projective A-homomorphism. If φ is surjective (injective) then V (W, respectively) has a projective non-trivial direct summand.*

Proof. Let P be a projective A-module and $\varphi = \sigma \circ \tau$ with $\sigma \in \mathrm{Hom}_A(P, W)$ and $\tau \in \mathrm{Hom}(V, P)$. If φ is surjective then σ is also surjective and by Lemma 1.6.20 factors through $P(W)$, so we may as well assume that $P = P(W)$. Then $P = \mathrm{im}\,\tau + \mathrm{Rad}(P)$ and so τ is surjective and V has a direct summand isomorphic to P. The other assertion follows by dualizing. $\qquad\square$

Theorem 1.6.31 *Let A be a symmetric algebra over a field K and let V, W be A-modules. Then*

$$\overline{\mathrm{Hom}}_A(V, W) \cong_K \overline{\mathrm{Hom}}_A(\Omega(V), \Omega(W)).$$

Proof. Let $\chi \in \mathrm{Hom}_A(V, W)$. Since $P(V)$ is projective there is a $\psi \in \mathrm{Hom}_A(P(V), P(W))$ making the following diagram commutative, where π, π' are the natural projections and ι, ι' inclusions:

$$
\begin{array}{ccccc}
\Omega(V) & \xrightarrow{\;\iota\;} & P(V) & \xrightarrow{\;\pi\;} & V \\
\big\downarrow{\scriptstyle \psi|_{\Omega(V)}} & & \big\downarrow{\scriptstyle \psi} & & \big\downarrow{\scriptstyle \varphi} \\
\Omega(W) & \xrightarrow{\;\iota'\;} & P(W) & \xrightarrow{\;\pi'\;} & W
\end{array}
$$

If $\psi' \colon P(V) \to P(W)$ is another lift of $\varphi \circ \pi$ then $\pi' \circ (\psi - \psi') = 0$, that is $\mathrm{im}(\psi - \psi') \subseteq \Omega(W)$. Thus $\psi|_{\Omega(V)} - \psi'|_{\Omega(V)}$ factors through $P(V)$ and we obtain a well-defined K-linear map $\Omega \colon \mathrm{Hom}_A(V, W) \to \overline{\mathrm{Hom}}_A(\Omega(V), \Omega(W))$

$$\varphi \mapsto \Omega(\varphi) := \psi|_{\Omega(V)} + \mathrm{pHom}_A(\Omega(V), \Omega(W)) \in \overline{\mathrm{Hom}}_A(\Omega(V), \Omega(W)).$$

If $\varphi \in \ker \Omega$ then $\psi|_{\Omega(V)}$ factors through a projective module, hence through $P(V)$ by Theorem 1.6.27(d). That is, there is a $\sigma \colon \Omega(V) \to P(V)$ with $\psi|_{\Omega(V)} = \sigma \circ \iota$. Then $(\psi - \iota' \circ \sigma) \circ \iota = 0$ and there is a unique $\tau \colon V \to W$ with $\psi - \iota' \circ \sigma = \tau \circ \pi$. It follows that $\pi' \circ \tau \circ \pi = \pi' \circ \psi - \pi' \iota' \sigma = \varphi \circ \pi$. Thus $\varphi = \pi' \circ \tau \in \mathrm{pHom}_A(V, W)$. Hence $\ker(\Omega) = \mathrm{pHom}_A(V, W)$. $\qquad\square$

Exercises

Exercise 1.6.1 Prove Lemma 1.6.16.

Exercise 1.6.2 Let V_1, V_2 be A-modules. Prove that

$$\mathrm{Rad}(V_1 \oplus V_2) \cong_A \mathrm{Rad}\,V_1 \oplus \mathrm{Rad}\,V_2.$$

Exercise 1.6.3 Let Ω be a finite G-set, and let $H \leq G$ and K be a commutative ring. Let $\mathcal{O}_1, \ldots, \mathcal{O}_r$ be the H-orbits on Ω. Show that a K-basis for $\mathrm{Inv}_H(K\Omega) = \{x \in K\Omega \mid hx = x \text{ for all } h \in H\}$ is $\{\mathcal{O}_1^+, \ldots, \mathcal{O}_r^+\}$ with $\mathcal{O}_i^+ = \sum_{x \in \mathcal{O}_i} x$. Also $H^+ K\Omega \subseteq \mathrm{Inv}_H(K\Omega)$ (see (1.1), p. 3) with equality if $|H|$ is invertible in K. In this case let $e_H := \frac{1}{|H|} H^+$ and show that for $g \in G$

$$e_H g e_H \,\mathcal{O}_j^+ = \sum_{i=1}^r c_{ij} \frac{1}{|\mathcal{O}_i|} \mathcal{O}_i^+$$

with $c_{ij} = |\{y \in \mathcal{O}_j \mid g \cdot y \in \mathcal{O}_i\}|$.

Exercise 1.6.4 Keep the notation of Exercise 1.6.3 and assume that $\Omega = \{1, \ldots, n\}$, so that $G \leq S_n$. Assume that p is a prime not dividing $|H|$ and q is a power of p. Show that the following GAP-program returns the matrix of the action of $e_H g e_H$ on $e_H \mathbb{F}_q \Omega$ with respect to the basis $(\mathcal{O}_1^+, \ldots, \mathcal{O}_r^+)$:

```
cond := function( H, n , g, q )
 local condmat, orbs;
 orbs := Orbits( H , [1..n] );
 condmat := List( orbs, Oi -> List( orbs, Oj -> 1/(Size(Oi)*Z(q)^0) *
                Size( Intersection( List(Oi, x -> x^g), Oj) ) ) );
 return condmat;
end;
```

Hint: Recall that GAP uses right action (see Example 1.1.49) and use the fact that $|\{x^g \mid x \in \mathcal{O}_i\} \cap \mathcal{O}_j| = |\mathcal{O}_i \cap \{y^{g^{-1}} \mid y \in \mathcal{O}_j\}|$.

Exercise 1.6.5 Let $G := S_4$ act naturally on $\Omega := \{1, 2, 3, 4\}$ and let K be a field with char $K \neq 3$. Let $V := K\Omega$ be the corresponding permutation module. Assume that $H := \langle (2, 3, 4) \rangle$ and $e := e_H$ as in Example 1.6.13. Compare the poset of $eKGe$-submodules of eV with the poset of KG-submodules of V and determine the composition factors. Distinguish the cases char $K \neq 2$ and char $K = 2$. In the latter case show that $eKGe$ is isomorphic to KV_4, the group algebra of the Klein 4-group, and thus has only one simple module.

Exercise 1.6.6 Let A be a ring and let e and f be idempotents in A. Show that the following statements are equivalent:
(a) $Ae \cong_A Af$;
(b) $eA \cong_A fA$;
(c) there are two elements $a \in eAf$ and $b \in fAe$ such that $ab = e$ and $ba = f$.

Exercise 1.6.7 Let A be a ring and let

$$1 = e_1 + \cdots + e_n = f_1 + \cdots + f_n$$

be two decompositions of unity as a sum of orthogonal idempotents of A with $Ae_i \cong_A Af_i$. Show that there is a unit $u \in A$ such that $e_i = u^{-1} f_i u$ for $i = 1, \ldots, n$.

Exercise 1.6.8 Let K be a field with char $K = p > 0$ and let G be a finite group of order $|G| = p^a q$ with q coprime to p. Assume that G has a subgroup H of order q. Show that KGe_H is a projective cover of the trivial KG-module K_G.

Exercise 1.6.9 Find the number of simple $K S_4$-modules (up to isomorphism) for a field K of characteristic two or three and also the dimensions of the projective indecomposable modules.

1.7 Blocks

In this section we assume that A is an algebra over a commutative ring K. We consider decompositions of A as a direct sum of (two-sided) ideals.

Lemma 1.7.1 *Let I_1, I_2 be non-zero ideals of A. Then $A = I_1 \oplus I_2$ if and only if $I_i = A\epsilon_i$ with $\epsilon_i^2 = \epsilon_i \in \mathbf{Z}(A)$ for $i = 1, 2$ and $\epsilon_1 + \epsilon_2 = 1_A$ (so $\epsilon_1 \epsilon_2 = 0$).*

Proof. If ϵ_1, ϵ_2 are as indicated, then $A = A\epsilon_1 \oplus A\epsilon_2$ by Lemma 1.6.2(b). Conversely, let $A = I_1 \oplus I_2$ and write $1_A = \epsilon_1 + \epsilon_2$ with $\epsilon_i \in I_i$. Then $\epsilon_i^2 = \epsilon_i$ as in the proof of Lemma 1.6.2(b) and $\epsilon_i \in \mathbf{Z}(A)$ because for any $a \in A$ we have $\epsilon_1 a + \epsilon_2 a = 1 \cdot a = a \cdot 1 = a\epsilon_1 + a\epsilon_2$ and $\epsilon_i a, a\epsilon_i \in I_i$. $\qquad \square$

Observe that in the above lemma I_i is a K-algebra with identity ϵ_i. If I_i is a direct sum of ideals then ϵ_i can also be written as a sum of orthogonal central idempotents.

Definition 1.7.2 A **block idempotent** (also called a **centrally primitive** idempotent) of A is an idempotent $\epsilon = \epsilon^2 \in \mathbf{Z}(A)$ which cannot be written as a sum of two orthogonal central idempotents.

Corollary 1.7.3 (a) *If A can be written as*

$$A = B_1 \oplus \cdots \oplus B_m \quad \text{with} \quad B_i \trianglelefteq A \quad (i = 1, \dots, m),$$

where the B_i cannot be written as a direct sum of non-zero ideals of A, and if

$$1 = \epsilon_1 + \cdots + \epsilon_m \quad \text{with} \quad \epsilon_i \in B_i \quad (i = 1, \dots, m),$$

*then the ϵ_i are block idempotents and the B_i are called **block ideals**.*
(b) *Conversely, if $\epsilon_1, \dots, \epsilon_m$ are pairwise orthogonal block idempotents with $1 = \epsilon_1 + \cdots + \epsilon_m$ then $A\epsilon_i$ is a block ideal for $i = 1, \dots, m$ and*

$$A = A\epsilon_1 \oplus \cdots \oplus A\epsilon_m.$$

*This is usually called the **block decomposition** of A.*

Proof. (a) follows immediately from Lemma 1.7.1.
(b) Conversely, since the ϵ_i are central, it is clear that the $A\epsilon_i$ are two-sided ideals. If $A\epsilon_i = I \oplus I'$ could be written as the sum of two non-zero ideals $I, I' \trianglelefteq A$ then $\epsilon_i = \epsilon + \epsilon'$ with $\epsilon \in I$ and $\epsilon' \in I'$ and ϵ, ϵ' would be non-zero central orthogonal idempotents, which is a contradiction. $\qquad \square$

Theorem 1.7.4 *If $(\epsilon_1, \dots, \epsilon_m)$ and $(\epsilon'_1, \dots, \epsilon'_n)$ are tuples of block idempotents of A with*

$$1 = \sum_{i=1}^{m} \epsilon_i = \sum_{j=1}^{n} \epsilon'_j \qquad (1.19)$$

then $m = n$ and $\{\epsilon_1, \ldots, \epsilon_m\} = \{\epsilon_1', \ldots, \epsilon_n'\}$. If V is any A-module then

$$V = \epsilon_1 V \oplus \cdots \oplus \epsilon_m V \qquad \text{with submodules} \quad \epsilon_i V;$$

in particular, if V is indecomposable, then there is exactly one ϵ_i with $\epsilon_i V = V$ (and $\epsilon_j V = \{0\}$ for all other ϵ_j) and V is then said to "belong to the block $A\epsilon_i$." If A is semisimple then $\epsilon_j V$ is a homogeneous component of V.

Proof. In fact, it follows from (1.19) that $\epsilon_i = \epsilon_i 1 = \sum_{j=1}^n \epsilon_i \epsilon_j'$, and this is a sum of orthogonal central idempotents, so there is exactly one $j = j(i)$ with $\epsilon_i = \epsilon_i \epsilon_j' \neq 0$. Interchanging the roles of ϵ_i and ϵ_j' one gets $\epsilon_i = \epsilon_i \epsilon_j' = \epsilon_j'$. The other assertions are obvious, since the ϵ_i are central and orthogonal. $\qquad\square$

Remark 1.7.5 If ϵ is a central idempotent of A then so is $1_A - \epsilon$ and $A = A\epsilon \oplus A(1_A - \epsilon)$. It follows that block idempotents exist, if A is noetherian or artinian or if 1_A is a sum of primitive idempotents in A. Also the number of block idempotents of A in Theorem 1.7.4 is clearly 2^m.

Corollary 1.7.6 Let K be a splitting field for the K-algebra A and let $\epsilon_1, \ldots, \epsilon_m$ be the block idempotents of A. Then there are exactly m K-algebra homomorphisms $\omega_i \colon \mathbf{Z}(A) \to K$ and $\omega_i(\epsilon_j) = \delta_{i,j}$.

Proof. We have

$$\mathbf{Z}(A) = \mathbf{Z}(A)\epsilon_1 \oplus \cdots \oplus \mathbf{Z}(A)\epsilon_m$$

and each $\mathbf{Z}(A)\epsilon_i$ is indecomposable as a $\mathbf{Z}(A)$-module and has a unique maximal submodule by Theorem 1.6.23. Thus there is at most one K-algebra homomorphism $\omega \colon \mathbf{Z}(A) \to K$ with $\omega(\epsilon_j) = 1$. On the other hand, if V is a simple A-module with $\epsilon_j V = V$ then left multiplication with $z \in \mathbf{Z}(A)$ is an A-endomorphism of V, and by our assumption $zv = \omega(z)v$ for $v \in V$ for some $\omega(z) \in K$. Obviously $\omega \colon z \mapsto \omega(z)$ is a K-algebra homomorphism $\mathbf{Z}(A) \to K$ with $\omega(\epsilon_j) = 1$. $\qquad\square$

The rest of this section will only be used in Chapter 4.

If A is of finite length then By Lemma 1.7.4 any projective indecomposable A-module Ae belongs to exactly one block $A\epsilon_i$. In fact, if

$$A\epsilon_i = Ae_{i,1} \oplus \cdots \oplus Ae_{i,n_i} \quad \text{with primitive idempotents} \quad e_{i,j}$$

then Ae belongs to $A\epsilon_i$ if and only if $Ae \cong_A Ae_{i,j}$ for some $j \in \{1, \ldots, n_i\}$. But it is also possible to decide whether or not two projective indecomposable A-modules belong to the same block, without knowing the block idempotents (or blocks). Since a block idempotent either annihilates an indecomposable module or acts as the identity on it, it is clear that two indecomposable modules must belong to the same block if they have a common composition factor. What we are going to see now is that the equivalence relation "belonging to the same block" for projective indecomposable modules is the transitive hull of the relation "having a common composition factor."

Theorem 1.7.7 *Let A be of finite length. Then two projective indecomposable modules Ae, Ae' belong to the same block if and only if the idempotents e, e' are "linked," that is if and only if there is a series of primitive idempotents $e = e^{(0)}, e^{(1)}, \ldots, e^{(k)} = e'$ such that $Ae^{(i-1)}$ and $Ae^{(i)}$ have a composition factor in common for $i = 1, \ldots, k$.*

Proof. The "only if" part was already shown in the discussion above. So assume that Ae and Ae' belong to the same block $A\epsilon$. Thus $e\epsilon = e$ and $e'\epsilon = e'$. If $ee' \neq 0$ or $e'e \neq 0$ then $\mathrm{Hom}_A(Ae, Ae') \cong eAe' \neq \{0\}$ or $\mathrm{Hom}_A(Ae', Ae) \neq \{0\}$ and hence Ae and Ae' have a composition factor in common (namely $Ae/\mathrm{Rad}(Ae)$ or $Ae'/\mathrm{Rad}(Ae')$). Otherwise $\epsilon - e - e'$ is also idempotent and hence can be written as a sum of pairwise orthogonal primitive idempotents. We may choose our notation so that $e_0 = e, e_n = e'$ and

$$\epsilon = e_0 + \cdots + e_r + e_{r+1} + \cdots + e_n,$$

where the e_j are pairwise orthogonal primitive idempotents ordered in such a way that e is linked to e_j if and only if $j \leq r$. In particular $e_i Ae_j = \{0\} = e_j Ae_i$ for $i \leq r$ and $j > r$. We put

$$\epsilon = e_0 + \cdots + e_r , \quad \epsilon' = e_{r+1} + \cdots + e_n.$$

Then $\epsilon A\epsilon' = \{0\} = \epsilon' A\epsilon$ and

$$A\epsilon \cdot A = A\epsilon\epsilon_i A = A\epsilon A\epsilon_i = A\epsilon A(\epsilon + \epsilon') = A\epsilon A\epsilon \subseteq A\epsilon,$$

and similarly $A\epsilon' \cdot A \subseteq A\epsilon'$ and thus both $A\epsilon, A\epsilon'$ are ideals with $A\epsilon_i = A\epsilon \oplus A\epsilon'$, which contradicts the indecomposability of $A\epsilon_i$ unless $A\epsilon'_i = 0$, hence $r = n$. $\qquad\blacksquare$

Definition 1.7.8 Let A be of finite length, let $V_1, \ldots, V_r$ be a complete set of representatives of simple A-modules and let $P_i = P(V_i)$ be the projective cover of V_i for $i = 1, \ldots, r$. Let c_{ij} be the number of composition factors in a fixed composition series of P_i which are isomorphic to V_j. Then the $r \times r$-matrix $[c_{ij}]$ is called the **Cartan matrix** of A.

Corollary 1.7.9 *If A is of finite length and the simple A-modules are sorted according to the blocks then the Cartan matrix is a block diagonal matrix*

$$C = \mathrm{diag}(C^{(1)}, \ldots, C^{(m)}),$$

where $C^{(i)}$ is the Cartan matrix of the ith block $A\epsilon_i$ for $i = 1, \ldots m$. Furthermore it is impossible to arrange the simple modules in a block $A\epsilon_i$ in such a way that

$$C^{(i)} = \left[\begin{array}{c|c} C_1^{(i)} & 0 \\ \hline 0 & C_2^{(i)} \end{array}\right]$$

with square matrices $C_1^{(i)}, C_2^{(i)}$.

If A is semisimple then the Cartan matrix is the identity matrix. Also it follows from the definition that the diagonal entries in a Cartan matrix are always positive.

Example 1.7.10 The Cartan matrix of $K\,S_3$ for a field K of characteristic two and three can be read off from Example 1.6.26. We get

$$C = \left[\begin{array}{c|c} 2 & 0 \\ \hline 0 & 1 \end{array}\right] \ (\mathrm{char}\,K = 2) \,, \qquad C = \left[\begin{array}{cc} 2 & 1 \\ 1 & 2 \end{array}\right] \ (\mathrm{char}\,K = 3).$$

So $K\,S_3$ has two, one (or three) blocks depending on the characteristic.   ◆

Exercises

Exercise 1.7.1 Let K be a field and let A be the subring of $K^{4\times4}$ defined by

$$A = \left\{ \begin{bmatrix} a & b & 0 & 0 \\ 0 & c & 0 & 0 \\ 0 & 0 & c & d \\ 0 & 0 & 0 & a \end{bmatrix} \mid a, b, c, d \in K \right\} \subseteq K^{4\times4}.$$

Find all simple A-modules up to isomorphism and their projective covers, and show that the Cartan matrix of A is

$$C = \begin{bmatrix} 1 & 1 \\ 1 & 1 \end{bmatrix}.$$

Exercise 1.7.2 (a) Verify that the following GAP-program computes the block idempotents of $\mathbb{F}_q G$ for "small" q and G:

```
gap> blockidemps  := function( G, q )
> local FG, ZFG, T, A, cs, o, ids;
> FG := GroupRing( GF(q), G ); o := Embedding( G, FG );
> cs := List( ConjugacyClasses(G),c -> Sum(List(Elements(c),x-> x^o)) );
> ZFG := Subalgebra( FG, cs, "basis" );
> T := StructureConstantsTable( Basis(ZFG) );
> A := AlgebraByStructureConstants( GF(q), T );
> ids := Filtered( Elements(A), x -> x*x = x );
> return( Filtered( ids, e -> Length( Set( List(ids, x->e*x) ) ) = 2 ) );
> end;;
```

(b) Apply the program to find the block idempotents e of $\mathbb{F}_q G$ for $q \leq 5$ and all symmetric and alternating groups G of degree at most six. Show that in all cases $e \in \langle\, \{\, (g^G)^+ \mid p \nmid |\langle g \rangle| \}\, \rangle_{\mathbb{F}_q}$, where $p = \mathrm{char}\,\mathbb{F}_q$.
Note: This observation will be proved in general in Theorem 4.4.7.

Exercise 1.7.3 Modify the GAP-program of Exercise 1.7.2 to compute for each block idempotent $e \in \mathbf{Z}(\mathbb{F}_q G)$ also the dimensions of $\mathbf{Z}(\mathbb{F}_q G)e$ and $\mathbb{F}_q G\, e$. Apply it to the symmetric and alternating groups of degree at most five for $q \leq 5$.

Exercise 1.7.4 Let $e = \sum_{C \in \mathrm{cl}(G)} a_C\, C^+ \in KG$ with $a_C \in K$ be a block idempotent. Show that $\sum_{C \in \mathrm{cl}(G)} |C| a_C \in \{0, 1\}$ and that there is exactly one block idempotent for which this sum is unity.

1.8   Changing coefficients

If R is a commutative ring and $K \geq R$ is a commutative ring extension then any representation $\boldsymbol{\delta} \colon G \to \mathrm{GL}_n(R)$ of a finite group G over R can be considered as a K-representation $^K\boldsymbol{\delta}$ using the embedding $\mathrm{GL}_n(R) \subseteq \mathrm{GL}_n(K)$. Conversely, given a K-representation of G one might ask whether or not it is equivalent to such a representation $^K\boldsymbol{\delta}$ for a suitable R-representation $\boldsymbol{\delta}$ of G. We have two special situations in mind.

- K is the quotient field of R, e.g. K an algebraic number field, and R is the ring of algebraic integers in K (or a local ring containing this ring). In this case the question is whether we can write a K-representation "integrally," i.e. to find a suitable K-basis for the underlying representation module such that all group elements are represented by matrices over R with respect to the basis.

- $R \subset K$ is a field extension, for instance $K = \mathbb{C}$ and $R = \mathbb{Q}$ or some algebraic number field, and the question is whether we can realize a complex (or K-) representation over a smaller field.

If V is an R-free RG-module affording the representation $\boldsymbol{\delta} \colon G \to \mathrm{GL}_n(R)$ then a module affording $^K\boldsymbol{\delta}$ is the free K-module $KV := K \otimes_R V$, which may be turned into a KG-module satisfying

$$g \cdot \alpha \otimes v = \alpha \otimes gv \quad \text{for} \quad g \in G , \ \alpha \in K , \ v \in V.$$

It follows from the standard properties of tensor products (see e.g. [40], Section 12) that this is indeed a KG-module which has K-basis $(1 \otimes v_1, \ldots, 1 \otimes v_n)$, provided that $(v_1, \ldots, v_n)$ is an R-basis of V.

Definition 1.8.1 Let $K \supseteq R$ be commutative rings and let A be an R-**order**, that is an R-algebra, which is free and finitely generated as an R-module (i.e. it has a finite R-basis). Then $KA := K \otimes_R A$ is a K-algebra with multiplication satisfying

$$(\alpha \otimes a) \cdot (\beta \otimes b) = \alpha\beta \otimes ab \quad \text{for} \quad \alpha, \beta \in K \quad \text{and} \quad a, b \in A.$$

We have an embedding of rings given by $\epsilon \colon A \to KA$, $a \mapsto 1 \otimes a$. Furthermore, if V is an A-module, then $KV := K \otimes_R V$ is a KA-module satisfying

$$(\alpha \otimes a) \cdot (\beta \otimes v) = \alpha\beta \otimes av \quad \text{for} \quad \alpha, \beta \in K , \ v \in V.$$

Conversely, for a KA-module W we write $W_A := \mathrm{Inf}_\epsilon W$ for the restricted A-module. We often identify A with $\epsilon(A)$ and embed V in KV.

Of course, to *define* the multiplication in KA and the action of KA on KV one may use K-bases (see [40], p. 72). For the case $A = RG$ we have a K-algebra isomorphism $KRG \cong KG$ with $1 \otimes g \mapsto g$ for $g \in G$.

In this section we only look at the case of a field extension. The other case will be addressed in Section 4.1.

Definition 1.8.2 Let $K \subseteq L$ be a field extension and let A be a finitely generated K-algebra. If $\gamma \in \mathrm{Gal}(L/K)$ then γ extends to a ring automorphism

$$\gamma \otimes \mathrm{id}_A \colon LA \to LA \;, \quad \alpha \otimes a \mapsto \gamma(\alpha) \otimes a \qquad (\alpha \in L \;,\; a \in A).$$

If W is an LA-module we define $^\gamma W := \mathrm{Inf}_{(\gamma^{-1} \otimes \mathrm{id}_A)} W$ (see Remark 1.1.21). Thus $^\gamma W = W$ as abelian group, but the action of LA on $^\gamma W$ is changed to satisfy

$$(\alpha \otimes a) \star_\gamma w = (\gamma^{-1}(\alpha) \otimes a) \cdot w \qquad \text{for} \qquad w \in W.$$

The module $^\gamma W$ is called **algebraically conjugate** to W.

Remark 1.8.3 Keeping the notation of the definition it is clear that $\dim_L LA = \dim_K A$ and $\dim_L LV = \dim_K V$ for an A-module V. If W affords the representation

$$\delta \colon LA \to L^{n \times n}, \quad x \mapsto \delta(x) = [\alpha_{ij}(x)] \qquad \text{for} \qquad x \in LA$$

then $^\gamma W$ affords the representation (with respect to the same basis)

$$^\gamma\delta \colon LA \to L^{n \times n}, \quad x \mapsto {}^\gamma\delta(x) := [\gamma(\alpha_{ij}((\gamma^{-1} \otimes \mathrm{id}_A)(x)))] \qquad \text{for} \qquad x \in LA.$$

In particular,

$$^\gamma\delta(a) = [\gamma(\alpha_{ij}(a))] \qquad \text{for} \qquad a \in A \subseteq LA.$$

Furthermore, it follows that $^{\gamma_1 \gamma_2} W = {}^{\gamma_1}({}^{\gamma_2} W)$ for $\gamma_1, \gamma_2 \in \mathrm{Gal}(L/K)$ and that W is simple if and only if $^\gamma W$ is simple.

Theorem 1.8.4 *Let A be a K-algebra for some field K and let $L \supseteq K$ be a finite Galois extension. If V is a simple A-module (with $\dim_K(V) < \infty$) then there is a simple LA-module W such that*

$$LV \cong_{LA} m \cdot ({}^{\gamma_1} W \oplus \cdots \oplus {}^{\gamma_k} W) \qquad \text{with} \qquad \gamma_i \in \mathrm{Gal}(L/K),$$

with $^{\gamma_i} W \not\cong_{LA} {}^{\gamma_j} W$ for $i \neq j$, Here $m \cdot W$ is short for $\underbrace{W \oplus \cdots \oplus W}_{m}$. Furthermore $m\,k \leq [L : K]$. If K is a finite field, then $m = 1$.

Proof. Let $\delta \colon A \to K^{n \times n}$ be a matrix representation afforded by V. Then by Corollary 1.3.7 $\delta(A) \cong D^{r \times r}$ with a division ring $D := (\mathrm{End}_A V)^{\mathrm{op}}$ and some $r \in \mathbb{N}$. The LA-module LV affords a matrix representation $^L\delta \colon LA \to L^{n \times n}$ with $^L\delta(\alpha \otimes a) = \alpha\,\delta(a)$ for $\alpha \in L$, $a \in A$, and we have

$$^L\delta(LA) \cong L \otimes_K D^{r \times r} \cong (L \otimes_K D)^{r \times r}.$$

We first show that the K-algebra $L \otimes_K D^{r \times r}$ is semisimple. By Corollary 1.6.12 it suffices to show that $L \otimes_K D$ is semisimple. If $(\alpha_1, \ldots, \alpha_s)$ is a K-basis of L we have

$$L \otimes_K D = \bigoplus_{i=1}^{s} \alpha_i \otimes D$$

as a D-space. Any two-sided ideal $I \trianglelefteq L \otimes_K D$ is also a D-subspace, invariant under inner automorphisms γ of D (with $\gamma \colon \sum_{i=1}^{s} \alpha_i \otimes d_i \mapsto \sum_{i=1}^{s} \alpha_i \otimes \gamma(d_i)$). By Exercise 1.8.2 I has a D-basis consisting of elements of the form $\sum_{i=1}^{s} \alpha_i \otimes d_i$ with $\alpha_i \in L$ and $d_i \in \mathbf{Z}(D)$. Hence I is generated by an ideal $I' \trianglelefteq L \otimes_K \mathbf{Z}(D) \subseteq L \otimes_K D$. If I is nilpotent, then I' must be nilpotent as well. But since $L \supseteq K$ is a finite separable field extension, we infer from Exercise 1.8.1 that $L \otimes_K \mathbf{Z}(D)$ is semisimple. From Corollaries 1.4.7 and 1.5.3 we conclude that $L \otimes_K D$ is semisimple, too.

In the following we will identify (see Remark 1.1.22)

$$V = \mathrm{Inf}_{\boldsymbol{\delta}}\, K^n \qquad \text{and} \qquad LV = \mathrm{Inf}_{L\boldsymbol{\delta}}\, L^n.$$

Then $\mathrm{Gal}(L/K)$ acts naturally on LV. Since ${}^L\boldsymbol{\delta}(LA)$ is semisimple LV is a direct sum of simple LA-modules. If W is a simple submodule of LV with L-basis $(w_1, \ldots, w_d)$ and $\gamma \in \mathrm{Gal}(L/K)$, then $\gamma W := \{\gamma w \mid w \in W\}$ is also a simple submodule of LV, isomorphic to ${}^\gamma W$. Let $W' := \sum_{\gamma \in \mathrm{Gal}(L/K)} \gamma W$ and $\tilde{w}_i := \mathrm{Tr}_{L/K}(w_i)$ for $1 \leq i \leq d$, where $\mathrm{Tr}_{L/K}(x) := \sum_{\gamma \in \mathrm{Gal}(L/K)} \gamma x$ for $x \in L$ or $x \in L^n$, the usual field trace. Then $\tilde{w}_i \in K^n \cap W'$ ([110], Theorem VI.5.1, p. 285). Since $\mathrm{Tr}_{L/K} \neq 0$ (see [110], Theorem VI.5.2, p. 286) we can, if need be, multiply w_i by some constant in L, so that $\tilde{w}_i \neq 0$. Then

$$\{0\} \neq \langle \tilde{w}_1, \ldots, \tilde{w}_d \rangle_A \leq_A V,$$

and, since V is simple, we have equality. Hence $W' = LV$ and thus LV is a direct sum of simple LA-modules algebraically conjugate to W. Since $\mathrm{Gal}(L/K)$ acts transitively on $\{\gamma W \mid \gamma \in \mathrm{Gal}(L/K)\}$, and hence on the set of isomorphism classes of these submodules, it is clear that each γW occurs (up to isomorphism) with the same multiplicity m in a direct sum decomposition of LV. Since $LV = W'$ we have $m\,k \leq [L:K]$.

If K is a finite field, then D is a finite division ring, so by a well-known theorem of Wedderburn ([32], p. 101) D is a field. Hence by Exercise 1.8.1 there are finite fields $L_1, \ldots, L_t$ with $L \otimes_K D \cong L_1 \oplus \cdots \oplus L_t$. Consequently

$$ {}^L\boldsymbol{\delta}(LA) \cong (L \otimes_K D)^{r \times r} \cong L_1^{r \times r} \oplus \cdots \oplus L_t^{r \times r} \cong_{LA} r \cdot (L_1^r \oplus \cdots \oplus L_t^r).$$

Here $W_i := L_i^r$, the simple $L_i^{r \times r}$-modules, are simple LA-modules via inflation (see Definition 1.1.21) and $W_i \not\cong_{LA} W_j$ for $i \neq j$, since they belong to different block ideals of ${}^L\boldsymbol{\delta}(LA)$. On the other hand,

$$L \otimes_K D^{r \times r} \cong_{LA} r \cdot (L \otimes_K V) = r \cdot LV$$

and consequently $LV \cong_{LA} W_1 \oplus \cdots \oplus W_t$. $\qquad\square$

Exercises

Exercise 1.8.1 Let L_1, L_2 be field extensions of K and $L_1 \cong K[X]/(f)$ for some irreducible polynomial $f \in K[X]$. Show that

$$L_1 \otimes_K L_2 \cong L_2[X]/(f) \qquad \text{as} \qquad K\text{-algebras}.$$

Deduce that $L_1 \otimes_K L_2$ is a semisimple K-algebra, provided that $L_1 \supseteq K$ is a finite separable extension.

Exercise 1.8.2 Let D be a division ring, let G be a set of automorphisms of D and $n \in \mathbb{N}$. Any automorphism σ of D acts on $V := D^n$ by $\sigma\,[d_1, \ldots, d_n]^{\mathrm{T}} := [\sigma(d_1), \ldots, \sigma(d_n)]^{\mathrm{T}}$ for $d_1, \ldots, d_n \in D$. Show that any D-subspace W of V with $\sigma W = W$ for all $\sigma \in G$ has a D-basis consisting of vectors w_i with $\sigma\,w_i = w_i$ for all $\sigma \in G$. (See [166], p. 221.)

Exercise 1.8.3 Let $Q_8 := \langle a, b \mid a^2 = b^2 = [a, b],\ a^4 = 1 \rangle$ be the quaternion group, let $K \subseteq \mathbb{R}$ be a field and let $L := K(i)$, where $i^2 = -1$. Using Exercise 1.1.5 show that there is an irreducible representation $\boldsymbol{\delta} \colon Q_8 \to \mathrm{GL}_4(K)$ with

$$\boldsymbol{\delta} \colon a \mapsto \begin{bmatrix} 0 & 0 & -1 & 0 \\ 0 & 0 & 0 & -1 \\ 1 & 0 & 0 & 0 \\ 0 & 1 & 0 & 0 \end{bmatrix}, \qquad b \mapsto \begin{bmatrix} 0 & -1 & 0 & 0 \\ 1 & 0 & 0 & 0 \\ 0 & 0 & 0 & 1 \\ 0 & 0 & -1 & 0 \end{bmatrix}.$$

Let $V := K^4$ be the corresponding representation module. Show that $W := \langle w_1, w_2 \rangle_L$ with $w_1 := [1, -i, 0, 0]^{\mathrm{T}}$ and $w_2 := [0, 0, 1, i]^{\mathrm{T}}$ is a simple $L\,Q_8$-submodule of LV. Using the notation of the proof of Theorem 1.8.4 compute γW, where $\mathrm{Gal}(L/K) = \langle \gamma \rangle$ and $\tilde{w}_1, \tilde{w}_2$. Prove that $W \cong_{L\,Q_8} {}^{\gamma}W$. Deduce that $\boldsymbol{\delta}(K\,Q_8)$ is a division ring. For $K = \mathbb{R}$ it is called the division ring of real quaternions.

2

Characters

2.1 Characters and block idempotents

Throughout this section we assume that G is a finite group and A is a finite-dimensional algebra over a field K. All A-modules considered are supposed to be finitely generated over K.

Definition 2.1.1 Let V be an A-module and let $\delta\colon A \to \mathrm{End}_K V$ be the corresponding representation. Then the function

$$\chi_V = \chi_\delta \colon A \to K, \ a \mapsto \mathrm{trace}\, \delta(a)$$

is called the **character** of V or of δ. A character of an irreducible representation (or equivalently of a simple module) is called **irreducible**. Let $\mathrm{Irr}_K(A)$ be the set of irreducible characters of A over K. If $A = KG$ then $\chi = \chi_V$ is often identified with its restriction

$$\chi|_G \colon G \to K, \ g \mapsto \mathrm{trace}\, \delta(g).$$

A character of a group G (over a field K) is a character of a representation of G (over K) and the set of all these characters will be denoted by $\mathrm{Char}_K(G)$. We put $\mathrm{Irr}_K(G) := \mathrm{Irr}_K(KG)$ and $\mathrm{Irr}(G) := \mathrm{Irr}_{\mathbb{C}}(G)$.

Obviously a character χ_V of an A-module V as above is always completely determined by its values on a fixed K-basis of V (in fact often an even smaller set suffices, as we will see shortly). So the above identification of group characters is harmless. If $\dim_K V = 1$ then χ_V is the same as the matrix representation corresponding to V; in this case χ_V is an algebra homomorphism and $\chi|_G \colon G \to K^\times$ (in the case $A = KG$) is a group homomorphism called a **linear character**. In particular we will denote the character of the trivial representation of G of degree one by 1_G, thus $1_G(g) = 1$ for all $g \in G$.

Example 2.1.2 The character of the regular representation $\delta_{KG}\colon G \to \mathrm{GL}(KG)$ is the **regular character** and will be denoted by ρ_G. Obviously

$$\rho_G(g) = \begin{cases} 0 & \text{for } 1 \neq g \in G, \\ |G| \cdot 1_K & \text{for } g = 1. \end{cases}$$

If char $K \mid |G|$ then ρ_G is the zero function.   ◆

Example 2.1.3 If the group G acts on the set Ω and δ is the corresponding permutation representation, then we have $\chi_\delta(g) = |\operatorname{Fix}_\Omega(g)| \cdot 1_K$. Such a character will be called a **permutation character**. Actually, Example 2.1.2 is a special case of this construction and so is the trivial character 1_G.   ◆

Lemma 2.1.4 (a) *If* $V \cong_A V'$ *are isomorphic A-modules, then* $\chi_V = \chi_{V'}$.
(b) *If* V *is an A-module and* $W \leq_{KG} V$ *then*

$$\chi_V = \chi_W + \chi_{V/W}.$$

In particular, every character is a sum of irreducible characters.
(c) *Group characters are class functions, i.e. if* $\chi \in \operatorname{Char}_K(G)$ *then*

$$\chi(g^{-1}hg) = \chi(h) \text{ for all } g, h \in G.$$

(d) *If* $\varphi\colon A \to A_1$ *is an epimorphism of K-algebras (or* $\varphi\colon G \to G_1$ *is a group epimorphism) and* $\chi \in \operatorname{Irr}_K(A_1)$ *(or* $\chi \in \operatorname{Irr}_K(G_1)$, *respectively) then* $\chi \circ \varphi$ *is in* $\operatorname{Irr}_K(A)$ *(or* $\operatorname{Irr}_K(G)$, *respectively), called the* **inflation** *of* χ.

Proof.   Recall that similar matrices have the same trace. For (b) observe that addition of characters is to be taken point-wise (in the K-vector space of K-valued functions). (d) follows from Definition 1.1.20 and Remark 1.1.21.   □

Lemma 2.1.5 *The irreducible characters of A are linearly independent over K, if K is a splitting field for A.*

Proof.   By Theorem 1.4.6 the irreducible characters of A are the inflations of the irreducible characters of $A/\operatorname{J}(A)$. Hence we may assume that A is semisimple. In this case the result follows from Wedderburn's Theorem 1.5.5: if $A = \bigoplus_{i=1}^r A_i$ with simple two-sided ideals A_i corresponding to the irreducible representations δ_i with characters χ_i, choose $e_i \in A_i$ such that $\delta_i(e_i)$ is the matrix with unity in position $(1,1)$ and zeros elsewhere. Recall that by our assumption the image of δ_i is a full matrix ring over K. Then $\chi_j(e_i) = \delta_{i,j}$. So from $0 = \sum_{i=1}^r a_i\chi_i$ with $a_1, \ldots, a_r$ in K we conclude, by inserting e_i, that $a_i = 0$.   □

The assertion of Lemma 2.1.5 holds also for non-splitting fields, provided that $\operatorname{J}(LA) = L \otimes_K \operatorname{J}(A)$ for any extension field $L \supseteq K$, which is the case if $A = KG$ for a finite group G, for instance (see [41], theorem (7.9), p. 146). See Exercise 2.1.1 for an example where $\operatorname{Irr}_K(A)$ is linearly dependent.

The following theorem gives an explicit formula for the central primitive idempotents in terms of the irreducible characters of a group.

Theorem 2.1.6 *Let K be a field with* char $K \nmid |G|$ *and* $\mathrm{Irr}_K(G) = \{\chi_1, \ldots, \chi_r\}$. *Let* $\chi_i = \chi_{L_i}$ *for a simple KG-module L_i with $D_i = \mathrm{End}_{KG} L_i$ and corresponding block idempotent ϵ_i. Then*

(a)

$$\epsilon_i = \frac{\chi_i(1)}{[D_i : K]} \frac{1}{|G|} \sum_{g \in G} \chi_i(g^{-1})g \quad and \quad \mathrm{char}\, K \nmid \frac{\chi_i(1)}{[D_i : K]} \in \mathbb{N};$$

(b) *for $h \in G$ we have*

$$\frac{1}{|G|} \sum_{g \in G} \chi_i(gh)\chi_j(g^{-1}) = \frac{[D_i : K]}{\chi_i(1)} \chi_i(h)\, \delta_{i,j}.$$

Proof. (a) We use Maschke's and Wedderburn's Theorem 1.5.5 and write

$$KG = \bigoplus_{i=1}^{r} KG\epsilon_i \quad \text{with } KG\epsilon_i = L_i \oplus \cdots \oplus L_i \quad \text{with } n_i \text{ summands } L_i.$$

Let $\epsilon_i = \sum_{g \in G} \alpha_g g$. We apply the regular character $\rho_G = \sum_{j=1}^{r} n_j \chi_j$ and get

$$|G|\alpha_g = \rho_G(\epsilon_i g^{-1}) = \sum_{j=1}^{r} n_j \chi_j(\epsilon_i g^{-1}) = n_i \chi_i(\epsilon_i g^{-1}) = n_i \chi_i(g^{-1}).$$

From Theorem 1.5.5 we have $n_i = \dim_{D_i} L_i = \frac{\dim_K L_i}{[D_i : K]} = \frac{\chi_i(1)}{[D_i:K]}$. Observe that $\epsilon_i \neq 0$, so $n_i = \frac{\chi_i(1)}{[D_i:K]}$ cannot be divisible by char K, so it is invertible in K.

(b) We use part (a) and get

$$\delta_{i,j}\epsilon_i = \epsilon_i \epsilon_j = \frac{\chi_i(1)}{[D_i : K]} \frac{\chi_j(1)}{[D_j : K]} \frac{1}{|G|^2} \sum_{x \in G} \sum_{y \in G} \chi_i(x^{-1})\chi_j(y^{-1})xy.$$

$$= \frac{\chi_i(1)}{[D_i : K]} \frac{\chi_j(1)}{[D_j : K]} \sum_{h \in G} \left(\frac{1}{|G|^2} \sum_{g \in G} \chi_i(gh^{-1})\chi_j(g^{-1}) \right) h.$$

For $i \neq j$ the result follows using the fact that the factors in front of the sum are non-zero in K by part (a). For $i = j$ we apply the formula for ϵ_i from (a) in the left hand side and compare coefficients. $\qquad\square$

The special case in which K is a splitting field for G and of characteristic not dividing $|G|$ deserves particular attention.

Corollary 2.1.7 *If K is a splitting field for G of characteristic not dividing $|G|$ then the block idempotent ϵ_{χ_i} corresponding to $\chi_i \in \mathrm{Irr}_K(G)$ is given by*

$$\epsilon_{\chi_i} := \frac{\chi_i(1)}{|G|} \sum_{g \in G} \chi_i(g^{-1})g.$$

*If V is any KG-module then $\epsilon_i V$ gives the L_i-homogeneous component of V,
where L_i is a KG-module with character χ_i.*

The above result can be generalized (as can others in this section) to semisimple symmetric algebras (see [63], Chap. 7). In view of Lemma 1.2.22 the following theorem could be subsumed under this more general setting. Observe that $(g^{-1})_{g \in G}$ is the basis contragredient to the natural basis G of the symmetric K-algebra KG. Similarly, if Ω is a transitive G-set, by Lemma 1.2.22 the basis contragredient to the Schur basis $(\theta_j)_{j=1}^m$ in the symmetric algebra $\mathrm{End}_{KG} K\Omega$ is $(\theta_{j'}/|\mathcal{O}_j|)_{j=1}^m$, where $\mathcal{O}_{j'} := \mathcal{O}_j'$, the paired orbit; see Definition 1.2.10.

Theorem 2.1.8 *Let* char $K \nmid |G|$ *and let* Ω *be a transitive G-set. Assume that*

$$K\Omega \cong_{KG} \bigoplus_{i=1}^m \underbrace{L_i \oplus \cdots \oplus L_i}_{n_i}, \qquad L_i \not\cong_{KG} L_j \quad for \quad i \neq j,$$

with simple KG-modules L_i. Put $D_i := \mathrm{End}_{KG} L_i$ and $z_i := \dim_{D_i} L_i$. Then

$$E := \mathrm{End}_{KG} K\Omega = \bigoplus_{i=1}^m E\epsilon_i , \qquad E\epsilon_i \cong D_i^{n_i \times n_i}$$

with centrally primitive idempotents ϵ_i. Let $(\theta_1 = 1, \ldots, \theta_m)$ be the Schur basis of E (with $\theta_i = \theta_{\mathcal{O}_i}$ corresponding to the orbits $\mathcal{O}_i$ of G on $\Omega \times \Omega$; see Lemma 1.2.15) and let ζ be the character of the E-module $K\Omega$. Assume that ζ_i is the character of the simple E-module in $E\epsilon_i$. Then $\zeta = \sum_{i=1}^m z_i \zeta_i$ and

$$\epsilon_i = z_i \sum_{j=1}^m \frac{\zeta_i(\theta_{j'})}{|\mathcal{O}_j|} \theta_j, \tag{2.1}$$

$$\sum_{k=1}^m \frac{1}{|\mathcal{O}_k|} \zeta_i(\theta_k) \zeta_j(\theta_{k'}) = \frac{\zeta_i(1)}{z_i} \delta_{i,j}. \tag{2.2}$$

Proof. By Theorem 1.5.4 $E\epsilon_i \cong \mathrm{End}_{KG} \mathrm{H}_{L_i}(K\Omega) \cong D_i^{n_i \times n_i}$ and $\zeta = \sum_{k=1}^m z_k \zeta_k$.

Observe that ζ is a symmetrizing trace for the symmetric algebra E, denoted by τ in Lemma 1.2.22, where we had shown that $\zeta(\theta_i \theta_{j'}) = \delta_{i,j} |\mathcal{O}_i|$. Thus writing $\epsilon_i = \sum_{k=1}^m \alpha_i \theta_i$ we get

$$\alpha_j |\mathcal{O}_j| = \zeta(\epsilon_i \theta_{j'}) = \sum_{k=1}^m z_k \, \zeta_k(\epsilon_i \theta_{j'}) = z_i \, \zeta_i(\theta_{j'}).$$

Observe that $|\mathcal{O}_j|$ divides $|G|$ and hence is invertible in K. Thus (2.1) follows.

Since the ϵ_i are orthogonal idempotents we get $\delta_{i,j} \epsilon_i = \epsilon_i \epsilon_j$, hence

$$\delta_{i,j} \epsilon_i = z_i z_j \sum_{k=1}^m \sum_{l=1}^m \frac{\zeta_i(\theta_{k'}) \zeta_j(\theta_{l'})}{|\mathcal{O}_k||\mathcal{O}_l|} \theta_k \theta_l = z_i z_j \sum_{k=1}^m \sum_{l=1}^m \sum_{u=1}^m \frac{\zeta_i(\theta_{k'}) \zeta_j(\theta_{l'})}{|\mathcal{O}_k||\mathcal{O}_l|} a_{klu} \theta_u.$$

We compare the coefficients of $\theta_1 = 1_E$ and get, by Theorem 1.2.20(b),

$$z_i \frac{\zeta_i(1)}{|\mathcal{O}_1|} \delta_{i,j} = z_i z_j \sum_{k=1}^{m} \frac{\zeta_i(\theta_{k'})\zeta_j(\theta_k)}{|\mathcal{O}_k|^2} \frac{|\mathcal{O}_k|}{|\mathcal{O}_1|}.$$

From this the result follows, since $z_i = \frac{\chi_i(1)}{[D_i:K]}$ (with $\chi_i = \chi_{L_i} \in \mathrm{Irr}_K(G)$) is invertible in K by Theorem 2.1.6(a). $\qquad\square$

Corollary 2.1.9 *It is standard to choose $L_1 := K_G$ the trivial KG-module in Theorem 2.1.8, so that ζ_1 is the principal character (see Theorem 1.2.20). Then*

$$\sum_{k=1}^{m} \zeta_i(\theta_k) = \delta_{i,1}|\Omega|.$$

Observe that by choosing the regular G-set $\Omega = G/\{1\}$ in Theorem 2.1.8 we get essentially Theorem 2.1.6 (with $h = 1$ in part (b)).

As an easy consequence we obtain an interesting relation between the dimensions of the irreducible constituents of a multiplicity-free permutation module $\mathbb{C}\Omega$ and the subdegrees.

Corollary 2.1.10 *Let Ω be a transitive G-set of rank r with suborbits of lengths $1 = d_1, \ldots, d_r$. Assume that $\mathbb{C}\Omega$ is multiplicity-free and let $1 = z_1, \ldots, z_r$ be the dimensions of the irreducible constituents. Then*

$$f_\Omega := |\Omega|^{r-2} \prod_{i=1}^{r} \frac{d_i}{z_i},$$

*called the **Frame quotient**, is an integer.*

Proof. Since $\mathbb{C}\Omega$ is multiplicity-free, $\mathrm{End}_{\mathbb{C}G}\,\mathbb{C}\Omega$ is commutative (see Exercise 1.5.2) and hence $\zeta_i(1) = 1$ for all i. We write (2.2) in Theorem 2.1.8 as a matrix equation, take determinants and use Remark 1.2.12:

$$\det[\zeta_i(\theta_k)]_{ik} \det[\zeta_i(\theta_{k'})]_{ik} = \prod_{i=1}^{r} \frac{|\mathcal{O}_i|}{z_i} = |\Omega|^r \prod_{i=1}^{r} \frac{d_i}{z_i}.$$

Observe that the $\zeta_i(\theta_k)$ are eigenvalues of the intersection matrices, hence algebraic integers. Corollary 2.1.9 shows that if we add the second, third, ..., rth column of $[\zeta_i(\theta_k)]_{ik}$ to the first one we obtain $[|\Omega|, 0 \ldots, 0]^{\mathrm{T}}$ as the first column. Hence $\frac{1}{|\Omega|} \det[\zeta_i(\theta_k)]_{ik}$ is an algebraic integer. $\qquad\square$

From the proof we see that if all $\zeta_i(\theta_k)$ are rational, then f_Ω is a square. It is not hard to see that this is the case if the irreducible constituents of $\mathbb{C}\Omega$ have rational characters. It had been conjectured that the Frame quotient is always a square. But this is not true; see for instance Exercise 2.1.11.

Table 2.1. Irreducible characters (representations) of E

$\lvert \mathcal{O}_i \rvert :$	$\binom{n}{2}$	$n(n-1)(n-2)$	$6\binom{n}{4}$
	θ_1	θ_2	θ_3
ζ_1	1	$2n-4$	$\binom{n-2}{2}$
ζ_2	1	$n-4$	$-n+3$
ζ_3	1	-2	1

Example 2.1.11 In Example 1.2.25 we computed the irreducible characters (representations) $\zeta_1, \zeta_2, \zeta_3$ of $E = \mathrm{End}_{K\,\mathrm{S}_n} K\binom{\Omega}{2}$, where $\Omega = \{1, \ldots, n\}$ for the case that char $K \nmid (n-2)\binom{n}{2}$ and $n \geq 4$ (see Table 2.1).

If char $K \nmid n!$ we can use the above "orthogonality relations," i.e. (2.2) in Theorem 2.1.8, to compute the z_i. Since $E \cong K \oplus K \oplus K$ we have $K\binom{\Omega}{2} \cong_{K\,\mathrm{S}_n} K \oplus L_2 \oplus L_3$ with $\mathrm{End}_{K\,\mathrm{S}_n} L_2 \cong \mathrm{End}_{K\,\mathrm{S}_n} L_3 \cong K$, so that L_2, L_3 are absolutely simple $K\,\mathrm{S}_n$-modules with K-dimensions z_2, z_3. We conclude that $z_1 = 1$ (of course), $z_2 = n - 1$ and $z_3 = \frac{1}{2}n(n-3)$. So this method has given us the degrees of two non-trivial irreducible characters of the symmetric group. Compare this result with Example 1.5.7. We also see that $f_\Omega = (n-1)^2$. ◆

Example 2.1.12 As in Example 1.2.26, let $G = \mathrm{HS}$ be the Higman–Sims group acting on $\Omega = \{1, \ldots, 100\}$. From the "character table" of $E := \mathrm{End}_{KG} K\Omega$ for a field K with char $K \neq 2, 5$, computed at the end of Example 1.2.26, we immediately find the degrees z_i of the irreducible constituents of the permutation representation of G on Ω. These are $z_1 = 1, z_2 = 22, z_3 = 77$, which in this case coincide with the subdegrees. Here $f_\Omega = \lvert \Omega \rvert = 10^2$. ◆

Definition 2.1.13 Let G be a finite group and K a field, then $\mathrm{cf}(G, K) := \{\psi \colon G \to K \mid \psi(h^{-1}gh) = \psi(g) \text{ for all } g, h \in G\}$ is called the **space of class functions** on G. Obviously $\mathrm{cf}(G, K)$ is a K-vector space of dimension equal to the number of conjugacy classes of G. If char $K \nmid \lvert G \rvert$ one defines a symmetric bilinear form on $\mathrm{cf}(G, K)$ in the following way: For $\psi, \varphi \in \mathrm{cf}(G, K)$ we put

$$(\varphi, \psi)_G := \frac{1}{\lvert G \rvert} \sum_{g \in G} \varphi(g)\psi(g^{-1}).$$

Note that $\mathrm{N}(\varphi) := (\varphi, \varphi)_G$ is often called the **norm** of φ.

Corollary 2.1.14 *Let K be a field with char $K \nmid \lvert G \rvert$. Then, using the notation of Theorem 2.1.6,*

(a) $(\chi_i, \chi_j)_G = [D_i : K]\delta_{i,j}$;

(b) *if V, W are KG-modules with characters φ, ψ, respectively, then*

$$(\varphi, \psi)_G = \dim \mathrm{Hom}_{KG}(V, W) \cdot 1_K;$$

(c) *if Ω is a G-set with permutation character π then*

$$(\pi, 1_G)_G = m \cdot 1_K, \quad \text{where } m \text{ is the number of orbits of } G \text{ on } \Omega;$$

(d) *if Ω is transitive of rank r then $(\pi, \pi)_G = r \cdot 1_K$. If $r = 2$, that is G is doubly transitive on Ω, then*

$$\pi = 1_G + \chi \quad with \quad \chi \in \mathrm{Irr}_K(G).$$

Proof. (a) This is just Theorem 2.1.6(b) with $h = 1$.

(b) Since KG is semisimple we have

$$V \cong_{KG} \bigoplus_{i=1}^{r} a_i L_i, \quad W \cong_{KG} \bigoplus_{i=1}^{r} b_i L_i,$$

where $L_1, \ldots, L_r$ are as in Theorem 2.1.6 and $a_i, b_i \in \mathbb{N}_0$ are the multiplicities of L_i in V, W, respectively, e.g.

$$a_i L_i = \underbrace{L_i \oplus \cdots \oplus L_i}_{a_i}.$$

Then

$$\mathrm{Hom}_{KG}(V, W) \cong \bigoplus_{i=1}^{r} a_i b_i \, \mathrm{Hom}_{KG}(L_i, L_i) \cong \bigoplus_{i=1}^{r} a_i b_i D_i.$$

From this the result follows using part (a) since, by Lemma 2.1.4, $\varphi = \sum_{i=1}^{r} a_i \chi_i$ and $\psi = \sum_{i=1}^{r} b_i \chi_i$.

(c) From Lemma 1.2.15 it follows immediately that $\dim_K \mathrm{Hom}_{KG}(K\Omega, K\{1\}) = m$, the number of orbits of G on Ω, where, of course $\{1\}$ is the trivial G-set, so that the character of $K\{1\}$ is 1_G. So the assertion follows from part (b).

(d) Corollary 1.2.16 just says that $\dim_K \mathrm{End}_{KG} K\Omega = r$, if Ω is a transitive G-set of rank r, hence $(\pi, \pi)_G = r$ in this case. Finally, if $r = 2$ then $\pi - 1_G = \chi$ must have norm one. $\square$

Of course, Corollary 2.1.14 part (a) gives another proof of Lemma 2.1.5 for the case that $\mathrm{char}\, K \nmid |G|$. Part (c) of the corollary can also be written as

$$\frac{1}{|G|} \sum_{g \in G} |\mathrm{Fix}_\Omega(g)| = m, \quad \text{the number of orbits of } G \text{ on } \Omega, \tag{2.3}$$

which is a well-known theorem of Cauchy and Frobenius that is often cited as a lemma of Burnside.

Theorem 2.1.15 (Orthogonality relations) *Let K be a splitting field for G with $\mathrm{char}\, K \nmid |G|$ and $\{g_1, \ldots, g_r\}$ representatives of the conjugacy classes of G. Then $\mathrm{Irr}_K(G)$ is an orthonormal basis of $\mathrm{cf}(G, K)$, and for $\chi, \chi' \in \mathrm{Irr}(G)$*

$$(\chi, \chi')_G = \sum_{k=1}^{r} \frac{1}{|\mathbf{C}_G(g_k)|} \chi(g_k) \chi'(g_k^{-1}) = \delta_{\chi, \chi'}, \tag{2.4}$$

$$\sum_{\chi \in \mathrm{Irr}(G)} \chi(g_i) \chi(g_j^{-1}) = |\mathbf{C}_G(g_i)| \, \delta_{i,j}. \tag{2.5}$$

Proof. The first formula follows from Corollary 2.1.14 and Lemma 2.1.4. Recall that the number of elements in G conjugate to g_k is $[G : \mathbf{C}_G(g_k)]$.

Written in matrix form (2.4) says

$$[\frac{\chi(g_k)}{|\mathbf{C}_G(g_k)|}]_{\chi,k} \cdot [\chi(g_k^{-1})]_{k,\chi} = \mathbf{I}_r.$$

It follows that

$$[\chi(g_k^{-1})]_{k,\chi} \cdot [\frac{\chi(g_k)}{|\mathbf{C}_G(g_k)|}]_{\chi,k} = \mathbf{I}_r,$$

which gives formula (2.5).   □

Observe that we obtain from (2.5) of Theorem 2.1.15 – usually called the "orthogonality relations for the columns" – for $g_i = g_j = 1$ the equation

$$\sum_{\chi \in \mathrm{Irr}(G)} \chi(1)^2 = |G|,$$

which we have already seen in Theorem 1.5.10, since obviously the $\chi(1)$ are just the dimensions of the simple modules.

Corollary 2.1.16 *We assume the notation of Theorem 2.1.15 and in addition that* char $K = 0$.

(a) *A KG-module V with character χ is simple if and only if $(\chi, \chi)_G = 1$.*

(b) *If V is an arbitrary KG-module with character χ_V, then*

$$\chi_V = \sum_{i=1}^{r} (\chi_V, \chi_i)_G \chi_i.$$

So the character of V provides convenient information about the composition factors of V and their multiplicities.

(c) *Two KG-modules are isomorphic if and only if their characters coincide.*

It follows from Theorem 1.5.11 and Lemma 2.1.4 that the irreducible characters of a group G can be given by a square matrix $[\chi_i(g_j)]_{i,j,=1,\dots,r}$, where the g_j are representatives of the conjugacy classes of G. This matrix is called the **character table** of G in the case that $K = \mathbb{C}$ – actually we would get the same matrix for any splitting field K of characteristic zero up to some identification, as we will see in Section 2.2. It is customary to include in the character table of a group more information than just the character values. For example, as column headings we usually state the names of the conjugacy classes. These consist of a number giving the element order of a representative of the class and a letter in order to distinguish between classes containing elements of the same order. The centralizer orders of representatives of the conjugacy classes are also usually given in the column headings. Whereas this is just for convenience – by Theorem 2.1.15 formula (2.5) it can be computed from the matrix

$[\chi_i(g_j)]$ – the element orders in general *cannot* be computed from this matrix alone (see Example 2.1.19 below); so the class names provide genuinely additional information in general. Other genuinely additional information usually included in the "table head" of a character table includes the so-called **power maps**. These give for each prime divisor p of the group order $|G|$ and each class name mX the name of the conjugacy class, which contains the pth power of a representative g_j of the class mX. We will see that this information (in conjunction with the values of the irreducible characters) is in fact sufficient to find the class name of the conjugacy class of *any* power g_j^m of g_j for $m \in \mathbb{Z}$ (see Corollary 2.2.10).

The famous ATLAS of finite groups [38] contains the character tables of all sporadic simple groups and many other simple groups with a large amount of further information, to which we will return at a later stage. These character tables, and many others, are also contained in the library of character tables of GAP.

Example 2.1.17 We show what a character table of the alternating group A_5 on five letters looks like when it is displayed from the library of GAP:

```
gap> t := CharacterTable("A5");;  Display(t);
A5

      2  2  2  .  .  .
      3  1  .  1  .  .
      5  1  .  .  1  1

         1a 2a 3a 5a 5b
      2P 1a 1a 3a 5b 5a
      3P 1a 2a 1a 5b 5a
      5P 1a 2a 3a 1a 1a

X.1      1  1  1  1  1
X.2      3 -1  .  A *A
X.3      3 -1  . *A  A
X.4      4  .  1 -1 -1
X.5      5  1 -1  .  .

A = -E(5)-E(5)^4
  = (1-ER(5))/2 = -b5
```

The first three lines contain the centralizer orders in a factorized form. Thus the orders of the centralizers of the representatives of the conjugacy classes are $2^2 \cdot 3 \cdot 5$, 2^2, 3, 5, 5 in that order. The three lines following the class names contain the second, third and fifth power map, telling us, for example, that the squares of the elements of class **5a** are contained in class **5b**. The matrix of character values follows. We see that it contains irrational values denoted by "A" and "$*A$," where "A" is defined in the footnote as (translating into our notation) $A = -\zeta_5 - \zeta_5^4 = \frac{1-\sqrt{5}}{2} = -b5$ (the latter, $-b5$, is the ATLAS notation for this irrationality) and $*A$ is obtained from A by applying the non-trivial Galois automorphism of $\mathbb{Q}(A)$, thus $*A = \frac{1+\sqrt{5}}{2}$ in our case. In Example 2.1.24 we will show how to compute the character table of A_5.   ♦

Example 2.1.18 The character table of a cyclic group $G = \langle g \rangle$ of order, say, n can easily be written down in a "generic form." With $\zeta_n = e^{\frac{2\pi i}{n}}$ as before, we get n pair-wise different linear characters χ_i defined by $\chi_i(g^j) = \zeta_n^{ij}$.   ♦

Example 2.1.19 Let $G = V_4$ be the Klein 4-group. Obviously there are four linear characters and we obtain the following character table:

V_4	1a	2a	2b	2c
λ_1	1	1	1	1
λ_2	1	−1	1	1
λ_3	1	1	−1	1
λ_4	1	1	1	−1

Now suppose that G is either the dihedral group D_8 or the quaternion group Q_8 of order eight. We know that there is a homomorphism $\varphi \colon G \to V_4$ with kernel of order two. By Lemma 2.1.4(d) the inflations $\lambda_i \circ \varphi$ are in $\mathrm{Irr}(G)$ for $i = 1, \ldots, 4$. Example 1.5.13 shows that there is just one irreducible character missing. This can be easily computed using the orthogonality relations for the columns (Theorem 2.1.15, (2.5)). So we see that the matrix of irreducible character values is the same for D_8 and for Q_8, although there are differences in the element orders. We present both character tables in one matrix with two different column headings:

D_8	1a	2a	2b	2c	4a
Q_8	1a	2a	4a	4b	4c
χ_1	1	1	1	1	1
χ_2	1	1	−1	1	1
χ_3	1	1	1	−1	1
χ_4	1	1	1	1	−1
χ_5	2	−2	0	0	0

♦

Example 2.1.20 It is well known that two elements in a symmetric group are conjugate if and only if they have the same "cycle type," that is the lengths of the cycles (in a decomposition into disjoint cycles) coincide. Also, if char $K \neq 2$ any symmetric group S_n for $n \geq 2$ has exactly two "linear characters," i.e. characters of degree one, namely the inflation of the characters of S_n / A_n, or, to put it differently, the trivial character and the sign character. Furthermore, since S_n is doubly transitive in its natural permutation representation on $\Omega = \{1, \ldots, n\}$, it follows from Corollary 2.1.14 that $\pi - 1_G = \chi$ is an absolutely irreducible character of degree $n-1$ if π is the permutation character of $K\Omega$, provided that char $K \nmid n!$. This is sufficient to write down the character table of S_3:

$\mathbf{G} = \mathbf{S_3}$	$\{1\}$	$(1,2)^G$	$(1,2,3)^G$
χ_1	1	1	1
χ_2	1	-1	1
χ_3	2	0	-1

◆

The following simple observation is often useful for the computation of character tables.

Lemma 2.1.21 *If $\chi \in \mathrm{Irr}_K(G)$ and $\lambda \in \mathrm{Irr}_K(G)$ is a linear character then $\lambda \cdot \chi \colon G \to K$, $g \mapsto \lambda(g)\chi(g)$, is in $\mathrm{Irr}_K(G)$.*

Proof. If χ is afforded by the representation $\delta \colon G \to \mathrm{GL}(V)$ then it is obvious that $\lambda \cdot \delta \colon G \to \mathrm{GL}(V)$, $g \mapsto \lambda(g)\delta(g)$, is an irreducible representation with character $\lambda \cdot \chi$. □

Example 2.1.22 The symmetric group $G = S_4$ of degree four has five conjugacy classes, which we write as follows: $\mathbf{1a} = \{1\}, \mathbf{2a} = ((1,2)(3,4))^G, \mathbf{3a} = (1,2,3)^G, \mathbf{2b} = (1,2)^G$, $\mathbf{4a} = (1,2,3,4)^G$. As in Example 2.1.20 we obtain two linear characters and $\chi_4 = \pi - 1_G$ as an irreducible character, where π is the natural permutation character:

$\mathbf{G} = \mathbf{S_4}$	1a	2a	3a	2b	4a
χ_1	1	1	1	1	1
χ_2	1	1	1	-1	-1
χ_4	3	-1	0	1	-1

We get from Lemma 2.1.21 a further irreducible character $\chi_5 = \chi_2 \cdot \chi_4$. At this point just one irreducible character is missing. This can be computed easily from the orthogonality relations. Alternatively one may use the fact that S_3 is a factor group of S_4, and one can therefore inflate the two-dimensional irreducible character of S_3 to S_4. The result is as follows, where we follow the custom of ordering the irreducible characters according to their degrees, if there is no good reason for a different ordering:

$\mathbf{G} = \mathbf{S_4}$	1a	2a	3a	2b	4a
χ_1	1	1	1	1	1
χ_2	1	1	1	-1	-1
χ_3	2	2	-1	0	0
χ_4	3	-1	0	1	-1
χ_5	3	-1	0	-1	1

◆

Example 2.1.23 Let $G = \mathrm{S}_5$ be the symmetric group of degree five. The conjugacy classes of G are $\mathbf{1a} = \{1\}$, $\mathbf{2a} = ((1,2)(3,4))^G$, $\mathbf{3a} = (1,2,3)^G$, $\mathbf{5a} = (1,2,3,4,5)^G$, $\mathbf{2b} = (1,2)^G$, $\mathbf{4a} = (1,2,3,4)^G$ and $\mathbf{6a} = ((1,2)(3,4,5))^G$. As in Example 2.1.22 we immediately obtain four irreducible characters from the sign character and the natural permutation representation on $\Omega = \{1,2,3,4,5\}$. We also easily compute the character θ of the action of G on $\binom{\Omega}{2}$:

	1a	2a	3a	5a	2b	4a	6a
χ_1	1	1	1	1	1	1	1
χ_2	1	1	1	1	-1	-1	-1
χ_3	4	0	1	-1	2	0	-1
χ_4	4	0	1	-1	-2	0	1
θ	10	2	1	0	4	0	1

We verify that $(\theta, \theta)_G = 3$ and $(\theta, \chi_1)_G = (\theta, \chi_3)_G = 1$. Actually, this could also be derived from Example 1.2.25. It follows from Corollary 2.1.16 that $\chi_5 = \theta - \chi_1 - \chi_3 \in \mathrm{Irr}(G)$. Also by Lemma 2.1.21 the product $\chi_2 \cdot \chi_5$ is irreducible. Now only one irreducible character is missing, and this character can easily be computed using the orthogonality relations. The result is as follows:

$|\mathbf{C}_G(g)|$:	120	8	6	5	12	4	6
$G = \mathrm{S}_5$	1a	2a	3a	5a	2b	4a	6a
χ_1	1	1	1	1	1	1	1
χ_2	1	1	1	1	-1	-1	-1
χ_3	4	0	1	-1	2	0	-1
χ_4	4	0	1	-1	-2	0	1
χ_5	5	1	-1	0	1	-1	1
χ_6	5	1	-1	0	-1	1	-1
χ_7	6	-2	0	1	0	0	0

♦

Example 2.1.24 We proceed with the alternating group $G = \mathrm{A}_5$. The conjugacy class $\mathbf{5a}$ of S_5 splits into two classes in A_5 because, for an element $h \in \mathbf{5a}$ in S_5, the centralizer $\mathbf{C}_{\mathrm{S}_5}(h)$ is contained in A_5. Thus G has five conjugacy classes: $\mathbf{1a} = \{1\}, \mathbf{2a} = ((1,2)(3,4))^G$, $\mathbf{3a} = (1,2,3)^G$, $\mathbf{5a} = (1,2,3,4,5)^G$ and $\mathbf{5b} = (1,3,5,2,4)^G$. Obviously, the restriction of a character of a group to a subgroup is a character of the subgroup. If we restrict the irreducible characters of S_5 just calculated (which we are denoting by $\chi_i(\mathrm{S}_5)$) to A_5 we get

$|\mathbf{C}_G(g)|$:	60	4	3	5	5
$G = \mathrm{A}_5$	1a	2a	3a	5a	5b
χ_1	1	1	1	1	1
$\chi_4 = \chi_4(\mathrm{S}_5)|_{\mathrm{A}_5}$	4	0	1	-1	-1
$\chi_5 = \chi_5(\mathrm{S}_5)|_{\mathrm{A}_5}$	5	1	-1	0	0
$\psi = \chi_7(\mathrm{S}_5)|_{\mathrm{A}_5}$	6	-2	0	1	1

We calculate $(\chi_i, \chi_i)_G = 1$ for $i = 1, 4, 5$ and $(\psi, \psi)_G = 2$; also $(\psi, \chi_i)_G = 0$ for $i = 1, 4, 5$. Thus ψ is the sum of two irreducible characters, say $\psi = \chi_2 + \chi_3$. It is clear from the orthogonality relations for the columns that χ_i for $i = 2, 3$ takes the values 3, -1, 0 on the classes **1a**, **2a**, **3a**, respectively. Let $g \in$ **5a**, then $g^{-1} \in$ **5a** and $\chi_2(g)$ and $\chi_3(g)$ are solutions of the equation $5 = 2 + x^2 + (1-x)^2$. Thus we get

$\lvert \mathbf{C}_G(g) \rvert$:	60	4	3	5	5	
$\mathbf{G} = \mathrm{A}_5$	1a	2a	3a	5a	5b	
χ_1	1	1	1	1	1	
χ_2	3	-1	0	α	β	$\alpha = \frac{1}{2}(1 - \sqrt{5})$
χ_3	3	-1	0	β	α	$\beta = \frac{1}{2}(1 + \sqrt{5})$
χ_4	4	0	1	-1	-1	
χ_5	5	1	-1	0	0	

$\blacklozenge$

Exercises

Exercise 2.1.1 Let $L \supseteq K$ be a finite field extension. Considering L as a K-algebra show that $\mathrm{Irr}_K(L) = \{\, \mathrm{Tr}_{L/K} \,\}$, where $\mathrm{Tr}_{L/K}$ is the usual field trace; see, for example, [72], p. 115. Prove that $\mathrm{Irr}_K(L)$ is linearly independent if and only if $L \supseteq K$ is separable.

Exercise 2.1.2 Let A be a finite abelian group and K be an algebraically closed field with $\mathrm{char}\, K \nmid \lvert A \rvert$. Show that $\mathrm{Irr}_K(A) = \mathrm{Hom}(A, K^\times)$ becomes a group $(\hat{A}, \cdot)$ with

$$\lambda \cdot \lambda' \colon A \to K^\times, \ a \mapsto \lambda(a)\lambda'(a), \qquad \text{for} \qquad \lambda, \lambda' \in \hat{A},$$

and that $A \cong \hat{A}$.
Hint: Let $A = \langle a_1 \rangle \times \cdots \times \langle a_r \rangle$ and $\lvert \langle a_i \rangle \rvert = n_i$, and let $\zeta_{n_i} \in K$ be a primitive n_ith root of unity $(i = 1, \ldots, r)$. For $a = a_1^{m_1} \cdots a_r^{m_r} \in A$ define

$$\lambda_a \colon A \to K^\times, \ a_1^{j_1} \cdots a_r^{j_r} \mapsto \zeta_{n_1}^{j_1 m_1} \cdots \zeta_{n_r}^{j_r m_r}.$$

Verify that $\lambda_a \in \hat{A}$ and that $a \to \lambda_a$ defines an isomorphism $A \to \hat{A}$.

Exercise 2.1.3 Let $\boldsymbol{\delta} \colon G \to K^{m \times m}$, $\boldsymbol{\delta}' \colon G \to K^{n \times n}$ be matrix representations.
(a) For any $\mathbf{a} \in K^{m \times n}$ put $\mathbf{a}_G := \sum_{g \in G} \boldsymbol{\delta}(g^{-1}) \mathbf{a} \, \boldsymbol{\delta}'(g)$. Show that $\mathbf{a}_G \cdot \boldsymbol{\delta}'(g) = \boldsymbol{\delta}(g) \cdot \mathbf{a}_G$.
(b) If $\mathrm{char}\, K \nmid \lvert G \rvert$ and $\boldsymbol{\delta}$ is absolutely irreducible show that

$$\sum_{g \in G} \boldsymbol{\delta}(g^{-1})_{ij} \boldsymbol{\delta}(g)_{kl} = \delta_{j,k} \delta_{i,l} \frac{\lvert G \rvert}{n}.$$

(c) Give a new proof of Theorem 2.1.15.

Exercise 2.1.4 Let K be a field with $\mathrm{char}\, K \nmid \lvert G \rvert$ and let V be a KG-module. Show that $(\chi_V, \mathbf{1}_G)_G = \dim_K \mathrm{Inv}_G(V) \cdot \mathbf{1}_K$.

Exercise 2.1.5 Let K be a finite field and let V be a KG-module. Consider V as a G-set. Using (2.3) show that the number of orbits of G on V equals the number of orbits of G on $V^\star$. Show that the Klein 4-group G has a three-dimensional $\mathbb{F}_2 G$-module V, such that the lengths of the orbits of G on V and $V^\star$ are different.

Exercise 2.1.6 Let $T = [\chi_i(g_j)]$ be the character table of a finite group G with conjugacy classes $C_1, \ldots, C_r$. Use the orthogonality relations to show that

$$\det(T) = c \left(\frac{|G|^r}{|C_1| \cdots |C_r|} \right)^{1/2} \quad \text{with } c \in \langle \zeta_4 \rangle.$$

Exercise 2.1.7 Let $T = [\chi_i(g_j)]$ be the character table of a finite group G. Show that the sum of the entries in each row is a non-negative integer.
Hint: Use the permutation action of G on itself by conjugation.

Exercise 2.1.8 Let $n > 1$ be a natural number and let ζ be a primitive nth root of unity in some field K of characteristic not dividing n. Use the orthogonality relations to show that that for $j \in \mathbb{Z}$

$$\sum_{i=0}^{n-1} \zeta^{ij} = \begin{cases} n \cdot 1_K & \text{if } j \equiv 0 \mod n, \\ 0 & \text{if } j \not\equiv 0 \mod n. \end{cases}$$

Exercise 2.1.9 Let $H \leq G$ and $\chi \in \mathrm{Char}_\mathbb{C}(G)$ with $\chi(h) = 0$ for all $h \in H \backslash \{1\}$. Show that $|H| \mid \chi(1)$.

Exercise 2.1.10 Denote the number of conjugacy classes of a group G by $\mathrm{k}(G)$ and let $N \trianglelefteq G$. Write $G = \dot{\bigcup}_{i=1}^{n} g_i N$ and $\bar{G} = G/N$. Prove the following.

(a) $|\mathbf{C}_G(g)| \leq |\mathbf{C}_{\bar{G}}(gN)| \cdot |\mathbf{C}_N(g)|$ with equality if $(N\,\mathbf{C}_G(g))/N = \mathbf{C}_{\bar{G}}(gN)$.

(b) $\sum_{g \in G} |\mathbf{C}_G(g)| \leq \sum_{i=1}^{n} |\mathbf{C}_{\bar{G}}(g_i N)| \sum_{h \in N} |\mathbf{C}_N(g_i h)|$.

(c) $\sum_{h \in N} |\mathbf{C}_N(g_i h)| = \sum_{u \in N} |\mathbf{C}_{g_i N}(u)| \leq \sum_{u \in N} |\mathbf{C}_N(u)|$.

(d) $\mathrm{k}(G) \leq \mathrm{k}(G/N) \cdot \mathrm{k}(N)$. **Hint:** Use (2.3) on p. 93.

Exercise 2.1.11 (See ref. [80].) Let $G := \mathrm{J}_1$ be the sporadic simple Janko group and Ω the G-set considered in Exercise 1.2.7. Use the irreducible representations of $\mathrm{End}_{\mathbb{C}G} \mathbb{C}\Omega$ computed in Exercise 1.2.7 to verify that $\mathbb{C}\Omega$ is a direct sum of five simple submodules with dimensions 1, 56, 56, 76, 77. Compute the Frame quotient (see Corollary 2.1.10) and show that it is not a square.

2.2 Character values

In this section we always assume that G is a finite group and K is a field.

Notation: For $m \in \mathbb{N}$ coprime to char K we define K_m to be a splitting field of $X^m - 1 \in K[X]$ over K.

Lemma 2.2.1 *Let* χ *be a character of* G *of degree* n *over* K *and let* $g \in G$ *be an element of order* m. *Then*

$$\chi(g) = \xi_1 + \cdots + \xi_n$$

with m *roots of unity* $\xi_i \in K_m$. *In particular, if* char $K = 0$ *then the character value* $\chi(g)$ *is always an algebraic integer. Also, for any* $j \in \mathbb{Z}$ *we have*

$$\chi(g^j) = \xi_1^j + \cdots + \xi_n^j.$$

Proof. Let $\delta \colon G \to \mathrm{GL}_n(K)$ be a representation affording χ. Then the minimal polynomial of $\delta(g)$ is a divisor of $X^m - 1 \in K[X]$, which splits into linear factors in $K_m[X]$. Hence $\delta(g)$ is trigonalizable over K_m. This means that there is a non-singular matrix $T \in \mathrm{GL}_n(K_m)$ such that

$$T^{-1}\delta(g)T = \begin{bmatrix} \xi_1 & & 0 \\ & \ddots & \\ * & & \xi_n \end{bmatrix},$$

where the ξ_i's are the eigenvalues of $\delta(g)$, hence mth roots of unity. Since $\chi(g) = \xi_1 + \cdots + \xi_n$ the first assertion follows. Also for any $j \in \mathbb{Z}$ we have

$$T^{-1}\delta(g^j)T = (T^{-1}\delta(g)T)^j = \begin{bmatrix} \xi_1^j & & 0 \\ & \ddots & \\ * & & \xi_n^j \end{bmatrix},$$

so $\chi(g^j) = \xi_1^j + \cdots + \xi_n^j$. $\qquad\square$

The following is essentially a corollary to Lemma 2.2.1.

Lemma 2.2.2 *Let* char $K = 0$ *and* $\chi \in \mathrm{char}_K(G)$. *Then* $\chi(g)$ *is an algebraic integer for any* $g \in G$. *If* $m \in \mathbb{N}$ *and* $\zeta_m \in K$ *is a primitive* mth *root of unity then*

$$\mathrm{Gal}(\mathbb{Q}(\zeta_m)/\mathbb{Q}) = \{\eta_i \mid 1 \le i \le m,\ \mathrm{g.c.d.}(i, m) = 1\},$$

where η_i *is defined by* $\eta_i(\zeta_m) = \zeta_m^i$ *and* g.c.d.(i, m) *stands for the greatest common divisor of* i *and* m *as usual. Also, since* $\mathbb{Q}(\zeta_m)$ *is normal over* $\mathbb{Q}$, *any* η_i *can be extended to an automorphism of* $K_m = K(\zeta_m)$. *If* $g \in G$ *has order* m *then*

(a) $\chi(g^j) = \eta_j(\chi(g))$ *for any* $j \in \mathbb{N}$ *with* g.c.d.$(j, m) = 1$;

(b) *for any prime* $p \in \mathbb{N}$ *we have* $\chi(g^p) \equiv \chi(g)^p$ mod p *in the ring of integers of* K;

(c) *if* $\chi(g) \in \mathbb{Z}$ *and* p *is a prime, then* $\chi(g^p) \equiv \chi(g)$ mod p *in* $\mathbb{Z}$.

Proof. The first assertions are well known from algebra (see e.g. [110], theorem VI.3.1, p. 278). Assertion (a) follows immediately from Lemma 2.2.1 since $\eta_j(\xi_1 + \cdots + \xi_n) = \xi_1^j + \cdots + \xi_n^j$ in the notation of this lemma. The second assertion follows on putting $j = p$ in Lemma 2.2.1 and observing that

$$\chi(g)^p = (\xi_1 + \cdots + \xi_n)^p \equiv \xi_1^p + \cdots + \xi_n^p \qquad \text{mod } p$$

by the binomial theorem. We use the fact that the binomial coefficients $\binom{p}{i}$ are divisible by p for $i = 1, \ldots, p - 1$. Finally we get (c) since by Fermat's little theorem $x^p \equiv x$ mod p holds for all $x \in \mathbb{Z}$. $\qquad \square$

Remark 2.2.3 The congruence relations of Lemma 2.2.2 and their refinements in Exercise 2.2.5 are extremely useful for computing $\chi \in \mathrm{Irr}(G)$, provided that the power maps and centralizer orders of G are known. Often it is possible to compute χ with minimal effort just from $\chi(1)$ (or some other known value $\chi(g_0)$) by simply forming the class functions $\eta \in \mathrm{cf}(G)$, with $\eta(1) = \chi(1)$ and $\eta(g)$ for $g \in G \setminus \{1\}$ being of minimal absolute value compatible with the congruence relations. If η is of norm one and η is unique, then $\chi = \eta$. If there are several solutions of norm one, often all but one can be excluded by considering the scalar product with the trivial or other known character.

Example 2.2.4 As a simple, albeit typical, example we consider again $G = S_5$:

$\mathbf{C}_G(g)\| :$	120	8	6	5	12	4	6
$G = S_5$	1a	2a	3a	5a	2b	4a	6a
$\chi_{5/6}$	5	1	-1	0	± 1	± 1	± 1
χ_7	6	± 2	0	1	0	0	0

From Exercise 2.2.5 we infer that $\chi(h) \equiv \chi(1)$ mod 4 for any $\chi \in \mathrm{Char}_{\mathbb{C}}(G)$ and $h \in$ **2a**, because g is a square (of an element in **4a**). In the table we have given the values $\eta(g)$ of minimal absolute value compatible with $\eta(1) = 5$ and $\eta(1) = 6$, respectively (with the notation of Remark 2.2.3). All these η have norm one. Thus if $\chi \in \mathrm{Irr}(G)$ has degree six then χ is uniquely determined and $\chi(h) = -2$, since otherwise $(\chi, 1_G)_G \neq 0$. For $\chi(1) = 5$ there seem to be more possibilities, but observe that the values on the elements of **2b** and **6a** must be congruent modulo three and hence equal. So orthogonality with 1_G shows that there are, in fact, only two solutions, which are the characters χ_5, χ_6 listed on p. 98. $\qquad \blacklozenge$

Remark 2.2.5 If char $K = 0$ and m is the least common multiple of the orders of the elements of G – usually called the **exponent** of G – then

$$\chi(g) \in K \cap \mathbb{Q}(\zeta_m) \qquad \text{for all} \qquad \chi \in \mathrm{Irr}_K(G), \; g \in G,$$

where ζ_m is a primitive mth root of unity, which we will identify with $\zeta_m := e^{\frac{2\pi i}{m}}$. This means that we will always use a fixed embedding of $K \cap \mathbb{Q}(\zeta_m)$ in $\mathbb{C}$. Then all character values are complex numbers. In particular we may look at absolute values and complex conjugates of character values.

Lemma 2.2.6 *If χ is the character of a representation δ of G over a field K of characteristic zero then we have the following.*

(a) $|\chi(g)| \leq \chi(1)$ *and, if equality holds, $\delta(g) \in \mathbf{Z}(\delta(G))$, the center of $\delta(G)$. If δ is absolutely irreducible the converse also holds.*

(b) $\chi(g) = \chi(1)$ *if and only if $\delta(g) = \mathrm{id}$. Thus $\ker \delta = \{g \in G \mid \chi(g) = \chi(1)\}$, and this is also denoted by $\ker \chi$.*

(c) $\chi(g^{-1}) = \overline{\chi(g)}$, *the complex conjugate of $\chi(g)$ for any $g \in G$.*

Proof. (a) If the eigenvalues of $\delta(g)$ are $\xi_1, \ldots, \xi_n$ then

$$|\chi(g)| = |\xi_1 + \cdots + \xi_n| \leq |\xi_1| + \cdots + |\xi_n| = n$$

by the triangle inequality. Equality holds if and only if $\xi_1 = \xi_2 = \cdots = \xi_n$, and this is equivalent to $\delta(g) = \xi_1 \cdot \mathrm{id}$. Then certainly $\delta(g) \in \mathbf{Z}(\delta(G))$. If δ is absolutely irreducible then by Schur's lemma any $\delta(g) \in \mathbf{Z}(\delta(G))$ must be a scalar multiple of id, so the converse holds also.

(b) follows from the proof of (a).

(c) If the eigenvalues of $\delta(g)$ are $\xi_1, \ldots, \xi_n$ then those of $\delta(g^{-1})$ are $\overline{\xi_1}, \ldots, \overline{\xi_n}$ because $\xi^{-1} = \overline{\xi}$ for ξ, a root of unity. $\qquad\square$

Definition 2.2.7 A character χ of a group G is called **faithful** if it is the character of a faithful representation.

If $\mathrm{char}\, K = 0$ then $\chi \in \mathrm{char}_K(G)$ is faithful if and only if $\ker \chi = \{1\}$.

Corollary 2.2.8 *Let V be a KG-module with character χ_V and let $V^\star$ be the contragredient module (see Definition 1.1.35). If $\mathrm{char}\, K = 0$ then*

$$\chi_{V^\star}(g) = \overline{\chi_V(g)} \qquad \text{for } g \in G.$$

So the complex conjugate of a character is again a character. In fact, one can apply other field automorphisms as well.

Remark 2.2.9 If m is the exponent of G as in Remark 2.2.5 then

$$H = \mathrm{Gal}(\mathbb{Q}(\zeta_m)/\mathbb{Q}) = \{\eta_j \mid 1 \leq j < m , \ (m, j) = 1\}$$

(with η_j being the field automorphism mapping ζ_m to ζ_m^j as before) acts on the conjugacy classes of G by putting

$$\eta_j(g^G) = (g^j)^G \qquad \text{for} \qquad g \in G.$$

If char $K = 0$ then H acts also on the characters of G over K, and in particular on $\mathrm{Irr}_K(G)$, in a natural way:

$$^\eta\chi(g) = \eta(\chi(g)) \qquad \text{for} \qquad \eta \in H, \ \chi \in \mathrm{Irr}_K(G), \ g \in G.$$

Of course we have to show that $^\eta\chi$ *is* a character of G over K, which is irreducible if χ is irreducible. To this end assume that χ is a character afforded by a matrix representation

$$\delta \colon G \to \mathrm{GL}_n(K) , \ g \mapsto [d_{ij}(g)] \quad \text{for} \quad g \in G$$

and let $\eta \in H$. We observe that $\tilde{K} = K \cap \mathbb{Q}(\zeta_m)$ is a normal extension of $\mathbb{Q}$, so we can choose an extension $\eta' \in \mathrm{Gal}(K/\mathbb{Q})$ of $\eta|_{\tilde{K}}$. We define $\eta'\delta \colon G \to \mathrm{GL}_n(K)$ by $\eta'\delta(g) = [\eta'(d_{ij}(g))]$. Obviously this is a representation of G over K, which is irreducible if and only if δ is irreducible. The character of $\eta'\delta$ is

$$g \mapsto \sum_{i=1}^{n} \eta'(d_{ii}(g)) = \eta(\sum_{i=1}^{n} d_{ii}(g)) = \eta(\chi(g)) = {}^\eta\chi(g).$$

Corollary 2.2.10 *If $j \in \mathbb{Z}$ is coprime to $|G|$ then $g \in G$ is conjugate in G to g^j if and only if $\eta_j(\chi(g)) = \chi(g)$ for all $\chi \in \mathrm{Irr}(G)$, where η_j is as in Remark 2.2.9. For any irreducible representation $\delta \colon G \to \mathrm{GL}_n(\mathbb{C})$ the eigenvalues of $\delta(g)$ can be computed from the character table (provided it contains the power maps).*

Proof. Since $\chi(g^j) = \eta_j(\chi(g))$ the first claim holds because two elements $g, h \in G$ are conjugate in G if and only if $\chi(g) = \chi(h)$ for all $\chi \in \mathrm{Irr}(G)$. From the power maps of a character table one knows in which conjugacy class g^p is contained for any prime divisor (and thus for any divisor) of G. Thus $\chi(g^j)$ can be found for any $\chi \in \mathrm{Irr}(G)$ and any $j \in \mathbb{Z}$. If χ is the character of the representation δ and $g \in G$ has order m then the eigenvalues of $\delta(g)$ are powers of a primitive mth root ζ of unity. From the orthogonality relations applied to the cyclic group $\langle g \rangle$ it follows that the multiplicity of ζ^i as an eigenvalue of $\delta(g)$ is

$$(\chi|_{\langle g \rangle}, \lambda_i)_{\langle g \rangle} = \frac{1}{|\langle g \rangle|} \sum_{j=0}^{m-1} \chi(g^j)\zeta^{ij},$$

where $\lambda_i \in \mathrm{Irr}(\langle g \rangle)$ is given by $g^j \mapsto \zeta^{ij}$ for $0 \le j \le m - 1$. $\square$

Lemma 2.2.11 (Gallagher) *Let $g \in G$ be such that $\chi(g) \ne 0$ for all $\chi \in \mathrm{Irr}(G)$ and $N := [\langle g \rangle, G]$, the smallest $N \trianglelefteq G$ such that $gN \in \mathbf{Z}(G/N)$. Then*

$$|\mathrm{Irr}(G)| \le |\mathbf{C}_G(g)| - (|G/N| - |\mathrm{Irr}(G/N)|).$$

Proof. By Lemma 2.2.6, $|\chi(g)| = \chi(1)$ if and only if $\ker\chi \ge N$. Since algebraically conjugate characters have the same kernel, we get

$$|\mathbf{C}_G(g)| = |G/N| + \sum_{\chi \in \mathrm{Irr}(G) \ \ker\chi \not\ge N} |\chi(g)|^2 \ge |G/N| + (|\mathrm{Irr}(G)| - |\mathrm{Irr}(G/N)|)$$

using the orthogonality relations (2.5) and Exercise 2.2.2. $\square$

Definition 2.2.12 Keeping the notation of Remark 2.2.9, characters of G which are in one orbit under H are called **algebraic** or **Galois conjugate**. Similarly conjugacy classes of G in one orbit under H are called **algebraic** or **Galois conjugate**. A conjugacy class C of G is called **rational** if it is fixed by H (which means that any element of C is conjugate in G to all its powers, that have the same order) and it is called **real** if each element of C is conjugate in G to its inverse.

If p is a prime, $m = p^r n$ with $p \nmid n$ and $H_p := \mathrm{Gal}(\mathbb{Q}_m/\mathbb{Q}_n)$, then characters of G which are in one orbit under H_p are called **p-conjugate** and those which are fixed under H_p are called **p-rational**.

Observe that evidently a conjugacy class $g^G = C$ is rational (resp. real) if and only if $\chi(g)$ is rational (resp. real) for all $\chi \in \mathrm{Irr}(G)$.

We note that according to Remark 2.2.9 $H = \mathrm{Gal}(\mathbb{Q}(\zeta_m)/\mathbb{Q})$ acts on the columns and also on the rows of the character table of G, and the actions are compatible in the sense that

$$^\eta\chi_i(C_j) = \chi_i(^\eta C_j) \qquad \text{for } \eta \in H, \ \chi_i \in \mathrm{Irr}(G), \ C_j \text{ a conjugacy class of } G.$$

In addition, we recall from Exercise 2.1.6 that the matrix $[\chi_i(C_j)]_{i,j=1}^r$ is invertible, so that the hypotheses of the following general and very useful lemma are satisfied.

Lemma 2.2.13 (Brauer's permutation lemma) *Let G be a finite group and let $A = [a_{ij}] \in K^{n \times n}$ be an invertible matrix. Suppose that G acts as a permutation group on the set I of row indices of A via $\star$ and also on the set J of column indices of A via $\diamond$ such that*

$$a_{g \star i, g \diamond j} = a_{i,j} \quad \text{for all } i \in I, \ j \in J, \ g \in G.$$

Then the number of fixed points of any $g \in G$ on I equals the number of fixed points of any g on J. Also the number of orbits of G on I equals the number of orbits of G on J.

Proof. Let $P(g)$ and $Q(g)$ be the permutation matrices corresponding to $g \in G$ in the action on I and J, respectively. Our assumption yields

$$P(g) \cdot A = [a_{g^{-1} \star i, j}] = [a_{i, g \diamond j}] = A \cdot Q(g).$$

Hence $P(g) = A \cdot Q(g) \cdot A^{-1}$ and $\mathrm{trace}\, P(g) = \mathrm{trace}\, Q(g)$, i.e. both permutation actions have the same permutation character. The second assertion follows from this using Corollary 2.1.14(c). $\qquad\qquad \square$

Corollary 2.2.14 *The number of algebraic conjugate classes of irreducible characters of G over $\mathbb{C}$ is the same as the number of algebraic conjugate classes of conjugacy classes of elements. Also the number of irreducible real-valued characters is equal to the number of real conjugacy classes.*

Example 2.2.15 Let G be a simple group of order $|G| = 168$. From the Sylow theorems it follows that the number of Sylow 7-subgroups must be eight and the order of the normalizer of a Sylow 7-subgroup P_7 must hence be 21. A well-known theorem of Burnside (see [88] Hauptsatz IV.2.6, p. 419) says that a finite group with a Sylow subgroup which is the center of its normalizer has a proper normal subgroup. Since we assume that G is simple we conclude that

$$P_7 = \mathbf{C}_G(P_7) < \mathbf{N}_G(P_7) \quad \text{and} \quad [\mathbf{N}_G(P_7) : \mathbf{C}_G(P_7)] = 3.$$

$\mathbf{N}_G(P_7)$ acts by conjugation on $P_7 \backslash \{1\}$ with two orbits $\{g, g^2, g^4\}$ and $\{g^3, g^6, g^5\}$ if $P_7 = \langle g \rangle$. Thus G has two non-real conjugacy classes of elements of order seven. Hence there must be a pair of non-real (complex) characters $\psi, \psi' = \overline{\psi}$. Since g is conjugate to g^2 and g^4, $\psi(g)$ must be invariant under the automorphism $\eta_2 \in \mathrm{Gal}(\mathbb{Q}(\zeta_7)/\mathbb{Q})$ with $\eta_2(\zeta_7) = \zeta_7^2$. Hence $\psi(g) = r + s\alpha$ with integers r, s and $\alpha = \zeta_7 + \zeta_7^2 + \zeta_7^4$, where $s \neq 0$ and $r \geq 0$. Observe that $\overline{\alpha} = \zeta_7^6 + \zeta_7^5 + \zeta_7^3 = 1 - \alpha$ and $\alpha\overline{\alpha} = 2$ (and hence $\alpha, \overline{\alpha} = \frac{1}{2}(-1 \pm \sqrt{-7})$). So

$$|\psi(g)|^2 = \psi(g)\psi(g^{-1}) = r^2 + rs(\alpha + \overline{\alpha}) + s^2\alpha\overline{\alpha} = r^2 - rs + 2s^2,$$

and, of course, we get the same value for $\psi'(g)\psi'(g^{-1})$. Using $|\mathbf{C}_G(g)| = 7$ we conclude from the orthogonality relations for the columns of the character table that $1 + 2(r^2 - rs + 2s^2) \leq 7$. This implies $s = \pm 1$ and $r \in \{0, 1\}$ and thus

$$\{\psi(g), \psi'(g)\} = \{\alpha, \overline{\alpha}\} \quad \text{and} \quad \psi(1) = 3 \quad \text{or}$$
$$\{\psi(g), \psi'(g)\} = \{1 + \alpha, 1 + \overline{\alpha}\} \quad \text{and} \quad \psi(1) = 4.$$

In both cases $1 + |\psi(g)|^2 + |\psi'(g)|^2 = 5 = |\mathbf{C}_G(g)| - 2$ so $\chi(g) \in \mathbb{Z}$ for all other $\chi \in \mathrm{Irr}(G) \setminus \{\psi, \psi'\}$ and there must be two irreducible characters, say χ_4, χ_6, with $\chi_i(g) = \pm 1$ for $i \in \{4, 5\}$, whereas $\chi(g) = 0$ for all $\chi \in \mathrm{Irr}(G) \setminus X$ for $X = \{1_G, \psi, \psi', \chi_4, \chi_6\}$. The congruence relations (Lemma 2.2.2) imply that

$$\chi(g) = 1 \iff \chi(1) \in \{1, 8\},$$
$$\chi(g) = -1 \iff \chi(1) = 6,$$
$$\chi(g) = 0 \iff \chi(1) = 7,$$

because the degree of an irreducible character is always bounded by $\sqrt{|G|}$, which is < 13 in our case. The orthogonality relations for the columns $\{1\}$ and g^G require

$$1 + \psi(1)(\psi(g) + \psi'(g)) + \chi_4(1)\chi_4(g) + \chi_6(1)\chi_6(g) = 0.$$

The unique solution is given by

$$\psi(1) = 3, \quad \{\chi_4(1), \chi_6(1)\} = \{6, 8\}.$$

Since $|G| - \sum_{\chi \in X} \chi(1)^2 = 7^2$ it is clear that there is exactly one irreducible character, say $\chi_5 \in \mathrm{Irr}(G)$ with $\chi_5(1) = 7$. We know that G must have elements of orders two and three. So we have found the following fragment of the character table, where we may assume that **2a** is a class of involutions (= elements of order two) which are in the center of a Sylow 2-subgroup:

	1a	2a	3a	??	7a	7b
χ_1	1	1	1	1	1	1
χ_2	3	e			α	$\overline{\alpha}$
χ_3	3	e			$\overline{\alpha}$	α
χ_4	6	x			-1	-1
χ_5	7	y			0	0
χ_6	8	z			1	1

with $e, x, y, z \in \mathbb{Z}$, since character values on involutions are always rational integers.

The elements of 3a must be self-centralizing; otherwise there would be elements of order six and thus 2a would be the only class containing 2-elements $\neq 1$ and the centralizer of an element of 2a would have order $2^3 \cdot 3$. This would imply that G has a normal (elementary abelian) Sylow 2-subgroup, which is absurd. Since χ_4, χ_5, χ_6 are rational, the congruence relations in conjunction with the orthogonality relations imply

$$\chi_5(h) = 1, \quad \chi_6(h) = -1 \quad \text{so} \quad \chi_2(h) = \chi_3(h) = \chi_4(h) = 0 \quad (h \in \mathbf{3a}).$$

The orthogonality relations yield $e = -1$, $x = 2$, $y = -1$, $z = 0$, and after that the missing column is determined by these relations as well. In general, the element orders are not determined by the character values, but in our case we can conclude from the centralizer order, which is four, that the order of the elements of the last class is either two or four. But two is impossible, since then a Sylow 2-subgroup would be elementary abelian, which contradicts the centralizer order. So the last class should be denoted by 4a. We display the completed character table as follows:

$|\mathbf{C}_G(g)|:$	168	8	3	4	7	7
	1a	2a	3a	4a	7a	7b
χ_1	1	1	1	1	1	1
χ_2	3	-1	0	1	α	$\overline{\alpha}$
χ_3	3	-1	0	1	$\overline{\alpha}$	α
χ_4	6	2	0	0	-1	-1
χ_5	7	-1	1	-1	0	0
χ_6	8	0	-1	0	1	1

Of course, we have not shown that a group with this character table actually exists, but it is well known that the group $GL_3(2) \cong L_2(7)$ is, in fact, a simple group of order 168. So the table given above is the character table of this group. ♦

Exercises

Exercise 2.2.1 Let $\chi \in \mathrm{Irr}(G)$ and $Z = \{g \in G \mid |\chi(g)| = \chi(1)\}$. Show that $Z/\ker\chi = \mathbf{Z}(G/\ker\chi)$.

Exercise 2.2.2 Assume that $I \subseteq \mathrm{Irr}(G)$ is invariant under Galois conjugation and $g \in G$ is such that $\chi(g) \neq 0$ for all $\chi \in I$. Show that $\sum_{\chi \in I} |\chi(g)|^2 \geq |I|$.

Hint: One may use the fact that the arithmetic mean of positive real numbers is not less than the geometric mean.

Exercise 2.2.3 Let G be a finite cyclic group and let Ω_1 and Ω_2 be two finite G-sets. Show that if the $\mathbb{C}G$-permutation modules $\mathbb{C}\Omega_1$ and $\mathbb{C}\Omega_2$ are isomorphic, then Ω_1 and Ω_2 are isomorphic G-sets. Use this statement to give an alternative proof of Brauer's permutation lemma, Lemma 2.2.13.

Exercise 2.2.4 Let $\chi = \chi_\delta \in \mathrm{Irr}(G)$ with $\chi(1) = 2$.

(a) Show that $|G|$ must be even.

(b) Let $g \in G$ have order two. Show that $\delta(g) \in \mathbf{Z}(\delta(G))$ or $g \notin G'$.

(c) Conclude that a simple group cannot have an irreducible representation of degree two over $\mathbb{C}$.

Hint: Let $\chi(g) = \xi_1(g) + \xi_2(g)$, where $\xi_i(g)$ are the eigenvalues of $\delta(g)$. Write $(\chi, \chi)_G$ in terms of $\xi_i(g)$ and use the fact that if $|G| = 2m + 1$ then G can be written as $G = \{1, g_1, \ldots, g_m, g_1^{-1}, \ldots, g_m^{-1}\}$. We will show in Theorem 2.3.4 that $\chi(1)$ divides $|G|$ for all $\chi \in \mathrm{Irr}(G)$, but this should, of course, not be used in this exercise.

Exercise 2.2.5 (See ref. [60].) Let $\chi \in \mathrm{Char}_\mathbb{C} G$ and $\chi(g) \in \mathbb{Q}$. Show that

$$\chi(g^{p^r}) \equiv \chi(g^{p^{r-1}}) \bmod p^r \qquad \text{for all primes } p.$$

Exercise 2.2.6 Assume that $(\chi(1) \mid \chi \in \mathrm{Irr}(G)) = (1, 1, 1, 1, 2, 8)$. Compute the values $\chi(g)$ for all $\chi \in \mathrm{Irr}(G)$ and $g \in G$. Deduce that $G \cong C_3^2 \rtimes Q_8$ or $G \cong C_3^2 \rtimes D_8$. (We will exclude the second possibility in Exercise 2.9.1.)
Note: In this example all non-linear irreducible characters have degrees divisible by a prime $p \, (= 2)$ and G has a normal p-complement. This is not a coincidence; see Exercise 4.3.6.

Exercise 2.2.7 Let $G := \mathrm{GL}_2(3)$.

(a) Using rational canonical forms find representatives for $\mathrm{cl}(G)$.

(b) Consider the action of G on the set of one-dimensional subspaces of $\mathbb{F}_3^2$ and show that G maps homomorphically onto S_4. Inflate the irreducible characters of S_4 to G.

(c) Use the orthogonality relations and Lemma 2.2.2 to complete the character table of G. Verify that it can be written as follows:

$|\mathbf{C}_G(g)|$	48	48	8	6	6	4	8	8	
g^G	1a	2a	4a	3a	6a	2b	8a	8b	
χ_1	1	1	1	1	1	1	1	1	
χ_2	1	1	1	1	1	-1	-1	-1	
χ_3	2	2	2	-1	-1	.	.	.	
χ_4	3	3	-1	.	.	-1	1	1	
χ_5	3	3	-1	.	.	1	-1	-1	
χ_6	2	-2	.	-1	1	.	α	$-\alpha$	$\alpha := \sqrt{-2}$
χ_7	2	-2	.	-1	1	.	$-\alpha$	α	
χ_8	4	-4	.	1	-1	.	.	.	

2.3   Character degrees

We assume in this section that G is a finite group and that K is a field of characteristic not dividing $|G|$.

Let $C_1, \ldots, C_r$ be the conjugacy classes of G. We recall from Chapter 1 (Lemma 1.1.9) that the "class sums" $C_i^+ = \sum_{y \in C_i} y$ $(i = 1, \ldots, r)$ form a K-basis of the center $\mathbf{Z}(KG)$ of the group algebra KG. In particular there must be elements $\beta_{ijk} \in K$ with

$$C_i^+ \cdot C_j^+ = \sum_{k=1}^{r} \beta_{ij}^k C_k^+ \qquad \text{for} \qquad 1 \le i, j \le r.$$

The β_{ij}^k are called the **structure constants** for $\mathbf{Z}(KG)$ with respect to the class sums basis. In fact, they are integers modulo the characteristic of K and can be computed from the "class multiplication coefficients" within the group G, which are introduced in the following.

Remark 2.3.1 Choose $g_k \in C_k$ and let

$$\alpha_{ij}^k := |\{(g, h) \mid g \in C_i \, , \, h \in C_j \, , \, gh = g_k\}| \in \mathbb{N}_0.$$

Then α_{ij}^k is independent of the choice of $g_k \in C_k$. The numbers α_{ij}^k are usually called the **class multiplication coefficients** of G. With the above notation $\beta_{ij}^k = \alpha_{ij}^k \cdot 1_K$.

This follows from a simple computation by simplifying the product $\sum_{g \in C_i} g \cdot \sum_{h \in C_j} h$.

Theorem 2.3.2 Let $\delta \colon G \to \mathrm{GL}(V)$ be an absolutely irreducible representation over K with character χ. Then we have for $z \in \mathbf{Z}(KG)$

$$\delta(z) = \omega(z) \, \mathrm{id}_V \qquad \text{with} \quad \omega(z) \in K,$$

and

$$\omega \colon \mathbf{Z}(KG) \to K, \quad z \mapsto \omega(z)$$

is an algebra homomorphism called the **central character** $\omega = \omega_\chi$ corresponding to δ or χ. Furthermore

(a) $\chi(1) \ne 0$ in K and

$$\omega(C^+) = \frac{|C| \chi(g)}{\chi(1)} \qquad \text{for any conjugacy class} \qquad C = g^G \quad \text{in } G;$$

(b) with the α_{ij}^k as defined in Remark 2.3.1 and $M_i^K = [\alpha_{ij}^k \cdot 1_K] \in K^{r \times r}$ with row index j and column index k we obtain

$$\omega(C_i^+) \begin{bmatrix} \omega(C_1^+) \\ \vdots \\ \omega(C_r^+) \end{bmatrix} = M_i^K \begin{bmatrix} \omega(C_1^+) \\ \vdots \\ \omega(C_r^+) \end{bmatrix} \qquad \text{for } i = 1, \ldots r.$$

Proof. Since $z \in \mathbf{Z}(KG)$ it follows that $\delta(z) \in \mathrm{End}_{KG} V = K \cdot \mathrm{id}_V$, because V is absolutely irreducible. So $\delta(z) = \omega(z) \mathrm{id}_V$ and

$$\omega(z_1 z_2) \mathrm{id}_V = \delta(z_1 z_2) = \delta(z_1)\delta(z_2) = \omega(z_1)\omega(z_2) \mathrm{id}_V,$$

and hence ω is an algebra homomorphism.

(a) Computing the trace of $\delta(C^+)$ in two different ways we get

$$\chi(C^+) = \chi(1)\omega(C^+) = \sum_{y \in C} \chi(y) = |C|\chi(g)$$

for any conjugacy class C. If $\chi(1)$ were zero in K then since $\mathrm{char}\, K \nmid |C|$ we would deduce $\chi(g) = 0$ for all $g \in G$, which contradicts Corollary 2.1.14. So we can divide by $\chi(1)$ and get (a).

(b) Since ω is a K-algebra homomorphism we get, from the definition of the class multiplication coefficients α_{ij}^k,

$$\omega(C_i^+) \cdot \omega(C_j^+) = \sum_{k=1}^{r} \alpha_{ij}^k \, \omega(C_k^+). \tag{2.6}$$

This is the same as asserted in matrix form. $\qquad\square$

Corollary 2.3.3 *If* $\mathrm{char}\, K = 0$ *and* δ *is an absolutely irreducible representation of* G *over* K *with central character* ω *then* $\omega(C^+)$ *is an algebraic integer for every conjugacy class* C *of* G. *Moreover* $\omega(C^+) \in \mathbb{Z}[\zeta_m]$, *where* m *is the order of an element in* C *and* ζ_m *is an* mth *root of unity in* K.

Proof. Equation (2.6) says that $\omega(C_i^+)$ is an eigenvalue of the integral matrix $M_i = [\alpha_{ij}^k]$ and thus a root of the monic characteristic polynomial of M_i, and hence it is an algebraic integer. By part (a) of Theorem 2.3.2 and Lemma 2.2.1 $\omega(C^+) \in \mathbb{Q}(\zeta_m)$. By Theorem 2.6 in [168] the ring of algebraic integers in $\mathbb{Q}(\zeta_m)$ is $\mathbb{Z}[\zeta_m]$. $\qquad\square$

Theorem 2.3.4 *If* K *is a splitting field for* G *of characteristic zero then the degrees of the irreducible representations of* G *over* K *are divisors of* $|G|$.

Proof. Let χ be in $\mathrm{Irr}_K(G)$ with corresponding central character ω_χ. We use the orthogonality relations, see Theorem 2.1.15, and obtain

$$|G| = \sum_{i=1}^{r} |C_i|\chi(g_i)\chi(g_i^{-1}) = \sum_{i=1}^{r} \omega_\chi(C_i^+)\chi(1)\chi(g_i^{-1}),$$

hence

$$\frac{|G|}{\chi(1)} = \sum_{i=1}^{r} \omega_\chi(C_i^+)\chi(g_i^{-1}).$$

All terms on the right hand side are algebraic integers by Lemma 2.2.2 and Corollary 2.3.3. Since the algebraic integers form a ring, the left hand side is an algebraic integer and (being in $\mathbb{Q}$) is even a rational integer. $\qquad\square$

Lemma 2.3.5 *Assume that K is a splitting field for G with* char $K = 0$ *and* $\chi \in \mathrm{Irr}_K(G)$. *If $g \in G$ is conjugate in G to zg for some $z \in \mathbf{Z}(G)$ then*

$$\chi(g) = 0 \qquad or \qquad \chi(z) = \chi(1).$$

Proof. If χ is the character of the representation δ with central character ω, then $\omega(z) = \frac{\chi(z)}{\chi(1)}$ and

$$\delta(zg) = \omega(z)\delta(g) \qquad \text{for} \quad g \in G.$$

Hence by assumption

$$\chi(g) = \chi(zg) = \omega(z)\chi(g),$$

from which the result follows. □

Theorem 2.3.6 *If K is a splitting field for G with* char $K = 0$ *and $\chi \in \mathrm{Irr}_K(G)$ then $\chi(1) \mid [G : \mathbf{Z}(G)]$.*

Proof. We use induction with respect to the order of G. First, assume that χ is faithful. Note that the center $\mathbf{Z}(G)$ acts on the set of conjugacy classes by left multiplication. If the conjugacy class $C = g^G$ is in an orbit of length $< |\mathbf{Z}(G)|$, then there is $1 \neq z \in \mathbf{Z}(G)$ such that g is conjugate to zg in G, so $\chi(g) = 0$ by Lemma 2.3.5. If $C_1, \ldots, C_s$ are representatives of the orbits of length $|\mathbf{Z}(G)|$ with $g_i \in C_i$ then

$$|G| = \sum_{g \in G} \chi(g)\overline{\chi(g)} = \sum_{i=1}^{s} \sum_{z \in \mathbf{Z}(G)} |g_i^G| \chi(g_i z)\overline{\chi(g_i z)}$$

$$= |\mathbf{Z}(G)| \sum_{i=1}^{s} |g_i^G| |\chi(g_i)|^2 = |\mathbf{Z}(G)| \chi(1) \sum_{i=1}^{s} \omega(C_i^+)\overline{\chi(g_i)}.$$

So $\frac{[G:\mathbf{Z}(G)]}{\chi(1)}$ is an algebraic integer.

In the case that χ is not faithful, then χ can be considered as the inflation of a faithful irreducible character $\chi' \in \mathrm{Irr}_K(G/N)$ with $N = \ker \chi$, and the result follows by induction from

$$\chi(1) = \chi'(1) \mid [G/N : \mathbf{Z}(G/N)] \mid [G : N\,\mathbf{Z}(G)] \mid [G : \mathbf{Z}(G)].$$

□

In Section 3.6 we will prove a much stronger result; see Theorem 3.6.5.

Theorem 2.3.7 (Burnside) *Let $\delta \colon G \to \mathrm{GL}(V)$ be an irreducible representation of G over $\mathbb{C}$ with character χ. If C is a conjugacy class of G with* g.c.d.$(\chi(1), |C|) = 1$ *then for $g \in C$ we have $\delta(g) \in \mathbf{Z}(\delta(G))$ or $\chi(g) = 0$.*

Proof. By assumption there are $a, b \in \mathbb{Z}$ with $a\chi(1) + b|C| = 1$. Multiplying this equation by $\frac{\chi(g)}{\chi(1)}$ and using Theorem 2.3.2(a) we get

$$a\chi(g) + b\omega_\chi(C^+) = \frac{\chi(g)}{\chi(1)},$$

where ω_χ is the central character corresponding to χ. The left hand side of this equation is an algebraic integer by Lemma 2.2.2 and Corollary 2.3.3. So $\frac{\chi(g)}{\chi(1)}$ is an algebraic integer. By Lemma 2.2.6 part (a) we have $\delta(g) \notin \mathbf{Z}(\delta(G))$ if and only if $|\frac{\chi(g)}{\chi(1)}| < 1$. If C is rational, we can conclude immediately that $\chi(g) = 0$. In general, let $m = |\langle g \rangle|$, so that $\chi(g) \in \mathbb{Q}(\zeta_m)$. We note that

$$c = \prod_{\gamma \in \mathrm{Gal}(\mathbb{Q}(\zeta_m)/\mathbb{Q})} \gamma\left(\frac{\chi(g)}{\chi(1)}\right)$$

is an algebraic integer and invariant under $\mathrm{Gal}(\mathbb{Q}(\zeta_m)/\mathbb{Q})$, hence an integer. By Lemma 2.2.6 part (a) the absolute value of each factor is at most one (observe that $\gamma(\chi(g)) = \chi(g^j)$ for some $j \in \mathbb{N}$). So, if $\delta(g) \notin \mathbf{Z}(\delta(G))$ the integer c must be zero and hence also $\chi(g) = 0$. □

Theorem 2.3.8 *If G is a non-abelian finite simple group then $\{1\}$ is the only conjugacy class of G that has prime power length.*

Proof. Suppose C is a conjugacy class of G with $|C| = p^a$ for some prime p and suppose $1 \neq g \in C$. Any non-trivial irreducible representation δ of G is faithful, since G is simple and satisfies $\mathbf{Z}(\delta(G)) = 1$ since G is non-abelian. By Theorem 2.3.7 we have for any $1_G \neq \chi \in \mathrm{Irr}(G)$ either $p \mid \chi(1)$ or $\chi(g) = 0$. From the orthogonality relations we obtain

$$0 = \sum_{\chi \in \mathrm{Irr}(G)} \chi(g)\chi(1) = 1 + \sum_{\chi \in \mathrm{Irr}(G), p \nmid \chi(1)} \chi(g)\chi(1).$$

We conclude that

$$-\frac{1}{p} = \sum_{\chi \in \mathrm{Irr}(G), p \mid \chi(1)} \chi(g)\frac{\chi(1)}{p}$$

is an algebraic integer, which is absurd. □

As a corollary we get a famous purely group theoretical theorem as follows.

Theorem 2.3.9 (Burnside's $p^a q^b$-theorem) *If p, q are primes and G is a group of order $p^a q^b$ then G is solvable.*

Proof. Since p-groups are solvable we may assume that $p \neq q$ and $a, b > 0$. The hypothesis is inherited by normal subgroups and factor groups, so, using induction, it is enough to show that G is not simple. Let $P \in \mathrm{Syl}_p(G)$ and

$1 \neq g \in \mathbf{Z}(P)$ (this can be done since a non-trivial p-group has a non-trivial center). Then $P \leq \mathbf{C}_G(g)$, hence $|g^G| = [G : \mathbf{C}_G(g)] \mid q^b$. By Theorem 2.3.8 G is not simple. □

Since the assertion of this theorem has nothing to do with representations of groups, it seems natural to look for a proof which avoids representation theory. Such a proof has indeed been found by Goldschmidt and Matsuyama ([68], [118]; see also [107]). But it should be noted that these proofs are far more involved.

Exercises

Exercise 2.3.1 Show that the sum of the entries in any column of the character table of a finite groups is an integer. Find an example where this integer is negative.
Remark: The smallest examples are two groups of order 96, but there are many examples to be found in the ATLAS ([38]).

Exercise 2.3.2 Using the assumption and notation of Theorem 2.3.2 show that

$$\sum_{i=1}^r \frac{1}{|C_i|} \omega(C_i^+) \omega(C_{i'}^+) = \frac{|G|}{\chi(1)^2}.$$

Exercise 2.3.3 Let G be a group with $\mathbf{Z}(G) = \{1\}$ and let $\chi \in \mathrm{Irr}(G)$ be faithful with $\chi(1) = p$, a prime. If $P \in \mathrm{Syl}_p(G)$ show that $P = \mathbf{C}_G(P)$ has order p. Conclude that a group with trivial center cannot have a faithful irreducible character of degree two (compare with Exercise 2.2.4).
Hint: Choose $g \in \mathbf{Z}(P)$ of order p and let δ be a representation affording χ.

(a) Use Theorem 2.3.7 to show that $\chi(g) = 0$.

(b) Consider $\chi(g) = 0$ as a polynomial equation in ζ_p and conclude that the eigenvalues of $\delta(g)$ are $1, \zeta_p, \ldots, \zeta_p^{p-1}$.

(c) Conclude that $\mathbf{C}_G(g)$ and so P is abelian.

(d) Show that $\chi(x) = 0$ for all $x \in \mathbf{C}_G(P) \setminus \{1\}$ and use Exercise 2.1.9.

2.4   The Dixon–Schneider algorithm

In this section we assume that G is a finite group with conjugacy classes $C_1 = g_1^G, \ldots, C_r = g_r^G$, where $g_1 = 1$. Let K be a splitting field for G with char $K \nmid |G|$ and $\mathrm{Irr}_K(G) = \{\chi_1, \ldots, \chi_r\}$. We put $\omega_i := \omega_{\chi_i}$ (see Theorem 2.3.2).
Theorem 2.3.2 shows that the central characters ω_i (more precisely, the vectors $[\omega_i(C_1^+), \ldots, \omega_i(C_r^+)]^T$) are common (column-)eigenvectors of the matrices

$$M_i^K := [\alpha_{ij}^k \cdot 1_K]_{j,k=1}^r \in K^{r \times r},$$

where the $\alpha_{ij}^k \in \mathbb{N} \cup \{0\}$ are the class multiplication coefficients (see Remark 2.3.1) defined by

$$\alpha_{ij}^k = |\{(g, h) \mid g \in C_i , \ h \in C_j , \ gh = g_k\}| \in \mathbb{N} \cup \{0\},$$

with $g_k \in C_k$. If one is able to compute within the group sufficiently well in order to find the conjugacy classes and to decide in which class a product of elements lies, one may compute these matrices M_i^K, and it is tempting to determine the ω_i's by solving the corresponding eigenvalue problems and observing that $\omega_i(1) = 1$. It is straightforward to compute the degrees of the irreducible characters χ_i from these (see Exercise 2.3.2) so the irreducible characters can be obtained from the ω_i's using the known class lengths $|C_i|$. In fact, the procedure to compute the character table of a group, which Burnside suggested in [24], sect. 223, and illustrated for the dihedral group D_{10}, is equivalent to this approach (see also [40], p. 238).

An alternative way to find the irreducible characters directly as row-eigenvectors of the M_i^K was proposed by Schneider ([156]). We first need a simple lemma.

Lemma 2.4.1 *For a conjugacy class C_i let $C_{i'} := \{g^{-1} \mid g \in C_i\}$. Then*

$$\alpha_{ij}^k |C_k| = \alpha_{i'k}^j |C_j|.$$

Proof. Let

$$X = \{(x, y) \mid x \in C_i , \ y \in C_j , \ xy \in C_k\} \quad \text{and}$$
$$Y = \{(x', z) \mid x' \in C_{i'} , \ z \in C_k , \ x'z \in C_j\}.$$

Then

$$X \to Y \quad (x, y) \mapsto (x^{-1}, xy) \qquad \text{and} \qquad Y \to X \quad (x', z) \mapsto (x'^{-1}, x'z)$$

are inverse bijections. Obviously the left hand side in the assertion is just $|X|$, whereas the right hand side is $|Y|$. $\square$

Let $\omega = \omega_\chi$ for $\varphi \in \mathrm{Irr}_K(G)$. Applying the above equation to $\omega(C_{i'}^+)\omega(C_j^+) = \sum_k \alpha_{i'j}^k \omega(C_k^+)$ we obtain

$$\frac{|C_{i'}|\chi(g_i^{-1})}{\chi(1)} \frac{|C_j|\chi(g_j)}{\chi(1)} = \sum_{k=1}^r \alpha_{i'j}^k \frac{|C_k|\chi(g_k)}{\chi(1)} = \sum_{k=1}^r \alpha_{ik}^j \frac{|C_j|\chi(g_k)}{\chi(1)}.$$

Hence $\frac{|C_{i'}|\chi(g_i^{-1})}{\chi(1)}\chi(g_j) = \sum_k \chi(g_k)\alpha_{ik}^j$, an equation, that can also be found in [24], sect. 229. Using $|C_{i'}| = |C_i|$ we get

$$\frac{|C_i|\chi(g_i^{-1})}{\chi(1)}[\chi(g_1), \ldots, \chi(g_r)] = [\chi(g_1), \ldots, \chi(g_r)]M_i^K. \qquad (2.7)$$

Notation: We will identify a class function χ of G with the row-vector $[\chi(g_1), \ldots, \chi(g_r)]$, and if $\chi(1) \neq 0$ we define

$$\chi^{(n)} := [1, \frac{\chi(g_2)}{\chi(g_1)}, \ldots, \frac{\chi(g_r)}{\chi(g_1)}].$$

In the following we will call row-vectors with first entry 1 "normalized."

Theorem 2.4.2 *The $\chi_j \in \mathrm{Irr}_K(G)$ are row-eigenvectors of the matrices M_i^K. In fact, $\{\chi_1^{(n)}, \ldots, \chi_r^{(n)}\}$ is the set of normalized common row-eigenvectors of all the M_i^K for $i = 1, \ldots, r$.*

Proof. Because of (2.7) only the second statement still needs to be proved. Let $v := \sum_{j=1}^r a_j \chi_j$ be a common eigenvector of all M_i^K. Then there exist $\lambda_i \in K$ with $v M_i^K = \lambda_i v$. Hence

$$\lambda_i \sum_j a_j \chi_j = \sum_j a_j \chi_j M_i^K = \sum_j a_j \frac{|C_i| \chi_j(g_i^{-1})}{\chi_j(1)} \chi_j.$$

Comparing coefficients we get

$$\lambda_i a_j = a_j \frac{|C_i| \chi_j(g_i^{-1})}{\chi_j(1)} \qquad \text{for all } i.$$

Hence for any j with $a_j \neq 0$ we obtain

$$\frac{\lambda_i}{|C_i|} = \frac{\chi_j(g_i^{-1})}{\chi_j(1)} \qquad \text{for all } i.$$

So, if $a_j \neq 0$ and $a_k \neq 0$ then $\chi_j(g_i) = \frac{\chi_j(1)}{\chi_k(1)} \chi_k(g_i)$ for all i. Thus χ_j, χ_k are linearly dependent and hence $j = k$. So v is a multiple of some χ_j, and if v is normalized it must be equal to $\chi_j^{(n)}$. $\qquad\square$

Corollary 2.4.3 *From the class multiplication matrices M_i the irreducible characters $\chi_j \in \mathrm{Irr}_K(G)$ can be computed up to signs. If $\mathrm{char}\, K = 0$ or $\mathrm{char}\, K > 2 \cdot \sqrt{|G|}$ then the characters themselves can be determined.*

Proof. By Theorem 2.4.2 the $\chi_j^{(n)}$ are uniquely determined by the matrices M_i. The scalar product $(\chi_j^{(n)}, \chi_j^{(n)})_G = \frac{1}{\chi_j(1)^2} \cdot 1_K$ can be computed to yield $\pm\chi_j(1)$ and thus we obtain $\pm\chi_j$. If $\mathrm{char}\, K = 0$ then χ_j can be identified since we know that $\chi_j(1) > 0$. If $\mathrm{char}\, K = p > 0$ then $\chi_j(1) = d_j \cdot 1_K$, where d_j is the degree of a representation with character χ_j. Since K is supposed to be a splitting field for G we know that $d_j \leq \sqrt{|G|}$. If $p > 2 \cdot \sqrt{|G|}$ then $p - d_j > \sqrt{|G|}$ and we can identify the correct value of $\chi_j(1)$. $\qquad\square$

Obviously, in the corollary $\sqrt{|G|}$ may be replaced by any other upper bound for the degrees of the irreducible representations of G over K.

Of course, the main interest is in computing the complex irreducible characters, and these are determined uniquely from the M_i by Corollary 2.4.3. On the other hand, solving eigenvalue problems in $\mathbb{C}$ is not easy. Mostly this is done numerically rather than symbolically. But representing character values as floating point numbers with rounding errors is usually of no use. So we intend to solve the eigenvalue problem over a suitable finite field F and transform the answer back to $\mathbb{C}$ using a discrete Fourier transform. This method was first proposed by Dixon ([48]).

We will assume now that F is a splitting field for G of characteristic $p > 2 \cdot \sqrt{|G|}$ not dividing $|G|$. We also assume that F contains a primitive eth root of unity ϵ, where e is the exponent of G, i.e. the l.c.m. of the orders of the elements of G. Let

$$\mathrm{Irr}(G) = \{\chi_1, \ldots, \chi_r\} \qquad \text{and} \qquad \mathrm{Irr}_F(G) = \{\varphi_1, \ldots, \varphi_r\}.$$

We define a ring homomorphism

$$\theta \colon \mathbb{Z}[\zeta_e] \to F \qquad \text{with} \quad \zeta_e \mapsto \epsilon; \tag{2.8}$$

θ extends the canonical epimorphism $\mathbb{Z} \to \mathbb{Z}/(p) = \mathbb{F}_p \subseteq F$. It follows that $\theta(M_i) = M_i^F$, and, since θ is a ring homomorphism, we get from equation (2.7) for all $i, j \in \{1, \ldots, r\}$

$$\theta(|C_i|\chi_j(g_i^{-1})) \cdot \theta(\chi_j) = \theta(\chi_j(1))\theta(\chi_j) \cdot M_i^F,$$

and hence

$$\frac{\theta(|C_i|\chi_j(g_i^{-1}))}{\theta(\chi_j(1))} \cdot \theta(\chi_j) = \theta(\chi_j) \cdot M_i^F.$$

So $\theta(\chi_j)$ is a common eigenvector for all M_i^F, and by Theorem 2.4.2 and its corollary we may choose our notation so that

$$\theta(\chi_j) = \varphi_j \qquad \text{for} \quad 1 \le j \le r.$$

By Theorem 2.4.2 and its corollary we may compute φ_j by solving the eigenvalue problems for the matrices M_i^F over the finite field $\mathbb{F}_p$. We will show now how to obtain χ_j from this. For simplicity of notation we will omit the index j.

Let $g \in G$ have order q. From Lemma 2.2.1 we get for $j \in \mathbb{Z}$

$$\chi(g) = \sum_{k=0}^{q-1} m_k \zeta_q^k \qquad \text{and} \qquad \chi(g^j) = \sum_{k=0}^{q-1} m_k \zeta_q^{kj}. \tag{2.9}$$

Multiplying the latter equation by $\frac{1}{q}\zeta_q^{-ij}$ for $i = 1, \ldots q-1$ and summing over j we get

$$\frac{1}{q}\sum_{j=0}^{q-1} \chi(g^j)\zeta_q^{-ij} = \frac{1}{q}\sum_{j=0}^{q-1}\sum_{k=0}^{q-1} m_k\zeta_q^{kj}\zeta_q^{-ij} = \sum_{k=0}^{q-1} m_k \frac{1}{q}\sum_{j=0}^{q-1} \zeta_q^{(k-i)j} = m_i,$$

where we have used Exercise 2.1.8. We put $\epsilon_q := \epsilon^{e/q}$. Then $\theta(\zeta_q) = \theta(\zeta_e^{e/q}) = \epsilon_q$ and we obtain

$$\frac{1}{\theta(q)}\sum_{j=0}^{q-1} \varphi(g^j)\epsilon_q^{-ij} = \theta(m_i) \qquad (0 \le i \le q-1). \tag{2.10}$$

Since $m_i \leq \chi(1) < p$ is a non-negative integer and $\theta|_{\{0,\ldots,p-1\}}$ is injective we may use (2.10) to compute the m_i and thus χ, once we know φ.

Dixon–Schneider algorithm

In order to compute the complex characters $\chi_1, \ldots, \chi_r$ of the group G with exponent e we do the following.

(a) Compute the integral matrices $M_i = [\alpha_{ij}^k]$.

(b) Choose a prime p with $p > 2 \cdot \max\{\chi(1) \mid \chi \in \mathrm{Irr}(G)\}$ – if one does not have a better estimate of the character degrees one might take $p > 2 \cdot \sqrt{|G|}$ – and, in addition $p \equiv 1 \mod e$. Choose a primitive eth root $\epsilon \in \mathbb{F}_p$.

(c) Find the common normed row-eigenvectors $\varphi^{(n)}$ of the $M_i^{\mathbb{F}_p}$ in $\mathbb{F}_p$. Find a square root $d \in \{\theta(1), \ldots, \theta(\frac{p-1}{2})\}$ of $(\varphi^{(n)}, \varphi^{(n)})_G^{-1}$ and compute $\varphi = d \cdot \varphi^{(n)} \in \mathrm{Irr}_F(G)$. Here θ is as in (2.8).

(d) For each $\varphi \in \mathrm{Irr}_F(G)$ compute the corresponding $\chi \in \mathrm{Irr}(G)$ using (2.10) and (2.9).

Remark 2.4.4 A famous theorem of Dirichlet (see e.g. [128], p. 469) asserts that there are in fact infinitely many primes p with $p \equiv 1 \mod e$ (for any e), and so we certainly can find one which is large enough as requested in (b). Since $p \equiv 1 \mod e$ the field $\mathbb{F}_p$ contains a primitive eth root ϵ of unity.

In step (a) of the algorithm it is usually not necessary to compute all the matrices M_i. In fact, it follows from Theorem 2.4.2 that an eigenspace of any M_i^F is spanned by irreducible F-characters; in particular, if an eigenspace of M_i is one-dimensional, then one irreducible character is already obtained. Otherwise these eigenspaces can possibly be split up by taking their intersections with eigenspaces of a different M_j^F. Methods that avoid computing too many class multiplication matrices can be found in [48]. We illustrate the procedure by means of an example.

Example 2.4.5 We take $G = A_6$ and take as representatives for the conjugacy classes $g_1 = 1$, $g_2 = (1,2)(3,4)$, $g_3 = (1,2,3)$, $g_4 = (1,2,3)(4,5,6)$, $g_5 = (1,2,3,4)(5,6)$, $g_6 = (1,2,3,4,5)$, $g_7 = (1,3,5,2,4)$, and write $C_i = g_i^G$. We compute M_2 and get, for example using the following straightforward GAP-code:

```
gap> G := AlternatingGroup(6);;
gap> g := [ (), (1,2)(3,4), (1,2,3), (1,2,3)(4,5,6), (1,2,3,4)(5,6),
>           (1,2,3,4,5), (1,3,5,2,4) ];;
gap> cl := List( g, x -> ConjugacyClass(G,x) );; M2:=[];;
gap> for j in [1..Length(cl)] do M2[j]:=[];
>      w := List( Cartesian(cl[2],cl[j]), x->x[1]*x[2] );
>      for k in [1..Length(cl)] do M2[j][k]:=Length(Positions(w,g[k]));od;
>    od;
```

$$M_2 := \texttt{M2} = \begin{bmatrix} 0 & 1 & 0 & 0 & 0 & 0 & 0 \\ 45 & 4 & 9 & 9 & 4 & 5 & 5 \\ 0 & 8 & 9 & 0 & 4 & 5 & 5 \\ 0 & 8 & 0 & 9 & 4 & 5 & 5 \\ 0 & 8 & 9 & 9 & 17 & 10 & 10 \\ 0 & 8 & 9 & 9 & 8 & 10 & 10 \\ 0 & 8 & 9 & 9 & 8 & 10 & 10 \end{bmatrix}.$$

The exponent of G is 60, so we choose $p = 61 > \sqrt{|G|}$ and put $F := \mathbb{F}_p$. We compute the eigenvalues and (bases of) the eigenspaces of $M_2^F = \theta(M_2) = \texttt{M2*e}$, where $\texttt{e}$ is the identity of F:

```
gap> p := 61;; F := GF(p);; e := Identity(F);; ev := Eigenvalues(F,M2*e);;
gap> evecs := List( Eigenspaces(F,M2*e), GeneratorsOfVectorSpace );;
```

The elements of F are represented as powers of a primitive element $\texttt{Z(61)}$. To write them as integers $(\bmod\ p)$ we use the function $\texttt{dom} := (\theta|_{\{-\frac{p-1}{2},\dots,\frac{p-1}{2}\}})^{-1}$:

```
gap> dom := function( p, x )
> return( Position(List([-(p-1)/2..(p-1)/2], i->i*e), x) - (p+1)/2 );end;;
```

The following GAP-code:

```
gap> for sp in evecs do
> for c in ev do
> if sp[1] * M2*e = sp[1]*c then Print("\n",dom(p,c),":  "); fi;
> od;
> for v in sp do Print( List(v , x -> dom(p,x)), " ," ); od;
> od;
```

shows that M_2^F has five eigenspaces U_λ^2, where λ denotes the corresponding eigenvalue and the entries have to be interpreted as numbers in F:

$$
\begin{aligned}
U_0^2 &= & & \langle\, [1,0,-23,-23,0,0,23]\,,\ [0,0,0,0,0,1,-1]\,\rangle_F, \\
U_{-16}^2 &= \langle\varphi_1^{(n)}\rangle_F &=& \langle\, [1,1,1,1,1,1,1]\,\rangle_F, \\
U_{-9}^2 &= \langle\varphi_7^{(n)}\rangle_F &=& \langle\, [1,12,-6,-6,0,0,0]\,\rangle_F, \\
U_9^2 &= & & \langle\, [1,-12,0,-12,12,0,0]\,,\ [0,0,1,-1,0,0,0]\,\rangle_F, \\
U_5^2 &= \langle\varphi_6^{(n)}\rangle_F &=& \langle\, [1,-27,0,0,-27,27,27]\rangle_F.
\end{aligned}
$$

By Theorem 2.4.2 the normed generators of U_{-16}^2, U_{-9}^2 and U_5^2 are of the form $\varphi_i^{(n)}$, with $\varphi_i \in \mathrm{Irr}_F(G)$ for some i which we have named 1, 7 and 6, respectively. For each of the eigenvectors of M_2^F we compute its norm $\texttt{a}$ and a square root $\texttt{d}$ of $\texttt{a}^{-1}$ in the range $1,\dots,(p-1)/2 \in F$, if $\texttt{a}$ is a square. Observe that by the proof of Corollary 2.4.3, $\texttt{d}$ is the degree of $\varphi \in \mathrm{Irr}_F(G)$ if the eigenvector is $\varphi^{(n)}$. We first write a function $\texttt{scp}$ which returns the scalar product of two class functions of G (using, for simplicity, the fact that all classes of G are real):

```
gap> scp := function (v,w)
> return(Sum( List([1..Length(cl)], i->Size(cl[i])*v[i]*w[i]) )/Size(G));
```

```
> end;;
gap> for v in Concatenation(evecs) do
>       d := Filtered( [1..(p-1)/2], x -> (x*e)^2 = scp(v,v)^-1 );
>       Print( [dom( p, scp(v,v)^-1 ), d] , "," );
>    od;
```

We obtain the scalar products $5 = 26^2$, -28, $1 = 1^2$, $-22 = 10^2$, $-16 = 17^2$, -26, $20 = 9^2$ in F. In particular, we get (the trivial character φ_1 and)

$$\varphi_6 = 9 \cdot \varphi_6^{(n)} = [9, 1, 0, 0, 1, -1, -1], \quad \varphi_7 = 10 \cdot \varphi_7^{(n)} = [10, -2, 1, 1, 0, 0, 0].$$

U_0^2 and U_9^2 have dimensions larger than one and could be split into one-dimensional spaces generated by characters by computing further matrices M_i and their eigenspaces (or their action on U_0^2 and U_9^2). But this is not really necessary here. Calling the generators of U_0^2 (or U_9^2) listed above v, w (with v having first entry 1) we know (from Theorem 2.4.2) that there must be $\varphi, \varphi' \in \mathrm{Irr}_F(G)$ with

$$\varphi^{(n)} = v + a \cdot w, \quad \varphi'^{(n)} = v + b \cdot w \quad \text{for} \quad a, b \in F.$$

We may test all pairs $a, b \in F$ whether or not they yield orthogonal class functions with "degrees" d1, d2 dividing $|G|$ and such that $d1^2 + d2^2 < |G| - 9^2 - 10^2$:

```
gap> for i in [1,4] do              # we consider the 1. and 4. eigenspace
> for a in [0..p-1] do for b in [a..p-1] do
> v := evecs[i][1] + a* evecs[i][2]; w := evecs[i][1] + b* evecs[i][2];
> if IsSubset( List([1..(p-1)/2], x -> (x*e)^2), [scp(v,v),scp(w,w)] )
> then d1 :=  Filtered( [1..(p-1)/2], x -> (x*e)^2 = scp(v,v)^-1)[1];
>        d2 :=  Filtered( [1..(p-1)/2], x -> (x*e)^2 = scp(w,w)^-1)[1];
>    if IsInt(Size(G)/d1) and IsInt(Size(G)/d2) and scp(v,w) = 0 * Z(p)
>       and d1^2 + d2^2 < Size(G) - 9^2 - 10^2 then
>              Print([a,b],",",List(d1*v,x-> dom(p,x)),",",
>              List(d2*w,x-> dom(p,x)),"\n");
>    fi;
>  fi;
> od;od;
>    od;
```

In both cases we obtain a single solution ($\{a, b\} = \{36, 48\}$ and $\{a, b\} = \{12, 37\}$, respectively) and irreducible characters

$$\varphi = [\, 8, 0, -1, -1, 0, -17, 18\,], \quad \varphi' = [\, 8, 0, -1, -1, 0, 18, -17\,] \in \mathrm{Irr}_F(G)$$

and

$$\psi = [\, 5, 1, -1, 2, -1, 0, 0\,], \quad \psi' = [\, 5, 1, 2, -1, -1, 0, 0\,] \in \mathrm{Irr}_F(G);$$

ψ, ψ' lift trivially to characters in $\mathrm{Irr}(G)$, since the first five classes are rational. Lifting φ, φ' to $\mathbb{C}$ we obtain, say, $\chi, \chi' \in \mathrm{Irr}(G)$ which agree on the rational classes of G. Using the orthogonality relations it is easy to obtain the values on the remaining two classes mapping to $-17, 18 \in F$ under θ. But here we will use the formulas (2.10) and (2.9), which are easily encoded in **GAP**:

```
gap> epsq := Z(p)^((p-1)/5);; phi := [ 8, 0, -1, -1, 0, -17, 18 ]*e;;
gap> for x in g{[6,7]} do
>   m := List( [0..4] , i -> dom( p, (5*e)^-1 * Sum( List( [0..4], j ->
>             phi[Position(cl,ConjugacyClass(G,x^j))] * epsq^(-i*j))) ) );
>   Print("chi(g_",Position(g,x),") = ", m*List([0..4],i->E(5)^i),"    ");
> od;
```

We obtain

$$\chi(g_6) = \alpha := -\zeta_5^2 - \zeta_5^3 = \frac{1 + \sqrt{5}}{2}, \qquad \chi(g_7) = \alpha' := -\zeta_5 - \zeta_5^4 = \frac{1 - \sqrt{5}}{2}.$$

We finish by displaying the character table, which we have computed completely:

	360	8	9	9	4	5	5
A_6	1a	2a	3a	3b	4a	5a	5b
χ_1	1	1	1	1	1	1	1
χ_2	5	1	2	-1	-1	0	0
χ_3	5	1	-1	2	-1	0	0
χ_4	8	0	-1	-1	0	α	α'
χ_5	8	0	-1	-1	0	α'	α
χ_6	9	1	0	0	1	-1	-1
χ_7	10	-1	1	1	0	0	0

◆

The Dixon–Schneider algorithm has been implemented in GAP by Hulpke (see [87]). It is the standard method in GAP to compute character tables. The main problem is to compute the class multiplication matrices, where a large amount of testing elements for conjugacy may be required. In [87] the character tables of some maximal subgroups of F_{23} were computed as examples having orders up to almost 10^{10} and about 200 conjugacy classes.

Exercises

Exercise 2.4.1 (a) Let $\rho\colon \mathbf{Z}(KG) \to \mathrm{End}_K \mathbf{Z}(KG)$ be the regular representation of the center of the group algebra KG. Show that

$$M_i^K = [\rho(C_i^+)]_B \qquad \text{with} \qquad B = (C_1^+, \ldots, C_r^+).$$

(b) Show that $M_{i_1}^K, \ldots, M_{i_k}^K$ for $\{i_1, \ldots, i_k\} \subseteq \{1, \ldots, r\}$ have exactly r normalized common row-eigenvectors (the vectors in $\mathrm{Irr}_K(G)$) if and only if

$$\mathbf{Z}(KG) = \langle C_{i_1}^+, \ldots, C_{i_k}^+ \rangle_{\mathrm{alg}}.$$

Exercise 2.4.2 Show that for the group $G := \mathrm{SL}_2(3)$ one class multiplication matrix M_i suffices in order to compute the character table of G using the Dixon–Schneider algorithm. Give an example of an abelian group where this not true. Show, however, that there is always a linear combination of the M_i whose eigenvalues have all multiplicity one.

The following exercise is part of an important observation by John McKay; see for example [120].

Exercise 2.4.3 Let M be the Cartan matrix of the affine Weyl group of type E_6 given as follows:

$$M := \begin{pmatrix} 2 & 0 & 0 & 0 & -1 & 0 & 0 \\ 0 & 2 & 0 & 0 & 0 & -1 & 0 \\ 0 & 0 & 2 & 0 & 0 & 0 & -1 \\ 0 & 0 & 0 & 2 & -1 & -1 & -1 \\ -1 & 0 & 0 & -1 & 2 & 0 & 0 \\ 0 & -1 & 0 & -1 & 0 & 2 & 0 \\ 0 & 0 & -1 & -1 & 0 & 0 & 2 \end{pmatrix}.$$

Show that the irreducible characters of $SL_2(3)$ are a complete set of eigenvectors of M by using the GAP character table of $SL_2(3)$. This observation can be generalized, see [120], and in the generalized form is called the McKay correspondence.

2.5 Application – generation of groups

We have seen in Section 2.4 that the class multiplication coefficients α_{ij}^k as defined in Remark 2.3.1 can be used to construct the character table of a group. However, the converse is also true, that is, the α_{ij}^k can be computed from the character table; in fact, this is a very important application of character tables in practice. At first sight this might look like a vicious circle, but one has to keep in mind that the Dixon–Schneider algorithm is applicable only to relatively small groups, whereas character tables of even larger groups can sometimes be computed from only a very limited amount of information on the group. We give some modest examples of how this can be done in Section 2.10 and in Chapter 4, Section 4.12. In fact, for some of the sporadic simple groups the character table had been computed using some (at that time hypothetical) information about the order and local structure of the group before even the existence of the group had been proved, let alone any permutation or matrix representation was known that could lend itself to any direct computation with the group elements. The most spectacular example of this approach is the largest sporadic simple group, the "Monster group" (also called the "friendly giant") for which the character table had been computed in 1981 by Fischer, Livingstone and Thorne before the existence was proved by Griess and long before the first element representation and multiplication of two elements was carried out by Wilson in 1997.

Definition 2.5.1 Let G be a finite group and let $\boldsymbol{C} := (C_1, \dots, C_m)$ with normal subsets C_i of G, that is, unions of conjugacy classes of elements of G. For $H \leq G$ we put

$$\Sigma_{\boldsymbol{C}}^H(G) := \{(g_1, \dots, g_m) \mid g_i \in C_i, \ g_1 \cdots g_m = 1, \ \langle g_1, \dots, g_m \rangle =_G H\}$$

and

$$\Sigma_{\boldsymbol{C}}^{\leq H}(G) := \{(g_1, \ldots, g_m) \mid g_i \in C_i, \; g_1 \cdots g_m = 1, \; \langle g_1, \ldots, g_m \rangle \leq_G H\},$$

where $U =_G H$ (resp. $U \leq_G H$) means that U is conjugate in G to H (resp. to a subgroup of H). Furthermore we abbreviate $\Sigma_{\boldsymbol{C}} := \Sigma_{\boldsymbol{C}}^{\leq G}(G)$ and put

$$\alpha_{\boldsymbol{C}} = \alpha_{\boldsymbol{C}}^G := |\Sigma_{\boldsymbol{C}}| \quad \text{and} \quad \gamma_{\boldsymbol{C}}^G := |\Sigma_{\boldsymbol{C}}^G(G)|.$$

Remark 2.5.2 If $\Sigma_{\boldsymbol{C}}^H(G) \neq \emptyset$ for some $H \leq G$ then G acts by conjugation on this set with stabilizers conjugate to $\mathbf{C}_G(H)$. In particular, if $\Sigma_{\boldsymbol{C}}^G(G) \neq \emptyset$ then

$$[G : \mathbf{Z}(G)] \mid \gamma_{\boldsymbol{C}}^G.$$

An m-tuple $\boldsymbol{C} := (C_1, \ldots C_m)$ of conjugacy classes of G is called **rigid** if $\gamma_{\boldsymbol{C}}^G = [G : \mathbf{Z}(G)]$. This is equivalent to saying that G acts transitively on $\Sigma_{\boldsymbol{C}}^G(G)$.

Also, if $\mathcal{L}(G)$ is a complete set of representatives of the set of conjugacy classes of subgroups of G then

$$\Sigma_{\boldsymbol{C}} = \dot{\bigcup}_{H \in \mathcal{L}(G)} \Sigma_{\boldsymbol{C}}^H(G). \tag{2.11}$$

Let $\boldsymbol{C} := (C_1, \ldots, C_m)$ be as in Definition 2.5.1. We will see shortly that $\alpha_{\boldsymbol{C}}$ can be computed from the character table of G, whereas $\gamma_{\boldsymbol{C}}^G$ is what is of interest in most cases. Obviously one has

$$\Sigma_{\boldsymbol{C}} = \dot{\bigcup}_{H \leq G} \Sigma_{(C_1 \cap H, \ldots, C_m \cap H)}^H (H), \tag{2.12}$$

so α can be considered as the "summatory function" for γ^G. If the subgroup lattice (i.e. the poset of subgroups – it is, in fact, a lattice in the sense of [115], p. 60) of G is known, one can compute γ^G from α using a "Möbius inversion." To do this we have to introduce the Möbius function of the subgroup lattice or better yet of a finite or locally finite poset. Recall that a poset $(P, \leq)$ is called locally finite if for every pair $x, y \in P$ the set $\{z \in P \mid x \leq z \leq y\}$ is finite.

Definition 2.5.3 The Möbius function of a locally finite poset $(P, \leq)$ is the function $\mu_P \colon P \times P \to \mathbb{Z}$ satisfying $\mu_P(x, y) = 0$ unless $x \leq y$, in which case it is defined recursively by $\mu_P(x, x) = 1$ and

$$\sum_{x \leq z \leq y} \mu_P(x, z) = 0 \quad \text{for} \quad x < y.$$

Example 2.5.4 Taking $(P, \leq) = (\mathbb{N}, |)$, the natural numbers ordered by divisibility, we readily find

$$\mu_{\mathbb{N}}(1, n) = \begin{cases} 1 & \text{if} \quad n = 1, \\ (-1)^r & \text{if} \quad n \text{ is the product of } r \text{ distinct primes,} \\ 0 & \text{if} \quad n \text{ is divisible by a square of a prime.} \end{cases}$$

Thus $\mu_{\mathbb{N}}(1, n) = \mu(n)$, where $\mu \colon \mathbb{N} \to \mathbb{Z}$ denotes the (arithmetical) Möbius function of elementary number theory (see [32], p. 55). Also $\mu_{\mathbb{N}}(m, n) = \mu(\frac{n}{m})$ if m divides n. ◆

The Möbius function is the inverse of the incidence function in the incidence algebra. Perhaps the easiest way to look at it is to consider it as a matrix. For simplicity we will restrict our attention to finite posets. So let $P = \{x_1, \ldots, x_n\}$, where we may assume that $x_i \leq x_j$ for $i \leq j$. Then the incidence matrix

$$[z_{ij}]_{1 \leq i,j \leq n} \quad \text{with} \quad z_{ij} = \begin{cases} 1 & \text{if } x_i \leq x_j, \\ 0 & \text{else} \end{cases}$$

is a lower triangular matrix with ones on the diagonal. Hence it is invertible, and it is immediate that its inverse is

$$[\mu_{ij}]_{1 \leq i,j \leq n} \quad \text{with} \quad \mu_{ij} = \mu_P(x_i, x_j).$$

Since this is also a right inverse we obtain

$$\sum_{x \leq z \leq y} \mu_P(z, y) = \delta_{x,y}. \tag{2.13}$$

Lemma 2.5.5 *Let μ be the Möbius function of the subgroup lattice of G and let $C := (C_1, \ldots, C_m)$ with normal subsets C_i of G. Then*

$$\gamma_C^G = \sum_{H \leq G} \mu(H, G) \cdot \alpha_{(C_1 \cap H, \ldots, C_m \cap H)}^H.$$

Proof. From (2.12) we get $\alpha_{(C_1 \cap H, \ldots, C_m \cap H)}^H = \sum_{U \leq H} \gamma_{(C_1 \cap U, \ldots, C_m \cap U)}^G$. Multiplying this equation by $\mu(H, G)$ and summing over all $H \leq G$ we get

$$\sum_{H \leq G} \mu(H, G) \cdot \alpha_{(C_1 \cap H, \ldots, C_m \cap H)}^H = \sum_{H \leq G} \sum_{U \leq H} \mu(H, G) \gamma_{(C_1 \cap U, \ldots, C_m \cap U)}^U$$

$$= \sum_{U \leq G} \Big(\sum_{U \leq H \leq G} \mu(H, G) \Big) \gamma_{(C_1 \cap U, \ldots, C_m \cap U)}^U$$

$$= \gamma_{(C_1, \ldots, C_m)}^G,$$

using (2.13). $\qquad\square$

Example 2.5.6 We specialize all C_i to G, that is, we consider $C := \underbrace{(G, \ldots, G)}_{m}$.

Then $\alpha_C = |G|^{m-1}$ and γ_C^G is the number of $m-1$-tuples of elements of G generating G or, in other words, the number of epimorphisms of the free group F_{m-1} on $m-1$ generators to G. Since $\text{Aut}\, G$ acts on this set of epimorphisms (from the left) without fixed points and two epimorphisms $\varphi_i : F_{m-1} \to G$ are in the same orbit if and only if they have the same kernel, we see that

$$d_{m-1}(G) := \frac{1}{|\text{Aut}\, G|} \cdot \gamma_C^G = \frac{1}{|\text{Aut}\, G|} \sum_{H \leq G} \mu(H, G) \cdot |H|^{m-1}$$

is the number of normal subgroups of F_{m-1} with factor group isomorphic to G. $\qquad\blacklozenge$

Remark 2.5.7 If μ is the Möbius function of the subgroup lattice of the finite group G then $\mu(H, G)$ depends only on the conjugacy class of the subgroup H in G. Also $\mu(H, G) = -1$ for every maximal subgroup $H \leq G$.

For the computation of the relevant values of the Möbius function for larger groups we refer to Section 3.5. Here we illustrate the above notions with the alternating group A_5 as a concrete example.

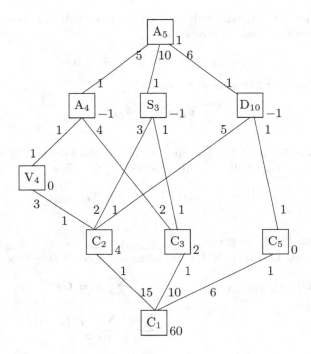

Figure 2.1. Subgroup pattern of A_5.

Example 2.5.8 The alternating group A_5 has exactly one conjugacy class of subgroups isomorphic to A_4, D_{10}, S_3, C_5, V_4, C_3, C_2 and C_1, respectively. We sketch the subgroup lattice in Figure 2.1 by drawing what is sometimes called the "subgroup pattern" (see [23]). Here any box stands for a conjugacy class of subgroups of A_5, and such boxes are connected by a line if a representative of one such class contains a conjugate of a representative of the other class as a maximal subgroup. The numerals next to the lines give the number of conjugates of a representative of the lower (resp. upper) conjugacy class contained in (resp. containing) a representative of the upper (resp. lower) conjugacy class. Thus, for example, any S_3 in A_5 contains three subgroups conjugate to C_2 and any such subgroup is contained in two subgroups conjugate to S_3. We have also displayed the values $\mu(H, G)$ as (lower right) indices at the box representing the conjugacy class of H. This value can be computed from the top using

(2.13) with $x = H$ and $y = G$ and the already computed values $\mu(U, G)$ for $H < U \leq G$. Here we find

$$d_2(\mathrm{A}_5) = \frac{1}{120}(-60 + 4 \cdot 15 \cdot 2^2 + 2 \cdot 10 \cdot 3^2 - 5 \cdot 12^2 - 10 \cdot 6^2 - 6 \cdot 10^2 + 60^2)$$
$$= 19.$$

Thus the free group F_2 has 19 normal subgroups with factor group isomorphic to A_5. Or, in other words, the direct product of 19 copies of A_5 can be generated by two elements, but not the direct product of 20 copies (see [75]). ◆

Theorem 2.5.9 *Assume that $C_1, \ldots, C_m$ are conjugacy classes (not necessarily distinct) of G and let $\alpha_{C_1,\ldots,C_m} := \alpha_{(C_1,\ldots,C_m)}$ (see Definition 2.5.1). Then*

$$\alpha_{C_1,\ldots,C_m} = \frac{|C_1| \cdots |C_m|}{|G|} \sum_{\chi \in \mathrm{Irr}(G)} \frac{\chi(g_1) \cdots \chi(g_m)}{\chi(1)^{m-2}} \qquad \text{with} \qquad g_j \in C_j.$$

Proof. For $C \in \mathrm{cl}(G)$ let $g_C \in C$ be a representative and $C' = \{g^{-1} \mid g \in C\}$. Also, as usual, $C^+ = \sum_{g \in C} g \in \mathbb{C}G$. Obviously

$$\alpha_{C_1,\ldots,C_m} = |C_m| \cdot |\{(x_1, \ldots, x_{m-1}) \mid x_j \in C_j \, , \ x_1 \cdots x_{m-1} = g_m^{-1}\}|.$$

Hence

$$C_1^+ \cdots C_{m-1}^+ = \sum_{C \in \mathrm{cl}(G)} \frac{\alpha_{C_1,\ldots,C_{m-1},C}}{|C|} C'^+.$$

Applying the central character ω_χ with $\chi \in \mathrm{Irr}(G)$ we get, using Theorem 2.3.2,

$$\frac{|C_1|}{\chi(1)}\chi(g_1) \cdots \frac{|C_{m-1}|}{\chi(1)}\chi(g_{m-1}) = \sum_{C \in \mathrm{cl}(G)} \alpha_{C_1,\ldots,C_{m-1},C} \frac{1}{\chi(1)}\chi(g_C^{-1}).$$

Multiplying both sides of this equation by $\frac{|C_m|}{|G|}\chi(1)\chi(g_m)$ and summing over all χ in $\mathrm{Irr}(G)$ we get

$$\frac{|C_1| \cdots |C_m|}{|G|} \sum_{\chi \in \mathrm{Irr}(G)} \frac{\chi(g_1) \cdots \chi(g_m)}{\chi(1)^{m-2}}$$

$$= \sum_{C \in \mathrm{cl}(G)} \alpha_{C_1,\ldots,C_{m-1},C} \frac{|C_m|}{|G|} \sum_{\chi \in \mathrm{Irr}(G)} \chi(g_m)\chi(g_C^{-1}) = \alpha_{C_1,\ldots,C_m},$$

by making use of the orthogonality relations (Theorem 2.1.15). □

The most important case of the theorem, and the only one usually treated in books on representation theory, is the one with $m = 3$. Observe that we have

$$\alpha_{C_i,C_j,C_k} = |C_k|\alpha_{ij}^{k'},$$

the latter being the class multiplication coefficients defined in Remark 2.3.1. We will call the $\alpha_{C_1,\ldots,C_m}$ **symmetric class multiplication coefficients**. The significance of the theorem can be seen in the following remark.

Remark 2.5.10 For $k_1, \ldots, k_m \in \mathbb{N} \setminus \{1\}$ we define the group $\Delta(k_1, \ldots, k_m)$ by generators and relations as follows:

$$\Delta(k_1, \ldots, k_m) := \langle\, x_1, \ldots, x_m \mid x_1^{k_1} = \cdots = x_m^{k_m} = x_1 \cdots x_m = 1 \,\rangle.$$

Suppose that the elements of the conjugacy class C_{i_j} of G have orders k_j for $1 \leq j \leq m$. Then $\alpha_{i_1, \ldots, i_m}$ is the number of group homomorphisms of $\Delta(k_1, \ldots, k_m)$ to G with the property that x_j is mapped onto an element of the class C_{i_j}:

$$\alpha_{i_1, \ldots, i_m} = |\{\varphi \colon \Delta(k_1, \ldots, k_m) \to G \mid \varphi(x_j) \in C_{i_j} \quad \text{for } 1 \leq j \leq m\}|.$$

Summing over all m-tuples of conjugacy classes of G such that the orders of the elements of C_{i_j} divide k_j we get the total number of homomorphisms from $\Delta(k_1, \ldots, k_m)$ to G.

Remark 2.5.11 $\Delta(k_1, \ldots, k_m)$ is finite for $1 < k_1, \ldots, k_m$ and $m \geq 3$ if and only if $\Delta(k_1, \ldots, k_m)$ is one of the following groups:

(a) $\Delta(2, 2, k) \cong D_{2k}$, the dihedral group of order $2k$,

(b) $\Delta(2, 3, 3) \cong A_4$,

(c) $\Delta(2, 3, 4) \cong S_4$,

(d) $\Delta(2, 3, 5) \cong A_5$.

Proof. See sect. 6.4 in [39]. □

Example 2.5.12 We consider the character table of the Mathieu group $G = M_{11}$ of degree 11 which is computed in Section 2.10 and printed on p. 177. We see that G has just one conjugacy class of elements of order two, three and five each, namely classes 2a, 3a, 5a. An easy computation (e.g. with GAP) shows that the symmetric class multiplication coefficient is $\alpha_{2a,3a,5a} = 15 \cdot |C_5| = 23760 = 198 \cdot 120$, and we deduce that M_{11} has exactly 198 subgroups isomorphic to A_5, for the number of non-trivial homomorphisms from A_5 to M_{11} is $\alpha_{2a,3a,5a}$. Since A_5 is simple, a non-trivial homomorphism from A_5 to M_{11} is an embedding; two embeddings with the same image differ just by an automorphism of A_5. So we have to divide $\alpha_{2a,3a,5a}$ by the order of $\mathrm{Aut}(A_5) \cong S_5$, which is 120. From the ATLAS ([38], p. 18) we get the information that M_{11} has a maximal subgroup isomorphic to S_5 with index 66; so we get one conjugacy class of subgroups isomorphic to A_5 consisting of 66 subgroups. Also the ATLAS states the existence of a maximal subgroup in M_{11} of index 11 isomorphic to $A_6 . 2_3$. This in turn contains an A_5 as a maximal subgroup of index 12 which is self-normalizing in M_{11}. So the 198 subgroups isomorphic to A_5 fall in just two conjugacy classes with lengths 66 and $132 = 11 \cdot 12$. ◆

In many applications it is important to find out whether or not a given group G is an epimorphic image of $\Delta(k_1, \ldots, k_m)$. This is the case if and only if there is an m-tuple of conjugacy classes $C := (C_1, \ldots, C_m)$ in G with k_j being a multiple of the order of the elements of C_j and $\Sigma_C^G(G) \neq \emptyset$.

In general it is not possible to compute $\gamma_C^G = |\Sigma_C^G(G)|$ from the character table alone. In addition one needs some information about the subgroup lattice as we have seen above. Nevertheless it is sometimes possible to conclude that $\gamma_C^G > 0$ just using information about maximal subgroups in cases where it would not be feasible to compute the full subgroup lattice.

Remark 2.5.2 yields a necessary condition for a given group G being an epimorphic image of $\Delta(k_1, \ldots, k_m)$. Namely, for some m-tuple $C := (C_1, \ldots, C_m)$ of conjugacy classes of G, with k_j being a multiple of the order of the elements of C_j, one must have $\alpha_C \geq [G : \mathbf{Z}(G)]$. Of course, this condition is not sufficient. But if the character tables of all the maximal subgroups of G (or just of those containing elements of orders $k_1, \ldots, k_m$) are known, one can sometimes use this information to give an upper bound for

$$|\{(g_1, \ldots, g_m) \in \Sigma_C \mid \langle g_1, \ldots, g_m \rangle \neq G\}|.$$

If this happens to be smaller than α_C then one can conclude that there must indeed be an epimorphism. We will give some examples below, but before we do this we shall give another necessary condition involving just the information contained in the character table for a finite group G being an epimorphic image of $\Delta(k_1, \ldots, k_m)$. This goes back to R. Brauer and was later refined by L. Scott and is often called the "Brauer trick." It is one of the rare cases where one can say something about the existence of subgroups using the character table alone. We will see in the next chapter that there are many necessary conditions for the existence of subgroups involving characters, so that it is not uncommon that deductions about the non-existence of subgroups of G can be drawn from the character table of G.

We first need a lemma recalling from Definition 1.1.18 that $\mathrm{Inv}^G(V)$ is the smallest submodule W of a KG-module V such that G acts trivially on V/W.

Lemma 2.5.13 *If $G = \langle g_1, \ldots, g_r \rangle$ (i.e. $g_1, \ldots, g_r$ are generators for the group G) and V is a KG-module for a field K then*

$$\mathrm{Inv}^G(V) = (1 - g_1)V + \cdots + (1 - g_r)V$$
$$= (1 - g_1)V + g_1(1 - g_2)V + \cdots + g_1 \cdots g_{r-1}(1 - g_r)V.$$

Proof. We start by proving the first equation. Since

$$g_i(1 - g_j)v = (1 - g_i)g_j v - (1 - g_i)v + (1 - g_j)v \quad \text{for} \quad 1 \leq i, j \leq r,$$

we know that $W_j := (1 - g_1)V + \cdots + (1 - g_j)V$ is invariant under the generators g_i for $1 \leq i \leq j$ and hence W_r is a KG-submodule of V. (In the case that G is infinite, observe that $(1 - g_i^{-1})v = -(1 - g_i)g_i^{-1}v$, hence $(1 - g_i^{-1})V = (1 - g_i)V$.) Obviously, all generators g_i act trivially on V/W_r and any submodule with trivial factor module must contain all $(1 - g_i)V$. Hence $\mathrm{Inv}^G(V) = W_r$.

We show that $W_j = (1 - g_1)V + g_1(1 - g_2)V + \cdots + g_1 \cdots g_{j-1}(1 - g_j)V$ holds for all j. This follows by induction from

$$(1 - g_j)v = (1 - g_1)v + g_1(1 - g_2)v + g_1 g_2(1 - g_3)v$$
$$+ \cdots + g_1 \cdots g_{j-1}(1 - g_j)v - (1 - g_1 \cdots g_{j-1})g_j v,$$

because

$$(1 - g_1 \cdots g_{j-1})g_j v = (1 - g_{j-1})g_j v + (1 - g_{j-2})g_{j-1}g_j v + \cdots + (1 - g_1)g_2 \cdots g_j v$$

is in W_{j-1} and thus

$$(1 - g_j)v - g_1 \cdots g_{j-1}(1 - g_j)v \in W_{j-1} \qquad \text{for all} \qquad v \in V.$$

$\square$

Theorem 2.5.14 (L. Scott) *Let $G = \langle g_1, \ldots, g_r \rangle$ be a group with $g_1 \cdots g_r = 1$ and let V be a KG-module over an arbitrary field K with $\dim_K V = n < \infty$. Then*

$$\sum_{i=1}^{r}(n - \dim_K \mathrm{Inv}_{\langle g_i \rangle}(V)) \geq 2n - \dim_K \mathrm{Inv}_G(V) - \dim_K \mathrm{Inv}_G(V^\star).$$

Proof. Let C be the K-subspace of $V^r = V \oplus \cdots \oplus V$ defined by

$$C = \{(v_1, \ldots, v_r) \mid v_i \in (1 - g_i)V \qquad (1 \leq i \leq r)\}.$$

We have K-linear maps

$$\beta\colon V \to C \qquad v \qquad \mapsto ((1 - g_1)v, \ldots, (1 - g_r)v),$$
$$\delta\colon C \to V \quad (v_1, \ldots, v_r) \mapsto v_1 + g_1 v_2 + \cdots + g_1 \ldots g_{r-1}v_r.$$

Since

$$0 = 1 - g_1 \cdots g_r = (1 - g_1) + g_1(1 - g_2) + \cdots + g_1 \cdots g_{r-1}(1 - g_r)$$

we have $\mathrm{im}\,\beta \subseteq \ker \delta$. Furthermore

$$\begin{aligned}
\mathrm{im}\,\delta &= (1 - g_1)V + g_1(1 - g_2)V + \cdots + g_1 \cdots g_{r-1}(1 - g_r)V \\
&= (1 - g_1)V + \cdots + (1 - g_r)V \\
&= \mathrm{Inv}^G(V)
\end{aligned}$$

by Lemma 2.5.13, the smallest submodule W of V such that G acts trivially on V/W. Hence $\dim_K(\mathrm{im}\,\delta) = n - \dim_K \mathrm{Inv}_G(V^\star)$ by duality (see Theorem 1.1.34). On the other hand it is obvious that $\ker \beta = \mathrm{Inv}_G(V)$ and

$(1 - g_i)V = \mathrm{Inv}^{\langle g_i \rangle}(V)$. By duality and Exercise 1.1.9 $\dim_K \mathrm{Inv}^{\langle g_i \rangle}(V) = n - \dim_K \mathrm{Inv}_{\langle g_i \rangle}(V)$. We thus obtain

$$\sum_{i=1}^{r}(n - \dim_K \mathrm{Inv}_{\langle g_i \rangle}(V)) = \dim_K C = \dim_K \mathrm{im}\,\delta + \dim_K \ker \delta$$

$$= \dim_K \mathrm{im}\,\delta + \dim_K \mathrm{im}\,\beta + \dim_K(\ker \delta/\mathrm{im}\,\beta)$$

$$\geq \dim_K \mathrm{Inv}^G(V) + (n - \dim_K \ker \beta)$$

$$= (n - \dim_K \mathrm{Inv}_G(V^*)) + (n - \dim_K \mathrm{Inv}_G(V)).$$

$\square$

Corollary 2.5.15 *Let $G = \langle g_1, \ldots, g_r \rangle$ be a finite group with $g_1 \cdots g_r = 1$ and $1_G \neq \chi \in \mathrm{Irr}(G)$. Then*

$$\sum_{i=1}^{r}(\chi_{\langle g_i \rangle}, 1_{\langle g_i \rangle})_{\langle g_i \rangle} \leq (r - 2)\chi(1).$$

Proof. If V is a $\mathbb{C}G$-module affording χ, then $\mathrm{Inv}_G(V) = \{0\}$ and $\mathrm{Inv}_G(V^*) = \{0\}$ because χ is irreducible and not the trivial character. Also

$$\dim \mathrm{Inv}_{\langle g_i \rangle}(V) = (\chi_{\langle g_i \rangle}, 1_{\langle g_i \rangle})_{\langle g_i \rangle}.$$

$\square$

Corollary 2.5.16 *For $g \in G$ let $d(g)$ denote the minimal number of conjugates of g in G which generate G. Then*

$$d(g) \geq \frac{\dim V}{\dim V - \dim \mathrm{Inv}_{\langle g \rangle}(V)}$$

for any KG-module on which G acts faithfully.

Proof. Let $d = d(g)$ and $G = \langle g_1, \ldots, g_d \rangle$, where g_i are all conjugates of g in G and $g_{d+1} = (g_1 \cdots g_d)^{-1}$. Then $\dim \mathrm{Inv}_{\langle g_i \rangle}(V) = \dim \mathrm{Inv}_{\langle g \rangle}(V)$ for $i = 1, \ldots, d$ and applying Theorem 2.5.14 we get

$$d \cdot (\dim V - \dim \mathrm{Inv}_{\langle g \rangle}(V)) + \dim V - \dim \mathrm{Inv}_{\langle g_{d+1} \rangle}(V) \geq 2 \dim V$$

since G acts faithfully. From this the result follows immediately. $\square$

Example 2.5.17 Again let $G = M_{11}$. Searching for candidates of rigid triples we list all triples of conjugacy classes C_i, C_j, C_k with $\alpha_{C_i, C_j, C_k} \leq 1$. Also, for each such triple we list in parentheses the value

$$n_\chi(C_i, C_j, C_k) = \sum_{l \in \{i,j,k\}}(\chi_{\langle g_l \rangle}, 1_{\langle g_l \rangle})_{\langle g_l \rangle},$$

with $\chi = \chi_5$ being the irreducible character of M_{11} of degree 11 (see the character table on p. 177):

$$\alpha_{2a,2a,2a} = \frac{|G|}{4} \ (21), \qquad \alpha_{2a,2a,3a} = \frac{2|G|}{3} \ (19),$$
$$\alpha_{2a,2a,4a} = \frac{|G|}{2} \ (17), \qquad \alpha_{2a,2a,5a} = |G| \ (17),$$
$$\alpha_{2a,2a,6a} = |G| \ (17), \qquad \alpha_{2a,2a,8a/b} = 0,$$
$$\alpha_{2a,2a,11a/b} = 0, \qquad \alpha_{2a,3a,3a} = |G| \ (17),$$
$$\alpha_{2a,3a,4a} = |G| \ (15), \qquad \alpha_{2a,3a,6a} = \frac{|G|}{2} \ (15),$$
$$\alpha_{2a,3a,8a/b} = \frac{|G|}{2} \ (13), \qquad \alpha_{2a,3a,11a/b} = |G| \ (13),$$
$$\alpha_{2a,4a,11a/b} = |G| \ (11), \qquad \alpha_{2a,8a,8b} = |G| \ (9),$$
$$\alpha_{2a,11a/b,11a/b} = |G| \ (9), \qquad \alpha_{3a,3a,11a/b} = |G| \ (11).$$

We can draw the following conclusions immediately: G has no dihedral subgroups of order 16 or 22, because $\alpha_{2a,2a,8a/b} = \alpha_{2a,2a,11a/b} = 0$ and it is also not an epimorphic image of $\Delta(2,3,6)$ or $\Delta(2,3,8)$ since $\alpha_{2a,3a,6a}, \alpha_{2a,3a,8a/b} < |G|$. In the cases where $\alpha_{C_i,C_j,C_k} = |G|$ we have to consider the following two alternatives: either $\Sigma^G_{(C_i,C_j,C_k)}(G) = \emptyset$ or $\Sigma^G_{(C_i,C_j,C_k)}(G) = \Sigma_{(C_i,C_j,C_k)}$ is one orbit under the conjugation action of G, i.e. we have a rigid triple.

From Corollary 2.5.15 we can conclude that

$$\Sigma^G_{(C_i,C_j,C_k)}(G) = \emptyset \quad \text{if} \quad n_\chi(C_i,C_j,C_k) > \chi(1) = 11.$$

Hence G can only be an epimorphic image of $\Delta(2,4,11), \Delta(2,8,8), \Delta(2,11,11)$ or $\Delta(3,3,11)$. The character tables of the maximal subgroups of M_{11} can be found in the **GAP** library of character tables together with the fusion maps into M_{11}, i.e. for each conjugacy class of a maximal subgroup the name of the conjugacy class of M_{11} is listed in which it is contained. We list this information:

$A_6.2_3$	1a	2a	3a	4a	5a	4a	8a	8b	
$L_2(11)$	1a	2a	3a	5a	5a	6a	11a	11b	
$3^2:Q_8.2$	1a	2a	2a	3a	4a	4a	6a	8a	8b
$A_5.2$	1a	2a	3a	5a	2a	4a	6a		
$2.S_4$	1a	2a	4a	2a	8a	8b	3a	6a	

In each line the conjugacy classes of the particular subgroup H are listed, not by giving the name in H but the name of the conjugacy class of G which contains it. For example, $L_2(11)$ contains two conjugacy classes of elements of order five, which both are contained in the conjugacy class 5a of G. Only the second maximal subgroup, $L_2(11)$, contains elements of order 11. We compute the (symmetric) class multiplication coefficients in this maximal subgroup of G and find $\alpha_{2a,11a,11a} = \alpha_{2a,11b,11b} = \alpha_{3a,3a,11a/b} = |L_2(11)|, \alpha_{2a,11a,11b} = 0$. This shows that the triples $(2a, 11a, 11a)$, $(2a, 11b, 11b)$ and $(3a, 3a, 11a/b)$ are not rigid, but that $(2a, 11a, 11b)$ is, in fact, a rigid triple. Furthermore, since no maximal subgroup of M_{11} contains elements of order 11 and elements of order four, it is clear that the triples $(2a, 4a, 11a/b)$ are also rigid.

The triples $(2a, 8a, 8a)$ and $(2a, 8b, 8b)$ can be excluded as rigid triples by looking at the maximal subgroup $A_6.2_3$, in which one can compute the relevant class multiplication coefficients $\alpha_{2a,8a,8a} \neq 0$ and $\alpha_{2a,8b,8b} \neq 0$. Thus in M_{11} we have $\Sigma^G_{(2a,8a,8a)}(G) \neq \Sigma_{(2a,8a,8a)}$ and $\Sigma^G_{(2a,8b,8b)}(G) \neq \Sigma_{(2a,8b,8b)}$.

To sum up: we have found exactly three rigid triples for M_{11}, namely $(2a, 11a, 11b)$ and $(2a, 4a, 11a/b)$. Also computing $n_\chi(2a, 3a, C)$ for all conjugacy classes C of M_{11} one can deduce readily from Corollary 2.5.15 that M_{11} cannot be generated by an element or order two and an element of order three. ◆

Why is it important to find rigid triples for a group? We give three reasons.

(1) Standard generators

It often happens that one has different sets of generators for the same group or for isomorphic groups and one wants to find explicitly an isomorphism in terms of the given generators, in order to compare some results or to make use of computations performed using one generating set for the isomorphic copy. Imagine, for example, that one has found inside some large group, such as the Fischer group Fi_{22}, a subgroup isomorphic to M_{11}. In practice it is often all but impossible to obtain one set of generators from another one even for the same group, unless special care has been taken in the choice of the generators. Wilson [171] has introduced **standard generators** for sporadic simple groups, which can, in fact, be constructed given an arbitrary generating set. They can be found in Wilson's online ATLAS of finite group representations ([170]), which can also be accessed via the GAP package `atlasrep`.

The idea is to characterize generators (in most cases one considers generating systems consisting of just two generators – all finite simple groups can be generated by a pair of elements) by certain equations they fulfill. For this, rigid triples are ideal: if C_i, C_j, C_k is a rigid triple for G and $(x, y, z) \in \Sigma^G(C_i, C_j, C_k)$ then $G = \langle x, y \rangle$, and if one somehow finds elements $x' \in C_i$, $y' \in C_j$ such that $(x'y')^{-1} \in C_k$ then there is an isomorphism φ such that $\varphi(x) = x', \varphi(y) = y'$. Of course one has to be able to decide whether some element is in C_i etc. This is easy if C_i is characterized by the orders of its elements; otherwise some further work has to be done.

Example 2.5.18 We again look at the example $G = M_{11}$. We choose the rigid triples $(2a, 4a, 11a)$, $(2a, 4a, 11b)$. Let $(x, y, z) \in \Sigma^G_{(2a,4a,11a)}(G)$ and $(x', y', z') \in \Sigma^G_{(2a,4a,11b)}(G)$. The classes $2a$, $4a$ are the only classes of elements of order two and four, respectively. So, whenever one finds in some group H isomorphic to M_{11} elements a, b of order two and four, respectively, such that ab has order 11 then there is an isomorphism $\varphi: H \to G$ such that $(\varphi(a), \varphi(b))$ is either (x, y) or (x', y'). The two cases can be distinguished by computing $abab^2ab^3$. This element has order five in one case – (a, b) are then called standard generators; see [170] – and three in the other. The probability of obtaining elements with the required properties using a random search is quite high and can be found in [170]. Here also representatives of the conjugacy classes of M_{11} are listed, written as products of the standard generators a, b:

1a : a^2	2a : a	3a : ab^2ab^2	4a : b	5a : $abab^2ab^{-1}$
6a : ab^2	8a : $abab^2ab^2$	8b : $ab^{-1}ab^2ab^2$	11a : ab	11b : ab^{-1}

Assume, for instance, that we are given permutations

$$a := (1,11)(2,7)(3,5)(4,6), \ c := (1,2,7,12,11,8,4,10,6,9,3) \in S_{12}.$$

It is easy to verify that they generate a simple group H of the same order as M_{11}. By [89], p. 314, we have $H \cong M_{11}$. To find an isomorphism and the corresponding permutation character pc we compute standard generators a, b for H. In GAP this can be done as follows:

```
gap> a := (1,11)(2,7)(3,5)(4,6);;  c:= (1,2,7,12,11,8,4,10,6,9,3);;
gap> H := Group(a,c);;   IsSimple(H) and Size(H) = 11*10*9*8;
true
gap> repeat b := Random(H); until Order(b) = 4 and Order(a*b) = 11 and
>                            Order(a*b*a*b^2*a*b^3)= 5;
gap> b;
(1,11,5,12)(2,10)(3,9,8,4)(6,7)
gap> cls := [ a^2, a, a*b^2*a*b^2, b, a*b*a*b^2*a*b^-1, a*b^2,
>                 a*b*a*b^2*a*b^2,a*b^-1*a*b^2*a*b^2, a*b, a*b^-1 ];;
gap> pc := List( cls, x -> 12 - NrMovedPoints(x) );
[ 12, 4, 3, 0, 2, 1, 0, 0, 1, 1 ]
gap> norm:=List(pc, x-> x^2) * List(cls, g -> 1/Size(Centralizer(H,g)));
2
gap> chi := pc - List( pc, x -> 1 );;
gap> Position( Irr( CharacterTable("M11") ), chi );
5
```

We have calculated that the norm of pc is two, so that chi := pc $- 1_H$ is an irreducible character, which we have identified as the fifth character in the character table of M_{11} (see p. 177).   ◆

Standard generators will also be used in Example 4.4.18 below.

(2) Inverse problem of Galois theory

Another important area where the above methods can be applied is the famous inverse problem of Galois theory, the question of whether every finite group is isomorphic to a Galois group over the rational numbers or at least over an abelian extension of $\mathbb{Q}$. This problem is still open, but in the 1980s considerable progress was made and the problem could in fact be decided for a number of specific groups using the criterion which we state below in a weak form (see [119], [160], [167]).

Theorem 2.5.19 (Belyi, Fried, Matzat, Thompson) *Suppose G is a finite group with trivial center $\mathbf{Z}(G) = \{1\}$ and that G has a rigid m-tuple of classes $C = (C_1, \ldots, C_m)$. Let $K = \mathbb{Q}(\{\chi(g_j) \mid \chi \in \mathrm{Irr}(G), 1 \leq j \leq m\})$, where as usual $g_j \in C_j$. Then G is isomorphic to a Galois group of a polynomial $f \in K[X]$. In particular, if the classes C_j are rational then G is a Galois group over $\mathbb{Q}$.*

Example 2.5.20 Let G be the sporadic simple Janko group J_4. We will use Theorem 2.5.19 to show that G is a Galois group over $\mathbb{Q}$ (see [135] and [116], p. 172). We consider $C := (2a, 4c, 11a)$ and compute $\alpha_C = \frac{3}{2}|G|$:

```
gap> ct:=CharacterTable("J4");;   C:= [ 2, 7, 19 ];; fus:=[];; ctm := [];;
gap> ClassNames(ct){C};
[ "2a", "4c", "11a" ]
gap> ClassStructureCharTable(ct,C)/Size(ct);
3/2
```

All maximal subgroups of G are known (see [101]), and their character tables can be accessed in GAP via the command Maxes. The fusions into G are also stored on the library tables. So we loop over all conjugacy classes of maximal subgroups $H < G$ and search for those H which contain elements of the three classes $C \in \mathbf{C}$. For these H we list the conjugacy classes intersecting the $C \in \mathbf{C}$ non-trivially and compute the corresponding class multiplication constants, printing them if they are not zero:

```
gap> for name in Maxes(ct) do
> ctm := CharacterTable( name );
> fus := Filtered(ComputedClassFusions(ctm),y-> y.name ="J4")[1].map;
>  if IsSubset( fus, C ) then
>  Print( name ); Cm := List( C , x -> Positions(fus,x) );
>  Print("\n",List(Cm, x->ClassNames(ctm){x}), "\n");
>   for Ch in Cartesian(Cm[1], Cm[2] ,Cm[3]) do
>    cs :=  ClassStructureCharTable(ctm, Ch);
>    if cs <> 0  then
>     Print( "alpha_",ClassNames(ctm){Ch}, " = ",cs/Size(ctm),"*|H| \n");
>    fi;
>   od;
>  fi;
> od;
c2aj4
[ [ "2a", "2b", "2d", "2f", "2h" ],
  [ "4c", "4g", "4l", "4o", "4r", "4u", "4x" ], [ "11a" ] ]
alpha_[ "2h", "4u", "11a" ] = 1/2*|H|
```

We see that the only maximal subgroups of G which contain elements of 2a, 4c and 11a are those isomorphic to $H := 2_{+}^{1+12} \cdot 3M_{22} : 2$, the centralizer of an element of 2a, with the GAP-name c2aj4. Also there is a single triple of conjugacy classes of H intersecting the $C \in \mathbf{C}$ non-trivially and having non-zero class multiplication constant $(\frac{1}{2}|H|)$, namely $(2h, 4u, 11a)$. Thus $\alpha_{\mathbf{C}}^{H} = \frac{1}{2}|G|$ and from Remark 2.5.2, (2.11), we conclude that $\gamma_{\mathbf{C}}^{G} = |G|$, so that $\mathbf{C}$ is, in fact, a rigid triple. Since the classes 2a, 4c, 11a are rational, G is a Galois group over $\mathbb{Q}$ by Theorem 2.5.19.

Incidentally, since H is the largest centralizer of a non-trivial element of G and $\mathbf{C}_G(H) \cong C_2$, we conclude from Remark 2.5.2 that

$$\Sigma_{\mathbf{C}}^{\leq G}(G) = \Sigma_{\mathbf{C}}^{G}(G) \cup \Sigma_{\mathbf{C}}^{H}(G)$$

and that $(2h, 4u, 11a)$ is a rigid triple of $H = 2_{+}^{1+12} \cdot 3M_{22} : 2$.   ♦

In Example 2.5.17 we saw that M_{11} has exactly three rigid triples up to conjugacy, none of which consists only of rational classes. So we can deduce only that M_{11} is a Galois group over $\mathbb{Q}(\sqrt{-11})$. This does not mean that M_{11} is not a Galois group over $\mathbb{Q}$. In fact, the following has been proved in a series of papers; see [116], sect. II.9.

Theorem 2.5.21 *All sporadic simple groups, with at most the exception the Mathieu group M_{23}, occur as Galois groups over $\mathbb{Q}$.*

Despite considerable efforts, the Mathieu group M_{23} has so far only been realized as a Galois group over $\mathbb{Q}(\sqrt{-23})$ and other quadratic extensions of $\mathbb{Q}$.

(3) Automorphism groups of compact Riemann surfaces
Another area where the above ideas have been applied is the study of groups of automorphisms of compact Riemann surfaces (see e.g. the survey by Jones [100]). A faithful action of a finite group G on a compact Riemann surface S of genus g corresponds to an epimorphism of

$$\Gamma = \langle a_1, b_1, \ldots, a_h, b_h, x_1, \ldots, x_m \mid x_i^{k_i} = \prod_{i=1}^{h}[a_i, b_i] \cdot \prod_{i=1}^{m} x_i = 1 \rangle$$

to G with torsion-free kernel, where

$$g = 1 + |G|(h - 1 + \frac{1}{2}\sum_{i=1}^{m}(1 - k_i^{-1})).$$

In the special and important case that the orbit space S/G is the Riemann sphere, we have $h = 0$ and hence $\Gamma = \Delta(k_1, \ldots, k_m)$ and the kernel is torsion-free if and only if the images of the generators have order precisely k_i. Our example above hence shows that M_{11} acts as a group of automorphisms on a Riemann surface of genus $g = 1 + 7920(-1 + \frac{1}{2}(1 - \frac{1}{2} + 1 - \frac{1}{4} + 1 - \frac{1}{11})) = 631$. There are numerous publications on determining the least genus of a compact Riemann surface on which a group G acts faithfully. This is called the **strong symmetric genus** of G. The strong symmetric genus of M_{11} is, in fact, 631; see [18], p. 63.

The order of the automorphism group G of a compact Riemann surface is bounded by Hurwitz's upper bound

$$|G| \leq 84(g - 1)$$

(see e.g. [18], theorem 3.17, p. 15). Those groups for which this bound is attained are called **Hurwitz groups**. They are precisely the finite non-trivial epimorphic images of $\Delta(2, 3, 7)$. Hurwitz groups are always perfect (see Exercise 2.5.3), and it is an attractive problem to find out which finite simple groups are Hurwitz groups. For instance, all but 64 of the alternating groups are Hurwitz groups (see [34]) and exactly 12 of the sporadic simple groups, including the Monster, are Hurwitz groups (see [100] and [172]). The most important methods for tackling this problem are those sketched in this section.

Exercises

Exercise 2.5.1 Let $C = (C_1, C_2, C_3)$ be a triple of conjugacy classes of the Janko group $G := J_4$ with $\alpha_C = |G|$ and let $(g_1, g_2, g_3) \in \Sigma_C^H(G)$ for some $H \leq G$. Show that H is a dihedral group, or $C = (2a, 4a, 11b)$ in the notation of the ATLAS, and H is a proper subgroup of the largest maximal subgroup $2^{11} : M_{24}$.

Exercise 2.5.2 Show that the Mathieu group M_{23} has no rigid triple of conjugacy classes.

Exercise 2.5.3 Let $p, q, r \in \mathbb{N}$ be pair-wise relatively prime. Show that $\Delta(p, q, r)$ is perfect.

Exercise 2.5.4 Compute standard generators for the Mathieu group M_{11} given in GAP as `MathieuGroup(11)`; see Example 1.6.14.

Exercise 2.5.5 Show that the following sporadic simple Janko groups are Hurwitz groups: J_1, J_2, J_4.

2.6 Character tables

We will now investigate the question of which group theoretical properties of a group are reflected in its character table and also its group algebra.

For finite abelian groups Exercise 2.1.2 implies that the character table of a finite abelian group G determines the group G up to isomorphism. On the other hand we have seen in Example 2.1.19 that the dihedral group D_8 and the quaternion group Q_8 both have the "same" character table. This requires a definition, since the ordering of the rows and columns in a character table is somewhat arbitrary.

Definition 2.6.1 (a) If $\mathbf{a} = [\alpha_{ij}] \in K^{m \times n}$ is a matrix over a commutative ring K and $\sigma \in S_n$ we put $^\sigma \mathbf{a} := [\alpha_{i\,\sigma(j)}]_{1 \leq i \leq m, 1 \leq j \leq n}$. If the sets of rows of $\mathbf{a}$ and $^\sigma \mathbf{a}$ coincide, we say that σ is an automorphism of $\mathbf{a}$. The group of automorphisms of $\mathbf{a}$ will be denoted by $\mathrm{Aut}(\mathbf{a})$.

(b) Let G, H be finite groups with conjugacy classes $g_1^G, \ldots, g_n^G$ and $h_1^H, \ldots, h_n^H$, respectively. Furthermore, let $\mathrm{Irr}(G) = \{\chi_1, \ldots, \chi_n\}$ and $\mathrm{Irr}(H) = \{\psi_1, \ldots, \psi_n\}$. The groups G, H are said to have the **same character table**, if there is $\sigma \in S_n$ such that the set of rows of $^\sigma[\chi_i(g_j)]$ and $[\psi_i(h_j)]$ coincide.

(c) The groups G, H as in (b) are said to form a **Brauer pair** if $G \not\cong H$ and σ as in (b) exists with the additional property that

$$h_i^k \in h_j^H \implies g_{\sigma(i)}^k \in g_{\sigma(j)}^G \quad \text{for} \quad 1 \leq i, j \leq n \quad \text{and} \quad k \in \mathbb{N}.$$

Thus G, H with $G \not\cong H$ form a Brauer pair if and only if they have the same character table and in addition the "same" power maps.

(d) If $[\chi_i(g_j)]_{1 \leq i, j \leq n}$ is a character table of G, then $\sigma \in \mathrm{Aut}[\chi_i(g_j)]$ is called a **character table automorphism** if

$$g_i^k \in g_j^G \implies g_{\sigma(i)}^k \in g_{\sigma(j)}^G \quad \text{for} \quad 1 \leq i, j \leq n \quad \text{and} \quad k \in \mathbb{N}.$$

The groups D_8 and Q_8 are by no means rare examples for non-isomorphic groups having the same character table. In fact, there are very many more examples of this sort, e.g among 2-groups, as Table 2.2 shows.

Table 2.2. Numbers of isomorphism classes and character tables of groups

Order	Number of isomorphism classes	Number of character tables
16	14	11
32	51	35
64	267	146
128	2328	904
256	56 092	9501

In fact, among these groups there are not just pairs, but also triples, quadruples or even larger families of groups having the same character tables. There are even families containing 256 pairwise non-isomorphic groups of order 256 (with rank five and elementary abelian commutator factor group and center of order eight) having the same character table (see [162]).

On the other hand there are many classes of non-solvable groups which are determined by their character tables. Nagao showed in [123] that the symmetric groups form such a class, that is, if a group G has the same character table as S_n then $G \cong S_n$. For a generalization of this result and other classes of groups with this property see e.g. [132] and [134].

Examples of Brauer pairs are not so common. Brauer asked in [13], Problem 4, whether or not such pairs would exist, and the first examples were given by Dade in [43]. These were certain p-groups of exponent p and order p^7 defined for $p \geq 5$. Observe that for such groups of exponent p the power maps do not give any further information which is not contained in the matrix of the character values.

A comprehensive search for Brauer pairs among the 2-groups of order up to 2^8 was carried out by Skrzipczyk in 1992 ([162]) using the library of 2-groups established by O'Brien (see [129]), which is available also in GAP. Perhaps surprisingly (in view of the numbers in the above table) no Brauer pairs were found among the groups of order up to 2^7. But for the groups of order 256 the search was successful, giving the first examples of Brauer pairs among 2-groups and the first examples of such pairs not consisting of groups of exponent p.

Theorem 2.6.2 *Among the* 56092 *groups of order* 2^8 *there are exactly ten Brauer pairs. These are*

$$(G_{1734}, G_{1735}), (G_{1736}, G_{1737}), (G_{1739}, G_{1740}), (G_{1741}, G_{1742}), (G_{3378}, G_{3380}),$$

$$(G_{3379}, G_{3381}), (G_{3678}, G_{3679}), (G_{4154}, G_{4157}), (G_{4155}, G_{4158}), (G_{4156}, G_{4159}),$$

where G_i *denotes the group of order* 2^8 *with number* i *in O'Brien's data base mentioned above.*

These are the smallest examples for Brauer pairs. For further examples see [53].

Let $\chi \in \mathrm{Irr}(G)$. Recall that $\ker \chi := \{g \in G \mid \chi(g) = \chi(1)\}$ is the kernel of a representation with character χ and that $|\chi(g)| \leq \chi(1)$ for all $g \in G$. Since every normal subgroup N of a finite group G is the intersection of kernels of irreducible representations of G – consider the inflations of all irreducible representations of G/N – it is clear that the character table of G determines the poset of normal subgroups of G together with their orders and the character tables of the factor groups. More precisely, it can be decided which unions of conjugacy classes of G form normal subgroups.

Lemma 2.6.3 *Given the character table of a finite group one can determine*

(a) *the poset of normal subgroups of G;*

(b) *the isomorphism type of the factor commutator subgroup G/G' of G;*

(c) *the isomorphism type of the center $\mathbf{Z}(G)$ of G;*

(d) *whether or not G is nilpotent or solvable.*

Brauer's Problem 10 in [13] asked if one can decide from the character table of a finite group G alone, whether or not a normal subgroup $N \trianglelefteq G$ (given by the columns of the character table corresponding to $\{C \in \mathrm{cl}(G) \mid C \subseteq N\}$) is abelian. Saksonov [154] showed that this is not the case, finding counter examples among extraspecial groups of odd order. In fact, Exercise 2.6.2 shows that even the number of abelian normal subgroups cannot be determined from the character table alone. This also shows that given the character table of G and a normal subgroup N (identified as above) it is not possible in general to find the character table of N.

Lemma 2.6.4 *An element $g \in G$ is a commutator in G if and only if*

$$\sum_{\chi \in \mathrm{Irr}(G)} \frac{\chi(g)}{\chi(1)} \neq 0.$$

Proof.   A commutator in G is of the form $g = x^{-1}y^{-1}xy = x^{-1}x^y$ with $x, y \in G$. If g is in the conjugacy class C_k of G, then g is of this form with $x \in C_i$ if and only if $xg = y^{-1}xy \in C_i$, i.e. if and only if $\alpha_{ik}^i = \alpha_{ki}^i \neq 0$ with the notation of Remark 2.3.1. Note that α_{ki}^i is the (diagonal) (i, i)-entry of the regular matrix representation $\rho(C_k^+)$ of $\mathbf{Z}(\mathbb{C}G)$ evaluated at C_k^+ with respect to the basis $(C_1^+, \ldots, C_r^+)$. Since all entries in this matrix are non-negative, the regular character at C_k^+ is positive if and only if g is a commutator in G. But

$$\mathrm{trace}\,\rho(C_k^+) = |C_k| \sum_{\chi \in \mathrm{Irr}(G)} \frac{\chi(g)}{\chi(1)}$$

by Theorem 2.3.2.   □

It is well known that not every element of the commutator subgroup of a finite group G is necessarily a commutator; see [90]. The preceding two lemmas give a convenient method to search for examples of this phenomenon. In fact, taking advantage of this method and the library of groups of small order in

GAP one finds that the smallest examples are two groups of order 96, a non-split extension $2^3 \cdot A_4$ of an elementary abelian group of order eight with A_4, and a split extension $Q_8 : A_4$ of a quaternion group with A_4. In both cases the commutator subgroup is a Sylow 2-subgroup and contains a class of involutions which are non-commutators. Many other examples may be found among the groups of order 132 and three examples among the groups of order 144.

There are also perfect groups having elements which are not commutators, the smallest examples being $2^4 : A_5$, the extension in which A_5 acts transitively on the non-trivial elements of the elementary abelian group of order 2^4 and $3 \cdot A_6$, the triple cover of A_6 (see [38]). But no example of a finite non-cyclic simple group having non-commutators is known; in fact, there is a conjecture often referred to as **Ore's conjecture** that no such group exists. Ore showed in [130] that every element of an alternating group is a commutator. There is a stronger conjecture as follows.

Conjecture 2.6.5 (J. Thompson) *If G is a non-abelian finite simple group there is a conjugacy class C in G such that $G = CC$.*

Obviously, if $G = CC$ then any element of G is of the form $g = [x, y]$ with $x \in C$ (see Exercise 2.6.3). Thus Thompson's conjecture implies Ore's conjecture. Both conjectures have been verified for a large variety of finite simple groups (see [12] and [54]), where the authors show that both conjectures hold for simple groups of Lie-type over finite fields of size greater than eight. For the Ore conjecture the only cases left to check are the orthogonal and unitary groups over fields of size at most eight; see [161].

Exercises

Exercise 2.6.1 Let $\mathrm{Irr}(G) = \{\chi_1, \ldots, \chi_9\}$ and $g_1, \ldots, g_9$ be representatives of the conjugacy classes of G. Assume that

$$[\chi_i(g_j)]_{i,j=1}^9 = \begin{bmatrix} 1 & 1 & 1 & 1 & 1 & 1 & 1 & 1 & 1 \\ 1 & 1 & -1 & -1 & 1 & 1 & 1 & 1 & 1 \\ 2 & 2 & . & . & 2 & -1 & -1 & -1 & 2 \\ 2 & 2 & . & . & -1 & 2 & -1 & -1 & -1 \\ 2 & 2 & . & . & -1 & -1 & 2 & -1 & -1 \\ 2 & 2 & . & . & -1 & -1 & -1 & 2 & -1 \\ 3 & -1 & -1 & 1 & 3 & . & . & . & -1 \\ 3 & -1 & 1 & -1 & 3 & . & . & . & -1 \\ 6 & -2 & . & . & -3 & . & . & . & 1 \end{bmatrix}.$$

Determine the poset of normal subgroups of G. For each proper normal subgroup N of G describe the structure of N and G/N.

Exercise 2.6.2 (See [134].) Let P_4 be the set of all 4×4 permutation matrices, $D_4 := \{\mathrm{diag}(a_1, \ldots, a_4) \mid a_i \in \{1, -1\}\}$ and $G := D_4 \cdot P_4 \leq \mathrm{GL}_4(\mathbb{Z})$. We put $G_1 := \{A \in G \mid \det(A) = 1\}$ and $G_2 := \{A \in G \mid \mathrm{per}(A) = 1\}$, where "per" denotes the "permanent" defined by $\mathrm{per}([a_{ij}]_{1 \leq i,j \leq n}) := \sum_{\sigma \in S_n} a_{1\,\sigma(1)} \cdots a_{n\,\sigma(n)}$. ($G_2$ is isomorphic to the Weyl group of type D_4.) Show that G_1 and G_2 have the same character table. Show also that G_1 has exactly three normal subgroups of order eight, and these are elementary abelian, whereas G_2 has only one abelian normal subgroup of order eight. Verify that G_1, G_2 do not form a Brauer pair.

Exercise 2.6.3 Let C be a conjugacy class of a finite group G. Show that $G = CC$ if and only if C is a real class and

$$\sum_{\chi \in \mathrm{Irr}(G)} \frac{|\chi(x)|^2 \overline{\chi(g)}}{\chi(1)} \neq 0 \qquad \text{for every } x \in C,\ g \in G.$$

Exercise 2.6.4 For the group $\mathrm{GL}_2(3)$ use the character table (computed in Exercise 2.2.7) to determine the poset of normal subgroups.

2.7 Products of characters

In this section K is an arbitrary field and G is a finite group. We continue to assume that all KG-modules considered are finite dimensional over K.

The K-vector space $\mathrm{cf}(G, K)$ of class functions introduced in Section 2.1 is a K-algebra with multiplication defined by

$$(\varphi \cdot \psi)(g) = \varphi(g)\psi(g) \qquad \text{for} \quad \varphi, \psi \in \mathrm{cf}(G, K)\, ,\ g \in G.$$

Using tensor products we see that the product of characters is again a character.

Lemma 2.7.1 *If V, W are KG-modules affording the characters χ_V and χ_W, respectively, then the tensor product (with diagonal action) $V \otimes_K W$ has character*

$$\chi_{V \otimes_K W} = \chi_V \cdot \chi_W.$$

Thus the product of characters is a character.

Proof. This follows immediately from Lemma 1.1.38. □

Example 2.7.2 Looking at the character table of $\mathrm{L}_2(7)$ (see Example 2.2.15)

$|\mathbf{C}_G(g)|$:	168	8	3	4	7	7
	1a	2a	3a	4a	7a	7b
χ_1	1	1	1	1	1	1
χ_2	3	−1	0	1	α	$\overline{\alpha}$
χ_3	3	−1	0	1	$\overline{\alpha}$	α
χ_4	6	2	0	0	−1	−1
χ_5	7	−1	1	−1	0	0
χ_6	8	0	−1	0	1	1

with $\alpha = \zeta_7 + \zeta_7^2 + \zeta_7^4 = \frac{1}{2}(-1 + \sqrt{-7})$, we can immediately verify that

$$\chi_2 \cdot \chi_2 = \chi_3 + \chi_4, \quad \chi_2 \cdot \chi_3 = \chi_1 + \chi_6, \quad \chi_2 \cdot \chi_4 = \chi_3 + \chi_5 + \chi_6. \qquad (2.14)$$

Suppose we know only the character χ_2 and consequently its algebraic conjugate χ_3. We could then obtain the other irreducible characters in the following simple way. One can compute the scalar product of $\chi_2 \cdot \chi_2$ and $\chi_2 \cdot \chi_3$ with the known irreducible characters χ_1, χ_2, χ_3 and conclude that $\chi_2 \cdot \chi_2 - \chi_3$ and $\chi_2 \cdot \chi_3 - \chi_1$ are ordinary characters of norm 1, hence irreducible. Calling these χ_4 and χ_6, respectively, one subsequently finds that $\chi_5 = \chi_2 \cdot \chi_4 - \chi_3 - \chi_6$ is again an irreducible character, the only one which was still missing.

In GAP the above calculations can be performed like this:

```
gap> t := CharacterTable("L2(7)");;
gap> prod := Tensored( Irr(t){[2]} , Irr(t){[2,3,4]} );;
gap> Display( MatScalarProducts( Irr(t) , prod ) );
[ [  0,  0,  1,  1,  0,  0 ],
  [  1,  0,  0,  0,  0,  1 ],
  [  0,  0,  1,  0,  1,  1 ] ]
```

We have first retrieved the character table of $G = L_2(7)$ from the library of character tables and named it `t`. Then using `Tensored` we calculated the products of the second irreducible character with the sublist `Irr(t){[2,3,4]})`, obtaining three reducible characters. Finally we have given the matrix of scalar products of all the irreducible characters with these reducible characters. The matrix contains the same information as (2.14).

Here we have used the complete character table of $L_2(7)$. Suppose now, as above, that only the first three irreducible characters were known, that is, those of degree ≤ 3. Then

```
gap> irr := Irr(t){[1,2,3]};
gap> prod := Tensored( irr{[2]}, irr{[2,3]} );;
gap> red := Reduced( irr, prod );;
```

yields two further irreducible characters (of degrees six and eight). In fact the command

$$\mathbf{Reduced}(\ [\chi_1, \ldots, \chi_m], \ [\psi_1, \ldots, \psi_n] \ )$$

for a list of irreducible characters $[\chi_1, \ldots, \chi_m]$ and a list of arbitrary characters $[\psi_1, \ldots, \psi_n]$ computes the characters

$$[\ \psi_i - \sum_{j=1}^{m} (\chi_j, \psi_j)_G \chi_j \ | \ 1 \leq i \leq n \ ]$$

(in other words, the projections of the ψ_i onto the orthogonal complement of the span of the known irreducibles $\chi_1, \ldots, \chi_m$ in $\mathrm{cf}(G, \mathbb{C})$) and returns those characters which have norm 1 in the record component **irreducibles** and the others in the component **remainders**. The first ones may be added to the irreducibles in the list `irr` and may be used in the following sequel:

```
gap> Append( irr, red.irreducibles );       # now irr contains 5 characters
gap> Append( prod , Tensored( irr{[2]} , irr{[4,5]} ));;
gap> red := Reduced( irr , prod );;
gap> Append( irr, red.irreducibles );
```

```
gap> Display( t, rec( chars:=irr , powermap:=false) );
L3(2)

      2  3  3  .  2  .  .
      3  1  .  1  .  .  .
      7  1  .  .  .  1  1

         1a 2a 3a 4a 7a 7b

Y.1      1  1  1  1  1  1
Y.2      3 -1  .  1  A /A
Y.3      3 -1  .  1 /A  A
Y.4      6  2  .  . -1 -1
Y.5      8  . -1  .  1  1
Y.6      7 -1  1 -1  .  .
```

$$A = E(7)+E(7)^2+E(7)^4$$
$$= (-1+ER(-7))/2$$
$$= b7$$

The following classical result is mainly of theoretical importance.

Theorem 2.7.3 (Burnside, Brauer) *Suppose that K is a splitting field for G of characteristic not dividing $|G|$ and that ψ is a faithful (but not necessarily irreducible) character of G over K that takes on exactly m different values $a_1 = \psi(1), a_2, \ldots, a_m$ on the elements of G. Then every irreducible character $\chi \in \mathrm{Irr}(G)$ is a constituent of one of the powers $\psi^0 = 1_G, \psi, \psi^2, \ldots, \psi^{m-1}$.*

Proof. (G. R. Robinson.) We consider the following class function on G:

$$\theta = (\psi - a_2 1_G) \cdots (\psi - a_m 1_G) = \sum_{j=0}^{m-1} b_j \psi^j$$

with certain $b_j \in K$. Then

$$\theta(g) = \begin{cases} 0 & \text{for} \quad g \neq 1, \\ \prod_{i=2}^{m}(a_1 - a_i) & \text{for} \quad g = 1, \end{cases}$$

because ψ is faithful and thus $\psi(g) = a_1$ if and only if $g = 1$. Thus $\theta = a \cdot \rho_G$, with $a = \frac{1}{|G|} \prod_{i=2}^{m-1}(a_1 - a_i) \neq 0$. So for any $\chi \in \mathrm{Irr}(G)$ we have

$$(\theta, \chi)_G = a(\rho_G, \chi)_G = a\chi(1) \neq 0.$$

Thus for every $\chi \in \mathrm{Irr}(G)$ there must be $j \in \{0, \ldots, m-1\}$ with $b_j \neq 0$ and $(\psi^j, \chi)_G \neq 0$. $\square$

The regular character ρ_G is a good example for Theorem 2.7.3: it takes exactly two values on G and every irreducible character is a constituent of ρ_G. This example also shows that the theorem is more of theoretical interest; one cannot possibly hope, in general, to obtain the irreducible characters just from the knowledge of a faithful character of a group. The situation might improve if one knows a faithful (irreducible) character of small degree compared to the degrees of the other irreducible characters. It should also be clear that the bound in Burnside's theorem is by no means sharp. For example, choosing $\psi = \sum_{\chi \in \mathrm{Irr}(G)} \chi$ and G non-abelian, one can show that $m > 2$. Or, looking at the character table of the Mathieu group $G = M_{11}$ (see Section 2.10, p. 177) one finds that there is a faithful irreducible character χ_8 of degree 44 which takes five values on G, but every irreducible character of G is a constituent of χ_8^2.

Of course, the product of irreducible characters will not be in general irreducible. In fact, the square of a character χ will never be irreducible unless χ is a linear character, as we will see below, and the same holds true for higher powers of characters.

Theorem 2.7.4 *Let K be a field with* char $K \neq 2$ *and V a KG-module of dimension $n > 1$ and with character χ. Then there are KG-submodules $V^{[2]}$ and $V^{[1^2]}$ of V with*

$$V \otimes_K V = V^{[2]} \oplus V^{[1^2]}$$

and characters given by

$$\chi^{[2]}(g) = \frac{\chi(g)^2 + \chi(g^2)}{2} \,, \qquad \chi^{[1^2]}(g) = \frac{\chi(g)^2 - \chi(g^2)}{2} \qquad \text{for} \quad g \in G.$$

In fact, $V^{[2]}$ resp. $V^{[1^2]}$, are the K-submodules of symmetric (resp. skew-symmetric) tensors in $V \otimes_K V$.

Proof. Let $B = (v_1, \ldots, v_n)$ be a K-basis of V. We define a K-linear map

$$\tau \colon {}_K V \otimes_K V \to V \otimes V, \qquad \tau \colon v_i \otimes v_j \mapsto v_j \otimes v_i;$$

then $\tau(v \otimes v') = v' \otimes v$ for all $v, v' \in V$. Furthermore $\tau \in \mathrm{End}_{KG} V \otimes_K V$ and $\tau^2 = \mathrm{id}_V$. Since char $K \neq 2$ the endomorphism τ is diagonalizable and $V \otimes_K V$ is the direct sum of the eigenspaces

$$V^{[2]} = \{x \in V \otimes_K V \mid \tau(x) = x\} \,, \qquad V^{[1^2]} = \{x \in V \otimes_K V \mid \tau(x) = -x\}.$$

Since $\tau \in \mathrm{End}_{KG} V \otimes_K V$ these are KG-submodules. To calculate the values of their characters $\chi^{[2]}$ and $\chi^{[1^2]}$ at an element $g \in G$ we may assume that K contains the eigenvalues of $\delta_V(g)$, for otherwise we might extend the field K, which does not affect χ_V. We choose a basis $B = (v_1, \ldots, v_n)$ such that

$$[\delta_V(g)]_B = \begin{bmatrix} \xi_1 & & * \\ & \ddots & \\ & & \xi_n \end{bmatrix}.$$

So $gv_i \in \xi_i v_i + \sum_{k<i} K v_k$. A K-basis for $V^{[1^2]}$ is $(w_{i,j} = v_i \otimes v_j - v_j \otimes v_i \mid i < j)$. We use the lexicographical ordering $<$ of the pairs (i,j). It follows that

$$g \cdot w_{i,j} \in (\xi_i \xi_j) w_{i,j} + \sum_{(k,l)>(i,j)} K \cdot w_{k,l}.$$

So

$$\chi^{[1^2]}(g) = \sum_{i<j} \xi_i \xi_j = \frac{1}{2}\left(\left(\sum_{i=1}^n \xi_i\right)^2 - \sum_{i=1}^n \xi_i^2\right) = \frac{1}{2}(\chi(g)^2 - \chi(g^2)).$$

The formula for $\chi^{[2]}$ follows from $\chi(g)^2 = \chi^{[2]}(g) + \chi^{[1^2]}(g)$. $\qquad \square$

Theorem 2.7.4 just deals with a particularly simple case of a much more general construction, which is usually called **symmetrization** of characters or representations.

Definition 2.7.5 Let K be a commutative ring and let V be a free K-module of rank n. For any integer m the symmetric group S_m acts on

$$\otimes^m V := \underbrace{V \otimes V \otimes \cdots \otimes V}_{m} \qquad \text{from the right by}$$

$$v_1 \otimes \cdots \otimes v_m \cdot \sigma := v_{\sigma(1)} \otimes \cdots \otimes v_{\sigma(m)} \qquad \text{for} \quad \sigma \in \mathrm{S}_m , \ v_i \in V.$$

Then $\mathrm{S}_K(V,m) := \mathrm{End}_{K\,\mathrm{S}_m} \otimes^m V$ is called the **Schur algebra** $\mathrm{S}_K(n,m)$.

In the following we assume that K is a field of characteristic zero and that V is an n-dimensional K-vector space. We write $E := \mathrm{S}_K(V,m)$. The $K\,\mathrm{S}_m$-right module $\otimes^m V$ can be considered as a $(E, K\,\mathrm{S}_m)$-bimodule with

$$\varphi \cdot t := \varphi(t) \quad \text{for} \quad \varphi \in E , \ t \in \otimes^m V.$$

For a simple $K\,\mathrm{S}_m$-module X we will denote

$$V_X := \mathrm{Hom}_{K\,\mathrm{S}_m}(X, \ \otimes^m V), \tag{2.15}$$

which is naturally an E-module. In fact, if $X_1, \ldots, X_r$ are representatives of the isomorphism classes of simple $K\,\mathrm{S}_m$-submodules of $\otimes^m V$ then by Exercise 1.5.7 $\{V_{X_i} \mid 1 \le i \le r\}$ is a set of representatives of the isomorphism classes of simple E-modules.

Lemma 2.7.6 *For any simple $K\,\mathrm{S}_m$-module X let $\mathrm{H}_X(\otimes^m V)$ be the X-homogeneous component (see Definition 1.4.1) of $\otimes^m V$ when we consider $\otimes^m V$ as a $K\,\mathrm{S}_m$-module. This is E-invariant and*

$$\mathrm{H}_X(\otimes^m V) \cong V_X \otimes_K X \qquad \text{as } (E, K\,\mathrm{S}_m)\text{-bimodules.}$$

Proof. We have a K-bilinear map $V_X \times X \to \mathrm{H}_X(\otimes^m V)$, $(\varphi, x) \mapsto \varphi(x)$. This induces a K-linear map

$$\Psi \colon V_X \otimes_K X \to \mathrm{H}_X(\otimes^m V) \qquad \text{with} \qquad \varphi \otimes x \mapsto \varphi(x),$$

which is readily seen to be a homomorphism of $(E, K\,\mathrm{S}_m)$-bimodules. If X is not isomorphic to a direct summand of $\otimes^m V$, then $V_X = \{0\}$ and $\mathrm{H}_X(\otimes^m V) = \{0\}$. Let $\mathrm{H}_X(\otimes^m V) = Y_1 \oplus \cdots \oplus Y_k$ with $Y_j \cong_{K\mathrm{S}_m} X$ and $\psi_j \colon Y_j \to X$ a $K\,\mathrm{S}_m$-isomorphism for $j = 1, \ldots, k$. Then every $x \in \mathrm{H}_X(\otimes^m V)$ can be written as $x = \sum_{j=1}^k y_j$ with

$$y_j = \psi_j^{-1}\psi_j(y_j) = \Psi(\psi_j^{-1} \otimes \psi_j(y_j)) \in Y_j,$$

since we may consider $\psi_j^{-1} \in V_X$ by abuse of notation. Thus Ψ is surjective. We use the fact (proven in Theorem 3.3.8) that all simple $K\,\mathrm{S}_m$-modules are absolutely simple. Hence

$$\dim_K V_X = \dim_K \mathrm{Hom}_{K\,\mathrm{S}_m}(X, \mathrm{H}_X(\otimes^m V)) = \dim_K \mathrm{Hom}_{K\,\mathrm{S}_m}(X, \oplus_{j=1}^k Y_j) = k.$$

Therefore

$$\dim_K(V_X \otimes_K X) = \dim_K V_X \dim_K X = k \dim_K X = \dim_K \mathrm{H}_X(\otimes^m V)$$

and Ψ is bijective. $\qquad\square$

We have a natural embedding

$$\mu \colon \mathrm{End}_K V \to E \qquad \text{with} \qquad \mu(\varphi) \colon v_1 \otimes \cdots \otimes v_m \mapsto \varphi(v_1) \otimes \cdots \otimes \varphi(v_m).$$

Thus every E-module becomes a $K\,\mathrm{GL}(V)$-module. In particular, if V is a KG-module and $X \leq_{K\mathrm{S}_m} \otimes^m V$ is simple, then the simple E-submodule V_X defined in (2.15) is a KG-submodule of $\otimes^m V$.

Lemma 2.7.7 *Let K be a field of characteristic zero, let V be a KG-module with character χ_V and let $\sigma = \sigma_1 \cdots \cdots \sigma_k \in \mathrm{S}_m$ be a product of disjoint s_j-cycles, for $j = 1, \ldots, k$ (so that $m = \sum_{j=1}^k s_j$). If $X_1, \ldots, X_r$ is a set of representatives of the simple $K\,\mathrm{S}_m$-modules with characters $\chi_1, \ldots, \chi_r$, and η_i is the character of the KG-module V_{X_i} for $i = 1, \ldots, r$, then*

$$\sum_{i=1}^r \eta_i(g)\,\chi_i(\sigma) = \prod_{j=1}^k \chi_V(g^{s_j}) \qquad \text{for} \qquad g \in G.$$

Proof. Consider the K-linear map

$$f \colon \otimes^m V \to \otimes^m V, \quad w \mapsto g \cdot w \cdot \sigma$$

using the (G, S_m)-bimodule structure of $\otimes^m V$. Since

$$\otimes^m V = \bigoplus_{i=1}^r \mathrm{H}_{X_i}(\otimes^m V) = \bigoplus_{i=1}^r V_{X_i} \otimes_K X_i$$

we get $\text{trace}(f) = \sum_{i=1}^{r} \eta_i(g)\chi_i(\sigma)$.

We want to show, that $\text{trace}(f) = \prod_{j=1}^{k} \chi_V(g^{s_j})$. For this, we may assume that K contains the eigenvalues of g on V. We choose a K-basis B of V consisting of eigenvectors, so that for $v \in B$ we have $gv = \alpha(v)v$ with $\alpha(v) \in K$. Also, since permutations with the same cycle type are conjugate, we can assume that σ_j is the s_j-cycle $(\sum_{i=1}^{j-1} s_i + 1, \ldots, \sum_{i=1}^{j} s_i)$. We write

$$\otimes^m V = \underbrace{V \otimes \cdots \otimes V}_{:=W_1} \otimes \cdots \otimes \underbrace{V \otimes \cdots \otimes V}_{:=W_k} \quad \text{with} \quad W_j = \underbrace{V \otimes \cdots \otimes V}_{s_j}.$$

Then

$$f = f_1 \otimes \cdots \otimes f_k \quad \text{with} \quad f_j \colon W_j \to W_j \, , \ w \mapsto g \cdot w \cdot \sigma_j \quad (1 \le j \le k).$$

Since $\text{trace}(f) = \prod_{j=1}^{k} \text{trace}(f_j)$, we may assume $k = 1$, that is $\sigma = (1, \ldots, m)$. A basis of $\otimes^m V$ is $\otimes^m B = \{v_1 \otimes \cdots \otimes v_m \mid v_i \in B, (1 \le i \le m)\}$. Then

$$f(v_1 \otimes \cdots \otimes v_m) = (\prod_{i=1}^{m} \alpha(v_i)) \, v_2 \otimes \cdots \otimes v_m \otimes v_1.$$

Hence $\text{trace}(f) = \sum_{v \in B} \alpha(v)^m = \chi_V(g^m)$. □

Definition 2.7.8 For $\chi \in \text{cf}(G, K)$ and $\psi \in \text{cf}(S_m, K)$ we define the class function $\chi \boxdot \psi \in \text{cf}(G, K)$ by

$$(\chi \boxdot \psi)(g) := \frac{1}{n!} \sum_{\sigma \in S_m} \psi(\sigma) \prod_{j=1}^{m} \chi(g^j)^{a_j(\sigma)} \quad \text{for} \quad g \in G,$$

where $a_j(\sigma)$ is the number of j-cycles of σ.

Theorem 2.7.9 *If V is a KG-module for a field K with* $\text{char} \, K = 0$ *and X is a simple $K S_m$-module and V_X is as defined in (2.15), then* $\chi_{V_X} = \chi_V \boxdot \chi_X$. *Thus $\chi_V \boxdot \chi_X$ is an ordinary character of G or the 0-function. We have*

$$\chi_V^m = \sum_{\psi \in \text{Irr}(S_m)} \psi(1) \, \chi_V \boxdot \psi.$$

Proof. We have for $g \in G$

$$\chi_{V_X}(g) = \sum_{\psi \in \text{Irr}(S_m)} \chi_{V_X}(g) \, (\chi_X, \psi)_{S_m}$$

$$= \sum_{\psi \in \text{Irr}(S_m)} \chi_{V_X}(g) \frac{1}{m!} \sum_{\sigma \in S_m} \chi_X(\sigma) \, \psi(\sigma^{-1})$$

$$= \frac{1}{m!} \sum_{\sigma \in S_m} \chi_X(\sigma) \sum_{\psi \in \text{Irr}(S_m)} \chi_{V_X}(g) \, \psi(\sigma)$$

$$= \frac{1}{m!} \sum_{\sigma \in S_m} \chi_X(\sigma) \prod_{j=1}^{m} \chi(g^j)^{a_j(\sigma)} = \chi_V \boxdot \chi_X(g),$$

where we have used Lemma 2.7.7 and the well-known fact that in a symmetric group every element is conjugate to its inverse. As mentioned before, the fact that K is a splitting field for S_m will be proved in Theorem 3.3.8. □

We will see in Section 3.3 that the irreducible characters of the symmetric group over a field K with char $K = 0$ can be labeled by **partitions**

$$\lambda = [\lambda_1, \ldots, \lambda_j] \quad \text{with} \quad \lambda_1 \geq \lambda_2 \geq \cdots \geq \lambda_j > 0 \quad \text{and} \quad \sum_i \lambda_i = m.$$

Here $\psi^{[m]} = \mathbf{1}_{S_m}$ and $\psi^{[1^m]}$ is the sign character of S_m, where $[1^m] := \underbrace{[1, \ldots, 1]}_{m}$.

If λ is a partition of m and $\psi^\lambda \in \mathrm{Irr}(S_m)$ is the corresponding irreducible character we will also write

$$\chi^\lambda := \chi \boxdot \psi^\lambda \quad \text{for} \quad \chi \in \mathrm{cf}(G, K),$$

which is in accordance with the notation used in Theorem 2.7.4.

Example 2.7.10 From the character table of S_3 (Example 2.1.20) we see that for any ordinary character χ of an arbitrary group G we have

$$\chi^{[3]}(g) = \frac{1}{6}(\chi(g)^3 + 3\chi(g^2)\chi(g) + 2\chi(g^3)),$$

$$\chi^{[1^3]}(g) = \frac{1}{6}(\chi(g)^3 - 3\chi(g^2)\chi(g) + 2\chi(g^3)),$$

$$\chi^{[2,1]}(g) = \frac{1}{3}(\chi(g)^3 - \chi(g^3)).$$

In particular, we see that $\chi^{[1^3]} = 0$ if $\chi(1) \leq 2$. ◆

Another way to see symmetrizations is the following: if $\delta\colon G \to \mathrm{GL}(V)$ is a representation of a finite group G and $\tau\colon \mathrm{GL}(V) \to \mathrm{GL}(W)$ is a representation of $\mathrm{GL}(V)$ then the concatenation

$$\tau \circ \delta\colon G \to \mathrm{GL}(W)$$

is again a representation of G. One obtains the symmetrizations with respect to the simple $K\,S_m$-module X by choosing τ to be the representation afforded by V_X.

If $K \subseteq \mathbb{R}$ and the KG-module V carries a G-invariant non-degenerate symmetric or symplectic form – we will deal with the question of when this is possible in Section 2.9 – then the image of δ will be in the orthogonal group $\mathrm{O}(V)$ or in the symplectic group $\mathrm{Sp}(V)$ and one can compose δ with irreducible representations of these groups, which in general will yield finer decompositions of $\otimes^m V$ than the one given in Theorem 2.7.9.

Example 2.7.11 Let $G = M_{12}$ be the sporadic Mathieu group of degree 12. It is a simple permutation group acting (sharply) 5-fold transitive on 12 letters (see [38]). By Corollary 2.1.14(d) G has an irreducible character of degree 11. Let us assume that we know this character, the lengths of the conjugacy classes (equivalently, the centralizer orders of the representatives of the conjugacy classes) and the power maps of G. We will show that using symmetrizations and products of known characters we can compute further irreducible characters of G and eventually (in Example 2.8.12) the complete character table of G. We start by retrieving the (complete) character table of G from the GAP library and copying the first two irreducible characters (the trivial character and one character of degree 11) to the list `irr`, which we then display together with the centralizer orders (in factored form) and the power maps:

```
gap>  t := CharacterTable("M12");;
gap>  irr := Irr(t){[1,2]};;
gap>  Display( t , rec(chars:=irr) );
M12
```

```
      2  6  4  6  1  2  5  5  1  2  1  3  3  1  .  .
      3  3  1  1  3  2  .  .  .  1  1  .  .  .  .  .
      5  1  1  .  .  .  .  .  1  .  .  .  .  1  .  .
     11  1  .  .  .  .  .  .  .  .  .  .  .  .  1  1

        1a 2a 2b 3a 3b 4a 4b 5a 6a 6b 8a 8b 10a 11a 11b
     2P 1a 1a 1a 3a 3b 2b 2b 5a 3b 3a 4a 4b  5a 11b 11a
     3P 1a 2a 2b 1a 1a 4a 4b 5a 2a 2b 8a 8b 10a 11a 11b
     5P 1a 2a 2b 3a 3b 4a 4b 1a 6a 6b 8a 8b  2a 11a 11b
    11P 1a 2a 2b 3a 3b 4a 4b 5a 6a 6b 8a 8b 10a  1a  1a

Y.1      1  1  1  1  1  1  1  1  1  1  1  1   1   1   1
Y.2     11 -1  3  2 -1 -1  3  1 -1  . -1  1  -1   .   .
```

We then compute the second symmetrizations of χ_2 (denoted by `Y.2` in the table), that is $\chi_2^{[2]}, \chi_2^{[1^2]}$, and reduce these with the list of the two irreducible characters in `irr`, thereby obtaining two further irreducible characters of G of degrees 54 and 55. In fact, it is not hard to see that $\chi_2^{[1^2]}$ and $\chi_2^{[2]} - \chi_1 - \chi_2$ are irreducible:

```
gap>  r := Symmetrizations( t, irr{[2]}, 2 );;
gap>  red := Reduced( t , irr , r );;
gap>  Append( irr, red.irreducibles );
gap>  Display( t , rec(chars:=irr,powermap:=false,centralizers:=false) );
M12
```

```
        1a 2a 2b 3a 3b 4a 4b 5a 6a 6b 8a 8b 10a 11a 11b

Y.1      1  1  1  1  1  1  1  1  1  1  1  1   1   1   1
Y.2     11 -1  3  2 -1 -1  3  1 -1  . -1  1  -1   .   .
Y.3     54  6  6  .  .  2  2 -1  .  .  .  .   1  -1  -1
Y.4     55 -5 -1  1  1 -1  3  .  1 -1  1 -1   .   .   .
```

One can obtain reducible characters by forming products of the known irreducibles and also higher symmetrizations, and one might try to obtain further irreducible characters by reducing these with the four irreducible characters in `irr` known so far. But the outcome is the following:

```
gap> r := Tensored( irr{[2]} , irr{[3,4]} );;
gap> Append( r , Tensored( irr{[3]} , irr{[4]} ) );
gap> Append( r , Symmetrizations( t , irr{[3,4]} , 2 ) );
gap> Append( r , Symmetrizations( t , irr{[2,3,4]} , 3 ) );
gap> red := Reduced( t , irr , r );; r := red.remainders;;
gap> red.irreducibles;
[ ]
gap> SortParallel( List(r , Norm) , r ); List( r , Norm );
[2, 2, 2, 4, 4, 22, 25, 26, 26, 87, 6261, 6741, 7500, 8493, 27041,
30265]
gap> List( r , x -> x[1] );
[154, 165, 320, 474, 485, 1376, 1375, 1257, 1366, 2740,
23425, 24393, 25478, 27358, 49017, 51874]
```

So no new irreducible characters were found, but instead we have a list of characters of rather large norm and degree, which does not seem to be very useful. We will see in Section 2.8, however, on continuing this example in Example 2.8.12, that we can, in fact, derive all the irreducible characters of G from this list.

                                                           ◆

Exercises

Exercise 2.7.1 Let $\mathrm{Irr}(G) = \{\chi_1, \ldots, \chi_r\}$ and let

$$\chi_i \cdot \chi_j = \sum_{k=1}^{r} d_{ijk}\, \chi_k, \qquad d_{ijk} \in \mathbb{N}_0 \qquad 1 \le i, j, k \le r.$$

Show that the characters $\chi_i \in \mathrm{Irr}(G)$ can be calculated given the numbers d_{ijk} for $1 \le i, j, k \le r$.

Exercise 2.7.2 Let $m, n \in \mathbb{N}$. The symmetric group S_m acts on $\mathbf{n}^{\mathbf{m}} := \{f\colon \{1, \ldots, m\} \to \{1, \ldots, n\}\}$ from the right by $f \cdot \sigma := f \circ \sigma$. Let K be a commutative ring and let V be a free K-module of rank n. Show that

$$S_K(V, m) \cong \mathrm{End}_{K\,S_m}\, K\mathbf{n}^{\mathbf{m}}.$$

Exercise 2.7.3 Let J_1 be Janko's first sporadic simple group (see [38]). Show that every $\chi \in \mathrm{Irr}(J_1) \setminus \{1_{J_1}\}$ has the property that $(\chi^2, \chi_i)_{J_1} > 0$ for all $\chi_i \in \mathrm{Irr}(J_1)$.

Exercise 2.7.4 Let V, W be KG-modules, K a field and $W' \le_{KG} W$. Show that

$$W' \otimes_K V \le_{KG} W \otimes_K V \quad \text{and} \quad (W \otimes_K V)/(W' \otimes_K V) \cong_{KG} W/W' \otimes_K V.$$

Note: It is important that K is a field or, for instance, that V is free as a K-module. Consider as an example $K = \mathbb{Z}$, $G = \{1\}$, $V = \mathbb{Z}/2\mathbb{Z}$, $W' = \mathbb{Z}$,

$W = \mathbb{Q}$. Here we have

$$W' \otimes_K V \cong V, \qquad \text{whereas} \qquad W \otimes_K V = 0.$$

Exercise 2.7.5 This is a continuation of Exercise 2.4.3. Using the GAP character table of $G := \mathrm{SL}_2(3)$ tensor the irreducible character χ_5 of degree two with all the irreducible characters $\chi_1, \ldots, \chi_7$ of G. Compute the matrix $M := [m_{ij}]$, where m_{ij} is the multiplicity of χ_j in $\chi_5 \cdot \chi_i$ for $i, j = 1, \ldots, 7$ and check that the matrix A given in Exercise 2.4.3 is just $2\mathbf{I}_7 - M$. This observation can be generalized; see [120].

Exercise 2.7.6 Let $\mathbf{Z}(G) = \{1\}$ and $\chi = \overline{\chi} \in \mathrm{Irr}(G)$ be faithful with $\chi(1) = 3$. Show that

(a) $|\mathbf{C}_G(P_3)| = 3$ for $P_3 \in \mathrm{Syl}_3(G)$ (use Exercise 2.3.3);

(b) $\chi^{[1^2]} = \chi$ and $\psi := \chi^{[2]} - \mathbf{1}_G \in \mathrm{Irr}(G)$ is faithful with $\psi(1) = 5$;

(c) $\varphi \colon G \to \mathbb{C}, \; g \mapsto \begin{cases} 1 & \text{if } g \text{ has order three} \\ \psi(g) - 1 & \text{else} \end{cases} \in \mathrm{Irr}(G)$ with $\varphi(1) = 4$;

(d) $\chi(g) = \begin{cases} -1 & \text{if } g \text{ has order two,} \\ \frac{1 \pm \sqrt{5}}{2} & \text{if } g \text{ has order five;} \end{cases}$

(e) G has exactly two irreducible characters χ, χ' of degree three and no elements of order p for a prime $p > 5$;

(f) G has no elements x of order four (show that $\chi \chi' = \varphi + \psi$);

(g) $\langle \mathbf{1}_G, \chi, \chi', \varphi, \psi \rangle_\mathbb{Z}$ is closed under multiplication;

(h) $G \cong \mathrm{A}_5$.

Note: If $\chi \neq \overline{\chi} \in \mathrm{Irr}(G)$ and $\chi(1) = 3$ then $G \cong \mathrm{L}_2(7)$.

2.8 Generalized characters and lattices

In this section K is always assumed to be a field of characteristic zero.

If one knows the centralizer orders and a few characters of a finite group G one can find further characters using products (or symmetrizations, if the power maps are also known). These characters usually have rather large norms even if one starts with irreducible ones. One might try to obtain characters of smaller norm by looking at differences of characters, but in general it is hard to decide whether a difference of characters is a character. The best one can state, if one just knows the scalar products of the characters in question, is the following lemma (see [37]).

Lemma 2.8.1 (Guy's lemma) *Assume that φ and ψ are characters of a group G with $a := (\varphi, \varphi)_G \leq b := (\psi, \psi)_G$, $c := (\varphi, \psi)_G$ and*

$$d := b - (a \cdot b - c^2) \geq 0.$$

Then $\psi - \varphi$ is a (proper) character of G or $d = 0$ and $\varphi - \frac{c}{b}\psi \in \mathrm{Irr}(G)$.

Proof. Let

$$\varphi = \sum_{i=1}^{r} a_i \chi_i, \qquad \psi = \sum_{i=1}^{r} b_i \chi_i \qquad (a_i, b_i \in \mathbb{N}_0),$$

where $\mathrm{Irr}(G) = \{\chi_1, \ldots, \chi_r\}$. Then

$$a := (\varphi, \varphi)_G = \sum_{i=1}^{r} a_i^2, \quad b := (\psi, \psi)_G = \sum_{i=1}^{r} b_i^2, \quad c := (\varphi, \psi)_G = \sum_{i=1}^{r} a_i b_i.$$

If $\psi - \varphi$ is not a character then for some k we must have $d_k = a_k - b_k > 0$. By the Schwarz inequality we have

$$(\varphi - a_k \chi_k , \psi - b_k \chi_k)_G^2 \leq ((\varphi, \varphi)_G - a_k^2) \cdot ((\psi, \psi)_G - b_k^2).$$

Hence

$$a \cdot b - c^2 \geq a_k^2 b + b_k^2 a - 2 a_k b_k c \geq a_k^2 b + b_k^2 a - (a + b) a_k b_k,$$

since $2c \leq a + b$. From this it follows from our assumption that

$$0 \geq ab - c^2 - b \geq (a_k b - b_k a) d_k - b = ((b - a) b_k + b d_k) d_k - b \geq b(d_k^2 - 1) \geq 0.$$

Hence $d_k = 1$ and $b_k = 0$, since $a = b$ is clearly impossible. All inequalities above must be equalities, including the one in the Schwarz inequality. From this we see that $\varphi - \chi_k$ and ψ must be proportional. Hence there is a positive $x \in \mathbb{Q}$ such that $\chi_k = \varphi - x\psi$. Taking norms we get $bx^2 - 2cx + a = 1$, hence $x = \frac{c}{b}$. $\square$

Of course, the converse in Guy's lemma does not hold, that is $\psi - \varphi$ can be a character even if $d < 0$; see Example 2.8.2. On the other hand, Guy's lemma is best possible in the sense that if $d := b - (a \cdot b - c^2) < 0$ then there is a group G with characters φ, ψ such that $a = (\varphi, \varphi)_G, b = (\psi, \psi)_G$ and $c = (\varphi, \psi)_G$, and $\psi - \varphi$ is not a character (see [37]).

Example 2.8.2 We continue with Example 2.7.11 and compute the matrix of scalar products of the first ten characters of **r**. (Since this is a symmetric matrix, only the lower part of it is shown.)

```
gap> Display( MatScalarProducts( t, r{[1..10]} ) );
[ [   2 ],
  [   0,   2 ],
  [   0,   0,   2 ],
  [   2,   0,   2,   4 ],
  [   0,   2,   2,   2,   4 ],
  [   2,   3,   5,   7,   8,  22 ],
  [   1,   4,   5,   6,   9,  22,  25 ],
  [   3,   1,   5,   8,   6,  18,  15,  26 ],
  [   3,   2,   5,   8,   7,  20,  18,  25,  26 ],
  [   5,   5,  10,  15,  15,  43,  44,  35,  39,  87 ] ]
gap> y := r[3] - 1/5*r[7];
  [ 45, 1, 5, -18/5, -18/5, -3/5, 1/5, 0, 2/5, -2/5, 1/5, -1/5, 0, 1, 1 ]
```

It is immediately seen that the norms of r[4] − r[1] − r[3] and r[5] − r[2] − r[3] are zero, so the characters r[4], r[5] can be discarded. Lemma 2.8.1 shows that r[6] − r[3] and r[10] − r[6] are characters and also that r[7] − r[3] is a character since r[3] − $\frac{5}{25}$ r[7] is not integral and hence not in Irr(G). Deleting r[4], r[5] and replacing r[6], r[7] and r[10] by the differences r[6] − r[3], r[7] − r[3] and r[10] − r[6], respectively, we obtain a new list r of characters with the following matrix of scalar products:

```
[ [   2 ],
  [   0,   2 ],
  [   0,   0,   2 ],
  [   2,   3,   3,  14 ],
  [   1,   4,   3,  14,  17 ],
  [   3,   1,   5,  13,  10,  26 ],
  [   3,   2,   5,  15,  13,  25,  26 ],
  [   3,   2,   5,  16,  17,  17,  19,  23 ] ]
```

Now the norms are sufficiently small such that one can make further conclusions (although Lemma 2.8.1 applies only to r[3] and r[8]). For instance, we can see that r[4] - r[2] - r[3] is a proper character because r[4] (of norm 14) cannot have any irreducible constituent with multiplicity ≥3, as can be seen from the scalar products with the three characters of norm two which are pair-wise "disjoint," that is have no constituent in common. Similarly we may conclude that r[2] and r[3] can be subtracted from r[5], and further reductions can be made (see also Example 2.9.4). But we will not pursue this further at this point, because we will see (in Example 2.8.12) that there is an easier way to finish off the character table of M$_{12}$. ◆

We have seen that it is in general not easy to decide whether or not a difference of two characters is a character. It appears to be useful to consider quite generally differences of characters, although some of the arguments used at the end of Example 2.7.11 are no longer valid in this more general context. But one has more freedom to decrease the norms. If one obtains a ℤ-linear combination ψ of characters which has norm one then ψ or $-\psi$ is an irreducible character, depending on whether $\psi(1)$ is positive or not. We will see that there are even cases where one can decide that a ℤ-linear combination θ of a given list of reducible characters must be a multiple of an irreducible character, which then can, of course, easily be computed by dividing θ by $\sqrt{(\theta,\theta)_G}$. So it is certainly of practical importance to consider not only characters, but also ℤ-linear combinations of characters.

Definition 2.8.3 If K is a field of characteristic zero and G is a finite group then an element of

$$\mathbb{Z}\,\mathrm{Irr}_K(G) = \{ \sum_{\chi \in \mathrm{Irr}_K(G)} a_\chi \chi \in \mathrm{cf}(G,K) \mid a_\chi \in \mathbb{Z} \}$$

is called a **generalized character** (or virtual character) of G over K.

The most important case is $K = \mathbb{C}$, in which case no reference to K will be given.

Remark 2.8.4 By Lemma 2.7.1 it is clear that $\mathbb{Z}\,\mathrm{Irr}_K(G)$ is a ring. It can also be considered as a $\mathbb{Z}$-lattice in the sense of the following definition (cf. [29]).

Definition 2.8.5 A $\mathbb{Z}$-lattice is a free $\mathbb{Z}$-module L of finite rank together with a positive definite quadratic form N on $\mathbb{R} \otimes L$. We write

$$(v, w) = \frac{1}{2}(\mathrm{N}(v + w) - \mathrm{N}(v) - \mathrm{N}(w)) \quad \text{for} \quad v, w \in \mathbb{R} \otimes L.$$

We embed L into $\mathbb{R} \otimes L$ and write $av = a \otimes v$ for $a \in \mathbb{R}$ and $v \in L$. In the case where L is a $\mathbb{Z}$-lattice of generalized characters we define

$$\mathrm{N}\Big(\sum_{\chi \in \mathrm{Irr}_K(G)} a_\chi \chi \Big) = \sum_{\chi \in \mathrm{Irr}_K(G)} a_\chi^2 \cdot (\chi, \chi)_G \quad \text{for} \quad a_\chi \in \mathbb{R}.$$

Any set of (generalized) characters of G generates a sublattice L of

$$\mathbb{Z}\,\mathrm{Irr}(G) := \langle\, \chi \mid \chi \in \mathrm{Irr}(G)\, \rangle_{\mathbb{Z}} := \Big\{ \sum_{\chi \in \mathrm{Irr}(G)} a_\chi \chi \mid a_\chi \in \mathbb{Z} \Big\}$$

and in view of the above remarks it is useful to find a shortest vector in $L \setminus \{0\}$, that is a $\psi \in L \setminus \{0\}$ with $\mathrm{N}(\psi)$ being minimal. But this is computationally a very difficult problem, except for lattices of rank two.

Reduction algorithm
Assume that $L = \langle v, w \rangle$ is a $\mathbb{Z}$-lattice of rank two and that $\mathrm{N}(v) \leq \mathrm{N}(w)$.
Input: (v, w)
Output: (v', w') with $L = \langle v', w' \rangle$ and $\mathrm{N}(v') = \min\{\mathrm{N}(x) \mid x \in L \setminus \{0\}\}$ and
$$\mathrm{N}(w') = \min\{\mathrm{N}(x) \mid x \in L \setminus \mathbb{Z}v'\}$$

    while $|(v, w)| > \frac{1}{2}\mathrm{N}(v)$ **do**
        choose $k \in \mathbb{Z}$ such that $-\frac{1}{2}\mathrm{N}(v) < (v, w - kv) \leq \frac{1}{2}\mathrm{N}(v)$
        and put $w := w - kv$
        if $\mathrm{N}(v) \leq \mathrm{N}(w)$ **then return** (v, w)
            else interchange v and w
      end if
    end while

It is not hard to see that this algorithm is indeed correct (see e.g. [28], p. 68). The output is usually called a **reduced basis** of L, or a **reduced pair**. For lattices L of larger rank it looks tempting to use the same idea in order to obtain a generating set of L which is pair-wise reduced. The following straightforward GAP program does exactly this for a list **red** of generalized characters of a character table **t**:

```
weakreduce := function ( t, red )
  local  r, normsum, a, b, i, j, k;
  r := ShallowCopy( red );
  repeat
    normsum := Sum( List( r, Norm ) );
    SortParallel( List( r, Norm ), r );
    for i  in [ 1 .. Length( r ) ]  do
      a := Norm( r[i] );
      if a <> 0  then
        for j  in [ i + 1 .. Length( r ) ]  do
          b := ScalarProduct( t, r[i], r[j] );
          k := BestQuoInt( b, a );
              if b - k * a  = - a/2  then  k := k - 1;  fi;
              r[j] := r[j] - k * r[i];
          od;
      fi;
    od;
    r := Filtered( r, x -> Norm( x ) <> 0 );
  until normsum = Sum( List( r, Norm ) );
  return r;
end;
```

But observe that this algorithm may fail badly in some examples to produce generalized characters of smallest norm in the lattice spanned by red, as can be seen in the following example.

Example 2.8.6 Assume that we have four vectors $v_1, \ldots, v_r$ with matrix of scalar products given by

$$
M = \begin{bmatrix} 2 & 1 & 1 & 0 \\ 1 & 2 & 0 & 1 \\ 1 & 0 & 2 & 1 \\ 0 & 1 & 1 & 3 \end{bmatrix}.
$$

Then $(v_1, \ldots, v_r)$ is pair-wise reduced. But it is easy to see that the vector $v_4 - v_3 - v_2 + v_1$ has norm one. ◆

Also it should be noted that the output of the above algorithm is not necessarily a $\mathbb{Z}$-basis of the $\mathbb{Z}$-span of red.

A much better algorithm to compute a basis of short vectors in a lattice is the famous LLL-algorithm by Lenstra, Lenstra and Lovász ([111]), which had been used by these authors to show that polynomial factorization over the integers can be computed in polynomial time. The algorithm has meanwhile found numerous other applications. We describe this algorithm in its simplest form, referring to [29] for more details.

We use the following notation familiar from the Gram–Schmidt orthogonalization process. If $v_1, \ldots, v_r$ are linearly independent vectors in a lattice L define recursively

$$v_i^* = v_i - \sum_{j=1}^{i-1} \mu_{i,j} v_j^* \qquad (1 \le i \le r), \tag{2.16}$$

where

$$\mu_{i,j} = \begin{cases} \frac{(v_i, v_j^*)}{N(v_j^*)} & \text{if} \quad v_j^* \ne 0 \\ 0 & \text{if} \quad v_j^* = 0 \end{cases} \qquad (1 \le j < i \le r). \tag{2.17}$$

Thus $v_1^* = v_1$, and for $i > 1$ the vector v_i^* is the orthogonal projection of v_i onto the orthogonal complement of $\langle v_1, \ldots, v_{i-1} \rangle_\mathbb{R}$ in $\langle v_1, \ldots, v_i \rangle_\mathbb{R}$ and $v_i^* = 0$ if $v_i \in \langle v_1, \ldots, v_{i-1} \rangle_\mathbb{R}$. The non-zero v_i^* form an orthogonal basis for $\langle v_1, \ldots, v_r \rangle_\mathbb{R}$. Of course, if one starts with linearly independent vectors v_i then no v_i^* will be zero. We will use the notation

$$\text{GS}(v_1, \ldots, v_r) := \left((v_1^*, \ldots, v_r^*) \,, \, [\mu_{i,j}]_{1 \le j < i \le r} \right).$$

Definition 2.8.7 A $\mathbb{Z}$-basis $(v_1, \ldots, v_r)$ of a lattice L is called **LLL-reduced** if

$$|\mu_{i,j}| \le \frac{1}{2} \qquad \text{for} \quad 1 \le j < i \le r$$

and

$$N(v_i^* + \mu_{i,i-1} v_{i-1}^*) \ge \frac{3}{4} N(v_{i-1}^*) \qquad \text{for} \quad 1 < i \le r.$$

Theorem 2.8.8 *Let* $(v_1, \ldots, v_r)$ *be an LLL-reduced* $\mathbb{Z}$*-basis of a lattice* L. *Then for every* $x \in L \setminus \{0\}$ *one has*

$$N(v_1) \le 2^{r-1} N(x);$$

more generally, if $x_1, \ldots, x_s \in L$ *are linearly independent, then*

$$N(v_j) \le 2^{r-1} \max(N(x_1), \ldots, N(x_s)) \quad \text{for} \quad 1 \le j \le s.$$

Proof. See [29], p. 85. $\qquad\qquad\qquad\qquad\qquad\qquad\qquad\qquad\qquad\qquad\qquad$ $\square$

Thus, although an LLL-reduced basis of L does not necessarily contain a shortest non-zero vector of L, the norm of its first vector v_1 differs from the norm of a shortest non-zero vector just by a fixed factor in the worst case. Most importantly, there is an efficient algorithm to compute an LLL-reduced basis. Also, in many examples this algorithm produces vectors which are much shorter than those guaranteed in Theorem 2.8.8. We formulate this algorithm in a simple form, which is computationally far from being optimal.

LLL-algorithm
Input: $v_1, \ldots, v_r$ spanning a lattice L of rank r

Output: An LLL-reduced basis $(w_1, \ldots, w_r)$ of L
(1) $((v_1^*, \ldots, v_r^*)\,,\,[\mu_{i,j}]_{1 \le j < i \le r}) := \mathrm{GS}(v_1, \ldots, v_r)$ (see (2.16) and (2.17))
     Put $i := 2$.
(2)   **while** $i \le r$ **do**
(3)   **for** $j = i - 1, \ldots, 1$ **do**
                     $a := \lfloor \mu_{i,j} + \tfrac{1}{2} \rfloor\;;\quad v_i := v_i - a \cdot v_j\;;\quad \mu_{i,j} := \mu_{i,j} - a$
                     for $k = 1, \ldots, j - 1$ **do**
                         $\mu_{i,k} := \mu_{i,k} - a \cdot \mu_{j,k}$
                     end for
                 end for
(4)   **if** $i > 1$ **and** $\mathrm{N}(v_i^* + \mu_{i,i-1} v_{i-1}^*) < \tfrac{3}{4}\mathrm{N}(v_{i-1}^*)$ **then**
                     interchange v_i and v_{i-1} (and update:)
(4a) $((v_1^*, \ldots, v_r^*)\,,\,[\mu_{i,j}]_{1 \le j < i \le r}) := \mathrm{GS}(v_1, \ldots, v_r)$
                     $i := i - 1$
                 else $i := i + 1$
                 end if
             end while
(5) **return** $(v_1, \ldots, v_r)$

Proof. Observe that whenever step (3) of the algorithm has been carried out, one has achieved $|\mu_{k,j}| < \tfrac{1}{2}$ for $1 \le j < k \le i$. Also none of the $v_1^*, \ldots, v_r^*$ has been changed during this step. It follows from this and the condition in step (4) that the algorithm produces an LLL-reduced basis of L if it terminates. But it is not obvious at all that it does terminate. To see this, it is essential to analyze how the value

$$D = \prod_{k=1}^{r} d_k \quad \text{with} \quad d_k = \prod_{i=1}^{k} \mathrm{N}(v_i^*)$$

changes in the course of the algorithm. In fact, it does not change at all in step (3), and it changes in step (4) only if the condition is fulfilled and v_i and v_{i-1} are interchanged. We look more closely at what is happening in step (4a) in this case and see that $(v_1^*, \ldots, v_r^*)$ is replaced by $(v_1^*, \ldots, v_{i-2}^*, w_{i-1}^*, w_i^*, v_{i+1}^*, \ldots, v_r^*)$, with $w_{i-1}^* = v_i^* + \mu_{i,i-1} v_{i-1}^*$ and

$$w_i^* = v_{i-1}^* - \mu_{i,i-1} \frac{\mathrm{N}(v_{i-1}^*)}{\mathrm{N}(w_{i-1}^*)} w_{i-1}^*.$$

Hence

$$\mathrm{N}(w_{i-1}^*) = \mathrm{N}(v_i^* + \mu_{i,i-1} v_{i-1}^*) < \frac{3}{4}\mathrm{N}(v_{i-1}^*),$$

and an easy computation shows that

$$\mathrm{N}(w_{i-1}^*)\mathrm{N}(w_i^*) = \mathrm{N}(v_{i-1}^*)\mathrm{N}(v_i^*).$$

Thus all d_k remain unchanged except for d_{i-1}, which decreases by a factor $< \tfrac{3}{4}$. Hence the algorithm must terminate. $\square$

Of course the algorithm as formulated above can be improved in order to avoid a lot of unnecessary calculations, for instance in step (1) and step (4a). It can be reformulated so that the computation and storage of the v_i^* are avoided. Also the algorithm can be modified in such a way that instead of the $v_1, \ldots, v_r$ the matrix $[(v_i, v_j)]$ is entered and the base-change matrix is given as output.

If one analyzes the algorithm closely one can see that it works as formulated even if $v_1, \ldots, v_r$ are not a basis of L but just a generating set. If the rank of L is $s < r$ then the output of the algorithm is $(0, \ldots, 0, w_{r-s+1}, \ldots, w_r)$, with $(w_{r-s+1}, \ldots, w_r)$ being an LLL-reduced basis of L. Observe that in the LLL-algorithm the swap-condition step (4) is fulfilled if $v_i^* = 0$ and $v_{i-1}^* \neq 0$, that is if v_i is a linear combination of $v_1, \ldots, v_{i-1}$. Of course, the proof given has to be modified for this case.

Example 2.8.9 Applying the LLL-algorithm to the vectors of Example 2.8.6 we get a new basis with Gram-matrix

$$M = \begin{bmatrix} 2 & -1 & -1 & 0 \\ -1 & 2 & 0 & 0 \\ -1 & 0 & 2 & 0 \\ 0 & 0 & 0 & 1 \end{bmatrix}.$$

Hence a vector of norm one has been found.   ◆

We return to the case of generalized characters. If the LLL-algorithm is applied to a list of reducible characters, the output will be a list of generalized characters, usually of small norm. We have total success if a generalized character ψ of norm one is produced, because then $\pm\psi$ is an irreducible character as explained earlier. But even if this is not the case, and even if the $\mathbb{Z}$-span of the given reducible characters does not contain an irreducible character, we might be able to obtain an irreducible character from the output of the LLL-algorithm in favorable cases. The following lemma describes such a situation.

Lemma 2.8.10 *Assume that $\psi_1, \ldots, \psi_n \in \mathbb{Z}\operatorname{Irr}(G)$ have the following (Gram-) matrices $M_n = [(\psi_i, \psi_j)_G]_{1 \leq i,j \leq n}$ of scalar products:*

$$M_4 = \begin{bmatrix} 2 & -1 & 0 & 0 \\ -1 & 2 & -1 & -1 \\ 0 & -1 & 2 & 0 \\ 0 & -1 & 0 & 2 \end{bmatrix}, \quad or \quad M_5 = \begin{bmatrix} 2 & -1 & 0 & 0 & 0 \\ -1 & 2 & -1 & -1 & 0 \\ 0 & -1 & 2 & 0 & 0 \\ 0 & -1 & 0 & 2 & -1 \\ 0 & 0 & 0 & -1 & 2 \end{bmatrix},$$

that is the generalized characters ψ_i span a $\mathbb{Z}$-lattice of type D_4 or D_5. Then

(a) *if $n = 5$, then $\pm\frac{1}{2}(\psi_1 + \psi_3) \in \operatorname{Irr}(G)$;*

(b) *if $n = 4$, then for some $(i,j) \in \{(1,3),(1,4),(3,4)\}$ one has*
     $\pm\frac{1}{2}(\psi_i + \psi_j) \in \operatorname{Irr}(G)$.

Proof.   Since the ψ_i all have norm two, they are all of the form $\pm\chi_k \pm \chi_l$ for some $\chi_k \neq \chi_l \in \operatorname{Irr}(G)$. Since ψ_1, ψ_3, ψ_4 are pair-wise orthogonal, but all have non-zero scalar product with ψ_2, it is not possible that they are all proper characters or negatives of proper characters; instead two of these three generalized characters, say ψ_i, ψ_j, must be of the form $\chi_k - \chi_l$ and $\pm(\chi_k + \chi_l)$, and hence their sum is $2\chi_k$ or $-2\chi_l$. If ψ is a generalized character of norm two with $(\psi, \chi_k - \chi_l)_G = -1$ then $(\psi, \chi_k + \chi_l)_G = \pm 1$. Thus, if $n = 5$ we must have $(i, j) = (1, 3)$. $\qquad\qquad\square$

Corollary 2.8.11 *If $n = 5$ in Lemma 2.8.10 then the irreducible constituents of $\psi_1, \ldots, \psi_5$ can be computed and thus five irreducible characters of G are obtained. If $n = 4$ and if there is just one pair $(i, j) \in \{(1, 3), (1, 4), (3, 4)\}$ for which $\pm\frac{1}{2}(\psi_i + \psi_j)$ has algebraic integral values, then likewise the irreducible constituents of $\psi_1, \ldots, \psi_4$ can be computed and four irreducible characters of G are obtained.*

Example 2.8.12 We continue with Example 2.7.11, where we constructed two irreducible characters of $G = M_{12}$ and a list `r` of reducible characters of rather large norm. We apply the LLL-algorithm to this list:

```
gap> ll := LLL( t , r );;
gap> ll.irreducibles;
[ ]
gap> ll.norms;
[ 2, 2, 2, 3, 2, 2, 2, 2, 2, 3 ]
```

The LLL-algorithm does not find vectors (generalized characters) of norm one, but a $\mathbb{Z}$-basis of the lattice generated by `r` consisting of vectors of small norm. We filter out those of norm two and form the matrix of scalar products in order to try to apply Lemma 2.8.10:

```
gap> r := Filtered( ll.remainders, x-> Norm(x) = 2 );;
gap> gram := MatScalarProducts( t, r, r ) ;; Display (gram) ;
[ [ 2,  0,  0,  0,  0,  1,  0,  0 ],
  [ 0,  2,  0,  1,  1,  0,  0,  1 ],
  [ 0,  0,  2,  0,  0,  1,  1,  0 ],
  [ 0,  1,  0,  2,  0,  0,  0,  1 ],
  [ 0,  1,  0,  0,  2,  0,  0,  1 ],
  [ 1,  0,  1,  0,  0,  2,  1,  0 ],
  [ 0,  0,  1,  0,  0,  1,  2,  0 ],
  [ 0,  1,  0,  1,  1,  0,  0,  2 ] ]
```

A few transformations are necessary in order to see that the $\mathbb{Z}$-span of `r` contains a D_4-lattice spanned by $(r[2], r[4], r[5], r[8])$:

```
gap> d := [ r[4]-r[2], r[2], r[5]-r[2], -r[8] ];;
gap> Display( MatScalarProducts( t, d, d ));
[ [  2, -1,  0,  0 ],
  [ -1,  2, -1, -1 ],
  [  0, -1,  2,  0 ],
  [  0, -1,  0,  2 ] ]
```

By Lemma 2.8.10 we know that $\pm\frac{1}{2}(\psi_i + \psi_j)$ is an irreducible character for some $(i,j) \in \{(1,3),(1,4),(3,4)\}$. In order to find out for which (i,j) this is true, we compute $\frac{1}{2}(\psi_i + \psi_j)(1)$:

```
gap> for x in [ [1,3] , [1,4] , [3,4] ] do
> Print( (d[x[1]][1] + d[x[2]][1]) / 2, " , ");
> od;
-45 , -121/2 , -99/2 ,
```

Fortunately we can rule out two of the possibilities, and hence we obtain a new irreducible character of degree 45 (and in addition three further irreducible characters according to Corollary 2.8.11). In GAP there is a program that searches for D_n-lattices in a given lattice, spanned by vectors of norm two. Thus instead of scrutinizing the matrix **gram** as above we might have used the following:

```
gap> dn := DnLattice( t , gram , r );;
gap> dn.irreducibles;
[ [ 120, 0, -8, 3, 0, 0, 0, 0, 0, 1, 0, 0, 0, -1, -1 ] ,
  [ 11, -1, 3, 2, -1, 3, -1, 1, -1, 0, 1, -1, -1, 0, 0 ] ,
  [ 45, 5, -3, 0, 3, 1, 1, 0, -1, 0, -1, -1, 0, 1, 1 ] ,
  [ 55, -5, -1, 1, 1, 3, -1, 0, 1, -1, -1, 1, 0, 0, 0 ] ]
```

These are the four irreducible characters found. We add them to our list **irr** of irreducibles and continue by reducing the reducible generalized characters **ll.remainders** found by the previous call of **LLL** and repeat the process:

```
gap> Append( irr, dn.irreducibles );
gap> red := Reduced( t, irr, ll.remainders );;
gap> ll := LLL( t , red.remainders );; ll.norms;
[ 2, 2, 2, 2, 2, 2 ]
gap> r := ll.remainders;;
gap> dn := DnLattice( t , MatScalarProducts(t,r,r), r);
rec( gram := [ [ 2 ] ],        remainders :=
    [ [ 32, 8, 0, -4, 2, 0, 0, 2, 2, 0, 0, 0, -2, -1, -1 ] ],
  irreducibles :=
    [ [ 99, -1, 3, 0, 3, -1, -1, -1, -1, 0, 1, 1, -1, 0, 0 ],
      [ 55, -5, 7, 1, 1, -1, -1, 0, 1, 1, -1, -1, 0, 0, 0 ] ),
      [ 66, 6, 2, 3, 0, -2, -2, 1, 0, -1, 0, 0, 1, 0, 0 ] ,
      [ 144, 4, 0, 0, -3, 0, 0, -1, 1, 0, 0, 0, -1, 1, 1 ] ,
      [ 176, -4, 0, -4, -1, 0, 0, 1, -1, 0, 0, 0, 1, 0, 0 ]  ]
```

This time the $\mathbb{Z}$-span of **r** contains a D_5-lattice and we obtain five new irreducible characters. Now we have found all but two irreducible characters of G and in addition the sum of the two missing characters (in **dn.remainders**). It is clear that the latter is the sum of two algebraically conjugate characters which differ only on the non-rational classes **11a** and **11b**. From the orthogonality relations we see that the missing values on these classes are $\frac{1}{2}(-1 \pm \sqrt{-11})$. We finish this example by displaying the complete character table of G with the characters sorted according to their degrees:

M12

2	6	4	6	1	2	5	5	1	2	1	3	3	1	.	.
3	3	1	1	3	2	.	.	.	1	1	.	.	.	.	.
5	1	1	.	.	.	.	.	1	.	.	.	.	1	.	.
11	1	.	.	.	.	.	.	.	.	.	.	.	.	1	1

	1a	2a	2b	3a	3b	4a	4b	5a	6a	6b	8a	8b	10a	11a	11b
2P	1a	1a	1a	3a	3b	2b	2b	5a	3b	3a	4a	4b	5a	11b	11a
3P	1a	2a	2b	1a	1a	4a	4b	5a	2a	2b	8a	8b	10a	11a	11b
5P	1a	2a	2b	3a	3b	4a	4b	1a	6a	6b	8a	8b	2a	11a	11b
11P	1a	2a	2b	3a	3b	4a	4b	5a	6a	6b	8a	8b	10a	1a	1a

X.1	1	1	1	1	1	1	1	1	1	1	1	1	1	1	1
X.2	11	-1	3	2	-1	-1	3	1	-1	.	-1	1	-1	.	.
X.3	11	-1	3	2	-1	3	-1	1	-1	.	1	-1	-1	.	.
X.4	16	4	.	-2	1	.	.	1	1	.	.	.	-1	A	/A
X.5	16	4	.	-2	1	.	.	1	1	.	.	.	-1	/A	A
X.6	45	5	-3	.	3	1	1	.	-1	.	-1	-1	.	1	1
X.7	54	6	6	.	.	2	2	-1	.	.	.	.	1	-1	-1
X.8	55	-5	7	1	1	-1	-1	.	1	1	-1	-1	.	.	.
X.9	55	-5	-1	1	1	3	-1	.	1	-1	-1	1	.	.	.
X.10	55	-5	-1	1	1	-1	3	.	1	-1	1	-1	.	.	.
X.11	66	6	2	3	.	-2	-2	1	.	-1	.	.	1	.	.
X.12	99	-1	3	.	3	-1	-1	-1	-1	.	1	1	-1	.	.
X.13	120	.	-8	3	.	.	.	.	.	1	.	.	.	-1	-1
X.14	144	4	.	.	-3	.	.	-1	1	.	.	.	-1	1	1
X.15	176	-4	.	-4	-1	.	.	1	-1	.	.	.	1	.	.

```
A = E(11)+E(11)^3+E(11)^4+E(11)^5+E(11)^9
  = (-1+ER(-11))/2 = b11
```

Since we started with a rational character, all generalized characters we produced by tensoring, symmetrizations, reductions and the LLL-algorithm are rational, and it is quite clear that there is no chance of obtaining the irreducible characters X.4 and X.5 by this method alone. We were just lucky that there were only two irrational irreducible characters, so that we could finish off the character table by using the orthogonality relations. In Section 3.2 we will study a method to find irrational characters provided that enough information about the power maps is known. There we will also show that it is possible to compute the character table of M_{12} just from the knowledge of the power maps and centralizer orders. ♦

Lemma 2.8.10 is very useful in many cases, but its underlying idea can be generalized. The problem it solves in a special case is the following. Assume that $M \in \mathbb{Z}^{n \times n}$ is a symmetric positive definite matrix. We want to find all matrices $X \in \mathbb{Z}^{n \times m}$ with $M = XX^{\mathrm{T}}$ with $m \in \mathbb{N}$. In the applications we have in mind that $M = [(\psi_i, \psi_j)_G]_{1 \leq i,j \leq n}$ is the matrix of scalar products of a linearly independent list $(\psi_1, \ldots, \psi_n)$ of generalized characters and $X' = [x'_{i,j}] =$

$[(\psi_i, \chi_j)_G]_{1 \le i \le n, \chi_j \in \mathrm{Irr}(G)}$ is the (unknown) "decomposition" matrix, satisfying

$$\psi_i = \sum_{j=1}^{m} x'_{i,j} \chi_j, \qquad (1 \le i \le n), \qquad \{\chi_1, \ldots, \chi_m\} \subseteq \mathrm{Irr}(G), \qquad (2.18)$$

which shows how the ψ_i decompose into the yet unknown irreducibles. Of course we may replace $\mathrm{Irr}(G)$ by the set of those irreducible characters which occur as a constituent in some ψ_i, then no column of X' is the 0-column. Obviously $X'(X')^{\mathrm{T}} = M$. If we know X' we can consider (2.18) as a system of linear equations for the χ_j. Even if (2.18) does not have a unique solution it might have a subsystem with non-singular matrix, in which case some of the χ_j can be uniquely determined. It is clear that X' is just defined up to the ordering of the columns, since there is no a priori ordering of $\mathrm{Irr}(G)$. Also, we may allow the $\chi_j \in \mathrm{Irr}(G)$ to be multiplied by -1, because if we obtain $-\chi_j$, we get χ_j as well, just by looking at $\chi_j(1)$.

Definition 2.8.13 Two solutions $X_1, X_2 \in \mathbb{Z}^{n \times m}$ of $M = XX^{\mathrm{T}}$ are called **equivalent** if the columns can be rearranged in such a way that they differ only by a sign.

We will assume that no column of X is zero. If we denote the columns of X by $x_1, \ldots, x_m$ we have

$$M = XX^{\mathrm{T}} = \sum_{j=1}^{m} x_j x_j^{\mathrm{T}}.$$

By assumption M is positive definite. But also every subsum is at least positive semidefinite, because if $I \subseteq \{1, \ldots, m\}$ and $y \in \mathbb{R}^{n \times 1}$ then

$$y^{\mathrm{T}}(\sum_{j \in I} x_j x_j^{\mathrm{T}})y = \sum_{j \in I}(y^{\mathrm{T}} x_j)^2 \ge 0.$$

Lemma 2.8.14 *Suppose $M \in \mathbb{Z}^{n \times n}$ be a symmetric positive definite matrix and $M = XX^{\mathrm{T}}$ with $X = [x_1, \ldots, x_m] \in \mathbb{Z}^{n \times m}$. Then*

$$x_j^{\mathrm{T}} M^{-1} x_j \le 1 \qquad \text{for all} \quad j.$$

Proof. Put $P = X^{\mathrm{T}} M^{-1} X$. Then $P^2 = P \in \mathbb{R}^{m \times m}$ and $XP = X$. Thus $\mathrm{rk}\, P = \mathrm{rk}\, X = n$ and P has eigenvalues one and zero with multiplicities n and $m - n$, respectively. Consequently, $\mathbf{I}_n - P$ has eigenvalues zero and one with multiplicities n and $m - n$, respectively, and hence is positive semidefinite. Therefore its diagonal elements, which are $1 - x_j^{\mathrm{T}} M^{-1} x_j$, must be non-negative. $\square$

Lemma 2.8.14 provides us with an opportunity to compute the finitely many integral vectors $x_j \in \mathbb{Z}^{n \times 1}$ which may occur as column vectors in a solution $X \in \mathbb{Z}^{n \times m}$ of $XX^{\mathrm{T}} = M$. We will not go into the details of how this can be done efficiently, but formulate only a straightforward algorithm to find all such solutions. For a better algorithm and more details we refer to [147].

Additive-decomposition algorithm
Input: A positive definite matrix $M \in \mathbb{Z}^{n \times n}$ with $n \in \mathbb{N}$.
Output: A system of representatives of equivalence classes of solutions
$\qquad X \in \mathbb{Z}^{n \times m} \, (m \in \mathbb{N})$ without 0-columns of $M = XX^{\mathrm{T}}$.
$C_1 := (x \in \mathbb{Z}^{n \times 1} \mid 0 < x^{\mathrm{T}} M^{-1} x \le 1$, first non-zero entry of $x > 0)$
$Sol := \emptyset$; $k := 1$
$\qquad$ **while** $C_1 \ne ()$ **do**
$\qquad\qquad$ **if** $C_k \ne ()$ **then** $x_k := C_k[1]$ (first element in C_k)
$\qquad\qquad\qquad$ **if** $M = \sum_{i=1}^{k} x_i x_i^{\mathrm{T}}$ **then** $Sol := Sol \cup \{[x_1, \dots, x_k]\}$;
$\qquad\qquad\qquad\qquad C_k := C_k \setminus \{x_k\}$
$\qquad\qquad\qquad$ **else if** $M - \sum_{i=1}^{k} x_i x_i^{\mathrm{T}}$ is positive semidefinite **then**
$\qquad\qquad\qquad\qquad C_{k+1} := C_k$; $k := k + 1$
$\qquad\qquad\qquad$ **else** $C_k := C_k \setminus \{x_k\}$
$\qquad\qquad\qquad$ **end if**
$\qquad\qquad$ **else**
$\qquad\qquad\qquad C_{k-1} := C_{k-1} \setminus \{x_{k-1}\}$; $k := k - 1$
$\qquad\qquad$ **end if**
$\qquad$ **end while**
return *Sol*

Example 2.8.15 Let
$$M = \begin{bmatrix} 2 & 1 & 1 \\ 1 & 2 & 1 \\ 1 & 1 & 2 \end{bmatrix}.$$
We find
$$M^{-1} = A^{\mathrm{T}} \begin{bmatrix} \frac{3}{4} & 0 & 0 \\ 0 & \frac{2}{3} & 0 \\ 0 & 0 & \frac{1}{2} \end{bmatrix} A \quad \text{with} \quad A = \begin{bmatrix} 1 & -\frac{1}{3} & -\frac{1}{2} \\ 0 & 1 & -\frac{1}{2} \\ 0 & 0 & 1 \end{bmatrix}.$$
We easily see that the list of representatives C_1 of candidates for the columns of a matrix $X \in \mathbb{Z}^{n \times m}$ of $XX^{\mathrm{T}} = M$ is as follows:
$$C_1 = (\begin{bmatrix} 1 \\ 1 \\ 0 \end{bmatrix}, \begin{bmatrix} 1 \\ 0 \\ 1 \end{bmatrix}, \begin{bmatrix} 0 \\ 1 \\ 1 \end{bmatrix}, \begin{bmatrix} 1 \\ 1 \\ 1 \end{bmatrix}, \begin{bmatrix} 1 \\ 0 \\ 0 \end{bmatrix}, \begin{bmatrix} 0 \\ 1 \\ 0 \end{bmatrix}, \begin{bmatrix} 0 \\ 0 \\ 1 \end{bmatrix}).$$
The algorithm produces the following solutions:
$$X_1 = \begin{bmatrix} 1 & 1 & 0 \\ 1 & 0 & 1 \\ 0 & 1 & 1 \end{bmatrix}, \ X_2 = \begin{bmatrix} 1 & 1 & 0 & 0 \\ 1 & 0 & 1 & 0 \\ 1 & 0 & 0 & 1 \end{bmatrix}.$$

If M is the matrix of scalar products of some (generalized) characters ψ_1, ψ_2, ψ_3, and it is known that they are $\mathbb{Z}$-linear combinations of just three unknown irreducibles (because perhaps all but three irreducibles are already known), then one can dismiss solution X_2. Since X_1 is non-singular we get

$$\chi = \frac{1}{2}(\psi_1 - \psi_2 + \psi_3) \in \pm\mathrm{Irr}(G) := \mathrm{Irr}(G) \cup \{-\chi \mid \chi \in \mathrm{Irr}(G)\},$$

and likewise $\psi_1 - \chi \in \pm\mathrm{Irr}(G)$ and $\psi_3 - \chi \in \pm\mathrm{Irr}(G)$. On the other hand, if X_2 cannot be excluded we are not able to obtain new irreducible characters. The system of linear equations corresponding to X_2 has no subsystem having a unique solution.   ♦

In practice the number $|C_1|$ of candidates and the number of solutions $X \in \mathbb{Z}^{n \times k}$ can be very large and the algorithm can be applied only if the size of the matrix M and the size of its entries are not too large. There are some restrictions which can be made in our applications. One may restrict the vectors of C_1 to elements of $\mathbb{N}_0^{n \times 1}$ in the case that one is dealing only with ordinary instead of generalized characters. In Example 2.8.15 this was no restriction at all. Often it is more important to be able to give a bound for k (as in Example 2.8.15). This is usually the number of unknown irreducible characters. Even if the number of solutions to be considered is small, the method may fail because the system of linear equations for the unknown irreducibles may have no subsystem with a unique solution. Nevertheless we will give some rather convincing examples of the application of the method, mainly in Section 3.2. We finish this section by giving an extremely simple GAP example.

Example 2.8.16 We compute again the characters of S_5 using the (constituents of) the natural permutation character and the sign character (compare with Example 2.1.23):

```
gap> t := CharacterTable("S5");; irr := Irr(t){[1,2,5]};;
gap> Display( t, rec(chars:=irr, centralizers:=false, powermap:=false) );
A5.2
        1a 2a 3a 5a 2b 4a 6a

Y.1      1  1  1  1  1  1  1
Y.2      1  1  1  1 -1 -1 -1
Y.3      4  .  1 -1 -2  .  1
gap> r := Tensored( irr, irr );; Append( r, Tensored(irr,r) );
gap> red := Reduced( irr, r );; Length( red.irreducibles );
1
gap> Append(irr,red.irreducibles); r := red.remainders;; l:=LLL(t,r);;
gap> r := l.remainders;;
gap> M := MatScalarProducts( t, r, r );; Display( M );
[ [  2,  1,  1 ],
  [  1,  2,  1 ],
  [  1,  1,  2 ] ]
```

So we have arrived at the Gram-matrix of Example 2.8.15. Also we know that there are exactly three irreducible characters missing, because Y.4 = Y.2 ·

Y.3 was already found by reducing **r**. The additive-decomposition algorithm is implemented in GAP under the name `OrthogonalEmbeddings`. It returns for a given positive definite matrix M a record `rec` with component `vectors` $= (x_i^{\mathrm{T}} \mid x_i \in C_1)$, with C_1 as in the algorithm and component `solutions` a list of length $|Sol|$ indicating which columns one has to choose for the ith solution. Thus the output Sol in the formulation of the algorithm is:

```
       List( rec.solutions , x -> TransposedMat(rec.vectors{x}) )
gap> oe := OrthogonalEmbeddings( M );
rec( vectors := [ [ 1, 1, 0 ], [ 1, 0, 1 ], [ 0, 1, 1 ], [ 1, 1, 1 ],
        [ 1, 0, 0 ], [ 0, 1, 0 ], [ 0, 0, 1 ] ],
   solutions := [ [ 1, 2, 3 ], [ 4, 5, 6, 7 ] ] )
```

If we had typed `OrthogonalEmbeddings( M , 3 );`, indicating the maximal k in our above discussion, only the first solution would have been given:

```
gap> x_1 := oe.vectors{ oe.solutions[1] } ;; Display(x_1) ;
[ [  1,   1,   0 ],
  [  1,   0,   1 ],
  [  0,   1,   1 ] ]
gap> ch := TransposedMat( x_1^-1 ) * r;
[  [ 6, -2, 0, 1, 0, 0, 0 ] ,
   [ 5, 1, -1, 0, -1, 1, -1 ] ),
   [ 5, 1, -1, 0, 1, -1, 1 ] ) ] ]
```

These are the three missing irreducible characters. The procedure could have been abbreviated by using instead of `OrthogonalEmbeddings( m )` the following:

```
gap> OrthogonalEmbeddingsSpecialDimension ( t, r, M, 3);
rec(
  irreducibles := [ [ 5, 1, -1, 0, -1, 1, -1 ],    [ 5, 1, -1, 0, 1, -1, 1 ] ,
      [ 6, -2, 0, 1, 0, 0, 0 ] ) ], remainders := [ ] )
```

◆

Exercises

Exercise 2.8.1 Compute the character table of the symmetric group S_{12} starting from the natural permutation character and the trivial and the sign character. Use only tensor products, symmetrizations up to degree three and the GAP commands `Reduce` and `LLL`. Try to do the same without using the LLL-algorithm.

Exercise 2.8.2 Let

$$M := \begin{bmatrix} 4 & 6 & 5 & 2 & 2 \\ 6 & 13 & 7 & 4 & 4 \\ 5 & 7 & 11 & 2 & 0 \\ 2 & 4 & 2 & 8 & 4 \\ 2 & 4 & 0 & 4 & 8 \end{bmatrix}.$$

Show that there are, up to equivalence, just two matrices (without 0-columns) $X \in \mathbb{Z}^{5 \times n}$ with $X X^{\mathrm{T}} = M$ for $n \leq 6$ but only one in $\mathbb{Z}_{\geq 0}^{5 \times n}$. Use GAP to find the number of solutions (up to equivalence) in $\mathbb{Z}_{\geq 0}^{5 \times n}$ for arbitrary n.

2.9   Invariant bilinear forms and the Schur index

In this section we assume that G is a finite group and that K is a field.

Applying Lemma 1.1.40 to a simple module V and its dual, one obtains

$$V \cong V^\star \qquad \text{if and only if} \qquad \mathrm{Inv}_G(V \otimes_K V) \neq 0.$$

Lemma 2.9.1 *Let* V, W *be* KG-modules *with characters* χ_V, χ_W *and let* $\mathrm{char}\, K \nmid |G|$. *Then*

(a) $\dim_K \mathrm{Hom}_{KG}(V, W) = (\chi_V, \chi_W)_G$ *in* K;

(b) $\dim_K \mathrm{Inv}_G(V \otimes_K V^\star) = (\chi_V, \chi_V)_G$ *in* K.

(c) *If in addition* V *is absolutely irreducible then*

$$(\chi_V^2, 1)_G = \begin{cases} 1 & \text{if } V \cong_{KG} V^\star, \\ 0 & \text{if } V \not\cong_{KG} V^\star. \end{cases}$$

Proof. (a) Using Lemma 1.1.40 and Exercise 2.1.4 we get

$$\begin{aligned} \dim_K \mathrm{Hom}_{KG}(V, W) &= \dim_K \mathrm{Inv}_G(\mathrm{Hom}_K(V, W)) \\ &= \dim_K \mathrm{Inv}_G(V^\star \otimes_K W) \\ &= (\chi_{V^\star} \cdot \chi_W, 1)_G = (\chi_V, \chi_W)_G. \end{aligned}$$

(b) follows from (a), setting $W = V^\star$ by Lemma 1.1.40, since $(\chi_V, \chi_V)_G = (\chi_{V^\star}, \chi_{V^\star})_G$.

As for part (c), observe that $(\chi_V^2, 1)_G = (\chi_V, \chi_{V^\star})_G$. Since V and $V^\star$ are absolutely simple, this scalar product is one or zero depending on V and $V^\star$ being isomorphic or not. $\qquad\square$

If $V \cong V^\star$ is absolutely simple, and in addition $\mathrm{char}\, K \neq 2$, then by the above

$$1 = (\chi_V^2, 1) = (\chi_V^{[2]}, 1)_G + (\chi_V^{[1^2]}, 1)_G.$$

So exactly one of the two summands on the right hand side is one. Observe

$$(\chi^{[2]}, 1)_G = \frac{1}{|G|} \sum_{g \in G} \frac{\chi(g)^2 + \chi(g^2)}{2} = \frac{1}{2}\left[(\chi^2, 1)_G + \frac{1}{|G|} \sum_{g \in G} \chi(g^2) \right].$$

Definition 2.9.2 Let $\mathrm{char}\, K \nmid |G|$ and let $\varphi \in \mathrm{cf}(G, K)$ be a class function. Then

$$\nu_k(\varphi) = \frac{1}{|G|} \sum_{g \in G} \varphi(g^k) \qquad \text{for} \qquad 2 \leq k \in \mathbb{N}$$

is called the kth **Frobenius–Schur indicator** of φ.

Obviously, $\nu_k \colon \mathrm{cf}(G, K) \to K$ is a K-linear map. From the above discussion we conclude the following.

Theorem 2.9.3 *If K is a field with $2 \neq \mathrm{char}\, K \nmid |G|$ and V is an absolutely simple KG-module with character $\chi = \chi_V \in \mathrm{Irr}_K(G)$, then $\nu_2(\chi) \in \{0, 1, -1\}$ and*

(a) $\nu_2(\chi) = 0$ *if and only if* $V \not\cong V^\star$;

(b) $\nu_2(\chi) = 1$ *if and only if* $0 \neq \mathrm{Inv}_G(V \otimes_K V) \leq V^{[2]}$;

(c) $\nu_2(\chi) = -1$ *if and only if* $0 \neq \mathrm{Inv}_G(V \otimes_K V) \leq V^{[1^2]}$.

Of course, in order to compute the kth Frobenius–Schur indicator one needs to know the kth power map. The following trivial observation is sometimes useful for analyzing reducible characters.

Example 2.9.4 Continuing with Example 2.8.2, we replace $\mathtt{r[4]}$ by the character $\mathtt{r[4]}$ - $\mathtt{r[2]}$ - $\mathtt{r[3]}$ and $\mathtt{r[5]}$ by $\mathtt{r[5]}$ - $\mathtt{r[2]}$ - $\mathtt{r[3]}$ and get the following matrix of scalar products:

$$[\,(\mathtt{r[i]}, \mathtt{r[j]})_G\,]_{1 \leq i,j \leq 5} = \begin{bmatrix} 2 & . & . & 2 & 1 \\ . & 2 & . & 1 & 2 \\ . & . & 2 & 1 & 1 \\ 2 & 1 & 1 & 6 & 5 \\ 1 & 2 & 1 & 5 & 7 \end{bmatrix}.$$

Using the GAP command `Indicator(t,r[1..5],2)` we find that the second Frobenius–Schur indicators are $2, 2, 2, 6, 5$. Thus $\nu_2(\mathtt{r[4]}) = 6 = \mathrm{N}(\mathtt{r[4]})$. By Theorem 2.9.3 this shows that $\mathtt{r[4]}$ is multiplicity-free, that is $|(\mathtt{r[4]}, \chi)_G| \leq 1$ for all $\chi \in \mathrm{Irr}(G)$. Hence $\mathtt{r[4]}$ - $\mathtt{r[1]}$ must be a proper character (of norm four). ◆

We give another interpretation for the cases (b) and (c) of Theorem 2.9.3. We recall from Lemma 1.1.42 that the K-vector space $\mathrm{Bifo}_K V$ of bilinear forms $\Phi \colon V \times V \to K$ is isomorphic to $(V \otimes_K V)^\star$. Hence, if $\mathrm{char}\, K \neq 2$, we obtain from Theorem 2.7.4

$$\mathrm{Bifo}_K V \cong_{KG} (V^{[1^2]} \oplus V^{[2]})^\star \cong_{KG} (V^{[2]})^\circ \oplus (V^{[1^2]})^\circ.$$

(cf. Theorem 1.1.34). It is easily seen that $\Phi \in \mathrm{Bifo}_K V$ is mapped under the composed isomorphism onto an element of $(V^{[2]})^\circ$ (resp. $(V^{[1^2]})^\circ$) if and only if Φ is *symmetric* (resp. *anti-symmetric* also called *symplectic*). The space $\mathrm{Inv}_G(V \otimes_K V)$ of G-invariant bilinear forms is isomorphic to $\mathrm{Hom}_{KG}(V, V^\star)$ (Lemma 1.1.42). From this we get the following corollary.

Corollary 2.9.5 *If V is an absolutely simple selfdual KG-module with character χ, where $\mathrm{char}\, K \nmid |G|$, then there is up to a scalar multiple exactly one G-invariant bilinear form $\Phi \neq 0$ on V. If, furthermore, $\mathrm{char}\, K \neq 2$ then this bilinear form Φ is* *symmetric if and only if $\nu_2(\chi) = 1$,*

anti-symmetric if and only if $\nu_2(\chi) = -1$.

In the case of complex characters χ we have another characterization for $\nu_2(\chi)$ as follows.

Theorem 2.9.6 *If $\chi \in \mathrm{Irr}(G)$ then $\nu_2(\chi) = 1$ if and only if χ is a character of a real representation.*

Proof. (a) Assume that χ is the character of the real matrix representation $D \colon G \to \mathrm{GL}_n(\mathbb{R})$. Then $0 \neq \sum_{g \in G} D(g)D(g)^{\mathrm{T}}$ is the matrix of a symmetric G-invariant bilinear form on the corresponding representation space. From Corollary 2.9.5 we conclude that $\nu_2(\chi) = 1$.

(b) Conversely, let $\nu_2(\chi) = 1$ and let D be a unitary matrix representation of G with character χ. Since χ is a real character there is a $P \in \mathrm{GL}_n(\mathbb{C})$ such that $P^{-1}D(g)P = \overline{D(g)} = (D(g)^{\mathrm{T}})^{-1}$ for all $g \in G$. So $D(g)PD(g)^{\mathrm{T}} = P$ for all $g \in G$, and by Corollary 2.9.5 we conclude that $P = P^{\mathrm{T}}$. Hence

$$\overline{P}^{-1}P^{-1}D(g)P\overline{P} = D(g) \qquad \text{for all} \qquad g \in G.$$

By Schur's lemma $P\overline{P} = \alpha E_n$ for some $\alpha > 0$. Hence $Q = \frac{1}{\sqrt{\alpha}}P$ is a unitary matrix, that is $Q \in \mathrm{U}_n(\mathbb{C})$. By the spectral theorem there is a $U \in \mathrm{U}_n(\mathbb{C})$ with

$$U^{-1}QU = A = \begin{bmatrix} a_1 & & 0 \\ & \ddots & \\ 0 & & a_n \end{bmatrix} = \begin{bmatrix} t_1^2 & & 0 \\ & \ddots & \\ 0 & & t_n^2 \end{bmatrix} = T^2,$$

with a diagonal matrix $T = \mathrm{diag}(t_1, \dots, t_n)$, where we can assume that $a_i = a_j$ if and only if $t_i = t_j$, so that any matrix commuting with A also commutes with T. We get

$$A = A^{\mathrm{T}} = (U^{-1}QU)^{\mathrm{T}} = \overline{U}^{-1}UAU^{-1}\overline{U}.$$

Hence $U^{-1}\overline{U}$ also commutes with T and we get a matrix S which will eventually transform our representation to a real one by defining

$$S = UTU^{-1} = \overline{U}\,T\overline{U}^{-1}.$$

We compute

$$S\overline{S}^{-1} = UTU^{-1}U\overline{T}^{-1}U^{-1} = UAU^{-1} = Q,$$

so

$$\overline{S}S^{-1}D(g)S\overline{S}^{-1} = \overline{D(g)} \qquad \text{for all} \qquad g \in G.$$

Hence $S^{-1}D(g)S = \overline{S^{-1}D(g)S}$ is real for all $g \in G$. $\qquad\square$

Example 2.9.7 Looking at the character table of the quaternion group (see Example 2.1.19) one finds $\nu_2(\chi_5) = -1$, where χ_5 is the irreducible character of degree two. Thus χ_5 is not the character of a real representation. $\blacklozenge$

Lemma 2.9.8 *For* $2 \leq k \in \mathbb{N}$ *define*

$$\theta_k(g) = |\{\, x \in G \mid x^k = g \}| \qquad \text{for} \quad g \in G.$$

Then θ *is a class function on* G *and*

$$\theta_k = \sum_{\chi \in \mathrm{Irr}(G)} \nu_k(\chi)\chi.$$

Proof. For $\chi \in \mathrm{Irr}(G)$ we find

$$(\chi, \theta_k)_G = \frac{1}{|G|} \sum_{g \in G} \chi(g)\overline{\theta}_k(g) = \frac{1}{|G|} \sum_{g \in G} \sum_{\substack{x \in G \\ x^k = g}} \chi(x^k) = \frac{1}{|G|} \sum_{x \in G} \chi(x^k).$$

$\square$

We get as a specialization for $k = 2$ the following.

Theorem 2.9.9 (Frobenius and Schur) *If* G *has* t *involutions then*

$$1 + t = \sum_{\chi \in \mathrm{Irr}(G)} \nu_2(\chi)\chi(1).$$

$\square$

This result is extremely useful for computational purposes. It also has a famous application as follows.

Theorem 2.9.10 (Brauer and Fowler)

(a) *If a finite group* G *has exactly* t *involutions then there is a real conjugacy class* $\{1\} \neq C \subset G$ *with*

$$|C| \leq \left(\frac{|G| - 1}{t} \right)^2.$$

(b) *For any* $n \in \mathbb{N}$ *there are only finitely many simple groups* G *with an involution* x *with* $|\mathbf{C}_G(x)| = n$.

Proof. (a) Let $R = \{\chi \in \mathrm{Irr}(G) | 1 \neq \chi = \overline{\chi}\}$ so $r' = |R|$ is the number of real-valued irreducible non-trivial characters and, by Corollary 2.2.14, is also equal to the number of real conjugacy classes $\neq \{1\}$ of G. By Theorem 2.9.9 we have $t \leq \sum_{\chi \in R} \chi(1)$. Using the well-known Schwarz inequality we conclude that

$$t^2 \leq \left(\sum_{\chi \in R} 1 \cdot \chi(1) \right)^2 \leq |R| \sum_{\chi \in R} \chi(1)^2 \leq r'(|G| - 1).$$

Hence

$$|G| - 1 \leq r'(\frac{|G| - 1}{t})^2.$$

So there must be at least one real conjugacy class $C \neq \{1\}$ of G with $|C| \leq (\frac{|G|-1}{t})^2$.

(b) If x is an involution in G with $|\mathbf{C}_G(x)| = n$, then there are at least $|G|/n$ involutions in G. By (a) there is an element $g \in G \setminus \{1\}$ such that $|g^G| = [G : \mathbf{C}_G(g)] < n^2$. If G is simple and non-abelian then G acts faithfully on the cosets of $\mathbf{C}_G(g)$, and hence is isomorphic to a subgroup of S_{n^2-1} (being simple, even of A_{n^2-1}) of which there are only finitely many. $\square$

Of course the bound given in the proof is very rough. For instance, for $n = 4$ it yields that G can be embedded in A_{15}, whereas we will show in the next example that $G \cong A_5$.

Example 2.9.11 Let G be a finite simple (or, more generally, perfect) group containing an involution $h \in G$ with $|\mathbf{C}_G(h)| = 4$. We will show that $G \cong A_5$. The orthogonality relations imply that there are exactly four $\chi \in \mathrm{Irr}(G)$ with $\chi(h) \neq 0$, and since $\sum_{\chi \in \mathrm{Irr}(G)} \chi(1)\chi(h) = 0$ we may choose our notation so that $\chi_1 = 1_G$, $\chi_2(h) = 1$, $\chi_3(h) = \epsilon = \pm 1$ and $\chi_4(h) = -1$. We have $1 + \chi_2(1) + \epsilon\chi_3(1) = \chi_4(1)$. Also we order the conjugacy classes so that $C_2 = h^G$ and compute the class multiplication coefficients:

$$\alpha_{2,2}^2 = |H_h| \quad \text{with} \quad H_h = \{(x,y) \mid x,y \in C_2 \,, xy = h\}.$$

If x, y are involutions such that xy has order two then they generate an abelian group ($\cong C_2 \times C_2$). Thus, if $(x, y) \in H_h$ then $x \neq y \in \mathbf{C}_G(h) \setminus \{h\}$. It follows under our hypothesis that $\alpha_{2,2}^2 \in \{0, 2\}$. On the other hand Theorem 2.5.9 yields

$$\alpha_{2,2}^2 = \frac{|C_2|^2}{|G|} \cdot c \quad \text{with} \quad c = 1 + \frac{1}{\chi_2(1)} + \frac{\epsilon}{\chi_3(1)} - \frac{1}{\chi_4(1)}.$$

Since $|C_2| = \frac{1}{4}|G|$ we get $|G| = 16c^{-1}\alpha_{2,2}^2$. It follows that $\alpha_{2,2}^2 = 2$, in particular $\mathbf{C}_G(h) \cong C_2 \times C_2$. The congruence relations (Lemma 2.2.2) imply that $\chi_i(1)$ is odd for $i = 2, 3, 4$ and $\neq 1$, because G is perfect. If $\epsilon = 1$ then $\chi_4(1) \geq 7$ and $c > 6/7$, which gives $|G| \leq 37 < \frac{32 \cdot 7}{6}$, which is absurd. Thus $\epsilon = -1$ and we get $c > 1/3$ and hence $|G| < 96$. Thus $|G|$ cannot have three different odd divisors, and it follows that the only possibility is $\chi_2(1) = 5$, $\chi_3(1) = \chi_4(1) = 3$, which gives $|G| = 60$.

$\blacklozenge$

Actually we have shown in the example that the centralizer of an involution in a finite simple group cannot be cyclic of order four. More generally, it can be proved that it cannot be cyclic of any order (see [88], Satz IV.2.8).

We have seen in Example 2.9.7 that the quaternion group Q_8 has a character $\chi_5 \in \mathrm{Irr}(Q_8)$ which is not the character of a real representation, although χ_5 takes only rational values; χ_5 is the character of the representation $\delta_{i,0} : Q_8 \to \mathbb{Q}(i)^{2\times 2}$ defined in Exercise 1.1.5. We will see shortly that there does exist a rational representation with character $2\chi_5$.

Remark 2.9.12 Let $K \subseteq L$ be a field extension of finite degree $[L : K] = m$. If V is an LG-module with $\dim_L V = n$ then the restricted module V_{KG} is a KG-module with $\dim_K V_{KG} = nm$. In fact, if $B = (v_1, \ldots, v_n)$ is an L-basis of V and $B_{L/K} = (\alpha_1, \ldots, \alpha_m)$ is a K-basis of L, then

$$\tilde{B} := (\alpha_1 v_1, \ldots, \alpha_m v_1 , \ldots , \alpha_1 v_n, \ldots, \alpha_m v_n)$$

is a K-basis of V_{KG}.

If V affords the matrix representation $\boldsymbol{\delta} \colon G \to \mathrm{GL}_n(L)$ with respect to the basis B and has character $\chi = \chi_V$, it is not hard to calculate the matrix representation $\tilde{\boldsymbol{\delta}} \colon G \to \mathrm{GL}_{mn}(K)$ afforded by V_{KG} with respect to the basis $\tilde{B}$ and the corresponding character $\tilde{\chi}$. Let $\rho_{L/K} \colon L \to \mathrm{End}_K L$ be the regular representation, that is $\rho_{L/K}(\alpha) \colon L \to L$, $x \mapsto \alpha x$ and let $\rho_{L/K} \colon L \to K^{m \times m}$ be the corresponding matrix representation with respect to the basis $B_{L/K}$. Thus for $\alpha \in L$

$$\rho_{L/K}(\alpha) = [a_{i,j}]_{1 \le i,j \le m} \quad \text{if} \quad \alpha \alpha_j = \sum_{i=1}^{m} a_{i,j} \alpha_i.$$

We may extend $\rho_{L/K}$ to a ring homomorphism

$$\rho_{L/K} \colon L^{n \times n} \to K^{mn \times mn}$$

by replacing any element α in $A \in L^{n \times n}$ by the block matrix $\rho_{L/K}(\alpha)$. Then

$$\tilde{\boldsymbol{\delta}} := \rho_{L/K} \circ \boldsymbol{\delta} \colon G \to \mathrm{GL}_{mn}(K) \qquad \text{and} \qquad \tilde{\chi} = \mathrm{Tr}_{L/K} \circ \chi,$$

where $\mathrm{Tr}_{L/K}$ is the usual field trace; see for example [72], p. 115.

Example 2.9.13 Applying the procedure to the representation

$$\boldsymbol{\delta} \colon Q_8 \to \mathrm{GL}_2(\mathbb{Q}(i)), \quad a \mapsto \begin{bmatrix} i & 0 \\ 0 & -i \end{bmatrix}, \quad b \mapsto \begin{bmatrix} 0 & 1 \\ -1 & 0 \end{bmatrix}$$

(from Example 2.9.7) using the $\mathbb{Q}$-basis $(1, i)$, we obtain

$$\tilde{\boldsymbol{\delta}} \colon Q_8 \to \mathrm{GL}_4(\mathbb{Q}), \quad a \mapsto \begin{bmatrix} 0 & -1 & 0 & 0 \\ 1 & 0 & 0 & 0 \\ 0 & 0 & 0 & 1 \\ 0 & 0 & -1 & 0 \end{bmatrix}, \quad b \mapsto \begin{bmatrix} 0 & 0 & 1 & 0 \\ 0 & 0 & 0 & 1 \\ -1 & 0 & 0 & 0 \\ 0 & -1 & 0 & 0 \end{bmatrix}.$$

The character of $\tilde{\boldsymbol{\delta}}$ is $2\chi_5$, where χ_5 was the character of $\boldsymbol{\delta}$. ◆

Definition 2.9.14 If K is a field and χ the character of a representation of G over an extension field of K, then

$$K(\chi) := K(\{\chi(g) \mid g \in G\}).$$

Observe that if χ is the character of a representation of G over $L \supseteq K$, then certainly $K(\chi) \subseteq L$.

Lemma 2.9.15 *Let χ be the character of an absolutely simple LG-module W, where $L \supseteq K$ is a field extension. Then there is a unique $\mathrm{m}_K(\chi) \in \mathbb{N}$ such that $\mathrm{m}_K(\chi) \chi$ is a character of a simple $K(\chi)G$-module V. The integer $\mathrm{m}_K(\chi)$ is called the* **Schur index** *of χ relative to K. Also, V is up to isomorphism the unique simple $K(\chi)G$-module, such that W is isomorphic to a submodule of LV.*

Proof. By Remark 1.5.9 there is a finite field extension $L' \supseteq K$ such that L' is a splitting field for G. Taking a common extension field $L'' \supseteq L$, L' we conclude from Exercise 1.5.8 that $L''W \cong L''W'$ for some simple $L'G$-module W'. Of course, W' also affords the character χ.

Let $\boldsymbol{\delta} \colon G \to \mathrm{GL}_n(L')$ be a representation with character χ. By Remark 2.9.12 $\tilde{\boldsymbol{\delta}} := \boldsymbol{\rho}_{L'/K(\chi)} \circ \boldsymbol{\delta} \colon G \to \mathrm{GL}_{kn}(K)$ is a representation with character $\tilde{\chi} = \mathrm{Tr}_{L'/K(\chi)} \circ \chi$. If $[L : K(\chi)] = k$ we have $\mathrm{Tr}_{L'/K(\chi)}(\chi(g)) = k\chi(g)$ for all $g \in G$, hence $\tilde{\chi} = k\chi$. Let V be a simple submodule of a $K(\chi)G$-module affording $\tilde{\chi}$. Then the character of V is $\chi_V = m\chi$ for some $m \leq k$.

Now let V' be any simple $K(\chi)G$-module with character $m'\chi$ for some $m' \in \mathbb{N}$. Then by Lemma 2.1.5 the composition factors of $L'V$ and $L'V'$ are all isomorphic to W'. Hence $\mathrm{Hom}_{L'G}(L'V, L'V') \neq \{0\}$ and from Exercise 1.3.2 we conclude that $V \cong V'$ and consequently $m = m'$. The same kind of argument also shows that if V' is a simple KG-module such that W is isomorphic to a submodule of LV' then $V \cong V'$. $\square$

Remark 2.9.16 We recall from Definition 1.8.2 that the automorphisms $\gamma \in \mathrm{Aut}(K)$ of a field K act on the KG-modules. If V is a KG-module which (relative to some K-basis) affords the matrix representation $\boldsymbol{\delta} \colon G \to \mathrm{GL}_n(K)$, then according to Remark 1.8.3 $^{\gamma}V$ affords the matrix representation $^{\gamma}\boldsymbol{\delta} := \gamma \circ \boldsymbol{\delta}$ (with respect to the same basis), where we have extended γ to an automorphism of $K^{n \times n}$. If χ is the character of V then the character of $^{\gamma}V$ is $^{\gamma}\chi = \gamma \circ \chi$.

Theorem 2.9.17 *Let V be a simple KG-module with character $\psi := \chi_V$ and let $L \supseteq K$ be a splitting field for G. Assume that W is a simple LG-submodule of LV with character $\chi := \chi_W$. If $\mathrm{Gal}(K(\chi)/K) = \{\gamma_1, \ldots, \gamma_r\}$, then $[K(\chi) : K] = r$ and*

$$\psi = \mathrm{m}_K(\chi)\,(^{\gamma_1}\chi + \cdots + {}^{\gamma_r}\chi).$$

Proof. Let e be the exponent of G and let K_e be the splitting field of the (separable) polynomial $X^e - 1 \in K[X]$. Then for all $g \in G$ we have $\chi(g) \in K_e$. Thus $K(\chi) \subseteq K_e$. Since $\mathrm{Gal}(K_e/K)$ is abelian, $K(\chi) \supseteq K$ is also a Galois extension and we have indeed $[K(\chi) : K] = r = |\mathrm{Gal}(K(\chi)/K)|$.

By assumption, W is a simple LG-submodule of $LV = L \otimes_K K(\chi)V$ and thus of a simple $K(\chi)G$- submodule W_1 of $K(\chi)V$. Recall that $K(\chi)V$ is semisimple by Theorem 1.8.4. Lemma 2.9.15 shows that the character of W_1 is $\chi_{W_1} := m_K(\chi)\,\chi$. Since an irreducible character is not zero, we conclude $\operatorname{char} K \nmid m_K(\chi)$. Therefore, if $\gamma \in \operatorname{Gal}(K(\chi)/K)$ fixes χ_{W_1}, then it fixes χ and hence $\gamma = \operatorname{id}_{K(\chi)}$. Applying Theorem 1.8.4 with $K(\chi) \supseteq K$ and the simple modules V and W_1 we get

$$K(\chi)V \cong \bigoplus_{\gamma \in \operatorname{Gal}(K(\chi)/K)} {}^\gamma W_1 \qquad \text{and} \qquad \psi = \sum_{\gamma \in \operatorname{Gal}(K(\chi)/K)} m_K(\chi) \cdot {}^\gamma\chi$$

and the result follows. □

Theorem 2.9.18 (a) *If K is a finite field then $m_K(\chi) = 1$ for any character χ of an absolutely simple LG-module for a field extension $L \supseteq K$.*

(b) *If e is the exponent of G and K is a splitting field of $X^e - 1 \in \mathbb{F}_p[X]$, then K is a splitting field for G.*

Proof. (a) By Lemma 1.5.9 there is a finite splitting field $L' \supseteq K$ for G. Then $L' \supseteq K$ is a Galois extension. Note that χ is afforded by a simple $L'G$-module, say W. By Lemma 2.9.15 there is a simple $K(\chi)G$-module V with character $m_K(\chi)\,\chi$ and W is a simple submodule of $L'V$. But by Theorem 1.8.4 $L'V \cong W$ because, for all $\gamma \in \operatorname{Gal}(L'/K(\chi))$, we have ${}^\gamma\chi = \chi$ and hence ${}^\gamma W \cong_{L'G} W$. Thus $m_K(\chi) = 1$.

(b) Let $L \supseteq K$ be a splitting field for G with $[L : K] < \infty$ and let W be a simple LG-module with character χ. Since $K = K(\chi)$ it follows as in part (a) that there is a simple KG-module V with $W \cong LV$. By Exercise 1.5.8 K is a splitting field for G. □

The assertion of part (b) of the theorem holds also if one replaces $\mathbb{F}_p$ by $\mathbb{Q}$; see Theorem 3.10.8.

Theorem 2.9.19 *Let $\chi \in \operatorname{Irr}(G)$ and let θ be a character of a KG-module for some $K \subseteq \mathbb{C}$. Then $m_K(\chi) \mid (\theta, \chi)_G$.*

Proof. We may assume that $(\theta, \chi)_G > 0$. Let W be a KG-module affording θ and let W_1 be a simple submodule of W with $(\chi_{W_1}, \chi)_G > 0$. If

$$W \cong \underbrace{W_1 \oplus \cdots \oplus W_1}_{k} \oplus U \qquad \text{with} \qquad W_1 \nmid U$$

then

$$(\theta, \chi)_G = k\,(\chi_{W_1}, \chi)_G = k\,m_K(\chi),$$

where the final equality follows from Theorem 2.9.17. □

Theorem 2.9.19 is extremely useful in practice in computing the Schur index of a character, or at least in giving an upper bound for it. Good candidates for θ are permutation characters, because these are realizable over any field K, that is they are characters of KG-modules. Thus it follows that

$$m_K(\chi) \mid \text{g.c.d.}\{(\eta,\chi)_G \mid \eta \text{ a permutation character of } G\}. \qquad (2.19)$$

Using this, we will compute the Schur indices of the irreducible characters of a number of simple groups in Example 3.5.23. They are all one or two, which is in accordance with the following conjecture formulated by several authors.

Conjecture 2.9.20 *The Schur index of every irreducible character of every finite simple group, moreover of every covering group (see Definition 3.7.9) of a finite simple group, is always one or two.*

The same is definitely not true more generally for finite perfect groups. In fact one has the following theorem.

Theorem 2.9.21 (A. Turull) *Given any positive integer n, there exists some finite perfect group G of chief length two and some irreducible character $\chi \in \text{Irr}(G)$ such that $m_{\mathbb{Q}}(\chi) = n$.*

For a proof see [164].

Another important application of Theorem 2.9.19 is the proof of Theorem 3.10.8.

Exercises

Exercise 2.9.1 Assume that $(\chi(1) \mid \chi \in \text{Irr}(G)) = (1,1,1,1,4,8)$. In Exercise 2.2.6 we computed $\chi(g)$ for all $\chi \in \text{Irr}(G)$ and $g \in G$. Use Theorem 2.9.9 to compute the character table of G (including the power maps) and deduce that $G \cong C_3^2 \rtimes Q_8$.

Exercise 2.9.2 Assume that $\text{Irr}(G)$ is as in Exercise 2.6.1. Compute the second Frobenius–Schur indicators for all $\chi \in \text{Irr}(G)$ and the power maps for G.

Exercise 2.9.3 Let $G := \text{HS}$ be the Higman–Sims group. In Example 1.2.26 we saw that G has a permutation representation of degree 100; see also Example 2.1.12. Use this and the character table of G (see e.g. [38], p. 81) to show that $m_{\mathbb{Q}}(\chi) = 1$ for all $\chi \in \text{Irr}(G)$.

Hint: Starting with the irreducible constituents of the permutation character of degree 100 produce (using tensor products) characters of rational representations of G and use Theorem 2.9.19.

Exercise 2.9.4 Let G be a perfect group and let $\chi \in \text{Irr}(G)$ be faithful with $\chi(1) = 2$. Show that

(a) $\mathbf{Z}(G) = \{g \in G \mid g^2 = 1\}$ has order two.

(b) $G/\mathbf{Z}(G) \cong A_5$ (consider $\chi^{[2]}$ and use Exercise 2.7.6).

(c) Use Theorem 2.9.9 to find the degrees of the faithful irreducible characters.

(d) Complete the character table of G.

2.10 Computing character tables – an example

The Dixon–Schneider algorithm presented in Section 2.4 can be used to compute the character table of a finite group G provided that one can compute with the elements of G in order to find its conjugacy classes and (at least some) class multiplication coefficients α_{ij}^k. For very large groups this is not feasible. As mentioned before, character tables of very large groups (such as the "Monster group") have been computed – using completely different methods – from a very limited portion of information about the group $|G|$, such as the order of G and the structure of some centralizers, even before the existence of the group had been proved. In this section we give a modest example of how the character table of a group can sometimes be computed in such a way.

The Mathieu group M_{11} is a sharply 4-fold transitive permutation group of degree 11 and the smallest sporadic simple group (see [88]). We will show that its character table can be computed using only the fact that M_{11} is a simple permutation group of degree 11 and of size $11 \cdot 10 \cdot 9 \cdot 8$. The idea is to look at elements of prime order and find some information about their centralizers. If these are small the values of the irreducible characters on the corresponding classes are rather restricted, being algebraic integers of small absolute values. The congruence relations (Lemma 2.2.2) allow only certain degrees for each of these values. The orthogonality relations, in conjunction with the congruence relations, are sometimes sufficient to exclude so many cases that the degrees of the irreducible characters, and later the characters themselves, can be found. In more complicated cases the theory of blocks with cyclic defect groups can be used, provided the group in question has cyclic Sylow p-subgroups; see Section 4.12. Observe that congruence relations with respect to a prime give strong information only if the prime is large. So one usually starts by looking at the largest prime divisors of the group order.

Since the symmetric group S_{11} has self-centralizing Sylow 11-subgroups, it is clear that the same holds true for M_{11}. Being a group of prime degree, M_{11} is primitive, and it follows from a theorem of Jordan ([88], p. 171) that a primitive group of degree 11 cannot contain a 5-cycle unless it is isomorphic to A_{11} or S_{11}. So a Sylow 5-subgroup of M_{11} is generated by a product of two 5-cycles which has centralizer order 25 in A_{11} and hence is self-centralizing in M_{11}.

So we assume for the rest of this section that G is a finite group with

$$|G| = 11 \cdot 10 \cdot 9 \cdot 8, \quad \mathbf{C}_G(P_{11}) = P_{11}, \quad \mathbf{C}_G(P_5) = P_5,$$

where P_p denotes a Sylow p-subgroup of G. Since

$$\mathbf{N}_G(P_{11}) / \mathbf{C}_G(P_{11}) \leq \mathrm{Aut}(P_{11}) \cong \mathbb{Z}/10\mathbb{Z}$$

it follows from Sylow's third theorem ([110], theorem I.6.4(iii), p. 35) that $|\mathbf{N}_G(P_{11})| = 11 \cdot 5$. This implies that the elements g of order 11 in G are not rational (not even real) but distribute into two conjugacy classes **11a** and **11b** with **11b** (**11a**) containing the squares and the inverses of the elements of **11a** (and **11b**, respectively). Observe that $g \mapsto g^2$ defines an automorphism of $\langle g \rangle = P_{11}$ of order ten. The same arguments give $|\mathbf{N}_G(P_5)| = 5 \cdot 4$, so that the elements of order five form a single conjugacy class **5a** of G. We thus have the following lemma.

Lemma 2.10.1 *All characters of G are rational on* **5a** *and have values in* $\mathbb{Q}(\sqrt{-11})$ *on* **11a**, *and there is at least one pair of characters whose values on* **11a** *are complex conjugate and non-real.*

Because of the orthogonality relations for columns **5a** and **1a** it follows that there are exactly five irreducible characters of G which do not vanish on the class **5a**, and the non-zero values on this class are 1 and -1. In particular, by the congruence relations all but five irreducible characters of G have degrees divisible by five and all irreducible character degrees of G must be congruent to ± 1 or 0 mod 5.

If c and $\bar{c}$ denote a pair of non-real values of irreducible characters on **11a** then $|c|^2 \leq 5$ by the orthogonality relations for the classes **11a** and **1a**. Hence it follows that $c \in \{\frac{1 \pm \sqrt{-11}}{2}, \frac{-1 \pm \sqrt{-11}}{2}\}$ and $|c|^2 = 3$, so that there is exactly one pair of irreducible characters of G having non-rational values on **11a** and $11 - 6 = 5$ irreducible characters, with value 1 or -1, or else one character with value ± 2 on **11a** and the trivial character. All other irreducible characters must vanish on **11a** and hence have degrees divisible by 11.

The following table gives for each of the above-mentioned possible values of an irreducible character χ on **11a** the degrees $\chi(1)$ and values $\chi(5a)$ allowed by the congruence relations. Of course we use the fact that $\chi(1)^2 \leq |G|$ and that $\chi(1) \equiv \pm 1, 0$ mod 5 as observed above.

$\chi(11a)$	$\chi(1)$	$\chi(1)$	$\chi(1)$	$\chi(1)$	$\chi(5a)$	$\chi(5a)$	$\chi(5a)$	$\chi(5a)$
$\frac{1\pm\sqrt{-11}}{2}$	6				1			
$\frac{-1\pm\sqrt{-11}}{2}$	5	16	60		0	1	0	
1	1	45			1	0		
-1	10				0	0		
2	24				-1			
-2	9	20			-1	0		
0	11	44	55	66	1	-1	0	1

The cases $\chi(11a) = \pm 2$ are easily seen to be impossible by looking at

$$\sum_{\chi \in \mathrm{Irr}(G)} \chi(1a) \cdot \chi(11a) = 0, \tag{2.20}$$

because there would be just four irreducible characters not vanishing on **11a**. (Block theory would also immediately exclude this case.) The orthogonality relation $\sum_{\chi \in \mathrm{Irr}(G)} \chi(5a) \cdot \chi(11a) = 0$ now shows that there must be a pair of complex conjugate irreducible characters of degree 16, since only such a pair can give a negative contribution to the above sum. Since not all characters can be non-negative on 5a there must be an irreducible character of degree 44, and the orthogonality relations for the classes 1a and 5a reveal that the last irreducible character not vanishing on 5a must have degree 11. From (2.20) it follows that there must be one irreducible character of degree 45 and three of degree ten. All further irreducible characters must have degree 55. Looking at the sum of the squares of the degrees found so far, we see that there is exactly one such character of degree 55. So we see that G has ten irreducible characters and the following portion of the character table is known at this stage; observe that we know that for an element g_5 of order five we have $\mathbf{N}_G(\langle g_5 \rangle) \cong C_5 \rtimes C_4$, the full holomorph of C_5, so there must be elements of order four in G.

	1a	2a	3a	4a	5a	??	??	??	11a	11b
χ_1	1	1	1	1	1	1	1	1	1	1
χ_2	10				0				-1	-1
χ_3	10				0				-1	-1
χ_4	10				0				-1	-1
χ_5	11	x			1				0	0
χ_6	16	0		0	1				c	$\bar{c}$
χ_7	16	0		0	1				$\bar{c}$	c
χ_8	44	y			-1				0	0
χ_9	45		0		0				1	1
χ_{10}	55				0				0	0

All the zeros in the table are explained by Theorem 3.10.10, which says that $\chi(g) = 0$ for $\chi \in \mathrm{Irr}(G)$ and $g \in G$ if there is a prime p such that $p \nmid \frac{|G|}{\chi(1)}$ but $p \mid |\langle g \rangle|$. As one can see this is a very powerful theorem. Furthermore by orthogonality $1 + x - y = 0$. To get the values of x and y we take advantage of the class multiplication coefficients (see Remark 2.3.1):

$$a_{ij}^k = \frac{|C_i||C_j|}{|G|} \sum_{\chi \in \mathrm{Irr}(G)} \frac{\chi(g_i)\chi(g_j)\chi(g_k^{-1})}{\chi(1)} \qquad \text{with} \qquad g_i \in C_i.$$

We use this for $C_i = C_j = $ 2a and $C_k = $ 5a. Here we may assume that 2a $= t^G$ is a class of "central" involutions, that is $t \in \mathbf{Z}(P_2)$ for some $P_2 \in \mathrm{Syl}_2(G)$. Since the elements of orders 5 and 11 are self-centralizing $|\mathbf{C}_G(t)| = 2^4 \cdot 3^m$ with $m \in \{1, 2\}$. Observe that by Theorem 2.9.9 there are at most $(\sum_{\chi \in \mathrm{Irr}(G)} \chi(1)) - 2 \cdot 16 - 1 = 186$ involutions in G, so that $|\mathbf{C}_G(t)| > 16$. We obtain

$$a_{ij}^k = \frac{|G|}{2^8 \cdot 3^{2m}}\left(1 + \frac{x^2}{11} - \frac{y^2}{44}\right) = \frac{5 \cdot 3^2}{2^6 \cdot 3^{2m}}(48 + 3y^2 - 8y).$$

Since this is an integer, y must be divisible by four and different from zero. Also $m = 1$, because otherwise y would have to be divisible also by three and $\sum_{\chi \in \mathrm{Irr}(G)} |\chi(t)|^2$ would be too large. So $(x, y) \in \{(3, 4), (-5, -4)\}$ and $|\mathbf{C}_G(t)| - (x^2 + y^2) \in \{23, 7\}$. To see that the second alternative is impossible

176 Characters

we observe that there must be a class **6a** of elements of order six with cubes in the class **2a** and squares in **3a**. Applying Theorem 3.10.10 again with $p = 3$ we see that $\chi_9(\mathbf{6a}) = 0$, and by the congruence relations $\chi_9(t)$ must be odd and divisible by three, thus $\chi_9(t)^2 \geq 9$. Hence $(x, y) = (3, 4)$ is the only possibility. The congruence relations in conjunction with the orthogonality relations for the classes **2a** and **1a** are now sufficient to find the remaining five values on the class **2a**. Since there are at least $165 = \frac{|G|}{243}$ involutions Theorem 2.9.9 shows that the Frobenius–Schur indicators of $\chi_1, \chi_5, \chi_8, \chi_9, \chi_{10}$ and, say, χ_2 are all one, while those of χ_3, χ_4 are either both zero, in which case they form a pair of complex conjugate characters, or 1 and -1. In any case **2a** is the only class of involutions. So at this point the status is as the first part of the following table shows, where a $+$ in the column **ind** means that the corresponding character has Frobenius–Schur indicator 1. We have added the values of $\chi^{[1^2]}$ for $\chi \in \{\chi_2, \chi_3, \chi_4\}$, which we can compute from the known values of χ:

	ind	1a	2a	3a	4a	5a	6a	??	??	11a	11b
χ_1	$+$	1	1	1	1	1	1	1	1	1	1
χ_2	$+$	10	2			0				-1	-1
χ_3		10	-2			0				-1	-1
χ_4		10	-2			0				-1	-1
χ_5	$+$	11	3			1				0	0
χ_6	0	16	0		0	1	0			c	$\bar{c}$
χ_7	0	16	0		0	1	0			$\bar{c}$	c
χ_8	$+$	44	4			-1				0	0
χ_9	$+$	45	-3	0		0	0			1	1
χ_{10}	$+$	55	-1			0				0	0
$\chi^{[1^2]}$		45	-3			0				1	1

Thus $\chi^{[1^2]} = \chi_9$. By Theorem 2.9.3 we have $\nu_2(\chi) \neq -1$ and χ_3, χ_4 must be a pair of complex conjugate characters. Since $\chi_9(\mathbf{3a}) = 0$ we find $\chi(\mathbf{3a}) \in \{0, 1, \zeta_3, \zeta_3^2\}$. The congruence relations eliminate the case 0. Hence $\chi_2(\mathbf{3a}) = 1$. The same arguments give $\chi_2(\mathbf{6a}) = -1$. Furthermore

$$\chi_9(\mathbf{4a}) = \frac{\chi_2(\mathbf{4a})^2 - 2}{2} = \frac{\chi_3(\mathbf{4a})^2 + 2}{2} = \frac{\overline{\chi_3(\mathbf{4a})}^2 + 2}{2},$$

from which we get $\chi_2(\mathbf{4a})^2 = \chi_3(\mathbf{4a})^2 + 4 = \overline{\chi_3(\mathbf{4a})}^2 + 4$. It follows that **4a** must be a real class. The orthogonality relations for the classes **4a** and **11a** yield

$$\chi_2(\mathbf{4a}) = 2, \quad \chi_3(\mathbf{4a}) = \chi_4(\mathbf{4a}) = 0, \quad \chi_9(\mathbf{4a}) = 1.$$

We find readily that $\chi_2^{[2]} = \chi_1 + \chi_2 + \chi_8$ and $\chi_3^{[2]} = \chi_5 + \chi_8$. So we get also the values $\chi_8(\mathbf{4a}) = 0$ and $\chi_5(\mathbf{4a}) = -1$. The last entry $\chi_{10}(\mathbf{4a}) = -1$ in the column of **4a** can be found using orthogonality. Since we now know all the Frobenius–Schur indicators we can count the elements of order four from the known character values on **2a**. We find that any involution has six square roots in G. Thus there are $\frac{|G|}{243} \cdot 6 = \frac{|G|}{2^3}$ elements of order four so that **4a** is the only class of elements of this order. Also we see that the elements of order four have two square roots each, so there are $\frac{|G|}{8} \cdot 2$ elements of order eight in G, which must distribute into two conjugacy classes. So we have found the missing classes 8a, 8b, which are non-real. We can now find all the character values on the rational classes 3a and 6a in exactly the same way as we did for those on 4a, and filling up the table is a very easy task. We finish by showing the complete table in GAP format:

```
M11        2   4   4   1   3   .   1   3   3   .   .
           3   2   1   2   .   .   1   .   .   .   .
           5   1   .   .   1   .   .   .   .   .   .
          11   1   .   .   .   .   .   .   .   1   1

              1a  2a  3a  4a  5a  6a  8a  8b 11a 11b
           2P 1a  1a  3a  2a  5a  3a  4a  4a 11b 11a
           3P 1a  2a  1a  4a  5a  2a  8a  8b 11a 11b
           5P 1a  2a  3a  4a  1a  6a  8b  8a 11a 11b
          11P 1a  2a  3a  4a  5a  6a  8a  8b  1a  1a

    X.1        1   1   1   1   1   1   1   1   1   1
    X.2       10   2   1   2   .  -1   .   .  -1  -1
    X.3       10  -2   1   .   .   1   A  -A  -1  -1
    X.4       10  -2   1   .   .   1  -A   A  -1  -1
    X.5       11   3   2  -1   1   .  -1  -1   .   .
    X.6       16   .  -2   .   1   .   .   .   B  /B
    X.7       16   .  -2   .   1   .   .   .  /B   B
    X.8       44   4  -1   .  -1   1   .   .   .   .
    X.9       45  -3   .   1   .   .  -1  -1   1   1
    X.10      55  -1   1  -1   .  -1   1   1   .   .
```

A = E(8)+E(8)^3 = ER(-2) = i2
B = E(11)+E(11)^3+E(11)^4+E(11)^5+E(11)^9
  = (-1+ER(-11))/2 = b11

Exercises

Exercise 2.10.1 Let $n \in \mathbb{N}$ and let p be a prime. Show that the following GAP function returns a list S such that the number of Sylow p-subgroups of any simple group of order n is in S:

```
gap> syl := function( n, p )
> local divs, lpd, S;
> divs := Factors( n ); lpd := divs[Length( divs )];
> divs := Combinations( divs );
> Add( divs[1], 1 ); divs := List( divs, Product );
> S := Filtered( divs, x-> x mod p = 1 and x > lpd );
> S := Filtered( S, x -> Gcd( n/x, p-1 ) <> 1 );
> return( S );
> end;;
```

Exercise 2.10.2 Let G be a simple group of order 660. The aim of this exercise is to construct the character table of G.

(a) Using the Sylow theorems and a well-known theorem of Burnside (see [88], Hauptsatz IV.2.6, p. 419) show that the Sylow 11-subgroups of G are self-centralizing and that the elements of order 11 of G are contained in a pair of mutually inverse conjugacy classes 11a, 11b. Use this to find all non-zero values of the irreducible characters on these classes.

(b) Using the congruence relations, find the possible degrees of the irreducible characters of G and then show that G has exactly eight conjugacy classes.

(c) Using Lemma 2.2.2 and the orthogonality relations show that a conjugacy class of elements of order five cannot be rational. Find the values of the irreducible characters on the elements of order five.

(d) Show that the normalizer of a Sylow 3-subgroup has order 12 and the centralizer must have order six. Complete the character table.

It is well known that there is, in fact, a simple group of order 660, namely $L_2(11)$. We reproduce the character table of this group in **GAP** format for convenience:

```
L2(11)     2  2  2  1  .  .  1  .  .
           3  1  1  1  .  .  1  .  .
           5  1  .  .  1  1  .  .  .
          11  1  .  .  .  .  .  1  1

             1a 2a 3a 5a 5b 6a 11a 11b
          2P 1a 1a 3a 5b 5a 3a 11b 11a
          3P 1a 2a 1a 5b 5a 2a 11a 11b
          5P 1a 2a 3a 1a 1a 6a 11a 11b
         11P 1a 2a 3a 5a 5b 6a  1a  1a

         X.1   1   1   1   1   1   1   1   1
         X.2   5   1  -1   .   .   1   B  /B
         X.3   5   1  -1   .   .   1  /B   B
         X.4  10  -2   1   .   .   1  -1  -1
         X.5  10   2   1   .   .  -1  -1  -1
         X.6  11  -1  -1   1   1  -1   .   .
         X.7  12   .   .   A  *A   .   1   1
         X.8  12   .   .  *A   A   .   1   1
```

A = E(5)+E(5)^4 = (-1+ER(5))/2 = b5
B = E(11)+E(11)^3+E(11)^4+E(11)^5+E(11)^9 = (-1+ER(-11))/2 = b11

3

Groups and subgroups

3.1 Restriction and fusion

Throughout this section K is a field and H is a subgroup of a finite group G. We repeat our convention that all KG- or KH-modules considered are finite dimensional over K.

Using the embedding $KH \subseteq KG$ any KG-module V can be considered as a KH-module in a natural way by restriction of the action (see Example 1.1.21). As such, the module is denoted by V_H and is called the restricted module. If χ is the character of V then the character of V_H is simply the restriction $\chi_{|H}$ of the function χ to H and will also be denoted by χ_H. If the character χ is given as usual by the list of its values on representatives $(g_i \mid 1 \leq i \leq r)$ of the conjugacy classes $C_1, \ldots, C_r$ of G, and if $(h_j \mid 1 \leq j \leq s)$ is a list of representatives of the conjugacy classes $C_1', \ldots, C_s'$ of H, then

$$\chi_H(h_j) = \chi(g_{\mathrm{fus}(j)}),$$

where $\mathrm{fus}\colon \{1, \ldots, s\} \to \{1, \ldots, r\}$ is defined by $\mathrm{fus}(j) = i$ if $h_j \in C_i$. This map is called the fusion map from H to G; often it is considered as a map

$$\mathrm{fus}\colon \{C_1', \ldots, C_s'\} \to \{C_1, \ldots, C_r\}.$$

Of course, χ_H must be an $\mathbb{N}_0$-linear combination of the irreducible characters of H. This very simple idea provides a necessary condition for a group H to be isomorphic to a subgroup of a group G, provided that the character tables of H and G are known. We illustrate this by an example as follows.

Example 3.1.1 The largest sporadic simple group M, the "Monster," has just two conjugacy classes of involutions 2a, 2b. We copy a small part of the ordinary character table (see e.g. [38]), containing just the first three classes and characters:

M

	1a	2a	2b
X.1	1	1	1
X.2	196883	4371	275
X.3	21296876	91884	-2324

Suppose we want to estimate the 2-rank of M. Recall that for a prime p the *p*-**rank** of a finite group G is the largest integer k such that G has an elementary abelian *p*-subgroup of order p^k.

If H is an elementary abelian subgroup of M of order 2^k, any non-trivial conjugacy class of H must fuse into 2a or 2b in M. Suppose that H contains x elements of 2a and hence $2^k - x - 1$ elements of 2b. Then

$$(\chi_H, 1)_H = \frac{1}{2^k}(\chi(1a) + x \cdot \chi(2a) + (2^k - x - 1) \cdot \chi(2b))$$

for any character χ of M. This leads to a congruence

$$\chi(1a) - \chi(2b) + x \cdot (\chi(2a) - \chi(2b)) \equiv 0 \qquad \mathrm{mod} \quad 2^k.$$

For $\chi = $ X.2 and $\chi = $ X.3 these congruences read

$$2^{16} \cdot 3 + x \cdot 2^{12} \equiv 0 \quad \mathrm{mod} \quad 2^k,$$
$$2^{16} \cdot 325 + x \cdot 2^{12} \cdot 23 \equiv 0 \quad \mathrm{mod} \quad 2^k.$$

Solving the first for x we get

$$x + 2^4 \cdot 3 \equiv 0 \quad \mathrm{mod} \quad 2^l \quad \text{with} \quad l = k - 12,$$

and inserting this in the second congruence we get

$$2^4 \cdot (325 - 3 \cdot 23) = 2^{4+8} \equiv 0 \quad \mathrm{mod} \quad 2^l.$$

Thus $l \le 12$ and hence $k \le 24$. Thus the 2-rank of M is at most 24 (compare [163], p. 498, where this result is referred to as being well known). ♦

In the above example the subgroup H of G had a rather trivial structure and a rather trivial character table. Similar questions might arise for other groups too. Suppose one knows the character table of the groups G and U, and one wants to know whether G might have a subgroup isomorphic to U. If this is the case, then there must be a fusion map

$$\mathrm{fus}_U^G \colon \{C_1', \dots, C_s'\} \to \{C_1, \dots, C_r\}$$

of the set $\{C_1', \dots, C_s'\}$ of conjugacy classes of U to the corresponding set $\{C_1, \dots, C_r\}$ of G defined by $\mathrm{fus}(C_i') = C_j$ if and only if $\eta(C_i') \subseteq C_j$, where $\eta \colon U \to H$ is an isomorphism. One obtains a number of necessary conditions for such a map, observing that for $h \in H \le G$ one must have $|\mathbf{C}_H(h)| \mid |\mathbf{C}_G(h)|$

and that the fusion map must also be compatible with the power maps of the corresponding character tables. Furthermore, the composition of the fusion map with the (irreducible) characters of G must have non-negative integral scalar products with the irreducible characters of U. Of course these are only necessary conditions for the existence of a subgroup $H \leq G$ isomorphic to U. Also the fusion maps will in general not be unique even when considered up to character table automorphisms. T. Breuer ([17]) has designed algorithms for finding possible fusions of conjugacy classes as above and implemented these in GAP. Figure 3.1 gives the result if one applies the algorithms to the list $(G_i)_{1\leq i\leq 24}$ of sporadic simple groups G_i excluding the Monster and Baby Monster. For each pair (G_i, G_j) of these simple groups (up to isomorphism) with $|G_j| \leq |G_i|$ the number $f(G_i, G_j)$ of possible class fusions (up to character table automorphisms) of G_j into G_i is listed (printing "." instead of "0" for better readability). If $f(G_i, G_j) = 0$ this shows that G_i has no subgroup isomorphic to G_j; of course the converse does not necessarily hold.

	M_{11}	M_{12}	J_1	M_{22}	J_2	M_{23}	HS	J_3	M_{24}	McL	He	Ru	Suz	ON	Co_3	Co_2	Fi_{22}	HN	Ly	Th	Fi_{23}	Co_1	J_4	Fi'_{24}
M_{11}	1																							
M_{12}	1	1																						
J_1	.	.	1																					
M_{22}	1	.	.	1																				
J_2	.	.	.	.	1																			
M_{23}	1	.	.	1	.	1																		
HS	2	.	.	1	.	.	1																	
J_3	.	.	.	.	.	.	.	1																
M_{24}	1	1	.	1	.	1	.	.	1															
McL	1	.	.	1	.	.	.	.	.	1														
He	.	.	.	.	.	.	.	.	.	.	1													
Ru	.	.	.	.	.	.	.	.	.	.	.	1												
Suz	1	1	.	.	1	.	.	.	.	.	.	.	1											
ON	1	.	1	.	.	.	.	.	.	.	.	.	.	1										
Co_3	2	1	.	1	.	1	1	.	.	1	.	.	.	.	1									
Co_2	2	.	.	2	.	1	1	.	.	1	.	.	.	.	.	1								
Fi_{22}	2	1	.	1	.	.	.	.	.	.	.	.	.	.	.	.	1							
HN	3	1	.	.	.	.	.	.	.	.	.	.	.	.	.	.	.	1						
Ly	4	1	.	.	.	.	.	.	.	.	.	.	.	.	.	.	.	.	1					
Th	.	.	.	.	.	.	.	.	.	.	.	.	.	.	.	.	.	.	.	1				
Fi_{23}	15	3	.	.	.	.	.	.	.	.	.	.	.	.	.	.	.	.	.	.	1			
Co_1	2	2	.	1	1	1	1	.	1	1	.	.	.	.	1	1	.	.	.	.	.	1		
J_4	5	6	.	2	.	2	.	.	2	.	.	.	.	.	.	.	.	.	.	.	.	.	1	
Fi'_{24}	13	47	.	1	.	.	.	.	.	1	.	.	.	.	.	.	.	.	.	.	1	.	.	1

Figure 3.1. $[f(G_i, G_j)]_{1\leq i\leq j\leq 24}$.

Exercises

Exercise 3.1.1 (a) Verify Figure 3.1 $[f(G_i, G_j)]_{1\leq i\leq j\leq 24}$, using the GAPcommands `PossibleClassFusions` and `RepresentativesFusions`.

(b) Compare this with the table on p. 238 in [38].

Exercise 3.1.2 Show that the 2-rank of M_{11} is two using only the character table printed on p. 177.

Exercise 3.1.3 See refs. [26] and [121]. Let $\chi \in \mathrm{Irr}(G)$ and assume that $\chi_H \in \mathrm{Irr}(H)$. Let $\delta \colon G \to K^{n \times n}$ be a matrix representation affording χ. Show that

$$\chi(g\,g') = \frac{\chi(1)}{|H|} \sum_{h \in H} \chi(g\,h^{-1})\chi(h\,g') \qquad \text{for} \qquad g, g' \in G,$$

$$\delta(g) = \frac{\chi(1)}{|H|} \sum_{h \in H} \chi(g\,h^{-1})\delta(h).$$

Hint: Use Exercise 2.1.3.

Exercise 3.1.4 Let $\chi \in \mathrm{Irr}(G)$ with $\chi_H \in \mathrm{Irr}(H)$ and $\nu_2(\chi_H) = 1$. Show that $\nu_2(\chi) \in \{0, 1\}$. Give examples for both possibilities.

3.2   Induced modules and characters

In this section let K be a commutative ring and let G be a finite group.

Let H be a subgroup of G and W a KH-module. Since KG can be considered as a (KG, KH)-bimodule, the tensor product $W^G = KG \otimes_{KH} W$ is a KG-module (see Remark 1.1.48), called the module **induced** from W.

Lemma 3.2.1 *Let $\{g_1, \ldots, g_m\}$ be a left transversal of the subgroup H in G, i.e. $G = \dot{\bigcup} g_i H$. Furthermore, let W be a KH-module which is free (and, as always, finitely generated) as a K-module. Then*

(a) $KG = \bigoplus_{i=1}^m g_i KH$ *as K-modules and*

$$W^G = \{\sum_{i=1}^m g_i \otimes w^i \mid w^i \in W\} = \bigoplus_{i=1}^m g_i \otimes W$$

as K-modules. If $B = (w_1, \ldots, w_n)$ is a K-basis of W then

$$B^G = (g_1 \otimes w_1, \ldots, g_1 \otimes w_n, \ldots, g_m \otimes w_1, \ldots, g_m \otimes w_n)$$

is a K-basis of W^G.

(b) *If $\delta \colon H \to \mathrm{GL}(W)$ is the representation afforded by W and*

$$\delta \colon H \to \mathrm{GL}_n(K) \,, \quad h \mapsto \delta(h) = [\delta(h)]_B$$

is the corresponding matrix representation with respect to the basis B, then the matrices of the "induced representation" $\delta^G \colon G \to \mathrm{GL}(W^G)$ afforded by the induced module corresponding to the above basis B^G are as follows:

$$[\delta^G(g)]_{B^G} = \begin{bmatrix} \dot{\delta}_{11}(g) & \dot{\delta}_{12}(g) & & \dot{\delta}_{1m}(g) \\ \dot{\delta}_{21}(g) & \dot{\delta}_{22}(g) & & \dot{\delta}_{2m}(g) \\ & & \ddots & \\ \dot{\delta}_{m1}(g) & \dot{\delta}_{m2}(g) & & \dot{\delta}_{mm}(g) \end{bmatrix} \qquad \text{for} \quad g \in G$$

with

$$\dot{\delta}_{ij}(g) = \begin{cases} \delta(g_j^{-1}gg_i) \in K^{n \times n} & if \quad g_j^{-1}gg_i \in H, \\ 0_n \in K^{n \times n} & else. \end{cases}$$

Proof. All this follows from a simple computation. $\qquad\square$

Remark 3.2.2 If W is a KH-module and $g \in G$ then the K-submodule $g \otimes W \leq W^G$ can be considered as a $K(gHg^{-1})$-module and will also be denoted by ${}^gW$. It is convenient to identify $1 \otimes W$ with W, so that $W = {}^1W \leq_{KH} W^G$. If W is simple and $g \in G$ then ${}^gW$ is also simple. If K is a field and W has character φ, the character of ${}^gW$ is

$${}^g\varphi: gHg^{-1} \to K, \quad x \mapsto \varphi(g^{-1}xg).$$

Proof. This just follows from $ghg^{-1} \cdot g \otimes w = g \otimes hw$ for $w \in W$ and $h \in H$. If U is a KH-submodule of W, then obviously $g \otimes U$ is a $K(gHg^{-1})$-submodule of ${}^gW$. $\qquad\square$

Remark 3.2.3 For the trivial KH-module K_H the induced module $(K_H)^G$ is isomorphic to the permutation module $K[G/H]$. Using the notation of Lemma 3.2.1, $g_i \otimes \alpha \mapsto \alpha g_i H$ for $\alpha \in K$ defines a KG-isomorphism $(K_H)^G \to K[G/H]$.

The image of a group under an induced representation can be embedded into a wreath product. We recall the definition of a wreath product.

Definition 3.2.4 If H is a group and $P \leq S_\Omega$ is a permutation group acting on a finite set Ω then the wreath product $H \wr P$ is defined as

$$H \wr P = \{(f, \sigma) \mid \sigma \in P, \ f: \Omega \to H\}$$

with multiplication

$$(f_1, \sigma_1) \cdot (f_2, \sigma_2) = (f_{1,2}, \sigma_1 \circ \sigma_2) \quad with \quad f_{1,2}(i) = f_1(i) f_2(\sigma_1^{-1}(i)).$$

Remark 3.2.5 With the notation of Lemma 3.2.1 let $\pi: G \to S_{G/H} \cong S_m$ be the permutation representation of G on the cosets of H. For $g \in G$ let $f_g: \{1, \ldots, m\} \to \delta(H)$ be defined by $f_g(i) = \delta(g_{\pi(g)(i)}^{-1} gg_i)$. Then

$$\delta^G(g) \mapsto (f_g, \pi(g)) \qquad \delta^G(G) \to \delta(H) \wr S_m$$

is an injective homomorphism into the wreath product.

Definition 3.2.6 If K is a field and $\varphi: H \to K$ is a class function on H and $G = \dot{\bigcup}_{i=1}^m g_i H$ then the **induced class function** $\varphi^G: G \to K$ is given by

$$\varphi^G: g \mapsto \sum_{i=1}^m \dot{\varphi}(g_i^{-1} gg_i) \quad with \quad \dot{\varphi}(x) = \begin{cases} \varphi(x) & if \ x \in H, \\ 0 & else. \end{cases}$$

It is immediate that φ^G is, in fact, a class function on G and does not depend on the choice of the transversal $(g_1, \ldots, g_m)$.

Lemma 3.2.7 *Let H be a subgroup of G and let K be a field.*

(a) *For $g \in G$, $\varphi, \varphi' \in \mathrm{cf}(H, K)$ and $\psi \in \mathrm{cf}(G, K)$ we have*

$$(^g\varphi)^G = \varphi^G, \qquad \varphi^G + \varphi'^G = (\varphi + \varphi')^G, \qquad \varphi^G \cdot \psi = (\varphi \cdot \psi_H)^G.$$

(b) *Let $\varphi \in \mathrm{cf}(H, K)$. For $g \in G$ let $g^G \cap H = h_1^H \dot\cup \cdots \dot\cup h_r^H$. Then*

$$\varphi^G(g) = \sum_{i=1}^{r} [\mathbf{C}_G(h_i) : \mathbf{C}_H(h_i)] \, \varphi(h_i).$$

(c) *If W is a KH-module with character φ then the induced module W^G has character φ^G.*

Proof. (a) is obvious from the definition.

(b) Let T be a left transversal of H in G and let $g^G \cap H = h_1^H \dot\cup \cdots \dot\cup h_r^H$. For $r = 0$, that is $g^G \cap H = \emptyset$, the result is true. Since φ and φ^G are class functions we may assume that $h_i = g^{t_i}$ with $t_i \in T$. Let $T_i := \{t \in T \mid g^t \in h_i^H\}$. Then $\varphi^G(g) = \sum_{j=1}^{r} |T_i| \cdot \varphi(h_i)$. But

$$t \in T_i \iff h_i^{t_i^{-1}t} \in h_i^H \iff t_i^{-1}t \in \mathbf{C}_G(h_i)H.$$

Hence $|T_i| = [\mathbf{C}_G(h_i) : \mathbf{C}_G(h_i) \cap H] = [\mathbf{C}_G(h_i) : \mathbf{C}_H(h_i)]$.

(c) follows immediately from Lemma 3.2.1. $\qquad\square$

Theorem 3.2.8 (Janusz) *Let $H \leq G$ and let $e \in \mathbb{C}H$ be a primitive idempotent such that $\mathbb{C}He$ has character φ. Assume that $\chi \in \mathrm{Irr}(G)$ satisfies $(\chi, \varphi^G)_G = 1$. Then $\mathbb{C}Ge\epsilon_\chi$ is a simple module with character χ, where ϵ_χ is the block idempotent (see Corollary 2.1.7).*

Proof. Obviously (see Exercise 3.2.1) $\mathbb{C}Ge \cong_{\mathbb{C}G} (\mathbb{C}He)^G$ has character φ^G and $\mathbb{C}Ge\epsilon_\chi$ is the (χ-) homogeneous component of $\mathbb{C}Ge$. Since $(\chi, \varphi^G)_G = 1$ the result follows. $\qquad\square$

Corollary 3.2.9 *Let $H_0 \leq H_1 \leq \ldots \leq H_r = G$ be a chain of subgroups and $\chi_i \in \mathrm{Irr}(H_i)$ with block idempotent $\epsilon_i := \epsilon_{\chi_i}$ be such that $(\chi_i, \chi_{i-1}^{H_i})_{H_i} = 1$ for $i = 1, \ldots, r$ and $\chi_0(1) = 1$. Then $\epsilon_0\epsilon_1 \cdots \epsilon_r$ is a primitive idempotent of $\mathbb{C}G$ affording φ_r.*

Proof. Use Theorem 3.2.8 repeatedly and note that ϵ_χ is a primitive idempotent if $\chi(1) = 1$. $\qquad\square$

The corollary gives (under its assumption) a method to construct primitive idempotents of the group algebra and hence irreducible representations of G over $\mathbb{C}$. For an algorithm to compute irreducible matrix representations along these lines, see [49]. The corollary also shows that a representation affording $\chi = \chi_r$ can be written over the field $\mathbb{Q}(\zeta_m)$, where m is the exponent of G (compare Theorem 3.10.8 below). The following question was raised by Janusz.

Question 3.2.10 *Given* $\chi \in \mathrm{Irr}(G)$ *is there a maximal subgroup* $H \leq G$ *and* $\varphi \in \mathrm{Irr}(H)$ *such that* $(\chi|_H, \varphi)_H = 1$?

It will be an immediate consequence of Corollary 3.6.15 that the answer to the question is "yes" for all solvable groups G (and all $\chi \in \mathrm{Irr}(G)$). The same is true for symmetric and alternating groups, for $\mathrm{L}_n(q)$ and some other classes of groups (see [99], [159]). But in general the answer to the question is "no."

Example 3.2.11 For a number of tables (including the tables of the sporadic simple groups with the exception of the "Baby Monster" and the "Monster") in the library of character tables in GAP, all the character tables of maximal subgroups are also stored together with corresponding fusion. For these groups Question 3.2.10 can be answered immediately. One finds the following counterexamples:

$$\chi_{50} \in \mathrm{Irr}(\mathrm{J}_4), \quad \chi_{37} \in \mathrm{Irr}(\mathrm{Ly}), \quad \chi_{29}, \chi_{30} \in \mathrm{Irr}(\mathrm{O'N}), \quad \chi_{31} \in \mathrm{Irr}(\mathrm{Th}).$$

Here the names refer to the standard names of sporadic simple groups (see [38]), and the characters χ_i (also in the notation of [38]) are such that all irreducible characters φ of maximal subgroups H have multiplicities $(\chi_i|_H, \varphi) \neq 1$. ◆

Induction and restriction are, in a certain sense, "adjoint operations."

Theorem 3.2.12 (Frobenius–Nakayama reciprocity) *If* W *is a* KH*-module and* V *is a* KG*-module, then*

$$\mathrm{Hom}_{KG}(W^G, V) \cong \mathrm{Hom}_{KH}(W, V_H),$$
$$\mathrm{Hom}_{KG}(V, W^G) \cong \mathrm{Hom}_{KH}(V_H, W)$$

as K*-modules.*

Proof. (a) Using the embedding $W \subseteq W^G$, $w \mapsto 1 \otimes w$, we get a K-linear map $\mathrm{Hom}_{KG}(W^G, V) \to \mathrm{Hom}_{KH}(W, V_H)$, $\varphi \mapsto \varphi|_W$. Conversely, if $\psi \in \mathrm{Hom}_{KH}(W, V_H)$ and $G = \bigcup_i g_i H$ with $g_1 = 1$ we define $\tilde{\psi} \colon W^G \to V$ by

$$\tilde{\psi}\left(\sum_i g_i \otimes w_i\right) = \sum_i g_i \psi(w_i) \qquad (w_i \in W).$$

One easily checks that $\tilde{\psi}$ is a KG-homomorphism and that $\varphi \mapsto \varphi|_W$ and $\psi \mapsto \tilde{\psi}$ are inverse K-linear maps.

(b) Keeping the notation of part (a) we define a KH-homomorphism

$$\tau \colon W^G \to W, \quad \sum_i g_i \otimes w_i \mapsto w_1$$

and get a K-linear map

$$\mathrm{Hom}_{KG}(V, W^G) \to \mathrm{Hom}_{KH}(V_H, W), \quad \varphi \mapsto \varphi' := \tau \circ \varphi.$$

For $\psi \in \mathrm{Hom}_{KH}(V_H, W)$ we define $\hat{\psi}\colon V \to W^G$ by

$$\hat{\psi}(v) = \sum_{i=1}^{m} g_i \otimes \psi(g_i^{-1}v).$$

It is readily checked that $\hat{\psi}$ is a KG-homomorphism and that $\varphi \mapsto \varphi'$ and $\psi \mapsto \hat{\psi}$ are inverse K-linear maps.

$\square$

There is a corresponding reciprocity theorem for class functions as follows.

Theorem 3.2.13 (Frobenius reciprocity) *If K is a field with* $\mathrm{char}\, K \nmid |G|$ *and* $\chi \in \mathrm{cf}(G, K)$, $\varphi \in \mathrm{cf}(H, K)$, *then*

$$(\chi, \varphi^G)_G = (\chi_H, \varphi)_H.$$

Proof. Let $G = \bigcup_{i=1}^{m} g_i H$ as before. Then

$$(\chi, \varphi^G)_G = \frac{1}{|G|} \sum_{g \in G} \chi(g) \varphi^G(g^{-1}) = \frac{1}{|G|} \sum_{g \in G} \sum_{i=1}^{m} \chi(g) \dot{\varphi}(g_i^{-1} g^{-1} g_i)$$

$$= \frac{1}{|G|} \sum_{i=1}^{m} \sum_{g \in G} \chi(g_i^{-1} g g_i) \dot{\varphi}(g_i^{-1} g^{-1} g_i) = \frac{1}{|G|} \sum_{i=1}^{m} \sum_{h \in H} \chi(h) \varphi(h^{-1})$$

$$= (\chi_H, \varphi)_H.$$

$\square$

We collect some basic facts about induced modules.

Lemma 3.2.14 *Let $H \leq G$ with $G = \bigcup_{i=1}^{m} g_i H$ and let W be a KH-module.*

(a) *If V is a KG-module such that $V = \bigoplus_{i=1}^{m} g_i W$ then $V \cong W^G$. In particular,*

$$(KH)^G \cong_{KG} KG \qquad \text{as } KG\text{-modules.}$$

(b) *(Transitivity of induction) Let $H \leq U \leq G$. Then*

$$(W^U)^G \cong_{KG} W^G \qquad \text{as } KG\text{-modules.}$$

(c) *If W' is another KH-module then*

$$(W \oplus W')^G \cong_{KG} W^G \oplus W'^G.$$

(d) *If V is a KG-module then*

$$(W \otimes_K V_H)^G \cong_{KG} W^G \otimes_K V \qquad \text{as } KG\text{-modules.}$$

Proof. (a) We have $W^G = \bigoplus_{i=1}^m g_i \otimes W$, and it is easily checked that

$$\varphi: W^G \to V, \quad \sum_{i=1}^m g_i \otimes w_i \mapsto \sum_{i=1}^m g_i w_i \quad \text{with} \quad w_i \in W$$

defines an isomorphism of KG-modules.

(b) and (c) follow immediately from the basic properties of the tensor product and are easily verified.

(d) If $(w_1, \ldots, w_m)$ is a K-basis of W then

$$(W \otimes_K V_H)^G = \{ \sum_{i,j} g_i \otimes_{KH} (w_j \otimes_K v_{i,j}) \mid v_{i,j} \in V \},$$

$$W^G \otimes_K V = \{ \sum_{i,j} (g_i \otimes_{KH} w_j) \otimes_K v_{i,j} \mid v_{i,j} \in V \}.$$

We leave it to the reader to check that

$$\sum_{i,j} g_i \otimes (w_j \otimes v_{i,j}) \mapsto \sum_{i,j} (g_i \otimes w_j) \otimes g_i v_{i,j}$$

defines a KG-isomorphism. $\qquad\square$

Observe that for $H = \{1\}$ and a KH-module W, which is free as a K-module (that is, $W \cong K \oplus \cdots \oplus K$), the induced module W^G is just a free KG-module (see Lemma 3.2.14(a) and (c)). Thus a KG-module V is projective if and only if it is isomorphic to a direct summand of W^G for some free $K\{1\}$-module W. This observation leads naturally to a generalization of the concept of projectivity for modules over group algebras. Of course, by Maschke's theorem (Theorem 1.5.6) this is only interesting if $\operatorname{char} K \mid |G|$.

Definition 3.2.15 If $H \le G$ is a subgroup of G, a KG-module V is called **H-projective** if

$$V \mid W^G \qquad \text{for some } KH\text{-module } W,$$

which is free (and finitely generated) as a K-module.

As mentioned above $\{1\}$-projective is the same as projective, if we stick to our convention that all KG-modules considered are free as K-modules.

Corollary 3.2.16 *Assume $U \le H \le G$.*

(a) *If W is a U-projective KH-module then W^G is a U-projective KG-module.*

(b) *If V_p is an H-projective KG-module and V is an arbitrary KG-module then $V_p \otimes_K V$ is an H-projective KG-module.*

Proof. (a) Assume that $W \mid W_1^H$ for some KU-module W_1. By Lemma 3.2.14(c) and (b) we get

$$W^G \mid (W_1^H)^G \cong_{KG} W_1^G.$$

(b) If $V_p \mid W^G$ for some KH-module W then we obtain from Lemma 1.1.39 and Lemma 3.2.14(e)

$$V_p \otimes_K V \mid W^G \otimes_K V \cong_{KG} (W \otimes_K V_H)^G.$$

$\square$

We will come back to the concept of relative projectivity in Chapter 4, where it will play an essential role (see Section 4.8).

Theorem 3.2.17 (Mackey) *Let H and U be subgroups of a finite group G. Let T be a complete set of (U, H)-double coset representatives in G, thus $G = \dot{\bigcup}_{g \in T} UgH$. If W is a KH-module then*

$$(W^G)_U = \bigoplus_{g \in T} W_g \quad with \quad W_g \cong_{KU} ((g \otimes W)_{gHg^{-1} \cap U})^U,$$

and the KU-module W_g depends (up to KU-isomorphism) only on the double coset UgH and not on the representative g.

Proof. Let $G = \dot{\bigcup}_{i \in I} g_i H$. The subgroup U permutes the cosets of H, and the orbits are $\{g_i H \mid i \in J_g\}$ for $g \in T$, where $J_g = \{i \in I \mid g_i \in UgH\}$. Every g_j for $j \in J_g$ can be written in the form $g_j = u_j g h_j$ with $u_j \in U$ and $h_j \in H$. Then

$$W^G = \bigoplus_{i \in I} g_i \otimes W = \bigoplus_{g \in T} \bigoplus_{j \in J_g} g_j \otimes W = \bigoplus_{g \in T} \bigoplus_{j \in J_g} u_j(g \otimes W)$$

and the $W_g = \bigoplus_{j \in J_g} g_j \otimes W$ are KU-modules, which clearly depend only on the double cosets UgH. Since

$$\mathrm{Stab}_U(gH) = gHg^{-1} \cap U \quad \text{we have} \quad U = \dot{\bigcup}_{j \in J_g} u_j(gHg^{-1} \cap U).$$

From Lemma 3.2.14(a) applied to the $K(gHg^{-1} \cap U)$-submodule $g \otimes W$ of W_g the assertion follows. $\square$

Corollary 3.2.18 *Let $N \trianglelefteq G$ with $G = \dot{\bigcup}_{i=1}^s g_i N$. Assume that W is a KN-module. Then*

$$(W^G)_N = \bigoplus_{i=1}^s g_i \otimes W.$$

In particular, if W is simple, $(W^G)_N$ is semisimple.

Proof. Take $U = H = N$ and $T = \{g_1, \ldots, g_s\}$ in Theorem 3.2.17. If W is simple, then $g_i \otimes W$ is also simple by Remark 3.2.2. $\square$

We combine Mackey's theorem with Theorem 3.2.12 to obtain Theorem 3.2.19.

Theorem 3.2.19 (Intertwining number theorem) *Let $U, H \leq G$ and D be a set of representatives of (U, H)-double cosets of U and H in G. If V is a KU-module and W is a KH-module then*

$$\operatorname{Hom}_{KG}(V^G, W^G) \cong_K \bigoplus_{g \in D} \operatorname{Hom}_{KH_g}(V_{H_g}, (g \otimes W)_{H_g})$$

as K-modules, where $H_g = gHg^{-1} \cap U$ for $g \in D$.

Proof. Using Theorem 3.2.12 we get

$$
\begin{aligned}
\operatorname{Hom}_{KG}(V^G, W^G) &\cong_K \operatorname{Hom}_{KU}(V, (W^G)_U) \\
&\cong_K \bigoplus_{g \in D} \operatorname{Hom}_{KU}(V, ((g \otimes W)_{H_g})^U \\
&\cong_K \bigoplus_{g \in D} \operatorname{Hom}_{KH_g}(V_{H_g}, (g \otimes W)_{H_g}).
\end{aligned}
\tag{3.1}
$$

$\square$

Corollary 3.2.20 *Let $U, H \leq G$ and D be a set of representatives of (U, H)-double cosets in G. If $\varphi \in \operatorname{Char}_{\mathbb{C}}(U)$ and $\psi \in \operatorname{Char}_{\mathbb{C}}(H)$ then*

$$(\varphi^G, \psi^G)_G = \sum_{g \in D} (\varphi_{gHg^{-1} \cap U}, {}^g\psi_{gHg^{-1} \cap U})_{gHg^{-1} \cap U}.$$

Lemma 3.2.21 *Let $H \leq G$ with $G = \dot{\bigcup}_{g \in T} gH$ and let V, W be KH-modules. For $\varphi \in \operatorname{Hom}_{KH}(V, W)$ define*

$$\varphi^G \left(\sum_{g \in T} g \otimes v_g \right) := \sum_{g \in T} g \otimes \varphi(v_g) \qquad (v_g \in V).$$

Then

$$\Phi : \operatorname{Hom}_{KH}(V, W) \to \operatorname{Hom}_{KG}(V^G, W^G), \quad \varphi \mapsto \varphi^G$$

is an injective K-homomorphism independent of the choice of the transversal T. For $V = W$ it is a K-algebra homomorphism. Furthermore we have the following.

(a) *φ is injective (surjective) if and only if φ^G is injective (surjective).*

(b) *A sequence of KH-homomorphisms*

$$\{0\} \longrightarrow V_1 \xrightarrow{\varphi} V \xrightarrow{\psi} V_2 \longrightarrow \{0\} \tag{3.2}$$

is exact if and only if the sequence of KG-modules

$$\{0\} \longrightarrow V_1^G \xrightarrow{\varphi^G} V^G \xrightarrow{\psi^G} V_2^G \longrightarrow \{0\} \tag{3.3}$$

is exact.

(c) *Sequence* (3.2) *is non-split if and only if* (3.3) *is non-split.*

(d) *If K is a field and $\mathrm{Hom}_{KH}(V, ((g \otimes V)_{gH\cap H})^H) = \{0\}$ for all $g \in G \setminus H$ then Φ is an isomorphism.*

Proof.   Obviously φ^G is KG-linear and independent of the choice of the transversal. We know that Φ is injective because $\varphi^G(1 \otimes v) = 1 \otimes \varphi(x)$ for $v \in V$.

(a) and (b) follow immediately from the definition of φ^G.

(c) If (3.2) splits then obviously (3.3) also splits. Conversely, assume that there is a $\sigma \in \mathrm{Hom}_{KG}(V_2^G, V^G)$ with $\psi^G \circ \sigma = \mathrm{id}_{V_2^G}$. We define KH-homomorphisms

$$\pi \colon W^G \to W \;,\; \sum_{g \in T} g \otimes w_g \mapsto w_1, \qquad \gamma \colon W_2 \to W \;,\; w_2 \mapsto \pi(\sigma(1 \otimes w_2)).$$

Then it is readily verified that $\psi \circ \gamma = \mathrm{id}_{V_2}$, so (3.2) splits also.

(d) follows by applying Theorem 3.2.19, more precisely by applying (3.1), with $U := H$.   □

We recall from Definition 1.6.29 that for A-modules V, W (A a K-algebra) $\overline{\mathrm{Hom}}_A(V, W)$ denotes the factor module of $\mathrm{Hom}_A(V, W)$ modulo the K-submodule of projective homomorphisms. The following extension of Frobenius Nakayama reciprocity will be of importance in Chapter 4.

Corollary 3.2.22 *For $H \leq G$ let V be a KG-module and let W be a KH-module. Then*

$$\overline{\mathrm{Hom}}_{KG}(W^G, V) \cong_K \overline{\mathrm{Hom}}_{KH}(W, V_H).$$

Proof.   From the proof of Theorem 3.2.12 we know a K-isomorphism $\mathrm{Hom}_{KH}(W, V_H) \to \mathrm{Hom}_{KG}(W^G, V)$, $\psi \mapsto \tilde{\psi}$, which is the inverse of the restriction map $\mathrm{Hom}_{KG}(W^G, V) \to \mathrm{Hom}_{KH}(W, V_H)$. It suffices to show that both isomorphisms map projective homomorphisms to projective homomorphisms.

If $\varphi \in \mathrm{Hom}_{KG}(W^G, V)$ factors through a projective KG-module P, then $\varphi|_W$ factors through P_H, which is a projective KH-module by Lemma 1.6.17. Conversely, if $\psi \in \mathrm{Hom}_{KH}(W, V_H)$ factors through the projective KH-module P_1, that is $\psi = \sigma \circ \tau$ with $\sigma \colon P_1 \to V_H$ and $\tau \colon W \to P_1$, then $\tilde{\psi} = \tilde{\sigma} \circ \tau^G$ with $\tau^G \colon W^G \to P_1^G$, as in Lemma 3.2.21, and

$$\tilde{\sigma} \colon P_1^G \to V_H, \; \sum_{i=1}^n g_i \otimes x_i \mapsto \sum_{i=1}^n g_i \sigma(x_i) \qquad (x_i \in P_1),$$

as in the proof of Theorem 3.2.12. By Corollary 3.2.16 P_1^G is projective and the proof is complete.   □

We close this section by giving a number of practical examples showing how induced characters can be used to compute irreducible characters of a group.

Example 3.2.23 In this example we show how induction may be used to construct the character table of $G = A_6$ using that of $H = A_5$. We start from the table of A_5 and (as in Example 2.1.23) from the irreducible characters of A_6 obtained from the natural permutation characters of A_6 of the actions on $\Omega = \{1, \ldots, 6\}$ and $\binom{\Omega}{2}$, namely $\theta_\Omega = \chi_1 + \chi_2$ and $\theta_{\binom{\Omega}{2}} = \chi_1 + \chi_2 + \chi_6$. The first line contains the centralizer orders and the next one the names of the classes. For A_6 these are $1a = \{1\}$, $2a = ((1,2)(3,4))^G$, $3a = (1,2,3)^G$, $3b = ((1,2,3)(4,5,6))^G$, $4a = ((1,2)(3,4,5,6))^G$, $5a = (1,2,3,4,5)^G$ and $5b = (1,3,5,2,4)^G$:

	60	4	3	5	5		360	8	9	9	4	5	5
A_5	1a	2a	3a	5a	5b	A_6	1a	2a	3a	3b	4a	5a	5b
φ_1	1	1	1	1	1	χ_1	1	1	1	1	1	1	1
φ_2	3	-1	0	α	α'	χ_2	5	1	2	-1	-1	0	0
φ_3	3	-1	0	α'	α	χ_6	9	1	0	0	1	-1	-1
φ_4	4	0	1	-1	-1								
φ_5	5	1	-1	0	0								
$(\chi_2)_H$	5	1	2	0	0	φ_2^G	18	-2	0	0	0	α	α'
$(\chi_9)_H$	9	1	0	-1	-1	φ_4^G	24	0	3	0	0	-1	-1
						φ_5^G	30	2	-3	0	0	0	0

with $\alpha, \alpha' = \frac{1}{2}(1 \pm \sqrt{5})$. One can see that $(\chi_2)_H = \varphi_1 + \varphi_4$ and $(\chi_6)_H = \varphi_4 + \varphi_5$, hence, by Theorem 3.2.13,

$$(\varphi_4^G, \chi_2)_G = (\varphi_4^G, \chi_6)_G = 1 \quad \text{and} \quad (\varphi_5^G, \chi_6)_G = 1.$$

A short calculation gives $(\varphi_2^G, \varphi_2^G)_G = 2$, $(\varphi_4^G, \varphi_4^G)_G = 3$, $(\varphi_5^G, \varphi_5^G)_G = 4$. Hence $\varphi_4^G - \chi_2 - \chi_6 \in \mathrm{Irr}(G)$, which we call χ_7 with values $10, -2, 1, 1, 0, 0, 0$. As one can see, $(\chi_7)_H = \varphi_2 + \varphi_3 + \varphi_4$, hence $(\varphi_2^G, \chi_7)_G = 1$. Thus we have $\varphi_2^G - \chi_7 = \chi_4 \in \mathrm{Irr}(G)$ with values $8, 0, -1, -1, 0, \alpha, \alpha'$ and $(\chi_4)_H = \varphi_2 + \varphi_5$. We denote the character which is algebraically conjugate to χ_4 by χ_5 and finally get $\varphi_5^G - \chi_6 - \chi_4 - \chi_5 = \chi_3 \in \mathrm{Irr}(G)$. The complete character table of A_6, which we had already computed using the Dixon–Schneider algorithm (see Example 2.4.5), can be found on p. 120 ◆

In order to compute the character of a group G induced from a known character of a subgroup H, it is necessary to know the fusion map. In the special case that $H = \langle g \rangle$ is cyclic, the fusion map of H into G can be obtained from – and is, in fact, equivalent to – that part of the power maps of G which give for each prime p with $p \leq |H|$ the conjugacy class $C_{i(p)}$ of G which contains g^p. Since the character tables of the cyclic groups are known, we obtain a good source of characters of G, provided we know its power maps. These induced characters can be particularly useful for finding character values on algebraically conjugate conjugacy classes. Of course they will usually have very large norms, so that in most cases the LLL-algorithm will be applied.

Example 3.2.24 We want to compute the character table of the sporadic simple Mathieu group M_{22}, which is a 3-fold transitive permutation group on 22 letters of order $443520 = 2^7 \cdot 3^2 \cdot 5 \cdot 7 \cdot 11$. We assume that we know the power maps of $G := \mathrm{M}_{22}$ and the centralizer orders.

Using the command `InducedCyclic` we induce up to G all linear characters of all the representatives of the conjugacy classes of cyclic subgroups of G to get:

```
gap> t := CharacterTable("M22");; ind := InducedCyclic( t, "all" );;
```

We obtain 26 characters including the regular one. Observe that the 11 characters of a cyclic subgroup of order 11 yield only three different characters when induced to G. We reduce these characters with the trivial character of G and use the LLL-algorithm in order to obtain one irreducible character (of degree 231) and a list **r** of ten generalized characters of norm two and three:

```
gap> irr := Irr(t){[1]};; red := Reduced( t, irr, ind );;
gap> l := LLL( t, red.remainders );;
gap> Append( irr, l.irreducibles ); # irr now contains 2 irreducibles
gap> r := l.remainders;;
```

Since we know that there are ten irreducible characters missing in our list of irreducibles, we apply the additive decomposition algorithm of Section 2.8 with the command `OrthogonalEmbeddings`; see Example 2.8.16:

```
gap> m := MatScalarProducts( t, r, r );;
gap> oe := OrthogonalEmbeddingsSpecialDimension( t, r, m, 10 );;
gap> Length( oe.irreducibles );
10
```

Thus all irreducible characters of G have been obtained. We display the irreducible characters:

```
M22
           1a 2a 3a 4a 4b 5a 6a 7a 7b 8a 11a 11b

X.1         1  1  1  1  1  1  1  1  1  1   1   1
X.2        21  5  3  1  1  1 -1  .  . -1  -1  -1
X.3        45 -3  .  1  1  .  .  A /A -1   1   1
X.4        45 -3  .  1  1  .  . /A  A -1   1   1
X.5        55  7  1  3 -1  .  1 -1 -1  1   .   .
X.6        99  3  .  3 -1 -1  .  1  1 -1   .   .
X.7       154 10  1 -2  2 -1  1  .  .  .   .   .
X.8       210  2  3 -2 -2  . -1  .  .  .   1   1
X.9       231  7 -3 -1 -1  1  1  .  . -1   .   .
X.10      280 -8  1  .  .  .  1  .  .  .   B  /B
X.11      280 -8  1  .  .  .  1  .  .  .  /B   B
X.12      385  1 -2  1  1  . -2  .  .  1   .   .
```

with $A = \frac{1}{2}(-1 + \sqrt{-7})$ and $B = \frac{1}{2}(-1 + \sqrt{-11})$.   ♦

To give a rough idea of the possible range of applications of the above method we look at a larger example.

Example 3.2.25 We want to construct the character table of the sporadic simple Suzuki group $G = \text{Suz}$ (see [38]) using only the power maps and centralizer orders. Suz is a simple group of order $448345497600 = 2^{13} \cdot 3^7 \cdot 5^2 \cdot 7 \cdot 11 \cdot 13$ with 43 conjugacy classes.

```
gap> t := CharacterTable("Suz");;
gap> irr := Irr(t){[1]};;              ind := InducedCyclic( t, "all");;
gap> red := Reduced( t, irr, ind );;   l := LLL( t, red.remainders );;
gap> r := l.remainders;;  List( r , Norm );
[ 2, 2, 2, 11, 10, 15, 14, 12, 12, 11, 11, 8, 6, 14, 14, 10, 14, 18, 13,
   14, 19, 10, 27, 29, 25, 21, 21, 24, 18, 22, 15, 19, 21, 19, 27, 25,
   24, 23, 30, 23, 23, 20 ]
```

The characters `ind` induced by the linear characters of all cyclic subgroups span a $\mathbb{Z}$-lattice. We have reduced `ind` with the trivial character; in other words, we have computed a generating system `red.remainders` for the orthogonal complement $L \le \mathbb{Z}\,\text{Irr}(G)$ of $\langle 1_G \rangle_{\mathbb{Z}}$ in $\langle \text{ind}, 1_G \rangle_{\mathbb{Z}}$ and have finally computed an LLL-reduced basis `r` for L. No irreducible character of G was found and `r` contains only few generalized characters of very small norm. Trying to use `OrthogonalEmbedding` at this stage looks hopeless. Instead, we generate more generalized characters of G by symmetrizing the elements of `r`, thereby enlarging the lattice L and fortunately obtaining "vectors" of smaller norm. One might be tempted to try to enlarge L by forming products, but L is, in fact, closed under multiplication by Lemma 3.2.7(a):

```
gap> for i in [2,3,4] do
>    Append(r , Symmetrizations(t,r,i) );
>    r := Reduced(t,irr,r).remainders;
>    r := LLL(t,r).remainders;
>    od;  List(r , Norm);
[ 2, 2, 3, 2, 4, 3, 3, 5, 5, 4, 8, 6, 7, 6, 14, 12, 12, 11, 11, 7, 10,
   5, 10, 11, 10, 12, 13, 8, 5, 12, 13, 9, 6, 15, 11, 12, 16, 9, 12, 14,
   12, 13 ]
gap> M := MatScalarProducts( t, r, r );;
gap> oe := OrthogonalEmbeddings( M, 42 );;
gap> Length( oe.solutions ); C := oe.vectors;;
12
```

This time `OrthogonalEmbedding` was successful (after an hour of CPU-time on a Pentium III, 1 GHz workstation) and produced all 12 solutions $X_i \in \mathbb{Z}^{42 \times 42}$ of $X\,X^{\mathrm{T}} = M$ (up to equivalence, see Definition 2.8.13). We get these X_i^{T} by `C{oe.solutions[i]}` for $i = 1, \ldots, 12$ (see Example 2.8.16). For each solution X_i the system of equations (2.18) has a unique solution consisting of a list $(\psi_{i,1}, \ldots, \psi_{i,n})$ of $n = 42$ class functions $\psi_{i,j}$ of norm one. Observe that we consider solutions up to sign, so we multiply those class functions $\psi_{i,j}$ with $\psi_{i,j}(1) < 0$ by -1:

```
gap> Xli := List( oe.solutions , x -> C{x} );;
gap> irrli := List( Xli , x -> ( TransposedMat( x ) )^-1 * r );;
gap> for i in [1..Length( irrli )] do
>        for j in [1..Length( irrli[i] )] do
```

```
>          if irrli[i][j][1] < 0 then irrli[i][j] := - irrli[i][j]; fi;
>       od;
>    od;
gap>
```

We have arrived at 12 different candidates for the character table of G, all of which, of course, satisfy the orthogonality relations. Inspecting the values $\psi_{i,j}[1]$ we find four of these candidates contain a "character" of degree zero. Leaving these aside we next test whether or not (tensor) products reduce properly:

```
gap> irrli := Filtered(irrli, x -> not  0 in List( x, x_i -> x_i[1] ) );;
gap> Length(irrli);
8
gap> for i in [1..8] do
> red:= Reduced( t, irrli[i], Tensored(irrli[i], irrli[i] ) );
> if red.remainders = [] then Print(i,","); fi;
> od;
5,6,7,8,
gap> irrli := irrli{[5,6,7,8]};;
```

Thus only four of our candidates pass the test. Next we check whether or not the nth symmetrizations reduce properly and find that all do for $n = 2$ but only one does for $n = 3$:

```
gap>    for i in [1..4] do
> red:= Reduced( t, irrli[i], Symmetrizations (t, irrli[i], 3) );
> if red.remainders = [] then Print(i,","); fi;
> od;
3,
gap> Append( irr, irrli[3] );
```

Thus we have computed all irreducibles. If instead of `OrthogonalEmbeddings` we had used `OrthogonalEmbeddingsSpecialDimension` (see Example 2.8.16) we would have obtained automatically a list of 18 irreducible characters which could then have been used to produce further reducible characters. But we intended to find all non-trivial irreducibles in one go.   ♦

In the above examples we were able to compute all irreducible characters of a group G just by using the power maps of G. We will give further examples for some sporadic simple groups (cf. [38]) where this approach is successful.

Definition 3.2.26 For a finite group G let

$$\mathrm{Ind}(G, \mathrm{Cyc}) = \langle\ \lambda^G \mid \lambda \in \mathrm{Irr}(\langle g \rangle),\ g \in G\ \rangle_{\mathbb{Z}}.$$

Remark 3.2.27 (a) $\mathrm{Ind}(G, \mathrm{Cyc})$ is an ideal in the ring $\mathbb{Z}\,\mathrm{Irr}(G)$ of generalized characters by Lemma 3.2.7 and $\mathrm{I}_1(G) = \mathrm{Ind}(G, \mathrm{Cyc}) + \mathbb{Z}\mathbf{1}_G$ is a subring. A $\mathbb{Z}$-basis of this can be computed from the power maps.

(b) The character tables of the following sporadic simple groups can be computed from the power maps and centralizer orders (see Table 3.1).

Table 3.1.

G	$|\mathrm{Irr}(G)|$	$\mathbb{Z}\,\mathrm{Irr}(G)/\mathrm{I}_1(G)$	$|\mathrm{Irr}(G) \cap \mathrm{I}_1(G)|$
M_{11}	10	$\{1\}$	$|\mathrm{Irr}(G)|$
M_{12}	15	C_2^4	2
J_1	15	$\{1\}$	$|\mathrm{Irr}(G)|$
M_{22}	12	C_2^3	2
J_2	21	$C_2^5 \times C_{10}$	2
M_{23}	17	C_2^2	5
HS	24	$C_2^5 \times C_8 \times C_{48}$	1
J_3	21	$C_2 \times C_6$	4
M_{24}	26	$C_2^5 \times C_8 \times C_{16}$	1
McL	24	$C_6^2 \times C_{18}$	1
He	33	$C_2^7 \times C_8 \times C_{336}$	4
Ru	36	$C_2^6 \times C_4^3 \times C_8 \times C_{16} \times C_{32}$	3
Suz	43	$C_2^7 \times C_6^4 \times C_{12} \times C_{24} \times C_{48}^2 \times C_{144} \times C_{288}$	1
O'N	30	$C_2 \times C_6 \times C_{24}$	4
Co_3	42	$C_2^4 \times C_6^2 \times C_{12} \times C_{24} \times C_{72}^2 \times C_{864}$	1

Observe that for a set $L = \{\psi_1, \ldots, \psi_s\}$ of generalized characters, the structure of $\mathbb{Z}\,\mathrm{Irr}(G)/\langle L\rangle_{\mathbb{Z}}$ can easily be found by computing the elementary divisors of the matrix of scalar products $M = [(\chi, \psi)_G]_{\chi \in \mathrm{Irr}(G), \psi \in L}$. In GAP this can be done in the following manner:

```
gap> t := CharacterTable( "M22" );;
gap> L := InducedCyclic( t, "all" );; Add( L, Irr(t)[1] );
gap> ElementaryDivisorsMat( MatScalarProducts( Irr(t), L ) );
[ 1, 1, 1, 1, 1, 1, 1, 1, 1, 2, 2, 2 ]
```

Example 3.2.28 We consider $G := \mathrm{SL}_2(q)$ for a prime power q. Let B be the subgroup of lower triangular matrices in G and let T be the subgroup of diagonal matrices in G. If $\mathbb{F}_q^\times = \langle \alpha \rangle$ then $T = \langle t \rangle$ with $t := \mathrm{diag}(\alpha, \alpha^{-1}) \in G$ and

$$B = \left\{ \begin{bmatrix} \alpha^k & 0 \\ \beta & \alpha^{-k} \end{bmatrix} \mid 1 \le k \le q-1,\ \beta \in \mathbb{F}_q \right\}.$$

If $w := \begin{bmatrix} 0 & 1 \\ -1 & 0 \end{bmatrix} \in G$ then it is easily seen that $\{1, w\}$ is a system of representatives of (B, B)-double cosets in G (i.e. $G = B \cup BwB$) and $B \cap wBw^{-1} = T$. Since $B/B' \cong T \cong \mathbb{F}_q^\times$, it is clear that B has exactly $q-1$ linear characters $\mathbf{1}_B = \lambda_0, \ldots, \lambda_{q-2}$ over $\mathbb{C}$ given by

$$\lambda_i \colon \begin{bmatrix} \alpha^k & 0 \\ \beta & \alpha^{-k} \end{bmatrix} \mapsto \zeta_{q-1}^{ik} \in \mathbb{C} \qquad (0 \le i \le q-1).$$

From Corollary 3.2.20 we get

$$(\lambda_i^G, \lambda_j^G)_G = (\lambda_i, \lambda_j)_B + ((\lambda_i)_T, {}^w(\lambda_j)_T)_T.$$

Since $w^{-1}tw = t^{-1}$ for $t \in T$ we get ${}^w(\lambda_i)_T = (\lambda_{q-1-i})_T$.

(a) If q is even we see that $\chi_i := \lambda_i^G = \lambda_{q-1-i}^G$ is an irreducible character of degree $(q+1)$ for $1 \leq i \leq \frac{q-2}{2}$. Furthermore $(\lambda_0^G, \lambda_0^G)_G = 2$ and hence $\lambda_0^G = \mathbf{1}_G + \chi_0$ with $\chi_0 \in \mathrm{Irr}(G)$ with $\chi_0(1) = q$.

(b) If q is odd then it follows that $\chi_i := \lambda_i^G = \lambda_{q-1-i}^G$ is irreducible of degree $(q+1)$ for $1 \leq i \leq \frac{q-3}{2}$. Furthermore as in (a) we have $\lambda_0^G = \mathbf{1}_G + \chi_0$ with $\chi_0 \in \mathrm{Irr}(G)$ with $\chi_0(1) = q$, and for $\lambda = \lambda_{\frac{q-1}{2}}$ we get $(\lambda^G, \lambda^G)_G = 2$, hence $\lambda^G = \chi' + \chi''$ with $\chi' + \chi'' \in \mathrm{Irr}(G)$.

We proceed to determine the conjugacy classes of G. If $X^2 + aX + b \in \mathbb{F}_q[X]$ is the minimal polynomial of $\gamma \in \mathbb{F}_{q^2}$ with $\mathbb{F}_{q^2}^\times = \langle\gamma\rangle$, then $s' := \begin{bmatrix} 0 & -a \\ 1 & -b \end{bmatrix} \in$ $\mathrm{GL}_2(q)$ has eigenvalues γ, γ^q and hence order $q^2 - 1$ and determinant γ^{q+1}. Consequently $s := (s')^{q-1} \in \mathrm{SL}_2(q)$ has order $q + 1$. We put $m := q/2$ if q is even and $m := (q-1)/2$ otherwise. Then $t^1, \ldots, t^{m-1}$ and $s^1, \ldots, s^m$ have pairwise different eigenvalues and hence are pair-wise non-conjugate in G. Also it is easily seen that $\mathbf{C}_G(t^k) = \langle t \rangle$ for $1 \leq k \leq m-1$ and $\mathbf{C}_G(s^l) = \langle s \rangle$ for $1 \leq l \leq m$.

(a) If q is even then the centralizer of $u := \begin{bmatrix} 1 & 0 \\ 1 & 1 \end{bmatrix}$ in G is

$$U := \{\begin{bmatrix} 1 & 0 \\ \beta & 1 \end{bmatrix} \mid \beta \in \mathbb{F}_q\} \cong (\mathbb{F}_q, +).$$

We find $|u^G| + (m-1)|t^g| + m|s^G| = q^3 - q - 1 = |G| - 1$. Hence we have found representatives for all conjugacy classes of G:

g	1	u	t^k	s^l	$1 \leq k \leq m-1$
$\lvert\mathbf{C}_G(g)\rvert$	$\lvert G\rvert$	q	$q-1$	$q+1$	$1 \leq l \leq m$

(b) If q is odd we put $z := t^m = \begin{bmatrix} -1 & 0 \\ 0 & -1 \end{bmatrix} \in \mathbf{Z}(G)$ and $u' := \begin{bmatrix} 1 & 0 \\ \alpha & 1 \end{bmatrix}$, which is conjugate to u in $\mathrm{GL}_2(q)$ but not in G, and we obtain the following representatives for $\mathrm{cl}(G)$:

g	1	z	u	zu	u'	zu'	t^k	s^l	$1 \leq k \leq m-1$
$\lvert\mathbf{C}_G(g)\rvert$	$\lvert G\rvert$	$\lvert G\rvert$	$2q$	$2q$	$2q$	$2q$	$q-1$	$q+1$	$1 \leq l \leq m$

To complete the character table we induce up the linear characters $\mu_j \in \mathrm{Irr}(\langle s \rangle)$ defined by $\mu_j(s^l) = \zeta_{q+1}^{jl}$. Observe that $\mu_j{}^G(s^l) = \zeta_{q+1}^{jl} + \zeta_{q+1}^{-jl}$ for $l = 1, \ldots, m$ and $\mu_j{}^G(t^k) = \mu_j{}^G(u) = \mu_j{}^G(u') = 0$ for $k = 1, \ldots, m-1$. We consider the generalized character $\theta_j := \chi_0 \cdot \lambda_j^G - \lambda_j^G - \mu_j^G$ and find that $(\theta_j, \theta_j)_G = 1$, $\theta_j(1) > 0$ and hence $\theta_j \in \mathrm{Irr}(G)$ for $j = 1, \ldots, m$.

(a) If q is even, we see that the number of irreducible characters found equals

$|\operatorname{cl}(G)|$, so the character table is as follows:

Character table of $L_2(q)$ for even q

$SL_2(q)$	1	u	t^k	s^l	
$|C_G(g)|$	$|G|$	q	$q-1$	$q+1$	
1_G	1	1	1	1	
χ_0	q	0	1	-1	
χ_i	$q+1$	1	$\zeta_{q-1}^{ik} + \zeta_{q-1}^{-ik}$	0	$1 \le i \le m-1$
θ_j	$q-1$	-1	0	$-(\zeta_{q+1}^{jl} + \zeta_{q+1}^{-jl})$	$1 \le j \le m$
			$1 \le k \le m-1$	$1 \le l \le m$	$m = \frac{q}{2}$

(b) If q is odd then it turns out that $(\theta_{m+1}, \theta_{m+1})_G = 2$. We list the (generalized) characters of norm two found so far, writing $\epsilon := (-1)^m$:

g	1	z	u	zu	u'	zu'	t^k	s^l
$(\chi + \chi')(g)$	$q+1$	$\epsilon(q+1)$	1	ϵ	1	ϵ	$2(-1)^k$	0
$\theta_{m+1}(g)$	$q-1$	$-\epsilon(q-1)$	-1	ϵ	-1	ϵ	0	$2(-1)^{l+1}$

To split these up into irreducibles we use the fact that u' is a power of u, so that the classes u^G and u'^G are algebraically conjugate. Also we first consider the non-faithful characters, that is, the characters of $L_2(q) = SL_2(q)/Z$ with $Z := \mathbf{Z}(SL_2(q))$. We see that a single pair of algebraically conjugate irreducible characters is missing, which must be (χ, χ') if $\epsilon = 1$ (that is $q \equiv 1 \mod 4$) or θ, θ' with $\theta + \theta' = \theta_{m+1}$ else. Here we denote characters of $L_2(q)$ and their inflations to G by the same symbol. The orthogonality relations are sufficient to find the values of χ, χ' or θ, θ', respectively. Observe that $t^k Z$ and $t^{m-k} Z$ are conjugate, as are $s^l Z$ and $s^{m-l} Z$.

Character table of $L_2(q)$ for odd q

	$1Z$	uZ	$u'Z$	$t^k Z$	$s^l Z$	
1_G	1	1	1	1	1	
χ_0	q	0	0	1	-1	
χ_{2i}	$q+1$	1	1	$\zeta_{q-1}^{2ik} + \zeta_{q-1}^{-2ik}$	0	$1 \le i \le \lfloor \frac{q-3}{4} \rfloor$
θ_{2j}	$q-1$	-1	-1	0	$-(\zeta_{q+1}^{2jl} + \zeta_{q+1}^{-2jl})$	$1 \le j \le \lfloor \frac{q-1}{4} \rfloor$
χ	$(q+1)/2$	$-b_q$	$-b'_q$	$(-1)^k$	0	if $q \equiv 1 \mod 4$
χ'	$(q+1)/2$	$-b'_q$	$-b_q$	$(-1)^k$	0	if $q \equiv 1 \mod 4$
θ	$(q-1)/2$	b_q	b'_q	0	$(-1)^{l+1}$	if $q \equiv -1 \mod 4$
θ'	$(q-1)/2$	b'_q	b_q	0	$(-1)^{l+1}$	if $q \equiv -1 \mod 4$
				$1 \le k \le \lfloor \frac{q-3}{4} \rfloor$	$1 \le l \le \lfloor \frac{q-1}{4} \rfloor$	

with $b_q := \frac{1}{2}(-1 + \sqrt{\epsilon q})$, $b'_q := \frac{1}{2}(-1 - \sqrt{\epsilon q})$. The completion of the character table of G is left to the exercises (Exercise 3.2.10). The character table of $L_2(q)$ was first computed by Schur (see [157], pp. 113–137). ◆

Exercises

Exercise 3.2.1 Let $H \le G$ and let $W \subseteq KG$ be a KH-module. Show that $W^G \cong_{KG} KG \cdot W$.

Exercise 3.2.2 Let $H \leq G$ and let W be a KH-module. Show that for any $g \in G$

(a) $W^G \cong_{KG} (g \otimes W)^G$,

(b) any H-projective KG-module is also H^g-projective.

Exercise 3.2.3 Let $P \in \mathrm{Syl}_p(G)$ and let K be a field of characteristic p. Assume that V is a projective KG-module of dimension $p^k r$ with $p \nmid r$. Show that

$$\underbrace{V \oplus \cdots \oplus V}_{r} \mid (K[G/P])^G.$$

Here $K[G/P]$ is the permutation module on the cosets of P.
Hint: Use Theorem 3.2.12, Lemma 1.6.17 and Corollary 1.6.25.

Exercise 3.2.4 Let $H \trianglelefteq G$ with finite index, $G = \dot{\bigcup}_{i=1}^s g_i H$ and let V be a KG-module. Prove that

$$\Phi \colon \sum_{i=1}^s g_i \otimes v_i \mapsto \sum_{i=1}^s g_i H \otimes g_i v_i \qquad \text{for } v_i \in V$$

defines a KG-isomorphism

$$\Phi \colon (V_H)^G \to K(G/H) \otimes_K V,$$

where the regular module $K(G/H)$ is considered as a KG-module by inflation.

Exercise 3.2.5 Let $H \leq G$ and let W be a KH-module with $W' \leq_{KH} W$. Show that $W'^G \leq_{KG} W^G$. Conclude that if W^G is simple then W must be simple.

Exercise 3.2.6 Let $N \trianglelefteq G$ with natural projection $\pi \colon G \to \bar{G} := G/N$ and $\chi \in \mathrm{Irr}(G)$. Prove the following.

(a) If $\bar{\psi} \in \mathrm{Irr}(\bar{G})$ then $\psi := \bar{\psi} \circ \pi \in \mathrm{Irr}(G)$ and $(\chi|_N)^G = \sum_{\bar{\psi} \in \mathrm{Irr}(\bar{G})} \psi(1) \psi \otimes \chi$. Furthermore $((\chi|_N)^G, (\chi|_N)^G)_G = [G : N] (\chi|_N, \chi|_N)_N$.

(b) If $\chi|_N$ is irreducible then

$$\{\psi \otimes \chi \mid \bar{\psi} \in \mathrm{Irr}(\bar{G})\} = \{\chi' \in \mathrm{Irr}(G) \mid \chi'|_N = a\chi_N \text{ for some } a \in \mathbb{N}\}.$$

Exercise 3.2.7 Let $Z \leq \mathbf{Z}(G) \leq G$ and let $\lambda \in \mathrm{Hom}(Z, \mathbb{C}^\times)$ be a linear character of Z. Let

$$\{\chi_1, \dots, \chi_r\} = \{\chi \in \mathrm{Irr}(G) \mid (\chi_Z, \lambda)_Z > 0\}.$$

Show that $\sum_{i=1}^r \chi_i(1)^2 = [G : Z]$ and $\sum_{i=1}^r \chi_i(1)\chi_i(g) = 0$ for $g \notin Z$.

Exercise 3.2.8 Assume that $A = [a_{ij}]_{i,j=1}^n \in \mathbb{C}^{n \times n}$ is a hermitian matrix of rank r and assume that the first r column vectors are linearly independent. Show that there is a matrix $P \in \mathbb{C}^{n \times n}$ such that

$$P = \left[\begin{array}{c|c} \mathbf{I}_r & C \\ \hline 0 & \mathbf{I}_{n-r} \end{array}\right] \quad \text{with} \quad P^{\mathrm{T}} A^{\mathrm{T}} \bar{P} = \left[\begin{array}{c|c} A_1^{\mathrm{T}} & 0 \\ \hline 0 & 0 \end{array}\right],$$

with $\bar{P}$ denoting the complex conjugate of P and $A_1 = [a_{ij}]_{i,j=1}^r$, and conclude that the submatrix A_1 of A is invertible.

Exercise 3.2.9 Compute the character table of the sporadic Hall–Janko group J_2, which is a simple group of order $604800 = 2^7 \cdot 3^3 \cdot 5^2 \cdot 7$ with 21 conjugacy classes (see [38]) using only the power maps and centralizer orders. Observe that there is not a unique solution (as in Example 3.2.25) but that there are two such tables which differ by a permutation of the conjugacy classes, which, however, can be compensated by a permutation of the irreducible characters; that is, the two tables differ by a character table automorphism.

Exercise 3.2.10 Complete the character table of $\mathrm{SL}_2(q)$ for odd q.

3.3   Symmetric groups

The representation theory of the symmetric groups is well developed and has a huge literature, starting with a fundamental paper [61] by Frobenius in 1900. We will hardly touch this theory, which has a strong combinatorial flavour, and refer instead to the standard work by James and Kerber [96]. In this section we just show how the irreducible characters of the symmetric groups over a field of characteristic zero can be labeled by partitions and how they can be computed recursively. First we have to introduce some combinatorial notions.

Definition 3.3.1 A **partition** of an integer n is a sequence $\lambda = (\lambda_1, \ldots, \lambda_r)$ of positive integers λ_i with

$$\lambda_1 \geq \lambda_2 \geq \cdots \geq \lambda_r \quad \text{with} \quad \sum_{i=1}^r \lambda_i = n.$$

We shall write $\lambda \vdash n$ to indicate that λ is a partition of n. It is standard to abbreviate repeated parts of a partition using exponents; for instance $(3^2, 2, 1^3) := (3, 3, 2, 1, 1, 1) \vdash 11$.

It is well known that the conjugacy classes of the symmetric group S_n can be labeled by the partitions of n. Any element $\sigma \in S_n$ can be written as a product of disjoint λ_i-cycles σ_i with $n \geq \lambda_1 \geq \cdots \geq \lambda_r \geq 1$ and thus determines a partition $\lambda = (\lambda_1, \ldots, \lambda_r) \vdash n$, which we call the **type** of σ. If $a_j(\sigma) \in \mathbb{N}_0$ is the number of j-cycles in this product decomposition of σ, then $a(\sigma) := (a_1(\sigma), \ldots, a_n(\sigma))$ is called the **cycle type** of σ. Two elements are conjugate in S_n if they have the same type, and this holds if and only if they have the same cycle type.

Definition 3.3.2 Let $\mathbf{n} = \{1, \ldots, n\} = \Lambda_1 \dot{\cup} \cdots \dot{\cup} \Lambda_r$ with all $\Lambda_i \neq \emptyset$. If $|\Lambda_1| \geq \cdots \geq |\Lambda_r|$ then $\Lambda = (\Lambda_1, \ldots \Lambda_r)$ is called a **set partition** of $\mathbf{n}$ of type λ, where $\lambda := (|\Lambda_1|, \ldots, |\Lambda_r|) \vdash n$. We define S_{Λ_i} to be the point-wise stabilizer of $\mathbf{n} \setminus \Lambda_i$. Then

$$S_\Lambda = S_{\Lambda_1} \times \cdots \times S_{\Lambda_r}$$

is called a **Young subgroup** of S_n.

If $\lambda \vdash n$ then S_n acts transitively on the set partitions of type λ and $S_\Lambda = \mathrm{Stab}_{S_n}(\Lambda)$. We single out one representative of Young subgroups for each partition $\lambda = (\lambda_1, \ldots, \lambda_r) \vdash n$: we put $\underline{\lambda}_i = \{\lambda_1 + \cdots + \lambda_{i-1} + 1, \ldots, \lambda_1 + \cdots + \lambda_i\}$ for $1 \leq i \leq r$ and

$$S_\lambda = S_{\underline{\lambda}_1} \times \cdots \times S_{\underline{\lambda}_r}.$$

This is the Young subgroup for the set partition $(\underline{\lambda}_1, \ldots, \underline{\lambda}_r)$.

Let K be a field with char $K \neq 2$ and $H \leq S_n$. The one-dimensional KH-module with character ϵ_H defined by

$$\epsilon_H(\sigma) = \begin{cases} 1 & \text{for } \sigma \in H \cap A_n, \\ -1 & \text{for } \sigma \in H \setminus A_n, \end{cases}$$

will be denoted by K_H^ϵ. Furthermore, for $\lambda \vdash n$ we put

$K_\lambda := K_{S_\lambda}$, the one-dimensional KS_λ-module with character $\mathbf{1}_\lambda := \mathbf{1}_{S_\lambda}$,
$K_\lambda^\epsilon := K_{S_\lambda}^\epsilon$, the one-dimensional KS_λ-module with character $\epsilon_\lambda := \epsilon_{S_\lambda}$.

From Theorem 3.2.19 we get for $\lambda, \mu \vdash n$

$$\dim_K \mathrm{Hom}_{KS_n}(K_\lambda^{S_n}, K_\mu^{\epsilon S_n}) = \sum_{S_n = \dot{\cup} S_\lambda\, g\, S_\mu} \dim_K \mathrm{Hom}_{KH_g}(K_{H_g}, K_{H_g}^\epsilon) \quad (3.4)$$

with $H_g = g\, S_\mu\, g^{-1} \cap S_\lambda$. If $\sigma \in H_g$ and $\sigma(i) = j \neq i$, then $(i,j) \in H_g$. Since char $K \neq 2$ we obtain

$$\mathrm{Hom}_{KH_g}(K_{H_g}, K_{H_g}^\epsilon) = \begin{cases} K & \text{if} & H_g = \{1\}, \\ 0 & \text{if} & H_g \neq \{1\}. \end{cases} \quad (3.5)$$

We also conclude that

$$\dim_K \mathrm{Hom}_{KS_n}(K_\lambda^{S_n}, K_\mu^{\epsilon S_n}) = \dim_K \mathrm{Hom}_{KS_n}(K_\mu^{S_n}, K_\lambda^{\epsilon S_n}). \quad (3.6)$$

Lemma 3.3.3 $S_\lambda\, g\, S_\mu = S_\lambda\, h\, S_\mu$ *if and only if* $|\underline{\lambda}_i \cap g\underline{\mu}_j| = |\underline{\lambda}_i \cap h\underline{\mu}_j|$ *for all* i, j.

Proof. (a) If $h = g_\lambda g g_\mu \in S_\lambda\, g\, S_\mu$ with $g_\lambda \in S_\lambda$, $g_\mu \in S_\mu$ then

$$\underline{\lambda}_i \cap h\underline{\mu}_j = \underline{\lambda}_i \cap g_\lambda g g_\mu \underline{\mu}_j = g_\lambda(\underline{\lambda}_i \cap g\underline{\mu}_j) \qquad \text{for all} \quad i, j.$$

(b) Conversely, let $\lambda = (\lambda_1, \ldots, \lambda_r)$, $\mu = (\mu_1, \ldots, \mu_s)$ and suppose that for all i, j we have $|\underline{\lambda}_i \cap g\underline{\mu}_j| = |\underline{\lambda}_i \cap h\underline{\mu}_j|$. Since

$$\underline{\lambda}_i = \dot{\bigcup}_{j=1}^s (\underline{\lambda}_i \cap g\underline{\mu}_j) = \dot{\bigcup}_{j=1}^s (\underline{\lambda}_i \cap h\underline{\mu}_j)$$

we can find for each i a permutation $g_i \in S_{\underline{\lambda}_i}$ with

$$g_i(\underline{\lambda}_i \cap g\underline{\mu}_j) = \underline{\lambda}_i \cap h\underline{\mu}_j \qquad \text{for all } j.$$

Putting $g_\lambda = g_1 \cdots g_s$ we see that $g_\lambda \in S_\lambda$ and $g_\lambda(g\underline{\mu}_j) = h\underline{\mu}_j$ for all j, so $h^{-1}g_\lambda g \in S_\mu$ and hence $h \in S_{\underline{\lambda}} g S_{\underline{\mu}}$. $\qquad \square$

So, assuming that the numbers of parts of λ and μ are r and s, respectively, every double coset $S_\lambda g S_\mu$ determines a matrix $A_{\lambda\mu} = [a_{ij}] = [|\underline{\lambda}_i \cap g\underline{\mu}_j|] \in \mathbb{N}_0^{r \times s}$ with row sums $\sum_{j=1}^s a_{ij} = \lambda_i$ and column sums $\sum_{i=1}^r a_{ij} = \mu_j$.

Let $\mu = (\mu_1, \ldots, \mu_s) \vdash n$ and $g \in S_n$. Then $g S_\mu g^{-1} = S_\Lambda$ for the set partition $\Lambda = (g\underline{\mu}_1, \ldots, g\underline{\mu}_s)$, where, of course, $g\underline{\mu}_i = \{gx \mid x \in \underline{\mu}_i\}$. In order to evaluate (3.5) we study the intersections $g S_\mu g^{-1} \cap S_\lambda$ and for this we need the notion of associated partitions, which can best be understood by looking at Young diagrams.

A partition λ can be visualized by a **Young diagram**, which consists of n boxes arranged in rows (left-aligned) with the ith row containing λ_i boxes. For example, the partition $(4,2,1)$ has the following Young diagram:

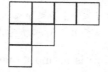

Obviously, the lengths of the columns of the Young diagram form another partition $\lambda' = (\lambda_1', \lambda_2', \ldots)$ with $\lambda_j' = |\{i \mid \lambda_i \geq j\}|$. This is called the partition **associated** with λ. Its Young diagram is obtained from that of λ just by transposing. Thus in our example $\lambda' = (3, 2, 1, 1)$ with Young diagram

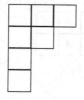

Definition 3.3.4 For $\lambda = (\lambda_1, \ldots, \lambda_r), \mu = (\mu_1, \ldots, \mu_s) \vdash n$ we define $\lambda \trianglelefteq \mu$ if and only if for all $i \in \mathbf{n}$ we have $\lambda_1 + \cdots + \lambda_i \leq \mu_1 + \cdots + \mu_i$. Here we put $\lambda_i = \mu_j = 0$ for $i > r$, $j > s$. This ordering of partitions is called the **dominance order**. Furthermore we write $\lambda \leq \mu$ if there is a $k \in \mathbb{N}$ such that $\lambda_i = \mu_i$ for $i < k$ and $\lambda_k \leq \mu_k$. This is the **lexicographical order**.

Unlike the lexicographical order the dominance order of partitions is not a total ordering; for example we have neither $(2, 2, 2) \trianglelefteq (3, 1, 1, 1)$ nor $(3, 1, 1, 1) \trianglelefteq (2, 2, 2)$. But obviously

$$\lambda \trianglelefteq \mu \qquad \text{implies} \qquad \lambda \leq \mu. \tag{3.7}$$

Lemma 3.3.5 *Let* $n \in \mathbb{N}$ *and* $\lambda = (\lambda_1, \ldots, \lambda_r)$, $\mu = (\mu_1, \ldots, \mu_s) \vdash n$.

(a) *If* $g\,\mathrm{S}_\mu\,g^{-1} \cap \mathrm{S}_\lambda = \{1\}$ *for some* $g \in \mathrm{S}_n$ *then* $\lambda \trianglelefteq \mu'$.

(b) *There is exactly one* $D = \mathrm{S}_\lambda\, h\, \mathrm{S}_{\lambda'}$ *such that* $g\,\mathrm{S}_{\lambda'}\,g^{-1} \cap \mathrm{S}_\lambda = \{1\}$ *for* $g \in D$.

Proof. Let $g \in \mathrm{S}_n$. Since $g\,\mathrm{S}_\mu\,g^{-1}$ is the stabilizer of $(g\underline{\mu}_1, \ldots, g\underline{\mu}_s)$ we have

$$g\,\mathrm{S}_\mu\,g^{-1} \cap \mathrm{S}_\lambda = \{1\} \quad \Longleftrightarrow \quad |\underline{\lambda}_i \cap g\underline{\mu}_j| \le 1 \qquad \text{for all} \quad i, j. \tag{3.8}$$

We write the elements of $g\underline{\mu}_j$ in the jth row of the Young diagram, using the natural ordering of the integers, obtaining an array $Y_{\mu,g}$ of numbers, usually called a Young tableau (so that $g\,\mathrm{S}_\mu\,g^{-1}$ is the set of $h \in \mathrm{S}_n$ stabilizing the rows of $Y_{\mu,g}$ set-wise). If (3.8) holds, the numbers $1, \ldots, \lambda_1$ must occur in different rows, hence in the first column of $Y_{\mu,g}$, and the numbers $1, \ldots, \lambda_1 + \lambda_2$ must appear in the first two columns of $Y_{\mu,g}$, and so on. Thus $\lambda_1 \le \mu_1'$, then $\lambda_1 + \lambda_2 \le \mu_1' + \mu_2'$, etc. Hence $\lambda \trianglelefteq \mu'$ and (a) is proved.

If $\mu = \lambda'$, then $\mu_i' = \lambda_i$ for all i and we can conclude from (3.8) that the ith column of $Y_{\mu,g}$ contains exactly $\underline{\lambda}_i$, so that

$$|\underline{\lambda}_i \cap g\underline{\mu}_j| = \begin{cases} 1 & \text{for } 1 \le i \le r,\ 1 \le j \le \lambda_i, \\ 0 & \text{for } 1 \le i \le r,\ \lambda_i < j \le \lambda_1. \end{cases} \tag{3.9}$$

From Lemma 3.3.3 we conclude that there is at most one double coset $D = \mathrm{S}_\lambda\, h\, \mathrm{S}_{\lambda'}$ such that (3.8) holds for $g \in D$. Conversely, transposing the Young tableau $Y_{\lambda,1}$ we obtain a Young tableau $Y_{\mu,g}$ with $g \in \mathrm{S}_n$ and (3.9) holds. □

Example 3.3.6 Let $\lambda := (4, 2, 1)$, $\mu := (3, 3, 1)$ and $g := (1, 2, 5, 4)(3, 6, 7) \in \mathrm{S}_7$. Then the Young tableau $Y_{\mu,g}$ is given by

$$Y_{\mu,g} =$$

2	5	6
1	4	7
3		

and $g\,\mathrm{S}_\mu\,g^{-1} \cap \mathrm{S}_\lambda = \langle (1,4), (5,6) \rangle$.   ◆

Corollary 3.3.7 *Let* K *be a field with* $\mathrm{char}\,K \ne 2$ *and* $\lambda, \mu \vdash n$.

(a) *If* $\mathrm{Hom}_{K\,\mathrm{S}_n}(K_\lambda{}^{\mathrm{S}_n}, K_\mu^{\epsilon\,\mathrm{S}_n}) \ne \{0\}$ *then* $\lambda \trianglelefteq \mu'$.

(b) $\dim_K \mathrm{Hom}_{K\,\mathrm{S}_n}(K_\lambda{}^{\mathrm{S}_n}, K_{\lambda'}^{\epsilon\,\mathrm{S}_n}) = 1$.

Proof. (a) If $\lambda \ntrianglelefteq \mu'$ then by Lemma 3.3.5(a) $g\,\mathrm{S}_\mu\,g^{-1} \cap \mathrm{S}_\lambda \ne 1$ for all $g \in \mathrm{S}_n$, so the result follows from (3.5) and (3.4).

(b) follows from Lemma 3.3.5(b) and (3.5) and (3.4). □

Theorem 3.3.8 *Let K be a field of characteristic zero. For any $\lambda \vdash n$ the $K S_n$-modules $K_\lambda^{S_n}$ and $K_{\lambda'}^{\epsilon, S_n}$ have exactly one irreducible constituent in common. This constituent is absolutely irreducible and occurs with multiplicity one in both modules; it will be denoted by $[\lambda]$ (identifying isomorphic modules). Then $\{[\lambda] \mid \lambda \vdash n\}$ is a complete set of representatives of the isomorphism classes of simple $K S_n$-modules. Let χ_λ be the character of $[\lambda]$. Then $\mathrm{Irr}(S_n) = \{\chi_\lambda \mid \lambda \vdash n\}$.*

Proof. Since $K S_n$ is semisimple we conclude from Corollary 3.3.7 (and Theorem 1.1.15) that $K_\lambda^{S_n}$ and $K_{\lambda'}^{\epsilon, S_n}$ indeed have exactly one irreducible constituent $[\lambda]$ in common and that $\dim_K \mathrm{End}_{K S_n}([\lambda]) = 1$. Hence $[\lambda]$ is absolutely irreducible (see Corollary 1.3.7). Let $\lambda, \mu \vdash n$ and assume that $[\lambda] \cong_{K S_n} [\mu]$. Since this is a common constituent of $K_\lambda^{S_n}$ and $K_{\mu'}^{\epsilon, S_n}$ we have

$$\mathrm{Hom}_{K S_n}(K_\lambda^{S_n}, K_{\mu'}^{\epsilon, S_n}) \neq \{0\}, \qquad \text{and hence} \qquad \lambda \trianglelefteq \mu$$

by Corollary 3.3.7. Interchanging λ, μ we also get $\mu \trianglelefteq \lambda$, thus $\lambda = \mu$. $\qquad\square$

Corollary 3.3.9 *Let $n \in \mathbb{N}$ and $\lambda \vdash n$. Then*

$$\chi_\lambda = \mathbf{1}_\lambda^{S_n} - \sum_{\mu > \lambda}(\mathbf{1}_\lambda^{S_n}, \chi_\mu)_{S_n} \cdot \chi_\mu, \tag{3.10}$$

$$\chi_{\lambda'} = \epsilon_{S_n} \cdot \chi_\lambda. \tag{3.11}$$

Proof. Suppose $\mu \vdash n$ and $(\mathbf{1}_\lambda^{S_n}, \chi_\mu)_{S_n} \neq 0$. Then the definition of χ_μ implies $(\mathbf{1}_\lambda^{S_n}, \epsilon_{\mu'}^{S_n})_{S_n} \neq 0$. From Corollary 3.3.7 we conclude that $\lambda \trianglelefteq (\mu')' = \mu$. So (3.10) follows from (3.7), since $(\mathbf{1}_\lambda^{S_n}, \chi_\lambda)_{S_n} = 1$.

From Lemma 3.2.14(d) we get $\epsilon_{S_n} \cdot \mathbf{1}_\lambda^{S_n} = (\epsilon_\lambda \cdot \mathbf{1}_\lambda)^{S_n} = \epsilon_\lambda^{S_n}$. Hence

$$(\epsilon_{S_n} \cdot \chi_\lambda, \mathbf{1}_{\lambda'}^{S_n})_{S_n} = (\chi_\lambda, \epsilon_{S_n} \cdot \mathbf{1}_{\lambda'}^{S_n})_{S_n} = 1$$

and

$$(\epsilon_{S_n} \cdot \chi_\lambda, \epsilon_\lambda^{S_n})_{S_n} = (\chi_\lambda, \mathbf{1}_\lambda^{S_n})_{S_n} = 1.$$

From this (3.11) follows. $\qquad\square$

As a special case of (3.10) and (3.11) we obtain

$$\chi_{(n)} = \mathbf{1}_{S_n} \qquad \text{and} \qquad \chi_{(1^n)} = \epsilon_{S_n}. \tag{3.12}$$

Of course, (3.10) would also hold with $\mu > \lambda$ replaced by $\mu \rhd \lambda$. But the formula is more useful as it stands, because $\leq$ is a total order and (3.10) suggests a way to compute the irreducible characters χ_λ in an inductive manner, starting with $\chi_{(n)} = \mathbf{1}_{S_n}$ and (proceeding in the anti-lexicographic order) from the corresponding $\mathbf{1}_\lambda^{S_n}$. Observe that these permutation characters are easily obtained; see Exercise 3.3.2. The second formula (3.11) shows that we need to compute only about half of the χ_λ in this way.

Example 3.3.10 We illustrate the procedure for $n = 5$. Since

$$(2^2, 1) = (3, 2)', \quad (2, 1^3) = (4, 1)', \quad (1^5) = (5)',$$

we compute $1_\lambda{}^{S_n}$ only for $\lambda = (5), (4, 1), (3, 2), (3, 1^2)$. The conjugacy class of S_n corresponding to the partition λ will be denoted by C_λ. We order the classes so that the first four are contained in A_n:

C_λ	$C_{(1^5)}$	$C_{(2^2,1)}$	$C_{(3,1^2)}$	$C_{(5)}$	$C_{(2,1^3)}$	$C_{(3,2)}$	$C_{(4,1)}$
$|C_\lambda|$	1	15	20	24	10	20	30
$1_{(5)}{}^{S_n}$	1	1	1	1	1	1	1
$1_{(4,1)}{}^{S_n}$	5	1	2	0	3	0	1
$1_{(3,2)}{}^{S_n}$	10	2	1	0	4	1	0
$1_{(3,1^2)}{}^{S_n}$	20	0	2	0	6	0	0

We know $\chi_{(5)} = 1_{S_n}$ from (3.12) and compute (or infer from Theorem 3.2.13) that $(1_\lambda{}^{S_n}, \chi_{(5)})_{S_n} = 1$ for all λ. From (3.10) we get $\chi_{(4,1)} = 1_{(4,1)}{}^{S_n} - 1 \cdot \chi_{(5)}$. We then compute

$$(1_{(3,2)}{}^{S_n}, \chi_{(4,1)})_{S_n} = 1, \quad (1_{(3,1^2)}{}^{S_n}, \chi_{(4,1)})_{S_n} = 2$$

and obtain from (3.10) that $\chi_{(3,2)} = 1_{(3,2)}{}^{S_n} - \chi_{(5)} - \chi_{(4,1)}$. Finally

$$(1_{(3,1^2)}{}^{S_n}, \chi_{(3,2)})_{S_n} = 1 \quad \text{yields} \quad \chi_{(3,1^2)} = 1_{(3,1^2)}{}^{S_n} - \chi_{(5)} - 2 \cdot \chi_{(4,1)} - \chi_{(3,2)}.$$

We thus obtain the character table of S_5, the last three lines being obtained from (3.11):

S_5	$C_{(1^5)}$	$C_{(2^2,1)}$	$C_{(3,1^2)}$	$C_{(5)}$	$C_{(2,1^3)}$	$C_{(3,2)}$	$C_{(4,1)}$
$\chi_{(5)}$	1	1	1	1	1	1	1
$\chi_{(4,1)}$	4	0	1	-1	2	-1	0
$\chi_{(3,2)}$	5	1	-1	0	1	1	-1
$\chi_{(3,1^2)}$	6	-2	0	1	0	0	0
$\chi_{(2^2,1)}$	5	1	-1	0	-1	-1	1
$\chi_{(2,1^3)}$	4	0	1	-1	-2	1	0
$\chi_{(1^5)}$	1	1	1	1	-1	-1	-1

◆

There is a more efficient recursive way to compute the character values for the irreducible characters of S_n. To describe it, we need some terminology. For each **node** (i, j) of the Young diagram of $\lambda = (\lambda_1, \ldots, \lambda_r) \vdash n$ (that is for each i, j with $1 \le i \le r$, $1 \le j \le \lambda_i$) there is a **hook** H_{ij}^λ which consists of the node (i, j), the nodes of λ to the right of (i, j) and the l_{ij}^λ nodes of λ below (i, j). The node (i, j) is called the **corner** of H_{ij}^λ. The number h_{ij}^λ of nodes of H_{ij}^λ is called the **hook length** and l_{ij}^λ is the **leg length** of H_{ij}^λ. Removing the nodes of H_{ij}^λ from the Young diagram of λ and sorting the rows, so that the lengths of the rows again have non-increasing lengths, one obtains an empty diagram i.e. [ ] (if $h_{ij}^\lambda = n$) or the Young diagram of a partition of $n - h_{ij}^\lambda$, which is denoted by $\lambda - R_{ij}^\lambda$. The Young diagram of this partition can also be obtained by removing the **rim** of H_{ij}^λ, as is illustrated in the following example.

Example 3.3.11 The following diagram shows the Young diagram of the partition $\lambda = (5, 4, 2)$. We have marked the nodes of the hook H_{12}^{λ} by $\times$ and the nodes of the corresponding rim by $\bigcirc$. We count $h_{12}^{\lambda} = 6$ and $l_{12}^{\lambda} = 2$. Furthermore $\lambda - R_{12}^{\lambda} = (3, 1, 1)$.

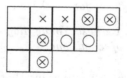

$\blacklozenge$

We can now state the recursion formula for the computation of the character χ^{λ} of $[\lambda]$ for $\lambda \vdash n$. For this we agree that $S_0 := \{1\}$ and $\chi^{[\,]} := 1$.

Theorem 3.3.12 (Murnaghan–Nakayama formula) *Let $\lambda \vdash n$ and let $\sigma \in S_n$ be of cycle type $a(\sigma)$ with $a_k(\sigma) > 0$ for some fixed $k \le n$. Let $\rho \in S_{n-k}$ be of cycle type $a(\rho)$ with*

$$a_i(\rho) = \begin{cases} a_k(\sigma) - 1 & \text{if } i = k, \\ a_i(\sigma) & \text{otherwise.} \end{cases}$$

Then

$$\chi^{\lambda}(\sigma) = \sum_{\substack{i,j \\ h_{i,j}^{\lambda}=k}} (-1)^{l_{ij}^{\lambda}} \chi^{\lambda-R_{ij}^{\lambda}}(\rho),$$

where the sum extends over all nodes (i, j) of the Young diagram of λ which are corners of hooks of length k.

Proof. See [96], p. 60. $\qquad\square$

Thus for $\sigma := (1, 2, 3)(4, 5)(6, 7, 8, 9, 10, 11) \in S_{11}$ and $\lambda := (5, 4, 2)$, as in the above example, we can compute the following (taking $k := 6$):

$$\chi^{\lambda}(\sigma) = \chi^{(3,1,1)}((1, 2, 3)(4, 5)) = -\chi^{(1,1,1)}((1, 2, 3)) + \chi^{(3)}((1, 2, 3)) = -1 + 1 = 0$$

or, using $k := 3$ and H_{14}^{λ}, the only hook of length three,

$$\chi^{\lambda}(\sigma) = -\chi^{(3,3,2)}((1, 2)(3, 4, 5, 6, 7, 8)) = 0,$$

because the Young diagram of $(3, 3, 2)$ has no hook of length six.

The theorem provides us with an efficient inductive method for computing character values for S_n. See [145] for a careful description of how this is implemented in GAP.

Corollary 3.3.13 *If λ is a partition of n and $\sigma = (1, \dots, n)$, then*

$$\chi^{\lambda}(\sigma) = \begin{cases} (-1)^r & \text{if } \lambda = [n - r, 1^r] \text{ for some } 0 \le r \le n - 1, \\ 0 & \text{otherwise.} \end{cases}$$

Proof.   Note that $a_i(\sigma) = \delta_{i,n}$, so we have to put $k := n$ in Theorem 3.3.12; $\rho := 1 \in S_0$ and $\chi^\lambda(\sigma) \neq 0$ only if the Young diagram of λ is a hook.   □

Corollary 3.3.14 *Embedding* S_{n-1} *in* S_n *in the usual way we have*

$$\chi^\lambda|_{S_{n-1}} = \sum \chi^\mu \qquad \text{for} \qquad \lambda \vdash n,$$

where we sum over all those $\mu \vdash n - 1$ *whose Young diagram is obtained from the diagram of* λ *by removing a hook of length one.*

Proof.   Apply Theorem 3.3.12 with $k := 1$, observing that $a_1(\sigma) > 0$ if $\sigma \in S_n$ is contained in S_{n-1}.   □

Thus, for instance,

$$\chi^{(3,3,2,2,1)}|_{S_{10}} = \chi^{(3,2,2,2,1)} + \chi^{(3,3,2,1,1)} + \chi^{(3,3,2,2)}.$$

Exercises

Exercise 3.3.1 Let K be a commutative ring, let $n \in \mathbb{N}$ and let $\Omega := \{1, \ldots, n\}$. Show that for $k \in \mathbb{N}$ with $1 \leq k \leq n/2$ the $K S_n$-module $K\binom{\Omega}{k}$ introduced in Example 1.2.25 is isomorphic to $K_{(n-k,k)}{}^{S_n}$.

Exercise 3.3.2 Let $\lambda = (\lambda_1, \ldots, \lambda_r)$, $\mu = (\mu_1, \ldots, \mu_s) \vdash n$ and $\sigma \in S_n$ have type σ. Show that $1_\lambda{}^{S_n}(\sigma)$ is the number of different set partitions $(\Lambda_1, \ldots, \Lambda_r)$ of type λ such that each Λ_i is a union of sets in $\{\underline{\mu}_1, \ldots, \underline{\mu}_s\}$.

Exercise 3.3.3 Let $\lambda \vdash n$. Show that $\chi_\lambda|_{A_n}$ is irreducible if and only if $\lambda \neq \lambda'$. Show also that $(\chi_\lambda|_{A_n}, \chi_\lambda|_{A_n})_{A_n} = 2$ if $\lambda = \lambda'$.

Exercise 3.3.4 Let $\lambda \vdash n$ and let r_λ be the number of parts of λ. Show that

$$r_\lambda = (\chi_{H_\lambda}, 1_{H_\lambda})_{H_\lambda},$$

where $\chi := \chi_{(n-1,1)}$ is the natural permutation character of S_n.

3.4   Permutation characters

In this section K is an arbitrary field and G is a finite group.

If G acts transitively on the finite set Ω and H is the stabilizer of some point $\omega \in \Omega$, then choosing coset representatives g_i with $G = \dot{\bigcup}_{i=1}^{n} g_i H$ we have $\Omega = \{g_i \cdot \omega \mid 1 \leq i \leq n\}$, and it is clear that we have a KG-isomorphism

$$K\Omega \rightarrow K_H{}^G \qquad \sum_{i=1}^{n} \alpha_i(g_i \cdot \omega) \mapsto \sum_{i=1}^{n} g_i \otimes \alpha_i \qquad (\alpha_i \in K),$$

where K_H denotes the trivial KG-module K. From this we conclude the following.

Remark 3.4.1 If G acts transitively on the finite set Ω and H is the stabilizer of some point $\omega \in \Omega$ then $\theta := 1_H^G$ is the permutation character of the corresponding permutation module $K\Omega$. In other words,

$$\theta(g) = |\text{Fix}_\Omega(g)| 1_K \qquad \text{for} \qquad g \in G.$$

From Lemma 3.2.7 we derive the useful formula

$$\theta(g) = |\mathbf{C}_G(g)| \cdot \frac{|g^G \cap H|}{|H|} \cdot 1_K.$$

It shows for instance how to compute $n(\theta, m) := |\{h \in H \mid |\langle h \rangle| = m\}|$ for $m \in \mathbb{N}$ given θ and the character table of G.

Lemma 3.4.2 If $H, U \leq G$ then $(1_H^G, 1_U^G)_G$ is the number of orbits of U on G/H and also equals the number of (U, H)-double cosets of U and H in G.

Proof. $\theta = 1_H^G$ is the permutation character of the action of G on G/H. By Frobenius reciprocity we have $(\theta, 1_U^G)_G = (\theta_U, 1_U)_U$, which by Corollary 2.1.14 is equal to the number of orbits of U on G/H. The orbits of U on G/H are of the form $\{ugH \mid u \in u\}$ for some $g \in G$ and correspond bijectively with the double cosets UgH of U and H in G. $\qquad \square$

Corollary 3.4.3 G acts doubly transitively on the finite set Ω if and only if the corresponding permutation character θ over $\mathbb{C}$ is of the form $\theta_\Omega = 1_G + \chi$ for some $\chi \in \text{Irr}(G)$.

Theorem 3.4.4 The following are necessary conditions for a character θ of a finite group G to be a transitive permutation character of G, i.e. the character of a transitive G-set:

(a) $\theta(1)$ divides $|G|$;

(b) $(\theta, \chi) \leq \chi(1)$ for each character χ of G;

(c) $(\theta, 1_G) = 1$;

(d) $\theta(g)$ is a non-negative integer for every $g \in G$;

(e) $\theta(g) \leq (|G| - \theta(1))/(|g^G| \cdot |\{g_1^G \in \text{cl}(G) \mid \langle g \rangle = \langle g_1 \rangle\}|)$ for $1 \neq g \in G$;

(f) $\theta(g) \leq \theta(g^m)$ for any $m \in \mathbb{Z}$;

(g) $\theta(g) = 0$ if the order of g does not divide $\frac{|G|}{\theta(1)}$;

(h) $\theta(1)$ divides $|g^G| \cdot \theta(g)$ for all $g \in G$;

(i) If a prime p divides $|G|/\theta(1)$ just once then, with the notation of Remark 3.4.1, $n(\theta, p)/(p - 1)$ divides $|G|/\theta(1)$ and $n(\theta, p)/(p - 1) \equiv 1 \mod p$.

Proof.   If $H \leq G$ is a stabilizer of a point of the corresponding G-set and $\theta = 1_H^G$, then clearly $\theta(1) = [G : H]$, hence (a) holds. Also

$$(\theta, \chi)_G = (1_H^G, \chi)_G = (1_H, \chi_H)_H \leq \chi(1)$$

by Frobenius reciprocity. (c) follows from Lemma 3.4.2. Conditions (d) and (f) are obvious from the definition. (e) follows easily from Remark 3.4.1 since for $g \neq 1$ one always has $g^G \cap H \subset H$. In our setting $\frac{|G|}{\theta(1)} = |H|$, so if the order of g does not divide this number, then certainly no conjugate of g can be in H, hence $g^G \cap H = \emptyset$, and (g) follows from Definition 3.2.6. Now, $[\mathbf{N}_G(\langle g \rangle) : \mathbf{C}_G(g)] = k$ is the number of different powers of g which are conjugate in G to g. Obviously k divides $|g^G \cap H|$. Since

$$|G| \cdot \theta(g) = [G : H] \cdot |\mathbf{C}_G(g)| \cdot |g^G \cap H| = \theta(1) \cdot |\mathbf{C}_G(g)| \cdot |g^G \cap H|$$

condition (h) follows. Note that under the assumption of (i) $n(\theta, p)/(p - 1)$ is the number of Sylow-p-subgroups of H. $\square$

In GAP a character satisfying the conditions of Theorem 3.4.4 is called a possible permutation character (see [62], sect. 70.13 and 70.14), and there are several algorithms implemented (see also [21]) to find all such characters of a group G (or all those of a fixed degree) using just the character table of G. Applying these programs to the Mathieu group M_{11} one obtains 39 possible permutation characters, and M_{11} has exactly 39 conjugacy classes of subgroups (see Section 3.5). But having a closer look – in Section 3.5 we will see how to find all transitive permutation characters of a group – one observes that there are three pairs of non-conjugate subgroups having the same permutation characters and there are also three possible permutation characters which are not permutation characters. This is not at all a rare phenomenon, as Table 3.2, comparing the number $np(G)$ of possible and the number $nt(G)$ of transitive permutation characters with the number $cls(G)$ of conjugacy classes of subgroups of some groups G, shows.

Table 3.2. Numbers of possible permutation characters

G	$np(G)$	$nt(G)$	$cls(G)$	G	$np(G)$	$nt(G)$	$cls(G)$
M_{12}	285	137	147	J_1	44	37	40
M_{22}	228	108	156	J_2	304	140	146
M_{23}	209	122	204	J_3	387	124	137

Thus the conditions given in Theorem 3.4.4 are far from sufficient to guarantee genuine permutation characters. This remains true even if one takes into account additional known necessary conditions for permutation characters; see, for instance [127], sect. 5.3. Hence given the character table of a finite group G, using this method it is only possible to exclude the existence of subgroups of G with certain properties, e.g. having a certain order or containing a specified number of elements of the conjugacy classes of G. It is more difficult to prove the existence of proper subgroups using just the character table of a group, apart from cyclic subgroups, dihedral subgroups or subgroups isomorphic to A_4, S_4 or A_5; see Section 2.5.

We note that finding all possible permutation characters is feasible only for groups of moderate size. The situation is better if one restricts oneself to multiplicity-free permutation characters. In [15] all such characters have been found for the sporadic simple groups and their automorphism groups.

Remark 3.4.5 In $G = L_2(29)$ there are two subgroups $H_1, H_2 \cong A_5$ which are not conjugate in G and for which not only the permutation characters $1_{H_i}{}^G$ are equal for $i = 1, 2$, but also the integral permutation modules $\mathbb{Z}\Omega_i$ with $\Omega_i = G/H_i$ are isomorphic as $\mathbb{Z}G$-modules for $i = 1, 2$ (see [158]).

Remark 3.4.6 There are infinitely many examples of finite groups G having subgroups $H_1, H_2 \le G$ with $1_{H_1}{}^G = 1_{H_2}{}^G$ and such that H_1 is a maximal subgroup of G and H_2 is not (see [16] and [74]). Thus one cannot decide from the permutation character of a transitive action whether or not it is primitive.

Remark 3.4.7 The sporadic simple group J_2 has two subgroups H_1, H_2 of order 16 with $1_{H_1}{}^G = 1_{H_2}{}^G$, but H_1 is abelian whereas H_2 is non-abelian.

Thus the information on fixed points of elements is not enough to characterize a permutation action. On the other hand, the permutation character in conjunction with the power maps of a given group G provides complete information on the cycle types of the permutations afforded on the G-set.

Lemma 3.4.8 *Let Ω be a finite G-set with permutation character θ and corresponding homomorphism $\pi\colon G \to S_\Omega$. For any $g \in G$ and $i \in \mathbb{N}$ let $\mathrm{cyc}_i(g)$ denote the number of i-cycles of $\pi(g)$. Then*

$$\mathrm{cyc}_i(g) = \frac{1}{i} \sum_{j \mid i} \mu(\frac{i}{j}) \, \theta(g^j), \qquad (3.13)$$

where μ denotes the arithmetical Möbius function (see Example 2.5.4).

Proof. Obviously $\mathrm{cyc}_1(g) = \theta(g)$ and, more generally, $\theta(g^j) = \sum_{i \mid j} i \cdot \mathrm{cyc}_i(g)$ for any $j \in \mathbb{N}$. From this the result follows using Möbius inversion. □

Note that (3.13) yields another necessary condition for a permutation character θ: the right hand side must be a non-negative integer for any $i \in \mathbb{N}$ and $g \in G$. Observe also that for a prime p and $r \in \mathbb{N}$ we get $\theta(g^{p^r}) - \theta(g^{p^{r-1}}) = p^r \cdot \mathrm{cyc}_{p^r}(g)$, a result that should be compared with Exercise 2.2.5.

Exercises

Exercise 3.4.1 Given the character table of a group G and the permutation character $1_H{}^G$ of a subgroup $H \le G$ write a GAP program that computes for characters χ, ψ of G the scalar product $(\chi_H, \psi_H)_H$.

Exercise 3.4.2 Let $G \cong C_n$ and let φ be the Eulerian function. Show that

(a) $\mathrm{Irr}_\mathbb{Q}(G) = \{\chi_d \mid d \text{ divides } n\}$ with $\chi_d(1) = \varphi(d)$ and $\chi_d \ne \chi_m$ for $d \ne m$;

(b) G has for each $d \mid n$ exactly one permutation character θ_d of degree d;

(c) $\theta_d = \sum_{m \mid d} \psi_m$ and $\psi_d = \sum_{m \mid d} \mu(\frac{d}{m}) \theta_m$;

(d) $\theta_d(g) = \frac{d}{n}\theta_n(g^d)$ for $g \in G$ and θ_n is the regular character of G.

Hint: For (a) see Exercise 1.5.6.

3.5   Tables of marks

As we have seen in Section 3.4, the permutation character of a transitive permutation representation of a finite group G is not sufficient to characterize the action of G. The permutation character just gives the number of fixed points of any cyclic subgroup of G. So it is natural to enhance this information. We recall from basic algebra that every transitive G-set is isomorphic to a coset space G/H for some subgroup $H \leq G$ and that G/H is isomorphic as a G-set to G/U if and only if H and U are conjugate in G.

Throughout this section G is assumed to be a finite group.

Definition 3.5.1 Let $\mathcal{L}(G) = \{H_1, \ldots, H_n\}$ be a complete set of representatives of the set of conjugacy classes of subgroups of a finite group G. For simplicity we will assume that $|H_i| \leq |H_j|$ for $i \leq j$.

(a) If Ω is any G-set then the function

$$m_\Omega \colon \mathcal{L}(G) \to \mathbb{N}_0, \ H \mapsto |\operatorname{Fix}_\Omega(H)|$$

is called the **mark** of Ω; it is often considered as a row-vector.

(b) The **table of marks** of G is the square matrix

$$M(G) = [\, m_{G/H_i}(H_j)]_{1 \leq i,j \leq n} \ .$$

Remark 3.5.2 It is clear that conjugate subgroups of G have the same number of fixed points on any G-set. In fact, it is easily seen that, for any finite G-set Ω and any subgroup $H \leq G$, one has $|\operatorname{Fix}_\Omega(H)| = |\operatorname{Hom}_G(G/H, \Omega)|$. Thus the mark of a G-set is independent of the choice of representatives in $\mathcal{L}(G)$ and likewise $M(G)$. Of course, $M(G)$ depends on the ordering of $\mathcal{L}(G)$.

Lemma 3.5.3 *If* $M(G) = [m_{ij}] = [\, m_{G/H_i}(H_j)]$ *is the table of marks of* G *then*

(a) $m_{ij} = [\mathbf{N}_G(H_i) : H_i] \cdot b_{ij}$, *where* b_{ij} *is the number of subgroups conjugate to H_i which contain H_j. In particular* $m_{ii} = [\mathbf{N}_G(H_i) : H_i]$.

(b) *If* $H_1 = \{1\}$ *then* $m_{i1} = [G : H_i]$.

(c) *The number* c_{ij} *of subgroups of H_i which are conjugate in G to H_j is equal to*

$$c_{ij} = \frac{m_{ij} \cdot m_{j1}}{m_{i1} \cdot m_{jj}} = |H_i| \cdot m_{ij} \cdot |\mathbf{N}_G(H_j)|^{-1}.$$

Proof. By definition

$$m_{ij} = |\{gH_i \mid g \in G, \; u \cdot gH_i = gH_i \quad \text{for all} \quad u \in H_j \}|$$
$$= |\{gH_i \mid g \in G, \; H_j \leq H_i^{g^{-1}} \}|$$
$$= [\mathbf{N}_G(H_i) : H_i] \cdot |\{H_i^g \mid g \in G, H_j \leq H_i^g \}|,$$

thus (a) holds, and (b) follows immediately from (a). For (c) observe that with the above notation

$$|\{(U, V) \mid U \leq V, \; U =_G H_j, \; V =_G H_i\}| = b_{ij} \cdot [G : \mathbf{N}_G(H_j)]$$
$$= c_{ij} \cdot [G : \mathbf{N}_G(H_i)],$$

from which the result follows using (a). $\qquad\square$

Corollary 3.5.4 *Let $M(G)$ be the table of marks of G.*

(a) $M(G)$ is invertible in $\mathbb{Q}^{n \times n}$.

(b) Two finite G-sets Ω and Ω' are isomorphic if and only if they have the same mark, i.e. if and only if $m_\Omega = m_{\Omega'}$.

(c) If Ω is a finite G-set with mark m_Ω having a_i orbits isomorphic to G/H_i then

$$[a_1, \ldots, a_n] = m_\Omega \cdot M(G)^{-1}.$$

Proof. (a) By Lemma 3.5.3(a) and our assumption on the ordering of $\mathcal{L}(G)$, $M(G)$ is a lower triangular matrix with $[\mathbf{N}_G(H_i) : H_i]_{1 \leq i \leq n}$ on the diagonal.

Any finite G-set can be decomposed into orbits which are isomorphic to G-sets of the form G/H_i. If a_i orbits of Ω are isomorphic to G/H_i then $m_\Omega = \sum a_i \, m_{G/H_i}$. By (a) the m_{G/H_i} are linearly independent, so the other results follow. $\qquad\square$

The table of marks of G contains some information about the Möbius function of the subgroup lattice of G (see Definition 2.5.3). In particular one can get the values $\mu_G(H_j, G)$ used in Section 2.5 (see Lemma 2.5.5 and Example 2.5.8).

Lemma 3.5.5 *Let μ_G be the Möbius function of the subgroup lattice of the finite group G. Then*

$$[M(G)^{-1}]_{i1} = |\mathbf{N}_G(H_i)|^{-1} \cdot \mu_G(\{1\}, H_i), \qquad (3.14)$$
$$[M(G)^{-1}]_{nj} = [\mathbf{N}_G(H_j) : H_j]^{-1} \cdot \mu_G(H_j, G). \qquad (3.15)$$

Proof. Using the notation of Lemma 3.5.3, let $[a_{ij}] = [c_{ij}]^{-1}$. Then

$$\delta_{i,k} = \sum_{j=1}^n c_{ij} a_{jk} = \sum_{U \leq H_i} a_{U H_k},$$

where $a_{UH_k} = a_{jk}$ if $U =_G H_j$. For $k = 1$ this implies that $a_{i1} = \mu_G(\{1\}, H_i)$. Since

$$M(G)^{-1} = \operatorname{diag}(|\mathbf{N}_G(H_1)|, \ldots, |\mathbf{N}_G(H_n)|)^{-1} \cdot [c_{ij}]^{-1} \cdot \operatorname{diag}(|H_1|, \ldots, |H_n|),$$

the first result follows. Similarly, let $[d_{ij}] = [b_{ij}]^{-1}$ with the notation of Lemma 3.5.3. Then

$$\delta_{i,k} = \sum_{j=1}^{n} d_{ij} b_{jk} = \sum_{H_k \leq U} d_{H_i U},$$

where $d_{H_i U} = d_{ij}$ if $U =_G H_j$. For $i = n$ this implies that $d_{ni} = \mu_G(H_i, G)$. Since

$$M(G)^{-1} = [b_{ij}]^{-1} \cdot \operatorname{diag}([\mathbf{N}_G(H_1) : H_1], \ldots, [\mathbf{N}_G(H_n) : H_n])^{-1}$$

the second equation follows.   □

Observe that in general the value $\mu_G(U, V)$ for subgroups $U, V \leq G$ does depend on U and V and not only on the conjugacy classes to which U and V belong. On the other hand, there is, of course, a Möbius function, sometimes denoted by λ_G for the poset of conjugacy classes of subgroups of G, writing $[H_i] \leq [H_j]$ if and only if $H_i \leq_G H_j$. If G is solvable there is a relation

$$\mu(H, G) = [\mathbf{N}_{G'}(H) : H \cap G'] \, \lambda([H], [G])$$

for $H \leq G$, but this does not hold in general; see [141].

For small groups G one can compute the table of marks of G by computing the whole subgroup lattice of G and using Lemma 3.5.3. This is implemented, for example, in GAP. In addition GAP contains a large library of table of marks of simple groups, which is almost complete up to order 10^{10}. For the larger groups their tables of marks have not been computed via the whole subgroup lattice, but in an inductive procedure; see e.g. [146].

Example 3.5.6 The table of marks of A_5 is as follows:

	C_1	C_2	C_3	V_4	C_5	S_3	D_{10}	A_4	A_5
A_5 / C_1	60								
A_5 / C_2	30	2							
A_5 / C_3	20	.	2						
A_5 / V_4	15	3	.	3					
A_5 / C_5	12	.	.	.	2				
A_5 / S_3	10	2	1	.	.	1			
A_5 / D_{10}	6	2	.	.	1	.	1		
A_5 / A_4	5	1	2	1	.	.	.	1	
A_5 / A_5	1	1	1	1	1	1	1	1	1
$H_j :=$	C_1	C_2	C_3	V_4	C_5	S_3	D_{10}	A_4	A_5

Here the zeros above the diagonal have been omitted and those below the diagonal been replaced by dots. Looking at the columns with numbers 1, 2, 3, 5 we get the full set of transitive permutation characters of A_5, keeping in mind that the generators of a C_5 in A_5 fall into two conjugacy classes of elements. Observe that the table of marks contains all the information encoded in the subgroup pattern (compare Example 2.5.8) but also additional information, which can, in general, not be deduced from the subgroup pattern. A further advantage of the table of marks is that there are good checks for its consistency, namely any (point-wise) product of two rows must be a non-negative integral linear combination of the rows, as will soon become obvious. ◆

Marks are also useful when one wants to decompose G-sets into orbits. To this end we introduce the following notions (see [139]).

Definition 3.5.7 We fix a G-set V. For $H \leq G$ and any $W \subseteq V$ we put $\mathbf{C}_G(W) := \{g \in G \mid g\,w = w \text{ for all } w \in W\}$ and define

$$\overline{H} := \mathbf{C}_G(\mathrm{Fix}_V(H)), \qquad \overline{W} := \mathrm{Fix}_V(\mathbf{C}_G(W)),$$

and call this the V-**closure** of H and and W, respectively. Note that H is called V-**closed** if $\overline{H} = H$.

Clearly we have $H \leq \overline{H}$ and $\overline{\overline{H}} = \overline{H}$ for $H \leq G$. Similarly we see $W \subseteq \overline{W}$ and $\overline{\overline{W}} = \overline{W}$ for $W \subseteq V$. Also any stabilizer $H = \mathrm{Stab}_G(v)$ for $v \in V$ is necessarily V-closed.

Example 3.5.8 Let Ω be a G-set and let $V := F\Omega$ be the corresponding permutation module over a finite commutative ring F. Then V is also a G-set, which we might want to decompose into G-orbits. Here we see that for $H \leq G$ we have

$$\mathrm{Fix}_V(H) = \langle \mathcal{O}_1^+, \ldots, \mathcal{O}_r^+ \rangle_F,$$

where $\mathcal{O}_1, \ldots, \mathcal{O}_r$ are the H-orbits in Ω (and $\mathcal{O}_i^+ = \sum_{\omega \in \mathcal{O}_i} \omega \in V$, as before). Clearly $g \in \overline{H}$ if and only if $g\mathcal{O}_i = \mathcal{O}_i$ for all $i = 1, \ldots, r$, so $\overline{H}$ is the largest subgroup of G having the same orbits on Ω as H has.

Consider, for example, $G := S_n$ with its natural action on $\Omega = \{1, \ldots, n\}$ and $V := F\Omega$. Then the V-closed subgroups of G are just the Young subgroups; see Definition 3.3.2. If S_Λ is a Young subgroup for the set partition $\Lambda = (\Lambda_1, \ldots, \Lambda_{r_\Lambda})$ of Ω then $|\mathrm{Fix}_V(S_\Lambda)| = q^{r_\Lambda}$, where $q = |F|$. Thus the mark of the G-set V is completely determined:

$$m_V(H) = \begin{cases} 0 & \text{if } H \text{ is not a Young subgroup,} \\ q^{r_\Lambda} & \text{if } H =_G S_\Lambda. \end{cases}$$

Note that $r_\Lambda = (\chi_H, \mathbf{1}_H)_H$, where $H = S_\Lambda$ and χ is the character of $\mathbb{C}\Omega$ (see Exercise 3.3.4). Let f_V^Λ be the number of orbits of G on V isomorphic to G/S_Λ. Write $\Lambda' \geq \Lambda$ if $\Lambda' = (\Lambda_1', \ldots, \Lambda_s')$ and every Λ_j' is contained in some Λ_i and thus $S_{\Lambda'} \geq S_\Lambda$. Then

$$q^{r_\Lambda} = \sum_{\Lambda' \geq \Lambda} f_V^{\Lambda'} \cdot [\mathbf{N}_G(S_{\Lambda'}) : S_{\Lambda'}].$$

Using Möbius inversion we obtain

$$f_V^\Lambda = [\mathbf{N}_G(S_\Lambda) : S_\Lambda]^{-1} \sum_{\Lambda' \geq \Lambda} \mu_n(\Lambda, \Lambda') \cdot q^{r_{\Lambda'}},$$

where μ_n is the Möbius function of the poset $(\mathcal{P}(n), \leq)$ of set partitions of $\{1, \ldots, n\}$. But $\{\Lambda' \in \mathcal{P}(n) \mid \Lambda' \geq \Lambda\}$ is isomorphic as a poset to $\mathcal{P}(r)$ with $r := r_\Lambda$, and $\sum_{\Lambda \in \mathcal{P}(r)} \mu_r(\{\{1\}, \ldots, \{r\}\}, \Lambda) \cdot q^{r_\Lambda}$ is just the number of injective maps from $\{1, \ldots, r\}$ to F. Hence

$$f_V^\Lambda = [\mathbf{N}_G(S_\Lambda) : S_\Lambda]^{-1} \prod_{i=0}^{r_\Lambda - 1} (q - i).$$

$\blacklozenge$

We will now generalize the ideas of Example 3.5.8.

Definition 3.5.9 Let $[\gamma_{ij}] := M(G)^{-1}$ and $H =_G H_j \in \mathcal{L}(G)$. Then for any $\chi \in \mathrm{Char}_{\mathbb{C}}(G)$ we call

$$f_\chi^H = \sum_{i=1}^n \gamma_{ij} X^{d_i} \in \mathbb{Q}[X] \qquad \text{with} \quad d_i := (\chi_{H_i}, \mathbf{1}_{H_i})_{H_i} \qquad (3.16)$$

the **character polynomial** of χ and H.

If $\chi = \chi_{\mathbb{C}\Omega}$ is the permutation character with some transitive G-set Ω and $\mathbb{F}_q$ is any finite field, then $\mathbb{F}_q\Omega$ is a G-set with mark $m_{\mathbb{F}_q\Omega} \colon H_j \mapsto q^{d_j}$ with $d_j := (\chi_{H_j}, \mathbf{1}_{H_j})_{H_j}$. From Corollary 3.5.4(c) we find that the number of orbits of G on $\mathbb{F}_q\Omega$ with stabilizers conjugate to H_j is precisely $f_\chi^{H_j}(q)$. From Example 3.5.8 we obtain for the natural permutation character χ of $G := S_n$ that $f_\chi^H = 0$ if H is not a Young subgroup of G and, for a Young subgroup S_Λ,

$$f_\chi^{S_\Lambda} = [\mathbf{N}_G(S_\Lambda) : S_\Lambda]^{-1} \prod_{i=0}^{r_\Lambda - 1} (X - i).$$

Note that S_n can be considered as an example of a real reflection group, that is a finite subgroup of $\mathrm{GL}(\mathbb{R}^n)$ generated by reflections (= elements g with $\mathrm{Fix}_{\mathbb{R}^n}(g)$ a hyperplane). The character polynomials for the natural characters of all real and complex reflection groups have been computed (see [131] and [139]). They all split into linear factors with non-negative integers as roots.

Now let V be an arbitrary $\mathbb{F}_q G$-module with character φ and assume that $q = p^r$ with $p \nmid |G|$. Applying Exercise 2.1.4 to $H \leq G$ we get $|\mathrm{Fix}_V(H)| = q^d$, where $(\varphi, \mathbf{1}_H)_H \equiv d \mod p$. Thus if $p > \dim V$ the mark m_V can be determined. Using Corollary 3.5.4(c), for each $H \leq G$, the number of orbits of G on V which are isomorphic to G/H can be found.

In fact, in Chapter 4 we will attach (in the above situation) an ordinary character $\chi \in \mathrm{Char}_{\mathbb{C}}(G)$ to V, called the "Brauer character" of V, such that $|\mathrm{Fix}_V(H)| = q^d$ with $\mathrm{d} = (\chi_H, \mathbf{1}_H)_H$. Then the number of orbits of G on V

which are isomorphic to G/H is again $f_\chi^H(q)$. Observe that if χ is faithful, there can only be $\chi(1)$ values of q such that χ is the Brauer character of an $\mathbb{F}_q G$-module without a regular G-orbit. We will see in Section 4.14 an example where it is important to know whether or not an $\mathbb{F}_q G$-module contains a regular G-orbit.

Example 3.5.10 As an example we list in Table 3.3 the polynomials $f_\chi^{H_j}$ for $\chi \in \{\chi_2, \chi_4\} \subseteq \mathrm{Irr}(A_5)$ (see Example 2.1.24) and H_j as in Example 3.5.6. Observe that $\chi_4 = \chi - \mathbf{1}_{A_5}$, where χ is the natural permutation character of A_5,

Table 3.3. List of polynomials $f_{\chi_2}^{H_j}$ and $f_{\chi_4}^{H_j}$

H_j	$f_{\chi_2}^{H_j}$	$f_{\chi_4}^{H_j}$
C_1	$\frac{1}{60}(X-5)(X-1)(X+6)$	$\frac{1}{60}(X-3)(X-2)(X-1)(X+6)$
C_2	$\frac{1}{2}(X-1)$	$\frac{1}{2}(X-2)(X-1)$
C_3	$\frac{1}{2}(X-1)$	$\frac{1}{2}(X-2)(X-1)$
V_4	0	0
C_5	$\frac{1}{2}(X-1)$	0
S_3	0	$X-1$
D_{10}	0	0
A_4	0	$X-1$
A_5	1	1

and its character polynomials should be compared with that of χ which is given for all n in Exercise 3.5.6. Note that in Table 3.3 $f_{\chi_2}^{C_1}(7)$ is not an integer. This is an indication that χ_2 is not a Brauer character of an $\mathbb{F}_7 A_5$-module, which will not be surprising. $\blacklozenge$

We recall that for two G-sets M, N the disjoint union $M \dot\cup N$ and the Cartesian product $M \times N$ are also G-sets in a natural way. Since

$$\mathrm{Fix}_{M \dot\cup N}(H) = \mathrm{Fix}_M(H) \dot\cup \mathrm{Fix}_N(H)$$

and

$$\mathrm{Fix}_{M \times N}(H) = \mathrm{Fix}_M(H) \times \mathrm{Fix}_N(H)$$

for G-sets M, N and any subgroup $H \leq G$, it follows that

$$m_{M \dot\cup N} = m_M + m_N, \quad m_{M \times N} = m_M \cdot m_N, \tag{3.17}$$

with the usual point-wise operations. Thus the $\mathbb{Z}$-linear combinations of marks of G form a ring which is isomorphic to the Burnside ring of G defined as follows.

Definition 3.5.11 The **Burnside ring** $\mathcal{B}(G)$ is the Grothendieck ring of the category of finite G-sets; i.e. as an abelian group $\mathcal{B}(G) = F/F_0$, where F is the free abelian group generated by the isomorphism classes (M) of finite G-sets and

$$F_0 = \langle\, (M \dot\cup N) - (M) - (N) \mid M, N \text{ finite } G\text{-sets} \,\rangle_{\mathbb{Z}}.$$

Defining $(M) \cdot (N) = (M \times N)$ on F as above, and extending $\mathbb{Z}$-linearly, F becomes a ring and F_0 an ideal, so $\mathcal{B}(G)$ also becomes a ring.

For any G-set M we will denote the image of (M) in $\mathcal{B}(G)$ by $[M]$. It is easily seen that $[M] = [N]$ if and only if M and N are isomorphic G-sets. Also it is clear that $\mathcal{B}(G)$ is a commutative ring with $\mathbb{Z}$-basis $([G/H_1], \ldots, [G/H_n])$, keeping our notation from Definition 3.5.1 with identity element $[G/G]$. It follows from the above and Corollary 3.5.4 that $\mathcal{B}(G)$ is, in fact, isomorphic to the ring of $\mathbb{Z}$-linear combinations of marks. Also putting

$$\varphi_H \colon [M] \mapsto m_M(H) \tag{3.18}$$

for any subgroup $H \leq G$, and extending $\mathbb{Z}$-linearly, we get a ring homomorphism

$$\varphi_H \colon \mathcal{B}(G) \to \mathbb{Z}.$$

Extending φ_H $\mathbb{Q}$-linearly, it is clear that for any $H \leq G$

$$\varphi_H \colon \mathbb{Q} \otimes \mathcal{B}(G) \to \mathbb{Q}$$

is an irreducible character of the $\mathbb{Q}$-algebra $\mathbb{Q} \otimes \mathcal{B}(G)$, so that the table of marks $M(G)$ can also be considered as the "character table" of this algebra with respect to the basis $\{[G/H_1], \ldots, [G/H_n]\}$, with the characters being the columns.

The product of marks m_{G/H_i} of transitive G-sets has a simple group-theoretical interpretation as follows.

Lemma 3.5.12 *For $H_i, H_j \in \mathcal{L}(G)$ let D_{ij} be a set of double coset representatives of H_i, H_j in G, i.e.*

$$G = \bigcup_{d \in D_{ij}}^{\cdot} H_i d H_j.$$

Then

$$m_{G/H_i} \cdot m_{G/H_j} = \sum_{d \in D_{ij}} m_{G/(H_i^d \cap H_j)}.$$

Proof. Since $G \cdot (gH_i, g'H_j) = G \cdot (H_i, dH_j)$ if and only if $g^{-1}g' \in H_i d H_j$, we have

$$G/H_i \times G/H_j = \bigcup_{d \in D_{ij}}^{\cdot} G \cdot (H_i, dH_j).$$

Using (3.17) the result follows immediately, because

$$\mathrm{Stab}_G((H_i, dH_j)) = H_i \cap H_j^{d^{-1}} =_G H_i^d \cap H_j.$$

$\square$

Restricting the marks to the set of cyclic subgroups we immediately obtain a result that might of course also have been deduced from Mackey's Theorem 3.2.17 and Lemma 3.2.14(d).

Corollary 3.5.13 *With the notation of Lemma 3.5.12 we have*

$$1_{H_i}{}^G \cdot 1_{H_j}{}^G = \sum_{d \in D_{ij}} (1_{H_i^d \cap H_j})^G.$$

Remark 3.5.14 Lemma 3.5.12 has many practical applications. For instance, it can be used to compute the conjugacy classes of subgroups which are intersections of two maximal subgroups of a group G, provided the table of marks of G is known. All the results of [105], which lists the intersections of the maximal subgroups of the simple groups of order less than 10^6, can be automatically computed using the GAP library of tables of marks. A similar application is the computation of *subdegrees* (see Definition 1.2.13) of a transitive G-set or, more generally, the lengths of the orbits of a subgroup of G on another transitive G-set.

Example 3.5.15 From the table of marks of A_5 (Example 3.5.6) we see, for instance,

$$[A_5 / V_4] \cdot [A_5 / V_4] = 3[A_5 / V_4] + 3[A_5 / \{1\}],$$
$$[A_5 / V_4] \cdot [A_5 / D_{10}] = 3[A_5 / C_2].$$

Thus there are six (V_4, V_4)-double cosets and three (V_4, D_{10})-double cosets in A_5. The subdegrees of A_5 / V_4 are $1, 1, 1, 4, 4, 4$, whereas the lengths of the D_{10}-orbits on A_5 / V_4 are $5, 5, 5$ (see Remark 1.2.12). Furthermore it follows for instance that different Klein 4-groups in A_5 intersect trivially. ♦

We recall that

$$m \colon \sum_{i=1}^{n} a_i [G/H_i] \mapsto \sum_{i=1}^{n} a_i m_{G/H_i} \quad (a_i \in \mathbb{Z})$$

gives an embedding

$$m \colon \mathcal{B}(G) \to \mathbb{Z}^{\mathcal{L}(G)} \cong \mathbb{Z}^n.$$

In this context $\mathbb{Z}^{\mathcal{L}(G)}$ is often called the **ghost ring** of G. From Lemma 3.5.3 we get the following.

Remark 3.5.16 $([N_G(H) : H]^{-1} m([G/H]))_{H \in \mathcal{L}(G)}$ is a $\mathbb{Z}$-basis of $\mathbb{Z}^{\mathcal{L}(G)}$. Thus

$$[\mathbb{Z}^{\mathcal{L}(G)} : m(\mathcal{B}(G))] = \prod_{H \in \mathcal{L}(G)} [N_G(H) : H].$$

Lemma 3.5.17 *With the above notation the image of m is given by*

$$m(\Omega(G)) = \{f \in \mathbb{Z}^{\mathcal{L}(G)} \mid \sum_{gH \in N_G(H)/H} f(\langle gH \rangle) \equiv 0 \quad \mathrm{mod} \quad [N_G(H) : H], \ H \le G\}.$$

Proof. (a) Let us denote the right hand side of the asserted equality by Y for the moment. If M is any G-set and $H \leq L \leq \mathbf{N}_G(H)$ then L/H acts on $\mathrm{Fix}_M(H)$ and $\mathrm{Fix}_{\mathrm{Fix}_M(H)}(L/H) = \mathrm{Fix}_M(\langle L, H \rangle)$. Hence, using the well known Cauchy–Frobenius–Burnside lemma (see Corollary 2.1.14)

$$\sum_{gH \in \mathbf{N}_G(H)/H} m_M(\langle gH \rangle) = \sum_{\overline{g} \in \mathbf{N}_G(H)/H} |\mathrm{Fix}_{\mathrm{Fix}_M(H)}(\overline{g})| = |\mathbf{N}_G(H)/H| \cdot r,$$

where r is the number of $\mathbf{N}_G(H)/H$-orbits in $\mathrm{Fix}_M(H)$. This proves that $m(\Omega(G)) \subseteq Y$.

(b) Conversely, it is clear that $[\mathbb{Z}^{\mathcal{L}(G)} : Y] \leq \prod_{H \in \mathcal{L}(G)} [\mathbf{N}_G(H) : H]$, since for any $H \in \mathcal{L}(G)$ we have $f_H \in \mathbb{Z}^{\mathcal{L}(G)}$ defined by $f_H(K) = \delta_{H,K} \cdot [\mathbf{N}_G(H) : H]$ is in Y. From Remark 3.5.16 we conclude that $m(\Omega(G)) = Y$. $\square$

What properties of subgroups can be read off from the table of marks? Obviously, if $M(G)$ is given then for every subgroup $H \leq G$ for which the mark $m_{G/H}$ is known the number of subgroups of each possible order can be computed using Lemma 3.5.3. Since cyclic groups are characterized by the fact that they have exactly one subgroup for each divisor of its order (see Exercise 3.5.1), it is clear that it can be decided whether or not H is cyclic. On the other hand, it is not possible in general to decide whether or not H is abelian.

Definition 3.5.18 For a finite group G and a prime p let $\mathbf{O}^p(G)$ be the intersection of all normal subgroups N of G with G/N being a p-group.

Obviously $\mathbf{O}^p(G)$ is a characteristic subgroup of G, the smallest normal subgroup of G with a p-group as factor group.

Theorem 3.5.19 (Dress) *Suppose that H and L are subgroups of G. Then* $\mathbf{O}^p(H) =_G \mathbf{O}^p(L)$ *if and only if* $\varphi_H \equiv \varphi_L \mod p$, *with φ_H as in (3.18).*

Proof. Of course the last assertion means $\varphi_H(x) \equiv \varphi_L(x) \mod p$ for all $x \in \mathcal{B}(G)$, but obviously it is sufficient to prove this congruence for a $\mathbb{Z}$-basis of $\mathcal{B}(G)$, e.g. of all $x = [G/U]$, $(U \leq G)$. So we have to show that $\mathbf{O}^p(H) =_G \mathbf{O}^p(L)$ if and only if

$$m_{G/U}(H) \equiv m_{G/U}(L) \mod p \qquad \text{for all subgroups } U \leq G. \qquad (3.19)$$

If M is any finite G-set and $H \trianglelefteq N \leq G$ then N and also N/H act on $\mathrm{Fix}_M(H)$. If, in addition, N/H is a p-group then all N/H-orbits on $\mathrm{Fix}_M(H)$ have p-power length, hence $|\mathrm{Fix}_M(H)| \equiv |\mathrm{Fix}_M(N)| \mod p$. In particular it follows that

$$m_{G/U}(H) \equiv m_{G/U}(\mathbf{O}^p(H)) \mod p \qquad \text{for all subgroups } U \leq G. \qquad (3.20)$$

This already proves one implication because $\mathbf{O}^p(H) =_G \mathbf{O}^p(L)$ implies that $m_{G/U}(\mathbf{O}^p(H)) = m_{G/U}(\mathbf{O}^p(L))$.

For the converse we assume (3.19) and let

$$H_p/\mathbf{O}^p(H) \in \mathrm{Syl}_p(\mathbf{N}_G(\mathbf{O}^p(H))/\mathbf{O}^p(H)).$$

Observe that $\mathbf{O}^p(H_p) = \mathbf{O}^p(H)$ and consequently $\mathbf{O}^p(H)$ is characteristic in H_p. Hence $\mathbf{N}_G(H_p) \le \mathbf{N}_G(\mathbf{O}^p(H))$, so that by the choice of H_p the prime p does not divide $[\mathbf{N}_G(H_p) : H_p] = m_{G/H_p}(H_p)$. Since by (3.20)

$$m_{G/H_p}(H) \equiv m_{G/H_p}(\mathbf{O}^p(H)) = m_{G/H_p}(\mathbf{O}^p(H_p)) \equiv m_{G/H_p}(H_p) \not\equiv 0 \bmod p$$

we get from our assumption ((3.19) with $U = H_p$) that $m_{G/H_p}(L_p) \ne 0$, where L_p is defined in the same way as H_p. By Lemma 3.5.3 L_p is contained in a conjugate of H_p. By reversing the roles of H and L we conclude that $H_p =_G L_p$ and hence also $\mathbf{O}^p(H) = \mathbf{O}^p(H_p) =_G \mathbf{O}^p(L_p) = \mathbf{O}^p(L)$. □

Corollary 3.5.20 *Given the table of marks $M(G)$ of a finite group G one can find the columns corresponding to subgroups $H \le G$ with $H = \mathbf{O}^p(H)$ for any prime p. In particular one can determine those columns corresponding to solvable subgroups.*

Proof. If $\mathcal{L}(G) = \{H_1, \ldots, H_n\}$ is as in Definition 3.5.1 it follows from Theorem 3.5.19 and Lemma 3.5.3 that $\mathbf{O}^p(H_i) =_G H_j$ with

$$j = \min\{k \mid m_{G/H_i}(H_k) \ne 0 \text{ and } \varphi_{H_i} \equiv \varphi_{H_k} \bmod p \}.$$

Furthermore it is well known that solvable groups are characterized by the fact that every non-trivial subgroup has a normal subgroup of index p for some prime p. □

Theorem 3.5.21 (Dress) *A finite group G is solvable if and only if $[G/G]$ is the only idempotent in the Burnside ring $\mathcal{B}(G)$. In general there is a bijection between the set or primitive idempotents of $\mathcal{B}(G)$ and the set of conjugacy classes of perfect subgroups of G.*

This is equivalent to saying that the connected components of the prime spectrum of $\mathcal{B}(G)$ are in bijection with the conjugacy classes of perfect subgroups (see [52]).

Proof. An element $e \in \mathcal{B}(G)\setminus\{0\}$ is an idempotent if and only if $\varphi_H(e) \in \{0, 1\}$ for all $H \le G$. Let $\sup(e) = \{H \le G \mid \varphi_H(e) \ne 0\}$. By Theorem 3.5.19 for any prime p and any subgroup $H \le G$ we have $H \in \sup(e)$ if and only if $\mathbf{O}^p(H) \in \sup(e)$. For any subgroup $U \le G$ let $U^{(s)} = \cap\{V \trianglelefteq U \mid U/V \text{ solvable}\}$ be the solvable residuum of U. We define $U \sim V$ if and only if $U^{(s)} = V^{(s)}$. Then obviously $U^{(s)}$ is a perfect subgroup of G, and it is easily seen that $U \sim H$ whenever $H \trianglelefteq U$ with U/H solvable. It follows from the above that for an idempotent $e \in \mathcal{B}(G)$ and $U \sim V$ we have $\varphi_U(e) = \varphi_V(e)$. Conversely, if we define for any perfect subgroup $H \in \mathcal{L}(G)$ an element $e_H \in \mathbb{Z}^{\mathcal{L}(G)}$ by

$$e_H(U) = \begin{cases} 1 & \text{if } U^{(s)} =_G H, \\ 0 & \text{else,} \end{cases}$$

then $e_H \in m(\mathcal{B}(G))$ by Lemma 3.5.17, which must be a primitive idempotent by the above.   □

Remark 3.5.22 At present the library of table of marks in GAP contains the table of marks of the following simple groups:

$$\begin{aligned}
&A_n & &\text{for} & &n = 5, \ldots, 12, \\
&L_3(q) & &\text{for} & &q \in \{3, 4, 5, 7, 8, 9, 11\}, \\
&U_3(q) & &\text{for} & &q \in \{3, 4, 5, 7, 8, 9, 11\},
\end{aligned}$$

$L_4(3)$, $L_5(2)$, $U_4(2)$, $U_4(3)$, $U_5(2)$, ${}^2F_4(2)'$, ${}^3D_4(2)$,

$G_2(3)$, $G_2(4)$, $O_8^+(2)$, $O_8^-(2)$, $S_4(4)$, $S_4(5)$, $S_6(2)$, $Sz(32)$, $Sz(8)$

M_{11}, M_{12}, M_{22}, M_{23}, M_{24}, J_1, J_2, J_3, HS, McL, Co_3.

Example 3.5.23 For the groups listed in Remark 3.5.22 we may compute all transitive permutation characters, and using Theorem 2.9.19 or (2.19) we obtain the following information about the Schur indices of the irreducible characters.

Let G be a group listed in Remark 3.5.22 and let $K \subseteq \mathbb{C}$ be a field. Then $m_K(\chi) \in \{1, 2\}$ for all $\chi \in \mathrm{Irr}(G)$; in fact, $m_K(\chi) = 1$ with the following possible exceptions:

$$\chi_2 \in \mathrm{Irr}(U_3(q)) \quad \text{for} \quad q \in \{3, 4, 5, 7, 8, 9, 11\},$$

$$\chi_{16} \in \mathrm{Irr}(U_4(3)), \quad \chi_2, \chi_{12}, \chi_{25} \in \mathrm{Irr}(U_5(2)), \quad \chi_{32} \in \mathrm{Irr}(S_4(5)),$$

$$\chi_{21} \in \mathrm{Irr}(J_2), \quad \chi_9 \in \mathrm{Irr}(J_3), \quad \chi_{11}, \chi_{13} \in \mathrm{Irr}(McL).$$

Here the numbering of the irreducible characters is as in the ATLAS [38]. Computing the Frobenius–Schur indicator one finds that in the above cases $m_K(\chi) = 2$ for $K \subseteq \mathbb{R}$, with the possible exceptions

$$\chi_{32} \in \mathrm{Irr}(S_4(5)), \quad \chi_{21} \in \mathrm{Irr}(J_2), \quad \chi_9 \in \mathrm{Irr}(J_3).$$

Concerning the final two characters it is shown in [57] that $m_{\mathbb{Q}}(\chi_{21}) = 2$ and $m_{\mathbb{Q}}(\chi_9) = 2$.   ♦

Example 3.5.24 The table of marks of $G := M_{11}$ is contained in the library of GAP. We reproduce it in Figure 3.2, as we would get it using the GAP command `Display( TableOfMarks("M11") );`. Figure 3.2 shows that G has 39 conjugacy classes of subgroups. In fact, representatives $H_1, \ldots, H_{39}$ of these conjugacy classes are stored together with the GAP library of table of marks of G and can be accessed for $1 \leq i \leq 39$ via `RepresentativeTom( TableOfMarks("M11"), i )`.

From Figure 3.2 one can immediately see that G has five conjugacy classes of maximal subgroups, with representatives $H_{38}, H_{37}, H_{35}, H_{34}, H_{27}$ and indices $11, 12, 55, 66, 165$. Also the cyclic subgroups among the H_i can easily be identified (namely H_i for $i \in \{1, 2, 3, 5, 6, 9, 11, 15\}$) and thus the values of the transitive permutation characters can be found (see Exercise 3.5.1). Looking at the corresponding columns of the table of marks we find that G has 36

```
7920
3960 24
2640  . 12
1980 36  .  6
1980 12  .  .  4
1584  .  .  .  .  4
1320 24  6  .  .  .  6
1320 24  6  .  .  .  .  2
1320  8  6  .  .  .  .  .  2
 990  6  .  6  .  .  .  .  6
 990  6  .  .  2  .  .  .  .  .  2
 990 30  .  6  2  .  .  .  .  .  2
 880  . 16  .  .  .  .  .  .  .  . 16
 792 24  .  .  .  2  .  .  .  .  .  .  2
 720  .  .  .  .  .  .  .  .  .  .  .  .  .  5
 660 12 12  2  .  .  .  .  .  .  .  .  .  2
 660 28  3  6  .  .  3  1  1  .  .  .  .  .  1
 495 15  .  3  3  .  .  .  .  3  1  1  .  .  .  .  1
 440 24  8  .  .  .  .  8  .  .  .  .  8  .  .  .  .  .  8
 440  8  8  .  .  .  2  .  2  .  .  .  8  .  .  .  .  .  .  2
 396 12  .  .  4  1  .  .  .  .  .  .  .  1  .  .  .  .  .  .  1
 330  2  6  .  2  .  .  .  2  2  .  .  .  .  .  .  .  .  .  .  .  2
 330 18  6  4  2  .  .  2  .  .  .  2  .  .  .  1  .  .  .  .  .  .  1
 220 12  4  .  4  .  .  4  .  .  .  4  .  .  .  .  4  .  .  .  .  4
 220 12  4  .  4  .  .  .  4  .  .  .  4  .  .  .  .  4  .  .  .  .  .  2
 220 20  4  6  .  .  2  4  2  .  .  .  4  .  .  .  2  .  4  2  .  .  .  .  .  2
 165 13  3  3  1  .  3  1  1  1  1  1  .  .  .  .  1  1  .  .  .  1  .  .  .  .  1
 144  .  .  .  .  4  .  .  .  .  .  .  1  .  .  .  .  .  .  .  .  .  .  .  1
 132 12  6  2  .  2  6  .  .  .  .  .  .  2  .  2  .  .  .  .  .  .  .  .  .  .  2
 132 12  6  2  .  2  .  2  .  .  .  .  2  .  2  .  .  .  .  .  .  .  .  .  .  .  1
 110  6  2  .  6  .  .  2  .  6  .  .  2  .  .  .  2  .  .  .  2  2  .  .  .  .  2
 110 14  2  6  2  .  2  2  2  .  .  2  2  .  .  .  2  .  2  2  .  .  .  2  .  2  .  .  .  .  2
 110  6  2  .  2  .  .  2  .  .  2  .  2  .  .  .  .  2  .  .  .  2  .  .  .  .  2
  66 10  3  4  2  1  3  1  1  .  .  2  .  1  .  1  1  1  .  .  .  1  .  1  .  .  .  .  1  .  .  1
  55  7  1  3  3  .  1  1  1  3  1  1  1  .  .  .  1  1  1  1  .  .  .  1  1  1  .  .  .  .  1  1  1  .  1
  22  6  4  2  2  2  .  4  .  .  .  2  4  2  .  .  4  .  .  .  2  .  2  .  .  .  .  2  .  .  .  .  2
  12  4  3  2  .  2  3  1  1  .  .  2  1  2  1  .  .  2  1  2  1  .  .  .  .  .  1  2  1  .  .  .  .  1
  11  3  2  1  3  1  .  2  .  3  1  1  2  1  .  1  .  1  2  .  1  .  1  2  2  .  .  .  .  1  2  .  .  .  .  1  .  1
   1  1  1  1  1  1  1  1  1  1  1  1  1  1  1  1  1  1  1  1  1  1  1  1  1  1  1  1  1  1  1  1  1  1  1  1  1  1  1
```

Figure 3.2. Table of marks of M_{11}.

transitive permutation characters and that $(G/H_7, G/H_8)$, $(G/H_{24}, G/H_{25})$ and $(G/H_{29}, G/H_{30})$ are pairs of G-sets yielding the same permutation character.

Abbreviating $m_i := m_{H_i}$ for the mark of the G-set G/H_i we may compute for instance:

$$m_{35} \cdot m_{35} = m_4 + m_{10} + m_{35}, \qquad m_{36} \cdot m_{36} = m_{19} + 2\,m_{36},$$
$$m_{37} \cdot m_{37} = m_{29} + m_{37}, \qquad m_{38} \cdot m_{38} = m_{31} + m_{38}.$$

This shows that the G-sets G/H_{35} and G/H_{36} have rank three and G/H_{37} and G/H_{38} have rank two.

Table 3.4 gives for each $i = 1, \ldots, 39$ the index $[M_{11} : H_i]$ and a description of the structure of a representative H_i of the ith conjugacy class of subgroups of M_{11}. This can be obtained via the GAP command

```
StructureDescription( RepresentativeTom(TableOfMarks("M11"),i)) );
```

Table 3.4. Subgroups of M_{11}

1	7920	C_1	2	3960	C_2	3	2640	C_3
4	1980	V_4	5	1980	C_4	6	1584	C_5
7	1320	S_3	8	1320	S_3	9	1320	C_6
10	990	Q_8	11	990	C_8	12	990	D_8
13	880	C_3^2	14	792	D_{10}	15	720	C_{11}
16	660	A_4	17	660	D_{12}	18	495	QD_{16}
19	440	$C_3^2 \rtimes C_2$	20	440	$C_3 \times S_3$	21	396	$C_5 \rtimes C_4$
22	330	$SL_2(3)$	23	330	S_4	24	220	$C_3^2 \rtimes C_4$
25	220	$C_3^2 \rtimes C_4$	26	220	$S_3 \times S_3$	27	165	$GL_2(3)$
28	144	$C_{11} \rtimes C_5$	29	132	A_5	30	132	A_5
31	110	$C_3^2 \rtimes Q_8$	32	110	$(S_3 \times S_3) \rtimes C_2$	33	110	$C_3^2 \rtimes C_8$
34	66	S_5	35	55	$(C_3^2 \rtimes C_8) \rtimes C_2$	36	22	A_6
37	12	$L_2(11)$	38	11	M_{10}	39	1	M_{11}

Here QD_{16} is a quasidihedral group (see[88], p. 91) of order 16 and M_{10}, a point stabilizer of M_{11} in its natural representation, is a non-split extension of A_6 by C_2 (see [38], p.4).   ◆

Exercises

Exercise 3.5.1 (a) Show that a finite group G is cyclic if and only if for each divisor $d \mid |G|$ there is exactly one subgroup $H \leq G$ with $|H| = d$. Use this to write a GAP program which, for a given table of marks $M(G)$ of a finite group G, finds the columns corresponding to the conjugacy classes of cyclic subgroups of G. This can be used to find a complete list of transitive permutation characters of G.

(b) Let $[\gamma_{ij}] = M(G)^{-1}$ with $M(G)$ as in Definition 3.5.1. Show that

$$\sum_{j=1}^{n} \gamma_{ij} = \begin{cases} \frac{\varphi(|H_i|)}{|N_G(H_i)|} & \text{if } H_i \text{ is cyclic,} \\ 0 & \text{otherwise.} \end{cases}$$

Here φ is the Eulerian function.

Exercise 3.5.2 We use the notation of Example 3.5.24. We saw there that the G-sets G/H_{24} and G/H_{25} yield the same permutation character. Find the orbit structure of G in $G/H_i \times G/H_i$ for $i \in \{24, 25\}$. Show that

$$\text{rk}_K \text{Hom}_{KG}(K(G/H_{24}), K(G/H_{25})) = 12$$

for any commutative ring K. Also show, using only the table of marks of G, that H_i has a normal Sylow 3-subgroup P and that H_i/P is cyclic for $i \in \{24, 25\}$. It will follow from Corollary 4.10.14 that $K(G/H_{24}) \not\cong K(G/H_{25})$ for a field K of characteristic three and hence $\mathbb{Z}(G/H_{24}) \not\cong \mathbb{Z}(G/H_{25})$, although the permutation characters of these modules coincide. Similarly, show that the permutation modules are not integrally equivalent for the other two pairs of transitive G-sets having the same permutation characters.

Exercise 3.5.3 (a) Assume that $M(G)$ is a table of marks of a group G such that representatives H_i $(1 \le i \le r)$ of the conjugacy classes of subgroups of G are either stored or can be computed. Write a GAP program to compute for $H \le G$, the fusion of $M(H)$ into $M(G)$, that is, a list f such that $U_j =_G H_{f(j)}$, where U_j $(1 \le j \le s)$ are representatives of the conjugacy classes of subgroups of H.

(b) Using the notation of Example 3.5.24 show that

$$m_{37} \cdot m_{26} = 2\, m_{17} + m_8 \quad \text{with} \quad H_{17} \cong D_{12}, \ H_8 \cong S_3 \, .$$

Conclude that $H := H_{26}$ has three orbits on $\Omega := G/H_{37}$ with stabilizers S_1, S_2, S_3 isomorphic to D_{12}, D_{12} and S_3. Using part (a) compute the mark of Ω considered as an H-set. Furthermore, show that S_1, S_2 are not conjugate in H and are centralizers of involutions in H and that there is an $h \in S_3$ with $|\mathbf{C}_H(h)| = 9$.

Exercise 3.5.4 We use the notation of Example 3.5.10. Show that

$$f_\chi^{H_j} = \frac{1}{[\mathbf{N}_G(H_j) : H_j]} \sum_{H_j \le H \le G} \mu(H_j, H) X^{d(H)} \quad \text{with} \quad d(H) = (\chi_H, \mathbf{1}_H)_H,$$

where μ denotes the Möbius function of the lattice of subgroups of G.

Exercise 3.5.5 Let $G := L_2(7)$ and let $\chi \in \mathrm{Irr}(G)$ with $\chi(1) = 3$. Show that $f_\chi^{\{1\}} = \frac{1}{168}(X - 1)(X^2 + X - 48)$.

Exercise 3.5.6 (See [136].) Consider the natural A_n-set $\Omega := \{1, \ldots, n\}$ and the permutation module $V := \mathbb{F}_q \Omega$. Show that the V-closed subgroups of G are of the form $S_\Lambda \cap A_n$ for a Young subgroup $S_\Lambda \le S_n$ and that $[S_\Lambda : S_\Lambda \cap A_n] = 2$ whenever $S_\Lambda \ne \{1\}$. Let $\chi := \chi_{\mathbb{C}\Omega}$. Show that $f_\chi^H = 0$ unless $H =_G S_\Lambda \cap A_n$, in which case

$$f_\chi^H = [\mathbf{N}_G(H) : H]^{-1} \prod_{i=0}^{r_\Lambda - 1} (X - i) \quad \text{if} \quad H \ne \{1\}$$

and

$$f_\chi^{\{1\}} = \frac{2}{n!} \left(X + \frac{(n-2)(n-1)}{2} \right) \prod_{i=0}^{n-2} (X - i).$$

3.6 Clifford theory

Throughout this section K is an arbitrary field and $N \trianglelefteq G$. We continue to assume that G is a finite group, although some of the results require only that $[G : N]$ is finite.

Remark 3.6.1 If W is a KN-module with representation δ and character φ then $g \otimes W$ ($\subseteq W^G$) is also a KN-module for every $g \in G$, which is called a module **conjugate** to W. Since $h \cdot g \otimes w = g \otimes (g^{-1}hg)w$ for $h \in N$, $w \in W$, this leads to a representation

$$^g\delta \colon N \to \mathrm{GL}(W) , \ h \mapsto \delta(g^{-1}hg),$$

and the character of $g \otimes W$ is $^g\varphi$ given by $^g\varphi(h) = \varphi(g^{-1}hg)$. Observe that $^g\delta(N) = \delta(N)$, and thus δ, is irreducible if and only if $^g\delta$ is irreducible.

If the KN-module W is contained in a KG-module V then obviously

$$g \otimes W \to gW \subseteq V, \quad g \otimes w \mapsto gw \quad \text{is a } KN\text{-isomorphism.}$$

If in addition W is stable under G, hence the restriction of a KG-module, then clearly $W = gW \cong_{KN} g \otimes W$, but otherwise $W \cong_{KN} g \otimes W$ need not hold.

Theorem 3.6.2 (Clifford) *Let V be a simple KG-module and let $N \trianglelefteq G$.*

(a) V_N *is a direct sum of simple KN-modules, which are conjugate in G.*

(b) *Let $W_1, \ldots, W_m \leq_{KN} V_N$ be representatives of the isomorphism classes of simple KN-submodules of V_N. For $i = 1, \ldots, m$ put $V_i := \mathrm{H}_{W_i}(V_N)$ and $T_i := T_G(W_i) := \{g \in G \mid W_i \cong g \otimes_{KN} W_i\}$, which is called the **inertia subgroup** of W_i. Then $N \trianglelefteq T_i \leq G$ with $[G : T_i] = m$ and V_i is a simple KT_i-module. There is an $e \in \mathbb{N}$ independent of i such that*

$$V \cong_{KG} (V_i)^G \qquad and \qquad V_N \cong_{KN} \bigoplus_{i=1}^{m} \underbrace{(W_i \oplus \cdots \oplus W_i)}_{e}.$$

There are $g_i \in G$, with $G = \dot{\bigcup}_{i=1}^{m} g_i T_1$, $V_i \cong_{KT_i} g_i V_1$ and $W_i \cong_{KN} g_i W_1$.

Proof. (a) Let $G = \dot{\bigcup}_{j=1}^{n} h_j N$. Then for $W \leq_{KN} V_N$ the sum $\sum_{j=n}^{m} h_j W \subseteq V$ is obviously a KG-submodule and, since V is supposed to be simple, we have $V = \sum_{j=1}^{m} g_j W$ provided that $W \neq \{0\}$. Choosing W simple, part (a) follows from Lemma 1.5.2.

(b) Obviously $N \leq T_i \leq G$ since $h \otimes W_i \cong_{KN} hW_i = W_i$ for $h \in N$. By (a) G acts transitively on the isomorphism classes $[W_i]$ of simple KN-submodules W_i of V and $\mathrm{Stab}_G([W_i]) = T_i$. Thus $[G : T_i] = m$ and $V_i = \mathrm{H}_{W_i}(V_N)$ is stable under the action of T_i, that is $V_i \leq_{KT_i} V_{T_i}$. If $V_1 = W_1^1 \oplus \cdots \oplus W_1^e$ with $W_1 \cong_{KN} W_1^j \leq V_N$ and $W_i = g_i W_1$ then $g_i V_1 = g_i W_1^1 \oplus \cdots \oplus g_i W_1^e = V_i$. Since $V = \oplus_{i=1}^{m} g_i V_i$ we have $V \cong_{KG} (V_i)^G$ by Lemma 3.2.14(a). Finally, since $(V_i)^G$ is simple, V_i must be a simple KT_i-module (see Exercise 3.2.5). $\qquad \square$

Corollary 3.6.3 *If V is a simple KG-module, char $K = p > 0$ and N is a normal p-subgroup of G, then $V_N = K \oplus \cdots \oplus K$ is a trivial KN-module.*

Proof. Use part (a) of Theorem 3.6.2 and Theorem 1.3.11. □

Corollary 3.6.4 *If $N \trianglelefteq G$ and W is a simple KN-module with inertia subgroup $T_G(W) = N$, then W^G is simple.*

Proof. Let V be a simple factor module of W^G. By Theorem 3.2.12

$$\{0\} \neq \mathrm{Hom}_{KG}(W^G, V) \cong \mathrm{Hom}_{KN}(W, V_N).$$

Thus V_N contains a simple submodule W' isomorphic to W. By Clifford's Theorem 3.6.2 and the assumption, V_N contains $[G : T_G(W')] = [G : T_G(W)] = [G : N]$ submodules conjugate to W' as direct summands. But then $\dim V \geq [G : N] \dim W = \dim W^G \geq \dim V$ and hence $V \cong W^G$. □

As an application of Clifford's theorem we present the following.

Theorem 3.6.5 (Ito) *Let V be a simple $\mathbb{C}G$-module, then $\dim V \mid [G : N]$ for every abelian normal subgroup $N \trianglelefteq G$.*

Proof. We use induction with respect to the group order $|G|$. Let $N \trianglelefteq G$ be an abelian normal subgroup and let W be a simple summand of V_N.

If $T = T_G(W) < G$ there is a simple $\mathbb{C}T$-module V_1 with $V \cong V_1^G$. By induction $\dim V_1 \mid [T : N]$, and so $\dim V = [G : T] \cdot \dim V_1 \mid [G : T] \cdot [T : N] = [G : N]$.

If $T = G$ then $V_N \cong W \oplus \cdots \oplus W$. Since N is abelian $\dim_{\mathbb{C}} W = 1$. Thus the elements of N are represented by scalar matrices in the representation δ corresponding to V, so $\delta(N) \leq \mathbf{Z}(\delta(G))$. By Theorem 2.3.6 we have $\dim V \mid [G : \mathbf{Z}(G)]$ so the result follows immediately if δ is faithful. Otherwise we may use the induction hypothesis for $G/\ker \delta$, since $(N \ker \delta)/\ker \delta \trianglelefteq G/\ker \delta$ is abelian. □

We note the following interesting consequence of Theorem 3.6.5, which will be used in Section 4.14.

Corollary 3.6.6 *Assume that G has an abelian normal Sylow-p-subgroup P and that $\mathbf{C}_G(g) = P$ for some $g \in P$. Then*

$$|\mathrm{Irr}(G)| \leq |P|$$

and $|\mathrm{Irr}(G)| < |P|$ unless G/P is abelian.

Proof. Theorem 3.6.5 implies that $p \nmid \chi(1)$ for all $\chi \in \mathrm{Irr}(G)$. From the congruence relations (Lemma 2.2.2(b)) we conclude that $\chi(g) \neq 0$ for any $\chi \in \mathrm{Irr}(G)$ and $g \in P$. Choosing $g \in P$ with $\mathbf{C}_G(g) = P$ and putting $N := [\langle g \rangle, G] \leq P$ we obtain from Lemma 2.2.11 that

$$|\mathrm{Irr}(G)| \leq |P| - (|G/N| - |\mathrm{Irr}(G/N)|).$$

Thus $|\mathrm{Irr}(G)| \leq |P|$, and equality implies that G/N and hence G/P is abelian. □

Theorem 3.6.7 *Assume that $W_1, \ldots, W_s$ are representatives of the G-conjugacy classes of simple KN-modules (up to isomorphism) and $T_i := T_G(W_i)$). Furthermore let V_{ij}, $1 \leq j \leq k_i$, be all simple KT_i-modules up to KT_i-isomorphism with $(V_{ij})_N \cong \underbrace{W_i \oplus \cdots \oplus W_i}_{e_{i,j}}$ (i.e. with $\mathrm{Hom}_{KN}((V_{ij})_N, W_i) \neq \{0\}$). Then*

$$\{V_{ij}^G \mid 1 \leq i \leq s \,, \ 1 \leq j \leq k_i\}$$

is a complete set of representatives of the isomorphism classes of simple KG-modules. If V is a simple KG-module such that V_{T_i} has V_{ij} as a composition factor, then $V \cong V_{ij}^G$ and the multiplicity of V_{ij} in V_{T_i} is one.

If $W_1 = K_N$, then $T_1 = G$ and $\{V_{i1} \mid 1 \leq j \leq k_1\}$ is just the set of simple KG/N-modules (up to isomorphism) inflated to KG-modules.

Proof.   By Theorem 3.6.2 every simple KG-module is isomorphic to one of the form V_{ij}^G. It suffices to prove that all V_{ij}^G are simple and pair-wise non-isomorphic. Obviously $V_{ij}^G \not\cong_{KG} V_{kl}^G$ for $i \neq k$, since $(V_{ij}^G)_N \not\cong_{KN} (V_{kl}^G)_N$.

We fix i and assume $G = \dot{\bigcup}_{k=1}^{m} g_k T_i$ with $g_1 = 1$. Since $g_k T_i = N g_k T_i$ we have, by Mackey's Theorem 3.2.17,

$$((V_{ij}^G)_{T_i})_N = (V_{ij}^G)_N \cong (V_{ij})_N \oplus \bigoplus_{k=2}^{m} (g_k \otimes V_{ij})_N.$$

Note that $(g_k \otimes V_{ij})_N$ is for $k \geq 2$ a direct sum of modules isomorphic to $g_k \otimes W_i \not\cong W_i$ since $g_k \in G \setminus T_i$. Thus V_{ij} is isomorphic to the only composition factor X of $(V_{ij}^G)_{T_i}$ with $X_N \cong W_i \oplus \cdots \oplus W_i$. In particular, $V_{ij}^G \not\cong V_{il}^G$ for $j \neq l$ and the last claim follows too.

Finally we show that every V_{ij}^G is simple. Let V be a simple factor module of V_{ij}^G. Then $0 \neq \mathrm{Hom}_{KG}(V_{ij}^G, V) \cong \mathrm{Hom}_{KT_i}(V_{ij}, V_{T_i})$. Since V_{ij} is simple V_{T_i} contains a module isomorphic to V_{ij} and V_N contains a submodule isomorphic to $(V_{ij})_N$. Since V_N contains all modules conjugate to W_i with equal multiplicity, by Theorem 3.6.2, we get $\dim V \geq [G : T_i]e_{i,j} \dim W_i = \dim(V_{ij}^G)$, so $V_{ij}^G = V$. $\qquad\square$

The mapping $V_{ij} \to V_{ij}^G$ described in Theorem 3.6.7 is sometimes referred to as the **Clifford correspondence**. It is often expressed in terms of characters as follows.

Corollary 3.6.8 *Suppose that G is a finite group and $N \trianglelefteq G$. For $\varphi \in \mathrm{Irr}(N)$ let $T_G(\varphi) := \{g \in G \mid {}^g\varphi = \varphi\}$ and for $N \leq H \leq G$ let*

$$\mathrm{Irr}(\, H \mid \varphi \,) := \{\psi \in \mathrm{Irr}(H) \mid (\psi_N, \varphi)_N \neq 0\}.$$

*Then for $T_G(\varphi) \leq T \leq G$ we have a bijection $\mathrm{Irr}(T \mid \varphi) \to \mathrm{Irr}(G \mid \varphi)$, $\psi \mapsto \psi^G$ called the **Clifford correspondence**, and*

$$\mathrm{Irr}(G) = \dot{\bigcup}\{\mathrm{Irr}(\, G \mid \varphi \,) \mid \varphi \in \mathrm{Irr}(N)\} = \dot{\bigcup}_{i=1}^{m} \mathrm{Irr}(\, G \mid \varphi_i \,)$$

if $\varphi_1, \ldots, \varphi_m$ are representatives of the G-conjugacy classes of irreducible characters of N.

Proof. For $T = T_G(\varphi)$ this is an immediate consequence of Theorem 3.6.7. If $T_G(\varphi) < T < G$ use the bijections $\mathrm{Irr}(T_G(\varphi) \mid \varphi) \to \mathrm{Irr}(T \mid \varphi)$, $\psi \mapsto \psi^T$ and $\mathrm{Irr}(T_G(\varphi) \mid \varphi) \to \mathrm{Irr}(G \mid \varphi)$, $\psi \mapsto \psi^G$ and the transitivity of induction. $\qquad\square$

Observe, however, that Theorem 3.6.7 is much more general than Corollary 3.6.8, since it applies to representations over arbitrary fields, in particular to representations over fields of characteristic $p > 0$, for which the language of (ordinary) characters is not adequate, as we will see in Section 4.2.

We illustrate the above theorem with a familiar example.

Example 3.6.9 Let $G = \mathrm{S}_4$ and $N = \mathrm{V}_4$, the Klein 4-group. We assume that char $K \neq 2$. Then N has four irreducible representations, which are linear characters $\lambda_1 = 1_V, \lambda_2, \lambda_3, \lambda_4$ (see Example 2.1.19); G has two orbits on these, $\{\lambda_1\}$ with inertia subgroup $T_1 = G$ and $\{\lambda_2, \lambda_3, \lambda_4\}$ with inertia subgroup $T_2 = T_G(\lambda_2) \cong \mathrm{D}_8$, a dihedral group. Let $\pi\colon G \to G/N \cong \mathrm{S}_3$ be the projection. Then

$$\mathrm{Irr}(G \mid \lambda_1) = \{\chi \circ \pi \mid \chi \in \mathrm{Irr}(G/N)\},$$
$$\mathrm{Irr}(G \mid \lambda_2) = \{\psi^G \mid \psi \in \mathrm{Irr}(T_2),\ \psi_N = \lambda_2\}.$$

So $\mathrm{Irr}(G \mid \lambda_2) = \{\chi_4, \chi_5\}$ in the notation of Example 2.1.22. $\qquad\blacklozenge$

The preceding result suggests the following strategy for finding a system of representatives of the isomorphism classes of simple KG-modules, provided that G has a normal subgroup N.

(1) Find representatives W_i of the G-conjugacy classes of simple KN-modules and their inertia groups T_i.

(2) Construct (up to isomorphism) all simple KT_i-modules V that satisfy $\mathrm{Hom}_{KN}(W_i, V_N) \neq \{0\}$.

(3) For all the KT_i-modules V constructed in step (2) compute the induced modules V^G.

Of course, the question arises of how to find the modules V in step (2). It turns out that the V's with the desired property are just the composition factors of $W_i^{T_i}$. We change the perspective slightly to simplify the notation and put $G = T_i$, i.e. we assume that $W = W_i$ is G-invariant.

Lemma 3.6.10 *If V is a simple KG-module with $V_N \cong W \oplus \cdots \oplus W$ for some simple KN-module W, then V is a factor module of W^G. Conversely, any composition factor V of W^G has the property that $V_N \cong W \oplus \cdots \oplus W$.*

Proof. The first assertion follows immediately from Theorem 3.2.12, since the assumption implies that $\mathrm{Hom}_{KN}(W, V_N) \neq 0$. Since $N \trianglelefteq G$ we have $NgN = gN$ for any $g \in G$, and Mackey's theorem yields

$$(W^G)_N \cong \sum_{G = \dot\bigcup gN} g \otimes W \cong W \oplus \cdots \oplus W.$$

Thus any constituent of W^G becomes homogeneous upon restriction to N. □

So the task is to find the simple factor modules of W^G for each G-invariant simple KN-module W. We first look at important special cases.

Theorem 3.6.11 *Let W be KG-module such that W_N is absolutely simple. Then for every simple $K(G/N)$-module V the KG-module $W \otimes_K V'$ is simple, where $V' := \mathrm{Inf}_\varphi V$ with $\varphi \colon KG \to K(G/N)$ extending K-linearly the canonical map $G \to G/N$. Furthermore, for simple $K(G/N)$-modules V_1, V_2 we have $W \otimes_K V_1' \cong_{KG} W \otimes_K V_2'$ if and only if $V_1 \cong_{K(G/N)} V_2$.*

Proof. Let $(w_1, \ldots, w_m)$ be a K-basis of W. Then $W \otimes_K V' = \bigoplus_{i=1}^m w_i \otimes V'$. Since W_N is absolutely simple, we deduce from Lemma 1.3.6(b) that there are elements $a_{ij} \in KN$ such that $a_{ij} \cdot w_k = 0$ for $k \neq i$ and $a_{ij} \cdot w_i = w_j$. Observe that

$$a \cdot (w \otimes v) = (a \cdot w) \otimes v \qquad \text{for} \qquad a \in KN, w \in W, v \in V'$$

because N acts trivially on V'. Let $0 \neq u = \sum_{i=1}^m w_i \otimes v_i$ with $v_i \in V'$ and, say, $v_1 \neq 0$. We show that $U := \langle u \rangle_{KG} = W \otimes_K V'$:

$$\left(\sum_{j=1}^m \alpha_j\, a_{1,j}\right) \cdot u = \left(\sum_{j=1}^m \alpha_j\, v_j\right) \otimes v_1 \in U \qquad \text{for all} \qquad \alpha_1, \ldots, \alpha_m \in K.$$

Thus $W \otimes_K v_1 \subseteq U$. Since V' is simple we get $V' = \langle \{g \cdot v_1 \mid g \in G\} \rangle_K$. Thus $W \otimes_K V' \subseteq U$ and $W \otimes_K V'$ is simple by Lemma 1.3.2.

Now let $\psi \colon W \otimes_K V_1' \to W \otimes_K V_2'$ be a KG-isomorphism and let $v \in V_1$. Since

$$a_{i,i} \cdot w_i \otimes v = w_i \otimes v = a_{j,i} \cdot w_j \otimes v \qquad \text{for} \qquad 1 \leq i, j \leq m$$

we see that $\psi(w_i \otimes v) = w_i \otimes \psi'(v)$ for a uniquely determined $\psi'(v) \in V_2$ independent of i. If $g \in G$ and $g \cdot w_i = \sum_j \alpha_{ji} w_j$ with $0 \neq \alpha_{ki} =: \alpha$ then $a_{i,k}\, g \cdot w_i = \alpha\, w_k$ and

$$\alpha\, w_k \otimes g \cdot \psi'(v) = (a_{i,k}\, g \cdot w_i) \otimes (g \cdot \psi'(v)) = a_{i,k}\, g \cdot \psi(w_i \otimes v) = \alpha\, w_k \otimes \psi'(g \cdot v).$$

Hence $\psi' \colon V_1' \to V_2'$ is a KG-isomorphism. □

Corollary 3.6.12 (Gallagher) *Let $\pi \colon G \to G/N$ $g \mapsto gN$ and $\chi \in \mathrm{Irr}(G)$. If $\chi_N \in \mathrm{Irr}(N)$ then $\chi \cdot (\psi \circ \pi) \in \mathrm{Irr}(G)$ for every $\psi \in \mathrm{Irr}(G/N)$ and*

$$(\chi_N)^G = \sum_{\psi \in \mathrm{Irr}(G/N)} \psi(1)\, \chi \cdot (\psi \circ \pi). \tag{3.21}$$

Proof. The first assertion is an immediate consequence of Theorem 3.6.11. Since $(\chi, \chi \cdot (\psi \circ \pi))_N = \psi(1)$, formula (3.21) follows comparing the degrees.
$\qquad\square$

Theorem 3.6.13 *Let G/N be cyclic of order n and let W be a G-invariant absolutely simple KN-module. Assume that (i) the map $x \mapsto x^n$, $K \to K$ is surjective and $X^n - 1 \in K[X]$ splits into linear factors or (ii) that K is a splitting field for G. Then the following hold.*

(a) *There is a KG-module V with $V_N \cong_{KN} W$.*

(b) *If X is a simple KG-module with $X_N \cong W \oplus \cdots \oplus W$ with, say, e summands isomorphic to W, then $e = 1$ and $X \cong V \otimes_K Z$ with V as in (a) and a KG/N-module Z of dimension one inflated to a KG-module. Conversely, for any such module Z, the KG-module $X = V \otimes_K Z$ is simple and satisfies $V_N \cong W$.*

Proof. (a) Assume first that K is a splitting field for G and let $\bar{K}$ be an algebraic closure of K. If there is a simple $\bar{K}G$-module $\bar{V}$ with $\bar{V}_N \cong_{\bar{K}N} \bar{K}W$, then by Exercise 1.5.8 $\bar{V} \cong \bar{K}V$, for a simple KG-module V and $V_N \cong_{KN} W$. So we may assume that K is algebraically closed. In this case, conditions (i) are satisfied. So we assume (i).

Choose an element $g \in G$ with $G = \langle N, g \rangle$, then $G = \bigcup_{i=0}^{n-1} g^i N$ with $n = [G : N]$. Let $\delta \colon N \to \mathrm{GL}(W)$ be the representation of N on W. Since $T_G(W) = G$ there is $\varphi \in \mathrm{GL}(W)$ with

$$\delta(g^{-1}hg) = \varphi^{-1}\delta(h)\varphi \qquad \text{for all} \quad h \in N.$$

Since $g^n \in N$ we have

$$\varphi^{-n}\delta(h)\varphi^n = \delta(g^{-n}hg^n) = \delta(g^n)^{-1}\delta(h)\delta(g^n).$$

Hence

$$\epsilon = \varphi^n \delta(g^n)^{-1} \in \mathrm{End}_{KN} W = K \cdot \mathrm{id}_W,$$

so there is $0 \neq \alpha \in K$ with $\varphi = \alpha \cdot \delta(g^n)$. By our assumption there is a $\gamma \in K$ with $\gamma^n = \alpha$. We extend δ to a mapping on all of G by defining

$$\delta(g^i h) = \gamma^{-i} \cdot \varphi^i \delta(h) \qquad \text{for} \quad h \in N , \ i \in \{0, \ldots, n-1\}.$$

An easy computation shows that $\delta \colon G \to \mathrm{GL}(W)$ is, in fact, a representation. Thus W becomes a KG-module, which we might call V, and $V_N = W$.

(b) By Exercise 3.2.4 we have $W^G \cong_{KG} K(G/N) \otimes_K V$. Let $n = p^a m$, where $p := \mathrm{char}\, K$, and let $\zeta \in K$ be a primitive mth root of unity. If $\bar{g} = gN$, so that $G/N = \langle \bar{g} \rangle$, we have by Example 1.1.23

$$K(G/N) \cong_{K(G/N)} \left(\bigoplus_{i=0}^{m-1} Z_i \right) \otimes U \quad \text{with} \quad Z_i = \langle v_i \rangle_K \quad \text{and} \quad \bar{g} \cdot v_i = \zeta^i v_i$$

and $U \cong K[X]/(X-1)^{p^a}$, a uniserial $K(G/N)$-module with trivial composition factors ($\cong Z_0$). By Theorem 3.2.12 $\mathrm{Hom}_{KG}(W^G, X) \cong_K \mathrm{Hom}_{KN}(W, X_N) \neq 0$, so X is a (top) composition factor of W^G and thus $X \cong Z_i \otimes_K V$, for some $i \in \{0, \ldots, m-1\}$.
$\qquad\square$

Corollary 3.6.14 *Let G/N be a p-group and let W be a simple FN-module for some $F = \mathbb{F}_{p^r}$. Assume that $\mathrm{End}_{FN} W = F \cdot \mathrm{id}_W$.*

(a) If W is G-invariant there is a unique FG-module V with $V_N = W$.

(b) There is up to isomorphism a unique simple FG-module V such that W is isomorphic to a direct summand of V_N and $V_N \cong_N W^{g_1} \oplus \cdots \oplus W^{g_m}$, where $\{g_1, \ldots, g_m\}$ is a transversal of $T_G(W)$ in G.

Proof.   (a) Let $N_0 := N \trianglelefteq N_1 \trianglelefteq \ldots \trianglelefteq N_n = G$ be a subnormal series of G with $[N_i : N_{i-1}] = p$ for all $i \in \{1, \ldots, n\}$. If V_i is a unique FN_i-module with $(V_i)_N = W$, then V_i is also invariant in G and $\mathrm{End}_{FN_i} V_i = F \,\mathrm{id}_{V_i}$. So, for (a) we may assume that $[G : N] = p$. Now $x \mapsto x^p$ is an automorphism of F and we may apply Theorem 3.6.13, from which the result follows, since the trivial module is the only simple $F(G/N)$-module.
  (b) follows from (a) and Corollary 3.6.8.   □

Corollary 3.6.15 *Assume $[G : N] = p$, a prime. Let K be a splitting field for N and G. If W is a simple KN-module then one of the following holds.*

(a) $T_G(W) = N$ and W^G is simple. There are, up to isomorphism, exactly p KN-modules $W = W_1, \ldots, W_p$ conjugate to W and $W_i^G = W^G$ for all i.

(b) $T_G(W) = G$ and W has an extension to a KG-module; i.e. there is a KG-module V with $V_N = W$. Furthermore one has

if $\mathrm{char}\,K \neq p$ there are exactly p such KG-modules $V_1, \ldots, V_p$ with $(V_i)_N = W$,

if $\mathrm{char}\,K = p$ then there is a unique KG-module V with $V_N = W$.

Proof.   This follows immediately from Theorem 3.6.2 and Theorem 3.6.13, since in the case considered here G/N has either p irreducible representations (if $\mathrm{char}\,K \neq p$) or just the trivial one by Example 1.1.23 if $\mathrm{char}\,K = p$.   □

Corollary 3.6.15 has many applications; for example, for a solvable group G, it can be applied to the terms of a composition series

$$1 = N_0 \trianglelefteq N_1 \trianglelefteq \ldots \trianglelefteq N_l = G,$$

since each factor N_i/N_{i+1} is cyclic of prime order. So it is possible to construct the irreducible representations or characters of N_{i+1} by extending or inducing those of N_i. We first illustrate this principle with a simple example.

Example 3.6.16 $G = \mathrm{GL}_2(3)$ has the following composition series, where we denote the subgroups by names identifying their isomorphism type:

$$1 \trianglelefteq C_2 \trianglelefteq V_4 \trianglelefteq Q_8 \trianglelefteq \mathrm{SL}_2(3) \trianglelefteq \mathrm{GL}_2(3).$$

We denote the irreducible representations by their degrees and join two of these by a line, if one is a constituent of the other upon restriction to the relevant subgroup. Note that the trivial character can always be extended, as well as any character which can be conjugate only to itself. So, much of Figure 3.3 is determined without even looking at the groups.

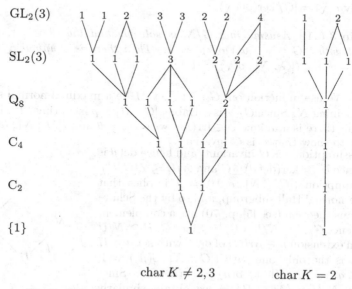

Figure 3.3. Degrees of irreducible representations.

Observe that we obtain a subgraph of the first graph for the case of char $K = 2$. There is a similar graph for the case char $K = 3$, which we leave as an exercise; see Exercise 3.6.7. ◆

Remark 3.6.17 Conlon has used the above ideas to develop very efficient algorithms to compute the degrees of the irreducible characters of a finite solvable group (see [36]) and the character tables of finite p-groups ([35]). The latter algorithm can be extended to finite super-solvable groups (see [11]) and is often called **Conlon's algorithm**. Observe that for such a group G the polycyclic structure can be used to compute very efficiently with the elements of G, see for example [84], so that the algorithm usually works, even for very large groups which are beyond the scope of the Dixon–Schneider algorithm.

We know from Corollary 3.6.14 that a G-invariant simple FN-module can be extended to an FG-module provided that G/N is a p-group and F is a splitting field for N with characteristic p. We will see that the same also holds under suitable assumptions if F is replaced by $\mathbb{C}$. For this we need a definition.

Definition 3.6.18 If $\delta\colon G \to \mathrm{GL}(V)$ is a representation over K with character χ, then

$$\det\chi\colon G \to K^{\times},\ g \mapsto \det(\delta(g))$$

is a linear character, and the order of $\det\chi$ is called the **determinantal order** of δ or χ and will be denoted by $o(\chi)$.

Obviously $o(\chi) = |G/\ker(\det\chi)|$.

Theorem 3.6.19 *Asusme that G/N is solvable and that $\theta \in \mathrm{Irr}(N)$ is G-invariant with $([G:N],\, \theta(1)o(\theta)) = 1$. Then there is a unique $\chi \in \mathrm{Irr}(G)$ with $\chi_N = \theta$ and $([G:N],\, o(\chi)) = 1$.*

Proof. We use induction on $[G:N]$. Let M be a maximal normal subgroup of G containing N. Since G/N is solvable, $[G:M] = p$ is a prime. By induction hypothesis there is a unique $\psi \in \mathrm{Irr}(M)$ with $\psi_N = \theta$ and $([M:N],\, o(\psi)) = 1$. We want to show that ψ is G-invariant.

By assumption θ is G-invariant, and hence $\det\theta$ is also. Hence $U := \ker(\det\theta) \trianglelefteq G$ and $N/U \leq \mathbf{Z}(G/U)$. The assumption $([G:N],\, o(\theta)) = 1$ implies that N/U is a normal Hall subgroup, and so by the Schur–Zassenhaus theorem (see [5], p. 70) has a complement H/U. Hence $G/U = N/U \times H/U$. Since $G/H \cong N/U$ we get an extension $\mu \in \mathrm{Irr}(G)$ of $\det\theta$ with $\ker\mu = H$ and this is the only one with $([G:N],\, o(\mu)) = 1$. We put $q := |N/U| = o(\mu) = o(\det\theta)$. Since $M/U \cong N/U \times (M \cap H)/U$ we obtain similarly that $\mu_M = \det\psi$ is the only extension of $\det\theta$ with $([M:N],\, o(\mu_M)) = 1$.

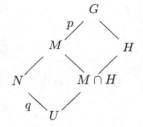

Let $g \in G$. Since $\det\psi^g|_N = \det\theta$ we have $\det\psi^g = \det\psi$. On the other hand, by Exercise 3.2.6 $\psi^g = \psi\lambda$ with λ a linear character of M/N inflated to M. Then

$$\det\psi = \det\psi^g = \lambda^{\theta(1)}\det\psi.$$

Since $o(\lambda)$ divides $[G:N]$ and thus $(o(\lambda),\, \theta(1)) = 1$, by our assumption we conclude that $\lambda = 1_M$ and $\psi^g = \psi$ is G-invariant.

By Theorem 3.6.13 ψ has p extensions $\chi_i \in \mathrm{Irr}(G)$ $(1 \leq i \leq p)$ to G; in fact, we may assume that $\chi_i = \chi_1\mu_i$ if $1_G = \mu_1, \ldots, \mu_p$ are the linear characters of G with kernel M. It follows that

$$X := \{\det\chi_i \mid 1 \leq i \leq p\} = \{\mu_i^{\theta(1)}\det\chi_1 \mid 1 \leq i \leq p\} = \{\mu_i\det\chi_1 \mid 1 \leq i \leq p\},$$

because $(p,\, \theta(1)) = 1$. Since $q = o(\psi) \mid o(\chi_1) \mid pq$ and X is the orbit of χ_1 under multiplication with the elements of a cyclic group of order p, there is exactly one element in X of order q. Hence there is exactly one $\chi := \chi_i$ with $o(\chi_i) = q$, while $o(\chi_j) = pq$ for $j \neq i$. $\qquad\square$

In fact, Theorem 3.6.19 holds without the assumption that G/N is solvable; see [92], corollary 8.16.

If N is a normal subgroup of G then G acts on the conjugacy classes and also on the irreducible characters of N. Here we have another instance, where Brauer's permutation lemma (Lemma 2.2.13) can be applied.

Corollary 3.6.20 *Let K be a splitting field for N with* char $K \nmid |N|$ *and* $N \trianglelefteq G$. *Then the number of G-conjugacy classes of irreducible characters of N is equal to the number of G-conjugacy classes of elements lying in N.*

Proof. Let A be the character table of N (over K). By Exercise 2.1.6 the matrix A is non-singular. The rows of A are naturally indexed by $\mathrm{Irr}_K(N)$, and the columns are naturally indexed by the conjugacy classes h^N of N, putting $\varphi(h^N) := \varphi(h)$ as usual for $\varphi \in \mathrm{Irr}_K(N)$ and $h \in N$. The group G acts on $\mathrm{Irr}_K(N)$ by ${}^g\varphi(h) = \varphi(g^{-1}hg)$ for $g \in G, \varphi \in \mathrm{Irr}_K(N), h \in N$, and on the conjugacy classes by $g * h^N = (ghg^{-1})^N$. By definition we have ${}^g\varphi(g * h^N) = \varphi(h^N)$. Thus the result follows immediately from Brauer's permutation lemma (Lemma 2.2.13). □

Although the number of orbits of G on $\mathrm{Irr}(N)$ and on the set $\mathcal{C}(N)$ of conjugacy classes of N coincide, the orbit lengths may be different; see Exercise 3.6.10.

Example 3.6.21 In Example 3.2.23 we have computed and displayed the character table of the alternating group A_6. We know that $A_6 \trianglelefteq S_6$ with $[S_6 : A_6] = 2$ and also that the classes **5a** and **5b** fuse to one class in S_6 because the centralizer of a 5-cycle in S_6 is contained in A_6. The other classes of A_6 are also conjugacy classes in S_6, so S_6 has six orbits on $\mathrm{Irr}(A_6)$. Hence there must be one pair of irreducible characters of A_6 conjugate in S_6. This can only be the pair of characters χ_2, χ_3 of degree five, so $\chi_2^{S_6} = \chi_3^{S_6} = \varphi \in \mathrm{Irr}(S_6)$ is an irreducible character whose values we can write down immediately, since

$$\varphi(g) = \begin{cases} \chi_2(g) + \chi_3(g) & \text{for } g \in A_6, \\ 0 & \text{for } g \notin A_6. \end{cases}$$

The remaining five irreducible characters of A_6 can each be extended in two ways differing just by the sign character, so $|\mathrm{Irr}(S_6)| = 11$, and hence there are $11 - 6 = 5$ conjugacy classes in $S_6 \setminus A_6$. In this example we know the conjugacy classes; those not contained in A_6 are **2b** $= (1,2)^G$, **2c** $= (1,2)(3,4)(5,6)^G$, **4a** $= (1,2,3,4)^G$, **6a** $= ((1,2)(3,4,5))^G$, **6b** $= (1,2,3,4,5,6)^G$, and we can immediately write down the values of the extensions of χ_4 using the natural permutation character of S_6 as in Example 3.2.23. Finally the values of the extensions of χ_5 on the "outer classes" **2b**, **2c**, **4a**, **6a**, **6b** can be computed using the orthogonality relations (Theorem 2.1.15, (2.5)) and the known values for the centralizer orders.

Observe that all one actually needs to compute the character table of G from the character table of a normal subgroup N with cyclic factor group G/N is the following:

(a) the list of "outer classes," i.e. the G-classes in $G \setminus N$;

(b) information which irreducible characters of N are conjugate in G – from this one finds which conjugacy classes of N fuse in G – , and

(c) one extension to G for each invariant $\chi \in \mathrm{Irr}(N)$. The others then can be simply obtained from this by multiplying with the linear characters of G/N.

In the ATLAS ([38]) exactly this information is provided for many simple groups and their cyclic non-trivial extensions. So the character tables of A_6 and S_6 are printed together somehow as follows:

A_6, S_6	1a	2a	3a	3b	4a	5a	5b	:	2b	2c	4b	6a	6b
χ_1	1	1	1	1	1	1	1	:	1	1	1	1	1
χ_2	5	1	2	−1	−1	0	0	:	3	−1	1	0	−1
χ_3	5	1	−1	2	−1	0	0	:	−1	3	1	−1	0
χ_4	8	0	−1	−1	0	α	α'	|	0	0	0	0	0
χ_5	8	0	−1	−1	0	α'	α	|	0	0	0	0	0
χ_6	9	1	0	0	1	−1	−1	:	3	3	−1	0	0
χ_7	10	−2	1	1	0	0	0	:	2	−2	0	−1	1

with α, α' as in Example 3.2.23. The colon signifies that the corresponding character extends with just one extension being printed. To get the other extension one just has to multiply the values to the right of the colon by -1. The conjugate characters are grouped together using the vertical bar $|$. It is apparent from this that 5a and 5b must fuse to one conjugacy class of S_6. ◆

Example 3.6.22 It is known that the symmetric group S_6 possesses an exceptional outer automorphism, so that $G = \mathrm{Aut}(S_6) = S_6 \cdot 2$. This information is sufficient to construct the character table of G. In fact, looking at the character table of S_6, reproduced below essentially in GAP-format, it is apparent that the characters marked by a colon must be extendible to G:

```
S_6    2  4  4  1  1  3  .  4  4  3  1  1
       3  2  .  2  2  .  .  1  1  .  1  1
       5  1  .  .  .  .  1  .  .  .  .  .

      1a 2a 3a 3b 4a 5a 2b 2c 4b 6a 6b

X.1    1  1  1  1  1  1  1  1  1  1  1 :
X.2    1  1  1  1  1  1 -1 -1 -1 -1 -1 :
X.3    5  1  2 -1 -1  .  3 -1  1  . -1
X.4    5  1  2 -1 -1  . -3  1 -1  .  1
X.5    5  1 -1  2 -1  . -1  3  1 -1  .
X.6    5  1 -1  2 -1  .  1 -3 -1  1  .
X.7   16  . -2 -2  .  1  .  .  .  .  . :
X.8    9  1  .  .  1 -1  3  3 -1  .  . :
X.9    9  1  .  .  1 -1 -3 -3  1  .  . :
X.10  10 -2  1  1  .  .  2 -2  . -1  1
X.11  10 -2  1  1  .  . -2  2  .  1 -1
```

It is also clear that the only pairs of conjugacy classes of S_6 which can fuse in G are 3a/b, 2b/c and 6a/b (and either none or all three pairs fuse in G, considering the values of X.3/5, X.4/6 and X.10/11 on these classes). Since all elements of order five are conjugate their centralizers must have order ten. Hence there is a class **10a**, and the values of all irreducible characters on this class are known (they are 0 or ± 1 by the congruence relations; see Lemma 2.2.2). The fifth powers of the elements of **10a** form a conjugacy class **2d** – so G is a split extension of S_6 – and for $z \in$ **2d** the values $\chi_i(z)$ are determined modulo 10 for all $\chi_i \in \mathrm{Irr}(G)$ (take congruences with the values on **10a** and **1a**). Furthermore, since z is not central, we conclude from Lemma 2.2.6 that $|\chi_i(z)| < \chi_i(1)$ for all non-linear χ_i. Thus the values of all $\chi_i \in \mathrm{Irr}(G)$ on **2d** are completely determined except for the irreducible characters of degree 16, where we have two possibilities, ± 4 or ± 14. But the last possibility is excluded, for instance, by the orthogonality relations (with the column corresponding to **10a**). It follows that the characters of degree five must be pair-wise conjugate and thus induce up to irreducible characters of degree ten of G, otherwise the congruence relations would be violated. Since by Lemma 3.6.20 the number of pairs of conjugate irreducible characters of S_6 must be the same as the number of pairs of conjugacy classes fusing in G, the characters of degree ten must also be conjugate. At this stage what is known about the character table of $G = S_6 .2$ is collated in Table 3.5.

Table 3.5. Character table of $G = S_6 .2$

$S_6 .2$	2	5	5	1	4	1	4	4	1	3	?	1	?	?
	3	2	.	2	.	.	1	.	1	.	.	.	.	.
	5	1	.	.	.	1	.	.	.	1	.	1	.	.
		1a	2a	3a	4a	5a	2b	4b	6a	2d	?	10a	?	?
χ_1		1	1	1	1	1	1	1	1	1	1	1	1	1
χ_2		1	1	1	1	1	1	1	1	-1	-1	-1	-1	-1
χ_3		1	1	1	1	1	-1	-1	-1	1		1		
χ_4		1	1	1	1	1	-1	-1	-1	-1		-1		
χ_5		10	2	1	-2	0	2	2	-1	0	0	0	0	0
χ_6		10	2	1	-2	0	-2	-2	1	0	0	0	0	0
χ_7		16	0	-2	0	1	0	0	0	4	0	-1	0	0
χ_8		16	0	-2	0	1	0	0	0	-4	0	1	0	0
χ_9		9	1	0	1	-1	3	-1	0	-1		-1		
χ_{10}		9	1	0	1	-1	3	-1	0	1		1		
χ_{11}		9	1	0	1	-1	-3	1	0	-1		-1		
χ_{12}		9	1	0	1	-1	-3	1	0	1		1		
χ_{13}		20	-4	2	0	0	0	0	0	0	0	0	0	0

Observe that χ_7, χ_8 must vanish outside the known classes, since otherwise their norm would not be one. Counting elements of order ≤ 2 in G known, so far one observes that this is $\sum_{\chi \in \mathrm{Irr}(G)} \chi(1)$. This implies that all second Frobenius–Schur indicators are one (see Theorem 2.9.9). We then find (using Lemma 2.9.8) that any element of class **4a** has four square roots in G and the involutions in **2a** have eight square roots in G, whereas they had only four square roots in S_6. We conclude that there are $4 \cdot \frac{|G|}{16}$ elements of order eight in G which have to distribute into at least two different conjugacy classes (since the centralizer orders must be at least eight). Also there must be $4 \cdot \frac{|G|}{32}$ elements of order four (with squares in **2a**) which are in $G \setminus S_6$. So we have found the missing conjugacy classes, centralizer orders (they are all eight) and incidentally also the second power map. All the missing character values must be ± 1, and it is an easy exercise to find the correct signs using the orthogonality relations. We present the complete table essentially in GAP format as follows.

```
S_6.2    2 5 5 1 4 1 4 4 1 3 3  1 3 3
         3 2 . 2 . . 1 . 1 .  . . .
         5 1 . . . 1 . . . 1  . 1 .

         1a 2a 3a 4a 5a 2b 4b 6a 2c 8a 10a 4c 8b
      2P 1a 1a 3a 2a 5a 1a 2a 3a 1a 4a 5a  2a 4a
      3P 1a 2a 1a 4a 5a 2b 4b 2b 2c 8a 10a 4c 8b
      5P 1a 2a 3a 4a 1a 2b 4b 6a 2c 8a 2c  4c 8b

X.1      1  1  1  1  1  1  1  1  1  1   1  1  1
X.2      1  1  1  1  1  1  1  1 -1 -1  -1 -1 -1
X.3      1  1  1  1  1 -1 -1 -1  1  1   1 -1 -1
X.4      1  1  1  1  1 -1 -1 -1 -1 -1  -1  1  1
X.5     10  2  1 -2  .  2  2 -1  .  .   .  .  .
X.6     10  2  1 -2  . -2 -2  1  .  .   .  .  .
X.7     16  . -2  .  1  .  .  .  4  .  -1  .  .
X.8     16  . -2  .  1  .  .  . -4  .   1  .  .
X.9      9  1  .  1 -1  3 -1  . -1  1  -1  1 -1
X.10     9  1  .  1 -1  3 -1  .  1 -1   1 -1  1
X.11     9  1  .  1 -1 -3  1  . -1  1  -1 -1  1
X.12     9  1  .  1 -1 -3  1  .  1 -1   1  1 -1
X.13    20 -4  2  .  .  .  .  .  .  .   .  .  .
```

◆

Example 3.6.23 Let G be a group which contains the Mathieu group M_{11} as a subgroup of index two. We shall conclude from the character table of M_{11} (computed in Section 2.10) that $G \cong C_2 \times M_{11}$. As in Example 3.6.22 we immediately see that there is a class 10a in G with squares lying in 5a (with centralizer order 10). All character values on the class 10a are integers and can be determined using the congruence relations. In particular, it follows that $\chi_6, \chi_7 \in \mathrm{Irr}(M_{11})$ of degree 16 are not conjugate in G but extend to irreducible characters of G. This implies that the conjugacy classes 11a and 11b do not fuse in G. Thus their centralizer orders are 22 and there must be an involution $z \in G$ commuting with some element $g \in$ 11a. The rational character $\chi_2 \in \mathrm{Irr}(M_{11})$ of degree ten extends to characters $\chi_2', \chi_2'' \in \mathrm{Irr}(G)$. If δ is a matrix representation with character χ_2' or χ_2'' then $\delta(g)$ is similar to $\mathrm{diag}(\zeta_{11}, \ldots, \zeta_{11}^{10})$ and hence $\delta(z) = \pm \mathbf{I}_{10}$. It follows that $z \in \mathbf{Z}(G)$ and $G \cong \langle z \rangle \times M_{11}$. We conclude that M_{11} has no outer automorphism of order two.   ◆

Remark 3.6.24 Let $N \trianglelefteq G$ with $[G : N] = 2$ and suppose that there are no conjugacy classes of N which fuse in G. Unlike in the above example one cannot conclude from this in general that $G \cong N \times C_2$. There might be automorphisms of N which leave invariant all conjugacy classes of N, without being inner automorphisms. The smallest example can be found in Exercise 3.7.3.

Exercises

Exercise 3.6.1 Let $N \trianglelefteq G$ and $\chi \in \mathrm{Irr}(G)$ such that $\varphi := \chi_N \in \mathrm{Irr}(N)$. Show that $(\chi^G, \varphi^G)_G = [G : N]$ and that

$$\mathrm{Irr}(G \mid \varphi) = \{\psi \cdot \chi \mid \psi \in \mathrm{Irr}(G), \ker \psi \supseteq N\}.$$

Hint: Use Lemma 3.2.14(d) to show that $\varphi^G = (1_N \cdot \varphi)^G = 1_N^G \cdot \chi$.

Exercise 3.6.2 Let P be a normal Sylow p-subgroup of G. Show that

$$P' = \bigcap \{ \ker \chi \mid \chi \in \mathrm{Irr}(G), \ p \nmid \chi(1) \}.$$

Exercise 3.6.3 Let p be a prime. Show that G has an abelian normal subgroup N such that G/N is a p-group if and only if $\chi(1)$ is a power of p for all $\chi \in \mathrm{Irr}(G)$.

Exercise 3.6.4 Let H be a (not necessarily finite) abelian group and let N be a subgroup of finite index. Assuming that K is an algebraically closed field, show that any $\lambda \in \mathrm{Hom}(N, K^\times)$ can be extended to a $\tilde{\lambda} \in \mathrm{Hom}(H, K^\times)$.
Hint: Consider first the case that H/N is cyclic and argue similarly as in the proof of Theorem 3.6.13.

Exercise 3.6.5 Let $N \trianglelefteq G$, with W_1 a simple KN-module and $T_G(W_1) \leq T \leq G$. Assume that W is a simple KT-module with $\mathrm{Hom}_{KN}(W_N, W_1) \neq \{0\}$ and that V is a simple KG-module. Prove that if W is isomorphic to a composition factor of V_T then its multiplicity is one and $V \cong W^G$.

Exercise 3.6.6 Let $\chi \in \mathrm{Irr}(G)$ and $\psi \in \mathrm{Irr}(H)$, where $H \trianglelefteq G$. Show that $(\chi_H, \psi)_H > 0$ if and only if $\omega_\chi(C^+) = \omega_\psi(C^+)$ for all $C = h^G$ with $h \in H$.
Hint: (a) Show that $\omega_\psi(C^+) = \omega_{\psi^g}(C^+)$ for all $g \in G$ and $C = h^G$ with $h \in H$.
(b) Show that $\sum_{g \in G} \epsilon_\psi^g \in \mathbf{Z}(\mathbb{C}G) \cap \mathbf{Z}(\mathbb{C}H) = \langle \{C^+ \mid C = h^G \text{ with } h \in H\} \rangle_{\mathbb{C}}$.
(c) Conclude that $\omega_{\psi'}(\sum_{g \in G} \epsilon_\psi^g) \neq 0$ for $\psi' \in \mathrm{Irr}(H)$ with $\omega_\psi(C^+) = \omega_{\psi'}(C^+)$ for all $C = h^G$ with $h \in H$. (Recall that ω_ψ is the central character and ϵ_ψ is the block idempotent corresponding to ψ; see Theorem 2.3.2 and Corollary 2.1.7, respectively.)

Exercise 3.6.7 Investigate Example 3.6.16 for the case that K is a field of characteristic three.

Exercise 3.6.8 Using only the character table of S_5, show that any group H that contains S_5 as a subgroup of index two is isomorphic to $S_5 \times C_2$. In particular, show that S_5 has no outer automorphisms of order two.
Hint: Consider an element $g \in H$ of order five. Show that $\mathbf{C}_H(g)$ has order ten and thus contains an involution z. Verify that $z \in \ker \chi$ for some $\chi \in \mathrm{Irr}(H)$ with $\chi(1) = 5$ and $\ker \chi \cap S_5 = \{1\}$.

Exercise 3.6.9 Let $[G : N] = 2$ and $N \cong M_{12}$. Assume that $G \not\cong N \times C_2$. Using the character table of M_{12} (see Example 2.8.12, p. 159) compute the character table of G.

Hint: Which conjugacy classes of N can fuse to classes of G and which irreducible characters of N may not be invariant in G? What are the possible Frobenius–Schur indicators $\nu_2(\chi)$ for $\chi \in \mathrm{Irr}(G)$? Exercise 3.1.4 is useful here. For each $g \in N$ find the possible number $x_2(g)$ of square roots of g in $G \setminus N$. Observe that $\sum_{g \in N} x_2(g)$ must be equal to $|N|$.

Note: It is known that $\mathrm{Out}(\mathrm{M}_{12}) \cong C_2$ (see [89], p. 312), hence $G \cong \mathrm{Aut}(\mathrm{M}_{12})$.

Exercise 3.6.10 Assume $N \trianglelefteq G$. Let G act on $\mathrm{Irr}(N)$ and on the set $\mathcal{C}(N)$ of conjugacy classes of N as in the proof of Corollary 3.6.20. Give an example for a group G where the lengths of the orbits of G on $\mathrm{Irr}(N)$ are different from the lengths of the orbits of G on $\mathcal{C}(N)$. One may use Exercise 2.1.5 to construct such an example with $|G| = 32$. Why? Show that for $|G| < 32$ the lengths of the orbits of G on $\mathrm{Irr}(N)$ and $\mathcal{C}(N)$ always coincide.

Exercise 3.6.11 Let K be a field and let $N \trianglelefteq G$ for a finite group G. Let V be a KG-module and $E = \mathrm{End}_{KN} V$. Show that G acts on E by

$$g \cdot \alpha(v) = g\alpha(g^{-1}v) \quad \text{for} \quad g \in G, \alpha \in E, v \in V$$

as ring automorphisms. Thus one obtains a homomorphism $\varphi \colon G \to \mathrm{Aut}(E)$. In the case that K is a finite field and V_N is simple, use Wedderburn's theorem on finite division rings ([32], p. 101) to show that E is a finite field extension of $L = \mathrm{End}_{KG} V$ and that $\mathrm{im}(\varphi) \subseteq \mathrm{Gal}(E/L)$. Furthermore, show that $N \mathbf{C}_G(N) \leq \ker \varphi$, so that $\mathrm{im}(\varphi)$ is isomorphic to a subgroup of $\mathrm{Out}(N)$.

We know that V can naturally be considered as an absolutely irreducible LG-module. Let $\mathrm{Gal}(E/L) = \{\gamma_1, \ldots \gamma_t\}$, so in particular $[E : L] = t$ and $\dim_L V = \frac{\dim_K(V)}{[L:K]}$. Extending scalars to E we let $EV = E \otimes_L V$ to obtain

$$(EV)_N \cong_{EN} {}^{\gamma_1} V_N \oplus \cdots \oplus {}^{\gamma_t} V_N$$

by Theorem 1.8.4, where the ${}^{\gamma_i} V_N$ are the modules algebraically conjugate to V_N considered as an EN-module.

3.7 Projective representations

In this section G is always a finite group and K is a field.

We have already encountered several examples where a G-invariant absolutely irreducible representation δ of a normal subgroup N of G can be extended to a representation of G. In general this will not be true, but we will show here that any such δ can in fact be extended to a "projective representation" of G, which we will define now.

Definition 3.7.1 A map $\delta \colon G \to \mathrm{GL}(V)$ for some K-vector space V is called a **projective representation** of G if $\delta(1) = \mathrm{id}_V$ and for all $g, g' \in G$

$$\delta(g)\delta(g') = \alpha(g, g') \cdot \delta(gg') \quad \text{for some} \quad \alpha(g, g') \in K^\times.$$

The projective representation δ is called irreducible if V has no proper nontrivial subspace invariant under all $\delta(g)$. The mapping $\alpha \colon G \times G \to K^\times$ is called the **factor set** corresponding to δ. Two projective representations $\delta\colon G \to$ GL(V) and $\delta'\colon G \to$ GL(V') are called **equivalent** if there is a K-isomorphism $\varphi\colon V' \to V$ such that

$$\delta'(g) = \gamma(g) \cdot \varphi^{-1}\delta(g)\varphi \qquad \text{for all} \quad g \in G \quad \text{with} \quad \gamma(g) \in K^\times.$$

In this chapter representations in the sense of Definition 1.1.1 are often called "*ordinary representations.*"

Remark 3.7.2 (a) Let $\pi\colon$ GL(V) $\to$ PGL(V) = GL(V)/$K^\times$ id$_V$ be the canonical projection. Then $\delta\colon G \to$ GL(V) is a projective representation if and only if $\delta(1) = \mathbf{1}_V$ and $\hat{\delta} = \pi \circ \delta$ is a group homomorphism. Hence the name "projective" representation; it has nothing to do with *projective modules* (i.e. direct summands of free modules). The requirement that $\delta(1) = \mathbf{1}_V$ in the definition is not really important (and is often omitted in the literature) but it sometimes simplifies matters. Obviously it is equivalent to require $\alpha(1,1) = 1$. As an immediate consequence of this statement we conclude that $\alpha(g,1) = \alpha(1,g) = 1$ holds for all $g \in G$.

(b) If δ, δ' are equivalent projective representations with factor sets α, α' then a simple calculation shows that, with the notation of the definition,

$$\alpha'(g,g') = \frac{\gamma(gg')}{\gamma(g)\gamma(g')}\alpha(g,g') \qquad (\gamma\colon G \to K^\times).$$

(c) If δ is a projective representation with factor set $\alpha\colon G \to K^\times$ then computing $\delta(g_1 g_2 g_3)$ in two different ways using the associativity in G one finds that

$$\alpha(g_1 g_2, g_3)\alpha(g_1, g_2) = \alpha(g_1, g_2 g_3)\alpha(g_2, g_3) \quad \text{for all } g_1, g_2, g_3 \in G. \tag{3.22}$$

Any function $\alpha\colon G \to K^\times$ with this property and $\alpha(g,1) = \alpha(1,g) = 1$ for all $g \in G$ is called a (normalized) "2-cocycle," and the set of these 2-cocycles from G to $K^\times$ forms a group $Z^2(G, K^\times)$ with point-wise multiplication. The set

$$B^2(G, K^\times) = \{(g,g') \mapsto \frac{\gamma(gg')}{\gamma(g)\gamma(g')} \mid \gamma\colon G \to K^\times , \; \gamma(1) = 1\}$$

is a subgroup – the subgroup of (normalized) "2-coboundaries" – and

$$H^2(G, K^\times) = Z^2(G, K^\times)/B^2(G, K^\times)$$

is called the second cohomology group. In the case that $K = \mathbb{C}$ it is also called the **Schur multiplier** of G and is usually denoted by $M(G)$.

(d) It follows from (b) that equivalent projective representations have factor sets which are in the same coset of $B^2(G, K^\times)$, so an equivalence class of projective representations determines a well-defined element of $H^2(G, K^\times)$. Conversely, if δ is a projective representation of G with factor set α and $\alpha B^2(G, K^\times) = \beta B^2(G, K^\times)$ for $\beta \in Z^2(G, K^\times)$ then there is a projective representation of G with factor set β which is equivalent to δ. In particular a projective representation is equivalent to an ordinary representation if and only if its factor set in in $B^2(G, K^\times)$.

Theorem 3.7.3 *Let $N \trianglelefteq G$ and let $\delta \colon N \to \mathrm{GL}(W)$ be an absolutely irreducible G-invariant representation. Then δ extends to a projective representation of G. In fact, there is an $\bar{\alpha} \in H^2(G/N, K^\times)$ such that for every $\alpha \in Z^2(G/N, K^\times)$ with $\bar{\alpha} = \alpha B^2(G/N, K^\times)$ a projective representation $\delta_\alpha \colon G \to \mathrm{GL}(W)$ exists with factor set $\tilde{\alpha} \colon (g, g') \mapsto \alpha(gN, g'N)$ and $\delta_\alpha|_N = \delta$. If $H^2(G/N, K^\times) = \{1\}$ then δ may be extended to an ordinary representation of G.*

Proof. Let $G = \bigcup_{i=1}^{n} g_i N$ with $g_1 = 1$. Since δ is G-invariant the representations δ and $^{g_i}\delta$ are equivalent and hence there are $\tilde{\delta}(g_i) \in \mathrm{GL}(W)$ with

$$\delta(g_i^{-1} h g_i) = {}^{g_i}\delta(h) = \tilde{\delta}(g_i)^{-1} \delta(h) \tilde{\delta}(g_i) \qquad \text{for all} \quad h \in N,\ 1 \le i \le n.$$

Observe that $\tilde{\delta}(g_i)$ is uniquely determined up to a non-zero scalar multiple, since δ is absolutely irreducible. We may choose $\tilde{\delta}(g_1) = \mathrm{id}_W$ and define

$$\tilde{\delta} \colon G \to \mathrm{GL}(W), \quad g_i u \mapsto \tilde{\delta}(g_i)\delta(u) \quad \text{for} \quad u \in N.$$

For $1 \le i, j \le n$ let $g_i g_j = g_k u_{i,j}$ with $k \in \{1, \ldots, n\}$ and $u_{i,j} \in N$. Then we get for arbitrary $h \in N$

$$\begin{aligned}
\tilde{\delta}(g_j)^{-1}\tilde{\delta}(g_i)^{-1}\delta(h)\tilde{\delta}(g_i)\tilde{\delta}(g_j) &= \delta((g_i g_j)^{-1} \cdot h \cdot (g_i g_j)) \\
&= \delta(u_{i,j}^{-1} g_k^{-1} h g_k u_{i,j}) \\
&= \delta(u_{i,j})^{-1}\tilde{\delta}(g_k)^{-1}\delta(h)\tilde{\delta}(g_k)\delta(u_{i,j}) \\
&= \tilde{\delta}(g_i g_j)^{-1}\delta(h)\tilde{\delta}(g_i g_j).
\end{aligned}$$

Hence $\tilde{\delta}(g_i g_j)\tilde{\delta}(g_j)^{-1}\tilde{\delta}(g_i)^{-1} \in \mathrm{End}_{KN} W = K \cdot \mathrm{id}_W$ and

$$\tilde{\delta}(g_i)\tilde{\delta}(g_j) = \alpha'(g_i N, g_j N)\tilde{\delta}(g_i g_j) \qquad \text{with} \quad \alpha'(g_i N, g_j N) \in K^\times,$$

where $\alpha'(1N, g_i N) = \alpha'(g_i N, 1N) = 1$ for all i. We get

$$\begin{aligned}
\tilde{\delta}(g_i u)\tilde{\delta}(g_j u') &= \tilde{\delta}(g_i)\delta(u)\tilde{\delta}(g_j)\delta(u') & = \tilde{\delta}(g_i)\tilde{\delta}(g_j)\delta(u^{g_j} u') \\
&= \alpha'(g_i N, g_j N)\tilde{\delta}(g_i g_j)\delta(u^{g_j} u') = \alpha'(g_i N, g_j N)\tilde{\delta}(g_k)\delta(u_{i,j} u^{g_j} u') \\
&= \alpha'(g_i N, g_j N)\tilde{\delta}(g_i u g_j u')
\end{aligned}$$

for arbitrary $u, u' \in N$. Thus $\tilde{\delta}$ is a projective extension of δ with factor set $\tilde{\alpha}' \colon (g, g') \mapsto \alpha'(gN, g'N)$. It follows that $\alpha' \in Z^2(G/N, K^\times)$. Assume that $\gamma \colon G/N \to K^\times$ with $\gamma(1N) = 1$ and

$$\alpha(gN, g'N) := \frac{\gamma(gN)\gamma(g'N)}{\gamma(gNg')}\, \alpha'(gN, g'N) \qquad \text{for} \quad g, g' \in G,$$

thus $\alpha B^2(G/N, K^\times) = \alpha' B^2(G/N, K^\times)$. We then define $\delta_\alpha \colon G \to \mathrm{GL}(W)$ by

$$\delta_\alpha(g_i u) := \gamma(g_i N)^{-1}\tilde{\delta}(g_i u) \qquad \text{for} \quad u \in N,\ 1 \le i \le n.$$

Clearly δ_α is a projective representation with factor set $\tilde{\alpha} \colon (g, g') \mapsto \alpha(gN, g'N)$ and $\delta_\alpha|_N = \delta$. $\qquad\qquad \square$

We have already seen in Theorem 3.6.13 that every G-invariant irreducible representation of a normal subgroup $N \trianglelefteq G$ can be extended to an ordinary representation of G, provided that G/N is cyclic and the ground field K is algebraically closed. For char $K = 0$ this can also be deduced from Theorem 3.7.3, since it is known that for a cyclic group C we have $H^2(C, K^\times) = \{1\}$ in this case (see [88], Satz V.25.3). Recall also that Theorem 3.6.19 gave another example where every G-invariant absolutely irreducible representation of a normal subgroup can be extended to an ordinary representation of G.

We add yet another particularly easy case, in which invariant irreducible representations of normal subgroups can be extended to ordinary representations.

Lemma 3.7.4 *Let $N \trianglelefteq G$ be an abelian normal subgroup having a complement $H \leq G$ in G. Then any absolutely irreducible representation δ of N which is invariant under G extends to an ordinary representation of G.*

Proof. Since δ is invariant under G it is clear that $U = \ker(\delta) \trianglelefteq G$. Furthermore δ must have degree one because it is absolutely irreducible, and we may consider δ as a faithful irreducible G/U-invariant character of N/U, hence $N/U \leq \mathbf{Z}(G/U)$. So $G/U \cong N/U \times HU/U$ and $\delta \times \mathbf{1}_{HU/U}$ defines an irreducible character of G/U. Inflating this to G we get an extension of δ to G. $\qquad \square$

Theorem 3.7.5 (Clifford) *Let $N \trianglelefteq G$ and W be an absolutely simple G-invariant KN-module with representation $\delta_W \colon N \to \mathrm{GL}(W)$. Assume that $\alpha B^2(G/N, K^\times) \in H^2(G/N, K^\times)$ and that $\delta_\alpha \colon G \to \mathrm{GL}(W)$ is a projective representation with factor set $(g, g') \mapsto \alpha(gN, g'N)$ as in Theorem 3.7.3 extending δ_W. If V is a simple KG-module with representation $\delta \colon G \to \mathrm{GL}(V)$ and*

$$V_N \cong W \oplus \cdots \oplus W \qquad \text{with } e \text{ summands } W,$$

then $V \cong_K W \otimes_K X$ with an e-dimensional K-vector space X and we can write

$$\delta(g) = \delta_{\alpha^{-1}}(gN) \otimes \delta_\alpha(g) \qquad \text{for} \qquad g \in G$$

with an irreducible projective representation $\delta_{\alpha^{-1}} \colon G/N \to \mathrm{GL}(X)$ with factor set α^{-1}.

Proof. $V_N = W_1 \oplus \cdots \oplus W_e$ with $W_i \cong_{KN} W$. We can find an adapted basis B of V such that for $h \in N$ and $g \in G$

$$[\delta(h)]_B = \begin{bmatrix} D_W(h) & & 0 \\ & \ddots & \\ 0 & & D_W(h) \end{bmatrix}, \quad [\delta(g)]_B = \begin{bmatrix} D_{11}(g) & & D_{1e}(g) \\ & \cdots & \\ D_{e1}(g) & & D_{ee}(g) \end{bmatrix},$$

where $D_W(h) = [\delta_W(h)]_{B_1}$ with an arbitrary basis B_1 of W and $D_{ij}(g) \in K^{m \times m}$ with $m = \dim W$. From $\delta(h)\delta(g) = \delta(g)\delta(g^{-1}hg)$ we obtain

$$D_W(h)D_{ij}(g) = D_{ij}(g)D_W(g^{-1}hg) \qquad \text{for} \quad h \in N , \ g \in G.$$

Put $D_\alpha(g) := [\delta_\alpha(g)]_{B_1}$, so that $D_\alpha(h) = D_W(h)$ for $h \in N$. Then, as in the proof of Theorem 3.7.3,

$$D_W(g^{-1}hg) = D_\alpha(g)^{-1}D_W(h)\,D_\alpha(g) \qquad \text{for} \quad h \in N, g \in G,$$
$$D_W(h)\,(D_{ij}(g)\,D_\alpha(g)^{-1}) = (D_{ij}(g)\,D_\alpha(g)^{-1})\,D_W(h) \quad \text{for} \quad h \in N, g \in G.$$

Since W is absolutely simple we conclude that

$$D_{ij}(g)D_\alpha(g)^{-1} = a_{ij}(g)\mathbf{I}_e \qquad \text{for some} \quad a_{ij}(g) \in K$$

with $a_{ij}(gh) = a_{ij}(g)$ for $g \in G, h \in N$. We define $D_{\alpha^{-1}}(gN) = [a_{ij}(g)]_{i,j=1}^e$. Then it follows that

$$[\delta(g)]_B = D_{\alpha^{-1}}(gN) \otimes D_\alpha(g) \qquad \text{for} \quad g \in G.$$

Since δ is an ordinary representation and D_α is a projective matrix representation of G, the multiplication rules for the Kronecker product yield that $D_{\alpha^{-1}}: g \mapsto D_{\alpha^{-1}}(gN)$ is also a projective matrix representation with factor set inverse to that of D_α. Since δ is irreducible, both D_α and $D_{\alpha^{-1}}$ must be irreducible. $\qquad\square$

We give a simple, but important, application in the following.

Corollary 3.7.6 *Let K be a field and let $G = G_1 \times G_2$ be the direct product of the subgroups G_1, G_2. Let $\pi_i: G \to G_i$ be the corresponding natural projections.*

(a) *If W_i is an absolutely simple KG_i-module for $i = 1, 2$, then $\mathrm{Inf}_{\pi_1} W_1 \otimes_K \mathrm{Inf}_{\pi_2} W_2$ is an absolutely simple KG-module. If K is a splitting field for G_1 and G_2, every simple KG-module is isomorphic to one of this form.*

(b) *For $\varphi_i \in \mathrm{cf}(G_i)$ $(i = 1, 2)$ we write $\varphi_1 \times \varphi_2: G \to \mathbb{C}$, $(g_1, g_2) \mapsto \varphi_1(g_1)\varphi_2(g_2)$. Then $\varphi_1 \times \varphi_2 \in \mathrm{cf}(G)$ and*

$$\mathrm{Irr}(G) = \{\, \chi_1 \times \chi_2 \mid \chi_1 \in \mathrm{Irr}(G_1)\,,\ \chi_2 \in \mathrm{Irr}(G_2) \,\}.$$

Proof. Obviously it suffices to prove (a). $\delta_i: KG_i \to \mathrm{End}_K W_i$ is absolutely irreducible if and only if it is surjective (by Corollary 1.3.7). The first assertion follows, since $\mathrm{End}_K W_1 \otimes_K W_2 \cong_K \mathrm{End}_K W_1 \otimes_K \mathrm{End}_K W_2$. Now let K be a splitting field for G_1 and G_2 and let V be a simple KG-module. Since G_1 and G_2 centralize each other, every KG_1-module is G-invariant and every representation $\delta_1: KG_1 \to \mathrm{End}_K W_1$ can be extended to an (ordinary) representation $\delta_1': KG \to \mathrm{End}_K W_1$ by putting $\delta_1'(g_1 g_2) := \delta_1(g_1)$ for $g_1 \in G_1, g_2 \in G_2$. So the result follows from Theorem 3.7.5 $\qquad\square$

Remark 3.7.7 (a) For a weakening of the hypothesis in part (a) of Corollary 3.7.6 see Exercise 3.7.7.

(b) If A_1, A_2 are K-algebras and W_1, W_2 are A_1, respectively, A_2-modules then $W_1 \otimes_K W_2$ is an $A_1 \otimes_K A_2$-module in a natural way, the "outer tensor product" of the modules. Observe that $K(G_1 \times G_2) \cong KG_1 \otimes_K KG_2$ as K-algebras.

But one cannot conclude in general that $W_1 \otimes_K W_2$ is simple if W_1 and W_2 are simple. Consider, for example, $A_1 = \mathbb{C}$ and $A_2 = \mathbb{Q} C_3$ with the cyclic group C_3 of order three and W_1 the trivial module $\mathbb{C}$, whereas $W_2 = \mathbb{Q}[X]/(X^2 + X + 1)$ (cf. Example 1.1.24 and Example 1.1.31). Then W_i are simple A_i-modules for $i = 1, 2$, but $W_1 \otimes_{\mathbb{Q}} W_2 = \mathbb{C}[X]/(X^2 + X + 1)$ is not a simple $\mathbb{C} C_3$-module, although $A_1 \otimes_{\mathbb{Q}} A_2 \cong \mathbb{C} C_3$.

There are two ways to "linearize" a projective representation of G with factor set α. One is to consider the "twisted group algebra" $K_\alpha G$ and the other is to use "central extensions" of G. We briefly mention the first alternative.

Definition 3.7.8 If $\alpha \colon G \times G \to K^\times$ is a factor set for the finite group over the field K, the K-vector space with basis G can be turned into an associative K-algebra by defining

$$g \star g' = \alpha(g, g') g g' \quad \text{for all} \quad g, g' \in G$$

and extending K-linearly. This K-algebra is called a **twisted group algebra** and will be denoted by $K_\alpha G$.

It is obvious that every (finitely generated) $K_\alpha G$-module V defines a projective representation of G with factor set α and conversely. Isomorphic $K_\alpha G$-modules lead to equivalent projective representations, but the converse does not necessarily hold ($K_\alpha G$-modules in the same orbit under the action of $\mathrm{Hom}(G, K^\star)$ also yield equivalent projective representations).

We proceed with the second option to linearize projective representations, which we will use almost exclusively.

Definition 3.7.9 Let $N \trianglelefteq G$ and let $\pi \colon G \to \overline{G} = G/N$, $g \mapsto \overline{g} = gN$ be the canonical projection; then a function $\tau \colon \overline{G} \to G$ with $\pi \circ \tau = \mathrm{id}_{\overline{G}}$ is called a **section** provided that $\tau(\overline{1}) = 1$ and τ maps conjugate elements of $\overline{G}$ to conjugate elements of G. The function $\eta_\tau \colon \overline{G} \times \overline{G} \to N$ defined by $\tau(\overline{g}) \tau(\overline{g}') = \tau(\overline{g}\,\overline{g}')\, \eta_\tau(\overline{g}, \overline{g}')$ is called the **factor set** corresponding to τ. If $N \leq \mathbf{Z}(G)$ we call the short exact sequence

$$\{1\} \xrightarrow{} N \xrightarrow{\ \iota\ } G \xrightarrow{\ \pi\ } \overline{G} \xrightarrow{} \{1\}$$

a **central extension** of $\overline{G}$, where ι is the embedding, and G is called a **covering group** of $\overline{G}$ provided that G and $\overline{G}$ are perfect.

The above definition is more special than the usual one. Often any function $\tau \colon \overline{G} \to G$ mapping each coset $\overline{g} = gN$ to a representative $g = \tau(gN) \in G$ is called a section (or it is just assumed in addition that $\tau(\overline{1}) = 1$). But it is clear that it is possible to choose the representatives in such a way that conjugate elements of $\overline{G}$ are represented by conjugate elements in G. And in practice this is quite useful. Obviously, N has a **complement** H in G (that is $H \leq G$ with $G = HN$ and $H \cap N = \{1\}$) if and only if there is a section $\tau \colon \overline{G} \to G$, which is a group homomorphism.

Remark 3.7.10 If $\tau \colon \overline{G} = G/N \to G$ is a section with factor set $\eta := \eta_\tau$ as in Definition 3.7.9 then the associativity implies

$$\eta(\overline{g}_1\overline{g}_2, \overline{g}_3)\eta(\overline{g}_1, \overline{g}_2)^{\tau(\overline{g}_3)} = \eta(\overline{g}_1, \overline{g}_2\overline{g}_3)\eta(\overline{g}_2, \overline{g}_3) \quad \text{for all} \quad \overline{g}_1, \overline{g}_2, \overline{g}_3 \in \overline{G}.$$

If $N \le \mathbf{Z}(G)$ this looks like the "2-cocycle" equation (3.22) on p. 239 and so $\lambda \circ \eta \in Z^2(\overline{G}, K^\times)$ for every $\lambda \in \mathrm{Hom}(N, K^\times)$. If $\delta \colon G \to \mathrm{GL}(V)$ is an ordinary representation over K, with $\delta(z) = \lambda(z)\,\mathrm{id}_V$ for $z \in N$, then $\delta \circ \tau \colon \overline{G} \to \mathrm{GL}(V)$ is a projective representation with factor set $\lambda \circ \eta$. Conversely, if $\overline{\delta} \colon \overline{G} \to \mathrm{GL}(V)$ is a projective representation with factor set $\lambda \circ \eta$ then we get an ordinary representation (assuming again that $N \le \mathbf{Z}(G)$)

$$\delta \colon G \to \mathrm{GL}(V), \quad g \mapsto \lambda(z_g)\overline{\delta}(\pi(g)) \quad \text{with} \quad z_g := g\,(\tau \circ \pi(g))^{-1} \in N$$

with $\delta \circ \tau = \overline{\delta}$. We say that "$\overline{\delta}$ is **lifted** to an ordinary representation δ of G."

Lemma 3.7.11 *Let $N \trianglelefteq G$ with natural projection $\pi \colon G \to G/N$ and let*

$$\{1\} \longrightarrow Z \overset{\iota}{\longrightarrow} H \overset{\pi'}{\longrightarrow} G/N \longrightarrow \{1\}$$

be a central extension with section $\tau \colon G/N \to H$ and corresponding factor set $\eta := \eta_\tau \colon G/N \times G/N \to Z$. Then

$$N \times Z \ \trianglelefteq\ G \curlywedge H := \{(g, h) \in G \times H \mid \pi(g) = \pi'(h)\} \ \le\ G \times H.$$

With $\tilde{\iota} \colon Z \to G \curlywedge H$, $z \mapsto (1, \iota(z))$ and $\tilde{\pi} \colon G \curlywedge H \to G$, $(g, h) \mapsto g$ we obtain a central extension

$$\{1\} \longrightarrow Z \overset{\tilde{\iota}}{\longrightarrow} G \curlywedge H \overset{\tilde{\pi}}{\longrightarrow} G \longrightarrow \{1\}$$

with section $\tilde{\tau} \colon g \mapsto (g, \tau(gN))$ and factor set $\eta_{\tilde{\tau}} \colon (g, g') \mapsto \eta(gN, g'N)$.

Proof. The verification is left to the exercises. Note that $G \curlywedge H$ is often called a "direct product with amalgamated factor groups;" see [88], Satz I.9.11, p. 50. $\qquad\square$

Example 3.7.12 Let N be an abelian normal subgroup of G and $\varphi \colon N \to K^\times$ a G–invariant linear character of N. Assume that $\tau \colon \overline{G} = G/N \to G$ is a section with factor set η. Then

$$\tilde{\varphi} \colon G \to K^\times, \qquad g \mapsto \varphi(\tau(\overline{g})^{-1} \cdot g)$$

is a projective representation of G with $\tilde{\varphi}|_N = \varphi$ with factor set $\alpha = \varphi \circ \eta$. $\qquad\blacklozenge$

Definition 3.7.13 A finite group H is called a **representation group** of G (over K) if there is $Z \le \mathbf{Z}(H)$ with $H/Z \cong G$ such that every irreducible projective representation δ' of G (over K) is equivalent to one of the form $\delta \circ \tau$ with an ordinary irreducible representation δ of H and a section $\tau \colon G \to H$.

If K is a splitting field for H, as we will assume most of the time, then $\delta|_Z = \lambda \colon Z \to K$ is a linear character, and the factor set of $\delta \circ \tau$ is $\alpha = \lambda \circ \eta$ with $\eta \colon G \times G \to Z$ being the factors set corresponding to τ.

Schur ([157]) has proved the existence of representation groups (over $\mathbb{C}$) for any finite group. We follow roughly the exposition given in [88].

Theorem 3.7.14 *Let H be a (not necessarily finite) group $Z \le \mathbf{Z}(H)$ with $H/Z \cong G$ and let $\tau \colon G \to H$ be a section with factor set $\eta := \eta_\tau \colon G \times G \to Z$. If $\lambda \in \hat{Z} = \operatorname{Hom}(Z, K^\times)$ then $\lambda \circ \eta \in Z^2(G, K^\times)$. The map*

$$\epsilon \colon \hat{Z} \to H^2(G, K^\times), \qquad \lambda \mapsto \lambda^\epsilon = (\lambda \circ \eta) B^2(G, K^\times)$$

is a group homomorphism. If K is algebraically closed then

$$\ker \epsilon = (Z \cap H')^\perp = \{ \lambda \in \hat{Z} \mid \lambda(z) = 1 \quad \textit{for all} \quad z \in Z \cap H' \}$$

and H is a representation group of G if and only if $Z \le \mathbf{Z}(H) \cap H'$ and ϵ is surjective.

Proof. The first assertion was already noted in Remark 3.7.10. Obviously ϵ is a homomorphism. By definition, $\lambda \in \ker \epsilon$ if and only if $\lambda \circ \eta \in B^2(G, K^\times)$, that is if and only if there is a $\gamma \colon G \to K^\times$ with $\gamma(1) = 1$ and $\lambda \circ \eta(g_1, g_2) = \gamma(g_1 g_2)^{-1} \gamma(g_1) \gamma(g_2)$ for all $g_1, g_2 \in G$. For such a γ we define

$$\mu = \mu_\gamma \colon H \to K^\times, \ \tau(g)z \mapsto \gamma(g)\lambda(z) \quad \text{for } z \in Z \ , \ g \in G.$$

Then $\mu|_Z = \lambda$. Furthermore for $g_1, g_2 \in G$, $z_1, z_2 \in Z$

$$\mu(\tau(g_1)z_1\tau(g_2)z_2) = \mu(\tau(g_1 g_2)\eta(g_1, g_2)z_1 z_2) = \gamma(g_1 g_2)\lambda(\eta(g_1, g_2))\lambda(z_1)\lambda(z_2)$$
$$= \mu(\tau(g_1)z_1)\mu(\tau(g_2)z_2);$$

i.e. μ is a homomorphism and hence $H' \le \ker \mu$, hence $Z \cap H' \le \ker \lambda$, that is $\lambda \in (Z \cap H')^\perp$.

Conversely, let $\lambda \in (Z \cap H')^\perp$. Then λ can be extended to an H-invariant linear character of ZH' having H' in its kernel and by Exercise 3.6.4 to a character $\tilde{\lambda}$ of H. Then $\lambda \circ \eta \in B^2(G, K^*)$ because

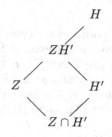

$$\tilde{\lambda}\tau(g_1)\tilde{\lambda}\tau(g_2) = \tilde{\lambda}(\tau(g_1)\tau(g_2)) = \tilde{\lambda}\tau(g_1 g_2)\lambda\eta(g_1, g_2).$$

If $Z \le \mathbf{Z}(H) \cap H'$ then ϵ is injective and by a theorem of Schur (see [88], Satz IV.2.3, p. 417) H is finite. So the last claim follows from Remark 3.7.10. $\square$

Theorem 3.7.15 *Assume that K is algebraically closed and char $K = 0$. Let F be a free group on $\{x_1, \ldots, x_n\}$ such that $F/R \cong G$ for some $R \trianglelefteq F$. Put*

$$\tilde{F} := F/[R, F] \quad \text{and} \quad \tilde{R} := R/[R, F].$$

Then the following holds.

(a) $\tilde{R} \leq \mathbf{Z}(\tilde{F})$ and $\tilde{T} := (R \cap F')/[R, F]$ is the torsion subgroup of $\tilde{R}$. Furthermore $\tilde{T}$ has a complement $\tilde{S} \cong \mathbb{Z}^n$ in $\tilde{R}$.

(b) $H^2(G, K^\times) \cong \tilde{T}$ is finite and $\tilde{F}/\tilde{S}$ is a representation group of G.

(c) If H is any representation group of G (over K) then there is an epimorphism

$$\tilde{\sigma}: \tilde{F} \to H \quad \text{with} \quad \tilde{\sigma}(\tilde{R}) = Z, \quad \tilde{\sigma}(\mathbf{Z}(\tilde{F})) = \mathbf{Z}(H), \quad \tilde{\sigma}|_{\tilde{F}'}: \tilde{F}' \cong H',$$

where Z is as in Definition 3.7.13. Also $Z \cong H^2(G, K^\times) \cong M(G)$.

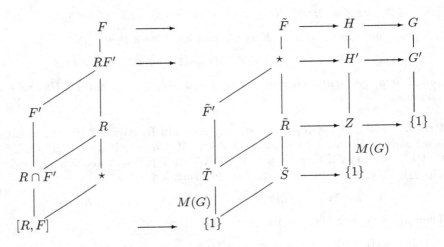

Figure 3.4. H as a homomorphic image of F.

Proof. See Figure 3.4. Clearly $\tilde{R} \leq \mathbf{Z}(\tilde{F})$. Since $[\tilde{F} : \tilde{R}] = |G| < \infty$ it follows from a theorem of Schur ([88], Satz IV.2.3, p. 417) that $\tilde{F}' = F'/[R, F]$ and hence $\tilde{T} := (R \cap F')/[R, F]$ is finite. Note that F/F' is a free abelian group of rank n and so is $(RF')/F' \cong R/(R \cap F') \cong \tilde{R}/\tilde{T}$, because $F/(RF')$ is finite. Hence $\tilde{R}$ is a finitely generated abelian group with torsion subgroup $\tilde{T}$, and by [110], Theorem III.7.3, p. 147, $\tilde{T}$ has a complement $\tilde{S} \cong \mathbb{Z}^n$.

(b) Let $\pi: F \to G$ be an epimorphism with $\ker \pi = R$ and $\tilde{\pi}: \tilde{F} \to G$, $x[R, F] \mapsto \pi(x)$. We apply Theorem 3.7.14 to the central extension (with ι the inclusion)

$$\{1\} \longrightarrow \tilde{R} \xrightarrow{\iota} \tilde{F} \xrightarrow{\tilde{\pi}} G \longrightarrow \{1\}.$$

Let $\tilde{\tau}: G \to \tilde{F}$ be a section and let $\tilde{\eta}: G \times G \to \tilde{R}$ be the corresponding factor set. To show that the map ϵ defined in Theorem 3.7.14 is surjective we take an arbitrary $\alpha \in Z^2(G, K^\times)$. We turn $G \times K^\times$ into a group $G(\alpha)$ by defining

$$(g_1, s_1) \cdot (g_2, s_2) := (g_1 g_2, \alpha(g_1, g_2) s_1 s_2) \quad \text{for} \quad g_1, g_2 \in G, \; s_1, s_2 \in K^\times.$$

Observe that the associativity follows, since $\alpha \in Z^2(G, K^\times)$ and the inverse of $(g, s) \in G(\alpha)$ is $(g^{-1}, \alpha(g^{-1}, g)^{-1} s^{-1})$. Clearly $K^\times \cong \{(1, s) \mid s \in K^\times\} \leq \mathbf{Z}(G(\alpha))$. Since F is free on $\{x_1, \ldots, x_n\}$ there is a homomorphism $\psi: F \to G(\alpha)$

mapping x_i to $(\pi(x_i), 1) \in G(\alpha)$ for $i = 1, \ldots, n$. Then

$$\psi: F \to G(\alpha), \; x \mapsto (\pi(x), \beta(x)) \qquad \text{with} \qquad \beta(x) \in K^\times \quad \text{and} \quad \beta|_R \in \hat{R}.$$

Now, $[R, F] \le \ker \psi$ since $\psi(R) \le \mathbf{Z}(G(\alpha))$, so ψ induces a group homomorphism $\tilde{\psi}: \tilde{F} \to G(\alpha)$, $\tilde{x} \mapsto (\tilde{\pi}(\tilde{x}), \tilde{\beta}(\tilde{x}))$ with $\tilde{\pi}(x[R, F]) := \pi(x)$ and $\tilde{\beta}(x[R, F]) := \beta(x)$. Then $\lambda := \tilde{\beta}|_{\tilde{R}} \in \hat{\tilde{R}}$. Using

$$\tilde{\psi}(\tilde{\tau}(g_1)\tau(g_2)) = \tilde{\psi}(\tilde{\tau}(g_1))\tilde{\psi}(\tilde{\tau}(g_2))$$

we get

$$\lambda \circ \tilde{\eta}(g_1, g_2) = \alpha(g_1, g_2) \, \tilde{\beta}(\tilde{\tau}(g_1)) \, \tilde{\beta}(\tilde{\tau}(g_2)) \, \tilde{\beta}(\tilde{\tau}(g_1 g_2))^{-1} \qquad \text{for} \qquad g_1, g_2 \in G.$$

Hence $\epsilon(\lambda) = \alpha B^2(G, K^\times)$ and ϵ is surjective. By Theorem 3.7.14 we have $H^2(G, K^\times) \cong \hat{\tilde{R}}/\tilde{T}^\perp$. Since any $\mu \in \hat{\tilde{T}}$ can be extended to a $\lambda \in \hat{\tilde{R}}$ with $\tilde{S} \le \ker \lambda$, the restriction map $\hat{\tilde{R}} \to \hat{\tilde{T}}$, $\lambda \mapsto \lambda|_{\hat{\tilde{T}}}$ is an epimorphism with kernel $\tilde{T}^\perp$. Hence $\hat{\tilde{R}}/\tilde{T}^\perp \cong \hat{\tilde{T}} \cong \tilde{T}$, where the last isomorphism follows from Exercise 2.1.2.

Now put $H := \tilde{F}/\tilde{S}$ and $Z := \tilde{R}/\tilde{S} \cong \tilde{T}$. Then $Z \le \mathbf{Z}(H) \cap H'$, and from Theorem 3.7.14 we conclude that H is a representation group of G.

(c) Let H be an arbitrary representation group of G with epimorphism $\rho: H \to G$ with $\ker \rho = Z$, where $Z \le \mathbf{Z}(H) \cap H'$ by Theorem 3.7.14. Choose $h_i \in H$ with $\rho(h_i) = \pi(x_i)$ for $i = 1, \ldots, n$. Then $H = \langle h_1, \ldots, h_n, Z \rangle$. By a theorem of Gaschütz (see [88], Satz III.3.12, p. 272) Z is contained in the Frattini subgroup $\Phi(H)$ and hence $H = \langle h_1, \ldots, h_n \rangle$ (see [88], Satz III.3.2, p. 268). Thus we obtain an epimorphism $\sigma: F \to H$ with $\sigma(x_i) = h_i$ for $i = 1, \ldots, n$ and therefore $\rho \circ \sigma = \pi$. Consequently $\sigma(R) = Z$ and thus $[R, F] \le \ker \sigma$, so that σ induces an epimorphism $\tilde{\sigma}: \tilde{F} \to H$ with $\rho \circ \tilde{\sigma} = \tilde{\pi}$, hence $\tilde{\sigma}(\tilde{R}) = Z$. Since $|Z| = |H^2(G, K^\times)| = [\tilde{R} : \ker \tilde{\sigma}]$ it follows that $\ker \tilde{\sigma}$ is a complement to $\tilde{T}$ in $\tilde{R}$. So $\tilde{\sigma}|_{\tilde{F}'}$ is injective and hence an isomorphism $\tilde{F}' \to H'$. If $\tilde{\sigma}(\tilde{x}) \in \mathbf{Z}(H)$ then $[\tilde{x}, \tilde{y}] \in \ker \tilde{\sigma} \cap \tilde{F}' = \{1\}$ for all $\tilde{y} \in \tilde{F}$, hence $\tilde{x} \in \mathbf{Z}(\tilde{F})$. Finally note that the definition of $\tilde{F}$, $\tilde{T}$ and $\tilde{S}$ is independent of the field K, in particular $H^2(G, K^\times) \cong M(G)$. $\square$

Corollary 3.7.16 *The degree of an irreducible projective representation of G over an algebraically closed field of characteristic zero divides $|G|$.*

Proof. An irreducible projective representation δ of G lifts to an ordinary irreducible representation δ_o of a representation group H of G as in Theorem 3.7.15. By Theorem 2.3.6 the degree of δ_o divides $[H : \mathbf{Z}(H)] = |G|/[\mathbf{Z}(H) : Z]$. $\square$

The construction of a representation group of G in Theorem 3.7.15 involves a choice of a complement of the torsion subgroup of a finitely generated abelian group (in addition to a choice of a finite presentation $G \cong F/R$). In general it is not determined up to isomorphism by G. For example the quaternion group Q_8 and the dihedral group D_8 of order eight are both representation groups of the Klein 4-group V_4 (see Exercise 3.7.2). On the other hand, if G is perfect (i.e. $G' = G$) then $H \cong F'/[R, F]$, and a representation group of G is unique up to isomorphism. It follows from Theorem 3.7.15 that in general the order of a representation group of G is determined; moreover it turns out that any two representation groups of a finite group G are isoclinic. To define this notion, first introduced by Hall [76], it is convenient to use the following convention.

Remark 3.7.17 The commutator map $H \times H \to H'$, $(g, h) \mapsto [g, h]$ is constant on cosets of the center $\mathbf{Z}(H)$. So we can define $[g \, \mathbf{Z}(H), h \, \mathbf{Z}(H)] := [g, h]$.

Definition 3.7.18 Two not necessarily finite groups H_1, H_2 are called **isoclinic** if there are isomorphisms $\varphi \colon H_1 / \mathbf{Z}(H_1) \to H_2 / \mathbf{Z}(H_2)$, $\psi \colon H_1' \to H_2'$ satisfying

$$\psi([x, y]) = [\varphi(x \, \mathbf{Z}(H_1)), \varphi(y \, \mathbf{Z}(H_1))] \qquad \text{for all} \qquad x, y \in H_1. \tag{3.23}$$

Since $H_i / \mathbf{Z}(H_i) \cong \operatorname{Inn}(H_i)$, the group of inner automorphisms H_1 acts on H_1' and via φ also on H_2'. The condition (3.23) says that ψ is an H_1-equivariant isomorphism of groups. Clearly isoclinism is an equivalence relation, and it is also obvious that any two abelian groups are isoclinic.

Corollary 3.7.19 *Any two representation groups of G are isoclinic.*

Proof. Let $\tilde{\sigma} \colon \tilde{F} \to H$ be as in Theorem 3.7.15(c). Clearly it suffices to show that $H_1 := \tilde{F}$ is isoclinic to H. We define

$$\varphi \colon \tilde{F} / \mathbf{Z}(\tilde{F}) \to H / \mathbf{Z}(H), \ x \, \mathbf{Z}(\tilde{F}) \mapsto \tilde{\sigma}(x) \, \mathbf{Z}(H) \qquad \text{and} \qquad \psi := \tilde{\sigma}|_{\tilde{F}'}.$$

By Theorem 3.7.15(c) φ and ψ are isomorphisms and (3.23) obviously holds. $\qquad \square$

Lemma 3.7.20 *Two finite groups H_1, H_2 are isoclinic if and only if there are embeddings $\iota_i \colon H_i \to G$ into a finite group G such that $G = \iota_i(H_i) \mathbf{Z}(G)$ for $i = 1, 2$.*

Proof. (See [78].) (a) If $G = H \mathbf{Z}(G)$ then $G' = H'$ and $\mathbf{Z}(H) = \mathbf{Z}(G) \cap H$. So $\psi := \operatorname{id}_{H'}$ and $\varphi \colon H / H' \to G / G'$, $h \, \mathbf{Z}(H) \mapsto h \, \mathbf{Z}(G)$ define an isoclinism from H to G.

(b) Conversely, let φ, ψ be as in Definition 3.7.18. Let $\tilde{H} := H_1 \ltimes H_2$ be the semidirect product using the action of H_1 on H_2 mentioned above, that is $x^g := x^h = h^{-1} x h$ for $g \in H_1$, $x, h \in H_2$ if $h \in \varphi(g \, \mathbf{Z}(H_1))]$. Then

$$H_1 \cong \tilde{H}_1 := \{(h, 1) \mid h \in H_1\} \leq \tilde{H} \quad \text{and} \quad H_2 \cong \tilde{H}_2 := \{(1, h) \mid h \in H_2\} \trianglelefteq \tilde{H}.$$

It is readily verified that $D := \{(h, \psi(h)^{-1}) \mid h \in H_1'\} \trianglelefteq \tilde{H}$. We put $G := \tilde{H} / D$. Since ψ is bijective, $\tilde{H}_i \cap D = \{1\}$ for $i = 1, 2$ and we get embeddings

$$\iota_1 \colon H_1 \to G, \ h \mapsto (h, 1) D \qquad \text{and} \qquad \iota_2 \colon H_2 \to G, \ h \mapsto (1, h) D.$$

Defining $\tilde{Z} := \{(y, z) \in \tilde{H} \mid z \in \varphi(y^{-1} \, \mathbf{Z}(H_1))\}$ we see that $D \leq \tilde{Z}$ and clearly $\tilde{H} = \tilde{H}_1 \tilde{Z} = \tilde{H}_2 \tilde{Z}$. So the claim follows, once we have seen that $\tilde{Z} / D \leq \mathbf{Z}(G)$, or, equivalently, $[\tilde{Z}, \tilde{H}] \leq D$. To this end, let $(y, z) \in \tilde{Z}$ and for $(g, 1) \in \tilde{H}_1$

choose $h \in \varphi(g\, \mathbf{Z}(H_1))$. Then $z^y = z$ and consequently

$$[(y,z),(g,1)] = (y^{-1},z^{-1})\,(g^{-1},1)\,(y,z)\,(g,1) = (y^{-1},z^{-1})\,(g^{-1}yg,z^g)$$

$$= ([y,g],(z^{-1})^{g^{-1}yg}z^g) = ([y,g],[z^{-1},h]^{-1}) \in D$$

since $\psi([y,g]) = [z^{-1},h]$. Hence $\tilde{Z}/D \le \mathbf{Z}(G)$ because for any $h_2 \in H_2$ we have

$$[(y,z),(1,h_2)] = (y^{-1},z^{-1}h_2^{-1})\,(y,z\,h_2) = (1,(z^{-1}h_2^{-1})^y z\,h_2) = (1,1).$$

$\square$

In Lemma 3.7.20 we may assume that $G = \iota_1(H_1)\iota_2(H_2)$.

Corollary 3.7.21 *Let H_1, H_2 be representation groups of G. Then there are bijections $\beta\colon \mathrm{Irr}(H_1) \to \mathrm{Irr}(H_2)$, $\chi \mapsto \chi^\beta$ and $\gamma\colon H_1 \to H_2$, $h \mapsto h^\gamma$ such that*

$$\chi^\beta(h^\gamma) = \lambda_h \cdot \chi(h) \qquad \text{with} \quad \lambda_h \in \mathbb{C}^\times \quad \text{and} \quad \lambda_h = 1 \quad \text{for} \quad h \in H_1'.$$

Proof. Clearly we may assume $H_1 \not\cong H_2$ and in addition (by Lemma 3.7.20) that H_1, H_2 can be embedded into a group H such that $H = H_1 H_2 = H_1\, \mathbf{Z}(H) = H_2\, \mathbf{Z}(H)$. Then there are $z_i, y_i \in \mathbf{Z}(H)$ with

$$H = \dot{\bigcup}_{i=1}^n z_i H_1 = \dot{\bigcup}_{i=1}^n y_i H_2, \quad H_1 = \dot{\bigcup}_{i=1}^n y_i (H_1 \cap H_2), \quad H_2 = \dot{\bigcup}_{i=1}^n z_i (H_1 \cap H_2).$$

We define $\gamma\colon H_1 \to H_2$, $y_i h \mapsto z_i h$, $(h \in H_1 \cap H_2)$. If $\chi \in \mathrm{Irr}(H_1 \mid \lambda_1)$ for $\lambda_1 \in \mathrm{Irr}(\mathbf{Z}(H_1))$ and $\lambda \in \mathrm{Irr}(\mathbf{Z}(H))$ with $\lambda_{\mathbf{Z}(H_1)} = \lambda_1$ (which exists by Exercise 3.6.4) then define

$$\chi^\beta\colon H_2 \to \mathbb{C}, \quad z_i h \mapsto \lambda(z_i)\delta(h), \qquad \text{for} \qquad h \in H_1 \cap H_2.$$

Then $\chi^\beta \in \mathrm{Irr}(H_2)$. Clearly $\chi^\beta((y_i h)^\gamma) = \lambda(z_i)\delta(h)$ and $\chi(y_i h) = \lambda(y_i)\delta(h)$. Finally observe that the assumptions imply that $H' = H_1' = H_2'$. $\square$

Examples of isoclinic groups can be constructed using the following.

Definition 3.7.22 *Let $Z_i \le \mathbf{Z}(G_i)$ for $i = 1,2$ and let $\alpha\colon Z_1 \to Z_2$ be an isomorphism. Then $D = \{(z,\alpha(z)^{-1}) \mid z \in Z_1\} \le \mathbf{Z}(G_1 \times G_2)$ and*

$$G_1 \curlyvee_\alpha G_2 = G_1 \curlyvee_{Z_1} G_2 := (G_1 \times G_2)/D$$

*is called the **direct product** of G_1 and G_2 **with amalgamated central subgroups**.*

There are obvious embeddings of G_1, G_2 into $G_1 \curlyvee_{Z_1} G_2$ and $G_1 \curlyvee_{Z_1} G_2 = G_1 G_2$ with $G_1 \cap G_2 = Z_1 = Z_2$ using these embeddings.

Example 3.7.23 Let H_0 be a finite group with $C_p \cong \mathbf{Z}(H_0) \le H_0'$ and $[H_0 : H_0'] = p$. Assume in addition that $H_0'/\mathbf{Z}(H_0)$ simple. Then we can draw the complete lattice of normal subgroups of $H_0 \times C_{p^2}$ and consequently of $H := H_0 \curlyvee_{C_p} C_{p^2}$. For simplicity we do this for $p = 2$ (see Figure 3.5). Here D is

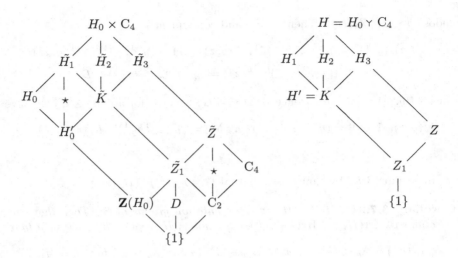

Figure 3.5. Direct product with amalgamated center.

the diagonal subgroup in $\tilde{Z}_1 := \mathbf{Z}(H_0) \times C_2 \cong V_4$. Also $\tilde{H}_i$, for $i = 1, 2, 3$, are the subgroups of index two in $H_0 \times C_4$ intersecting in $\tilde{K}$ and $H_i = \tilde{H}_i/D$. Furthermore $K := \tilde{K}/D = H'$. Finally

$$\tilde{Z} := \mathbf{Z}(H_0 \times C_4) = \mathbf{Z}(H_0) \times C_4 \qquad \text{and} \qquad Z := \tilde{Z}/D = \mathbf{Z}(H).$$

By Lemma 3.7.20 H_1 and H_2 are isoclinic, $H_1' = H_2' = H'$ and $\mathbf{Z}(H_1) = \mathbf{Z}(H_2) = Z_1$. Furthermore $H_1 \cong H_0$ and the character tables of H can be automatically produced from the character table of H_0 (using the known character table of C_4 and Corollary 3.7.6; cf. Exercise 3.7.3). In fact, if $Z = \langle z \rangle$ then

$$H_2 = H' \, \dot{\cup} \, \{h_1 z \mid h_1 \in H_1 \setminus H'\}.$$

For $\chi_1 \in \mathrm{Irr}(H_1)$ we have $\chi_1^H = \chi + \chi'$ with $\chi \neq \chi' \in \mathrm{Irr}(H)$. Then $\chi_2 := \chi|_{H_2}$ and $\chi_2' := \chi'|_{H_2}$ are irreducible and

$$\chi_2 + \chi_2' = (\chi_1^H)|_{H_2} = (\chi_1|_{H'})^{H_2}.$$

It is clear that $\{\chi_2(h_1 z), \chi_2'(h_1 z)\} = \{i \, \chi_1(h_1), -i \, \chi_1(h_1)\}$ for $h_1 \in H \setminus H'$.

If H_1 is a representation group of $G := H_1/Z_1 \cong H/Z \cong H_2/Z_2$, then so is H_2 and clearly $H_1 \not\cong H_2$.

Conversely, suppose now that G is a finite group with $[G : G'] = 2$ and G' is simple with multiplier $M(G) \cong C_2$. (In the ATLAS ([38]) one can find many examples; also one may take $G = S_n$ for $n = 5$ or $n > 7$.) Let $H_1 \not\cong H_2$ be representation groups of G. We use the notation of the proof of Corollary 3.7.21. In particular $H = H_1 H_2 = H_i \mathbf{Z}(H)$, $H' = H_i'$ and $[H : H_i] = n$ for $i = 1, 2$. Since $2 = [G : G'] = [H_i : H_i']$ for $i = 1, 2$ we have $n = 2$. Since $H_i'/\mathbf{Z}(H_i) \cong G'$ we conclude from Theorem 3.7.14 that $\mathbf{Z}(H_i)$ embeds into $M(G')$. Thus $\mathbf{Z}(H_1) = \mathbf{Z}(H_2) \cong C_2$ and $\mathbf{Z}(H)$ has order four. In fact, since

$H_1 \cap \mathbf{Z}(H) = H_2 \cap \mathbf{Z}(H)$ it follows that $\mathbf{Z}(H) \cong C_4$. Now it is easily seen that $H \cong H_1 \curlyvee C_4$, and we conclude that G has exactly two representation groups up to isomorphism. ♦

A representation group H can be used to lift all projective representations (up to equivalence) to ordinary representations. When dealing with a single projective representation a factor group of H is often more appropriate.

Corollary 3.7.24 *Assume that K is algebraically closed with* char $K = 0$. *For $\bar{\alpha} \in H^2(G, K^\times)$ there is a central extension with cyclic kernel C_α*

$$\{1\} \longrightarrow C_\alpha \overset{\iota}{\longrightarrow} H_\alpha \overset{\pi}{\longrightarrow} G \longrightarrow \{1\},$$

a section $\tau_\alpha\colon G \to H_\alpha$ and a faithful $\lambda_\alpha \in \mathrm{Irr}(C_\alpha)$ so that $(\lambda_\alpha \circ \eta_{\tau_\alpha})B^2(G, K^\times) = \bar{\alpha}$. Thus every projective representation of G with factor set $\alpha \in \bar{\alpha}$ is equivalent to one which can be lifted to an ordinary representation of H_α.

Proof. Assume that H is a representation group of G and that the notation is as in Theorem 3.7.14. Abbreviate $Z_\alpha := \ker \lambda$ and let $H_\alpha := H/Z_\alpha$ and $C_\alpha := Z/Z_\alpha$. Finally, put $\tau_\alpha\colon G \to H_\alpha$, $g \mapsto \tau(g)Z_\alpha$ and $\lambda_\alpha\colon C_\alpha \to K^\times$, $zZ_\alpha \mapsto \lambda(z)$. Then $\lambda \circ \eta = \lambda_\alpha \circ \eta_{\tau_\alpha}$. Observe that C_α is cyclic, because λ_α is faithful. □

Our motivation to introduce projective representations was to be able to extend a G-invariant absolutely irreducible representation δ or character φ of a normal subgroup $N \trianglelefteq G$ to G. We now combine such an extension with a lifting to an ordinary representation or character as in Corollary 3.7.24 in order to obtain a concrete description of $\mathrm{Irr}(G \mid \varphi)$.

Theorem 3.7.25 *Let $N \trianglelefteq G$ and let $\varphi \in \mathrm{Irr}(N)$ be G-invariant. Then there is a central extension*

$$\{1\} \longrightarrow Z \overset{\iota}{\longrightarrow} H \overset{\pi}{\longrightarrow} G \longrightarrow \{1\}$$

with section $\tau\colon G \to H$, a faithful $\lambda \in \mathrm{Irr}(Z)$ and a $\chi_o \in \mathrm{Irr}(H)$ so that

$$\pi^{-1}(N) = N \times Z, \quad \text{and} \quad \chi_o|_{N \times Z} = \varphi \times \lambda, \tag{3.24}$$

$$\mathrm{Irr}(G \mid \varphi) = \{ (\chi_o \cdot \chi) \circ \tau \mid \chi \in \mathrm{Irr}(H \mid \mathbf{1}_N \times \lambda^{-1}) \}. \tag{3.25}$$

Observe that in (3.25) $(\chi_o \cdot \chi)|_{N \times Z} = \chi(1) \cdot \varphi \times \mathbf{1}_Z$, so that $Z \leq \ker(\chi_o \cdot \chi)$ (for $\chi \in \mathrm{Irr}(H \mid \mathbf{1}_N \times \lambda^{-1})$).

Proof. Let φ be afforded by $\delta\colon N \to \mathrm{GL}(W)$, where W is a simple $\mathbb{C}N$-module. By Theorem 3.7.3 there is $\bar{\alpha} \in H^2(G/N, \mathbb{C}^\times)$ such that for every $\alpha \in Z^2(G/N, \mathbb{C}^\times)$ with $\bar{\alpha} = \alpha B^2(G/N, \mathbb{C}^\times)$ there is a projective representation $\delta_\alpha\colon G \to \mathrm{GL}(W)$ with factor set $\tilde{\alpha}\colon (g, g') \mapsto \alpha(gN, g'N)$ and $\delta_\alpha|_N = \delta$. By Corollary 3.7.24 there is a group H_α with $H_\alpha/Z_\alpha \cong G/N$ for $Z_\alpha \leq \mathbf{Z}(H_\alpha) \cap H_\alpha'$, a faithful linear character $\lambda\colon Z_\alpha \to K^\times$ and a section $\tau = \tau_\alpha\colon G/N \to H_\alpha$ with

factor set $\eta\colon G/N\times G/N \to Z_\alpha$ such that $\alpha := \lambda\circ\eta$ satisfies $\alpha B^2(G/N, K^\times) = \bar\alpha$.
From Lemma 3.7.11 we obtain a central extension

$$\{1\} \longrightarrow Z_\alpha \xrightarrow{\ \iota\ } G \curlywedge H_\alpha \xrightarrow{\ \pi\ } G \longrightarrow \{1\}$$

with section $\tilde\tau\colon g \mapsto (g, \tau(gN))$ and factor set $\eta_{\tilde\tau}\colon (g, g') \mapsto \eta(gN, g'N)$. Then δ_α has factor set $\tilde\alpha\colon (g, g') \mapsto \lambda(\eta_{\tilde\tau}(g, g'))$.

It follows that

$$\delta_o\colon G \curlywedge H_\alpha \to \mathrm{GL}(W),\ (g,z) \mapsto \lambda(z)\delta(\tau(g))$$

is an ordinary representation with $\delta_o|_{N\times Z_\alpha} = \delta \times \lambda$. We put $H := G \curlywedge H_\alpha$, $Z := \iota(Z_\alpha)$ and denote the character of δ_o by χ_o. Then $\chi_o|_{N\times Z} = \varphi \times \lambda$.

Let $(\mathbf{1}_N\times\lambda^{-1})^H = \sum_{\chi\in\mathrm{Irr}(H)} e_\chi\chi$. By Exercise 3.6.1

$$\sum_{\chi\in\mathrm{Irr}(H)} e_\chi^2 = [H : N \times Z] = [G : N].$$

The following diagram appears to the right:

$$
\begin{array}{ccccc}
H_\alpha & \longleftarrow & H = G \curlywedge H_\alpha & \xrightarrow{\ \pi\ } & G \\
\downarrow & & \downarrow & & \downarrow \\
Z_\alpha & \longleftarrow & N \times Z & \longrightarrow & N \\
\downarrow & & & & \downarrow \\
\{1\} & N & & Z & \{1\} \\
& & \{1\} & &
\end{array}
$$

Then $(\varphi\times\mathbf{1}_Z)^H = ((\varphi\times\lambda)\cdot(\mathbf{1}_N\times\lambda^{-1}))^H = \chi_o\cdot(\mathbf{1}_N\times\lambda^{-1})^H = \sum_{\chi\in\mathrm{Irr}(H)} e_\chi\chi_o\cdot\chi$.
Again by Exercise 3.6.1 $((\varphi \times \mathbf{1}_Z)^H, (\varphi \times \mathbf{1}_Z)^H)_H = [G : N]$, and it follows that

$$(\chi_o\cdot\chi\ ,\ \chi_o\cdot\chi')_H = \delta_{\chi,\chi'} \qquad \text{for} \qquad \chi, \chi' \in \mathrm{Irr}(H \mid \mathbf{1}_N \times \lambda^{-1}).$$

Thus $\mathrm{Irr}(H \mid \varphi \times \mathbf{1}_Z) = \{\chi_o\cdot\chi \mid \chi \in \mathrm{Irr}(H \mid \mathbf{1}_N \times \lambda^{-1})\}$. $\qquad\square$

We proceed to discuss characters of projective representations and their orthogonality relations. The basic lemma is the following.

Lemma 3.7.26 *Let $Z \le \mathbf{Z}(G)$, $\overline{G} := G/Z$ and let $\lambda \in \mathrm{Irr}(Z)$ be faithful.*

(a) *If $g \in G$ satisfies $\mathbf{C}_{\overline{G}}(gZ) > \mathbf{C}_G(g)/Z$ then $\chi(g) = 0$ for all $\chi \in \mathrm{Irr}(G \mid \lambda)$.*

(b) *For $g, h \in G$ we have*

$$\sum_{\chi\in\mathrm{Irr}(G\mid\lambda)} \chi(g)\chi(h^{-1}) = \begin{cases} |\mathbf{C}_{\overline{G}}(gZ)| & \text{if } h \in g^G \text{ and } \mathbf{C}_{\overline{G}}(gZ) = \mathbf{C}_G(g)/Z, \\ 0 & \text{else.} \end{cases}$$

Proof. (a) $\mathbf{C}_{\overline{G}}(gZ) \ne \mathbf{C}_G(g)/Z$ holds if and only if there is an $x \in G$ with $1 \ne z = [g, x] \in Z$, that is $g^x = gz$. In this case Lemma 2.3.5 implies $\chi(g) = 0$ for all $\chi \in \mathrm{Irr}(G, \lambda)$, because λ is faithful.

(b) We use induction on $n := |Z|$. For $n = 1$ see the orthogonality relations (Theorem 2.1.15). By assumption Z is cyclic of order, say, n. For each divisor d of n there is exactly one subgroup C_d of order n/d, which is the kernel of λ^d.

If $j \in \mathbb{N}$ is coprime to n/d we obtain $\mathrm{Irr}(G \mid \lambda^{dj})$ from $\mathrm{Irr}(G \mid \lambda^d)$ by applying a cyclotomic field automorphism η_j. Hence we have

$$a_{g,h} := \sum_{\chi \in \mathrm{Irr}(G)} \chi(g)\chi(h^{-1}) = \sum_{d \mid n} \sum_{(j, \frac{n}{d})=1} \Big(\sum_{\chi \in \mathrm{Irr}(G \mid \lambda^d)} \chi(g)\chi(h^{-1}) \Big)^{\eta_j}.$$

If g, h are not conjugate in G then $a_{g,h} = 0$, and by induction we get

$$\sum_{\chi \in \mathrm{Irr}(G \mid \lambda)} \chi(g)\chi(h^{-1}) = 0. \tag{3.26}$$

Clearly (3.26) also holds by part (a) if $h \in g^G$ and $\mathbf{C}_{\overline{G}}(gZ) > \mathbf{C}_G(g)/Z$. So finally we may assume that $h = g$ and $\mathbf{C}_{\overline{G}}(gZ) = \mathbf{C}_G(g)/Z$. Then, by the orthogonality relations,

$$|\mathbf{C}_G(g)| = a_{g,g} = \sum_{d \mid n} \varphi(n/d) \sum_{\chi \in \mathrm{Irr}(G \mid \lambda^d)} |\chi(g)|^2, \tag{3.27}$$

with φ being the Eulerian function. On the other hand, by our assumption,

$$|\mathbf{C}_G(g)| = n \, |\mathbf{C}_{\overline{G}}(gZ)| = \sum_{d \mid n} \varphi(n/d) \, |\mathbf{C}_{\overline{G}}(gZ)|. \tag{3.28}$$

Now, λ^d may be considered as a faithful character of $Z_d = Z/C_d$, which is a central subgroup of $G_d = G/C_d$. Our assumption implies for $\tilde{g} = gC_d$ that $\mathbf{C}_{\overline{G}_d}(\tilde{g}Z_d) = \mathbf{C}_{G_d}(\tilde{g})/Z_d$, where $\overline{G}_d := G_d/Z_d \cong G/Z = \overline{G}$. Hence by induction $\sum_{\chi \in \mathrm{Irr}(G \mid \lambda^d)} |\chi(g)|^2 = |\mathbf{C}_{\overline{G}}(gZ)|$ for $d > 1$. Inserting this in (3.27) and comparing with (3.28) we obtain $\sum_{\chi \in \mathrm{Irr}(G \mid \lambda)} |\chi(g)|^2 = |\mathbf{C}_{\overline{G}}(gZ)|$. $\qquad\square$

Definition 3.7.27 $g \in G$ is called α**-regular** for an $\alpha \in Z^2(G, K^\times)$ if

$$\alpha(g, x) = \alpha(x, g) \qquad \text{for all} \quad x \in \mathbf{C}_G(g).$$

Lemma 3.7.28 *Let G be a finite group, let K be a field and $\alpha \in Z^2(G, K^\times)$.*

(a) *If $g \in G$ and $\mathrm{Trace}(\delta(g)) \neq 0$ for some projective representation over K, then g is α-regular.*

(b) *If $\alpha' \in \alpha B^2(G, K^\times)$ then $g \in G$ is α-regular if and only if g is α'-regular and this holds if and only if any element in g^G is α-regular.*

(c) *If K is algebraically closed of characteristic zero then $g \in G$ is α-regular if and only if $\mathrm{Trace}(\delta(g)) \neq 0$ for some projective representation with factor set α. If H_α, Z_α and τ_α are as in Corollary 3.7.24 then g is α-regular if and only if $\tau(g)$ is not conjugate in H_α to any $z \tau(g)$ with $1 \neq z \in Z_\alpha$. In particular, if the order of g is coprime to the order of $\alpha B^2(G, K^\times)$ in $H^2(G, K^\times)$ then g is α-regular.*

Proof. (a) For $x \in \mathbf{C}_G(g)$ we get $\alpha(x,g) \cdot \delta(x)\delta(g)\delta(x)^{-1} = \alpha(g,x) \cdot \delta(g)$ and consequently $(\alpha(x,g) - \alpha(g,x)) \cdot \mathrm{Trace}(\delta(g)) = 0$.

(b) The first part of (b) is obvious. The second part follows from a lengthy calculation, which is not worth reproducing. It follows also immediately from (c) in this more special situation, which, however, is the only one we are interested in.

(c) Let $g \in G$ be α-regular. We may use Corollary 3.7.24 to assume that $\alpha = \lambda \circ \eta$ with $G \cong H_\alpha/Z_\alpha$, $(Z_\alpha \le \mathbf{Z}(H_\alpha))$ a section $\tau\colon G \to H_\alpha$, corresponding factor set $\eta\colon G \times G \to Z_\alpha$ and a faithful $\lambda \in \mathrm{Irr}(Z_\alpha)$. Since λ is faithful, Definition 3.7.27 says that $\eta(g,x) = \eta(x,g)$ for all $x \in \mathbf{C}_G(g)$. This is equivalent to $\tau(g)\tau(x) = \tau(x)\tau(g)$ for all $x \in \mathbf{C}_G(g)$ and therefore to $\mathbf{C}_G(g) \cong \mathbf{C}_{H_\alpha}(\tau(g))/Z_\alpha$. Thus no two elements in the coset $\tau(g)Z_\alpha$ can be conjugate in H_α, and, at the same time, Lemma 3.7.26 shows that there is an irreducible projective representation δ with $\mathrm{Trace}(\delta(g)) \neq 0$. Clearly, if the order of g, say m, is coprime to the order of $\alpha B^2(G, K^\times)$, which is $|Z_\alpha|$, then we may choose $\tau(g)$ to have order m, and $\tau(g)$ is not conjugate to any other element in the coset $\tau(g)Z_\alpha$. $\qquad\square$

If $\alpha \in Z^2(G, K^\times)$ and $U \le G$ then $\alpha' := \alpha|_{U \times U} \in Z^2(U, K^\times)$. Obviously, if $g \in U$ is α-regular then g is also α'-regular. We give a typical example, where this simple idea can be used to find non-α-regular classes.

Example 3.7.29 $G := \mathrm{L}_2(11)$ has a representation group $H := \mathrm{SL}_2(11) \cong 2 \cdot G$ and maximal subgroups U isomorphic to A_5 with index 11 (see [38], p. 7). Thus H has maximal subgroups V containing $Z := \mathbf{Z}(H)$ with $V/Z \cong U$. If Z had a complement in V then by a theorem of Gaschütz (see [88], Satz I.17.4) Z would have a complement in H (since $(|Z|, [H : U]) = 1$), which is impossible because $Z \le H'$. Hence $V \cong 2 \cdot \mathrm{A}_5$, using the well known fact that the Schur multiplier of A_5 has order two. Let $\alpha \in Z^2(G, \mathbb{C}^\times)$ be in a cohomology class of order two and let $\alpha' := \alpha|_{U \times U}$. In Example 3.7.33 we will show that the involutions in A_5 are not α'-regular. So we may conclude that the class of involutions in G is not α-regular, or in other words that all faithful irreducible characters of H vanish on the elements of order four (which map onto involutions under the natural map $H \to G$). $\qquad\blacklozenge$

If $\delta\colon G \to \mathrm{GL}(V)$ and $\delta\colon G \to \mathrm{GL}(V')$ are projective representations over K with factor sets α, α' then

$$\delta \otimes \delta'\colon G \to \mathrm{GL}(V \otimes V'), \quad g \mapsto \delta(g) \otimes \delta'(g)$$

defines a projective representation with factor set $\alpha\alpha'$. In particular tensoring α-projective representations with ordinary ones yields α-projective representations. Naturally the trace function $\chi_\delta\colon G \to K$, $h \mapsto \mathrm{trace}(\delta(h))$ is called an α-**projective character**. In general this need not be a class function (see also Exercise 3.7.1), and certainly equivalent projective representations need not have the same projective characters. It follows from Corollary 3.7.24 that every irreducible projective representation of a finite group G over $\mathbb{C}$ is equivalent to one for which the corresponding projective character is a class function.

Definition 3.7.30 A factor set $\alpha\colon G \times G \to K^\times$ will be called normalized if there is a central extension $H/Z \cong G$ with $Z \leq \mathbf{Z}(H)$ and a section $\tau\colon G \to H$ (with corresponding factor set η_τ) such that $\alpha = \lambda \circ \eta_\tau$ for some faithful $\lambda \in \mathrm{Hom}(Z, K^\times)$. If $\alpha\colon G \times G \to \mathbb{C}^\times$ is a normalized factor set we define $\mathrm{Irr}_\alpha(G)$ to be the set of irreducible α-projective characters of G over $\mathbb{C}$. The K-vector space of class functions vanishing on all conjugacy classes of G which are not α-regular will be denoted by $\mathrm{cf}_\alpha(G, K)$.

Theorem 3.7.31 (Orthogonality relations) *Let* $\alpha\colon G \times G \to \mathbb{C}$ *be a normalized factor set and let* $g, h \in G$ *be in* α-regular classes. Then

$$\sum_{\chi \in \mathrm{Irr}_\alpha(G)} \chi(g)\chi(h^{-1}) = \begin{cases} |\mathbf{C}_G(g)| & \text{if } g \text{ and } h \text{ are conjugate in } G, \\ 0 & \text{else.} \end{cases} \tag{3.29}$$

$\mathrm{Irr}_\alpha(G)$ *is an orthonormal basis for* $\mathrm{cf}_\alpha(G, \mathbb{C})$ *with respect to the bilinear form* $(\varphi, \psi)_G := \frac{1}{|G|} \sum_{g \in G} \varphi(g)\psi(g^{-1})$. *In particular,* $k_\alpha := |\mathrm{Irr}_\alpha(G)|$ *is equal to the number* r_α *of* α-regular classes of G.

Proof. Let $H/Z \cong G$, with $Z \leq \mathbf{Z}(H)$ and $\tau\colon G \to H$, be a section such that $\alpha = \lambda \circ \eta_\tau$ with $\lambda \in \mathrm{Irr}(Z)$ faithful. Then $\mathrm{Irr}_\alpha(G) = \{\psi \circ \tau \mid \psi \in \mathrm{Irr}(H \mid \lambda)\}$ and (3.29) follows immediately from Lemma 3.7.26. Let $g_1, \ldots, g_{r_\alpha}$ be representatives of the α-regular classes of G and $X := [\chi(g_j)]_{\chi \in \mathrm{Irr}_\alpha(G), j}$. Then (3.29) can be written as $X^{\mathsf{T}}\bar{T} = \mathrm{diag}(|\mathbf{C}_G(g_1)|, \ldots, |\mathbf{C}_G(g_{r_\alpha})|)$. Thus $r_\alpha \leq k_\alpha$.

Let $\psi_1, \psi_2 \in \mathrm{Irr}(H \mid \lambda)$ and $\chi_i = \psi_i \circ \tau$ for $i = 1, 2$. Then

$$(\psi_1, \psi_2)_H = \frac{1}{|H|} \sum_{g \in G} \sum_{z \in Z} \psi_1(\tau(g)z)\psi_2(\tau(g)^{-1}z^{-1})$$

$$= \frac{1}{|G|} \frac{1}{|Z|} \sum_{g \in G} \sum_{z \in Z} \chi_1(g)\lambda(z)\chi_2(g^{-1})\lambda(z^{-1}) = (\chi_1, \chi_2)_G.$$

So the characters in $\mathrm{Irr}_\alpha(G)$ are indeed orthonormal and a basis of $\mathrm{cf}_\alpha(G, \mathbb{C})$ because $r_\alpha \leq k_\alpha$. $\qquad\square$

As a consequence we obtain the following important result.

Theorem 3.7.32 *Let* $N \trianglelefteq G$ *and* $\varphi \in \mathrm{Irr}(N)$.

(a) *If* $\chi \in \mathrm{Irr}(G \mid \varphi)$ *then* $\frac{\chi(1)}{\varphi(1)} \mid [G : N]$.

(b) $|\mathrm{Irr}(G \mid \varphi)| = k_\alpha(T_G(\varphi)/N)$ *for some normalized factor set* α *of* $T_G(\varphi)/N$.

Proof. (a) If $\chi \in \mathrm{Irr}(G \mid \varphi)$ and $T := T_G(\varphi)$ then $\chi = \psi^G$ for some $\psi \in \mathrm{Irr}(T \mid \varphi)$ and $\frac{\chi(1)}{\varphi(1)} = [G : T]\frac{\psi(1)}{\varphi(1)}$. So we may assume that $T = G$. In this case we conclude from Theorem 3.7.5 that $\chi(1) = \varphi(1)\,d$, where d is the degree of an irreducible projective representation of G/N. So the result follows from Corollary 3.7.16.

(b) Because of the Clifford correspondence (Corollary 3.6.8) we may assume that $T_G(\varphi) = G$. Using Theorem 3.7.25 and the notation of its proof we get

$$|\operatorname{Irr}(G \mid \varphi)| = |\operatorname{Irr}(H_\alpha \mid \lambda^{-1})| = |\operatorname{Irr}_{\alpha^{-1}}(G/N)|,$$

where $\lambda \in \operatorname{Irr}(Z_\alpha)$, $Z_\alpha \leq \mathbf{Z}(H_\alpha)$, $H_\alpha/Z_\alpha \cong G/N$ with section τ and $\alpha^{-1} = \lambda^{-1} \circ \eta_\tau$. Since a class is α-regular if and only if it is α^{-1}-regular; the claim follows from Theorem 3.7.31. $\qquad\square$

Note that on choosing N abelian in part (a) of Theorem 3.7.32 we recover Theorem 3.6.5.

Example 3.7.33 We have already mentioned that $M(A_5) \cong C_2$. We will compute the character table of a representation group $H = 2 \cdot A_5 \cong SL_2(5)$ (see [38], p. 2) or, in other words, compute $\operatorname{Irr}_\alpha(A_5)$, where $\langle \alpha B^2(A_5, \mathbb{C}^\times) \rangle = H^2(A_5, \mathbb{C}^\times)$. By Lemma 3.7.28(c) all elements of odd order in A_5 are α-regular, and using Theorem 3.7.31 we get $|\operatorname{Irr}_\alpha(A_5)| \in \{4, 5\}$ and $\sum_{\chi \in \operatorname{Irr}_\alpha(A_5)} \chi(1)^2 = 60$. By Exercise 3.7.5 $\chi(1)$ is even for all $\chi \in \operatorname{Irr}_\alpha(A_5)$. Since the only way to write $15 = 60/4$ as a sum of four or five squares is $15 = 1 + 1 + 4 + 9$ we find $\operatorname{Irr}_\alpha(A_5) = \{\psi_1, \ldots, \psi_4\}$ with $\psi_{1,2}(1) = 2, \psi_3(1) = 4, \psi_4(1) = 6$ and the involutions in A_5 are not α-regular. Note that $\psi_i \cdot \psi_j$ is an ordinary character for $1 \leq i, j \leq 4$, hence $\{\psi_1^2, \psi_2^2\} = \{1_{A_5} + \chi_2, 1_{A_5} + \chi_3\}$ in the notation of Example 2.1.24. Thus the values of $\psi_{1,2}$ are determined up to the signs, which are easily found using the congruence relations (see Lemma 2.2.2). Theorem 3.7.31 also implies that $\chi_3 \cdot \psi_1$ is irreducible and thus equal to ψ_4 and that $\psi_3 = \chi_2 \cdot \psi_1 - \psi_4$. Inflating the χ_i $(1 \leq i \leq 5)$ and lifting the ψ_j $(1 \leq j \leq 4)$ to characters of H, we get the character table of H:

2.A5

2	3	3	2	1	1	1	1	1	1
3	1	1	.	1	1	.	.	.	.
5	1	1	.	.	.	1	1	1	1

	1a	2a	4a	3a	6a	5a	10a	5b	10b
2P	1a	1a	2a	3a	3a	5b	5b	5a	5a
3P	1a	2a	4a	1a	2a	5b	10b	5a	10a
5P	1a	2a	4a	3a	6a	1a	2a	1a	2a

X.1	1	1	1	1	1	1	1	1	1
X.2	3	3	-1	.	.	A	A	*A	*A
X.3	3	3	-1	.	.	*A	*A	A	A
X.4	4	4	.	1	1	-1	-1	-1	-1
X.5	5	5	1	-1	-1	.	.	.	.
X.6	2	-2	.	-1	1	-A	A	-*A	*A
X.7	2	-2	.	-1	1	-*A	*A	-A	A
X.8	4	-4	.	1	-1	-1	1	-1	1
X.9	6	-6	.	.	.	1	-1	1	-1

A = -E(5)-E(5)^4 = (1-ER(5))/2 = -b5

That a preimage $h \in H$ of an involution of A_5 has indeed order four (as indicated by the class name 4a) follows from $\psi^{[2]} = X.2$ for $\psi := X.6$, since this implies $\psi(h^2) = 2 \cdot X.2(h) - \psi(h)^2 = -2$ and thus $h^2 \in 2a$.

In the ATLAS ([38]) this table is abbreviated by just listing $\mathrm{Irr}(A_5)$ and $\mathrm{Irr}_\alpha(A_5)$ and (in between) the element orders of the preimages (in H) of the conjugacy classes of A_5. So here the table looks approximately like Table 3.6:

Table 3.6.

	ind	1a	2a	3a	5a	5b
χ_1	+	1	1	1	1	1
χ_2	+	3	−1	0	$-b5$	*
χ_3	+	3	−1	0	*	$-b5$
χ_4	+	4	0	1	−1	−1
χ_5	+	5	1	−1	−2	0
	ind	1	4	3	5	5
		2		6	10	10
χ_6	−	2	0	−1	$b5$	*
χ_7	−	2	0	−1	*	$b5$
χ_8	−	4	0	1	−1	−1
χ_9	−	6	0	0	1	−1

Example 3.7.34 We saw in Example 3.7.23 that $M(A_5) \cong C_2$ implies that $M(S_5) \cong C_2$ and that $G = S_5$ has exactly two representation groups H_i up to isomorphism. We have now two methods at our disposal to compute the character table of H_i, namely extending the characters of $2.A_5$ as in Example 3.6.22 or computing the projective characters of S_5 as in Example 3.7.33. One of the tables in GAP notation is the following:

```
2.A5.2
        2   4  4  3  2  2  1    1  2  3  3  2  2
        3   1  1  .  1  1  .    .  1  .  .  1  1
        5   1  1  .  .  .  1    1  .  .  .  .  .

            1a 2a 4a 3a 6a 5a 10a 2b 8a 8b 6b 6c
        2P  1a 1a 2a 3a 3a 5a  5a 1a 4a 4a 3a 3a
        3P  1a 2a 4a 1a 2a 5a 10a 2b 8a 8b 2b 2b
        5P  1a 2a 4a 3a 6a 1a  2a 2b 8b 8a 6c 6b

X.1   +   1   1  1  1  1  1    1  1  1  1  1  1
X.2   +   1   1  1  1  1  1    1 -1 -1 -1 -1 -1
X.3   +   6   6 -2  .  .  1    1  .  .  .  .  .
X.4   +   4   4  .  1  1 -1   -1  2  .  . -1 -1
X.5   +   4   4  .  1  1 -1   -1 -2  .  .  1  1
X.6   +   5   5  1 -1 -1  .    .  1 -1 -1  1  1
X.7   +   5   5  1 -1 -1  .    . -1  1  1 -1 -1
X.8   -   4  -4  . -2  2 -1    1  .  .  .  .  .
X.9   0   4  -4  .  1 -1 -1    1  .  .  .  B -B
X.10  0   4  -4  .  1 -1 -1    1  .  .  . -B  B
X.11  0   6  -6  .  .  .  1   -1  .  A -A  .  .
X.12  0   6  -6  .  .  .  1   -1  . -A  A  .  .
```

$A = E(8)+E(8)^3 = ER(-2) = i2, \qquad B = E(3)-E(3)^2 = ER(-3) = i3$

Table 3.7.

χ	ind	1A	2A	3A	5A	5B	fus	ind	2B	4A	6A
		60	4	3	5	5			6	2	3
p power			A	A	A	A			A	A	AB
p' part			A	A	A	A			A	A	AB
χ_1	+	1	1	1	1	1	:	++	1	1	1
χ_2	+	3	-1	0	-b5	*	|	+	0	0	0
χ_3	+	3	-1	0	*	-b5	|				
χ_4	+	4	0	1	-1	-1	:	++	2	0	-1
χ_5	+	5	1	-1	-2	0	:	++	1	-1	1
	ind	1	4	3	5	5	fus	ind	2	8	6
		2		6	10	10				8	6
χ_6	−	2	0	-1	b5	*	|	−	0	0	0
χ_7	−	2	0	-1	*	b5	|				
χ_8	−	4	0	1	-1	-1	:	00	0	0	i3
χ_9	−	6	0	0	1	-1	:	00	0	i2	0

In the ATLAS this is given as shown in Table 3.7. How to obtain the character table of the other representation group from this was discussed in Example 3.7.23. One just has to multiply i2, i3 (or A, B) by $i = \zeta_4$ and adjust the element orders and power maps. In the GAP version the first means that the classes 2b, 8a, 8b, 6b, 6c should be replaced by 4b, 8a, 8b, 12a, 12b. Also, of course, the Frobenius–Schur indicators of the last four irreducible characters change from 0 to −1. ◆

Exercises

Exercise 3.7.1 Let $\rho : G \to \mathrm{GL}(V)$ be a projective representation with factor set α. Show that
$$\rho(h)^{-1}\rho(g)\rho(h) = \alpha(g,h)\alpha(h, h^{-1}gh)\rho(h^{-1}gh) \quad \text{for all} \quad g,h \in G.$$

Exercise 3.7.2 Let F be a free group on $\{x,y\}$ and let R be the normal subgroup of F generated by $\{x^2, y^2, [x,y]\}$, so that $F/R \cong \mathrm{V}_4$. Using the notation of Theorem 3.7.15 and writing $\tilde{a} := a[R,F]$ for $a \in F$ show that

(a) $\tilde{F}' = \langle [\tilde{x}, \tilde{y}] \rangle$ has order two;

(b) $\tilde{S}_1 := \langle \tilde{x}^2, \tilde{y}^2 \rangle$ and $\tilde{S}_2 := \langle \tilde{x}^2\tilde{y}^2, [\tilde{x}, \tilde{y}]\tilde{x}^2 \rangle$ are both complements of $\tilde{F}'$ in $\tilde{R}$;

(c) $\tilde{F}/S_1 \cong \mathrm{D}_8$ and $\tilde{F}/S_2 \cong \mathrm{Q}_8$.

Conclude that D_8 and Q_8 are representation groups of V_4.

Exercise 3.7.3 Let $\langle z \rangle = \mathrm{C}_4$ and $\alpha\colon \langle z^2 \rangle \to \mathrm{Z}(\mathrm{D}_8)$ be the unique isomorphism. Define $H := \mathrm{D}_8 \curlyvee_\alpha \mathrm{C}_4$.

(a) Compute the character table of H.

(b) Verify that H has three normal subgroups isomorphic to D_8 and one isomorphic to Q_8 and that all irreducible characters of these normal subgroups are invariant in H.

Hint: For (a) one may use the GAP commands `CharacterTable("Cyclic", 4)`, `CharacterTable("Dihedral", 8)`, `CharacterTableDirectProduct` and `CharacterTableFactorGroup`.

Exercise 3.7.4 Let H be a finite group with cyclic Sylow p-subgroup P. Use the transfer homomorphism $H \to P$ (see [88], IV.2.2, p. 444) to show that $H' \cap \mathbf{Z}(H)$ is a p'-group. Deduce from this that p does not divide $|M(G)|$ if the Sylow p-subgroups of a finite group G are cyclic.

Exercise 3.7.5 Let H be a finite group with $Z \le H' \cap \mathbf{Z}(H)$ and let $\chi \in \mathrm{Irr}(H)$ be faithful. Show that $\mathbf{Z}(H)$ is cyclic and $|\mathbf{Z}(H)| \mid \chi(1)$.
Hint: Consider $\det D(z)$ for $z \in Z$ and D a representation with character χ.

Exercise 3.7.6 Use the character table of $L_2(11)$ (see Exercise 2.10.2) and Example 3.7.29 to compute the character table of $H := 2 \cdot L_2(11)$.
Hint: Observe that the square of a faithful character can be considered as a character of $L_2(11)$ and must vanish on the involutions. Use this and Exercise 3.7.5 to find the degrees of the faithful irreducible characters of H and the square of the faithful irreducible characters of smallest degree.

Exercise 3.7.7 Using the notation of Corollary 3.7.6 show the following.

(a) If $G_1 \cong G_2 \cong Q_8$ then G has an absolutely simple $\mathbb{R}G$ module which is not a tensor product of two absolutely simple $\mathbb{R}\,Q_8$-modules.

(b) If K is a finite field then every absolutely simple KG-module is isomorphic to $\mathrm{Inf}_{\pi_1} W_1 \otimes_K \mathrm{Inf}_{\pi_2} W_2$ with absolutely simple KG_i-modules W_i, $(i = 1, 2)$.
Hint: Use Theorem 1.8.4 and Theorem 3.6.2.

3.8 Clifford matrices

Let G be a finite group with a normal subgroup N and let $\overline{G} = G/N$ with $\pi \colon G \to \overline{G}$, $g \mapsto \overline{g}$ being the canonical epimorphism. In this section we will always write $\overline{U} = \pi(U)$ for any subset (or element) U of G. The aim is to construct the character table of G from the character tables of $\overline{G}$ and of suitable subgroups of $\overline{G}$ (the inertia factors). The method described in this section is due to Fischer (see [58]). In this section the ground field is always $\mathbb{C}$ and character means character over $\mathbb{C}$.

We summarize Corollary 3.6.8 and Theorem 3.7.5 in the language of characters as follows.

Theorem 3.8.1 *Let $\varphi_1 = 1_N, \varphi_2, \ldots, \varphi_t$ be representatives of the G-conjugacy classes of irreducible characters of N. Let $T_m = T_G(\varphi_m)$ be the inertia subgroup of φ_m for $m = 1, \ldots, t$. Then*

$$\mathrm{Irr}(G) = \dot{\bigcup}_{m=1}^{t} \{\psi^G \mid \psi \in \mathrm{Irr}(\,T_m \mid \varphi_m\,)\}.$$

Furthermore, every $\psi \in \mathrm{Irr}(\,T_m \mid \varphi_m\,)$ is of the form $\tilde{\varphi}_m \cdot \hat{\psi}$, where $\tilde{\varphi}_m \in \mathrm{Irr}_\alpha(T_m)$ is fixed with $\tilde{\varphi}_m|_N = \varphi_m$ and $\hat{\psi} \in \mathrm{Irr}_{\alpha^{-1}}(\overline{T_m})$ is inflated to T_m. Here α is a normalized factor set (see Definition 3.7.30) constant on cosets of N, so that it can also be considered as an element of $Z^2(\overline{T_m}, \mathbb{C}^\times)$.

We will use the notation of Theorem 3.8.1 throughout this section. In addition we fix the notation for the conjugacy classes.

- Let $\overline{g_1} = g_1 N, \ldots, \overline{g_r} = g_r N$ be representatives for the conjugacy classes of $\overline{G} = G/N$, thus $g_i \in G$ for $1 \le i \le r$ and we assume that $g_1 = 1$.

- $\pi^{-1}(\overline{g_i}^{\overline{G}}) = \dot{\bigcup}_{j=1}^{s_i} g_{ij}^G$ and we may assume that $\pi(g_{ij}) = \overline{g_i}$ and $g_{i1} = g_i$. Of course, s_1 is the number of G-conjugacy classes in N.

- $(g_{ij})^G \cap T_m = \dot{\bigcup}_{k=1}^{s(i,j,m)} (g_{ijk}^m)^{T_m}$ with $g_{ijk}^m \in T_m$. We have $g_{111}^m = 1$ and $s(1,1,m) = 1$ for all m.

- $\overline{g_i}^{\overline{G}} \cap \overline{T_m} = \dot{\bigcup}_{l=1}^{t(i,m)} \overline{y_{il}^m}^{\overline{T_m}}$ with $y_{il}^m \in T_m$, and we may assume that $y_{i1}^m = g_i$ if $\overline{g_i} \in \overline{T_m}$. Of course, $t(1,m) = 1$; more generally, $t(i,m) \in \{0,1\}$ if $\overline{g_i} \in \mathbf{Z}(\overline{G})$. We always have $\pi((g_{ijk}^m)^{T_m}) = \overline{y_{il}^m}^{\overline{T_m}}$ for some $l = l(j,k)$.

Remark 3.8.2 If $g_i \in \mathbf{Z}(G)$ we have $s_i = s_1$ and we may choose $g_{ij} := g_i g_{1j}$ for $j = 1, \ldots, s_1$.

We assume that the character tables of $\overline{G}$ and the inertia factor groups $\overline{T_m} = T_m/N$ are known, as are the corresponding fusion maps of $\overline{T_m}$ into $\overline{G}$, which means that for every conjugacy class $\overline{g_i}^{\overline{G}}$ we know the classes $\overline{y_{il}^m}^{\overline{T_m}}$ or, to be more precise, the values of the irreducible (ordinary or projective) characters of $\overline{T_m}$ on these classes. By our convention ($\varphi_1 = 1_N$) we always have $T_1 = G$, so $t(i,1) = 1, s(i,j,1) = 1$, and we may assume that $g_{ij1}^1 = g_{ij}$ and $y_{i1}^1 = g_i$ for $1 \le i \le r$.

Keeping the above notation we get for $\psi \in \mathrm{Irr}(\,T_m \mid \varphi_m\,)$, using Lemma 3.2.7,

$$\begin{aligned}
\psi^G(g_{ij}) &= \sum_{k=1}^{s(i,j,m)} \frac{|\mathbf{C}_G(g_{ij})|}{|\mathbf{C}_{T_m}(g_{ijk}^m)|} \psi(g_{ijk}^m) \\
&= \sum_{k=1}^{s(i,j,m)} \frac{|\mathbf{C}_G(g_i)|}{|\mathbf{C}_{T_m}(g_{ijk}^m)|} \tilde{\varphi}_m(g_{ijk}^m) \hat{\psi}(\overline{g_{ijk}^m}) \\
&= \sum_{l=1}^{t(i,m)} c_{(m,l),j}^i \hat{\psi}(\overline{y_{il}^m}),
\end{aligned} \tag{3.30}$$

with

$$c^i_{(m,l),j} = \sum_{k,l(j,k)=l} \frac{|\mathbf{C}_G(g_{ij})|}{|\mathbf{C}_{T_m}(g^m_{ijk})|} \tilde{\varphi}_m(g^m_{ijk})).$$

Let

$$I_i = \{(m,l) \mid 1 \le m \le t, \ 1 \le l \le t(i,m), \ \overline{y^m_{il}}^{\overline{T_m}} \ \text{is} \ \alpha_m\text{-regular} \}.$$

We order I_i lexicographically.

Definition 3.8.3 The matrix

$$\mathrm{Fi}_i = \mathrm{Fi}_{G,N}(\overline{g_i}^{\overline{G}}) = [c^i_{(m,l),j}]_{(m,l)\in I_i, 1 \le j \le s_i}$$

is called the **Clifford matrix** (or **Fischer matrix**) corresponding to the class of $\overline{g_i} \in \overline{G}$.

Observe that the Clifford matrices depend on the choice of the projective extensions $\tilde{\varphi}_m$ of the irreducible characters φ_m of N. We will always assume that the $\tilde{\varphi}_m$ are chosen as ordinary characters whenever possible. We also can assume that the values of $\tilde{\varphi}_m$ are algebraic integers. Then $\mathrm{Fi}(\overline{g_i}^{\overline{G}})$ is a matrix with algebraic integers as entries with s_i columns, where s_i is the number of conjugacy classes of G into which the preimage $\pi^{-1}\overline{g_i}^{\overline{G}}$ splits. It will soon turn out to be a square matrix. The rows are divided into blocks, each block corresponding to an inertia group T_m and containing as many rows as the number of α_m-regular $\overline{T_m}$-conjugacy classes there are in $\overline{g_i}^{\overline{G}} \cap \overline{T_m}$. By our convention the first block corresponds to $T_1 = G$ and thus contains just one row, and from $\tilde{\varphi}_1 = \mathbf{1}_G$ we get $c^i_{(1,1),j} = 1$ for all j.

Also for $i = 1$ we get, since $g_1 = g_{11} = g^m_{111} = 1$ and $t(1,m) = s(1,1,m) = 1$, that

$$c^1_{(m,1),j} = \frac{\psi^G(g_{1j})}{\psi(1)\varphi_m(1)} \qquad \text{for} \quad \psi \in \mathrm{Irr}(\,T_m \mid \varphi_m\,). \qquad (3.31)$$

In particular, $c^1_{(m,1)1} = [G : T_m]\varphi_m(1)$, and, if N is abelian, the first column of Fi_1 contains just the orbit lengths of G on $\hat{N}$.

Since the notation is somewhat complicated, let us first explain it in a very simple example, although this is somewhat misleading. In fact, in practice one will *never* compute the g^m_{ijk} or try to compute the irreducible characters of the T_m, simply because the character tables of these inertia subgroups are usually much larger and more complicated to compute than the character table of G. Instead, the idea is to use only the character tables of the inertia factor groups $\overline{T_m}$ and arithmetical properties of the Clifford matrices.

Example 3.8.4 Let $N = \mathrm{V}_4 \trianglelefteq G = \mathrm{S}_4$ so that $\overline{G} = G/N \cong \mathrm{S}_3$; for simplicity of notation we identify $\overline{G}$ with S_3. We have three conjugacy classes: $C_1 = \overline{1}^{\overline{G}}, C_2 = \overline{(1,2)}^{\overline{G}}, C_3 = \overline{(1,2,3)}^{\overline{G}}$ in $\overline{G}$, and hence there are three Clifford matrices, $\mathrm{Fi}_1 =$

$\mathrm{Fi}(C_1), \mathrm{Fi}_2 = \mathrm{Fi}(C_2), \mathrm{Fi}_3 = \mathrm{Fi}(C_3)$. Also G has two orbits on $\mathrm{Irr}(N)$, the trivial character $\mathbf{1}_N$ forms one orbit with inertia group $T_1 = G$ and the other three linear characters form another one, with inertia groups T_2 of index three in G, so $T_2 \cong D_8$ and $\overline{T_2} \cong C_2$. We may choose φ_2 to have kernel $\langle (1,3)(2,4) \rangle$, hence $T_2 = \langle (1,2)(3,4), (1,2,3,4) \rangle$ and (omitting the trivial cases $g_{ij1}^1 = g_{ij}$)

$$
\begin{aligned}
&g_1 = g_{11} = 1, &&g_{12} = (1,2)(3,4), &&y_{11}^2 = 1 \\
&g_{111}^2 = 1, &&g_{121}^2 = (1,2)(3,4), \quad g_{122}^2 = (1,3)(2,4), \\
&g_2 = g_{21} = (1,2), &&g_{22} = (1,2,3,4), &&y_{21}^2 = (1,2) \\
&g_{211}^2 = (1,3), &&g_{221}^2 = (1,2,3,4), \\
&g_3 = g_{31} = (1,2,3), &&g_{31}^G \cap T_2 = \emptyset.
\end{aligned}
$$

We conclude that

$$
c_{(1,1)1}^1 = \frac{|\mathbf{C}_G(g_{11})|}{|\mathbf{C}_{T_2}(g_{111}^2)|} \tilde{\varphi}_2(g_{111}^2) = \frac{24}{8} \cdot 1 = 3,
$$

$$
c_{(1,1)2}^1 = \frac{|\mathbf{C}_G(g_{12})|}{|\mathbf{C}_{T_2}(g_{121}^2)|} \tilde{\varphi}_2(g_{121}^2) + \frac{|\mathbf{C}_G(g_{12})|}{|\mathbf{C}_{T_2}(g_{122}^2)|} \tilde{\varphi}_2(g_{122}^2) = \frac{8}{4}(-1) + \frac{8}{8} = -1,
$$

$$
c_{(2,1)1}^2 = \frac{|\mathbf{C}_G(g_{21})|}{|\mathbf{C}_{T_2}(g_{211}^2)|} \tilde{\varphi}_2(g_{211}^2) = \frac{4}{4} \cdot 1 = 1,
$$

$$
c_{(2,1)2}^2 = \frac{|\mathbf{C}_G(g_{22})|}{|\mathbf{C}_{T_2}(g_{221}^2)|} \tilde{\varphi}_2(g_{221}^2) = \frac{4}{4} \cdot (-1) = -1.
$$

Thus

$$
\mathrm{Fi}(C_1) = \begin{bmatrix} 1 & 1 \\ 3 & -1 \end{bmatrix}, \quad \mathrm{Fi}(C_2) = \begin{bmatrix} 1 & 1 \\ 1 & -1 \end{bmatrix}, \quad \mathrm{Fi}(C_3) = [1].
$$

To construct the character table of G we just have to compose these matrices with the character tables of the inertia factor groups S_3 and S_2 in the right way. The result is as follows:

S_4	g_{11}	g_{12}	g_{21}	g_{22}	g_{31}
χ_1	1	1	1	1	1
χ_2	1	1	-1	-1	1
χ_3	2	2	0	0	-1
χ_4	3	-1	1	-1	0
χ_5	3	-1	-1	1	0

with the two blocks of rows corresponding to the two inertia groups and the three blocks of columns corresponding to the Clifford matrices. In fact, the matrix X of characters values of G can be written as a product of two matrices, the first being the block diagonal matrix with the character tables of the inertia factor groups on the diagonal and the second factor being formed by the Clifford matrices and zeros, arranged suitably:

$$
X = \left[\begin{array}{ccc|cc}
1 & 1 & 1 & & \\
1 & 1 & -1 & & \\
2 & 0 & -1 & & \\
\hline
& & & 1 & 1 \\
& & & 1 & -1
\end{array}\right] \cdot \left[\begin{array}{cc|cc|c}
1 & 1 & \cdot & & \cdot & \cdot \\
\cdot & \cdot & 1 & 1 & \cdot \\
\cdot & \cdot & \cdot & \cdot & 1 \\
\hline
3 & -1 & \cdot & \cdot & \cdot \\
\cdot & \cdot & 1 & -1 & \cdot
\end{array}\right].
$$

This holds in general and shows that the Clifford matrices are unique, once the (projective) character tables of the inertia factors have been selected. ◆

Let us summarize.

Lemma 3.8.5 *With the above notation the character table of G can be written as a product of two matrices, the first one being a block diagonal matrix with (projective) character tables of the inertia factors $\overline{T_m}$ on the diagonal and the second factor being built up by the Clifford matrices and zeros, arranged suitably. In particular the Clifford matrices are uniquely determined up to the ordering of the columns and rows, provided the tables for the $\overline{T_m}$ have been selected.*

We leave the proof (which is essentially transforming equation (3.30) into a matrix equation) to the reader (Exercise 3.8.1). For the uniqueness observe that (projective) character tables are invertible, so the second factor in the asserted product decomposition is unique. Of course, a permutation of the columns of the Clifford matrices will result in a corresponding permutation of the columns of the character table of G. We will always assume that the (projective) character tables of the $\overline{T_m}$ have been chosen in advance, although in practice making the right choice might be difficult unless the Schur multipliers of the $\overline{T_m}$ are trivial or there are other reasons which guarantee that one may work with ordinary characters. Sometimes it is necessary to test all possible choices, and one is successful if all but one lead to a contradiction.

As mentioned above, the idea is not to compute Clifford matrices as in the above example, but to use properties of these matrices in order to determine them. The Clifford matrices satisfy orthogonality relations, which follow from the usual orthogonality relations for character tables. Keeping the notation as above, let

$$m_j^i = [\mathbf{N}_G(g_i N) : \mathbf{C}_G(g_{ij})] = |N| \frac{|\mathbf{C}_{\overline{G}}(\overline{g_i})|}{|\mathbf{C}_G(g_{ij})|} \in \mathbb{N}, \qquad (3.32)$$

$$c_j^i = |\mathbf{C}_G(g_{ij})|, \qquad (3.33)$$

$$b_{(m,l)}^i = |\mathbf{C}_{\overline{T_m}}(\overline{y_{il}^m})|; \qquad (3.34)$$

so, in particular we obtain

$$b_{(1,1)}^i = |\mathbf{C}_{\overline{G}}(\overline{g_i})| \qquad \text{and}$$
$$m_j^i \cdot c_j^i = |N|\, b_{(1,1)}^i \qquad \text{independent of } j.$$

We now can state the orthogonality relations that hold for Clifford matrices.

Theorem 3.8.6 (Orthogonality relations) *With the notation introduced above, the following statements hold.*

(a)

$$\sum_{\rho \in I_i} b_\rho^i c_{\rho j}^i \overline{c_{\rho k}^i} = \delta_{j,k} c_j^i.$$

(b) *For $\rho, \sigma \in I_i$ the following equation holds:*

$$\sum_j m^i_j c^i_{\rho j} \overline{c^i_{\sigma j}} = \delta_{\rho \sigma} \frac{b^i_{(1,1)}}{b^i_\rho} |N|.$$

(c) *The* Fi_i *are square matrices.*

Proof. (a) By Theorem 2.1.15 we have

$$\sum_{\chi \in \mathrm{Irr}(G)} \chi(g_{ij}) \overline{\chi(g_{ij'})} = c^j_i \cdot \delta_{j,j'}.$$

Furthermore

$$\sum_{\chi \in \mathrm{Irr}(G)} \chi(g_{ij}) \overline{\chi(g_{ij'})} = \sum_{m=1}^{t} \sum_{\psi \in \mathrm{Irr}(T_m | \varphi_m)} \psi^G(g_{ij}) \overline{\psi^G(g_{ij'})}$$

$$= \sum_{m=1}^{t} \sum_{\psi \in \mathrm{Irr}(T_m | \varphi_m)} \left(\sum_l c^i_{(m,l),j} \tilde{\psi}(y^m_{il} N) \right) \overline{\left(\sum_{l'} c^i_{(m,l'),j} \tilde{\psi}(y^m_{il'} N) \right)}$$

$$= \sum_{m=1}^{t} \sum_{l,l'} c^i_{(m,l),j} \overline{c^i_{(m,l),j'}} \sum_{\psi \in \mathrm{Irr}(T_m | \varphi_m)} \tilde{\psi}(y^m_{il} N) \overline{\tilde{\psi}(y^m_{il'} N)}$$

$$= \sum_{m=1}^{t} \sum_{l,l'} c^i_{(m,l),j} \overline{c^i_{(m,l),j'}} b^i_{(m,l)} \delta_{(i,l),(i,l')} = \sum_{m=1}^{t} \sum_{l} c^i_{(m,l),j} \overline{c^i_{(m,l),j'}} b^i_{(m,l)}$$

using the orthogonality relations for projective characters.

(c) Writing $B_i = \mathrm{diag}(b^i_\rho \mid \rho \in I_i)$ and $C_i = \mathrm{diag}(c^i_1, \ldots, c^i_{s_i})$ we can write the orthogonality relations proved in part (a) as a matrix equation:

$$\mathrm{Fi}_i^{\mathrm{T}} B_i \overline{\mathrm{Fi}_i} = C_i,$$

or

$$(\mathrm{Fi}_i^{\mathrm{T}} B_i) \cdot (\overline{\mathrm{Fi}_i} \, C_i^{-1}) = \mathbf{I}_{s_i}. \tag{3.35}$$

Hence it follows that the number $|I_i|$ of rows of Fi_i is greater than or equal to the number s_i of columns. On the other hand,

$$\sum_{i=1}^{r} s_i = |\mathrm{Irr}(G)|,$$

whereas

$$\sum_{i=1}^{r} |I_i| = \sum_{i=1}^{r} \sum_{m=1}^{t} |\{\alpha_m\text{-regular classes of } \overline{T_m} \text{ in } \overline{g_i}^G\}|$$

$$= \sum_{m=1}^{t} |\{\alpha_m\text{-regular classes of } \overline{T_m}\}|$$

$$= \sum_{m=1}^{t} |\{\alpha_m^{-1} \text{ projective irreducible characters of } \overline{T_m}\}|$$

$$= \sum_{m=1}^{t} |\operatorname{Irr}(T_m \mid \varphi_m)| = |\operatorname{Irr}(G)|.$$

Hence $s_i = |I_i|$.

(b) Since Fi_i is a square matrix it follows with the notation introduced above from (3.35) that

$$(\operatorname{Fi}_i C_i^{-1}) \cdot (\operatorname{Fi}_i^{\mathrm{T}} B_i) = \mathbf{I}_{s_i}.$$

$\square$

Corollary 3.8.7 $\sum_j m_j^i = |N|.$

Proof. This follows from Theorem 3.8.6(b) using $\rho = \sigma = (1,1)$, since $c_{(1,1),j}^i = 1$. $\square$

Theorem 3.8.8 *Let* $M_i = \pi^{-1} \mathbf{C}_{\overline{G}}(\overline{g_i}) = \mathbf{N}_G(g_i N)$. *Then, using a suitable choice of the projective extensions of the irreducible representations of N to the respective inertia subgroups and an appropriate ordering of the rows and columns, one has*

$$\operatorname{Fi}_{G,N}(\overline{g_i}^G) = \operatorname{Fi}_{M_i,N}(\overline{g_i}^{M_i}).$$

The choice of the projective extensions and the orderings of the rows and columns are specified in the proof.

Proof. Since $g_{ij} \in g_i N$, by our general assumption it is clear that $g_{ij}^G \cap g_i N$ is a conjugacy class of M_i, since any element conjugating $g_i u$ into $g_i u'$ for $u, u' \in N$ must be in M_i. This means that $g_{ij}^G \mapsto g_{ij}^G \cap g_i N$ gives a bijection of the set of G-conjugacy classes of $\pi^{-1}\overline{g_i}^G$ to the set of M_i-conjugacy classes of $\pi^{-1}\overline{g_i}^{M_i}$ and we have a natural bijection between the columns of the two Clifford matrices in question. In particular these Clifford matrices have the same size.

Concerning the rows, we observe that any G-orbit φ_m^G for $\varphi_m \in \operatorname{Irr}(N)$ splits into M_i-orbits $\varphi_{(m,\mu)}^{M_i}$ with $\varphi_{(m,\mu)} = {}^{x_\mu}\varphi_m$ for some representatives $x_\mu \in G$. Then the inertia subgroups in M_i are

$$T_{(m,\mu)} = T_{M_i}(\varphi_{(m,\mu)}) = \{g \in M_i \mid {}^g\varphi_{(m,\mu)} = \varphi_{(m,\mu)}\} = {}^{x_\mu}T_m \cap M_i.$$

We choose $^{x_\mu}\tilde{\varphi}_m|_{T_{(m,\mu)}}$ as a projective extension of $\varphi_{(m,\mu)}$. The corresponding factor set is $\alpha_{(m,\mu)}$ with

$$\alpha_{(m,\mu)}(g,h) = \alpha_m(g^{x_\mu}, h^{x_\mu}) \qquad \text{for} \qquad g, h \in T_{(m,\mu)}.$$

Since $g_i N \in \mathbf{Z}(\overline{M_i})$ we have $t(i,(m,\mu)) \leq 1$, and we may label the rows of $\mathrm{Fi}_{M_i,N}(\overline{g_i}^{M_i})$ by (m,μ) instead of $((m,\mu),1)$, if $g_i \in T_{(m,\mu)}$ is $\alpha_{(m,\mu)}$-regular.

If $g_i \in T_{(m,\mu)}$ then $g_i^{x_\mu} \in T_m$, hence $\overline{g_i^{x_\mu}} \in \overline{g_i}^G \cap \overline{T_m}$, so there is exactly one l such that $\overline{g_i^{x_\mu}} \in \overline{y_{il}^m}^{T_m}$. If $t \in T_m$ and $(\overline{g_i^{x_{\mu'}}})^{\bar t} = \overline{g_i^{x_\mu}}$ then $x_{\mu'} t x_\mu^{-1} \in M_i$ and $^{x_{\mu'} t x_\mu^{-1}}\varphi_{(m,\mu)} = {}^{x_{\mu'} t}\varphi_m = \varphi_{(m,\mu')}$ and hence $\mu = \mu'$. Also if $g \in G$ satisfies $\overline{g_i^g} = \overline{y_{il}^m}$ for some $l \in \{1,\ldots,t(i,m)\}$, then $^g\varphi_m = {}^{hx_\mu}\varphi_m$ for some $h \in M_i$, hence $g^{-1} h x_\mu = t \in T_m$. Thus $\overline{g_i^g} = \overline{g_i^{hx_\mu t}} \in \overline{g_i^{x_\mu}}^{T_m}$. Thus the map

$$\eta_m \colon (m,\mu) \mapsto (m,l) \qquad \text{if} \qquad \overline{g_i^{x_\mu}} \in \overline{y_{il}^m}^{T_m}$$

is bijective. Obviously, if $g^{x_\mu} \in T_m$ is α-regular then $g \in T_{(m,\mu)}$ is $\alpha_{(m,\mu)}$-regular, and on restricting η_m to those (m,μ) which are $\alpha_{(m,\mu)}$-regular we get a bijection of the labels for the rows of $\mathrm{Fi}_{M_i,N}(\overline{g_i}^{M_i})$ to those of $\mathrm{Fi}_{G,N}(\overline{g_i}^G)$. We may assume that the notation has been chosen so that $\eta_m = \mathrm{id}$, i.e. we will identify $(m,\mu) = (m,l)$ henceforth.

We now have to compare the corresponding entries in the Clifford matrices. Let $g_{ij\kappa}^{(m,l)}$ be representatives of the $T_{(m,l)}$-classes in $g_{ij}^{M_i}$. Then

$$(g_{ij\kappa}^{(m,l)})^{x_l} \in g_{ij}^G \cap T_m = \bigcup_{k=1}^{\cdot \, s(i,j,m)} (g_{ijk}^m)^{T_m}$$

and we get a bijective map

$$\eta \colon \{\kappa \mid g_{ij\kappa}^{(m,l)} \in g_i N\} \;\rightarrow\; \{k \mid g_{ijk}^m \in y_{il}^m N\}$$
$$g_{ij\kappa}^{(m,l)} \;\mapsto\; g_{ijk}^m \quad \text{with} \quad (g_{ij\kappa}^{(m,l)})^{x_l} \in (g_{ijk}^m)^{T_m},$$

as is easily verified. Then

$$\tilde{\varphi}_{(m,l)}(g_{ij\kappa}^{(m,l)}) = {}^{x_l}\tilde{\varphi}_m(g_{ij\kappa}^{(m,l)}) = \tilde{\varphi}_m((g_{ij\kappa}^{(m,l)})^{x_l}) = \tilde{\varphi}_m(g_{ijk}^m).$$

Observe that $\mathbf{C}_G(g_{ij}) \leq M_i$ and also $\mathbf{C}_G(g_{ij\kappa}^{(m,l)}) \leq M_i$ because $g_{ij} N = g_i N = g_{ij\kappa}^{(m,l)} N$. Hence $|\mathbf{C}_G(g_{ij})| = |\mathbf{C}_{M_i}(g_{ij})|$. Furthermore if $(g_{ij\kappa}^{(m,l)})^{x_l} \in (g_{ijk}^m)^{T_m}$ then

$$|\mathbf{C}_{T_{(m,l)}}(g_{ij\kappa}^{(m,l)})| = |\mathbf{C}_{{}^{x_l}T_m}(g_{ij\kappa}^{(m,l)}) \cap M_i| = |\mathbf{C}_{T_m}((g_{ij\kappa}^{(m,l)})^{x_i})| = |\mathbf{C}_{T_m}(g_{ijm}^m)|.$$

Thus the corresponding entries in the Clifford matrices agree. It also follows that the column weights m_i^j coincide and obviously also $b_{(1,1)}^i = |\mathbf{C}_{\overline{G}}(\overline{g_i})| = |\mathbf{C}_{M_i/N}(\overline{g_i})|$. From Theorem 3.8.6(b) it follows that the weights b_ρ^i agree as well. $\qquad\square$

In the following we will assume that the normal subgroup N is abelian.

Lemma 3.8.9 *Let N be abelian, let $g_i \in \mathbf{Z}(G)$ and let g_{ij} be as in* Remark 3.8.2. *Then we may choose projective extensions $\tilde{\varphi}_m \colon T_m \to \mathbb{C}$ of the $\varphi_m \in \mathrm{Irr}(N)$ such that*

(a) $\quad \mathrm{Fi}_i = \mathrm{Fi}_1 \quad$ *and* $\quad c^i_{\rho,j} = \sum_{g \in G/T_m} {}^g \varphi_m(g_{1,j});$

(b) $\quad c^i_{\rho,1} = b^i_{(1,1)}(b^i_\rho)^{-1} \quad (\rho \in I_i);$

(c) $\quad \sum_{\rho \in I_i} c^i_{\rho,j} = 0 \quad$ *for* $\quad 2 \leq j \leq s_i.$

Proof. We choose a section $\tau \colon \bar{G} \to G$ with $\tau(g_i) = \overline{g_i}$ and extend each $\varphi_m \in \mathrm{Irr}(N)$ to a projective character $\tilde{\varphi}_m \colon T_m \to \mathbb{C}$ as in Example 3.7.12. By assumption, $g_i \in T_m$ for all m and we have

$$\tilde{\varphi}_m(g_{ij}) = \varphi_m(g_i^{-1} g_{ij}) = \varphi_m(g_{1j}).$$

Since $t(i,m) = 1$ and $\overline{y_{i1}^m} = \overline{g_i}$ we get from (3.30) that $\psi^G(g_{ij}) = c^i_{(m,1),j} \hat{\psi}(\overline{g_i})$. On the other hand,

$$\psi^G(g_{ij}) = \frac{1}{|T_m|} \sum_{g \in G} {}^g \psi(g_{ij}) = \Big(\frac{1}{|T_m|} \sum_{g \in G} {}^g \tilde{\varphi}_m(g_{ij}) \Big) \hat{\psi}(\overline{g_i}) = \frac{|N|}{|T_m|} \varphi^G_m(g_{1j}) \hat{\psi}(\overline{g_i}).$$

Now (a) follows from Corollary 3.2.18. For $j = 1$ we obtain $c^i_{\rho,1} = |G||T_m|^{-1} = b^i_{(1,1)}(b^i_\rho)^{-1}$. Finally, (c) follows from (b) and Theorem 3.8.6(a) applied to $k = 1$. $\qquad \square$

As in Theorem 3.8.8 we put $M_i = \mathbf{N}_G(g_i N)$, thus $M_i/N = \mathbf{C}_{\overline{G}}(\overline{g_i})$. Furthermore we define

$$N_i = N_{g_i} = [g_i, N] = \langle [g_i, u] \mid u \in N \rangle.$$

Lemma 3.8.10 *With the above notations we have for abelian $N \trianglelefteq G$ the following.*

(a) *For any $g \in G$ we have a homomorphism $\kappa_g \colon N \to [g, N]$* $\quad u \mapsto [g, u]$.

(b) *For any $g \in g_i N$ one has $N_i = [g, N] = \{[g, u] \mid u \in N\}$.*

(c) *$N_i \trianglelefteq M_i$ and $N_i \leq N$.*

(d) *If $\varphi \in \mathrm{Irr}(N)$ then $N_i \leq \ker(\varphi)$ or $T_G(\varphi) \cap g_i N = \emptyset$.*

Proof. (a) $[g, u][g, u'] = (u^{-1})^g u (u'^{-1})^g u' = (u^{-1} u'^{-1})^g u u' = [g, uu']$.

(b) If $u \in N$ and $g = g_i u \in g_i N$, then for $v \in N$

$$[g, v] = [g_i u, v] = [g_i, v]^u [u, v] = [g_i, v]$$

because N is abelian and $[g_i, v] \in N$ since $N \trianglelefteq G$.

(b) Obviously $N_i \leq N$. Furthermore, for $y \in M_i$ and $u \in N$ we have (because N is abelian) $[g_i, u]^y = [g_i^y, u^y] = [g_i, u^y] \in N_i$.

(c) g_i is in $T_G(\varphi)$ if and only if $\varphi((u^{-1})^{g_i}) = \varphi(u^{-1})$ for all $u \in N$ hence if and only if $N_i \leq \ker(\varphi)$. Since $N \leq T_G(\varphi)$ the result follows. $\qquad \square$

Corollary 3.8.11 *With the above notation one has*

$$\mathrm{Fi}_{G,N}(\overline{g_i}^G) = \mathrm{Fi}_{\widetilde{M_i/N_i}}(\widetilde{g_iN_i}^{\widetilde{M_i/N_i}}),$$

with $\widetilde{M_i/N_i} = (M_i/N_i)/(N/N_i)$ *and* $M_i/N_i \to \widetilde{M_i/N_i}$, $x \mapsto \tilde{x}$ *denoting the canonical epimorphism.*

Proof. By Theorem 3.8.8 we may assume that $G = M_i$, so that $N_i \trianglelefteq G$. The result then follows by applying the natural isomorphism $G/N \to (G/N_i)/(N/N_i)$. Observe that by Lemma 3.8.10(c) $\ker \varphi_m \leq N_i$ unless $t(i,m) = 0$ (in which case Fi_i contains no row corresponding to the inertia group T_m). $\qquad\square$

Definition 3.8.12 $g_iN \in \overline{G}$ *is called a* **split coset** *if there is a* $g_i' \in g_iN$ *such that* $M_i = N \cdot \mathbf{C}_G(g_i')$.

Remark 3.8.13 (a) If g_iN is a split coset we will always assume that $g_i = g_{i,1}$ has been chosen such that $M_i = N \cdot \mathbf{C}_G(g_i)$.

(b) If G is a split extension of N, i.e. if there is a complement $H \leq G$ of N in G, then every coset of N in G is a split coset.

Proof. (b) Obviously $M_i = N \cdot (M_i \cap H)$. If we choose $g_i = g_i' \in H$ then $\mathbf{C}_G(g_i) \subseteq M_i \cap H$, because $g_i^{M_i} \cap H = \{g_i\}$. $\qquad\square$

Corollary 3.8.14 *If N is abelian and g_iN is a split coset in G, then the rows of $\mathrm{Fi}_{G,N}(\overline{g_i}^G)$ are M_i-orbit sums of characters of N/N_i, where $M_i = \mathbf{N}_G(g_iN)$. We have, using the convention of Remark 3.8.13(a),*

(a) $\mathrm{Fi}_{G,N}(\overline{g_i}^G) = \mathrm{Fi}_{M_i/N_i,N/N_i}(\{1\})$;

(b) $c_{\rho,1}^i = b_{(1,1)}^i (b_\rho^i)^{-1}$ $\quad (\rho \in I_i)$;

(c) $|c_{\rho,j}^i| \leq c_{\rho,1}^i$ $\quad (1 \leq j \leq s_i, \ \rho \in I_i)$;

(d) *if $c_{\rho_1,1}^i = 1$ for some $\rho_1 \in I_i$ then for every $\rho \in I_i$ there is a $\rho' \in I_i$ with*

$$c_{\rho',j}^i = c_{\rho,j}^i \cdot c_{\rho_1,j}^i \qquad for \qquad 1 \leq j \leq s_i;$$

(e) $\sum_{\rho \in I_i} c_{\rho,j}^i = 0$ $\qquad for \qquad 2 \leq j \leq s_i$;

(f) *if N is an elementary abelian p-group $c_{\rho,j}^i \in \mathbb{Z}[\zeta_p]$ for all j.*

Proof. By Theorem 3.8.8 and Corollary 3.8.11 we may assume that $G = M_i$ and $N_i = \{1\}$. This means that $\overline{g_i} \in \mathbf{Z}(\overline{G})$ and g_i acts trivially on N. Since g_iN is a split coset, we have $g_i \in \mathbf{Z}(G)$. Then the result follows from Lemma 3.8.9. For (d) observe that the assumption means that $[c_{\rho_1,j}^i]_{1 \leq j \leq s_i}$ is a linear character of N invariant under M_i. $\qquad\square$

The method of Clifford matrices works particularly smoothly for abelian N and split extensions (i.e. if N has a complement in G). Then by Lemma 3.7.4 one has to deal only with ordinary characters instead of projective ones. Also by Corollary 3.8.14 the first column of all Fi_i is known.

Example 3.8.15 Let us consider a split extension $G := C_2^4 \rtimes A_5$ of $N = C_2^4$ by $H = A_5$ with non-trivial action of H on N. In fact, it can be shown that any such extension will be split (see e.g. [85]), so our assumption is not restrictive. Since A_5 has no non-trivial representation of degree less than four over $\mathbb{F}_2$, the action of H on N is irreducible and it follows immediately that

H has one orbit (of length 15) on $N\setminus\{1\}$ or

H has two orbits (of lengths five and ten) on $N\setminus\{1\}$.

Let us consider the second case, the first one being almost trivial. It follows that H has three orbits on $\mathrm{Irr}(N)$ with stabilizers $\overline{T_1} = H = A_5$, $\overline{T_2} = A_4$, $\overline{T_3} = S_3$. The first point is, to write down the character tables of these groups and determine the fusion maps of T_i in H (see Table 3.8).

Table 3.8. Character table of $\overline{T_1}$, $\overline{T_2}$ and $\overline{T_3}$

		g_1 1a	g_2 2a	g_3 3a	g_4 5a	g_5 5b	representative
$\overline{T_1}$		60	4	3	5	5	centralizer order
$\hat{\psi}_1^{(1)}$		1	1	1	1	1	
$\hat{\psi}_2^{(1)}$		3	-1	0	A	A^*	$A = \frac{1}{2}(1 + \sqrt{5})$
$\hat{\psi}_3^{(1)}$		3	-1	0	A^*	A	$A^* = \frac{1}{2}(1 - \sqrt{5})$
$\hat{\psi}_4^{(1)}$		4	0	1	-1	-1	
$\hat{\psi}_5^{(1)}$		5	1	-1	0	0	

$\overline{T_2}$	y_{11}^2 1a	y_{21}^2 2a	y_{31}^2 3a	y_{32}^2 3b	representative		$\overline{T_3}$	y_{11}^3 1a	y_{21}^3 2a	y_{31}^3 3a
	12	4	3	3	centralizer order			6	2	3
$\hat{\psi}_1^{(2)}$	1	1	1	1			$\hat{\psi}_1^{(3)}$	1	1	1
$\hat{\psi}_2^{(2)}$	1	1	B	$\overline{B}$	$B = \zeta_3$		$\hat{\psi}_2^{(3)}$	1	-1	1
$\hat{\psi}_3^{(2)}$	1	1	$\overline{B}$	B			$\hat{\psi}_3^{(3)}$	2	0	-1
$\hat{\psi}_4^{(2)}$	3	-1	0	0						

In the tables given in Table 3.8, $g_i N = y_{i1}^1 N$ and $y_{il}^m N$ have been replaced by g_i and y_{il}^m, respectively. The fusion maps are given by the names of the representatives: remember $y_{il}^m N$ of $\overline{T_m}$ always fuses to $g_i N$ of H. Thus one has five Clifford matrices of sizes 3, 3, 4, 1 and 1. The b_ρ^i can be copied from the character tables of the $\overline{T_m}$ and the first columns are given by Corollary 3.8.14(b). By (e), (c) and (f) of the same corollary the second or third column of Fi$_1$ is of the form $[1, x, -1 - x]^T$ with $x \in \mathbb{Z}$ and $-5 \leq x \leq 5$. The corresponding centralizer order is $c_j^i = 60 + 12x^2 + 6(-1-x)^2$ and must be a divisor of $|G| = 960$. Hence $x \in \{-3, 1\}$. The values for Fi$_2$ and Fi$_3$ are obtained even more easily. The result is listed below, with the first column always giving the weights b_ρ^i

and the first row the centralizer orders c_j^i:

c_j^1	960	192	96
	g_{11}	g_{12}	g_{13}
60	1	1	1
12	5	−3	1
6	10	2	−2

c_j^2	16	16	8
	g_{21}	g_{22}	g_{23}
4	1	1	1
4	1	1	−1
2	2	−2	0

c_j^3	12	12	12	12
	g_{31}	g_{32}	g_{33}	g_{34}
3	1	1	1	1
3	1	1	−1	−1
3	1	−1	1	−1
3	1	−1	−1	1

The character table (Table 3.9) of $G = C_2^4 \rtimes A_5$ is now obtained by (3.30), where $C = \sqrt{-3}$ and $A = \frac{1}{2}(1 + \sqrt{5})$ as before.

Table 3.9. Character table of $C_2^4 \rtimes A_5$

c_j^i	960	192	96	16	16	8	12	12	12	12	5	5
	g_{11}	g_{12}	g_{13}	g_{21}	g_{22}	g_{23}	g_{31}	g_{32}	g_{33}	g_{34}	g_{41}	g_{51}
	1a	2a	2b	2c	4a	4b	3a	6a	6b	6c	5a	5b
χ_1	1	1	1	1	1	1	1	1	1	1	1	1
χ_2	3	3	3	−1	−1	−1	0	0	0	0	A	A^*
χ_3	3	3	3	−1	−1	−1	0	0	0	0	A^*	A
χ_4	4	4	4	0	0	0	1	1	1	1	−1	−1
χ_5	5	5	5	1	1	1	−1	−1	−1	−1	0	0
χ_6	5	−3	1	1	1	−1	2	0	0	−2	0	0
χ_7	5	−3	1	1	1	−1	−1	C	−C	1	0	0
χ_8	5	−3	1	1	1	−1	−1	−C	C	1	0	0
χ_9	15	3	3	−1	−1	1	0	0	0	0	0	0
χ_{10}	10	2	−2	2	−2	0	1	−1	−1	1	0	0
χ_{11}	10	2	−2	−2	2	0	1	−1	−1	1	0	0
χ_{12}	20	4	−4	0	0	0	−1	1	1	−1	0	0

We have already included class names and thus given the orders of the representatives of the conjugacy classes. Since N is elementary abelian we know that $o(g_{1j}) = 2$ for $j = 2, 3$. Hence $\chi_6^{[1^2]} \in \{\chi_{10}, \chi_{11}\}$ and we see that one of g_{21} or g_{22} has order two. We may arrange the classes so that $o(g_{21}) = 2$ (and hence $\chi_6^{[1^2]} = \chi_{10}$). This determines the second power map and the element orders.

♦

In the above example it was very easy to see that a character $(\chi_6^{[1^2]})$ was induced from a character of one of the inertia subgroups, namely $(\tilde{T}_3)$. In general it is very useful that one can decompose a (reducible) character into a sum of characters which are induced up from inertia subgroups, by forming scalar products of a character χ of G with the rows $c_{(m,l)}^i$ of the Clifford matrices F_i. These are defined by

$$(\chi, c_{(m,l)}^i) := \frac{b_{(m,l)}^i}{b_{(1,1)}^i |N|} \sum_j m_j^i \chi(g_{ij}) \overline{c_{(m,l),j}^i}. \tag{3.36}$$

Lemma 3.8.16 *Let* $\chi = \sum_{m=1}^{t} (\chi^{(m)})^G$ *be a generalized character of* G *with*

$$\chi^{(m)} = \sum_k a_{\chi,k}^{(m)} \psi_k^{(m)}, \qquad \psi_k^{(m)} \in \mathrm{Irr}(\ T_m \mid \varphi_m\ ), \qquad a_{\chi,k}^{(m)} \in \mathbb{Z}$$

then

$$(\chi, c_{(m,l)}^i) = \sum_k a_{\chi,k}^{(m)} \hat{\psi}_k^{(m)} (\overline{y_{il}^m}),$$

where $\psi_k^{(m)} = \tilde{\varphi}_m \cdot \hat{\psi}_k^{(m)}$ *with the notation of Theorem 3.8.1.*

Proof. By (3.30) and the orthogonality relations (Theorem 3.8.6) one gets

$$(\chi, c_{(m,l)}^i) = \frac{b_{(m,l)}^i}{b_{(1,1)}^i |N|} \sum_j m_j^i (\sum_{m'=1}^{t} \sum_k (a_{\chi,k}^{(m')} \psi_k^{(m')})^G (g_{ij})) \overline{c_{(m,l),j}^i}$$

$$= \frac{b_{(m,l)}^i}{b_{(1,1)}^i |N|} \sum_j m_j^i \sum_{m'=1}^{t} \sum_k a_{\chi,k}^{(m')} (\sum_{l'} c_{(m',l'),j}^i \hat{\psi}_k^{(m')} (\overline{y_{il'}^{m'}})) \overline{c_{(m,l),j}^i}$$

$$= \sum_k a_{\chi,k}^{(m)} \hat{\psi}_k^{(m)} (\overline{y_{il}^m}). \qquad \square$$

Observe that the $\hat{\psi}_k^{(m)} (\overline{y_{il}^m})$ are known. Thus, if all the Clifford matrices $c_{(m,l)}^i$ are given then by computing the scalar products (3.36) one can decompose a (generalized) character χ of G into its constituents belonging to the various inertia groups. What is more, Lemma 3.8.16 can also be used if only some of the $c_{(m,l)}^i$ are known in order to obtain linear equations for the integers $a_{\chi,k}^{(m)}$. In some cases (for instance, if one knows that χ is of small norm) these equations can be sufficient to determine the $a_{\chi,k}^{(m)}$. On the other hand, if all $a_{\chi,k}^{(m)}$ are known then the equation

$$\chi(g_{ij}) = \sum_m \sum_l c_{(m,l),j}^i \sum_k a_{\chi,k}^{(m)} \hat{\psi}_k^{(m)} (\overline{y_{il}^m}) \tag{3.37}$$

obtained from (3.30) gives a linear equation for the $c_{(m,l),j}^i$.

Example 3.8.17 We end this section with a somewhat larger and more realistic, although very simple, example. For many applications it is useful to have not only the character table of a group, but also the character tables of the maximal subgroups. This is particularly important for the sporadic simple groups. In fact, at the time of writing, the character tables of the maximal subgroups of all sporadic simple groups are known except for some of the maximals of the two largest sporadic simple groups, the "Baby Monster" B and the "Monster" M. In the example given here, we consider the case of the sporadic simple Conway group Co_2, which has a maximal subgroup $G = 2_+^{1+8} : \mathrm{S}_6(2)$ (see the ATLAS [38]). Note that G is the centralizer of an involution (in class 2a) in Co_2. We put $N = 2_+^{1+8}$, the extraspecial group of order 2^9 with automorphism group isomorphic to $\mathrm{O}_8^+(2)$ (see [88], p. 357), and $\tilde{N} = N/\mathbf{Z}(N)$, which can be considered as an eight-dimensional $\mathbb{F}_2 S_6(2)$-module. Also, $\tilde{N}$ affords the unique irreducible representation of $S_6(2)$ of degree eight over $\mathbb{F}_2$ (up to isomorphism [98]).

The table of marks of $\bar{G} := S_6(2)$ is known and available in the GAP library of table of marks. Using the table of marks one observes that there are exactly nine conjugacy classes of subgroups of $\bar{G}$ of index less than 2^8. Below, we display the corresponding permutation characters on the first 14 conjugacy classes of $\bar{G}$. Actually one would get the same permutation characters quickly from the character table of $S_6(2)$ by searching for possible non-trivial transitive permutation characters of degree less than 2^8 as in Section 3.4.

$S_6(2)$	1a	2a	2b	2c	2d	3a	3b	3c	4a	4b	4c	4d	4e	5a
θ_1	28	16	4	8	4	10	1	1	4	2	6	.	2	3
θ_2	36	16	12	8	4	6	.	3	.	6	2	4	2	1
θ_3	56	.	8	16	.	20	2	2	8	.	.	.	4	6
θ_4	63	31	15	15	7	15	.	3	3	7	7	3	3	3
θ_5	72	.	24	16	.	12	.	6	.	.	.	8	4	2
θ_6	120	.	24	.	8	.	3	6	12	.	.	4	.	.
θ_7	126	2	6	26	6	30	.	6	6	12	.	2	4	6
θ_8	135	15	39	15	7	.	.	9	3	3	3	11	3	.
θ_9	240	.	48	.	.	.	6	12	8	.	.	8	.	.

The permutation character θ of the action of $\bar{G}$ on the non-trivial elements of $\tilde{N}$ (and also on the non-trivial linear characters of N) must be a sum of some of these permutation characters, and for any $\bar{g} \in \bar{G}$ we must have

$$\theta(\bar{g}) = 2^k - 1 \quad \text{for some} \quad 0 \le k \le 7,$$

since we have a linear action. There are just two such characters, namely

$$\theta_1 + \theta_2 + \theta_3 + \theta_4 + \theta_5 \quad \text{and} \quad \theta_6 + \theta_8.$$

But as shown in Section 4.2 (Exercise 4.2.4), we can see from [98] that $\theta(5a) = 0$ and hence $\theta = \theta_6 + \theta_8$. Thus $\bar{G}$ has three orbits on $\tilde{N}$ and the same number of orbits on the linear characters of N with stabilizers $\bar{T}_1, \bar{T}_2, \bar{T}_3$ with indices 1, 135 and 120. Looking at the ATLAS ([38]) we see that $\bar{T}_2 \cong 2^6.L_3(2)$ and $\bar{T}_3 \cong U_3(3).2$. For an alternative way of finding these stabilizers, see Exercise 3.8.2. Since N has a unique faithful irreducible character φ_4 (of degree 16) there are in total four orbits $\{1_N\}, \varphi_2^G, \varphi_3^G, \{\varphi_4\}$ of irreducible characters of N under the action of $\bar{G}$, with stabilizers

$$\begin{aligned}
\bar{T}_1 &= S_6(2), & [S_6(2) : \bar{T}_1] &= 1, \\
\bar{T}_2 &\cong 2^6.L_3(2), & [S_6(2) : \bar{T}_2] &= 135, \\
\bar{T}_3 &\cong U_3(3).2, & [S_6(2) : \bar{T}_3] &= 120, \\
\bar{T}_4 &= S_6(2), & [S_6(2) : \bar{T}_4] &= 1.
\end{aligned}$$

Thus in our notation $s_1 = 4$ and the elements $1 = g_{1,1}, z = g_{1,2}, g_{1,3}, g_{1,4}$, with z being the central involution of N, are representatives of the conjugacy classes of G contained in N. These have lengths

$$1, \ 1, \ 2 \cdot 135, \ 2 \cdot 120.$$

We know that N contains $2^8 - 2^4 = 240$ elements of order four; these must be in $g_{1,4}^G$ and so $g_{1,3}$ is an involution. Since $|\mathbf{C}_G(g_{1,3})| = 2^{17} \cdot 3 \cdot 7$ and $|\mathbf{C}_G(g_{1,4})| = 2^{14} \cdot 3^3 \cdot 7$ we see that $g_{1,3}$ is in class $\mathbf{2a}$ or $\mathbf{2b}$ of Co_2 and $g_{1,4} \in \mathbf{4a}$, because this is the only class of elements of order four in Co_2 with centralizer order divisible by seven. So we have two possibilities for the fusion of N into Co_2. Using both possibilities and computing the scalar product of the restriction of the second irreducible character of Co_2 with $\mathbf{1}_N$, we get $-1/2$ and 7, respectively. It follows that $g_{1,3} \in \mathbf{2b}$ and the fusion of N into Co_2 is determined.

It is known that Co_2 has a unique (candidate for a) transitive permutation character of degree $[\mathrm{Co}_2 : G] = 56925$, which in GAP can be found by

```
gap>  PermChars( CharacterTable("Co2"), 56925 );;
```

Using this permutation character we can compute the scalar products of the restrictions of characters of Co_2 to G; see Exercise 3.4.1. We will write $\mathrm{Irr}(\mathrm{Co}_2) = \{\chi_1, \ldots, \chi_{60}\}$ using the ordering of the ATLAS to get

$$[(\chi_i|_G , \chi_j|_G)_G]_{1 \leq i,j \leq 4} = \begin{bmatrix} 1 & . & . & 1 \\ . & 2 & . & . \\ . & . & 3 & 1 \\ 1 & . & 1 & 4 \end{bmatrix}. \tag{3.38}$$

We will write

$$\chi_n|_G = \sum_{m=1}^{4} (\chi_n^{(m)})^G \qquad \text{for} \quad \chi_n \in \mathrm{Irr}(\mathrm{Co}_2) \tag{3.39}$$

with $\chi_n^{(m)}$ being a character of T_m or zero. Since we know the fusion of $N = 2_+^{1+8}$ and $\mathbf{Z}(N)$ into Co_2 we can easily derive $\chi_n^{(1)}(1)$ and $\chi_n^{(4)}(1)$:

$$a_n := \chi_n^{(1)}(1) = (\chi_n|_N, \mathbf{1}_N)_N,$$

$$b_n := \chi_n^{(7)}(1) = (\chi_n|_{\mathbf{Z}(N)}, -\mathbf{1}_{\mathbf{Z}(N)})_{\mathbf{Z}(N)} = \frac{1}{2}(\chi_n(1) - \chi_n(2a)).$$

Here $-\mathbf{1}_{\mathbf{Z}(N)}$ is the non-trivial linear character of $\mathbf{Z}(N)$. We list the relevant values for $n = 2, 3, 4$ in Table 3.10.

From the first line of Table 3.10 we see that $\eta := \chi_2^{(4)} \in \mathrm{Irr}(G)$ is an *ordinary* character which extends φ_4, the unique faithful irreducible character of N. Furthermore $\eta^2|_N = \varphi_4^2$ is the inflation of the regular character of $N/\mathbf{Z}(N) \cong 2^8$ and hence the sum of the orbits of $\varphi_1, \ldots, \varphi_3$ under G. From Clifford correspondence (Corollary 3.6.8) we conclude that there are uniquely determined linear characters $\tilde{\varphi}_m \in \mathrm{Irr}(T_m)$ with

$$\eta^2 = \sum_{m=1}^{3} \tilde{\varphi}_m^G \qquad \text{and} \qquad \tilde{\varphi}_m|_N = \varphi_m, \quad (1 \leq m \leq 3). \tag{3.40}$$

Table 3.10.

n	$\chi_n(1)$	$(\chi_n|_G, \chi_n|_G)_G$	a_n	b_n	$\chi_n(1) - a_n - b_n$
2	23	2	7	2^4	0
3	253	3	21	$7 \cdot 2^4$	120
4	275	4	28	$7 \cdot 2^4$	135

Thus all the φ_m extend to *ordinary* characters of their respective inertia subgroups T_m and we do not have to consider projective characters of $\bar{T}_m$. We will use the extensions $\tilde{\varphi}_m$ defined by (3.40) for the definition of the Clifford matrices. Thus

$$\mathrm{Irr}(T_m \mid \varphi_m) = \{\psi_j^{(m)} := \tilde{\varphi}_m \cdot \hat{\psi}_j^{(m)} \mid \hat{\psi}_j^{(m)} \in Irr(\bar{T}_m)\} \qquad \text{for} \qquad m = 1, \ldots, 4.$$

In particular, $\psi_j := \psi_j^{(1)}$ is the inflation of the jth irreducible character of $S_6(2)$ to G and $\psi_j^{(4)} = \eta \cdot \psi_j$.

The above table also gives us the decomposition of $\chi_n|_G$ for $n = 2, 3, 4$:

$$
\begin{aligned}
\chi_2|_G &= \psi_2 & & & + & \eta, \\
\chi_3|_G &= \psi_i & & + \ (\psi_j^{(3)})^G & + & \psi_2^{(4)}, \\
\chi_4|_G &= 1_G + \psi_6 & + \ (\psi_1^{(2)})^G & & + & \psi_2^{(4)},
\end{aligned}
$$

with $i \in \{4, 5\}$ and $j \in \{1, 2\}$, because ψ_4, ψ_5 are the irreducible characters of degree 21 of $\bar{T}_1$ and $\psi_1^{(3)}, \psi_2^{(3)}$ are the linear characters of $\bar{T}_3$. Also ψ_2 is the unique character in $\mathrm{Irr}(\bar{T}_1)$ of degree seven. Observe that we have used the information from (3.38). From equation (3.31) we the first rows $\mathbf{r}_k$ of Fi_1, namely

$$\mathbf{r}_4 = \chi_2' - 7 \cdot \mathbf{r}_1, \quad \mathbf{r}_3 = \chi_3' - 21 \cdot \mathbf{r}_1 - 7 \cdot \mathbf{r}_4, \quad \mathbf{r}_2 = \chi_4' - 21 \cdot \mathbf{r}_1 - 7 \cdot \mathbf{r}_4,$$

where $\chi_n' := [\chi_n(1), \chi_n(2a), \chi_n(2b), \chi_n(4a)]$. Hence we have the first Clifford matrix

g	$g_{1,1}$	$g_{1,2}$	$g_{1,3}$	$g_{1,4}$
$|g^G|$	1	1	270	240
in Co_2	1a	2a	2b	4a
$(1,1)$	1	1	1	1
$(2,1)$	135	135	7	-9
$(3,1)$	120	120	-8	8
$(4,1)$	16	-16	0	0

Instead of the $c_j^1 = |\mathbf{C}_G(g_{1,j})|$ (which we need for the orthogonality relations) we have listed the class lengths $|g_{1,j}^G|$, because they are smaller. The third line contains the names of the conjugacy classes of B which contain $g_{1,j}$. Also we have omitted the $b_{m,1}^1 = |\bar{T}_m|$, which can be obtained by dividing $|S_6(2)|$ by the first entry in row m, except for $m = 4$, where it is, of course, equal to $|S_6(2)|$.

The character tables of the inertia factor groups $\bar{T}_i$ are all contained in the GAP library of character tables together with their fusions into $\bar{G} = S_6(2)$. They can be accessed by

```
gap>  Maxes( CharacterTable( "S6(2)" ) );
[ "U4(2).2", "A8.2", "2^5:S6", "U3(3).2", "2^6:L3(2)", "2.[2^6]:(S3xS3)",
  "S3xS6", "L2(8).3" ]
```

We find $[\,|\operatorname{Irr}(\bar{T}_m)|\,]_{1 \le m \le 4} = [30, 24, 16, 30]$, hence $|\operatorname{Irr}(G)| = 100$.

From the character tables of the inertia factor groups $\bar{T}_i$ and the corresponding fusions we immediately obtain all the b_ρ^i (see (3.34)) and in particular the sizes of all the Clifford matrices Fi_i for the 30 conjugacy classes $\bar{g}_i{}^G$. The result is the following list, which gives for each conjugacy class $\bar{g}_i{}^{S_6(2)}$ the number s_i of conjugacy classes g_{ij}^G of $G = 2^{1+8} S_6(2)$ into which $\pi^{-1}(\bar{g}_i{}^{S_6(2)})$ splits:

1a	4	2a	3	2b	5	2c	4	2d	5	3a	2	3b	3	3c	4
4a	5	4b	3	4c	3	4d	6	4e	4	5a	2	6a	2	6b	2
6c	3	6d	2	6e	3	6f	3	6g	4	7a	5	8a	4	8b	4
9a	2	10a	2	12a	2	12b	2	12c	5	15a	2				

Thus G has 100 conjugacy classes and one can see that the task to compute the character table, a 100×100-matrix, has been reduced to the task to compute 30 square matrices of sizes ranging from two to six, most of which can be written down immediately.

Omitting the last row and the first column of the Clifford matrices Fi_i of G, we obtain the corresponding Clifford matrices $\overline{\mathrm{Fi}}_i$ for $2^8{:}S_6(2)$. Since this is a split extension, we know from Remark 3.8.13 that every coset $g_i N$ is a split coset in G. From Corollary 3.8.14 we get

$$c_{(m,l),1}^i = c_{(m,l),2}^i = \frac{b_{(1,1)}^i}{b_{(m,l)}^i} \qquad 1 \le m \le 3, \ 1 \le l \le t(i,m),$$

where we have chosen $g_{i,2} = z\, g_{i,1}$. From the orthogonality relations we see that the last row of Fi_i has the form $[q_i, -q_i, 0, \dots, 0]$, where

$$q_i = |\,\mathbf{C}_G(g_i)|\,|\,\mathbf{C}_{\bar{G}}(\bar{g}_i)|^{-1},$$

a power of two. We thus know the first two columns and the first and last row of all Clifford matrices. Also we conclude from Corollary 3.8.14(f) that the entries in all Clifford matrices are rational integers. Thus

$$\mathrm{Fi}_i = \begin{bmatrix} 1 & 1 \\ 1 & -1 \end{bmatrix} \quad \text{for} \quad i \in \{6, 14, 15, 16, 18, 25, 26, 27, 28, 30\}$$

and

$$\mathrm{Fi}_2 = \begin{bmatrix} 1 & 1 & 1 \\ 15 & 15 & -1 \\ 4 & -4 & 0 \end{bmatrix}, \ \mathrm{Fi}_i = \begin{bmatrix} 1 & 1 & 1 \\ 3 & 3 & -1 \\ 2 & -2 & 0 \end{bmatrix} \quad \text{for} \quad i \in \{7, 10, 11, 17, 19, 20\}$$

From Corollary 3.8.14(c) and (e) we see that Clifford matrices with first column $[1, 1, 2, 2]^T$ or $[1, 1, 1, 1, 2]^T$ are uniquely determined by the orthogonality relations, and we obtain

$$\mathrm{Fi}_i = \begin{bmatrix} 1 & 1 & 1 & 1 \\ 1 & 1 & 1 & -1 \\ 2 & 2 & -2 & 0 \\ 2 & -2 & 0 & 0 \end{bmatrix}, \ \mathrm{Fi}_j = \begin{bmatrix} 1 & 1 & 1 & 1 & 1 \\ 1 & 1 & 1 & -1 & -1 \\ 1 & 1 & -1 & 1 & -1 \\ 1 & 1 & -1 & -1 & 1 \\ 2 & -2 & 0 & 0 & 0 \end{bmatrix}$$

for $i \in \{13, 21, 23, 24\}$ and $j \in \{22, 29\}$.

At this point we have computed all Clifford matrices Fi_i except those for $i \in \{3, 4, 5, 8, 9, 12\}$, for which we know only the first two columns. This means that all but 17 columns of the character table of G are known. Since we know the decomposition of $\chi_2|_G$ and $\chi_4|_G$ we can evaluate these characters on all but 17 conjugacy classes and can determine the fusion of 74 conjugacy classes into Co_2.

To compute the remaining Clifford matrices, we investigate the decomposition of the restrictions $\chi_n|_G$. Since we know Fi_1 and the fusion of the conjugacy classes $g_{1,j}^G$ into Co_2, we can use Lemma 3.8.16 to find $d_n^m := \chi_n^{(m)}(1)$ for $n = 2, \dots, 6$ and $m = 1, \dots, 4$:

n	d_n^1	d_n^2	d_n^3	d_n^4	$\|\|\chi_n\|_G\|\|$
2	7	0	0	1	2
3	21	0	1	7	3
4	28	1	0	7	4
5	35	0	7	56	4
6	63	7	1	56	7

We use equations (3.37) for χ_n and for the 74 indices i, j for which we know the fusion of $g_{i,j}$ into Co_2 and the jth column of Fi_i restricting the (m, k) to those for which $\hat{\psi}_k^{(m)}(1) \le d_n^m$. For each of the above n we obtain a system of linear equations in the unknowns $a_{\chi_n, k}^{(m)}$ which has a unique solution.

Having determined the $a_{\chi_n, k}^{(m)}$ for $n = 2, \dots, 6$ (and all m, k) we consider equations (3.37) for χ_n with $n = 2, \dots, 6$ as a system of linear equations for the unknown elements $c_{[m,l],j}^i$ in the jth column of Fi_i, $i \in \{3, 4, 5, 8, 9, 12\}$. There is a problem though. Since we do not know the fusion of the g_{ij} into Co_2 yet we cannot evaluate the left hand side of (3.37). But there are restrictions for the fusion. If $\overline{g_i}$ has order o, then the order of g_{ij} must be in $\{o, 2\,o, 4\,o\}$. Also we can eliminate all classes of Co_2 for which the corresponding system of equations does not have a solution with first component 1 and last component 0. We demonstrate this for $i = 3$.

Here the jth column of Fi_i is a solution of

$$
\begin{bmatrix}
-1 & 0 & 0 & 0 & 1 \\
-3 & 0 & 0 & 1 & -1 \\
4 & 1 & 1 & 0 & -1 \\
3 & 0 & 0 & -1 & 0 \\
15 & 7 & 3 & 1 & -8
\end{bmatrix}
\cdot
\begin{bmatrix}
x_1 \\
x_2 \\
x_3 \\
x_4 \\
x_5
\end{bmatrix}
=
\begin{bmatrix}
\chi_2(g_{ij}) \\
\chi_3(g_{ij}) \\
\chi_4(g_{ij}) \\
\chi_5(g_{ij}) \\
\chi_6(g_{ij})
\end{bmatrix}.
$$

We have 16 possible conjugacy classes of Co_2 (with elements of order two, four or eight) which can contain g_{ij}, but for only six of these (for the classes 2c, 4d, 4f, 4g, 8a, 4g) do we get a solution x with $x_1 = 1$ and $x_5 = 0$. From these the last three conjugacy classes can be excluded, since the corresponding solutions would lead to $|\,\mathbf{C}_G(g_{ij})|\nmid|\mathbf{C}_{\mathrm{Co}_2}(g_{ij})|$, a contradiction. In exactly the same way the remaining Clifford matrices can be obtained, except for $i = 12$, for which one should use also the decomposition of $\chi_n|_G$ for $n = 7, 8, 9$. In

this case Corollary 3.8.14(d) may also be applied, which immediately gives the
second row and last column of Fi_{12}:

$$
Fi_3 = \begin{bmatrix}
1 & 1 & 1 & 1 & 1 \\
3 & 3 & 3 & 3 & -1 \\
36 & 36 & 4 & -12 & 0 \\
24 & 24 & -8 & 8 & 0 \\
8 & -8 & 0 & 0 & 0
\end{bmatrix}, \quad
Fi_{12} = \begin{bmatrix}
1 & 1 & 1 & 1 & 1 & 1 \\
1 & 1 & 1 & 1 & 1 & -1 \\
2 & 2 & 2 & 2 & -2 & 0 \\
8 & 8 & -8 & 0 & 0 & 0 \\
4 & 4 & 4 & -4 & 0 & 0 \\
4 & -4 & 0 & 0 & 0 & 0
\end{bmatrix}.
$$

$\blacklozenge$

Using similar techniques the character table of $2_+^{1+22}.Co_2$, one of the largest
maximal subgroups of BM, has been computed in [137] using GAP. All GAP programs used can be found in [138]. The group $2_+^{1+22}.Co_2$ has 448 conjugacy
classes, and the largest Clifford matrix is a 20×20-matrix.

Exercises

Exercise 3.8.1 Prove Lemma 3.8.5.

Exercise 3.8.2 Listed in [170] is an eight-dimensional irreducible representation $\delta \colon S_6(2) \to GL_8(2)$, given by $\delta(a)$ and $\delta(b)$ for standard generators of
$S_6(2)$. Find stabilizers T_i for representatives of the orbits of $S_6(2)$ acting on $\mathbb{F}_2^8$
and compute the character tables of the T_i.

Exercise 3.8.3 Compute the Clifford matrices Fi_i in Example 3.8.17 for $i \in \{4, 5, 8, 9\}$.

3.9   M-groups

Whereas permutation characters of a group can only be irreducible in the trivial case, it may very well happen that some non-trivial linear characters of a
subgroup induce up to irreducible characters. There is even an important class
of groups, called M-groups, where all irreducible characters may be obtained in
this simple way.

We keep the assumption that all groups considered in this section are finite.

Definition 3.9.1 (a) A representation $\delta \colon G \to GL(V)$ over a field K is called
monomial if it is induced from a degree one representation of a subgroup of G.

(b) A group G is called an **M-group** if all irreducible representations of G over
$\mathbb{C}$ are monomial.

Let $\delta \colon G \to GL(V)$ be induced from a representation λ of degree one of
$H \leq G$. Choosing coset representatives $G = \dot{\bigcup}_{i=1}^{n} g_i H$, any $g \in G$ permutes

the cosets, so $g \cdot g_i = g_{\pi_g(i)} h_i(g)$ with $h_i(g) \in H$, where $\pi \colon G \to \mathrm{S}_n$, $g \mapsto \pi_g$ is the corresponding permutation representation with stabilizer H. Then $V = KG \otimes_{KH} K$, $B = (g_1 \otimes 1, \ldots, g_n \otimes 1)$ is a K-basis of V and

$$[\delta(g)]_B = [\delta_{\pi_g(i),j} \cdot \lambda(h_i(g))]_{i,j} = P_{\pi_g} \cdot \mathrm{diag}(\lambda(h_1(g)), \ldots, \lambda(h_n(g))) \quad (3.41)$$

is a monomial matrix, i.e. one with exactly one non-zero entry in each row and column. Here P_{π_g} is the permutation matrix corresponding to the permutation π_g. Obviously

$$\ker \pi = \{g \in G \mid [\delta(g)]_B = \text{diagonal matrix}\}$$

and $\ker \pi / \ker \delta$ is abelian.

Conversely, it is not hard to see (Exercise 3.9.1) that for every irreducible matrix representation $\delta \colon G \to \mathrm{GL}_n(K)$ mapping all $g \in G$ to monomial matrices there is a monomial representation $\delta \colon G \to \mathrm{GL}(V)$ and a basis B of V such that $\delta(g) = [\delta(g)]_B$ for every $g \in G$.

Observe that in (3.41) the field elements $\lambda(h_i(g))$ are mth roots of unity for some $m \in \mathbb{N}$ dividing the exponent of G. Choosing a suitable fixed root of unity $\zeta \in K$ we have $\lambda(h_i(g)) = \zeta^{m_i}$ with integers m_i and the matrix $[\delta(g)]_B$ may be stored just by giving $\pi_g \in \mathrm{S}_n$ and $[m_1, \ldots, m_n] \in \mathbb{N}_0^n$. Since for the multiplication of monomial matrices no field addition is needed, the multiplication (and inversion) of the representing matrices can carried out with just n modular additions of integers.

The following theorem shows that any M-group is solvable.

Theorem 3.9.2 (Taketa) *If G is an M-group and $1 = n_1 \leq n_2 \leq \cdots \leq n_k$ are the degrees of the irreducible $\mathbb{C}$-representations of G, then for every irreducible representation δ of G of degree n_i one has*

$$G^{(i)} \leq \ker \delta,$$

where $G^{(i)}$ is the ith commutator subgroup of G. In particular, every M-group is solvable.

Proof. We use induction on $\deg \delta$. If $\deg \delta = 1$ it is clear that $G' = G^{(1)} \leq \ker \delta$. So let δ be a monomial representation of degree $n = n_i$ and let, as above, $\pi \colon G \to \mathrm{S}_n$ be the corresponding permutation representation. So as in the foregoing discussion $\ker \pi / \ker \delta$ is abelian. Since π considered as a representation is reducible (cf. Lemma 1.2.4) all irreducible constituents have degree $< n_i$. Hence, by induction, $G^{(i-1)} \leq \ker \pi$. Since $\ker \pi / \ker \delta$ is abelian, $G^{(i)} \leq \ker \delta$. $\qquad \square$

On the other hand, we will convince ourselves that the class of M-groups contains the following class of groups.

Definition 3.9.3 A group is **super-solvable** if it has a normal series

$$G = N_0 > N_1 > \cdots > N_m = \{1\}$$

with N_{i-1}/N_i cyclic for $i = 1, \ldots, m$.

Actually we will not only give a constructive proof that super-solvable groups are M-groups, but also we will show that there is an effective algorithm to compute in the semisimple case the irreducible representations up to equivalence for a class of groups, slightly larger than the class of super-solvable groups. This is due to Baum and Clausen (see [7], [9] and [11]). To describe the idea of this algorithm we need the following definition.

Definition 3.9.4 Let $\mathcal{C} = (\ \{1\} = G_0 < G_1 < \cdots < G_n = G\ )$ be a series of subgroups. A matrix representation $\mathbf{D}\colon G \to \mathrm{GL}_m(K)$ over a field K is called $\mathcal{C}$-adapted if the following conditions are satisfied:

(a) $\mathbf{D}|_{G_i} = \mathbf{D}_1^{(i)} \oplus \cdots \oplus \mathbf{D}_{r_i}^{(i)}$ with irreducible matrix representations $\mathbf{D}_j^{(i)}$ of G_i for $1 \leq j \leq r_i$;

(b) if $\mathbf{D}_j^{(i)}$ and $\mathbf{D}_k^{(i)}$ are equivalent $(1 \leq j, k \leq r_i)$, then $\mathbf{D}_j^{(i)} = \mathbf{D}_k^{(i)}$.

It is easy to see that any representation $\mathbf{D}\colon G \to \mathrm{GL}_m(K)$ is equivalent to a $\mathcal{C}$-adapted representation if $\mathrm{char}\,K \nmid |G|$.

The algorithm is based on the following lemma.

Lemma 3.9.5 *Let*

$$\mathcal{C} := (\ \{1\} = G_0 \lhd G_1 \lhd \cdots \lhd G_n = G\ ) \tag{3.42}$$

be a normal series of G with G_1 abelian and G_{i+1}/G_i of prime order for $i = 1, \ldots, n-1$. Assume that K is a field of characteristic not dividing $|G|$, which is a splitting field for all the G_i. If $\mathbf{D}, \mathbf{D}'\colon G \to \mathrm{GL}_m(K)$ are irreducible equivalent $\mathcal{C}$-adapted representations, then the intertwining space

$$\mathrm{Int}(\mathbf{D}, \mathbf{D}') := \{X \in K^{m \times m} \mid X\mathbf{D}(g) = \mathbf{D}'(g)X \quad \text{for all} \quad g \in G\} = \langle Y \rangle_K$$

is generated by a monomial matrix Y.

Proof. We give a proof which is constructive and uses induction (recursion) on n. For $n = 1$ we may put $Y = [1]$. Now assume $n > 1$. Let $\mathbf{D}, \mathbf{D}'$ be as indicated and $0 \neq X \in \mathrm{Int}(\mathbf{D}, \mathbf{D}')$. If $\mathbf{D}|_{G_{n-1}}$ is irreducible then so is $\mathbf{D}'|_{G_{n-1}}$, and by induction we have a monomial $Y \in K^{m \times m}$ with

$$X \in \mathrm{Int}(\mathbf{D}|_{G_{n-1}}, \mathbf{D}'|_{G_{n-1}}) = \langle Y \rangle_K.$$

Otherwise $\mathbf{D}|_{G_{n-1}} = \mathbf{D}_1 \oplus \cdots \oplus \mathbf{D}_p$ and $\mathbf{D}'|_{G_{n-1}} = \mathbf{D}_1' \oplus \cdots \oplus \mathbf{D}_p'$ with irreducible representations $\mathbf{D}_i$ and $\mathbf{D}_i'$ of G_{n-1} for $1 \leq i \leq p := [G : G_{n-1}]$. Replacing $\mathbf{D}'$ by $\mathbf{D}''\colon g \mapsto P^{-1}\mathbf{D}'(g)P$ with a suitable permutation matrix P (and observing that $\mathrm{Int}(\mathbf{D}, \mathbf{D}'') = P \cdot \mathrm{Int}(\mathbf{D}, \mathbf{D}')$) if need be, we may assume that $\mathbf{D}_i$ is equivalent to $\mathbf{D}_i'$ for $i = 1, \ldots, p$. By induction, $\mathrm{Int}(\mathbf{D}_i, \mathbf{D}_i') = \langle Y_i \rangle_K$ with monomial $Y_i \in K^{m/p \times m/p}$. Then $X = s_1 Y_1 \oplus \cdots \oplus s_p Y_p$ with $s_1, \ldots, s_p \in K \setminus \{0\}$. Thus X is monomial and may be obtained by solving the equation $X\mathbf{D}(h) = \mathbf{D}'(h)X$ for some $h \in G \setminus G_{n-1}$, considered as a system of linear equations in $s_1, \ldots, s_p$. $\square$

We remark that Lemma 3.9.5 holds true in greater generality (see [8]). But the constructive proof we have given depends on the special structure of $\mathcal{C}$.

Theorem 3.9.6 *Let $\mathcal{C}$ and K be as in Lemma 3.9.5. Then there is a complete system of representatives $(\mathbf{D}_1, \ldots, \mathbf{D}_r)$ of the equivalence classes of irreducible representations of G over K such that all $\mathbf{D}_i$ are monomial and $\mathbf{D}_1 \oplus \cdots \oplus \mathbf{D}_r$ is $\mathcal{C}$-adapted.*

Proof. We use induction on n, the case $n = 1$ being trivial. We put $N := G_{n-1}$ and $\mathcal{C}' := (\ \{1\} = G_0 \lhd G_1 \lhd \ldots \lhd G_{n-1} = N\ )$. By induction we have a set of monomial representatives $(\mathbf{D}'_1, \ldots, \mathbf{D}'_{r'})$ of the equivalence classes of irreducible representations of N such that $\mathbf{D}'_1 \oplus \cdots \oplus \mathbf{D}'_{r'}$ is $\mathcal{C}'$-adapted. We will show how to obtain the desired matrix representations $(\mathbf{D}_1, \ldots, \mathbf{D}_r)$ from these in a constructive manner initializing our list by $\mathcal{D} := ()$.

We choose a fixed $g \in G \setminus N$ and use the coset representatives $1, g, \ldots, g^{p-1}$ of N in G for the definition of the induced matrix representations $\mathbf{D}'^G_i$; see Lemma 3.2.1. For all $\mathbf{D}' := \mathbf{D}'_i$ with $i \in \{1, \ldots, r'\}$ we have to distinguish two cases (according to Corollary 3.6.15) as follows.

(a) $\mathbf{D}'^G$ is irreducible. Then we have a family $\{^{g^j}\mathbf{D}' \mid 0 \le j \le p - 1\}$ of p conjugate pair-wise non-equivalent representations of N. It is easy to see that they all are $\mathcal{C}'$-adapted, and by Lemma 3.9.5 there are monomial matrices X_j such that

$$X_j^{-1}\, {}^{g^j}\mathbf{D}'(x)\, X_j = \mathbf{D}'_{i_j}(x) \quad \text{for all} \quad x \in N$$

for some $i_j \in \{1, \ldots, r'\}$. We put $X := \operatorname{diag}(X_0, \ldots, X_{p-1})$ and obtain a monomial $\mathcal{C}$-adapted representation $\mathbf{D}$ of G by putting $\mathbf{D}(y) := X^{-1}\mathbf{D}'^G(y)X$ for $y \in G$. We add $\mathbf{D}$ to our list $\mathcal{D}$.

(b) $\mathbf{D}'$ and $^g\mathbf{D}'$ are equivalent and $\mathbf{D}'$ can be extended to a representation of G. Using Lemma 3.9.5 we obtain a monomial matrix Y such that $Y^{-1}\mathbf{D}'(x)Y = {}^g\mathbf{D}'(x) = \mathbf{D}'(g^{-1}xg)$ for all $x \in N$. Then $\mathbf{D}'(g^p)\, Y^{-p} \in \operatorname{Int}(\mathbf{D}, \mathbf{D})$. By our assumptions the equation $c^p Y^p = \mathbf{D}'(g^p)$ has exactly p solutions $c_1, \ldots, c_p \in K$. So we obtain p monomial matrix representations $\mathbf{D}_i$ $(i = 1, \ldots p)$ by

$$\mathbf{D}_i(g) := c_i Y \qquad \text{and} \qquad \mathbf{D}_i|_N = \mathbf{D}',$$

which we add to our list $\mathcal{D}$.

Having thus processed all $\mathbf{D}'_i$ it follows from Theorem 3.6.7 and Corollary 3.6.15 that $(\mathbf{D}_1, \ldots, \mathbf{D}_r) := \mathcal{D}$ is a complete set of representatives of the irreducible representations of G. Since all irreducible constituents of all $\mathbf{D}_i|_N$ are in $\{\mathbf{D}'_1, \ldots, \mathbf{D}'_{r'}\}$, $\mathbf{D}_1 \oplus \cdots \oplus \mathbf{D}_r$ is $\mathcal{C}$-adapted. $\qquad\square$

Remark 3.9.7 Theorem 3.9.6 in conjunction with Lemma 3.9.5 leads to a very efficient algorithm often called the **Baum–Clausen algorithm** for computing the irreducible representations (and characters) of a finite group which is an extension of an abelian group by a super-solvable one. In such a group, computations can be carried out very efficiently using a power-commutator presentation (see e.g. [84]).

Corollary 3.9.8 (a) *An extension of an abelian group by a super-solvable group is an M-group.*

(b) *Every nilpotent group, so in particular any p-group, is an M-group.*

Proof. (a) follows directly from Theorem 3.9.6 because of Exercise 3.9.1.

(b) Since a non-trivial nilpotent group has a non-trivial central (cyclic) subgroup it follows by induction that a nilpotent group is super-solvable. Hence (a) implies (b). □

The converse of Corollary 3.9.8(a) does not hold; see Exercise 3.9.2. It is clear that factor groups of M-groups are M-groups. But normal subgroups of M-groups are not necessarily M-groups; see [45]. A group-theoretical characterization of M-groups has been given by Parks [144].

Exercises

Exercise 3.9.1 (a) Let $\delta\colon G \to \mathrm{GL}_n(K)$ be an irreducible matrix representation of G such that $\delta(g)$ is a monomial matrix for every $g \in G$. Show that there is a monomial representation $\delta\colon G \to \mathrm{GL}(V)$ and a basis B of V such that $\delta(g) = [\delta(g)]_B$ for every $g \in G$.

(b) Show that (a) does not hold if one omits the hypothesis that δ is irreducible.

Exercise 3.9.2 (a) Find a solvable group that is not an M-group and an M-group that is not super-solvable.
Hint: It suffices to look at groups of order 24.

(b) Find an M-group which is not an extension of an abelian group by a super-solvable group.
Hint: Consider an extension of an extraspecial group of order 32 by a cyclic group of order three.

3.10   Brauer's induction theorem

In this section we will present a crucial result, Brauer's induction theorem, which has many important applications. It shows that any ordinary character of a group G can be written as a $\mathbb{Z}$-linear combination of characters induced from certain subgroups of G. In the following, let G be a finite group and let K be a field of characteristic zero.

Definition 3.10.1 Let $\mathbb{Z}\,\mathrm{Irr}_K(G)$ be the ring of generalized characters of G over K (see Definition 2.8.3). If $\mathcal{H}$ is a family of subgroups of G then let

$$\mathcal{I}(G, \mathcal{H}) := \sum_{H \in \mathcal{H}} \sum_{\psi \in \mathrm{Irr}_K(H)} \mathbb{Z}\psi^G.$$

We are looking for a "small" family $\mathcal{H}$ such that $\mathcal{I}(G, \mathcal{H}) = \mathbb{Z} \operatorname{Irr}_K(G)$. We need the following obvious fact, which should be compared with Lemma 3.2.14.

Lemma 3.10.2 *For $H \le G$, $\chi \in \operatorname{cf}(G, K)$, and $\psi \in \operatorname{cf}(H, K)$ we have*

$$\chi \cdot \psi^G = (\chi_H \cdot \psi)^G.$$

In particular $\mathcal{I}(G, \mathcal{H})$ is an ideal in $\mathbb{Z} \operatorname{Irr}_K(G)$.

We now define two families of subgroups as follows.

Definition 3.10.3 (a) A group H is called **(p-)elementary** if $H = \langle h \rangle \times P$ with a p-group P (for some $h \in H$ and a prime p). Let

$$\mathcal{E} := \mathcal{E}(G) := \{ H \le G \mid H \ p\text{-elementary for some prime } p \}.$$

(b) H is called **(p-)quasi-elementary** if H contains a cyclic normal subgroup $\langle h \rangle$ with factor group $H/\langle h \rangle$ a p-group. Let

$$\mathcal{Q} := \mathcal{Q}(G) := \{ H \le G \mid H \ p\text{-quasi-elementary for some prime } p \}.$$

Obviously $\mathcal{E} \subseteq \mathcal{Q}$. Observe that one may assume, without loss of generality in (a), that $P \in \operatorname{Syl}_p(H)$ and in (b) that $H = \langle h \rangle \rtimes P$ with $P \in \operatorname{Syl}_p(H)$. It is also clear that elementary groups are nilpotent.

We will first show that $\mathcal{I}(G, \mathcal{Q}) = \mathbb{Z} \operatorname{Irr}_K(G)$ and later improve this to $\mathcal{I}(G, \mathcal{E}) = \mathbb{Z} \operatorname{Irr}_K(G)$. But first we need the following two lemmas.

Lemma 3.10.4 *If $\mathcal{H}$ is a family of subgroups of G which is closed under conjugation (in G) and intersections then*

$$\mathcal{A}_{\mathcal{H}} := \{ \sum_{H \in \mathcal{H}} a_H \mathbf{1}_H{}^G \mid a_H \in \mathbb{Z} \}$$

is multiplicatively closed, and is hence a subring of $\operatorname{cf}(G, K)$, possibly without identity.

Proof. This follows immediately from the formula

$$\mathbf{1}_{H_i}{}^G \cdot \mathbf{1}_{H_j}{}^G = \sum_{d \in D_{ij}} (\mathbf{1}_{H_i^d \cap H_j})^G$$

established in Corollary 3.5.13. $\square$

Lemma 3.10.5 *Let $\mathcal{A}$ be a ring (possibly without identity) of $\mathbb{Z}$-valued functions on a finite set X (with point-wise addition and multiplication). If the constant function $\mathbf{1}_X \colon X \to \mathbb{Z}$, $x \mapsto 1$ is not in $\mathcal{A}$ then there is an $x \in X$ and a prime p with $p \mid f(x)$ for all $f \in \mathcal{A}$.*

Proof. For $x \in X$ consider the additive subgroup $I_x := \{f(x) \mid f \in \mathcal{A}\}$ of $\mathbb{Z}$. If $I_x \neq \mathbb{Z}$ then I_x is contained in some maximal subgroup $p\mathbb{Z}$, so $p \mid f(x)$ for all $f \in \mathcal{A}$. If, on the other hand, $I_x = \mathbb{Z}$ for all $x \in X$, choose for every $x \in X$ an element $f_x \in \mathcal{A}$ with $f_x(x) = 1$. Then

$$\prod_{x \in X} (\mathbf{1}_x - f_x) = 0.$$

Expanding the left hand side we get $\mathbf{1}_X$ as a $\mathbb{Z}$-linear combination of products of some f_x, hence $\mathbf{1}_X \in \mathcal{A}$, contrary to the hypothesis. $\qquad\square$

We first show that for the family $\mathcal{Q}$ the following statements hold.

Theorem 3.10.6 *Let K be a field of characteristic zero. Then*

(a) $\mathcal{I}(G, \mathcal{Q}) = \mathbb{Z}\operatorname{Irr}_K(G)$;

(b) $\mathbf{1}_G = \sum_{H \in \mathcal{Q}} a_H \mathbf{1}_H^G$ *with integers $a_H \in \mathbb{Z}$.*

Proof. (a) follows from (b) since $\mathcal{I}(G, \mathcal{Q})$ is an ideal in $\mathbb{Z}\operatorname{Irr}_K(G)$. Since $\mathcal{Q}$ is a family of subgroups of G which is closed under conjugation and taking intersections, Lemmas 3.10.4 and 3.10.5 show that it is sufficient to prove the following: for each prime p and every element $g \in G$ there is a subgroup $H \in \mathcal{Q}$ such that $p \nmid \mathbf{1}_H{}^G(g)$.

So we choose a prime p and an arbitrary element $g \in G$ and write $\langle g \rangle = P_1 \times \langle y \rangle$ with $P_1 \in \operatorname{Syl}_p(\langle g \rangle)$. Let $N = \mathbf{N}_G(\langle y \rangle)$ and $P_1 \leq P_2 \in \operatorname{Syl}_p(N)$. Then $H = P_2\langle y \rangle \in \mathcal{Q}$ and

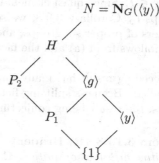

$$\mathbf{1}_H{}^G(g) = |\{xH \mid x \in G \, , \, gxH = xH\}|$$
$$= \mathbf{1}_H{}^N(g),$$

because $gxH = xH$ implies $x^{-1}gx \in H$, hence $x^{-1}yx \in H$ and $x \in N$. Observe that $\langle y \rangle = \{h \in H \mid \text{g.c.d.}(p, |\langle h \rangle|) = 1\}$.

Since $\langle y \rangle \trianglelefteq N$ the action of $\langle y \rangle$ on N/H is trivial, so the p-group $\langle g \rangle / \langle y \rangle \cong P_1$ is acting on N/H. Hence all orbits of $\langle g \rangle$ on N/H have lengths which are powers of p. It follows that the number of fixed points is given by

$$|\operatorname{Fix}_{N/H}(g)| = \mathbf{1}_H{}^N(g) \equiv [N : H] \mod p.$$

As $p \nmid [N : H]$ the assertion follows. $\qquad\square$

Finally, we can state Brauer's induction theorem.

Theorem 3.10.7 (Brauer's induction theorem)

(a) $\mathcal{I}(G, \mathcal{E}) = \mathbb{Z}\operatorname{Irr}(G)$.

(b) *Every generalized character of G over $\mathbb{C}$ is a $\mathbb{Z}$-linear combination of characters induced from linear characters of elementary subgroups of G.*

Proof. (a) We use induction on $|G|$. It suffices to show that $1_G \in \mathcal{I}(G, \mathcal{E})$. By Theorem 3.10.7 we already know that $1_G \in \mathcal{I}(G, \mathcal{Q})$. If $G \notin \mathcal{Q}$ then 1_G is a $\mathbb{Z}$-linear combination of characters ψ_i which are induced from characters of proper subgroups of G, namely of those in $\mathcal{Q}$. By induction the ψ_i are $\mathbb{Z}$-linear combination of characters which are induced from characters of elementary subgroups of G. Transitivity of induction (Lemma 3.2.14) yields the result in this case.

So we may assume that $G \in \mathcal{Q}$, that is $G = C \cdot P$, with C a cyclic p'-subgroup which is normal in G and P a p-group. Let $Z = \mathbf{C}_C(P)$. If $Z = C$ then $G = C \times P$ is elementary and the claim holds trivially. So we assume that $Z \neq C$. Let $H = Z \times P$, so $H \in \mathcal{E}$. Let

$$I = \{\chi \in \mathrm{Irr}(G) \mid 1_G \neq \chi\, , (1_H{}^G, \chi)_G > 0\} \quad \text{thus} \quad 1_H{}^G = 1_G + \sum_{\chi \in I} a_\chi \chi$$

with $a_\chi \in \mathbb{N}$. The trivial character 1_H is a constituent of every $\chi \in I$ by Frobenius reciprocity. Thus if $\chi(1) = 1$ for some $\chi \in I$ then $\chi_H = 1_H$ and $G = \ker\chi \cdot \mathbf{N}_G(P) = \ker\chi \cdot H = \ker\chi$ by the Frattini argument (see [88], Satz I.7.8) and $\chi = 1_G$, which contradicts the definition of I. Hence $\chi(1) > 1$ for all $\chi \in I$. Since quasi-elementary groups are super-solvable, and hence are M-groups by Corollary 3.9.8, we see that all $\chi \in I$ are induced from (linear) characters of proper subgroups, and the claim follows by induction.

(b) follows from (a) and the fact that elementary groups are M-groups. $\quad\square$

The preceding result has many important consequences. The first one we list, often called "Brauer's splitting field theorem" was an open problem for almost 50 years, the result being conjectured by Maschke around 1900.

Theorem 3.10.8 (R. Brauer) *If G is a finite group of exponent m and ζ is a primitive mth root of unity in $\mathbb{C}$ then $\mathbb{Q}(\zeta)$ is a splitting field for G.*

Proof. Let $K := \mathbb{Q}(\zeta)$ and $\chi \in \mathrm{Irr}(G)$. By Theorem 3.10.7

$$\chi = \sum_{i \in I} a_i \lambda_i^G \qquad a_i \in \mathbb{Z},$$

where the λ_i are linear characters of certain subgroups E_i for some index set I. Since $\lambda_i(g) \in K$ for all $g \in E_i$, all the λ_i and hence all the λ_i^G are characters of KG-modules. By Theorem 2.9.19 we have $\mathrm{m}_K(\chi) \mid (\lambda_i^G, \chi)_G$. Consequently

$$\mathrm{m}_K(\chi) \mid \sum_{i \in I} a_i(\lambda_i^G, \chi)_G = (\chi, \chi)_G = 1.$$

Hence $\mathrm{m}_K(\chi) = 1$, and by Lemma 2.9.15 χ is the character of a simple KG-module. By Exercise 1.5.8 K is a splitting field for G. $\qquad\square$

Theorem 3.10.9 (Brauer's characterization of characters) *A class function $\eta \in \mathrm{cf}(G,\mathbb{C})$ is a generalized character if and only if η_E is a generalized character for all elementary subgroups $E \leq G$. It is an irreducible character if and only if moreover $(\eta,\eta)_G = 1$ and $\eta(1) > 0$.*

Proof. Of course, a restriction of a generalized character is always a generalized character. On the other hand, assume that $\eta \in \mathrm{cf}(G,\mathbb{C})$ restricts to a generalized character on every elementary subgroup of G. By Theorem 3.10.7 $1_G = \sum_i a_i \lambda_i^G$ with integers a_i and linear characters $\lambda_i \in \mathrm{Irr}(E_i)$ of some elementary subgroups $E_i \leq G$. We conclude (using Lemma 3.2.7(a)) that

$$\eta = \eta \cdot 1_G = \sum_i a_i \eta \cdot \lambda_i^G = \sum_i a_i (\eta_{E_i} \cdot \lambda_i)^G \in \mathbb{Z}\,\mathrm{Irr}(G).$$

$\qquad\square$

Theorem 3.10.9 can be used to verify that certain class functions are indeed characters. As an application we give a simple proof of the following important result, which has already been utilized in Section 2.10. Actually the result is a special case of much more general results, which will be derived in Chapter 4, also using Theorem 3.10.9.

Theorem 3.10.10 *Let $\chi \in \mathrm{Irr}(G)$ and let p be a prime with $p \nmid \frac{|G|}{\chi(1)}$. Then $\chi(g) = 0$ for every $g \in G$ with $p \mid |\langle g \rangle|$.*

Proof. We define a class function $\eta \colon G \to \mathbb{C}$ by

$$\eta(g) = \begin{cases} \chi(g) & \text{for } p \nmid |\langle g \rangle|, \\ 0 & \text{else.} \end{cases}$$

We first show that η is a generalized character. To this end let $H \leq G$ be an elementary subgroup. Since H is nilpotent we may write $H = P \times Q$ with P a Sylow p-subgroup of H. Let $\psi \in \mathrm{Irr}(H)$ be arbitrary. Since η_H vanishes outside of Q we have

$$(\eta_H, \psi)_H = \frac{1}{|H|} \sum_{h \in Q} \eta(h)\psi(h^{-1}) = \frac{1}{|P|}(\chi_Q, \psi_Q)_Q.$$

Hence $|P|(\eta_H, \psi)_H \in \mathbb{Z}$. On the other hand, using Theorem 2.3.2 we get

$$(\eta_H, \psi)_H = \frac{1}{|H|} \sum_{h \in Q} \omega_\chi((h^G)^+) \frac{\chi(1)}{|G|} |\,\mathbf{C}_G(h)| \psi(h^{-1})$$

$$= \frac{\chi(1)}{|G||Q|} \sum_{h \in Q} \omega_\chi((h^G)^+) [\mathbf{C}_G(h) : P] \psi(h^{-1})$$

since $P \subseteq \mathbf{C}_G(h)$ for $h \in H$. Since $\omega_\chi((h^G)^+)$ is an algebraic integer (see Corollary 2.3.3) we get that $|Q|\frac{|G|}{\chi(1)}(\eta_H, \psi)_H \in \mathbb{Z}$. Our assumption implies that $|P|$ and $|Q|\frac{|G|}{\chi(1)}$ are coprime, hence $(\eta_H, \psi)_H \in \mathbb{Z}$ and $\eta \in \mathbb{Z}\,\mathrm{Irr}(G)$ by Theorem 3.10.9. Now it is obvious that

$$0 < (\eta, \eta)_G = (\eta, \chi)_G \leq (\chi, \chi)_G = 1.$$

Hence $\eta = \chi$ and χ vanishes on all elements of order divisible by p. $\square$

Theorem 3.10.9 has often been used for the construction of character tables. If some information about subgroups of a finite group G is available, including their character tables, it is sometimes possible to guess some candidate for an irreducible character of G and then use Theorem 3.10.9 to prove that it is indeed an irreducible character.

Example 3.10.11 Let $G := \mathrm{Fi}_{22}$ be the smallest of the sporadic simple Fischer groups; G has maximal subgroups $M_2 \cong \mathrm{O}_7(3)$ and $M_5 \cong 2^{10} : \mathrm{M}_{22}$, a split extension of an elementary abelian group of order 2^{10} by M_{22} (see [38], p. 163). For both groups the smallest degree of a faithful character is 78, and both groups have a unique faithful character ψ_2 and ψ_5, respectively, of this degree. In fact, $\psi_2 \in \mathrm{Irr}(M_2)$, while $\psi_5 = 1_{M_5} + \chi_5'$ with $\chi_5' \in \mathrm{Irr}(M_5)$.

So all irreducible characters of G must have degree ≥ 78 and one might guess that there is an irreducible character of G of this degree. Computing the class fusions of M_2 and M_5 into G, this guess is supported by the fact that $\psi_2(g_2) = \psi_5(g_5)$ whenever $g_2 \in M_2$ is conjugate in G to $g_5 \in M_5$, so that there are indeed class functions χ of G such that $\chi|_{M_i} = \psi_i$ for $i \in \{2, 5\}$. Any such class function is, of course, uniquely determined on all conjugacy classes of G which intersect non-trivially with M_2 or M_3, that is on all conjugacy classes of G except for

$$12a, 12b, 12c, 12h, 21a, 22a, 22b, 24a, 24b, 30a.$$

Assuming that χ is a character, the congruence relations (see Lemma 2.2.2) determine the values of χ on 21a modulo 21, on 22a, 22b modulo 22 on 30a modulo 30 and on the remaining classes modulo 6. Testing the norm of χ and $\chi^{[1^2]}$ for integrality, and whether or not $(\chi, 1_G)_G = 0$, we find that there is a unique solution, that is, there is exactly one class function $\chi \in \mathrm{cf}(G, \mathbb{C})$ with $\chi|_{M_i} = \psi_i$ for $i \in \{2, 5\}$, which possibly is a character.

To prove that this χ is a character we have to show that $\chi|_E$ is a generalized character for every elementary subgroup $E \leq G$. Of course, it suffices to consider only maximal elementary subgroups E, that is, those which are not contained in a larger elementary subgroup. Knowing the power maps of G, it is a simple matter to check whether or not $\chi|_E$ is a (generalized) character in the case that E is cyclic. The non-cyclic maximal elementary subgroups of G are in $\mathrm{Syl}_p(G)$ for $p \in \{2, 3, 5\}$ or of the form $\mathbf{C}_G(g)_p := \langle g \rangle \times P$ with $P \in \mathrm{Syl}_p(\mathbf{C}_G(g))$, where $g \in G$ has order coprime to p. From the orders and centralizer orders of the elements of G, we see that G has 15 conjugacy classes of non-cyclic maximal elementary subgroups. These are the Sylow p-subgroups for $p \in \{2, 3, 5\}$, and in addition

$$\mathbf{C}_G(g)_3 \quad \text{for} \quad g \in 2a \cup 2b \cup 2c \cup 4a \cup 4b \cup 4d$$

and

$$\mathbf{C}_G(g)_2 \quad \text{for} \quad g \in 3a \cup 3b \cup 3c \cup 3d \cup 5a \cup 9a.$$

Since $\chi|_U$ is a character for $U \leq M_2$ or $U \leq M_5$ we only have to consider non-cyclic maximal elementary subgroups which are not contained in M_2 or M_5. These are

$$P \in \mathrm{Syl}_5(G), \qquad \mathbf{C}_G(g)_2 \quad \text{for} \quad g \in 3a \cup 3b \quad \text{and} \quad \mathbf{C}_G(g)_3 \quad \text{for} \quad g \in 4a;$$

$P \cong C_5 \times C_5$ is trivial to check. For the other subgroups (of order $3 \cdot 2^8$, $3 \cdot 2^7$ and $4 \cdot 3^3$) one may compute the character tables and fusions into G using a permutation representation of G as in Example 3.10.12 below. Alternatively one may use the fact that these subgroups are contained in maximal subgroups $2.U_6(2)$ or $2^6 : S_6(2)$ of G.

Having checked that χ is indeed a character, it is a routine matter to calculate the complete character table of G using symmetrizations of χ, the characters induced from cyclic subgroups of G and the methods described in Section 2.8.

♦

Theorem 3.10.7 can also be used to compute the character table of a group G, provided that one can find the conjugacy classes and also sufficiently many elementary subgroups of G. In [165] an algorithm using this idea is described, which has been implemented in MAGMA and which for many interesting groups seems to perform better than the Dixon–Schneider algorithm presented in Section 2.4.

Example 3.10.12 Again we take $G := \mathrm{Fi}_{22}$. Using a permutation representation of G of degree 3510 (obtained from [170]) it is quite fast to compute the conjugacy classes and also the power maps for G. Hence one can immediately find the characters induced from linear characters of cyclic subgroups. Computing representatives of the 15 conjugacy classes of the non-cyclic maximal elementary subgroups of G as described in Example 3.10.11 is immediate. The computation of the character tables of these subgroups (with the Baum–Clausen algorithm, see Remark 3.9.7) and the determination of the fusions of their conjugacy classes into G, in order to induce their irreducible characters up to G, takes only a few minutes. Using the LLL-algorithm all the irreducible characters of G are found. A direct application of the Dixon–Schneider algorithm to G does not seem to be feasible.

♦

Exercises

Exercise 3.10.1 Let $G := \mathrm{McL}$ be the sporadic simple McLaughlin group. Use the table of marks of G (in the GAP library) to find all elementary subgroups of

G up to conjugacy in G. Show that all but three elementary subgroups (up to conjugacy) are cyclic or p-groups for some prime p. Verify that every elementary subgroup of G, which is not cyclic of order $11, 14, 15$ or 30, is conjugate to a subgroup of a maximal subgroup of G isomorphic to $U_4(3)$ or $U_3(5)$.

Exercise 3.10.2 Let $\mathcal{H}$ be a family of subgroups of G such that $\mathcal{I}(G, \mathcal{H}) = \mathbb{Z}\operatorname{Irr}_K(G)$. Show that every elementary subgroup $E \leq G$ is conjugate in G to a subgroup of H for some $H \in \mathcal{H}$.

Hint: Assume that $E = \langle h \rangle \times P$ with $P \in \operatorname{Syl}_p(E)$ is an elementary subgroup of G which is not conjugate in G to a subgroup of $H \leq G$. Let $\chi \in \operatorname{Irr}(H)$. Show that $\chi^G(h) \in p\,\mathbb{Z}[\zeta_m]$, where $m = |\langle h \rangle|$.

Exercise 3.10.3 Show that a finite group G is quasi-elementary if and only if $1_G \notin \langle\{1_H^G \mid H < G\}\rangle_\mathbb{Z}$.

Hint: Let $G = C \rtimes P$ with $P \in \operatorname{Syl}_p(G)$ and C cyclic. Assume $1_G = \sum_{i=1}^n a_i 1_{H_i}^G$ with $a_i \in \mathbb{Z} \setminus \{0\}$ and $G \neq H_i \neq_G H_j$ for $1 \leq i \neq j \leq n$. Choose the notation such that $p \nmid a_i [G : H_i]$ for $1 \leq i \leq m$ and $|H_1| \leq |H_i|$ for $i = 1, \ldots, m$. Show that $G = H_i C$ and $|H_i| \neq |H_j|$ for $1 \leq i \neq j \leq m$. Verify that there is a $\chi \in \operatorname{Irr}(G)$ with $(1_{H_1}^G, \chi)_G > 0$ and $\ker \chi \cap C = H_1 \cap C$ and that $(1_{H_i}^G, \chi)_G = 0$ for $2 \leq i \leq m$ and any such χ.

4

Modular representations

4.1 p-modular systems

In the preceding chapters we studied representations of a group G over a fixed field K, where most of the time the characteristic of K was assumed not to be a divisor of $|G|$, so that the group algebra KG was semisimple. Often the study of representations of a finite group G over a field of characteristic p dividing $|G|$ is called "modular representation theory." But this captures only a narrow aspect of the theory as it was developed by Richard Brauer. In fact, an important part of this theory is the interplay between representations of G in characteristic zero and in characteristic p.

The simplest example of this kind is to relate rational representations of G to representations over $\mathbb{F}_p$. If $\delta \colon G \to \mathrm{GL}_n(\mathbb{Q})$ is a representation, it is not hard to see (see Theorem 4.1.4 below) that δ is equivalent to a representation δ' with $\delta'(G) \subseteq \mathrm{GL}_n(\mathbb{Z})$. Extending the natural projection $\mathrm{pr}_p \colon \mathbb{Z} \to \mathbb{F}_p = \mathbb{Z}/p\mathbb{Z}$ to the matrix ring $\mathbb{Z}^{n \times n}$ we may then form

$$\overline{\delta'} := \mathrm{pr}_p \circ \delta' \colon G \to \mathrm{GL}_n(\mathbb{F}_p),$$

which is a representation of G over $\mathbb{F}_p$ and we may associate it to δ. As one can see here, representations over $\mathbb{Z}$ (or, equivalently, $\mathbb{Z}G$-lattices, that is $\mathbb{Z}G$-modules which are free and finitely generated as $\mathbb{Z}$-modules) form the connecting link between "ordinary representations" (over $\mathbb{Q}$) and "modular representations" (over $\mathbb{F}_p$). In particular we have an embedding and a projection

$$\mathbb{Q}G \hookleftarrow \mathbb{Z}G \longrightarrow \mathbb{F}_p G \tag{4.1}$$

of the relevant group algebras. We would like to generalize this idea by replacing $\mathbb{Q}$, for instance by an algebraic number field K, and $\mathbb{Z}$ by a suitable ring R with quotient field K and a maximal ideal P such that $F \cong R/P$ has characteristic $p > 0$. Again it is useful to consider more generally algebras and orders (see Definition 1.8.1), their modules and "lattices."

Definition 4.1.1 Let R be an integral domain with quotient field K and let A be an R-order. An A-**lattice** M is an A-module which has a finite R-basis. We will embed $M \subseteq KM := K \otimes_R M$.

For the above construction it is essential that every representation over $K := \mathbb{Q}$ is equivalent to one with entries in $R := \mathbb{Z}$. This will not hold for an arbitrary integral domain R and its quotient field K; for instance, it will generally not hold for the ring R of algebraic integers in an algebraic number field K, which is not always a principal ideal domain. But there is a good substitute, the important class of "valuation rings."

Definition 4.1.2 A **valuation ring** is an integral domain with quotient field K such that
$$K = R \cup \{x^{-1} \mid x \in R \setminus \{0\}\} \neq R.$$

Simple examples are $\mathbb{Z}_{(p)} := \{\frac{a}{b} \mid a, b \in \mathbb{Z}, \ p \nmid b\} \subseteq \mathbb{Q}$ for a prime p. In fact, it is an easy exercise to see that these are the only valuation rings in $\mathbb{Q}$ (see e.g. [32], p. 268).

We will need the following basic facts about valuation rings.

Lemma 4.1.3 *Let R be a valuation ring with quotient field K. Then we have the following.*

(a) *R is a local ring, that is $P := R \setminus R^\times \trianglelefteq R$ and $P = \mathrm{J}(R)$ is the only maximal ideal of R.*

(b) *R is integrally closed in K, that is any root $\alpha \in K$ of a monic polynomial in $R[X]$ is in R.*

(c) *$W := K^\times / R^\times$ is a (multiplicative) ordered abelian group with $\alpha R^\times \leq \beta R^\times$ for $\alpha, \beta \in K^\times$ if and only if $\beta \alpha^{-1} \in R$. The canonical map $\nu = \nu_R \colon K^\times \to W$ satisfies $\nu(\alpha + \beta) \geq \min(\nu(\alpha), \nu(\beta))$ for $\alpha, \beta \in K^\times$.*

(d) *Any finitely generated torsion-free R-module is free.*

Proof. (a) If $a \in P$ and $x \in R$ then clearly $ax \in P$. Let $a, b \in P \setminus \{0\}$. We may assume $\frac{a}{b} \in R$, interchanging a, b if need be. Then $a + b = b(\frac{a}{b} + 1) \in P$.

(b) If $\alpha \in K \setminus R$ and $\sum_{j=0}^{n} a_j \alpha^j = 0$ with $a_0, \ldots, a_n \in R$ and $a_n = 1$ then, since $\alpha^{-1} \in R$, we get $\alpha = -\sum_{j=0}^{n-1} a_j \alpha^{j-n} \in R$, a contradiction.

(c) The first assertion is immediate from the definition of a valuation ring. Let $\alpha, \beta \in K^\times$ and, say, $\frac{\beta}{\alpha} \in R$, that is $\nu(\alpha) \leq \nu(\beta)$. Then $(\alpha + \beta)\alpha^{-1} \in R$, hence the second assertion.

(d) As in (a) let $P := R \setminus R^\times$ be the maximal ideal of R. Let $V = \langle v_1, \ldots, v_n \rangle_R$ be a torsion-free R-module (that is $av = 0$ with $a \in R, v \in V$ implies $a = 0$ or $v = 0$). Then V/PV is an R/P-vector space generated by $\{\hat{v}_i := v_i + PV \mid 1 \leq i \leq n\}$. We may choose the notation so that $(\hat{v}_1, \ldots, \hat{v}_m)$ is a basis for V/PV. By Nakayama's lemma (Exercise 1.4.1) $V = \langle v_1, \ldots, v_m \rangle_R$. Assume that

$$\sum_{j=1}^{m} a_j v_j = 0 \qquad \text{with} \qquad a_j \in R \qquad \text{not all zero.} \tag{4.2}$$

We choose the notation so that $a_1 \neq 0$ and $\nu(a_1) \leq \nu(a_j)$ for all j with $a_j \neq 0$. Then $a_j a_1^{-1} \in R$ for all j. Multiplying equation (4.2) by a_1^{-1} and taking residues modulo PV we obtain a contradiction to the linear independence of the $\hat{v}_j$. $\qquad \square$

Theorem 4.1.4 *If R is a principal ideal domain or a valuation ring with quotient field K and A is an R-order then, for any finitely generated KA-module V, there is an "R-form," that is an A-submodule $M \leq_A V_A$ having an R-basis which is at the same time a K-basis for V. Thus M is an A-lattice with $KM = V$.*

Proof. Let $(v_1, \ldots, v_n)$ be a K-basis of V and let $(a_1, \ldots, a_m)$ be an R-basis of A. We put

$$M = \sum_{i=1}^{m} \sum_{j=1}^{n} R\, a_i\, v_j \subsetneq V.$$

It is easily verified that M is a finitely generated A-module with $KM = V$. It is R-torsion-free because it is contained in the K-vector space V, and hence is free; if R is a valuation ring the latter follows from Lemma 4.1.3(d), and for R a principal ideal domain it follows from the classification of finitely generated modules over such rings (see [110], Theorem III.7.3, p. 147). Any R-basis of M is obviously also a K-basis for V. $\qquad \square$

Corollary 4.1.5 *If R and K are as in the last theorem, any matrix representation $\delta \colon G \to \mathrm{GL}_n(K)$ of a finite group G over K is equivalent to a representation $\delta' \colon G \to \mathrm{GL}_n(K)$ with $\delta'(g) \in \mathrm{GL}_n(R)$ for all $g \in G$.*

Proof. We just apply the theorem with the R-order RG. $\qquad \square$

We note that an R-form M as in Theorem 4.1.4 is not uniquely determined up to A-isomorphism, as is seen in the following example.

Example 4.1.6 Let $A = \mathbb{Z}G$ with $G = \{1, g\} \cong C_2$, the cyclic group of order 2, and let $K = \mathbb{Q}$ so $KA = \mathbb{Q}G$. The regular module $V = \mathbb{Q}G$ has two $\mathbb{Z}$-forms $M_1 = \mathbb{Z}G$ and M_2 with $\mathbb{Z}$-bases

$$B_1 = (1 + g\,,\; 1) \quad \text{and} \quad B_2 = \left(\frac{1}{2}(1 + g)\,,\; \frac{1}{2}(1 - g)\right), \quad \text{respectively.}$$

The corresponding matrix representations with respect to the bases B_1, B_2 are given by

$$\delta_1(g) = \begin{bmatrix} 1 & 1 \\ 0 & -1 \end{bmatrix} \quad \text{and} \quad \delta_2(g) = \begin{bmatrix} 1 & 0 \\ 0 & -1 \end{bmatrix}.$$

Under the natural homomorphism $\mathbb{Z}^{2 \times 2} \to \mathbb{F}_2^{2 \times 2}$ these matrices map to $\begin{bmatrix} 1 & 1 \\ 0 & 1 \end{bmatrix}$ and $\begin{bmatrix} 1 & 0 \\ 0 & 1 \end{bmatrix}$, so it is clear that $\delta_1(g)$ and $\delta_2(g)$ cannot be conjugate in $\mathrm{GL}_2(\mathbb{Z})$. Hence $M_1 \not\cong_A M_2$. $\qquad \blacklozenge$

The advantage of choosing valuation rings for our construction can already be seen in the following lemma.

Lemma 4.1.7 *Let R be a valuation ring with quotient field K and let A be an R-order. Then $F := R/\operatorname{J}(R)$ is a field (of characteristic p if $p \in \operatorname{J}(R)$), $\hat{A} := A/\operatorname{J}(R)A$ is a finite dimensional F-algebra and*

$$\operatorname{J}(R)A \leq \operatorname{J}(A), \qquad hence \qquad A/\operatorname{J}(A) \cong \hat{A}/\operatorname{J}(\hat{A}). \qquad (4.3)$$

Thus every simple A-module is the inflation of a simple $\hat{A}$-module.

Proof.   Clearly $\operatorname{J}(R)A \trianglelefteq A$. We have to show that $\operatorname{J}(R)A \subseteq M$ for any maximal left ideal $M \leq_A A$. Assume on the contrary that $\operatorname{J}(R)A \not\subseteq M$ for such an M. Then $\{0\} \neq \operatorname{J}(R)(A/M)$, hence $\operatorname{J}(R)(A/M) = A/M$, because A/M is simple and $\operatorname{J}(R)(A/M) \leq_A A/M$. Since A is finitely generated over R, so is A/M. From Nakayama's lemma (Exercise 1.4.1) we conclude that $A/M = \{0\}$, a contradiction. $\square$

Remark 4.1.8 Using the assumptions of Lemma 4.1.7, $KA := K \otimes_R A$ is a finite dimensional K-algebra. Similarly as in (4.1) we have an embedding and a projection

$$KA \hookleftarrow A \longrightarrow \hat{A} = A/\operatorname{J}(R)A, \quad a \mapsto \hat{a} := a + \operatorname{J}(R)A.$$

By our embedding pair-wise orthogonal (central) idempotents in A are also pairwise orthogonal (central) idempotents in KA. Since KA has finite length it follows that

$$A = Ae_1 \oplus \cdots \oplus Ae_r \qquad \text{with primitive idempotents} \qquad e_i \in A,$$
$$A = A\epsilon_1 \oplus \cdots \oplus A\epsilon_m \qquad \text{with block idempotents} \qquad \epsilon_i \in \operatorname{Z}(A).$$

Observe that the e_i need not be primitive in KA and the ϵ_i need not be block idempotents of KA. Applying $A \to \hat{A}$ we get

$$\hat{A} = \hat{A}\hat{e}_1 \oplus \cdots \oplus \hat{A}\hat{e}_r \qquad \text{with} \qquad \hat{e}_i^2 = \hat{e}_i \in \hat{A},$$
$$\hat{A} = \hat{A}\hat{\epsilon}_1 \oplus \cdots \oplus \hat{A}\hat{\epsilon}_m \qquad \text{with} \qquad \hat{\epsilon}_i^2 = \hat{\epsilon}_i \in \operatorname{Z}(\hat{A}),$$

and again in general the $\hat{e}_i$ will not be primitive (and the $\hat{\epsilon}_i$ need not be block) idempotents in $\hat{A}$. But we shall eventually impose further conditions on R in order to achieve that one can find primitive mutually orthogonal idempotents $e_i \in A$ (block idempotents $\epsilon_i \in \operatorname{Z}(A)$) such that the $\hat{e}_i$ ($\hat{\epsilon}_i$) are also primitive (respectively block) idempotents in $\hat{A}$.

Furthermore, if M is an A-lattice, $KM := K \otimes_R M$ is a KA-module and $\hat{M} := M/(\operatorname{J}(R)M)$ is an $\hat{A}$-module. Conversely, if V is a KA-module with $\dim_K V < \infty$ and representation $\delta \colon KA \to \operatorname{End}_K V$ then by Theorem 4.1.4 there is an A-lattice M with R-basis $B = (v_1, \ldots, v_n)$ and $V = KM$. If $\delta \colon KA \to$

$K^{n \times n}$ is the matrix representation afforded by δ with respect to the basis B, then $\boldsymbol{\delta}(a) = [a_{ij}] \in R^{n \times n}$ for all $a \in A$ and $\hat{M}$ affords the matrix representation

$$\hat{\boldsymbol{\delta}} \colon \hat{A} \to F^{n \times n}, \quad \hat{a} \mapsto [\eta(a_{ij})]$$

with respect to the F-basis $(\hat{v}_1, \ldots, \hat{v}_n)$, where $\eta \colon R \to F$ is the natural map. We call $\hat{\boldsymbol{\delta}}$ and $\hat{M}$ a **p-modular reduction** (with respect to η) of $\boldsymbol{\delta}$ and V, respectively. Observe that these are *not* uniquely determined up to equivalence or isomorphism, but depend on the choice of M. In Example 4.1.6 both the trivial two-dimensional representation and the representation given by $g \mapsto \begin{bmatrix} 1 & 1 \\ 0 & 1 \end{bmatrix} \in \mathbb{F}_2^{2 \times 2}$ are 2-modular reductions of the regular representation of $\langle g \rangle \cong C_2$ over $\mathbb{Q}$. However, in Theorem 4.1.23 we will see that under suitable conditions on R the composition factors of a p-modular reduction do not depend on the choice of the R-form M up to isomorphism.

We will now show that for any prime p and any field K of characteristic zero there is a valuation ring R with quotient field K, such that $\mathrm{J}(R)$ contains p (and hence $\mathrm{char}(R/\mathrm{J}(R)) = p$). For this we first need a simple lemma.

Lemma 4.1.9 *Let I be a proper ideal in the subring R of a field K. Then for every $x \in K$ one has*

$$I\,R[x] \neq R[x] \qquad or \qquad I\,R[x^{-1}] \neq R[x^{-1}].$$

Here $R[x], R[x^{-1}]$ denote the subrings of K generated by R and x or x^{-1}, respectively.

Proof. (See [94], p. 561.) Assume that $I\,R[x] = R[x]$ and $I\,R[x^{-1}] = R[x^{-1}]$. Then we have $a_i, b_j \in I$ with

$$1 = \sum_{i=0}^{n} a_i\,x^i \qquad and \qquad 1 = \sum_{j=0}^{m} b_j\,x^{-j}.$$

Since $I \neq R$ we have $m + n \geq 2$, and by symmetry we may assume that $n \geq m$. Also we may assume that we have chosen the a_i, b_j so that n is minimal. Multiplying the first equation by $1 - b_0$ and the second one by $a_n\,x^n$ and subtracting we get the following equation:

$$1 = \sum_{k=0}^{n-1} c_k\,x^k \qquad with \qquad c_k \in I,$$

which contradicts the minimality of n. $\qquad\qquad\square$

Theorem 4.1.10 *If S is a subring of a field K and $\{0\} \neq I \lhd S$ then there is a valuation ring $R \supseteq S$ with quotient field K and $\mathrm{J}(R) \supseteq I$. In particular, if I is a maximal ideal in S then $\mathrm{J}(R) \cap S = I$.*

Proof. Let

$$\mathcal{M} := \{R' \mid S \subseteq R' \subseteq K, \ R' \text{ ring}, \ IR' \neq R'\}.$$

Clearly $S \in \mathcal{M}$. If $\mathcal{M}' \subseteq \mathcal{M}$ is a chain (a totally ordered subset w.r.t $\subseteq$) we put $R_1 := \bigcup \{R' \mid R' \in \mathcal{M}'\}$. It is easy to see that $R_1 \in \mathcal{M}$, hence by Zorn's lemma $\mathcal{M}$ has a maximal element, which we call R.

We apply Lemma 4.1.9 to $IR \lhd R \subseteq K$ and see that for any $x \in K$ we have $IR[x] \neq R[x]$ or $IR[x^{-1}] \neq R[x^{-1}]$. Hence $R[x] \in \mathcal{M}$ or $R[x^{-1}] \in \mathcal{M}$. From the maximality of R we conclude that $R[x] = R$ or $R[x^{-1}] = R$, that is $x \in R$ or $x^{-1} \in R$. Thus R is indeed a valuation ring (with quotient field K) containing S, and the elements of I being non-units are in $J(R)$. $\qquad\square$

Corollary 4.1.11 *If K is a field of characteristic zero and p is a prime, then there is a valuation ring R with quotient field K and $\mathrm{char}(R/J(R)) = p$.*

Proof. Apply the theorem with $S := \mathbb{Z}$ and $I := p\mathbb{Z}$. Since $J(R) \supseteq p\mathbb{Z}$ the prime field $\mathbb{Z}/p\mathbb{Z}$ can be embedded into $R/J(R)$. $\qquad\square$

Lemma 4.1.12 *Let $\bar{\mathbb{Q}}$ be the algebraic closure of $\mathbb{Q}$ in $\mathbb{C}$ and let p be a prime.*

(a) *There is a valuation ring R in $\bar{\mathbb{Q}}$ with $J(R) \cap \mathbb{Z} = p\mathbb{Z}$. Let*

$$\mathcal{U}_{p'} := \{\zeta \in \mathbb{C} \mid \zeta^m = 1 \ \text{ for some } \ m \ \text{ with } \ p \nmid m\}$$

be the multiplicative group of roots of unity of order coprime to p. Then $\mathcal{U}_{p'} \subseteq R$, and if $\theta : R \to F$ is a ring homomorphism onto a field F, then $F \cong R/J(R)$ is an algebraic closure of the prime field $\mathbb{F}_p$ and

$$\theta|_{\mathcal{U}_{p'}} : \mathcal{U}_{p'} \to F^{\times}$$

is an isomorphism of multiplicative groups.

(b) *Let $m \in \mathbb{N}$ and let $\theta : \mathbb{Z}[\zeta_m] \to F$ be a ring homomorphism into a field of characteristic p. Then θ may be extended to a ring homomorphism $\theta' : R \to \bar{F}$, where R is a valuation ring in $\bar{\mathbb{Q}}$ and $\bar{F}$ is an algebraic closure of $\mathrm{im}(\theta)$. Also θ induces a group isomorphism*

$$\theta|_{\mathcal{U}_{p',m}} : \mathcal{U}_{p',m} \to \{\alpha \in F \mid \alpha^m = 1\},$$

where $\mathcal{U}_{p',m} := \{\zeta \in \mathcal{U}_{p'} \mid \zeta^m = 1\}$.

Proof. (a) The first assertion follows from Theorem 4.1.10. By Lemma 4.1.3(b) we have $\mathcal{U}_{p'} \subset R$. If $\zeta \in \mathcal{U}_{p'}$, $\zeta \neq 1$, has multiplicative order m, then

$$f = 1 + X + \cdots + X^{m-1} = \frac{X^m - 1}{X - 1} = \prod_{i=1}^{m-1} (X - \zeta^i) \in R[X].$$

Hence $1 - \zeta$ divides $m = f(1)$ in R. Therefore, if $\theta(\zeta) = 1$ then $m \in \mathrm{J}(R) \cap \mathbb{Z} = p\mathbb{Z}$, which contradicts the definition of $\mathcal{U}_{p'}$. Thus $\theta|_{\mathcal{U}_{p'}}$ is injective. On the other hand, it is easily seen that every element of F is algebraic over $\mathbb{F}_p$ because every element of R is algebraic over $\mathbb{Z}$. Any element $\alpha \neq 0$ algebraic over F is also algebraic over $\mathbb{F}_p$; hence it is in a finite field and is thus an mth root of unity for some $m \in \mathbb{N}$ coprime to p. Thus $\alpha \in \theta(\mathcal{U}_{p'})$.

(b) Put $R' := \mathbb{Z}[\zeta_m]$. Using Theorem 4.1.10 we choose a valuation ring R in $\bar{\mathbb{Q}}$ such that $R' \subset R$ and $R' \cap \mathrm{J}(R) = \ker(\theta)$, which is certainly a maximal ideal in R'. The homomorphism θ can be extended to $\theta' : R' + \mathrm{J}(R) \to F$, mapping $r + x$ with $r \in R'$ and $x \in \mathrm{J}(R)$ to $\theta(r)$. Let $\theta_{\mathrm{nat}} : R \to R/\mathrm{J}(R)$ be the natural homomorphism. Then $\theta' = \tau \circ \theta_{\mathrm{nat}}|_{R'+\mathrm{J}(R)}$ with a field isomorphism $\tau : (R' + \mathrm{J}(R))/\mathrm{J}(R) \to \theta(R')$. Clearly τ can be extended to an isomorphism $\bar{\tau} : R/\mathrm{J}(R) \to \bar{F}$ (see [110], theorem V.2.8, p. 233). Then $\theta' := \bar{\tau} \circ \theta_{\mathrm{nat}}$ is the desired extension of θ. By part (a) we know that $\theta|_{\mathcal{U}_{p',m}} = \theta'|_{\mathcal{U}_{p',m}}$ is injective, and hence an isomorphism. $\qquad\square$

We remark in passing that if $K \subseteq L$ is a field extension and R is a valuation ring with quotient field L, then obviously $R \cap K$ is a valuation ring with quotient field K and maximal ideal $\mathrm{J}(R) \cap K$.

The proofs of Theorem 4.1.10 and Corollary 4.1.11 were non-constructive. This can be remedied if one restricts oneself to algebraic number fields, as we will frequently do. In this case valuation rings are always principal ideal domains.

Definition 4.1.13 A valuation ring which is also a principal ideal domain is called a **discrete valuation ring**.

Lemma 4.1.14 *If R is a discrete valuation ring with quotient field K and maximal ideal πR, then any element $\alpha \in K^\times$ has a unique representation in the form*

$$\alpha = \pi^n u \qquad \text{with} \qquad n \in \mathbb{Z} \, , \ u \in R^\times.$$

We define $\nu = \nu_\pi : K^\times \to \mathbb{Z}$, $\pi^n u \mapsto n$ with $n \in \mathbb{Z}$, $u \in R^\times$, and call it the **valuation** *corresponding to R.*

Proof. Since R is a principal ideal domain it is factorial ([110], theorem II.5.2, p. 112) and π is the only prime element up to equivalence. This implies the result using Definition 4.1.2. $\qquad\square$

We state without proofs a few basic facts about discrete valuation rings and their completions, which can be found in many textbooks on algebra.

Lemma 4.1.15 *Any valuation ring in an algebraic number field is discrete.*

Proof. Since the valuation rings $\mathbb{Z}_{(p)}$ in $\mathbb{Q}$ are obviously discrete, the result follows from [32], prop. 8.5.3, p. 297. $\qquad\square$

Under the assumptions of Lemma 4.1.14 ν_π obviously induces an order iso-morphism from the ordered abelian group $W := K^\times/R^\times = \{\pi^n R^\times \mid n \in \mathbb{Z}\}$ introduced in Lemma 4.1.3(c) to $(\mathbb{Z}, \leq)$. The "discrete valuation" ν_π can easily be used to define a norm on K in the topological sense by choosing $c \in \mathbb{R}$ with $0 < c < 1$ and putting

$$|\alpha|_\pi := \begin{cases} 0 & \text{for } \alpha = 0, \\ c^{\nu_\pi(\alpha)} & \text{for } \alpha \in K^\times. \end{cases}$$

This turns K also into a metric space, and it makes sense to talk about conver-gence and Cauchy sequences in K. But we can define these notions directly and in somewhat greater generality, without referring to metric spaces as follows.

Definition 4.1.16 Let R be a ring and V be an R-module.

(a) We call $\mathcal{F} = (V_n)_{n \in \mathbb{Z}}$ a **filtration** of V if

$$V_{n+1} \leq_R V_n \quad \text{for all} \quad n \in \mathbb{Z}, \qquad \bigcap_{n \in \mathbb{Z}} V_n = \{0\} \quad \text{and} \qquad \bigcup_{n \in \mathbb{Z}} V_n = V.$$

For $v \in V$ we put

$$\nu_\mathcal{F}(v) = \begin{cases} n & \text{if } v \in V_n \setminus V_{n+1}, \\ \infty & \text{if } v = 0. \end{cases}$$

(b) A sequence $(v_n)_{n \in \mathbb{N}}$ is called an $\mathcal{F}$-**Cauchy sequence** or is said to be $\mathcal{F}$-**convergent** with limit $v \in V$ (in symbols $v = \lim_{n \to \infty} v_n$) if

$$\lim_{n \to \infty} \nu_\mathcal{F}(v_n - v_{n-1}) = \infty \qquad \text{or} \qquad \lim_{n \to \infty} \nu_\mathcal{F}(v_n - v) = \infty \quad \text{in } \mathbb{R}, \text{ respectively.}$$

(c) V is called $\mathcal{F}$-**complete** if every $\mathcal{F}$-Cauchy sequence in V is $\mathcal{F}$-convergent.

(d) If R is a discrete valuation ring with quotient field K and maximal ideal πR we have a filtration $\mathcal{F} := (\pi^n R)_{n \in \mathbb{Z}}$ of K; R is called a **complete discrete valuation ring** if K is $\mathcal{F}$-complete.

Obviously in (d) we have $\nu_\mathcal{F}(\alpha) = \nu_\pi(\alpha)$ for $\alpha \in K^\times$.

Theorem 4.1.17 *If R is a discrete valuation ring with quotient field K and maximal ideal πR, there is a field extension $\tilde{K} \supseteq K$ and a complete discrete valuation ring $\tilde{R} \supseteq R$ with quotient field $\tilde{K}$ and maximal ideal $\pi \tilde{R}$ such that $\tilde{R}/\pi\tilde{R} \cong R/\pi R$, and every element of $\tilde{K}$ is a $\tilde{\nu}_\pi$-limit of a Cauchy sequence in K ("K is dense in $\tilde{K}$"); $\tilde{R}$ is usually called the **completion** of R.*

Proof. The proof is similar to the usual construction of real numbers as the "completion" of the rational numbers. See, for instance, [32], p. 272. □

Definition 4.1.18 If p is a prime, (K, R, F, η) is called a p-**modular system** if R is a complete discrete valuation ring with quotient field K of characteristic zero and $\eta: R \to F$ is a ring epimorphism onto a field F of characteristic p. A generator for $\ker \eta \trianglelefteq R$ will always be denoted by π. It is called a p-**modular**

splitting system for a finite group G if K contains a splitting field of $X^m - 1 \in$ $\mathbb{Q}[X]$, where m is the exponent of G. If $F = R/\pi R$ and $\eta\colon R \to R/\pi R$ is the canonical projection, we write (K, R, F) instead of (K, R, F, η).

In the literature it is common practice to consider only p-modular systems (K, R, F), but for practical purposes it is useful to have a homomorphism η mapping R onto a concrete field F, in which it is easier to compute. Observe that F can be considered as an R-module $(\mathrm{Inf}_\eta F)$ and $\hat{A} \cong F \otimes_R A$.

Remark 4.1.19 (a) If (K, R, F, η) is a p-modular splitting system for G then by Theorem 3.10.8 and Theorem 2.9.18 K and F are splitting fields for G.

(b) If G is a finite group with exponent m, we obtain a p-modular splitting system for G in the following way. Choose a ring homomorphism $\theta\colon \mathbb{Z}[\zeta_m] \to F$ onto a field F of characteristic p (see Lemma 4.1.12 or Exercise 4.1.2). By Theorem 4.1.10 and Lemma 4.1.15 there is a discrete valuation ring $R_1 \supset \mathbb{Z}[\zeta_m]$ in $\mathbb{Q}_m := \mathbb{Q}(\zeta_m)$ with maximal ideal πR_1 satisfying $\pi R_1 \cap \mathbb{Z}[\zeta_m] = \ker(\theta)$. We take the completion $R := \tilde{R}_1$ with quotient field K and maximal ideal πR. Similarly as in the proof of Lemma 4.1.12(b) one can extend θ first to a ring epimorphism $\theta'\colon R_1 \to F_1 \supseteq F$, where $F_1 \cong R_1/\pi R_1 \cong R/\pi R$ is a finite field extension of F and then to $\eta\colon R \to F_1$.

The completeness of R in a p-modular system (K, R, F, η) was not so important for the discussion in the preceding remark, but it is essential for the proof of Theorem 4.1.21. For this we also need the following simple observation.

Lemma 4.1.20 *If R is a complete discrete valuation ring with maximal ideal πR and V is a finitely generated R-module, then putting $V_n := \pi^n V$ for $n \geq 0$ and $V_n := \{0\}$ for $n < 0$ we get a filtration $\mathcal{F} := (V_n)_{n \in \mathbb{Z}}$ and V is $\mathcal{F}$-complete.*

Proof. This follows directly from the classification of the finitely generated modules over a principal ideal domain ([110], theorem III.7.3, p. 147, theorem III.7.5, p. 149). $\qquad\square$

Theorem 4.1.21 *Let (K, R, F, η) be a p-modular system and let A be an R-order and $\hat{A} := A/\pi A$ as above.*

(a) *If $\hat{x}$ is an idempotent in the F-algebra $\hat{A}$, then there is an idempotent $e \in A$ with $\hat{x} = \hat{e}$. "Idempotents in $\hat{A}$ can be lifted to A."*

(b) *$A^\times = \{a \in A \mid \hat{a} \in \hat{A}^\times\}$. The ring A is local if and only if $\hat{A}$ is local.*

(c) *If e_1 and e_2 are idempotents in A with $\hat{e}_1 = \hat{e}_2$ then there is a unit $u \in A^\times$ with $e_2 = u^{-1} e_1 u$.*

(d) *$\hat{}$ induces a bijection between the central idempotents of A and those of $\hat{A}$.*

(e) *If an A-lattice V is indecomposable then $\mathrm{End}_A V$ is a local ring.*

Proof. (a) We put $x_0 := x$ and define $x_n \in A$ for $n > 0$ recursively by $x_{n+1} := 3\,x_n^2 - 2\,x_n^3$. We show that $y_n := x_n^2 - x_n \in \pi^{2^n} A$ for all n. By assumption this holds for $n = 0$, and we assume inductively that it holds also for some $n \geq 0$. Then

$$x_{n+1}^2 - x_{n+1} = 9\,x_n^4 - 12\,x_n^5 + 4\,x_n^6 - 3\,x_n^2 + 2\,x_n^3 = 4\,y_n^3 - 3\,y_n^2 \in \pi^{2^{n+1}} A.$$

We also obtain $x_{n+1} - x_n = y_n\,(1 - 2\,x_n) \in \pi^{2^n} A$, which means that $(x_n)_{n \in \mathbb{N}}$ is a Cauchy sequence in A. Since A is a finitely generated R-module, we deduce from Lemma 4.1.20 that there is $e \in A$ with $e = \lim_{n \to \infty} x_n$. We also get $e^2 - e = \lim_{n \to \infty}(x_n^2 - x_n) = 0$ by the above. Finally, since $x_n - x_0 \in \pi A$ for all n we also have $e - x_0 \in \pi A$ and thus $\hat{x} = \hat{e}$.

(b) Let $u \in A$ and let $\hat{u} \in \hat{A}^\times$ with inverse $\hat{u}'$. Then $y := 1 - uu' \in \pi A$. Hence the series $\sum_{n=0}^\infty y^n$ is convergent in A and

$$u\,u' \sum_{n=0}^\infty y^n = (1 - y) \sum_{n=0}^\infty y^n = 1,$$

therefore u has a right inverse. Similarly u has a left inverse and so $u \in A^\times$. If $\hat{A}$ is local, then $A \setminus A^\times$ is the preimage of the ideal $\hat{A} \setminus \hat{A}^\times$ under the natural map $\hat{\ }$, hence an ideal. So A is local. The converse is clear.

(c) Put $u := 1 - e_1 - e_2 + 2e_1e_2$. Then $\hat{u} = 1$ and by part (b) we have $u \in A^\times$. Also $e_1u = e_1e_2 = ue_2$, hence the result.

(d) If e is a central idempotent in A, then obviously $\hat{e}$ is a central idempotent in $\hat{A}$. If e_1 and e_2 are central idempotents in A with $\hat{e}_1 = \hat{e}_2$, we see from (c) that $e_1 = e_2$. Now assume that $\hat{e}$ is a central idempotent in $\hat{A}$. By (a) we may assume that $e^2 = e \in A$. To show that $e \in \mathbf{Z}(A)$ we consider

$$A = eAe \oplus (1 - e)Ae \oplus eA(1 - e) \oplus (1 - e)A(1 - e).$$

Under the natural map $\hat{\ }$, the second and third summand are mapped to $\{0\}$ because $\hat{e} \in \mathbf{Z}(\hat{A})$. Hence $(1 - e)Ae = \pi(1 - e)Ae = \mathrm{J}(R)(1 - e)Ae$, and from Nakayama's lemma (Exercise 1.4.1) we get $(1 - e)Ae = \{0\}$ and similarly $eA(1 - e) = \{0\}$. Hence any element $x \in A$ can be written in the form $x = eae + (1 - e)b(1 - e)$, with $a, b \in A$, and thus it commutes with e.

(e) If V is indecomposable, $E := \mathrm{End}_A V$ contains only the trivial idempotent; E is an R-order since it is a subalgebra of the R-order $\mathrm{End}_R V$. By (a) the F-algebra $\hat{E} := E/\pi E$ has no non-trivial idempotent either. From Exercise 4.1.1 we see that the semisimple F-algebra $\hat{E}/\mathrm{J}(\hat{E})$ has no non-trivial idempotent. Thus $\hat{E}/\mathrm{J}(\hat{E})$ is a division ring (see Theorem 1.5.5). Then $\hat{E}$ is local and by (b) E is also. $\qquad\square$

It follows from the Krull–Schmidt theorem (Theorem 1.6.6) that the indecomposable direct summands of an A-lattice are uniquely determined up to isomorphism.

Corollary 4.1.22 (a) *Let (K, R, F, η) be a p-modular system and let A be an R-order. If $\epsilon_1, \ldots, \epsilon_m$ are block idempotents of A, then*

$$\hat{A} = \hat{A}\hat{\epsilon}_1 \oplus \cdots \oplus \hat{A}\hat{\epsilon}_m$$

is the block decomposition of $\hat{A}$. For any indecomposable $\hat{A}$-module Y there is exactly one block idempotent $\hat{\epsilon}_i$ such that $Y = \hat{\epsilon}_i Y$. Likewise, for any simple KA-module V there is exactly one block idempotent ϵ_i such that $V = \epsilon_i V$. In this case Y and V are said to **belong to the block** $B_i := A\epsilon_i$.

If $B_i(KA)$ is a set of representatives of the isomorphism classes of simple KA-modules V belonging to B_i, and if $\epsilon(V)$ is the block idempotent of KA with $\epsilon(V)V = V$, then

$$\epsilon_i = \sum_{V \in B_i(KA)} \epsilon(V).$$

(b) *If W is a projective $\hat{A}$-module, then there is a (projective) A-lattice M with $W \cong_{\hat{A}} M/\pi M$. The KA-module KM is sometimes called p-***projective**.

Proof.   (a) This follows from part (d) of Theorem 4.1.21, see also Corollary 1.7.3. Observe that we embed A into KA and that the block idempotents ϵ_i of A are central idempotents of KA but not necessarily centrally primitive in KA.

(b) Any projective $\hat{A}$-module is of the form (see Remark 1.6.22)

$$W = \hat{A}\hat{e}_1 \oplus \cdots \oplus \hat{A}\hat{e}_n$$

with idempotents $\hat{e}_i$ in $\hat{A}$. By part (a) of Theorem 4.1.21 we may assume that the e_i are idempotents in A for $1 \leq i \leq n$. Putting $M := Ae_1 \oplus \cdots \oplus Ae_n$ we get $W \cong_{\hat{A}} M/\pi M$.   □

Theorem 4.1.23 (R. Brauer) *Let A be an R-order and let (K, R, F, η) be a p-modular splitting system for A. Assume that $V_1, \ldots, V_k$ are representatives of the simple KA-modules and that $Y_1, \ldots, Y_l$ are representatives of the simple $\hat{A}$-modules.*

(a) *If Y is a simple $\hat{A}$-module, V a KA-module and M any R-form of V then the number of composition factors of $\hat{M}$ which are isomorphic to Y does not depend on the choice of M but only on V and hence may be denoted by $d_{V,Y}$. The matrix*

$$D := [d_{ij}] \in \mathbb{N}_0^{k \times l} \qquad with \qquad d_{ij} := d_{V_i, Y_j} \ (1 \leq i \leq k \ , \ 1 \leq j \leq l)$$

*is called the p-***decomposition matrix** *and its entries are called the p-***decomposition numbers** *of A.*

(b) (Brauer reciprocity) *If $\hat{A}\hat{e}_j$ is a projective cover of Y_j $(1 \leq j \leq l)$ with $e_j^2 = e_j \in A$ and $P(V_i)$ a projective cover of V_i $(1 \leq i \leq k)$ then*

$$KAe_j \cong \bigoplus_{i=1}^{k} d_{ij} P(V_i). \tag{4.4}$$

(c) *If $C = [c_{ij}] \in \mathbb{N}_0^{l \times l}$ is the Cartan matrix of the F-algebra $\hat{A}$ and $C' = [c'_{ij}]$ is the Cartan matrix of KA then*

$$C = D^T C' D. \qquad (4.5)$$

Of course, if KA is semisimple are given by as it will be in our applications, then (4.4) and (4.5)

$$KAe_j \cong \bigoplus_{i=1}^{k} d_{ij} V_i \quad \text{and} \quad C = D^T D. \qquad (4.6)$$

Proof.   (a) Let $P(Y)$ be a projective cover of Y. By Theorem 4.1.21(a) we may assume that $P(Y) = \hat{e}\hat{A}$ with an idempotent $e \in A$. Let

$$N_0 := \hat{M} > N_1 > \ldots > N_r := \{0\}$$

be a composition series. Since $P(Y)$ is projective, the sequence

$$\{0\} \to \text{Hom}_{\hat{A}}(P(Y), N_j) \to \text{Hom}_{\hat{A}}(P(Y), N_{j-1})$$
$$\to \text{Hom}_{\hat{A}}(P(Y), N_{j-1}/N_j) \to \{0\}$$

is exact for $j = 1, \ldots, r$. Since F is a splitting field for $\hat{A}$ we get

$$\text{Hom}_{\hat{A}}(P(Y), N_{j-1}/N_j) \cong \begin{cases} F & \text{if } N_{j-1}/N_j \cong Y, \\ \{0\} & \text{else (see (1.18) on p. 76).} \end{cases}$$

Hence

$$\dim_F \text{Hom}_{\hat{A}}(P(Y), \hat{M}) = \sum_{j=1}^{r} \dim_F \text{Hom}_{\hat{A}}(P(Y), N_{j-1}/N_j) = d(M, Y),$$

the number of composition factors of $\hat{M}$ isomorphic to Y. But by Theorem 1.6.8

$$\text{Hom}_{\hat{A}}(P(Y), \hat{M}) \cong_F \hat{e}\hat{M} \cong eM/\pi eM \qquad (i = 1, 2).$$

Since R is a principal ideal domain, eM is a free R-module, and we see that

$$d(M, Y) = \dim_F(\hat{e}\hat{M}) = \text{rk}(eM) = \dim_K eV$$

is independent of the choice of M.

(b) Let $KAe_j \cong \bigoplus_{i=1}^{k} d'_{ij} P(V_i)$ and let M_i be an R-form of V_i. Then

$$d'_{ij} = \dim_K \text{Hom}_{KA}(KAe_j, V_i) = \dim_K e_j V_i = \text{rk}_R e_j M_i = \dim_F \hat{e}_j \hat{M}_i$$

$$= \dim_F \text{Hom}_{\hat{A}}(\hat{A}\hat{e}_j, \hat{M}_i) = \sum_{m=1}^{l} d_{i,m} \dim_F \text{Hom}_{\hat{A}}(\hat{A}\hat{e}_j, Y_m) = d_{ij}.$$

(c) follows from (a) and (b). Namely, by part (a) of the proof,

$$c_{ij} = \dim_F \text{Hom}_{\hat{A}}(\hat{A}\hat{e}_j, \hat{A}\hat{e}_i) = \dim_F \hat{e}_j \hat{A}\hat{e}_i$$

$$= \dim_K e_j KAe_i = \dim_K \text{Hom}_{KA}(KAe_j, KAe_i) = \sum_{l=1}^{k} \sum_{l'=1}^{k} d_{l,j} d_{l',i} c'_{l',l}.$$

$\square$

Corollary 4.1.24 *If in Theorem 4.1.23 A and $\hat{A}$ are semisimple, then the simple modules may be labeled so that the decomposition matrix is the identity matrix.*

Proof. By our assumption the simple module Y_j coincides with its projective cover $\hat{A}\hat{e}_j$. Thus

$$\dim_F(Y_j) = \dim_K(KAe_j) = \sum_{i=1}^{k} d_{ij} \dim_K(V_i) \geq \sum_{i=1}^{k} d_{ij}^2 \dim_F(Y_j).$$

$\square$

Exercises

Exercise 4.1.1 Let A be a finite dimensional F-algebra and let $I \trianglelefteq A$ be nilpotent. Show that for every idempotent $\epsilon \in A/I$ there is an $e = e^2 \in A$ with $\epsilon = e + I$.
Hint: Mimic the proof of Theorem 4.1.21(a) and use induction on $\dim_F I$.

Exercise 4.1.2 Construct explicitly a ring homomorphism from $\mathbb{Z}[\zeta_m]$ onto a field of characteristic $p > 0$.

Exercise 4.1.3 Let (K, R, F, η) be a p-modular system. Show that $\mathbf{Z}(RG)$ is an R-order in $\mathbf{Z}(KG)$. Extend η to a ring homomorphism $\eta \colon \mathbf{Z}(RG) \to \mathbf{Z}(FG)$ in the natural way and show that η is an epimorphism with kernel $\pi \, \mathbf{Z}(RG)$.

Exercise 4.1.4 Let (K, R, F, η) be a p-modular system, let Ω_1, Ω_2 be G-sets, and let $R\Omega_i$, $K\Omega_i$ and $F\Omega_i$ for $i = 1, 2$ be the corresponding permutation modules over R, K, F.
(a) Show that $\mathrm{Hom}_{FG}(F\Omega_1, F\Omega_2) \cong \mathrm{Hom}_{RG}(R\Omega_1, R\Omega_2)/\pi \, \mathrm{Hom}_{RG}(R\Omega_1, R\Omega_2)$, i.e. any FG-homomorphism from $F\Omega_1$ to $F\Omega_2$ can be lifted to an RG-homomorphism from $R\Omega_1$ to $R\Omega_2$.
(b) For $\Omega := \Omega_1 = \Omega_2$ show that $\mathrm{Hom}_{RG}(R\Omega)$ is an R-order in $\mathrm{End}_{KG}(K\Omega)$. Moreover show that $\mathrm{End}_{FG}(F\Omega) \cong \mathrm{End}_{RG}(R\Omega)/\pi \, \mathrm{End}_{RG}(R\Omega)$ as F-algebras.

4.2 Brauer characters

Throughout this section G is a finite group and p is a prime. If K is a field of characteristic zero, then KG-modules are isomorphic if and only if their characters are equal (Corollary 2.1.16); also, the character of a KG-module provides complete information about its composition factors, including multiplicities, provided that the irreducible characters are known. All this does not hold for fields F of characteristic $p > 0$, regardless of whether or not p divides the group order $|G|$. For instance, if W is a vector space over F on which G acts trivially and if $\dim_F W = kp + 1$ for some $k \in \mathbb{N}$, then the character χ_W of W is the trivial character $\mathbf{1}_G$ over F. Obviously a character can only give information about multiplicities of composition factors of a module modulo p. Also, if p divides $|G|$ there are non-conjugate elements $g, g' \in G$ with $\chi(g) = \chi(g')$ for all characters $\chi \in \mathrm{Irr}_K(G)$. We need a definition.

Definition 4.2.1 An element $g \in G$ is called p-**regular** for a prime p if its order m is coprime to p and p-**singular** if it is not p-regular. If $m := |\langle g \rangle| = p^k q$ with $p \nmid q$ and $1 = a p^k + b q$ with $a, b \in \mathbb{Z}$ we call

$$g_{p'} := g^{a\,p^k} \quad \text{and} \quad g_p := g^{b\,q} \qquad (\text{so} \quad g = g_{p'}g_p = g_p g_{p'})$$

the p-**regular part** and the p-**part** of g, respectively. Let $G_{p'}$ be the set of p-regular elements of G. A conjugacy class is called p-regular or p-singular if its elements are p-regular or p-singular, respectively. The set of p-regular conjugacy classes will be denoted by $\mathrm{cl}(G_{p'})$.

Observe that in the definition a, b are uniquely determined modulo $p^k q$ and so the p-regular part and the p-part are well defined.

Lemma 4.2.2 *If F is a field of characteristic $p > 0$ then*

$$\chi(g) = \chi(g_{p'}) \qquad \text{for all} \qquad \chi \in \mathrm{Char}_F(G).$$

Proof.   We may assume that $G = \langle g \rangle$ and that F is algebraically closed. Let δ be a representation with character χ. Then $\delta(g) = \delta(g_{p'})\delta(g_p)$. By Example 1.1.23 we may choose a basis such that the matrices of $\delta(g_{p'})$ $\delta(g_p)$ are triangular matrices and the one for $\delta(g_p)$ is unipotent, and thus has only 1's on the diagonal. So the diagonal elements of $\delta(g)$ and $\delta(g_{p'})$ coincide. $\square$

From Lemma 4.1.12 we recall that for a prime p and a number $m \in \mathbb{N}$ we define

$$\mathcal{U}_{p'} = \langle \{ \, \zeta_k \mid p \nmid k \, \} \rangle \quad \text{and} \quad \mathcal{U}_{p',m} := \{ \zeta \in \mathcal{U}_{p'} \mid \zeta^m = 1 \}.$$

Definition 4.2.3 Let G be a finite group of exponent $m = p^r q$ with $p \nmid q$ and let F be a field of characteristic $p > 0$ with algebraic closure $\bar{F}$. Assume that W is an FG-module with $\dim_F W = n$ and representation $\delta \colon G \to \mathrm{GL}(W)$. Let $\theta \colon \mathbb{Z}[\zeta_m] \to \bar{F}$ be a ring homomorphism. For $g \in G_{p'}$ the eigenvalues of $\delta(g)$ are mth roots of unity in $\bar{F}$ and thus, by Lemma 4.1.12, are of the form

$$\theta(\zeta_1(g)), \ \ldots, \ \theta(\zeta_n(g)) \qquad \text{with uniquely determined} \qquad \zeta_i(g) \in \mathcal{U}_{p',m}.$$

We then define

$$\varphi_W(g) := \zeta_1(g) + \cdots + \zeta_n(g)$$

and call

$$\varphi_W \colon G_{p'} \to \mathbb{C} \qquad g \mapsto \zeta_1(g) + \cdots + \zeta_n(g)$$

the **Brauer character** of G afforded by W or δ with respect to θ. If W is simple then φ_W is called **irreducible**. The set of irreducible Brauer characters of G with respect to θ over a splitting field of characteristic p will be denoted by $\mathrm{IBr}_\theta(G)$. If (K, R, F, η) is a p-modular splitting system for G we put $\theta := \eta|_{\mathbb{Z}[\zeta_m]}$ and write $\mathrm{IBr}_\eta(G)$ (or simply $\mathrm{IBr}(G)$ if η is known) for $\mathrm{IBr}_\theta(G)$.

Obviously this definition of a Brauer character depends on θ. In order to avoid this ambiguity – which is particularly important if one is dealing with Brauer characters of different groups at the same time – we will often make a definite choice for an algebraic closure $\mathcal{F}_p$ of $\mathbb{F}_p$ and a ring homomorphism

$$\theta_m \colon \mathbb{Z}[\zeta_m] \to \mathcal{F}_p.$$

In fact, θ_m will be the restriction to $\mathbb{Z}[\zeta_m]$ of a surjective ring homomorphism $\theta \colon \mathcal{R} \to \mathcal{F}_p$, where $\mathcal{R}$ is a valuation ring in $\bar{\mathbb{Q}}$.

We first deal with the question of how to define $\mathcal{F}_p$ and how to compute in this field. Of course, a finite field $\mathbb{F}_{p^n}$ may be realized as $\mathbb{F}_p[X]/(f_n)$ for any irreducible polynomial $f_n \in \mathbb{F}_p[X]$ of degree n. We will make suitable choices for these f_n.

Definition 4.2.4 (a) A polynomial $f_n \in \mathbb{F}_p[X]$ of degree n is called **primitive** if it is the minimal polynomial of a primitive element α_n of $\mathbb{F}_{p^n}$, that is of a generator of the multiplicative group $\mathbb{F}_{p^n}^\times$.
(b) Let $n \in \mathbb{N}$ or $n = \infty$. A sequence $(f_m)_{m=1}^n$ of primitive polynomials $f_m \in \mathbb{F}_p[X]$ of degree m is called **compatible** if $(X + (f_m))^{(p^m-1)/(p^d-1)}$ is a root of f_d whenever $d \mid m$.

If $f_n \in \mathbb{F}_p[X]$ is primitive of degree n, then

$$\mathbb{F}_p[X]/(f_n) = \{\, X^i + (f_n) \mid i = 1, \ldots, p^n - 1 \,\} \cup \{0\}.$$

If $(f_m)_{m=1}^\infty$ is a compatible sequence of polynomials in $\mathbb{F}_p[X]$ and $d \mid m$ we have an embedding

$$\varphi_{d,m} \colon \mathbb{F}_p[X]/(f_d) \to \mathbb{F}_p[X]/(f_m) \quad \text{with} \quad X + (f_d) \mapsto X^{(p^m-1)/(p^d-1)} + (f_m).$$

Obviously we have

$$d \mid m \quad \text{and} \quad m \mid n \quad \Longrightarrow \quad \varphi_{m,n} \circ \varphi_{d,m} = \varphi_{d,n}. \tag{4.7}$$

Using these embeddings (thus identifying x with $\varphi_{d,m}(x)$ for $x \in \mathbb{F}_p[X]/(f_d)$ and $d \mid m$) we may form

$$\mathcal{F}_p := \bigcup_{m \in \mathbb{N}} \mathbb{F}_p[X]/(f_m),$$

and we see that this is an algebraic closure of $\mathbb{F}_p[X]/(f_1) \cong \mathbb{F}_p$.

More formally, $\mathcal{F}_p$ is a direct limit:

$$\mathcal{F}_p := \varinjlim(\mathbb{F}_p[X]/(f_m), \varphi_{d,m}) := (\bigsqcup_{m \in \mathbb{N}} \mathbb{F}_p[X]/(f_m)) / \sim, \tag{4.8}$$

where $\bigsqcup$ stands for the disjoint union (coproduct) and the equivalence relation $\sim$ is defined as follows. If $x \in \mathbb{F}_p[X]/(f_m)$ and $y \in \mathbb{F}_p[X]/(f_n)$ then

$$x \sim y \quad \text{if and only if} \quad \varphi_{m,r}(x) = \varphi_{n,r}(y)$$

for some $r \in \mathbb{N}$ with $m \mid r$ and $n \mid r$.

Remark 4.2.5 If $(f_m)_{m=1}^{\infty}$ is a compatible sequence of polynomials in $\mathbb{F}_p[X]$ and $\mathcal{F}_p$ is as in (4.8), then we have a group isomorphism

$$\eta\colon \mathcal{F}_p^{\times} \to \mathcal{U}_{p'}\,, \quad [X^j + (f_m)]_{\sim} \mapsto \zeta_{p^m-1}^j \quad (1 \le j \le p^m - 1).$$

It readily follows from Lemma 4.1.12 that infinite compatible sequences of polynomials do exist. In fact, if θ is as in that lemma and $f_m \in \mathbb{F}_p[X]$ is the minimal polynomial of $\theta(\zeta_{p^m-1})$, it is easily seen that $(f_m)_{m=1}^{\infty}$ is a compatible sequence. In order to define such a unique sequence for each prime p, we introduce a total ordering on the set of monic polynomials in $\mathbb{F}_p[X]$.

Definition 4.2.6 (a) On $\mathbb{F}_p$ we use the ordering $0 < 1 < \cdots < p - 1$. For

$$f = \sum_{i=0}^{n}(-1)^{n-i}a_i X^i \quad \text{and} \quad g = \sum_{i=0}^{n}(-1)^{n-i}b_i X^i, \quad \text{with} \quad a_i, b_i \in \mathbb{F}_p,$$

and $a_n = b_n = 1$, we define $f \prec g$ if and only if there is a $j > 0$ such that $a_i = b_i$ for $j \le i \le n$ and $a_j < b_j$.

(b) The **Conway polynomial** $f_{p,n} \in \mathbb{F}_p[X]$ is defined inductively as the smallest polynomial (with respect to the above ordering $\prec$) satisfying the following:

(i) $f_{p,n}$ is primitive of degree n and
(ii) if d divides n then $(X + (f_{p,n}))^{(p^n-1)/(p^d-1)}$ is a root of $f_{p,d}$.

Theorem 4.2.7 *For each prime p and each $n \in \mathbb{N}$ there is a unique Conway polynomial $f_{p,n} \in \mathbb{F}_p[X]$ of degree n.*

Proof. (a) Let $\bar{\mathbb{F}}_p$ be an algebraic closure of $\mathbb{F}_p$. We first show the following.

Claim: If $(f_m)_{m=1}^{n-1}$ is a compatible sequence in $\mathbb{F}_p[X]$ and roots $\alpha_m \in \bar{\mathbb{F}}_p$ of f_m have been chosen for $1 \le m \le n - 1$ such that

$$\alpha_m^{(p^m-1)/(p^d-1)} = \alpha_d \quad \text{whenever} \quad d \mid m, \tag{4.9}$$

then we can find $\alpha_n \in \bar{\mathbb{F}}_p$ with minimal polynomial $f_n \in \mathbb{F}_p[X]$ of degree n such that $\alpha_n^{(p^n-1)/(p^d-1)} = \alpha_d$ whenever $d \mid n$ and $(f_m)_{m=1}^{n}$ is a compatible sequence. If $n = 1$ we may choose any primitive element $\alpha_1 \in \mathbb{F}_p$ and $f_1 := X - \alpha_1$.
Now let $1 < n$ and $\alpha_m \in \bar{\mathbb{F}}_p$ with minimal polynomial $f_m \in \mathbb{F}_p[X]$ satisfy (4.9) for $1 \le m \le n - 1$. Let $n_1, \ldots, n_s$ be maximal among the proper divisors of n and let

$$n_{i,j} := \text{g.c.d.}(n_i, n_j) \quad \text{for} \quad 1 \le i < j \le s.$$

We choose any primitive element β in the unique subfield $\mathbb{F}_{p^n}$ of order p^n in $\bar{\mathbb{F}}_p$. Then $\beta_i := \beta^{\frac{p^n-1}{p^{n_i}-1}}$ is a primitive element of $\mathbb{F}_p(\beta_i) = \mathbb{F}_{p^{n_i}}$. Since α_{n_i} is also a primitive element in $\mathbb{F}_{p^{n_i}}$, there is a k_i coprime to $p^{n_i} - 1$ with $\beta_i^{k_i} = \alpha_{n_i}$ for

$i = 1, \ldots, s$. Then g.c.d.$(p^{n_i} - 1, p^{n_j} - 1) = p^{n_{i,j}} - 1$ and $\beta_{i,j} := \beta^{\frac{p^n - 1}{p^{n_{i,j}} - 1}}$ is a primitive element in $\mathbb{F}_{p^{n_{i,j}}}$. Furthermore by our assumption (4.9)

$$\gamma_{i,j} := \alpha_{n_i}^{\frac{p^{n_i} - 1}{p^{n_{i,j}} - 1}} = \alpha_{n_j}^{\frac{p^{n_j} - 1}{p^{n_{i,j}} - 1}} \qquad \text{for} \qquad 1 \le i < j \le s.$$

Hence

$$\gamma_{i,j} = (\beta_i^{k_i})^{\frac{p^{n_i} - 1}{p^{n_{i,j}} - 1}} = (\beta^{\frac{p^n - 1}{p^{n_i} - 1} k_i})^{\frac{p^{n_i} - 1}{p^{n_{i,j}} - 1}} = \beta^{\frac{p^n - 1}{p^{n_{i,j}} - 1} k_i} = \beta_{i,j}^{k_i}$$
$$= (\beta_j^{k_j})^{\frac{p^{n_j} - 1}{p^{n_{i,j}} - 1}} = (\beta^{\frac{p^n - 1}{p^{n_j} - 1} k_j})^{\frac{p^{n_j} - 1}{p^{n_{i,j}} - 1}} = \beta^{\frac{p^n - 1}{p^{n_{i,j}} - 1} k_j} = \beta_{i,j}^{k_j}.$$

Consequently, $p^{n_{i,j}} - 1 \mid k_i - k_j$ for $1 \le i < j \le s$. Let u be the product of all prime divisors of $p^n - 1$ which are coprime to $\prod_{i=1}^{s}(p^{n_i} - 1)$. Using a slight extension of the Chinese remainder theorem (see [112], theorem 3.16, p. 62) we conclude that there is a $k \in \mathbb{N}$ with

$$k \equiv 1 \mod u \qquad \text{and} \qquad k \equiv k_i \mod (p^{n_i} - 1) \qquad (1 \le i \le s).$$

It follows that k is coprime to all $p^{n_i} - 1$ and to u. Hence $\alpha_n := \beta^k$ is a primitive element of $\mathbb{F}_{p^n}$ and

$$\alpha_n^{\frac{p^n - 1}{p^{n_i} - 1}} = \beta_i^k = \beta_i^{k_i} = \alpha_{n_i} \qquad \text{for} \qquad 1 \le i \le s.$$

Choosing $f_n \in \mathbb{F}_p[X]$ as the minimal polynomial of α_n, and using the isomorphism

$$\psi_n \colon \mathbb{F}_p(\alpha_n) \to \mathbb{F}_p[X]/(f_n), \quad \alpha_n \mapsto X + (f_n),$$

we see that if d divides n_i for some $i \in \{1, \ldots, s\}$ we have

$$0 = \psi_n(f_d(\alpha_d)) = \psi_n(f_d(\alpha_n^{(p^n - 1)/(p^d - 1)})) = f_d(X^{(p^n - 1)/(p^d - 1)} + (f_n)).$$

Hence $(f_m)_{m=1}^{n}$ is compatible.

(b) It follows from part (a) that any compatible sequence $(f_m)_{m=1}^{n-1}$ of polynomials can be extended to a compatible sequence $(f_m)_{m=1}^{n}$. From this the assertion of the theorem follows by induction:

If α_1 is the smallest primitive element in $\mathbb{F}_p$, then $f_{p,1} := X - \alpha_1$ is the unique Conway polynomial of degree one.

Now assume that for $m \le n - 1$ we have a unique Conway polynomial $f_m = f_{p,m} \in \mathbb{F}_p[X]$ of degree m; then $f_{p,n}$ is the smallest f_n such that $(f_m)_{m=1}^{n}$ is compatible. $\square$

For methods that compute Conway polynomials, and for a data base containing Conway polynomials for a large number of finite fields, see [113] (see also [77], and [98]). At present some of the Conway polynomials of the largest degrees in this data base are $f_{2,409}, f_{3,263}, f_{5,251}$ and those which took the most amount of CPU time (years!) are $f_{53,15}$, computed by R.A. Parker, and $f_{2,92}$, computed by K. Minola and D. Berechung.

Example 4.2.8 Here is a list of some Conway polynomials over small fields:

$$f_{2,1} = X + 1, \qquad f_{2,2} = X^2 + X + 1, \quad f_{2,3} = X^3 + X + 1,$$
$$f_{2,4} = X^4 + X + 1, \qquad f_{2,5} = X^5 + X^2 + 1, \quad f_{2,6} = X^6 + X^4 + X^3 + X + 1,$$
$$f_{2,7} = X^7 + X + 1,$$
$$f_{3,1} = X + 1, \qquad f_{3,2} = X^2 + 2X + 2, \quad f_{3,3} = X^3 + 2X + 1,$$
$$f_{3,4} = X^4 + 2X^3 + 2, \quad f_{3,5} = X^5 + 2X + 1, \quad f_{3,6} = X^6 + 2X^4 + X^2 + 2X + 2,$$
$$f_{5,1} = X + 3, \qquad f_{5,2} = X^2 + 4X + 2, \quad f_{5,3} = X^3 + 3X + 3.$$

◆

Remark 4.2.9 In the following we will identify

$$\mathcal{F}_p = \bigcup_{n \in \mathbb{N}} \mathbb{F}_p[X]/(f_{p,n}) = \varinjlim \mathbb{F}_p[X]/(f_{p,n}).$$

Obviously we have a ring homomorphism

$$\mathbb{Z}[\zeta_{p^n-1}] \to \mathbb{F}_p[X]/(f_{p,n}) \quad \text{with} \quad \zeta_{p^n-1} \mapsto X + (f_{p,n}).$$

This can be extended to a ring homomorphism

$$\theta \colon \mathbb{Z}[\mathcal{U}_{p'}] = \bigcup_{n \in \mathbb{N}} \mathbb{Z}[\zeta_{p^n-1}] \to \mathcal{F}_p.$$

By Lemma 4.1.3, $\mathbb{Z}[\mathcal{U}_{p'}]$ is contained in any valuation ring in $\bar{\mathbb{Q}}$, and we may choose one $\mathcal{R}$ with maximal ideal $\mathcal{P}$ containing $\ker(\theta)$. Then $\mathcal{R} = \ker(\theta) + \mathcal{P}$, and we may extend the above θ to a ring epimorphism (again denoted by θ):

$$\theta \colon \mathcal{R} \to \mathcal{F}_p. \qquad (4.10)$$

In particular we have $\theta(\zeta_{p^n-1}) = X + (f_{p,n})$. We put $\theta_m := \theta|_{\mathbb{Z}[\zeta_m]}$.

Definition 4.2.10 Let G be a finite group of exponent m. Then (K, R, F, η) is called a **standard p-modular system** for G if R with quotient field K is the completion of a valuation ring R_1 in $\mathbb{Q}_m = \mathbb{Q}(\zeta_m)$ and $\eta \colon R \to F$ is a surjective homomorphism with $\eta|_{\mathbb{Z}[\zeta_m]} = \theta_m$. In particular, $F \subseteq \mathcal{F}_p$. We put

$$\mathrm{IBr}_p(G) := \mathrm{IBr}_{\theta_m}(G).$$

Remark 4.2.11 With this definition, $\mathrm{IBr}_p(G)$ is uniquely defined by G and the prime p. All the irreducible Brauer characters of finite simple groups G (or their covers or automorphism groups) published in [98] or [82] are in the so defined $\mathrm{IBr}_p(G)$. The uniqueness of $\mathrm{IBr}_p(G)$ is particularly useful when one is considering Brauer characters of a group and a subgroup at the same time, for instance if one wants to use induction and restriction.

  Assumption For the rest of this section let (K, R, F, η) be a p-modular splitting system for our finite group G and let Brauer characters of G be defined via η.

Example 4.2.12 Let $G = \langle g \rangle$ be a cyclic group of order eight. It is easily checked that we have a matrix representation $\delta \colon G \to \mathrm{GL}_5(\mathbb{F}_3)$, $g^i \mapsto \mathbf{a}^i$ with

$$\mathbf{a} := \begin{bmatrix} 1 & . & 1 & 2 & 2 \\ 2 & 1 & 1 & 2 & 2 \\ 1 & . & 2 & 1 & . \\ 1 & 1 & 1 & 1 & . \\ 2 & 1 & 1 & . & 2 \end{bmatrix} \in \mathbb{F}_3^{5 \times 5}.$$

The characteristic polynomial of $\mathbf{a}$ is

$$g_{\mathbf{a}} := (X + 1)(X^2 + X - 1)(X^2 + 1) \in \mathbb{F}_3[X].$$

Using the Conway polynomial $f_{3,2} = X^2 + 2X + 2 \in \mathbb{F}_3[X]$ and abbreviating $\alpha := X + (f_{3,2}) \in F := \mathbb{F}_3[X]/(f_{3,2})$, we see that

$$g_{\mathbf{a}} = (X + 1)(X + \alpha)(X + \alpha^3)(X + \alpha^2)(X + \alpha^6) \in F[X].$$

We abbreviate $\zeta := \zeta_8$ and see that, with the notation of Remark 4.2.9,

$$\theta_8 \colon \mathbb{Z}[\zeta] \to \mathcal{F}_3, \qquad \zeta \mapsto \alpha.$$

Thus we obtain for the Brauer character $\varphi := \varphi_\delta$ of δ

$$\varphi(g) = -1 - \zeta - \zeta^3 - \zeta^2 - \zeta^6 = -1 - \zeta - \zeta^3 = -1 - \sqrt{2}\,i \in \mathbb{C}.$$

The GAPcommand `BrauerCharacterValue( a )` gives exactly this result.
    On the other hand, we also have a ring homomorphism

$$\theta \colon \mathbb{Z}[\zeta] \to \mathcal{F}_3, \qquad \zeta \mapsto -\alpha,$$

and for the the Brauer character φ' of δ with respect to θ we have

$$\varphi'(g) = -1 + \zeta + \zeta^3 - \zeta^2 - \zeta^6 = -1 + \zeta + \zeta^3 = -1 + \sqrt{2}\,i \in \mathbb{C}.$$

$\blacklozenge$

We note the following basic properties of Brauer characters.

Lemma 4.2.13 *Let F be a field of characteristic $p > 0$ and let φ be a Brauer character of G afforded by the FG-module W. Then we have the following.*

(a) *φ is a class function on $G_{p'}$, in short $\varphi \in \mathrm{cf}(G_{p'}, \mathbb{C})$.*

(b) *For $g \in G_{p'}$ one has $\varphi(g^{-1}) = \overline{\varphi(g)}$, the complex conjugate of $\varphi(g)$.*

(c) *$\overline{\varphi} \colon G_{p'} \to \mathbb{C}$ defined by $\overline{\varphi}(g) = \overline{\varphi(g)}$ for $g \in G_{p'}$ is the Brauer character afforded by W^*.*

(d) *If $V \leq_{FG} W$ then*

$$\varphi_W = \varphi_V + \varphi_{W/V}.$$

(e) *If two FG-modules have isomorphic composition factors (counting multi-plicities) then they have the same Brauer characters.*

(f) *Let (K, R, F, η) be a p-modular system for G and let V be a KG-module with character $\chi = \chi_V$. If M is an R-form and $\hat{M}$ is a p-modular reduction of V, then the Brauer character of $\hat{M}$ (with respect to η) is*

$$\varphi_{\hat{M}} = \chi|_{G_{p'}}.$$

Proof. Parts (a) to (e) follow easily from the definitions. We consider part (f). Let $\delta\colon G \to \mathrm{GL}_n(K)$ be a matrix representation afforded by V. By Corollary 4.1.5 we may assume that $\delta(g) = [a_{ij}(g)] \in \mathrm{GL}_n(R)$ for all $g \in G$. Then $\hat{M}$ affords the matrix representation

$$\hat{\delta}(g)\colon G \to \mathrm{GL}_n(F),\ g \mapsto [\eta(a_{ij}(g))].$$

Now let $g \in G_{p'}$. Then by assumption the eigenvalues $\zeta_1, \ldots, \zeta_n$ of $\delta(g)$ are in $\mathcal{U}_{p',m}$. Since η maps the characteristic polynomial of $\delta(g)$ to the characteristic polynomial of $\hat{\delta}(g)$, it follows that the eigenvalues of $\hat{\delta}(g)$ are $\eta(\zeta_1), \ldots, \eta(\zeta_n)$. Thus by definition $\varphi_{\hat{M}}(g) = \zeta_1 + \cdots + \zeta_n = \chi(g)$. $\square$

Theorem 4.2.14 (a) $\mathrm{IBr}(G)$ *is linearly independent over* $\mathbb{C}$.

(b) *Two FG-modules have the same Brauer characters if and only if they have isomorphic composition factors (including multiplicities).*

(c) *For $\chi \in \mathrm{Irr}(G)$ and $\varphi \in \mathrm{IBr}(G)$ let V_χ be a simple KG-module affording χ and let Y_φ be a simple FG-module affording φ. Let $d_{\chi\varphi} := d_{VY}$ be the corresponding decomposition number (see Theorem 4.1.23). Then*

$$\chi|_{G_{p'}} = \sum_{\varphi \in \mathrm{IBr}(G)} d_{\chi\varphi}\,\varphi.$$

Proof. (a) Since the Brauer characters have their values in the cyclotomic field $\mathbb{Q}_m$, with m being the exponent of G, it is sufficient to prove that $\mathrm{IBr}(G)$ is linearly independent over $\mathbb{Q}_m$. Suppose that

$$\sum_{\varphi \in \mathrm{IBr}(G)} a_\varphi \varphi = 0 \qquad \text{for} \qquad a_\varphi \in \mathbb{Q}_m \tag{4.11}$$

with some $a_\varphi \neq 0$. Since $R' := R \cap \mathbb{Q}(\zeta_m)$ is a discrete valuation ring in $\mathbb{Q}_m$ with maximal ideal, say $\pi'R'$, multiplying (4.11) by a suitable power of π', if need be, we may assume that all a_φ are in R' and not all are divisible by π'. Applying η we obtain a non-trivial linear relation

$$\sum_{\varphi \in \mathrm{IBr}(G)} \eta(a_\varphi)(\eta \circ \varphi)(g) = 0 \qquad \text{for all} \qquad g \in G_{p'}$$

with not all $\eta(a_\varphi) \in F$ being zero. If $\varphi \in \mathrm{IBr}(G)$ is the Brauer character of the representation δ_φ and $\chi_\varphi \colon G \to F$ is the character of δ_φ, then $\chi_\varphi(g) = (\eta \circ \varphi)(g)$ for $g \in G_{p'}$. By Lemma 4.2.2 we have

$$\sum_{\varphi \in \mathrm{IBr}(G)} \eta(a_\varphi)\chi_\varphi(g) = 0 \qquad \text{for all} \qquad g \in G,$$

which contradicts Lemma 2.1.5.

(b) follows from (a) and Lemma 4.2.13(d) and (e).

(c) is an immediate consequence of Lemma 4.2.13(f) and (d). □

Corollary 4.2.15 (a) *If $p \nmid |G|$ then* $\mathrm{Irr}(G) = \mathrm{IBr}(G)$.

(b) *If $H \leq G$ and $p \nmid |H|$ and φ is a Brauer character of G then $\varphi|_H$ is an ordinary character of H.*

Proof. (a) follows from Corollary 4.1.24, because, by Maschke's theorem (Theorem 1.5.6), FG is semisimple. (b) follows from (a). □

In practice the irreducible Brauer characters of G will be given by a matrix $[\varphi_i(g_j)]_{1 \leq i \leq l, 1 \leq j \leq s}$, where $\mathrm{IBr}(G) = \{\varphi_1, \ldots, \varphi_l\}$ and $g_1, \ldots, g_s$ are representatives of the p-regular conjugacy classes of G. This matrix is called the **(p-)Brauer character table** of G. It is easy to see that this is a square matrix.

Theorem 4.2.16 (a) *The number $\mathrm{l}(G) := |\mathrm{IBr}(G)|$ of irreducible Brauer characters of G equals the number of p-regular conjugacy classes of G.*

(b) *There are rational integers $a_{\varphi\psi}$ such that*

$$\varphi = \sum_{\chi \in \mathrm{Irr}(B)} a_{\varphi\psi}\, \chi|_{G_{p'}} \qquad \text{for} \qquad \varphi \in \mathrm{IBr}(G).$$

(c) *The decomposition matrix $D = [d_{\chi\varphi}]_{\chi \in \mathrm{Irr}(G), \psi \in \mathrm{IBr}(G)}$ has rank $\mathrm{l}(G)$. There are unimodular matrices $U_1 \in \mathbb{Z}^{k \times k}$, $U_2 \in \mathbb{Z}^{l \times l}$ such that*

$$U_1 D U_2 = \begin{bmatrix} \mathbf{I}_l \\ \mathbf{0} \end{bmatrix} \in \mathbb{N}_0^{k \times l}.$$

Proof. (a) Let $g_1, \ldots, g_s$ be representatives of the p-regular conjugacy classes of G. Assume $\mathrm{Irr}(G) = \{\chi_1, \ldots, \chi_k\}$, $\mathrm{IBr}(G) = \{\varphi_1, \ldots, \varphi_l\}$ and let

$$\Phi := [\varphi_i(g_j)]_{1 \leq i \leq l, 1 \leq j \leq s} \quad \text{and} \quad X := [\chi_i(g_j)]_{1 \leq i \leq k, 1 \leq j \leq s}. \qquad (4.12)$$

Then $X = D\,\Phi$. Since the character table is invertible (see e.g. Exercise 2.1.6) we have $\mathrm{rk}\, X = s \leq \mathrm{rk}\, \Phi = l \leq s$, because by Theorem 4.2.14 the l rows of Φ are linearly independent. Thus $\mathrm{l}(G) = l = s$.

(b) We extend each $\varphi \in \mathrm{IBr}(G)$ to a class function $\breve{\varphi} \colon G \to \mathbb{C}$ by

$$\breve{\varphi}(g) := \varphi(g_{p'}) \qquad \text{for} \qquad g \in G. \qquad (4.13)$$

We show that $\breve{\varphi}$ is a generalized character. To this end we take an arbitrary p-elementary subgroup $H = P \times Q$ with a p-subgroup P and a p'-subgroup Q. By Corollary 4.2.15 $\breve{\varphi}|_Q$ is an ordinary character and by construction $\breve{\varphi}|_H$ is just the unique extension of this character with P in its kernel. By Brauer's characterization of characters (Theorem 3.10.9), $\breve{\varphi}$ is indeed a generalized character, that is there are $a_{\varphi\chi} \in \mathbb{Z}$ with $\breve{\varphi} = \sum_{\chi \in \mathrm{Irr}(G)} a_{\varphi\chi}\, \chi$. Restricting to $G_{p'}$ we obtain (b).

(c) Let Φ and X be as in (4.12). If $a_{\varphi\chi}$ are as in (b) and

$$A := [a_{\varphi_i \chi_j}]_{1 \le i \le l,\, 1 \le j \le k},$$

we conclude from (b) and the definition of D that $\Phi = AX = AD\Phi$. Hence $AD = \mathbf{I}_l$ because Φ is non-singular. From this the result follows by elementary linear algebra. $\square$

The following properties of Brauer characters are generalizations of analogous statements that hold for ordinary characters. They are very useful for constructing the irreducible Brauer characters of a group.

Remark 4.2.17 Let H be a subgroup of G and let (K, R, F, η) be a p-modular system. Furthermore, let V, W be FG-modules with Brauer characters φ_V, φ_W. Then the following statements hold.

(a) The Brauer character $\varphi_{V \otimes W}$ of $V \otimes W$ is given by $\varphi_V \varphi_W$, i.e. $\varphi_{V \otimes W}(g) = \varphi_V(g) \cdot \varphi_W(g)$ for all $g \in G_{p'}$.

(b) The Brauer character φ_{V_H} of the restriction of V to H is given as follows: $\varphi_{V_H}(h) = \varphi_V(h)$ for all $h \in H_{p'}$. If φ is a Brauer character of G, then $\varphi_H := \varphi|_{H_{p'}}$ is a Brauer character of H.

Proof.   We leave the proof to the reader; see Exercise 4.2.3. $\square$

In (4.13) we associated to an irreducible Brauer character φ a generalized (ordinary) character $\breve{\varphi}$. There is a similar construction which is often useful. Before defining this we introduce some convenient notation.

For $x \in \mathbb{Q} \setminus \{0\}$ we write $\nu_p(x) = k$ if $x = p^k \frac{a}{b}$ with $a, b \in \mathbb{Z}$ with $p \nmid a, b$; see Lemma 4.1.14. To simplify the notation we will also write

$$|G|_p := p^{\nu_p(|G|)} \qquad \text{and} \qquad |G|_{p'} := \frac{|G|}{|G|_p}, \tag{4.14}$$

and $[G : H]_p$ and $[G : H]_{p'}$ are used in the same way for $H \le G$.

Definition 4.2.18 For $\theta \in \mathrm{cl}(G) \cup \mathrm{cl}(G_{p'})$ we define $\widehat{\theta}, \tilde{\theta} \in \mathrm{cf}(G, \mathbb{C})$ by

$$\widehat{\theta}(g) := \begin{cases} \theta(g) & \text{for } g \in G_{p'}, \\ 0 & \text{else,} \end{cases}$$

and $\tilde{\theta} := |G|_p\, \widehat{\theta}$.

Lemma 4.2.19 (a) $\tilde{\theta}$ *is a generalized character for any* $\theta \in \mathrm{Irr}(G) \cup \mathrm{IBr}(G)$.
(b)

$$\frac{(\chi, \tilde{\theta})_G}{\chi(1)} \in R \qquad \text{for every} \quad \chi \in \mathrm{Irr}(G).$$

(c) *If* $\chi \in \mathrm{Irr}(G)$ *and* $a := \nu_p(\chi(1))$ *then* $\frac{1}{p^a}\tilde{\chi}$ *is a generalized character, while* $\frac{1}{p^{a+1}}\tilde{\chi}$ *is not.*

Proof.   (a) As in the proof of Theorem 4.2.16, we consider an arbitrary (p)-elementary subgroup $H = P \times Q$ with a p-subgroup P and a p'-subgroup Q. Then $\tilde{\theta}|_H = \frac{|G|_p}{|P|}\rho_P \times \theta_Q$, with ρ_P the regular character of P. Since θ_Q is a character (Corollary 4.2.15) the result follows again from Theorem 3.10.9.
    (b) Let $G_{p'} = g_1{}^G \dot\cup \ldots \dot\cup g_l{}^G$. Then

$$(\chi, \tilde{\theta})_G \;=\; \frac{|G|_p}{|G|} \sum_{i=1}^l |g_i{}^G| \chi(g_i)\theta(g_i^{-1}) \tag{4.15}$$

$$\;=\; \frac{\chi(1)}{|G|_{p'}} \sum_{i=1}^l \omega_\chi((g_i{}^G)^+)\theta(g_i^{-1}). \tag{4.16}$$

By Corollary 2.3.3 the summands are all algebraic integers, hence $(\chi, \tilde{\theta})_G \in R$.
    (c) We have

$$\tilde{\chi} = \sum_{\xi \in \mathrm{Irr}(G)} (\tilde{\chi}, \xi)_G\, \xi \qquad \text{with} \qquad (\tilde{\chi}, \xi)_G \in \mathbb{Z}$$

by (a). Using (b) we see

$$\frac{(\tilde{\chi}, \xi)_G}{\chi(1)} = \frac{(\chi, \tilde{\xi})_G}{\chi(1)} \in R,$$

hence $\nu_p((\tilde{\chi}, \xi)_G) \geq \nu_p(\chi(1)) = a$ and $\frac{1}{p^a}\tilde{\chi}$ is a generalized character. On the other hand, if $P \in \mathrm{Syl}_p(G)$ then $\tilde{\chi}|_P = \chi(1)\rho_P$ and $(\tilde{\chi}|_P, 1_p)_P = \chi(1)$. So if $\frac{1}{n}\tilde{\chi}$ is a generalized character, then $\frac{\chi(1)}{n} \in \mathbb{Z}$. $\square$

When one is computing decomposition matrices the following theorems are sometimes useful. We first state a lemma that we need.

Lemma 4.2.20 *Let* V *be an irreducible, selfdual* FG-*module, i.e.* $V \cong_{FG} V^\star$. *Then there exists up to scalars a unique non-degenerate* G-*invariant bilinear form* Ψ *on* V.

Proof.   We choose an FG-isomorphism $\mu \colon V \to V^\star$. By Lemma 1.1.42 we have an isomorphism

$$\mathrm{Hom}_F(V, V^\star) \cong_{FG} \mathrm{Bifo}_F\, V, \qquad \mu \mapsto ((v, w) \mapsto \mu(v)(w)).$$

Because of our general assumption, F is a splitting field for G. Thus V is absolutely irreducible and we have, by Schur's lemma, $\mathrm{Hom}_{FG}(V, V^\star) \cong F$, hence $\mathrm{Bifo}_{FG}\, V := \mathrm{Inv}_{FG}(\mathrm{Bifo}_F\, V) = \langle\, \Psi\, \rangle_F$, for a G-invariant bilinear form Ψ on V. Obviously $\Psi' \colon (v, w) \mapsto \Psi(w, v)$ is also G-invariant, hence

$$\Psi' = \alpha\Psi \qquad \text{for} \qquad \alpha \in F.$$

Hence

$$\Psi(v, w) = \alpha\Psi(w, v) = \alpha^2\Psi(v, w) \qquad \text{for all} \qquad v, w \in V$$

and $\alpha^2 = 1$. Since F is perfect (in fact finite by our assumption) α is 1 or -1, which means that Ψ is either a symmetric or a skew symmetric form on V, if F has odd characteristic, and symmetric otherwise. Obviously Ψ (corresponding to an isomorphism $V \to V^\star$) is non-degenerate. $\qquad\square$

Theorem 4.2.21 (Fong) *If $p = 2$ and $\varphi \in \mathrm{IBr}(G)$ is real-valued then $\varphi(1)$ is even or $\varphi = 1_{G_{p'}}$.*

Proof. Let $\varphi \in \mathrm{IBr}(G)$ be real-valued and $\varphi \neq 1_{G_{p'}}$, and let V be an FG-module with Brauer character φ. It follows from Lemma 4.2.13 that V is selfdual and from Lemma 4.2.20 that there is a G-invariant, non-degenerate symmetric bilinear form Ψ defined on V.

The subspace $U := \{v \in V \mid \Psi(v, v) = 0\}$ of isotropic vectors is G-invariant. Since V is a simple module we have either $U = V$ or $U = \{0\}$. If $U = V$ then Ψ is a non-degenerate symplectic form on V, hence $\dim_F V$ is even by linear algebra. If $U = \{0\}$ then it also follows from linear algebra that $\dim_F V = 1$. Then for $0 \neq v \in V$ and $g \in G$ we have $gv = \alpha v$, and from

$$\Psi(v, v) = \Psi(gv, gv) = \alpha^2\Psi(v, v)$$

we conclude that $\alpha^2 = 1$, hence $gv = v$ and V is the trivial FG-module. $\qquad\square$

In odd characteristic we use Lemma 4.2.20 to define the notion of a modular Frobenius–Schur indicator of an irreducible Brauer character.

Definition 4.2.22 Let $\varphi \in \mathrm{IBr}(G)$. If φ is not real-valued we define the Frobenius–Schur indicator of φ to be zero. If φ is real-valued then, according to Lemma 4.2.20, the corresponding simple FG-module V carries a non-degenerate G-invariant bilinear form Ψ. For p odd and Ψ being symmetric we define the Frobenius–Schur indicator of φ to be 1 and if Ψ is skew symmetric we define the indicator of φ to be -1. For p even, Ψ is always (skew) symmetric, so we define the Frobenius–Schur indicator of φ to be 1 if there is a non-degenerate G-invariant quadratic form on V and -1 otherwise.

As the following surprising theorem by Thompson, see [163], shows, the modular indicators for p odd can be derived from the decomposition numbers.

Theorem 4.2.23 *Let G be a finite group and let p be an odd prime, and let $\varphi \in$ IBr(G) be real-valued. Then there is a $\chi \in$ Irr(G), again real-valued, such that the decomposition number $d_{\chi,\varphi}$ is odd, and for all such χ the modular Frobenius–Schur indicator ind$_p(\varphi)$ is the same as the (ordinary) Frobenius–Schur indicator of χ.*

Proof. For a proof, see [163]. □

The situation in even characteristic, however, is much more complicated; see [169].

Example 4.2.24 We calculate the 2-Brauer character table for $G = A_5$. For convenience we reproduce the ordinary character table from Example 2.1.24:

$\lvert \mathbf{C}_G(g) \rvert :$	60	4	3	5	5	
$G = A_5$	1a	2a	3a	5a	5b	
χ_1	1	1	1	1	1	
χ_2	3	-1	0	α	β	$\alpha = \frac{1}{2}(1 - \sqrt{5})$
χ_3	3	-1	0	β	α	$\beta = \frac{1}{2}(1 + \sqrt{5})$
χ_4	4	0	1	-1	-1	
χ_5	5	1	-1	0	0	

For simplicity we abbreviate $\chi_i' := \chi_i|_{G_{2'}}$. Then $\varphi_i := \chi_1'$ is the trivial Brauer character. Furthermore χ_4 is a "defect zero character" for the prime 2, that is $2 \nmid \frac{|G|}{\chi_4(1)}$, and we will see shortly (Theorem 4.4.14) that this implies that $\varphi_4 := \chi_4' \in$ IBr(G). We find that

$$\chi_5' = -\chi_1' + \chi_2' + \chi_3'. \tag{4.17}$$

Thus $\mathbb{Z}\,$IBr$(G) := \langle$IBr$(G)\rangle_{\mathbb{Z}} = \langle \chi_1', \chi_2', \chi_3', \chi_4' \rangle$ and $\chi_2' + \chi_3' = \varphi_1 + \chi_5'$. This means that φ_1 must be a constituent of χ_2' or χ_3' (or both). Since no non-trivial representation of A_5 can have only trivial composition factors – the image would be a 2-group – we see that χ_2' or χ_3' has a constituent of degree two. To see that this actually holds for both of these characters, one may observe that the characters χ_2 and χ_3 are conjugate by an automorphism of A_5 induced by an inner automorphism of S_5. (For an alternative argument, one may use Theorem 4.2.21 or Exercise 4.2.1.) Finally, (4.17) shows us how to obtain the row of the 2-decomposition matrix corresponding to χ_5. The result is as follows:

	φ_1	φ_2	φ_3	φ_4
χ_1	1	·	·	·
χ_2	1	1	·	·
χ_3	1	·	1	·
χ_5	1	1	1	·
χ_4	·	·	·	1

(Why we have chosen the last line corresponding to χ_4 will become apparent in Section 4.3.) From this we obtain the 2-Brauer character table of G:

$G = A_5$	1a	3a	5a	5b
φ_1	1	1	1	1
φ_2	2	-1	$\frac{1}{2}(-1-\sqrt{5})$	$\frac{1}{2}(-1+\sqrt{5})$
φ_3	2	-1	$\frac{1}{2}(-1+\sqrt{5})$	$\frac{1}{2}(-1-\sqrt{5})$
φ_4	4	1	-1	-1

$\blacklozenge$

Exercises

Exercise 4.2.1 Let (K, R, F, η) be a p-modular splitting system and let $\sigma_p \colon F \to F$, $\sigma_p(\alpha) := \alpha^p$ for $\alpha \in F$ be the Frobenius automorphism of F. Let W be an n-dimensional FG-module with Brauer character φ_W and let $^{\sigma_p}W$ be the algebraically conjugate FG-module; see Remark 1.8.3.

(a) Show that if $\gamma_p \colon \mathbb{Z}\mathcal{U}_{p'} \to \mathbb{Z}\mathcal{U}_{p'}$ is the extension of the automorphism $\gamma_p \colon \mathcal{U}_{p'} \to \mathcal{U}_{p'}$ to $\mathbb{Z}\mathcal{U}_{p'}$ with $\gamma_p(u) := u^p$ for $u \in \mathcal{U}_{p'}$, then $\varphi_{\sigma_p W} = \gamma_p \circ \varphi_W$.

(b) Suppose now that $|F| = p^{2n} = q^n$ and let $\sigma_q := (\sigma_p)^n$. Show that if $\varphi = \overline{\varphi^{\sigma_q}}$ then we can choose a basis of W such that the image of G is a subgroup of the unitary group $U_n(q)$.
Hint: try to imitate the proof of Lemma 4.2.20.

The following exercise demonstrates that a Galois conjugate of an irreducible Brauer character need not be a Brauer character.

Exercise 4.2.2 Let G be $\mathrm{SL}_2(7)$ and let $p = 7$.

(a) Use Exercise 1.3.4 to show that G has exactly seven irreducible representations in characteristic p with dimensions $1, 2, 3, 4, 5, 6, 7$. Determine the corresponding Brauer characters $\varphi_1, \varphi_2, \ldots, \varphi_7$ with respect to a standard p-modular splitting system (K, R, F, η).

(b) Let σ be a field automorphism of K mapping ζ_8 to ζ_8^3. Show that φ_2^σ, where $\varphi_2^\sigma(g) := \sigma(\varphi_2(g))$ for $g \in G_{p'}$ is not a Brauer character.

(c) Generalize the previous result to $\mathrm{SL}_2(p)$ for p a prime greater than five.

Exercise 4.2.3 Prove Remark 4.2.17.

Exercise 4.2.4 Let V be an $\mathbb{F}_{p^k}G$-module with Brauer character φ. Then G acts on $V \setminus \{0\}$ and we obtain an ordinary permutation character θ of G of degree $p^{k \dim V} - 1$. Show that for a $g \in G_{p'}$ of order m, one has

$$\theta(g) = p^{nk} - 1 \qquad \text{with} \qquad n = \frac{1}{m} \sum_{j=1}^{m} \varphi(g^j).$$

In particular, using [98] show that an element of 5a of $G := S_6(2)$ has no fixed points on $V \setminus \{0\}$, where V is a simple $\mathbb{F}_{2^k}G$-module of dimension eight, as asserted in Example 3.8.17.

4.3   *p*-projective characters

Throughout this section (K, R, F, η) will denote a p-modular splitting system for our finite group G. By definition the p-decomposition matrix D of G gives the multiplicities of the composition factors of the p-modular reductions of the simple KG-modules. By Brauer reciprocity (Theorem 4.1.23(b)) D has another interpretation which will appear to be most useful for computing D; namely D also describes how the projective indecomposable FG-modules, when lifted to KG-modules, decompose as direct sums of irreducible KG-modules.

Definition 4.3.1 For $\varphi \in \mathrm{IBr}(G)$ the (ordinary) character

$$\Phi_\varphi := \sum_{\chi \in \mathrm{Irr}(G)} d_{\chi,\varphi} \chi$$

is called the **projective indecomposable character** associated to φ.

By Theorem 4.1.23, $\Phi_\varphi|_{G_{p'}}$ is the Brauer character of a projective cover $P(Y)$ of a simple FG-module Y with Brauer character φ, and Φ itself is the character of the KG-module eKG, where $e^2 = e \in RG$ is an idempotent with $FG\hat{e} = P(Y)$. More generally, if W is a projective FG-module then, by Corollary 4.1.22(b), there is a projective RG-lattice M with $W \cong_{FG} M/\pi M$. The character ψ of KM is called p-**projective** and $\psi|_{G_{p'}}$ is the Brauer character of W. Thus $\psi \in \mathrm{Char}_K(G)$ is p-projective if and only if

$$\psi = \sum_{\varphi \in \mathrm{IBr}(G)} a_\varphi \Phi_\varphi \quad \text{with} \quad a_\varphi \in \mathbb{N}_0 \quad \text{for all} \quad \varphi \in \mathrm{IBr}(G).$$

So, obviously sums of p-projective characters are p-projective. The elements of $\langle \Phi_\varphi \mid \varphi \in \mathrm{IBr}(G) \rangle_{\mathbb{Z}}$ are sometimes called generalized p-projective characters.

Lemma 4.3.2 *If $P \in \mathrm{Syl}_p(G)$ and ψ is a p-projective character then*

$$|P| \mid \psi(1) \quad \text{and} \quad \psi(g) = 0 \quad \text{for all} \quad g \in P \setminus \{1\}.$$

Proof.   By Lemma 1.6.17 the restriction of a p-projective character to P is p-projective, hence by Corollary 1.6.25 it is a multiple of the regular character ρ_P. $\qquad\square$

Actually, p-projective characters do not only vanish on non-trivial p-elements, but also on p-singular elements, as Theorem 4.3.3 shows.

For class functions φ, ψ defined on G or $G_{p'}$ we set

$$(\varphi, \psi)_{G_{p'}} := \frac{1}{|G|} \sum_{g \in G_{p'}} \varphi(g)\psi(g^{-1}).$$

Theorem 4.3.3 *The set $\{\Phi_\varphi \mid \varphi \in \mathrm{IBr}(G)\}$ is a basis of*

$$\mathrm{cf}_{p'}(G, K) := \{\psi \in \mathrm{cf}(G, K) \mid \psi(g) = 0 \quad \text{for all} \quad g \in G \setminus G_{p'}\}.$$

Furthermore

$$(\varphi, \Phi_\psi)_{G_{p'}} = (\Phi_\varphi, \psi)_{G_{p'}} = \delta_{\varphi,\psi} \qquad for \qquad \varphi, \psi \in \mathrm{IBr}(G)$$

and

$$[(\varphi, \psi)_{G_{p'}}]_{\varphi,\psi \in \mathrm{IBr}(G)} = C^{-1}, \tag{4.18}$$

where $C = [c_{\varphi,\psi}]_{\varphi,\psi \in \mathrm{IBr}(G)}$ is the Cartan matrix of FG.

Proof. Let $g_1, \ldots, g_r$ be representatives of the conjugacy classes of G with $g_1, \ldots, g_l$ being p-regular, $g_{l+1}, \ldots, g_r$ being p-singular, and let $i \leq l$. Inserting

$$\chi(g_i) = \sum_{\varphi \in \mathrm{IBr}(G)} d_{\chi\varphi}\varphi(g_i)$$

in the orthogonality relations (2.5) we get

$$|\mathbf{C}_G(g_i)| \cdot \delta_{i,j} = \sum_{\chi \in \mathrm{Irr}(G)} \chi(g_i)\chi(g_j^{-1}) = \sum_{\varphi \in \mathrm{IBr}(G)} \Phi_\varphi(g_j^{-1})\varphi(g_i). \tag{4.19}$$

For $j > l$ we get $\sum_{\varphi \in \mathrm{IBr}(G)} \overline{\Phi_\varphi(g_j)}\varphi = 0$; thus $\Phi_\varphi(g_i) = 0$, since $\mathrm{IBr}(G)$ is linearly independent. Now (4.19) says that the matrices

$$[\Phi_\varphi(g_i^{-1})]^{\mathrm{T}}_{\varphi \in \mathrm{IBr}(G), 1 \leq i \leq l} \qquad \text{and} \qquad [\varphi(g_i)\frac{1}{|\mathbf{C}_G(g_i)|}]_{\varphi \in \mathrm{IBr}(G), 1 \leq i \leq l}$$

are inverses of each other. Hence we get

$$(\varphi, \Phi_\psi)_{G_{p'}} = \sum_{i=1}^{l} \varphi(g_i)\Phi_\psi(g_i^{-1})|\mathbf{C}_G(g_i)| = \delta_{\varphi,\psi}.$$

Similarly, $(\Phi_\varphi, \psi)_{G_{p'}} = \delta_{\varphi,\psi}$. This means in effect that $\{\Phi_\varphi|_{G_{p'}} \mid \varphi \in \mathrm{IBr}(G)\}$ and $\mathrm{IBr}(G)$ are dual bases of the vector space of class functions on $G_{p'}$ with respect to the bilinear form $(\ ,\ )_{G_{p'}}$. Finally (4.18) follows, since $(\Phi_\varphi)|_{G_{p'}} = \sum_{\psi \in \mathrm{IBr}(G)} c_{\varphi,\psi}\,\psi$. $\qquad\qquad\square$

Corollary 4.3.4 *Assume that $\psi \in \mathrm{Char}_K(G)$ vanishes on all p-singular elements of G. Then ψ is a generalized p-projective character. If in addition*

$$(\varphi, \psi)_{G_{p'}} \geq 0 \qquad for\ all \qquad \varphi \in \mathrm{IBr}(G),$$

then ψ is a p-projective character.

Proof. By Theorem 4.3.3 we may write

$$\psi = \sum_{\varphi \in \mathrm{IBr}(G)} \alpha_\varphi \Phi_\varphi \qquad \text{with} \qquad \alpha_\varphi \in K.$$

By Theorem 4.2.16(a) there are $a_{\varphi,\chi} \in \mathbb{Z}$ with $\varphi = \sum_{\chi \in \mathrm{Irr}(G)} a_{\varphi,\chi} \, \chi|_{G_{p'}}$. Hence

$$\alpha_\varphi = (\varphi, \psi)_{G_{p'}} = \sum_{\chi \in \mathrm{Irr}(G)} a_{\varphi,\chi}(\chi, \psi)_{G_{p'}} = \sum_{\chi \in \mathrm{Irr}(G)} a_{\varphi,\chi}(\chi, \psi)_G \in \mathbb{Z}.$$

$\square$

There is a similar statement for class functions on the p-regular classes, which follows immediately from the fact that the projective indecomposable characters and the irreducible Brauer characters of G are dual bases with respect to the bilinear form $(\ ,\ )_{G_{p'}}$.

Corollary 4.3.5 *Assume that φ is a class function on the p-regular classes of G. Then φ is a Brauer character if and only if $(\varphi, \Phi_\psi)_{G'_p}$ is a nonnegative integer for all $\psi \in \mathrm{IBr}(G)$.*

The following properties of p-projective characters are useful for computing p-decomposition matrices.

Lemma 4.3.6 *Let H be a subgroup of G.*

(a) *If ψ is a p-projective character of G, then the restricted character ψ_H is a p-projective character of H.*

(b) *If ψ is a p-projective character of H, then the induced character ψ^G is a p-projective character of G.*

(c) *If ψ is a p-projective character of G, and φ is a Brauer character or an ordinary character of G, then $\psi \cdot \varphi$ is a p-projective character of G. Here, if φ is a Brauer character we define*

$$(\psi \cdot \varphi)(g) := \begin{cases} 0 & \text{for } g \notin G_{p'}, \\ \psi(g)\varphi(g) & \text{for } g \in G_{p'}. \end{cases}$$

Proof. This follows from Lemma 1.6.17 in case (a) and Corollary 3.2.16 for the remaining cases. $\square$

Theorem 4.3.7 *Let H be a subgroup of G and let $G = \dot\cup_{i=1}^m g_i H$. If W is an FH-module with Brauer character ψ, then the Brauer character of the induced module W^G is given by*

$$\psi_{W^G}(g) = \sum_{i=1}^m \dot\psi(g_i^{-1} g g_i) \qquad \text{with} \quad \dot\psi(x) := \begin{cases} \psi(x) & \text{if } x \in H, \\ 0 & \text{else,} \end{cases}$$

*for $g \in G_{p'}$. We will call $\psi^G := \psi_{W^G}$ the **induced** Brauer character.*

Proof. We define the class function $\psi_0 \in \mathrm{cf}(H, K)$ by

$$\psi_0(h) := \begin{cases} \psi(h) & \text{if } h \in H_{p'} \,, \\ 0 & \text{else,} \end{cases}$$

and have to show that the Brauer character of W^G is $\psi_0^G|_{G_{p'}}$, where ψ_0^G is the induced class function introduced in Definition 3.2.6.

Assume that W^G has Brauer character $\theta := \sum_{\varphi \in \mathrm{IBr}(G)} a_\varphi \varphi$. By Theorem 4.3.3 it suffices to show that $a_\varphi = (\theta, \Phi_\varphi)_{G_{p'}} = (\psi_0^G, \Phi_\varphi)_{G_{p'}}$ for every $\varphi \in \mathrm{IBr}(G)$. Obviously we may assume that W is simple. For $\varphi \in \mathrm{IBr}(G)$ let V_φ be a simple FG-module with Brauer character φ and projective cover $P(V_\varphi)$. By the proof of Theorem 4.1.23(a) and by Theorem 3.2.12 we have

$$a_\varphi = \dim_F \mathrm{Hom}_{FG}(P(V_\varphi), W^G) = \dim_F \mathrm{Hom}_{FH}(P(V_\varphi)|_H, W),$$

and this is the multiplicity m_φ of a projective cover $P(W)$ of W as a direct summand of $P(V_\varphi)|_H$. Using Theorem 4.3.3 again and Theorem 3.2.13 we get

$$m_\varphi = (\psi, \Phi_\varphi|_H)_{H_{p'}} = (\psi_0, \Phi_\varphi|_H)_H = (\psi_0^G, \Phi_\varphi)_G = (\psi_0^G, \Phi_\varphi)_{G_{p'}}. \qquad \square$$

Using the above notion of induced Brauer characters we obtain from Theorem 3.6.7 a Clifford correspondence for Brauer characters as follows.

Corollary 4.3.8 *Suppose* $N \trianglelefteq G$. *For* $\varphi \in \mathrm{IBr}(N)$ *let* $T_G(\varphi) := \{g \in G \mid {}^g\varphi = \varphi\}$, *and for* $N \leq H \leq G$ *define*

$$\mathrm{IBr}(\,H \mid \varphi\,) := \{\psi \in \mathrm{IBr}(H) \mid (\psi|_{N_{p'}}, \Phi_\varphi)_{N_{p'}} \neq 0\}.$$

If $T_G(\varphi) \leq T \leq G$ *we have a bijection* $\mathrm{IBr}(T \mid \varphi) \to \mathrm{IBr}(G \mid \varphi)$, $\psi \mapsto \psi^G$ *called the* **Clifford correspondence,** *and*

$$\mathrm{IBr}(G) = \dot\bigcup \{\,\mathrm{IBr}(\,G \mid \varphi\,) \mid \varphi \in \mathrm{IBr}(N)\,\}.$$

Proof. This is an immediate consequence of Theorem 3.6.7, Theorem 4.2.14 and Theorem 4.3.3. $\qquad \square$

The Clifford correspondence extends to projective indecomposable characters as we see in the following.

Theorem 4.3.9 *Let* $N \trianglelefteq G$ *and* $\theta \in \mathrm{IBr}(N)$. *If* $T_G(\theta) \leq T \leq G$ *and* $\psi \in \mathrm{IBr}(T \mid \theta)$ *then*

$$(\Phi_\psi)^G = \Phi_{\psi^G}.$$

Proof. By Lemma 4.3.2 and Definition 3.2.6, $(\Phi_\psi)^G \in \mathrm{cf}_{p'}(G, K)$, so

$$(\Phi_\psi)^G = \sum_{\varphi \in \mathrm{IBr}(G)} a_\varphi \Phi_\varphi \quad \text{with} \quad a_\varphi = ((\Phi_\psi)^G, \varphi)_{G_{p'}} = (\Phi_\psi, \varphi_T)_{T_{p'}}$$

because of Theorem 4.3.3 and Exercise 4.3.1. Let W be an FT-module with Brauer character ψ and let V be a simple FG-module with Brauer character φ. If $(\Phi_\psi, \varphi_T)_{T_{p'}} > 0$, then V_T has a composition factor isomorphic to W. By Exercise 3.6.5, $V \cong W^G$, and hence $\varphi = \psi^G$ and $a_\varphi = 1$. □

Example 4.3.10 We calculate the 3-Brauer character table of $G = A_5$. Similarly as in Example 4.2.24, we abbreviate $\chi_i' := \chi_i|_{G_{3'}}$ and we use the same numbering of the $\chi_i \in \mathrm{Irr}(G)$ as there. As always, $\varphi_1 := \chi_1'$. Now, χ_2, χ_3 are defect-zero characters for $p = 3$, and using Theorem 4.4.14 we infer that $\varphi_3 := \chi_2'$ and $\varphi_4 := \chi_3'$ are irreducible Brauer characters. Note that A_5 has a maximal subgroup $H \cong D_{10}$ isomorphic to a dihedral group. Since $3 \nmid |H|$, every character of H is 3-projective, and in particular, by Lemma 4.3.6, $(1_H)^G = \chi_1 + \chi_5$ is projective, and in fact, is a projective indecomposable, and hence is equal to Φ_{φ_1}. This gives us the first column of the decomposition matrix. Now χ_4' can have only the remaining irreducible Brauer character φ_2 as a constituent, and $\chi_4' = \varphi_2$, because $\frac{1}{2}\chi_4'$ has values which are not integers. Finally we observe that

$$\chi_5' = \chi_1' + \chi_4',$$

which yields the remaining row of the decomposition matrix D. Thus the 3-decomposition matrix is as follows:

	φ_1	φ_2	φ_3	φ_4
χ_1	1	.	.	.
χ_4	.	1	.	.
χ_5	1	1	.	.
χ_2	.	.	1	.
χ_3	.	.	.	1

The 3-Brauer character table is, of course, as follows:

$G = A_5$	1a	2a	5a	5b
φ_1	1	1	1	1
φ_2	4	0	-1	-1
φ_3	3	-1	$\frac{1}{2}(1 - \sqrt{5})$	$\frac{1}{2}(1 + \sqrt{5})$
φ_4	3	-1	$\frac{1}{2}(1 + \sqrt{5})$	$\frac{1}{2}(1 - \sqrt{5})$

♦

Exercises

Exercise 4.3.1 Let G be a group, H be a subgroup of G, φ a Brauer character of G, Φ a projective character of G, ψ a Brauer character of H and Ψ a projective character of H. Show that the following statements hold:

(a) $(\varphi, \Psi^G)_{G_{p'}} = (\varphi_H, \Psi)_{H_{p'}}$,

(b) $(\Phi, \psi^G)_{G_{p'}} = (\Phi_H, \psi)_{H_{p'}}$.

Exercise 4.3.2 Let $G := A_5$ and $H := A_4$. Verify for $p = 2$ that $\mathrm{IBr}(H) = \{\lambda_1 := 1_H|_{H_{2'}}, \lambda_2, \lambda_3\}$ with $\lambda_i(1) = 1$ for $i = 1, 2, 3$. Using the notation of Example 4.2.24, let a_{ij} and b_{ij} for $i = 1, \ldots, 4$ and let $j = 1, \ldots, 3$ be defined by $(\varphi_i)_H = \sum_{j=1}^3 a_{ij}\lambda_j$ and $(\lambda_j)^G = \sum_{i=1}^4 b_{ij}\varphi_i$. Show that

$$[a_{ij}] = \begin{bmatrix} 1 & 0 & 0 \\ 0 & 1 & 1 \\ 0 & 1 & 1 \\ 2 & 1 & 1 \end{bmatrix}, \qquad [b_{ij}] = \begin{bmatrix} 1 & 1 & 1 \\ 0 & 1 & 1 \\ 0 & 1 & 1 \\ 1 & 0 & 0 \end{bmatrix}.$$

Compare with Theorem 3.2.13.

Exercise 4.3.3 Show that the 5-decomposition matrix for A_5 is as follows:

	φ_1	φ_2	φ_3
χ_1	1	.	.
χ_2	.	1	.
χ_3	.	1	.
χ_4	1	1	.
χ_5	.	.	1

Exercise 4.3.4 Let V be an FG-module with Brauer character ψ on which G acts faithfully and let $m := |\{\psi(g) \mid g \in G_{p'}\}|$. Show that every irreducible FG-module is isomorphic to a composition factor of

$$\underbrace{V \otimes \cdots \otimes V}_{j} \text{ for some } 0 \le j \le m - 1.$$

Hint: Adapt the proof of Theorem 2.7.3.

Exercise 4.3.5 Let $\varphi \in \mathrm{IBr}(G)$ and $\lambda \in \mathrm{Irr}(G)$ with $\lambda(1) = 1$. Show that $(\Phi_\varphi, \lambda)_G = a \ge 1$ implies $a = 1$ and $\varphi = \lambda|_{G_{p'}}$. Conclude that any p-projective indecomposable character Φ has at most one (ordinary) linear constituent λ if $O^p(G) = G$ (see Definition 3.5.18).

Exercise 4.3.6 (a) Let $\lambda \in \mathrm{Irr}(G)$ with $\lambda(1) = 1$ and $\Phi := \Phi_{\lambda|_{G_{p'}}}$. Assume that all non-linear (ordinary) constituents of Φ have degrees divisible by p. Show that G has a normal p-complement, that is a normal subgroup N such that $G/N \cong P \in \mathrm{Syl}_p(G)$ (see [140]).
Hint: Let $N := O^p(G)$. Put $\varphi := \lambda|_{N_{p'}} \in \mathrm{IBr}(N)$ and $\Phi' := \Phi_\varphi$. Show that Φ' is a G-invariant character of N and that $\Phi|_N = \Phi' + \Phi''$ for some $\Phi'' \in \mathrm{Char}_K(N)$. Use Exercise 4.3.5, Clifford's theorem and Theorem 3.6.19 to show that $\Phi'(1) \equiv 1 \mod p$.
(b) Conversely, show that if G has a normal p-complement, then p divides the degree of any non-linear character in $\mathrm{Irr}(G)$.

4.4 Characters in blocks

As in Section 4.3 we fix a p-modular splitting system (K, R, F, η) for our finite group G.

By Corollary 4.1.22 every simple KG-module V and every simple FG-module Y belong to a particular block $B = RG\epsilon$ of RG, where ϵ is the block idempotent of RG with $\epsilon V = V$ (or $\hat{\epsilon}Y = Y$, respectively). Clearly Y belongs to B if and only if the projective cover $P(Y)$ belongs to B. So V belongs to B (and Y belongs to B) if and only if $\chi_V(\epsilon) \neq 0$ (and $\Phi_{\varphi_Y}(\epsilon) \neq 0$, respectively). We use the notation

$$\mathrm{Irr}(B) := \{\chi \in \mathrm{Irr}(G) \mid \chi(\epsilon) \neq 0\}, \qquad \mathrm{IBr}(B) := \{\varphi \in \mathrm{IBr}(G) \mid \Phi_\varphi(\epsilon) \neq 0\}.$$

Also, we will call B a p-**block** of G and write

$$\mathrm{Bl}_p(G) := \{B \mid B \text{ is a } p\text{-block of } G\} \qquad \text{and} \qquad \epsilon_B := \epsilon.$$

Then $\hat{\epsilon}_B \in \mathbf{Z}(FG)$ is the block idempotent of the block ideal $FG\hat{\epsilon}_B$ of the F-algebra FG.

In order to investigate $\mathrm{Irr}(B)$ for $B \in \mathrm{Bl}_p(G)$ the following notion is useful.

Definition 4.4.1 The p-**Brauer graph** of G is a graph with vertex set $\mathrm{Irr}(G)$, where $\chi, \chi' \in \mathrm{Irr}(G)$ are linked by an edge if and only if $d_{\chi\varphi} \neq 0 \neq d_{\chi'\varphi}$ for some $\varphi \in \mathrm{IBr}(G)$.

Theorem 4.4.2 *Let* $\mathrm{Bl}_p(G) = \{B_1, \ldots, B_m\}$, $\chi \in \mathrm{Irr}(G)$ *and* $\varphi \in \mathrm{IBr}(G)$.

(a) *If* $d_{\chi\varphi} \neq 0$ *then* $\chi \in \mathrm{Irr}(B_i)$ *if and only if* $\varphi \in \mathrm{IBr}(B_i)$. *Ordering* $\mathrm{Irr}(G)$ *and* $\mathrm{IBr}(G)$ *according to the blocks the p-decomposition matrix of G takes the form*

$$D = \begin{bmatrix} D_{B_1} & \mathbf{0} & & \mathbf{0} \\ \mathbf{0} & D_{B_2} & & \mathbf{0} \\ & & \ddots & \\ \mathbf{0} & \mathbf{0} & & D_{B_m} \end{bmatrix} \quad \text{with} \quad D_{B_i} = [d_{\chi\varphi}]_{\chi \in \mathrm{Irr}(B_i)\ \varphi \in \mathrm{IBr}(B_i)}.$$

The Cartan matrix of the ith block is $C(B_i) = D_{B_i}^{\mathrm{T}} D_{B_i}$.

(b) $\mathrm{Irr}(B_i)$ *is a connected component in the p-Brauer graph* $\Gamma_p(G)$ *for* $1 \leq i \leq m$.

Proof. (a) $\varphi \in \mathrm{IBr}(B)$ if and only if $\Phi_\varphi(1) = \Phi_\varphi(\epsilon_B)$. By Definition 4.3.1 this holds if and only if $\chi(1) = \chi(\epsilon_B)$ for all $\chi \in \mathrm{Irr}(G)$ with $d_{\chi\varphi} \neq 0$.

(b) By (a) each connected component of $\Gamma_p(G)$ is contained in $\mathrm{Irr}(B_i)$ for some i. Conversely, let $X \subseteq \mathrm{Irr}(B_i)$ be a connected component of $\Gamma_p(G)$ and let $Y := \{\varphi \in \mathrm{IBr}(G) \mid d_{\chi\varphi} \neq 0 \text{ for some } \chi \in X\}$. Put $X' := \mathrm{Irr}(B_i) \setminus X$ and $Y' := \mathrm{IBr}(B_i) \setminus Y$. We may order $\mathrm{Irr}(B_i)$ and $\mathrm{IBr}(B_i)$ so that

$$D_{B_i} = \begin{bmatrix} D_{XY} & \mathbf{0} \\ \mathbf{0} & D_{X'Y'} \end{bmatrix} \quad \text{with} \quad D_{XY} = [d_{\chi\varphi}]_{\chi \in X\ \varphi \in Y}.$$

Then

$$C(B_i) = \begin{bmatrix} D_{XY}^{\mathrm{T}} D_{XY} & 0 \\ 0 & D_{X'Y'}^{\mathrm{T}} D_{X'Y'} \end{bmatrix},$$

which contradicts Corollary 1.7.9 unless $\mathrm{Irr}(B_i) = X$. $\qquad\qquad\square$

The following orthogonality relation for Brauer characters is a simple consequence of the block decomposition of the Cartan matrix and its inverse.

Lemma 4.4.3 *If $\varphi, \psi \in \mathrm{IBr}(G) \cup \mathrm{Irr}(G)$ belong to different p-blocks, then*

$$(\varphi, \psi)_{G_{p'}} = 0.$$

Proof. If $B \in \mathrm{Bl}_p(G)$ and $\chi \in \mathrm{Irr}(B)$ then $\chi|_{G_{p'}} \in \langle \mathrm{IBr}(B) \rangle_{\mathbb{N}_0}$. Thus it suffices to consider irreducible Brauer characters. For these the claim follows from Theorem 4.3.3, (4.18), and the fact that $C^{-1} = \mathrm{diag}(\,(C(B_i)^{-1}, \ldots, C(B_m)^{-1})$ if $B_1, \ldots, B_m$ are the p-blocks of G. $\qquad\qquad\square$

Definition 4.4.4 If $\theta \in \mathrm{cf}(G, K)$ and $B \in \mathrm{Bl}_p(G)$ we call

$$\theta_B := \sum_{\chi \in \mathrm{Irr}(B)} (\theta, \chi)_G \, \chi$$

the *B-part* of θ.

Lemma 4.4.5 *If ψ is a p-projective character and B is a block then the B-part ψ_B is a p-projective character or zero.*

Proof. Let M be a projective RG-lattice such that ψ is the character of KM and let ϵ_B be the block idempotent of the block B of RG. Then $\epsilon_B M$ is a direct summand of M and hence the zero-module or projective. Since ϵ_B annihilates $(1 - \epsilon_B)M$, the trace of $g \in G$ on $\epsilon_B M$ equals $\psi(g\,\epsilon_B)$, the trace of $g\,\epsilon_B$ on M. The same argument and Corollary 4.1.22 also give for $\chi \in \mathrm{Irr}(G)$

$$\chi(g\,\epsilon_B) = \begin{cases} 0 & \text{for } \chi \notin \mathrm{Irr}(B), \\ \chi(g) & \text{for } \chi \in \mathrm{Irr}(B). \end{cases}$$

Thus

$$\psi(g\,\epsilon_B) = \sum_{\chi \in \mathrm{Irr}(G)} (\psi, \chi)_G\, \chi(g\,\epsilon_B) = \sum_{\chi \in \mathrm{Irr}(B)} (\psi, \chi)_G\, \chi(g) = \psi_B(g)$$

and ψ_B is the character of $\epsilon_B KM$ and hence a p-projective character. $\qquad\square$

Lemma 4.4.6 *Let B be a p-block of G and set*

$$\mathrm{k}(B) = |\,\mathrm{Irr}(B)| \qquad \text{and} \qquad \mathrm{l}(B) = |\,\mathrm{IBr}(B)|$$

(which is a standard notation). Then $\mathrm{l}(B) \le \mathrm{k}(B)$ and the decomposition matrix D_B has rank $\mathrm{l}(B)$. Moreover, there is a matrix $A_B \in \mathbb{Z}^{\mathrm{l}(B) \times \mathrm{k}(B)}$ with $A_B D_B = \mathbf{I}_{\mathrm{l}(B)}$.

Proof. This follows from Theorem 4.2.16. $\qquad\square$

Theorem 4.4.7 (Osima) *If B is a p-block of G and*

$$\epsilon_B = \sum_{g \in G} a_g g \in RG$$

is its block idempotent, then $a_g = 0$ for all $g \in G \setminus G_{p'}$.

Proof. By Corollary 4.1.22 and Corollary 2.1.7 we have

$$\epsilon_B = \sum_{\chi \in \mathrm{Irr}(B)} \epsilon_\chi = \sum_{\chi \in \mathrm{Irr}(B)} \frac{\chi(1)}{|G|} \sum_{g \in G} \chi(g^{-1})g. \qquad (4.20)$$

Thus

$$a_g = \frac{1}{|G|} \sum_{\chi \in \mathrm{Irr}(B)} \chi(1)\chi(g^{-1}).$$

But $\sum_{\chi \in \mathrm{Irr}(B)} \chi(1)\chi$ is the character of the projective module $RG\epsilon_B$, so by Lemma 4.3.2 $a_g = 0$ if g is not p-regular. $\qquad\square$

For every $\chi \in \mathrm{Irr}(G)$ there is by Theorem 2.3.2 a central character

$$\omega_\chi \colon \mathbf{Z}(KG) \to K \qquad \text{with} \quad C^+ \mapsto \omega_\chi(C^+) = \frac{|C|\chi(g)}{\chi(1)}$$

for any conjugacy class $C = g^G$. By Corollary 2.3.3 $\omega_\chi(C^+)$ is an algebraic integer, and hence is contained in R, because R is integrally closed (Lemma 4.1.3(b)) in K. Hence we obtain an F-algebra homomorphism

$$\hat{\omega}_\chi \colon \mathbf{Z}(FG) \to F \qquad \text{with} \quad C^+ \mapsto \eta(\omega_\chi(C^+)) \quad \text{for} \quad C \in \mathrm{cl}(G).$$

Theorem 4.4.8 *Two irreducible characters $\chi, \chi' \in \mathrm{Irr}(G)$ belong to the same block if and only if the following congruences hold in $\mathbb{Z}[\zeta_m]$, where m is the p'-part of the exponent of G:*

$$\frac{|g^G|\chi(g)}{\chi(1)} \equiv \frac{|g^G|\chi'(g)}{\chi'(1)} \qquad \mathrm{mod}\ p \qquad \text{for all} \quad g \in G_{p'}.$$

Proof. By the orthogonality relations (Theorem 2.1.15) we have for $\chi, \chi' \in \mathrm{Irr}(G)$

$$\omega_\chi(\epsilon_{\chi'}) = \delta_{\chi,\chi'}.$$

Let $\epsilon_B \in \mathbf{Z}(RG)$ be the block idempotent of $B \in \mathrm{Bl}_p(G)$. Since $\epsilon_B = \sum_{\chi \in \mathrm{Irr}(B)} \epsilon_\chi$ we have $\chi \in \mathrm{Irr}(B)$ if and only if $\omega_\chi(\epsilon_B) = 1$ and then $\hat{\omega}_\chi(\hat{\epsilon}_B) = 1$. But F being a splitting field for G, there is exactly one algebra homomorphism $\hat{\omega} \colon FG \to F$ with $\hat{\omega}(\hat{\epsilon}_B) = 1$ (see Corollary 1.7.6). Thus χ, χ' belong to the same p-block if and only if $\hat{\omega}_\chi = \hat{\omega}_{\chi'}$, and this holds if and only if these central characters agree on $\hat{\epsilon}_B$. By Theorem 4.4.7 $\hat{\omega}_\chi(\hat{\epsilon}_B) = \hat{\omega}_{\chi'}(\hat{\epsilon}_B)$ if and only if $\hat{\omega}_\chi(C^+) = \hat{\omega}_{\chi'}(C^+)$ for all $C \in \mathrm{cl}(G_{p'})$. This holds if and only if for all $g \in G_{p'}$

$$\alpha_g := \frac{|g^G|\chi(g)}{\chi(1)} - \frac{|g^G|\chi'(g)}{\chi'(1)} \in \ker(\eta) \cap \mathbb{Q}_m = p\,R \cap \mathbb{Q}_m.$$

Observe that $pR \cap \mathbb{Q}_m$ is a maximal ideal in $R \cap \mathbb{Q}_m$. This follows since p does not ramify in $\mathbb{Z}[\zeta_m]$, see Proposition 2.3, p. 10, in [168]. From Exercise 4.4.3(c) we conclude that for all $g \in G_{p'}$ we have

$$\gamma(\alpha_g) \in p\,R \cap \mathbb{Q}_m \qquad \text{for all} \qquad \gamma \in \mathrm{Gal}(\mathbb{Q}_m/\mathbb{Q}).$$

Hence, the coefficients of the minimal polynomial of $\frac{1}{p}\alpha_g$ over $\mathbb{Q}$ are in the ring $R \cap \mathbb{Q} = \mathbb{Z}_{(p)}$. If $f_g = \sum_{i=0}^r a_i X^i \in \mathbb{Q}[X]$ is the minimal polynomial of α_g then $\sum_{i=0}^r \frac{a_i}{p^{r-i}} X^i$ is the minimal polynomial of $\frac{1}{p}\alpha_g$. Since α_g is an algebraic integer and $\frac{a_i}{p^{r-i}} \in \mathbb{Z}_{(p)}$ for all $i = 0, \ldots, r$, we conclude that $\frac{1}{p}\alpha_g$ must be an algebraic integer too and hence $\alpha_g \in \mathbb{Z}[\zeta_m]$; see Theorem 2.6 in [168]. $\qquad\square$

For $B \in \mathrm{Bl}_p(G)$ we let $\hat{\omega}_B \colon \mathbf{Z}(FG) \to F$ be the unique algebra homomorphism with $\hat{\omega}_B(\hat{\epsilon}_B) = 1$. Then $\hat{\omega}_B$ is also called the **central character** of B. By the above,

$$\hat{\omega}_B((g^G)^+) = \eta(\omega_\chi((g^G)^+)) = \eta\Big(\frac{|g^G|\chi(g)}{\chi(1)}\Big) \qquad \text{for} \qquad \chi \in \mathrm{Irr}(B). \qquad (4.21)$$

Thus, given the character table of a finite group G it is easy to decide whether or not two ordinary irreducible characters belong to the same block. Also the central characters $\hat{\omega}_B \colon \mathbf{Z}(FG) \to F$ and the block idempotent $\epsilon_B \in \mathbf{Z}(RG)$ (or $\hat{\epsilon}_B \in \mathbf{Z}(FG)$) can readily be computed for any p-block B using (4.21) and (4.20), respectively. In particular, the number of p-blocks can easily be found. There seems to be no group theoretical interpretation of the number of p-blocks.

Example 4.4.9 Let $G = S_5$. From the character table computed in Example 2.1.23 we find that there are two 2-blocks B_1, B_2. We have $\mathrm{Irr}(B_1) = \{\chi_1, \chi_2, \chi_5, \chi_6, \chi_7\}$ with degrees $1, 1, 5, 5, 6$ and $\mathrm{Irr}(B_2) = \{\chi_3, \chi_4\}$ with degrees $4, 4$. We list the corresponding central characters and block idempotents (in $\mathbf{Z}(FG)$):

$\|\mathbf{C}_G(g)\|:$	$2^3 3\,5$	2^3	$2\,3$	5	$2^2 3$	2^2	$2^2 3$
$\mathbf{G} = \mathrm{S}_5$	1a	2a	3a	5a	2b	4a	6a
$\hat{\omega}_{B_1}$	1	1	0	0	0	0	0
$\hat{\omega}_{B_2}$	1	0	1	0	1	0	1
$\hat{\epsilon}_{B_1}$	1	0	1	1	0	0	0
$\hat{\epsilon}_{B_2}$	0	0	1	1	0	0	0

Here the block idempotents are given by their coefficients, that is

$$\hat{\epsilon}_{B_i} = \sum_{C \in \mathrm{cl}(G)} \hat{\epsilon}_{B_i}(C)\, C^+.$$

Note that $\hat{\epsilon}_{B_i}(C) = 0$ for all classes C with $C \not\subseteq G_{p'}$, as follows from Theorem 4.4.7. As a consequence we see that $\hat{\epsilon}_{B_j} \subseteq \mathbf{Z}(F\,A_5)$ for $j = 1, 2$. Since we know already from Example 4.2.24 (and Theorem 1.7.7) that A_5 has two 2-blocks, say b_1, b_2, these $\hat{\epsilon}_{B_j}$ are actually block idempotents in $F\,A_5$. To see that this does not hold in general, see Exercise 4.4.2. For completeness we list the central characters and block idempotents of A_5:

$\|\mathbf{C}_G(g)\|$:	$2^2 3\ 5$	2^2	3	5	5
$G = A_5$	1a	2a	3a	5a	5b
$\hat{\omega}_{b_1}$	1	1	0	0	0
$\hat{\omega}_{b_2}$	1	0	1	1	1
$\hat{\epsilon}_{b_1}$	1	0	1	1	1
$\hat{\epsilon}_{b_2}$	0	0	1	1	1

$\blacklozenge$

Definition 4.4.10 The p-block B of G containing the trivial character is called the **principal (p-)block** and will be denoted by $B_0(G)$.

Example 4.4.11 The simple Mathieu groups M_{22} and M_{24} have just one 2-block, the principal 2-block. $\qquad\blacklozenge$

Definition 4.4.12 Let $\chi \in \mathrm{Irr}(G)$ and let B be a p-block of G. Then

$$\mathrm{d}_p(\chi) := \nu_p(|G|) - \nu_p(\chi(1)) \qquad \text{and} \qquad \mathrm{d}(B) := \max_{\chi \in \mathrm{Irr}(B)} \mathrm{d}_p(\chi)$$

are called the **(p-)defect** of χ and B, respectively; $\chi \in \mathrm{Irr}(G)$ is called a **defect zero character** if $\mathrm{d}_p(\chi) = 0$. Furthermore, if $\chi \in \mathrm{Irr}(B)$ then

$$\mathrm{ht}_p(\chi) := \mathrm{d}(B) - \mathrm{d}_p(\chi)$$

is called the **(p-)height** of χ.

Example 4.4.13 In Example 4.4.9 we see $\mathrm{d}(B_1) = 3$ with characters of height $0, 0, 0, 0, 1$ and $\mathrm{d}(B_2) = 1$ with two characters of height zero. $\qquad\blacklozenge$

The following theorem, which we already have used in a couple of examples, shows incidentally that all characters in a block of defect one have height zero.

Theorem 4.4.14 *Let B be a p-block of G and let $\chi \in \mathrm{Irr}(B)$. Then the following assertions are equivalent:*

(a) $\mathrm{d}_p(\chi) = 0$;

(b) $\epsilon_\chi = \frac{\chi(1)}{|G|} \sum_{g \in G} \chi(g^{-1})g \in RG$;

(c) $\mathrm{Irr}(B) = \{\chi\}$;

(d) *the decomposition matrix of B is $D_B = [1]$;*

(e) χ *is the character of a projective RG-lattice;*

(f) $\chi(g) = 0$ *for all p-singular $g \in G$;*

(g) $\mathrm{d}(B) = 0.$

Proof. (a) $\Rightarrow$ (b); $\mathrm{d}_p(\chi) = 0$ means $\frac{\chi(1)}{|G|} \in R.$

  (b) $\Rightarrow$ (c) By assumption ϵ_χ is the block idempotent of B.
  (c) $\Rightarrow$ (d) This follows from Lemma 4.4.6.
  (d) $\Rightarrow$ (e) This is Brauer reciprocity (Theorem 4.1.23).
  (e) $\Rightarrow$ (f) By Theorem 4.3.3
  (f) $\Rightarrow$ (a), (g) By Corollary 4.3.4 χ is a generalized p-projective character, and is hence p-projective since $\chi \in \mathrm{Irr}(G)$ by assumption. From Lemma 4.3.2 we see that $\mathrm{d}_p(\chi) = 0$ holds, hence (a). Because of (c) we get $\mathrm{d}(B) = 0$ as well.
  (g) $\Rightarrow$ (a) This is by definition. $\qquad\qquad\qquad\qquad\qquad\qquad\qquad\quad\square$

Lemma 4.4.15 *Let $B \in \mathrm{Bl}_p(G)$ and $\theta \in \langle \mathrm{Irr}(B) \rangle_{\mathbb{Z}} \cup \langle \mathrm{IBr}(B) \rangle_{\mathbb{Z}}.$*

(a) $\tilde{\theta} \in \langle \mathrm{Irr}(B) \rangle_{\mathbb{Z}}$, *where $\tilde{\theta}$ is as in Definition* 4.2.18.

(b) *For $\chi, \chi' \in \mathrm{Irr}(B)$ one has*

$$\frac{(\chi, \tilde{\theta})_G}{\chi(1)} \equiv \frac{(\chi', \tilde{\theta})_G}{\chi'(1)} \quad \mathrm{mod}\ p.$$

Proof. (a) By Lemma 4.2.19 $\tilde{\theta}$ is a generalized character. Since $\tilde{\theta}$ vanishes on $G \setminus G_{p'}$ it is a generalized p-projective character, thus

$$\tilde{\theta} = \sum_{\varphi \in \mathrm{IBr}(G)} a_\varphi \Phi_\varphi \qquad \text{with} \qquad a_\varphi = (\varphi, \tilde{\theta})_{G_{p'}}.$$

But $(\varphi, \tilde{\theta})_{G_{p'}} = (\varphi, \theta)_{G_{p'}} = 0$ for all $\varphi \in \mathrm{IBr}(G) \setminus \mathrm{IBr}(B)$ by assumption and Lemma 4.4.3.
  (b) Let $G_{p'} = g_1{}^G \dot{\cup} \cdots \dot{\cup} g_l{}^G.$ Then, using (4.16) and Theorem 4.4.8, we have

$$\frac{(\chi, \tilde{\theta})_G}{\chi(1)} = \frac{1}{|G|_{p'}} \sum_{i=1}^{l} \omega_\chi((g_i{}^G)^+)\, \theta(g_i^{-1})$$

$$\equiv \frac{1}{|G|_{p'}} \sum_{i=1}^{l} \omega_{\chi'}((g_i{}^G)^+)\, \theta(g_i^{-1}) \equiv \frac{(\chi', \tilde{\theta})_G}{\chi'(1)} \quad \mathrm{mod}\ p.$$

$$\qquad\qquad\qquad\qquad\qquad\qquad\qquad\qquad\qquad\qquad\qquad\qquad\qquad\square$$

Theorem 4.4.16 *Let $\chi \in \mathrm{Irr}(B)$ for $B \in \mathrm{Bl}_p(G)$. Then the following are equivalent:*

(a) $\mathrm{ht}_p(\chi) = 0$;

(b) *there is a $\xi \in \mathrm{Irr}(B)$ such that $\nu_p((\tilde{\chi}, \xi)_G) = \nu_p(\xi(1))$;*

(c) $\nu_p((\tilde{\chi}, \xi)_G) = \nu_p(\xi(1))$ *for every $\xi \in \mathrm{Irr}(B)$.*

Proof. By Lemma 4.4.15(a) we may write

$$\tilde{\chi} = \sum_{\xi \in \mathrm{Irr}(B)} (\tilde{\chi}, \xi)_G \, \xi.$$

By Lemma 4.2.19(b) we have for all $\xi \in \mathrm{Irr}(B)$

$$\nu_p((\tilde{\chi}, \xi)_G) = \nu_p((\chi, \tilde{\xi})_G) \geq \nu_p(\chi(1)) \geq a - d$$

with $a := \nu_p(|G|)$ and $d := \mathrm{d}(B)$.

Using Lemma 4.2.19(c) we see that $\mathrm{ht}_p(\chi) = 0$ if and only if there is a $\xi \in \mathrm{Irr}(B)$ with $\frac{(\tilde{\chi}, \xi)_G}{p^{a-d+1}} \notin R$ or, equivalently, $\nu_p((\tilde{\chi}, \xi)_G) = a - d$. Since we also have $\nu_p((\tilde{\chi}, \xi)_G) \geq \nu_p(\xi(1)) \geq a - d$, this implies that $\nu_p((\tilde{\chi}, \xi)_G) = \nu_p(\xi(1))$. So (b) follows from (a).

Part (b) of Lemma 4.4.15 shows that (b) implies (c).

Finally, if (c) holds then

$$0 = \nu_p((\tilde{\chi}, \xi)_G) - \nu_p(\xi(1)) \geq \nu_p(\chi(1)) - \nu_p(\xi(1)) \qquad \text{for all} \qquad \xi \in \mathrm{Irr}(B)$$

and χ must be of height zero. $\qquad\square$

In the following we see that p-blocks behave well with respect to direct products of groups.

Lemma 4.4.17 *Let $G = G_1 \times G_2$ and $\mathrm{Bl}_p(G_1) = \{B_1, \ldots, B_r\}$, $\mathrm{Bl}_p(G_2) = \{B_1', \ldots, B_s'\}$. Then $\mathrm{Bl}_p(G) = \{B_{ij} \mid 1 \leq i \leq r, \ 1 \leq j \leq s\}$ with*

$$\mathrm{Irr}(B_{ij}) = \{\chi \times \chi' \mid \chi \in \mathrm{Irr}(B_i), \ \chi' \in \mathrm{Irr}(B_j')\},$$

$$\mathrm{IBr}(B_{ij}) = \{\varphi \times \varphi' \mid \varphi \in \mathrm{IBr}(B_i), \ \varphi' \in \mathrm{IBr}(B_j')\}.$$

Proof. By Corollary 3.7.6 $\mathrm{Irr}(G) = \{\chi \times \chi' \mid \chi \in \mathrm{Irr}(G_1), \ \chi' \in \mathrm{Irr}(G_2)\}$ and $\mathrm{IBr}(G) = \{\varphi \times \varphi' \mid \varphi \in \mathrm{IBr}(G_1), \ \varphi' \in \mathrm{IBr}(G_2)\}$. Since

$$\omega_{\chi \times \chi'}(((g\,g')^G)^+) = \omega_\chi((g^{G_1})^+)\, \omega_{\chi'}((g'^{G_2})^+) \quad \text{for} \quad g \in G_1, g' \in G_2,$$

it follows from Theorem 4.4.8 that $\chi \times \chi'$ and $\psi \times \psi'$ belong to the same block, say $B_{i,j} \in \mathrm{Bl}_p(G)$, if and only if χ, ψ belong to the same block, say $B_i \in \mathrm{Bl}_p(G_1)$, and χ', ψ' belong to the same block, say $B_j' \in \mathrm{Bl}_p(G_2)$. Let $\varphi \in \mathrm{IBr}(B_i)$ and $\varphi' \in \mathrm{IBr}(B_j')$. From Corollary 4.3.4 we may conclude that $\Phi_\varphi \times \Phi_{\varphi'}$ is a projective character, because obviously $G_{p'} = \{g\,g' \mid g \in (G_1)_{p'}, \ g' \in (G_2)_{p'}\}$. Since

$$(\Phi_\varphi \times \Phi_{\varphi'}, \varphi \times \varphi')_{G_{p'}} > 0 \quad \text{and} \quad (\Phi_\varphi \times \Phi_{\varphi'}, \chi \times \chi')_G = 0$$

for all $\chi \times \chi' \in \mathrm{Irr}(G) \setminus \mathrm{Irr}(B_{ij})$, we see that $\mathrm{IBr}(B_{ij})$ is as indicated. $\qquad\square$

Example 4.4.18 We compute the 3-decomposition matrix for $G = M_{11}$ by first determining the irreducible representations of G and their 3-Brauer characters. First we obtain the permutation representation on $\Omega := \{1, 2, \ldots, 11\}$ of M_{11} from the GAP package `atlasrep` data base. Note that `atlasrep` gives us access to the data collected about many finite simple groups in Wilson's online AT-LAS of finite group representations (see [170]). For example, the GAP command `DisplayAtlasInfo` supplied with the name of a group in the data base displays the information that can be retrieved. The command `AtlasGenerators` supplied with the name of a group in the data base and a number i, returns a record that contains the permutations (or the matrices) for the standard generators in the ith representation stored in the data base. So the following code will retrieve the permutation representation of M_{11} on Ω:

```
gap> LoadPackage("atlasrep","1.3.1",false);;
gap> m11 := Group(AtlasGenerators("M11",1).generators);
Group([ (2,10)(4,11)(5,7)(8,9), (1,4,3,8)(2,5,6,9) ])
```

In the next lines of code we consider the action of M_{11} on $V := \mathbb{F}_3\binom{\Omega}{2}$, a permutation module of dimension 55. We also compute the composition factors of the permutation module and verify that they are all absolutely irreducible.

```
gap> orb :=  Orbit( m11 , [1,2], OnSets );;
gap> g := Image( ActionHomomorphism( m11 , orb, OnSets ) );;
gap> V := PermutationGModule( g, GF(3) );;
gap> comps := Set( MTX.CompositionFactors(V) );;
gap> List(comps, W -> W.dimension);
[ 1, 5, 5, 10, 10, 24 ]
gap> List( comps, W -> MTX.IsAbsolutelyIrreducible(W) );
[ true, true, true, true, true, true ]
```

We use the representations of representatives of the conjugacy classes of G as words in the standard generators as given in Example 2.5.18 in order to compute the Brauer characters of the composition factors. The command `BrauerCharacterValue` was explained in Example 4.2.12.

```
gap> 3regclassreps := function( stgens )
> local a,b ; a:=stgens[1]; b:=stgens[2];
> return( [ a^2,  a,   b, a*b*a*b^2*a*b^-1,  a*b*a*b^2*a*b^2,
>                         a*b^-1*a*b^2*a*b^2, a*b, a*b^-1 ] ); end;;
gap> brauchars := List( comps,  W ->
>      List( 3regclassreps( W.generators ), BrauerCharacterValue ) );;
gap> brauchars := Set( brauchars );;   List( brauchars, y -> y[1] );
[ 1, 5, 5, 10, 24 ]
```

So we have found five irreducible Brauer characters, and incidentally we see (using Theorem 4.2.14(b)) that the two composition factors of degree ten are isomorphic.

This leaves us with the task of constructing three more irreducible Brauer characters to complete the Brauer character table of G. This can be achieved by finding the composition factors of tensor products of the representations we have already constructed. One way to do this is to look at the tensor product V26 of one of the irreducible modules of dimension five with the irreducible module of dimension 24:

```
gap> V26 := TensorProductGModule( comps[2] , comps[6]);;
gap> comps26 := Set( MTX.CompositionFactors(V26) );;
gap> ForAll( comps26, W -> MTX.IsAbsolutelyIrreducible(W) );
true
gap> brauchars26 := List( comps26,  W ->
> List( 3regclassreps( W.generators ), BrauerCharacterValue ) );;
gap> brauchars26 := Set( brauchars26 );; List( brauchars26, y -> y[1] );
[ 1, 5, 5, 10, 10, 10, 24, 45 ]
```

Thus it turns out that every absolutely irreducible $\mathbb{F}_3 G$-module occurs up to isomorphism as a composition factor of V26, and we have found all irreducible Brauer characters of G.

Of course, for larger examples it would be better to start analyzing tensor products of non-trivial known modules of smallest possible dimensions. Here one would start with comps[2] $\otimes$ comps[2]. This would yield a non-real irreducible Brauer character of degree ten, not occurring in the list brauchars. Adding its complex conjugate and the restriction of the ordinary irreducible defect-zero character of degree 45 would also finish the job.

We display the Brauer characters of G computed above as follows:

$G = M_{11}$	1a	2a	4a	5a	8a	8b	11a	11b
φ_1	1	1	1	1	1	1	1	1
φ_2	5	1	-1	.	α	$\bar{\alpha}$	γ	$\bar{\gamma}$
φ_3	5	1	-1	.	α	$\bar{\alpha}$	γ	$\bar{\gamma}$
φ_4	10	2	2	.	.	.	-1	-1
φ_5	10	-2	.	.	β	$-\beta$	-1	-1
φ_6	10	-2	.	.	$-\beta$	β	-1	-1
φ_7	24	.	.	-1	2	2	2	2
φ_8	45	-3	1	.	-1	-1	1	1

with

$$\alpha := -1 + \sqrt{-2}, \quad \beta := \sqrt{-2}, \quad \gamma := \frac{-1 + \sqrt{-11}}{2}.$$

From this we obtain the 3-decomposition matrix of M_{11}:

$\varphi_i(1):$	1	5	5	10	10	10	24	45
	φ_1	φ_2	φ_3	φ_4	φ_5	φ_6	φ_7	φ_8
χ_1	1	.	.	.	.	.	.	.
χ_2	.	.	.	1	.	.	.	.
χ_3	.	.	.	.	1	.	.	.
χ_4	.	.	.	.	.	1	.	.
χ_5	1	1	1	.	.	.	.	.
χ_6	1	1	.	.	.	1	.	.
χ_7	1	.	1	.	1	.	.	.
χ_8	.	1	1	1	.	.	1	.
χ_{10}	1	1	1	.	1	1	1	.
χ_9	.	.	.	.	.	.	.	1

♦

   The Brauer characters of M_{11} and the other Mathieu groups was first computed by James [95], except for one 2-Brauer character of M_{24}. Using the Meataxe Parker determined this missing Brauer character.

Exercises

Exercise 4.4.1 Let $B \in \text{Bl}_p(G)$. Show that

$$d(B) = \nu_p(|G|) - \min_{\varphi \in \text{IBr}(B)} \nu_p(\varphi(1)).$$

Hint: Use Theorem 4.2.14(c).

Exercise 4.4.2 Compute the central characters and block idempotents of A_6 and S_6 in characteristic two and show that the block idempotent of the non-principal block of $F S_6$ is a sum of two block idempotents of $F A_6$.

Exercise 4.4.3 Let $\epsilon_B = \sum_{g \in G} a_g g$ be a block idempotent in RG and let m be the p'-part of the exponent of G.

(a) Show that $a_g \in R \cap \mathbb{Q}_m$ for all $g \in G$.

(b) Let $\gamma \in \text{Gal}(\mathbb{Q}_m/\mathbb{Q})$. Show that $^\gamma \epsilon_B := \sum_{g \in G} \gamma(a_g) g$ is a block idempotent in RG of a block which we denote by $^\gamma B$.

(c) Show that $\chi \in \text{Irr}(B)$ implies $^\gamma \chi \in \text{Irr}(^\gamma B)$.

Exercise 4.4.4 Let (K, R, F, η) be a p-modular system for G and let B be a p-block of G with central character $\hat{\omega}_B \colon \mathbf{Z}(FG) \to F$ and $\chi \in \text{Irr}(B)$. Show that for $z \in \mathbf{Z}(RG)$ one has

$$\eta(\omega_\chi(z)) = \hat{\omega}_B(\eta(z)),$$

where η is extended to a ring homomorphism $\eta \colon \mathbf{Z}(RG) \to \mathbf{Z}(FG)$ in a natural way; see also Exercise 4.1.3.

Exercise 4.4.5 Using the assumptions of Lemma 4.4.17, let V_i for $i = 1, 2$ be a simple FG_i-module with projective cover $P(V_i)$. Show that $P(V_1 \otimes V_2) \cong_{FG} P(V_1) \otimes P(V_2)$. Deduce that $\Phi_{\varphi \times \varphi'} = \Phi_\varphi \times \Phi_{\varphi'}$ for $\varphi \in \mathrm{IBr}(G_1)$, $\varphi' \in \mathrm{IBr}(G_2)$.

Exercise 4.4.6 Let p be a prime and Z be a central p'-subgroup of G. Show that if $\chi, \chi' \in \mathrm{Irr}(G)$ belong to the same p-block then $\chi, \chi' \in \mathrm{Irr}(G \mid \lambda)$ for some $\lambda \in \mathrm{Irr}(Z)$. That is, one obtains a partition

$$\mathrm{Bl}_p(G) = \dot{\bigcup}_{\lambda \in \mathrm{Irr}(Z)} \{ B \in \mathrm{Bl}_p(G) \mid \mathrm{Irr}(B) \subseteq \mathrm{Irr}(G \mid \lambda) \}.$$

Exercise 4.4.7 For $\chi, \chi' \in \mathrm{Irr}(G)$ define $a_{\chi,\chi'} := (\chi, \chi')_{G_{p'}}$. Prove the following.

(a) If χ, χ' belong to different p-blocks, then $a_{\chi,\chi'} = 0$.

(b) If $B \in \mathrm{Bl}_p(G)$ with decomposition matrix D_B and Cartan matrix $C(B)$ then

$$[a_{\chi,\chi'}]_{\chi,\chi' \in \mathrm{Irr}(B)} = D_B \, C(B)^{-1} \, D_B^{\mathrm{T}}.$$

Exercise 4.4.8 Let $N \trianglelefteq G$, $b \in \mathrm{Bl}_p(N)$ and $\theta \in \mathrm{Irr}(b)$. Assume that $\chi \in \mathrm{Irr}(G \mid \theta)$ belongs to the block $B \in \mathrm{Bl}_p(G)$. Show that for any $\theta' \in \mathrm{Irr}(b)$ there is a $\chi' \in \mathrm{Irr}(B)$ with $\chi \in \mathrm{Irr}(G \mid \theta')$.
Hint: Show that if θ, θ' are connected in the Brauer graph $\Gamma_p(N)$ then there is a $\chi' \in \mathrm{Irr}(G \mid \theta')$ which is connected in $\Gamma_p(G)$ to χ. Then use Theorem 4.4.2.

4.5 Basic sets

Before introducing defect groups and Brauer's main theorems on blocks, we pause and discuss how to find information on the decomposition numbers using ordinary characters.

Definition 4.5.1 Let B be a p-block of G. A **basic set of Brauer characters** (a **basic set of p-projective characters**) of B is a $\mathbb{Z}$-basis of $\langle \mathrm{IBr}(B) \rangle_{\mathbb{Z}}$ (respectively of $\langle \Phi_\varphi \mid \varphi \in \mathrm{IBr}(B) \rangle_{\mathbb{Z}}$) consisting of Brauer characters (respectively p-projective characters).

The definition is not quite standard. In the literature, basic sets are often allowed to consist of generalized characters. We, however, follow [83].

Lemma 4.5.2 *A subset $S \subseteq \langle \mathrm{IBr}(B) \rangle_{\mathbb{N}_0}$ is a basic set of Brauer characters of B if and only if S is linearly independent over $\mathbb{Z}$ and $\{ \chi|_{G_{p'}} \mid \chi \in \mathrm{Irr}(B) \} \subseteq \langle S \rangle_{\mathbb{Z}}$.*

Proof. By Lemma 4.4.6, $\{ \chi|_{G_{p'}} \mid \chi \in \mathrm{Irr}(B) \} \subseteq \langle S \rangle_{\mathbb{Z}}$ if and only if $\mathrm{IBr}(B) \subseteq \langle S \rangle_{\mathbb{Z}}$. From this the result follows. $\qquad \square$

In order to find a basic set one might start with a set of Brauer characters, respectively p-projective characters, that spans but is not necessarily $\mathbb{Z}$-linear independent, which, for the case of Brauer characters, can be chosen to be $\{\chi|_{G_{p'}} \mid \chi \in \mathrm{Irr}(B)\}$. For the case of p-projective characters the set of induced p-projective indecomposable characters from all maximal subgroups restricted to B satisfies the requirement; see Exercise 4.5.7.

In both cases one might proceed by sorting the characters with respect to increasing degrees or norms and choose the first $l(B)$ $\mathbb{Z}$-linearly independent characters. In general, this simple procedure might not find a basic set (see Exercise 4.5.6), and even a basic set consisting of restricted ordinary characters is not known to exist in general. However, for all blocks of sporadic simple groups the existence can be easily verified using GAP. The following criterion can be used to verify that one indeed has found basic sets.

Lemma 4.5.3 *Let* $S^b \subseteq \langle \mathrm{IBr}(B) \rangle_{\mathbb{N}_0}$ *and* $S^p \subseteq \langle \Phi_\varphi \mid \varphi \in \mathrm{IBr}(B) \rangle_{\mathbb{N}_0}$ *with* $|S^b| = |S^p| = l(B)$. *Then* S^b *and* S^p *are basic sets if and only if*

$$U := [\, (\varphi, \Phi)_{G_{p'}} \,]_{\varphi \in S^b, \Phi \in S^p} \tag{4.22}$$

is invertible over $\mathbb{Z}$.

Proof. For $\psi \in S^b$ and $\Psi \in S^p$ we have

$$\psi = \sum_{\varphi \in \mathrm{IBr}(B)} u^1_{\psi,\varphi} \varphi, \qquad \Psi = \sum_{\varphi \in \mathrm{IBr}(B)} u^2_{\Psi,\varphi} \Phi_\varphi,$$

with $u^1_{\psi,\varphi}, u^2_{\Psi,\varphi} \in \mathbb{N}_0$, and S^b and S^p are basic sets if and only if the matrices

$$U_1 = [u^1_{\psi,\varphi}]_{\psi \in S^b, \varphi \in \mathrm{IBr}(B)} \quad \text{and} \quad U_2 = [u^2_{\Psi,\varphi}]_{\Psi \in S^p, \varphi \in \mathrm{IBr}(B)}$$

are unimodular. But from Theorem 4.3.3 we see that $U = U_1 U_2^{\mathrm{T}}$ and the lemma follows.

$\square$

Having found basic sets S^b, S^p of Brauer and p-projective characters, one may compute U as in (4.22). Determining $\mathrm{IBr}(B)$ is then equivalent to finding the unimodular matrix U_1 introduced in the proof of Lemma 4.5.3. Thus one method of attack is to find all solutions

$$X_1, X_2 \in \mathbb{N}_0^{l \times l} \qquad \text{of} \qquad X_1 X_2^{\mathrm{T}} = U \tag{4.23}$$

up to equivalence, where we call two solutions (X_1, X_2) and (X_1', X_2') equivalent if $X_1' = X_1 P$ and $X_2' = X_2 P$ for a permutation matrix $P \in \mathbb{N}_0^{l \times l}$. Each solution leads to a system of generalized Brauer characters which is a candidate for $\mathrm{IBr}(B)$ and may be tested, for instance by checking whether or not the scalar products with all known projective characters are non-negative.

Note that the problem is much harder than the one we have dealt with in Section 2.8, where the additive decomposition algorithm produced "all" solutions of $XX^{\mathrm{T}} = U$ for a positive definite matrix U.

In general it will not be feasible to solve equation (4.23) because the number of solutions will be much too large to be of any practical use. Instead the main idea is to improve gradually the basic sets leading to a matrix U with smaller entries. How this is done in practice, with the help of methods of integer linear programming, is described in [83]. This book describes the concepts and underlying methods of the computer algebra system MOC, which has been used to compute the Brauer character tables of a large number of simple groups.

We do not pursue the methods further, which are rather technical in nature, but present a comparatively simple example to illustrate the problems.

Example 4.5.4 We consider $G := J_1$, the smallest Janko group for the prime $p := 2$: G has five 2-blocks of defect zero; one, say B_1, of defect one; and the principal block $B := B_0(G)$ (of defect three). The block B_1 is easily described: $\mathrm{Irr}(B_1) = \{\chi_4, \chi_5\}$ and $\chi_4|_{G_{p'}} = \chi_5|_{G_{p'}}$. Hence

$$D_{B_1} = \begin{bmatrix} 1 \\ 1 \end{bmatrix}.$$

Turning to $B := B_0(G)$, it is straightforward to check that

$$\mathrm{Irr}(B) = \{\chi_1, \chi_6, \chi_7, \chi_8, \chi_{12}, \chi_{13}, \chi_{14}, \chi_{15}\}.$$

In GAP we get this by

```
gap> ct := CharacterTable( "J1" );; bl := PrimeBlocks( ct, 2 );;
gap> irrB := Positions( bl.block, 1 );
[ 1, 6, 7, 8, 12, 13, 14, 15 ]
```

Looking at the character table we see, abbreviating $\chi_i' := \chi_i|_{G_{p'}}$, that

$$\chi_{13}' = \chi_6' - \chi_7' + \chi_{12}' , \quad \chi_{14}' = \chi_6' - \chi_8' + \chi_{12}' , \quad \chi_{15}' = -\chi_1' + \chi_6' + \chi_{12}'.$$

So $S^b = \{\psi_1, \dots \psi_5\} := \{\chi_1', \chi_6', \chi_7', \chi_8', \chi_{12}'\}$ is a basic set of Brauer characters. We now produce projective characters:

```
gap> basm:=[1,6,7,8,12];;rest := Difference([1..Length(Irr(ct))],basm);;
gap> sb := Irr(ct){basm};;
gap> def0 := List( Positions(bl.defect,0), i -> Position(bl.block,i) );
[ 2, 3, 9, 10, 11 ]
gap> proj := Irr(ct){ def0 };; Add( proj , Sum( Irr(ct){[4,5]} ) );
```

So far, we have six projective indecomposable characters, five being of defect zero and $\chi_4 + \chi_5$, the projective indecomposable character of the block B_1. To obtain more 2-projective characters of G we take advantage of the fact that the 2-decomposition matrices of the maximal subgroups of G (see Maxes(ct)) are easily computed – see Exercise 4.5.4 and Corollary 4.13.6 – and can be found in the GAP library. We may thus induce up the indecomposable 2-projective characters of these subgroups and obtain further projectives:

```
gap> Maxes(ct);
[ "L2(11)", "2^3.7.3", "2xA5", "19:6", "11:10", "D6xD10", "7:6" ]
gap> for max in Maxes(ct) do
> ctu := CharacterTable( max ) ; d := DecompositionMatrix(ctu mod 2);;
> pimsmax := TransposedMat(d)*Irr(ctu);;
> Append( proj, InducedClassFunctions( pimsmax, ct ) );
> od;
gap> proj := Set(proj);; Length(proj);
30
```

We tensor these 30 characters with all characters in $\mathrm{Irr}(G)$ and compute the B-part (in GAP this means that one reduces the characters with $\mathrm{Irr}(G) \setminus \mathrm{Irr}(B)$). The result is a list of 198 projective characters in $\langle \Phi_\varphi \mid \varphi \in \mathrm{IBr}(B) \rangle_{\mathbb{N}_0}$:

```
gap> tens := Tensored( Irr(ct), proj );;
gap> otherblocks := Difference( [1..Length(Irr(ct))] , irrB );
[ 2, 3, 4, 5, 9, 10, 11 ]
gap> projectives := Reduced( ct,Irr(ct){otherblocks}, tens ).remainders;;
gap> Length( projectives );
198
```

We sort these characters according to their norms and retain only the first nine. Next, we test all $\binom{9}{5} = 126$ subsets of size five whether or not they form a basic set using Lemma 4.5.3:

```
gap> SortParallel( List(projectives, Norm), projectives );
gap> smallpro := projectives{[1..9]};;
gap> 5sets := Filtered( Combinations(smallpro), x -> Length(x) = 5 );;
gap> basicsets := Filtered( 5sets , x ->
>                          Determinant(MatScalarProducts(ct,sb,x)) in [-1,1] );;
gap> Length( basicsets );
17
```

Choosing a basic set $S^p = \{\Psi_1, \ldots, \Psi_5\} := \mathrm{sp}$ with minimal sum of norms we compute the matrix $A := [(\Psi, \chi)_G]_{\Psi \in S^p, \chi \in \mathrm{Irr}(B)}$. It is easy to identify these projective characters Ψ_i; see Exercise 4.5.2:

```
gap> SortParallel( List( basicsets,s -> Sum(List(s,Norm)) ), basicsets );
gap>  A := MatScalarProducts( ct, Irr(ct){irrB}, basicsets[1] );;
```

$$
A = \begin{bmatrix}
1 & 3 & 1 & 1 & 1 & 3 & 3 & 3 \\
1 & 2 & 1 & 2 & 2 & 3 & 2 & 3 \\
1 & 2 & 2 & 1 & 2 & 2 & 3 & 3 \\
0 & 1 & 2 & 2 & 3 & 2 & 2 & 4 \\
0 & 2 & 2 & 2 & 3 & 3 & 3 & 5
\end{bmatrix}, \quad \text{so} \quad U = \begin{bmatrix}
1 & 1 & 1 & 0 & 0 \\
3 & 2 & 2 & 1 & 2 \\
1 & 1 & 2 & 2 & 2 \\
1 & 2 & 1 & 2 & 2 \\
1 & 2 & 2 & 3 & 3
\end{bmatrix}
$$

in the notation of Lemma 4.5.3. Also the transposed rows of A are sums of columns of the (yet unknown) decomposition matrix D_B. The matrix U_1 (in the notation of the proof of Lemma 4.5.3) is in our case the submatrix of D_B consisting of the first five rows.

Since $\psi_1 = \varphi_1$, the trivial Brauer character $\Phi_1 := \Phi_{\varphi_1}$ is a summand in any 2-projective character Ψ with $(\Psi, \chi_1)_G > 0$. Hence we may find candidates for the first column of D_B by searching through the common p-subsums of all such Ψ. Here a "p-subsum" of a character $\chi = \sum_{\chi \in \mathrm{Irr}(G)} a_\chi \chi$ is any generalized p-projective character of the form $\sum_{\chi \in \mathrm{Irr}(G)} b_\chi \chi$, with $0 \le b_\chi \le a_\chi$ for all $\chi \in \mathrm{Irr}(G)$. It is easy to write a GAP program subs (see Exercise 4.5.3) which computes all p-subsums of a given Ψ. For practical reasons it is convenient to give Ψ by the vector of multiplicities of the irreducible constituents (thus, for instance, Ψ_1 by $A[1]$, the first row of A).

```
gap> li := [];;
gap> for sp in basicsets do
>       A1 := MatScalarProducts( ct, Irr(ct){irrB}, sp );
>       A1 := Filtered( A1 , x -> x[1] > 0);
>       ss := Intersection( List(A1 , y ->  subs(ct,basm,y,irrB,2)) );;
>       ss := Filtered( ss, y -> y[1] = 1 );  Add( li, ss );
>    od;
gap>  c1 := Intersection(li);
[ [ 1, 0, 0, 0, 1, 1, 1, 0 ], [ 1, 0, 0, 1, 1, 1, 0, 0 ],
  [ 1, 0, 1, 0, 1, 0, 1, 0 ], [ 1, 0, 1, 1, 1, 0, 0, 0 ],
  [ 1, 1, 0, 0, 0, 1, 1, 0 ], [ 1, 1, 0, 1, 0, 1, 0, 0 ],
  [ 1, 1, 1, 0, 0, 0, 1, 0 ], [ 1, 1, 1, 1, 0, 0, 0, 0 ],
  [ 1, 1, 1, 1, 1, 1, 1, 1 ] ]
```

Altogether we get only nine candidates for the first column of D_B.

Let x1 $\in$ c1. If the first column of D_B is x1$^\mathrm{T}$, and so $\Phi_1 =$ x1 * Irr(ct){irrB}, we see from A that $\Psi_i - \Phi_1$ is a p-projective character for $i = 1, \ldots, 3$ and $\Psi_1 - \Phi_1$ contains a projective indecomposable, say Φ_2, containing χ_6. This means that we may subtract x1 from the first three rows of A and that then the first row of A contains a p-subsum x2 with $\Phi_2 =$ x2* Irr(ct){irrB}. Similarly, if y2 is a row of A which is not in $\langle$x2$\rangle_{\mathbb{N}_0}$ then it contains a p-subsum x3 yielding another projective indecomposable. In this way we may enumerate all possibilities for the projective indecomposable characters or, equivalently, for the decomposition matrix. In each case we test the resulting matrix U_1 consisting of the first five rows for unimodularity and then whether the resulting candidates for the irreducible Brauer characters (obtained by multiplying the inverse of U_1 with $[\psi_i(g_j)]_{1 \le i \le 5, 1 \le j \le l}$, where $g_1, \ldots, g_l$ are representatives of the 2-regular classes) have non-negative integral scalar products with the projective characters we have selected. It is easy to write a one-line program isinspan := function(v , lv) which tests whether or not a vector $v \in \mathbb{N}_0^{1 \times n}$ is in the $\mathbb{N}_0$-span of a list lv of vectors in $\mathbb{N}_0^{1 \times n}$; see Exercise 4.5.1:

```
gap> cand := [];;
gap> for x1 in c1 do
>    a := ShallowCopy(A);    a{[1,2,3]}:= [ a[1]-x1, a[2]-x1, a[3]-x1 ];
>    for x2 in Filtered( subs(ct,basm,a[1],irrB,2), x -> x[2] > 0 ) do
>      y2 := First( a, x -> not isinspan( x, [x2]) );
>      for x3 in subs( ct,basm,y2,irrB,2 ) do
>        y3 := First( a ,  x -> not isinspan( x, [x2,x3] ) );
>        for x4 in subs( ct,basm,y3,irrB,2 ) do
>          y4 := First( a ,  x -> not isinspan(x, [x2,x3,x4]) );
>          for x5 in subs( ct,basm,y4,irrB,2 ) do
```

```
>                u := [ x1, x2, x3, x4, x5 ]; u1 := TransposedMat(u){[1..5]};
>                if Rank(u) = 5 and Determinant(u1) in [1,-1] then
>                    m := MatScalarProducts( ct, u1^-1 * sb, smallpro );
>                    if ForAll( m, y -> ForAll(y, x-> x >=0 and IsInt(x) ))
>                        then Add( cand, u );
>                    fi;
>                fi;
>            od;
>          od;
>        od;
>      od;
> od;
gap> for mat in cand do Sort(mat); od;
gap> cand:= Set( cand );; cand := List(cand, Reversed);; Length( cand );
5
```

There are only five candidates for the decomposition matrix D_B. These differ only in their first column:

$\chi(1)$	χ	φ_1	φ_1	φ_1	φ_1	φ_1	φ_2	φ_3	φ_4	φ_5
1	χ_1	1	1	1	1	1	0	0	0	0
77	χ_6	1	0	1	1	1	1	0	0	0
77	χ_7	1	1	0	0	1	0	1	1	0
77	χ_8	1	1	0	1	0	0	1	0	1
133	χ_{12}	1	1	0	0	0	0	1	1	1
133	χ_{13}	1	0	1	1	0	1	0	0	1
133	χ_{14}	1	0	1	0	1	1	0	1	0
209	χ_{15}	1	0	0	0	0	1	1	1	1

As one can see, one obtains the possibilities for the column corresponding to φ_1 from the first one (the all-1-solution) by subtracting one of the other columns (corresponding to $\varphi_2, \ldots, \varphi_5$). This is quite a common phenomenon: typically the decomposition matrix is determined up to some ambiguities resulting from the fact that it is not obvious whether or not some columns of the given matrix have to be subtracted from others in order to get the decomposition matrix; see for example Exercise 4.5.5. Often it is hard, or even impossible, to decide when one is working – as we are in this section – solely on the level of characters.

For the example above, Theorem 4.2.21 comes to our rescue. Since all ordinary characters of G are real-valued the same is true for the Brauer characters (see Theorem 4.2.16). So by Theorem 4.2.21 the degrees $\varphi_i(1)$ must be even for $i = 2, \ldots, 5$. Since $\chi(1)$ is odd for all $\chi \in \mathrm{Irr}(B)$, it is clear that all entries in the first column must be odd, hence the first column of D_B is the all-1-column.

The decomposition matrix of J_1 was first computed by Fong (see [59]) without the use of a computer.   ◆

In the above example the Brauer characters were determined just by analyzing basic sets. This is quite exceptional. In general further methods have to be used in conjunction with basic sets. We will demonstrate some of these, including the method of condensation, in the following example.

Example 4.5.5 We consider the Mathieu group $G := M_{22}$ over $F := \mathbb{F}_4$. We start similarly as in Example 4.5.4 – to which we refer for some of the notation – by producing 2-projective characters. Note that G has only one 2-block and is a subgroup of M_{23}, which has two complex conjugate defect-zero characters ψ_{12}, ψ_{13}, which we restrict to G to obtain 2-projective characters. Also we will use the known decomposition matrices of the maximal subgroups of G in order to obtain further 2-projective characters of G:

```
gap> ct := CharacterTable( "M22" );; irrB := [1..Length(Irr(ct))];;
gap> ctm23 := CharacterTable( "M23" );;
gap> proj := RestrictedClassFunctions( Irr(ctm23){[12,13]}, ct );;
gap> for max in Maxes(ct) do
>      ctu := CharacterTable( max ); d := DecompositionMatrix( ctu mod 2 );;
>      pimsmax := TransposedMat(d) * Irr(ctu);;
>      Append( proj, InducedClassFunctions( pimsmax, ct ) );
> od;
gap> projectives := Set( Tensored(Irr(ct), proj) );;
gap> SortParallel( List(projectives, Norm), projectives );
gap> smallpro := projectives{[1..10]};;
gap> basm := [ 1, 2, 3, 5, 6, 9, 10 ];; sb := Irr(ct){basm};;
gap> 7sets := Filtered( Combinations(smallpro), x -> Length(x) = 7 );;
gap> basicsets := Filtered( 7sets , x ->
>                   Determinant( MatScalarProducts(ct,sb,x) ) in [-1,1] );;
gap> SortParallel( List( basicsets, s -> Sum(List(s,Norm)) ), basicsets );
gap> A := MatScalarProducts( ct, Irr(ct), basicsets[1] );;
gap> Sort(A); A := Reversed(A);; Display(A);
[ [  1,  1,  1,  1,  1,  1,  2,  4,  5,  5,  5,  7 ],
  [  0,  1,  1,  0,  1,  1,  2,  3,  1,  3,  3,  3 ],
  [  0,  1,  0,  1,  1,  1,  2,  3,  1,  3,  3,  3 ],
  [  0,  0,  1,  1,  1,  2,  3,  4,  4,  5,  5,  7 ],
  [  0,  0,  0,  0,  0,  1,  1,  1,  0,  1,  1,  1 ],
  [  0,  0,  0,  0,  0,  0,  0,  0,  1,  1,  0,  1 ],
  [  0,  0,  0,  0,  0,  0,  0,  1,  0,  1,  1,  1 ] ]
gap> List( A, y -> Length( subs(ct, basm, y, [1..Length(Irr(ct))],2) ) );
[ 391, 19, 19, 121, 1, 1, 1 ]
```

Note that $\mathtt{A} = [a_{ij}]$ is the transpose of our first approximation to the 2-decomposition matrix D of G. In fact, we see that the last three transposed rows of $\mathtt{A}$ are columns of D because there are no proper 2-subsums. If

$$\Psi_i := \sum_{i=1}^{12} a_{ij}\, \chi_j \quad \text{then} \qquad \Psi_5 = (\theta_0)^G, \quad \Psi_6 = (\psi_{12})_G, \quad \Psi_7 = \overline{\Psi_6},$$

where θ_0 is a defect-zero character of $M_{21} \cong L_3(4)$. Note that Ψ_2 and Ψ_3 must contain a projective indecomposable character containing χ_2, so we search the subsums of $\mathtt{A[2]}$, $\mathtt{A[3]}$ for vectors with second entry "1":

```
gap> c2 := Filtered( subs(ct,basm,A[2],irrB,2), y -> y[2]=1 );;
gap> c3 := Filtered( subs(ct,basm,A[3],irrB,2), y -> y[2]=1 );;
gap> Set( List( c2, x -> x{[1..5]}) ); Set( List( c3, x -> x{[1..5]}) );
```

```
[ [ 0, 1, 1, 0, 1 ] ]
[ [ 0, 1, 0, 1, 1 ] ]
```

We see that the first five entries in A[2] and A[3] are uniquely determined. Since $\chi_4 = \overline{\chi}_3$ we must have a pair of complex conjugate 2-Brauer characters φ_2, φ_3, and we conclude that

$$\chi_2|_{G_{p'}} = \varphi_1 + \varphi_2 + \varphi_3, \quad \chi_3|_{G_{p'}} = \varphi_1 + \varphi_2 + \varphi_4, \quad \chi_4|_{G_{p'}} = \varphi_1 + \varphi_3 + \varphi_4.$$

Observe that φ_4 must be real and, by Theorem 4.2.21, of even degree. It follows that $\varphi_2(1) = \varphi_3(1) = 10$ and $\varphi_4(1) = 34$, and $\chi_5|_{G_{p'}} = \varphi_1 + \varphi_2 + \varphi_3 + \varphi_4$.

Since $((\chi_6)_{M_{21}}, \theta_0)_{M_{21}} = 1$, we see that χ_6 must have a (real) constituent $\varphi_5 \in \mathrm{IBr}(G)$ with $\varphi_5(1) \geq \theta_0(1) = 64$. Hence $\chi_6|_{G_{p'}} = \varphi_1 + \varphi_2 + \varphi_3 + \varphi_5$ or $\chi_6|_{G_{p'}} = \varphi_1 + \varphi_4 + \varphi_5$ or $\chi_6|_{G_{p'}} = \varphi_1 + \varphi_5$, implying that $\varphi_5(1)$ is 78, 64 or 98, respectively. The first possibility is easily excluded considering the possible 2-subsums, but not the second one. Therefore we use the Meataxe to find out how $\chi_6|_{G_{p'}}$ decomposes into Brauer characters. We easily see that

$$\varphi_2 \cdot \varphi_3 = \chi_6|_{G_{p'}} + \varphi_1.$$

Also φ_2, φ_3 are composition factors of the natural permutation module V:

```
gap> V := PermutationGModule( MathieuGroup(22), GF(4) );;
gap> comps := Set( MTX.CompositionFactors(V) );;
gap> ibr := List( comps, W -> W.dimension );
[ 1, 1, 10, 10 ]
gap> V100 := TensorProductGModule( comps[3] , comps[4] );;
gap> compsV100 := Set( MTX.CompositionFactors(V100) );;
gap> List( compsV100, W -> W.dimension );
[ 1, 1, 98 ]
```

Thus $\chi_6|_{G_{p'}} = \varphi_1 + \varphi_5$ and $\varphi_5(1) = 98$. We refine our matrix A:

```
gap> A{[2,3,4]} := [A[2]-A[5], A[3]-A[5], A[4] - 2*A[5]];; Display(A);
[ [ 1, 1, 1, 1, 1, 1, 2, 4, 5, 5, 5, 7 ],
  [ 0, 1, 1, 0, 1, 0, 1, 2, 1, 2, 2, 2 ],
  [ 0, 1, 0, 1, 1, 0, 1, 2, 1, 2, 2, 2 ],
  [ 0, 0, 1, 1, 1, 0, 1, 2, 4, 3, 3, 5 ],
  [ 0, 0, 0, 0, 0, 1, 1, 1, 0, 1, 1, 1 ],
  [ 0, 0, 0, 0, 0, 0, 0, 0, 1, 1, 0, 1 ],
  [ 0, 0, 0, 0, 0, 0, 0, 0, 1, 0, 1, 1 ] ]
```

There is a single pair φ_6, φ_7 of complex conjugate irreducible Brauer characters missing. Also at this stage all columns of D are known except for the first and fourth ones. There still are nine possibilities for D or its transpose A: we might subtract A[6] + A[7] once or twice from A[1] or from A[4].

In order to decrease the possibilities we induce up $\psi \in \mathrm{IBr}(M_{21})$ with $\psi(1) = 8$ to G and compute the scalar products with $\Psi_i := \sum_{j=1}^{12} a_{ij} \chi_j$ $(1 \leq i \leq 7)$:

```
gap> ct1 := CharacterTable(Maxes(ct)[1]);;
```

```
gap> psi :=  InducedClassFunction( Irr(ct1 mod 2)[2], ct1 );;
gap> psiG := InducedClassFunction( psi, ct );;  Psis := A * Irr(ct);;
gap> MatScalarProducts(ct , Psis, [psiG] );
[ [ 4, 0, 0, 3, 0, 1, 1 ] ]
```

Table 4.1.

x_1	x_2	$\varphi_6(1) = \varphi_7(1)$	Φ_1	Φ_4
4	3	35	Ψ_1	Ψ_4
4	1	69	Ψ_1	$\Psi_4 - (\Psi_6 + \Psi_7)$
2	3	36	$\Psi_1 - (\Psi_6 + \Psi_7)$	Ψ_4
2	1	70	$\Psi_1 - (\Psi_6 + \Psi_7)$	$\Psi_4 - (\Psi_6 + \Psi_7)$
0	3	37	$\Psi_1 - 2(\Psi_6 + \Psi_7)$	Ψ_4
0	1	71	$\Psi_1 - 2(\Psi_6 + \Psi_7)$	$\Psi_4 - (\Psi_6 + \Psi_7)$

We conclude that $\psi^G = x_1 \varphi_1 + x_2 \varphi_4 + \varphi_6 + \varphi_7$ with $x_1 \in \{4, 2, 0\}$ and $x_2 \in \{3, 1\}$. Thus there are only six possibilities for D or, equivalently, for the projective indecomposables Φ_i. Since $\psi^G(1) = 8 \cdot 22 = 176$, we obtain Table 4.1. So D is known if we know $\varphi_6(1)$. This can be easily decided using the Meataxe:

```
gap> W := TensorProductGModule( comps[3], compsV100[3] );;
gap> compsW := Set( MTX.CompositionFactors(W) );;
gap> ForAll( compsW , U -> MTX.IsAbsolutelyIrreducible(U) );
true
gap> Set( List( compsW , U -> U.dimension) );
[ 1, 10, 34, 70, 98 ]
```

Thus $70 \in \{\varphi(1) \mid \varphi \in \mathrm{IBr}(G)\}$, and D^{T} is obtained from the above A by subtracting the last two rows from the first and fourth ones:

$$D^{\mathrm{T}} = \begin{bmatrix}
1 & 1 & 1 & 1 & 1 & 1 & 2 & 4 & 1 & 3 & 3 & 3 \\
\cdot & 1 & 1 & \cdot & 1 & \cdot & 1 & 2 & 1 & 2 & 2 & 2 \\
\cdot & 1 & \cdot & 1 & 1 & \cdot & 1 & 2 & 1 & 2 & 2 & 2 \\
\cdot & \cdot & 1 & 1 & 1 & \cdot & 1 & 2 & 2 & 2 & 2 & 3 \\
\cdot & \cdot & \cdot & \cdot & \cdot & 1 & 1 & 1 & \cdot & 1 & 1 & 1 \\
\cdot & \cdot & \cdot & \cdot & \cdot & \cdot & \cdot & \cdot & 1 & 1 & \cdot & 1 \\
\cdot & \cdot & \cdot & \cdot & \cdot & \cdot & \cdot & 1 & \cdot & 1 & 1 & 1
\end{bmatrix}$$

Here we could finish the decomposition matrix, because we were able to construct a module V having a composition factor with one of the missing Brauer characters. We now want to describe the condensation method which would come to our rescue if such a module V and its composition factors could not be computed directly. So we go back to the point where we still had six possibilities for D^{T}, which we now collect in a list cand:

```
gap> cand:= [];;
gap> for x in Cartesian([0,1,2], [0,1]) do
>       a := ShallowCopy(A);
>       a[1] := a[1] - x[1]*(a[6]+a[7]);  a[4] := a[4] - x[2]*(a[6]+a[7]);
>       Add(cand, a);
>    od;
```

We start by finding a permutation FG-module V of smallest possible dimension having composition factors with Brauer characters φ_6, φ_7. For this we make use of the table of marks of G which is contained in the GAP library. Let $\mathcal{L}(G) = \{H_1, \ldots, H_n\}$ be representatives of conjugacy classes of subgroups of G, as in Definition 3.5.1. We compute $\texttt{perm} := ((1_{H_i})^G)_{1 \leq i \leq n}$ and the list $\texttt{permbrau}$ of their Brauer characters, which we write as linear combinations of the candidates for the irreducible Brauer characters of G:

```
gap> t := TableOfMarks( "M22" );;
gap> preg := Filtered( [1..Length(Irr(ct))],
>                i -> not IsInt( OrdersClassRepresentatives(ct)[i]/2 ) );;
gap> perm := PermCharsTom(ct,t);; permbrau := List(perm, y -> y{preg});;
gap> for A in cand do
>       ibr := TransposedMat( A{[1..7]}{basm} )^-1 * sb;;
>       ibr := List( ibr, y -> y{preg} );
>       x := List( permbrau, y -> SolutionMat( ibr, y ) );;
>       f := Filtered([1..Length(x)], i -> x[i][6] <>0 and x[i][7] <> 0);;
>       Print( f[Length(f)]," , ", x[f[Length(f)]], "\n" );
>    od;
149 , [ 10, 4, 4, 6, 1, 1, 1 ]
149 , [ 10, 4, 4, 4, 1, 1, 1 ]
149 , [ 8, 4, 4, 6, 1, 1, 1 ]
149 , [ 8, 4, 4, 4, 1, 1, 1 ]
149 , [ 6, 4, 4, 6, 1, 1, 1 ]
149 , [ 6, 4, 4, 4, 1, 1, 1 ]
```

We conclude that in each case $V = (F_{H_{149}})^G$ is the permutation module we were looking for. Its dimension is 462. We have also given the multiplicities of the composition factors. We see that these multiplicities vary only for factors with Brauer character φ_1 and φ_4. Hence we choose an idempotent e_H with $e_H V_4 \neq \{0\}$, where V_j is a simple FG-module with Brauer character φ_j. Of course, H has to have odd order. So we search for a subgroup H of odd order such that $n_4 \neq 0$, with $n_j := \dim_F(e_H V_j)$. Since $|H|$ is odd, $n_j = (\varphi_{jH}, 1_H)_H = (\varphi_j, \theta)_{G_{p'}}$, (see Exercise 4.3.1), where $\theta = (1_H)^G$. This is a p-projective character and is thus of the form $\theta = \sum_{j=1}^7 x_j \Phi_j$. By Theorem 4.3.3, $x_j = n_j$, and we can compute n_j by solving a system of linear equations, which we do for each of the six candidates for the $(\Phi_1, \ldots, \Phi_7)$:

```
gap> odds := Filtered([1..Length(perm)], i-> OrdersTom(t)[i] mod 2 <> 0);
[ 1, 3, 10, 13, 26, 28, 57, 94 ]
gap> thetas := perm{odds};;
gap> Filtered( [1..Length(odds)], i -> ForAll(cand ,
>                A -> SolutionMat(A *Irr(ct), thetas[i])[4] <> 0) );
[ 1, 2, 3, 4, 5, 6 ]
gap> Display( List( cand, A -> SolutionMat(A *Irr(ct), thetas[6]) ) );
[ [  1,  0,  0,  4,  8,  0,  0 ],
  [  1,  0,  0,  4,  8,  4,  4 ],
  [  1,  0,  0,  4,  8,  1,  1 ],
  [  1,  0,  0,  4,  8,  5,  5 ],
```

```
[  1,  0,  0,  4,  8,  2,  2 ],
[  1,  0,  0,  4,  8,  6,  6 ] ]
```

There are eight conjugacy classes of subgroups of odd order, and we see that $H := H_{28}$ (of order 11) is the group with maximal order we were looking for and which we choose as the condensation subgroup. Furthermore, $e_H V_j = \{0\}$ for $j = 2, 3$ in all cases and $e_H V_j \neq \{0\}$ for $j \neq 2, 3$ in all cases except for $A = \texttt{cand[1]}$. In view of Theorem 1.6.11, we expect the multiplicities of the composition factors for the $e_H A e_H$-module $e_H V$ as shown in Table 4.2.

Table 4.2. Multiplicities of the composition factors

A	multiplicities
cand[1]	10,6,1
cand[2]	10,4,1,1,1
cand[3]	8,6,1,1,1
cand[4]	8,4,1,1,1
cand[5]	6,6,1,1,1
cand[6]	6,4,1,1,1

We now construct the action of G on the cosets of H_{149} in GAP:

```
gap> s := RepresentativeTom( t, 149 );;
gap> G := UnderlyingGroup( t );;
gap> trans := RightTransversal( G, s );;
gap> g := Action( G, trans, OnRight );;
```

We finish by computing the matrices of the action of a few randomly selected elements of the form $e_H g e_H$ on the condensed module $e_H V$, where we make use of the GAP code from Exercise 1.6.4 stored in the file `cond.gap`, which we read in first. Finally, we use the GAP Meataxe to find the composition factors and their multiplicities in the condensed module for the algebra $CA \subseteq e_H F G e_H$ generated by the selected elements. Since GAP works with right modules, we have transposed the matrices in order to translate to our situation. Note that the dimension of the condensed module $e_H V$ is 42, the number of orbits of H, whereas $\dim V = 462$.

```
gap> Read("cond.gap");
gap> H := RepresentativeTomByGenerators( t, 28, GeneratorsOfGroup(g) );;
gap> mats:= List( [g.1, g.2, g.1*g.2, g.1*g.2^2*g.1*g.2, g.2^2*g.1*g.2^2],
>                      g -> TransposedMat( cond( H, 462, g, 2 ) ) );;
gap> m := GModuleByMats( mats, GF(4) );;
gap> List( MTX.CollectedFactors( m ), x -> [x[1].dimension, x[2]] );
[ [ 1, 8 ], [ 4, 4 ], [ 5, 1 ], [ 5, 1 ], [ 8, 1 ] ]
```

We see that the CA-module $e_H V$ has no composition factor with multiplicity greater than eight and only one composition factor with multiplicity greater than four. This is compatible only with $A \in \{\texttt{cand[4]}, \texttt{cand[6]}\}$, thus $\varphi_6(1) \in \{70, 71\}$. This last ambiguity could be resolved by showing that $CA = e_H F G e_H$.

The Brauer character table of M_{22} was first computed by James in [95] using only characters. In order to determine $\varphi_5(1)$ and $\varphi_6(1)$ he first computed some Brauer characters of $3 \cdot M_{22}$ and M_{23}. ◆

A further application of condensation is the determination of the structure of the projective indecomposable FG-modules; see [114]. Moreover, Exercise 4.5.8 shows that if we look at a condensed module $e_H V$, then the trace of an element $e_H g e_H$ in the condensed module $e_H V$ is just $\frac{1}{|H|} \sum_{h \in H} \mathrm{trace}_V(hg)$. This has been applied to determine the irreducible Brauer characters of the sporadic simple Lyons group for $p = 37$ and 67; see [122]. Historically one of the first uses of condensation consisted in finding vectors in a given FG-module that lie in a proper FG-submodule, thereby avoiding the step of computing the null space of matrices, as usually done in the Meataxe algorithm; see Section 1.3.

Exercises

Exercise 4.5.1 Let $v \in \mathbb{N}_0^{1 \times n}$ and let $\mathtt{lv}$ be a list of vectors in $\mathbb{N}_0^{1 \times n}$. Show that the following GAP program returns $\mathtt{true}$ if $v \in \langle \mathtt{lv} \rangle_{\mathbb{N}_0}$ and $\mathtt{false}$ otherwise:

```
isinspan := function( v , lv )
return( v in List (Cartesian(List(lv, x-> [0..Maximum(v)])),x -> x*lv) );
end;
```

Try to write a more efficient program, which also works for larger n.

Exercise 4.5.2 Using the notation of Example 4.5.4 and writing $\mathrm{Irr}(L_2(11)) = \{\psi_1, \ldots, \psi_8\}$, show that the projective characters $\Psi_1, \ldots, \Psi_5$ are the B-parts of

$$\psi_1^G + \psi_6^G, \quad {\chi_2}^2, \quad {\chi_3}^2, \quad \chi_2 \cdot \chi_3, \quad \chi_2 \cdot \chi_4.$$

Hint: Consider

```
gap> List(A, x -> Position( MatScalarProducts(t,Irr(t){irrB},tens), x ));
[ 9, 13, 26, 14, 37 ]
```

and examine how the characters in $\mathtt{tens}$ were formed.

Exercise 4.5.3 Show that the following GAP program computes the p-subsums needed in Example 4.5.4. Here we assume that $\mathtt{ct}$ is the character table of G and $B \in \mathrm{Bl}_p(G)$. Furthermore $\mathtt{irrB}$, $\mathtt{basm}$ and $\mathtt{vec}$ are lists such that $\mathrm{Irr}(B) = \mathrm{Irr}(\mathtt{ct})\{\mathtt{irrB}\}$, $\mathrm{Irr}(\mathtt{ct})\{\mathtt{basm}\}$ is a basic set of Brauer characters of B, and $\mathtt{vec} * \mathrm{Irr}(\mathtt{ct})\{\mathtt{irrB}\}$ is a p-projective character of G:

```
subs := function( ct, basm, vec, irrB, p )
local nullv, x, y, v, cands, psing, preg, rest, relations;
cands := [];
psing := Filtered( [1..Length(Irr(ct))],
                i->IsInt( OrdersClassRepresentatives(ct)[i]/p ) );
preg := Difference( [1..Length(Irr(ct))], psing );
nullv := List( [1..Length(psing)], x -> 0 );
rest := Difference( irrB, basm );
relations := List( rest, i -> SolutionMat( Irr(ct){basm}{preg},
                Irr(ct)[i]{preg} ) );;
```

```
for x in Cartesian ( List( vec{List(basm, i -> Position(irrB,i) )} ,
                            c -> [0..c] ) ) do
    v:= [] ; v{List( basm, i -> Position(irrB,i) )} := x;
    v{List( rest, i -> Position(irrB,i) )} := List([1..Length(rest)],
                                                   i-> x*relations[i]);
    y := v * Irr(ct){irrB};;
    if ForAll( v, c -> c >= 0 ) and  y{psing} = nullv
           and ForAll( [1..Length(v)] , i -> v[i] <= vec[i] ) and Sum(v) > 0
           then Add( cands, v );
    fi;
od;
return(cands);
end;;
```

Exercise 4.5.4 Show that the 2-decomposition matrix for the simple group $L_2(11)$ is given as below (the ordinary irreducible characters are labeled as in GAP).

	φ_1	φ_2	φ_3	φ_4	φ_5	φ_6
χ_1	1	.	.	.	.	.
χ_2	.	1	.	.	.	.
χ_3	.	.	1	.	.	.
χ_4	.	.	.	1	.	.
χ_5	.	.	.	1	.	.
χ_6	1	1	1	.	.	.
χ_7	.	.	.	.	1	.
χ_8	.	.	.	.	.	1

Hint: Induce the 2-projective indecomposable characters of the maximal subgroup 11: 5 to $L_2(11)$.

Exercise 4.5.5 (a) Verify that the sporadic simple Mathieu group $M_{11} = \langle x, y \rangle$ with $x := (2,10)(4,11)(5,7)(8,9)$ and $y := (1,4,3,8)(2,5,6,9)$ has two 2-blocks of defect zero with ordinary characters χ_5, respectively χ_6, and that the only 2-block not of defect zero is the principal block B_0. Show that the restrictions of the ordinary irreducible characters $\varphi_1 := \chi_1|_{2'}$, $\varphi_2 := \chi_2|_{2'}$ and $\varphi_3 := \chi_8|_{2'}$ to the 2-regular classes form a basic set for B_0.

(b) By restricting the basic set to the maximal subgroup $L_2(11)$, show that φ_1 and φ_2 are irreducible and that φ_3 is either irreducible or the sum of φ_2 and an irreducible Brauer character of degree 34.

(c) Let e_H be the fix idempotent of a Sylow 3-subgroup H. Verify that if S is a simple $F_2 M_{11}$-module in the principal block, then $e_H S$ is non-zero for both cases in (b).

(d) Use the results of Example 1.6.14 to prove that φ_3 is indeed the Brauer character of an irreducible $\mathbb{F}_2 M_{11}$-module. Thus the 2-decomposition matrix of

M_{11} is as follows:

$\varphi_i(1):$	1	10	44	16	16
	φ_1	φ_2	φ_3	φ_4	φ_5
χ_1	1	.	.	.	.
χ_2	.	1	.	.	.
χ_3	.	1	.	.	.
χ_4	.	1	.	.	.
χ_5	1	1	.	.	.
χ_8	.	.	1	.	.
χ_9	1	.	1	.	.
χ_6	.	.	.	1	.
χ_7	.	.	.	.	1

Exercise 4.5.6 (a) Show that the sporadic simple Mathieu group M_{22} has exactly seven irreducible Brauer characters in characteristic two all lying in the principal block.

(b) Using **GAP** show that the restrictions to the 2-regular classes of the ordinary irreducible characters $\chi_1, \chi_2, \chi_3, \chi_4, \chi_6, \chi_9, \chi_{10}$ are $\mathbb{Z}$-linearly independent.

(c) Show that the characters do not $\mathbb{Z}$-span $\mathbb{Z}\,\mathrm{IBr}_2(M_{22})$.

Exercise 4.5.7 Let G be a finite group and let p be a prime. Show that a generalized p-projective character Φ can be written as $\mathbb{Z}$-linear combination of projective characters induced from the p-elementary subgroups of G. Conclude that the set of all induced projective indecomposable characters from all maximal subgroups of G spans $\langle \Phi_\varphi | \varphi \in \mathrm{IBr}(B)\rangle_{\mathbb{Z}}$.
Hint: Use Brauer's induction Theorem 3.10.7.

Exercise 4.5.8 Let G be a finite group, p a prime and H be a subgroup of G of order prime to p and $e_H := \frac{1}{H}\sum_{h \in H} h$. Show that for a given FG-module V and $g \in G$ the following equation holds:

$$\mathrm{trace}_{e_H V}(e_H g e_H) = \frac{1}{|H|} \sum_{h \in H} \mathrm{trace}_V(hg).$$

4.6 Defect groups

In this section we will assign to each block a p-subgroup called a defect group of the block. We will see how the order of the defect group is closely related to the defect of the block, and we will study in detail the structure of blocks where the defect group is in the center. We retain the assumption that (K, R, F, η) is a p-modular splitting system for the finite group G.

Definition 4.6.1 For $g \in G$ any element of

$$\mathrm{Def}_p(g^G) := \{P \mid P \in \mathrm{Syl}_p(\mathbf{C}_G(h)),\ h \in g^G\}$$

is called a (p-)**defect group** of $C := g^G \in \mathrm{cl}(G)$.

Remark 4.6.2 If D is a p-defect group of $C \in \mathrm{cl}(G)$ then $\mathrm{Def}_p(C) = \{D^h \mid h \in G\}$. If $|D| = p^d$ then d is called the p-**defect** of C.

We will use the following notation for subgroups Q, P of G:

$$Q \leq_G P \quad \text{means} \quad \text{there is } g \in G \text{ with } g^{-1}Qg \leq P.$$

The following theorem shows that in characteristic p one can construct an ideal in $\mathbf{Z}(FG)$ given a p-subgroup P of G.

Theorem 4.6.3 *If P is a p-subgroup of G then*

$$\mathbf{Z}_P(FG) := \langle\, C^+ \mid C \in \mathrm{cl}(G) \ \text{ with } \ Q \leq_G P \ \text{ for } \ Q \in \mathrm{Def}_p(C) \,\rangle_F$$

is an ideal in $\mathbf{Z}(FG)$.

Proof. Let C_i, C_j be conjugacy classes of G and $D_i \in \mathrm{Def}_p(C_i)$ and $D_j \in \mathrm{Def}_p(C_j)$. By Remark 2.3.1, $C_i^+ C_j^+ = \sum_{x \in G} \alpha_x x$, with

$$\alpha_x = |\Omega| 1_F \quad \text{with} \quad \Omega := \{(g,h) \mid g \in C_i,\ h \in C_j \text{ with } gh = x\}.$$

Let $\alpha_x \neq 0$ and $Q \in \mathrm{Def}_p(x^G)$. The centralizer $\mathbf{C}_G(x)$, and hence also Q, acts on Ω by conjugation. Since $\alpha_x \neq 0$, the p-group Q must fix some $(g,h) \in \Omega$, and hence $Q \leq_G D_i$ and $Q \leq_G D_j$. $\qquad\square$

Definition 4.6.4 Let B be a p-block of G with block idempotent ϵ_B. A p-subgroup D of G is called a **defect group** for B if $\hat{\epsilon}_B \in \mathbf{Z}_D(FG)$ but $\hat{\epsilon}_B \notin \mathbf{Z}_Q(FG)$ for any proper subgroup Q of D. We will denote the set of defect groups of B by $\mathrm{Def}(B)$.

Theorem 4.6.5 *If B is a p-block of G with block idempotent ϵ_B and $D \in \mathrm{Def}(B)$ then*

$$\hat{\epsilon}_B \in \mathbf{Z}_Q(FG) \qquad \text{if and only if} \qquad D \leq_G Q.$$

In particular, any two defect groups of B are conjugate.

Proof. Firstly, if $D \leq_G Q$ then, by definition, $\hat{\epsilon}_B \in \mathbf{Z}_D(FG) \leq \mathbf{Z}_Q(FG)$.

Secondly, $D \in \mathrm{Def}(B)$ implies that $\hat{\epsilon}_B = \sum_{g \in G} \alpha_g g$ with $\alpha_g \in F$ and $0 \neq \alpha_g$ implies that $H \leq_G D$ for $H \in \mathrm{Def}_p(g^G)$. Also $\hat{\epsilon}_B \notin \mathbf{Z}_H(FG)$ for any proper subgroup H of D, and by Rosenberg's lemma (Lemma 1.6.10)

$$\hat{\epsilon}_B \notin \sum_{H < D} \mathbf{Z}_H(FG).$$

Hence there must be some g with $\alpha_g \neq 0$ and $D \in \mathrm{Def}_p(g^G)$. So if $\hat{\epsilon}_B \in \mathbf{Z}_Q(FG)$ then $D \leq_G Q$. $\qquad\square$

Theorem 4.6.6 (Min–max) *Let B be a p-block with defect group D and let $\hat{\epsilon}_B = \sum_{C \in \mathrm{cl}(G)} \alpha_C \, C^+ \in \mathbf{Z}(FG)$. Assume that $C \in \mathrm{cl}(G)$ and $Q \in \mathrm{Def}_p(C)$. Then*

(a) *if $\hat{\omega}_B(C^+) \neq 0$ then $D \leq_G Q$;*

(b) *if $\alpha_C \neq 0$ then $Q \leq_G D$;*

(c) *there is a conjugacy class C of G with $\hat{\omega}_B(C^+) \neq 0$ and $\alpha_C \neq 0$, and for any such class – usually called* **defect class** *of B – one has $D \in \mathrm{Def}_p(C)$.*

Proof. (a) Since $\hat{\omega}_B(C^+) \neq 0$ it follows that $C^+ \in \mathbf{Z}_Q(FG)$ is not contained in the maximal ideal $\ker(\hat{\omega}_B) \trianglelefteq \mathbf{Z}(FG)$. Hence $\mathbf{Z}(FG) = \ker(\hat{\omega}_B) + \mathbf{Z}_Q(FG)$. By Rosenberg's lemma (Lemma 1.6.10) $\hat{\epsilon}_B \in \mathbf{Z}_Q(FG)$, because $\hat{\epsilon}_B \notin \ker(\hat{\omega}_B)$. Now (a) follows from Theorem 4.6.5.

(b) holds because $\hat{\epsilon}_B \in \mathbf{Z}_D(FG)$.

(c) We have

$$1_F = \hat{\omega}_B(\hat{\epsilon}_B) = \hat{\omega}_B\Big(\sum_{C \in \mathrm{cl}(G)} \alpha_C \, C^+ \Big) = \sum_{C \in \mathrm{cl}(G)} \alpha_C \, \hat{\omega}_B(C^+).$$

Hence there is a class C with the properties claimed. The rest follows from (a) and (b) because $D \leq_G Q \leq_G D$ implies that D and Q are conjugate. $\qquad\square$

Example 4.6.7 In Example 4.4.9 (where $G = S_5$) we immediately see from the table of central characters and block idempotents that $B_1 = B_0(G)$ and B_2 each have just one defect class, namely **1a** and **3a**, respectively. From this we see that $P \in \mathrm{Syl}_p(G)$ is a defect group of $B_0(G)$ – obviously, this holds for the principal p-block $B_0(G)$ of any group G – and $\langle (1,2) \rangle \in \mathrm{Def}(B_2)$. On the other hand, we see in the same example that the non-principal 2-block b_2 of A_5 of defect zero has three defect classes, namely **3a, 5a** and **5b**. $\qquad\blacklozenge$

Theorem 4.6.8 *If D is a defect group of the p-block B then $|D| = p^{\mathrm{d}(B)}$.*

Proof. Let $|D| = p^d$ and let $C_d = g_d^G$ be a defect class of B. Let $\chi \in \mathrm{Irr}(B)$ with $h_p(\chi) = 0$. Since $\hat{\omega}_B(C_d^+) = \widehat{\omega_\chi(C_d^+)} \neq 0$, it follows that

$$\omega_\chi(C^+) = \frac{\chi(g_d)|C_d|}{\chi(1)} \in R \setminus \pi R.$$

Assume that $\nu_\pi(p) = e$, so that $\nu_p(a) = e\nu_\pi(a)$ for $a \in \mathbb{Q} \setminus \{0\}$. Then putting $\nu_p(|G|) = a$, so that $\nu_p(|C_d|) = a - d$, we obtain

$$
\begin{aligned}
0 &= \nu_\pi(\omega_\chi(C^+)) = \nu_\pi(\chi(g_d)) + \nu_\pi(|C_d|) - \nu_\pi(\chi(1)) \\
&= \nu_\pi(\chi(g_d)) + e(a - d) - e(a - \mathrm{d}(B)) = \nu_\pi(\chi(g_d)) + e(\mathrm{d}(B) - d) \\
&\geq \mathrm{d}(B) - d,
\end{aligned}
$$

since $\nu_\pi(\chi(g_d)) \geq 0$. Hence $\mathrm{d}(B) \leq d$.

On the other hand, let $\epsilon_B = \sum_{C \in \mathrm{cl}(G)} a_C \, C^+ \in \mathbf{Z}(RG)$ be the block idempotent of B. Then $a_{C_d} \notin \pi R$, that is $\nu_\pi(a_{C_d}) = 0$. We have by Corollary 2.1.7

$$
\begin{aligned}
a_{C_d} &= \frac{1}{|G|} \sum_{\chi \in \mathrm{Irr}(B)} \chi(1)\chi(g_d^{-1}) \\
&= \frac{1}{|G|} \sum_{\chi \in \mathrm{Irr}(B)} \sum_{\varphi \in \mathrm{IBr}(B)} \mathrm{d}_{\chi,\varphi}\, \chi(1)\varphi(g_d^{-1}) \\
&= \frac{1}{|G|} \sum_{\varphi \in \mathrm{IBr}(B)} \Phi_\varphi(1)\varphi(g_d^{-1}).
\end{aligned}
$$

Since by Lemma 4.3.2 $\frac{\Phi_\varphi(1)}{|G|} \in R$, there must be a $\varphi \in \mathrm{IBr}(B)$ with $\varphi(g_d^{-1}) \notin \pi R$. By Lemma 4.4.6, $\varphi \in \langle \chi|_{G'} \mid \chi \in \mathrm{Irr}(B)\rangle_{\mathbf{Z}}$, so there is a $\chi' \in \mathrm{Irr}(B)$ with $\chi'(g_d^{-1}) \notin \pi R$. Since

$$
\frac{\overline{\chi'(g_d)}|C_d|}{\chi'(1)} = \overline{\omega_{\chi'}(C_d^+)} \in R,
$$

we get $\nu_\pi(\frac{|C_d|}{\chi'(1)}) \geq 0$ and thus $\nu_p(\chi'(1)) \leq \nu_p(|C_d|)$. Hence

$$
\nu_p(\chi'(1)) = a - \mathrm{d}(B) + h_p(\chi') \leq \nu_p(|C_d|) = a - d
$$

and $d \leq \mathrm{d}(B) - h_p(\chi') \leq \mathrm{d}(B)$, as claimed. $\qquad \square$

We recall some standard notations from group theory.

Definition 4.6.9 If p is a prime $\mathbf{O}_p(G)$ denotes the largest normal p-subgroup of G. Obviously $\mathbf{O}_p(G)$ is the intersection of all Sylow p-subgroups of G. Similarly, $\mathbf{O}_{p'}(G)$ is the largest normal p'-subgroup of G.

Theorem 4.6.10 *If B is a p-block of G and C is a conjugacy class of G with $C \cap \mathbf{C}_G(\mathbf{O}_p(G)) = \emptyset$ then $\hat\omega_B(C^+) = 0$. In particular,*

$$
\mathbf{O}_p(G) \leq D \qquad \text{for every} \qquad D \in \mathrm{Def}(B).
$$

Proof. Put $P := \mathbf{O}_p(G)$ and let C be a class of G with $C \cap \mathbf{C}_G(P) = \emptyset$. By Corollary 1.4.7 it suffices to show that $C^+ \in \mathrm{J}(FG)$. Let $\delta\colon FG \to \mathrm{End}_F V$ be an irreducible representation. Then $P \subseteq \ker \delta|_G$ (see Corollary 3.6.3), and

$$
C = h_1^P \,\dot\cup\, \cdots \,\dot\cup\, h_m^P \qquad \text{for some} \qquad h_i \in C
$$

and $\delta(h_i^g) = \delta(h_i)\delta([h_i,g]) = \delta(h_i)$, since the commutator $[h_i,g] \in P$. Our assumption implies that each orbit h_i^P is non-trivial and has length n_i divisible by p. Thus $\delta(C^+) = n_1\delta(h_i) + \cdots + n_m\delta(h_m) = 0$. By Theorem 1.4.6 we have $C^+ \in \mathrm{J}(FG)$.

For the last assertion, observe that $D \in \mathrm{Def}_p(C_d)$ for a defect class C_d of B and $\hat\omega_B(C_d^+) \neq 0$. Hence $C_d \leq \mathbf{C}_G(\mathbf{O}_p(G))$, because $\mathbf{C}_G(\mathbf{O}_p(G)) \trianglelefteq G$. $\qquad \square$

We continue by giving a description of the relationship between the blocks of a group G and those of a factor G/N.

Lemma 4.6.11 *Let $N \trianglelefteq G$ with canonical homomorphism $\gamma\colon G \to \bar{G} := G/N$ and $\bar{B} \in \mathrm{Bl}_p(\bar{G})$. Then there is a unique block $B \in \mathrm{Bl}_p(G)$ such that*

$$\{\bar{\chi} \circ \gamma \mid \bar{\chi} \in \mathrm{Irr}(\bar{B})\} \subseteq \mathrm{Irr}(B) \qquad and \qquad \{\bar{\varphi} \circ \gamma \mid \bar{\varphi} \in \mathrm{IBr}(\bar{B})\} \subseteq \mathrm{IBr}(B).$$

We will write $\mathrm{Inf}_\gamma(\bar{B}) := B$. Now assume that N is a p-subgroup and $G = N\,\mathbf{C}_G(N)$. Then $\mathrm{Bl}_p(\bar{G}) \to \mathrm{Bl}_p(G), \bar{B} \mapsto \mathrm{Inf}_\gamma(\bar{B})$ is bijective and $\mathrm{IBr}(\mathrm{Inf}_\gamma(\bar{B})) = \{\bar{\varphi} \circ \gamma \mid \bar{\varphi} \in \mathrm{IBr}(\bar{B})\}$. The Cartan matrices are related by

$$C(\mathrm{Inf}_\gamma(\bar{B})) = |N| \cdot C(\bar{B}) \quad and \quad \mathrm{d}(\mathrm{Inf}_\gamma(\bar{B})) = \mathrm{d}(\bar{B}) + \nu_p(|N|).$$

Proof. If $\bar{\chi} \in \mathrm{Irr}(\bar{G})$, then $\chi := \bar{\chi} \circ \gamma \in \mathrm{Irr}(G)$, and for $C = g^G \in \mathrm{cl}(G)$

$$\omega_\chi(C^+) = [\gamma^{-1}(\mathbf{C}_{\bar{G}}(\gamma(g))) : \mathbf{C}_G(g)]\,\omega_{\bar{\chi}}(\bar{C}^+),$$

where $\bar{C} = \gamma(C)$. By Theorem 4.4.8 (and its proof) it follows that if $\bar{\chi}$ and $\bar{\chi}'$ belong to the same p-block of $\bar{G}$ then χ and χ' belong to the same p-block of G. Thus there is a unique $B \in \mathrm{Bl}_p(G)$ with $\{\bar{\chi} \circ \gamma \mid \bar{\chi} \in \mathrm{Irr}(\bar{B})\} \subseteq \mathrm{Irr}(B)$. If $\bar{\chi} \in \mathrm{Irr}(\bar{B})$ and $g \in G_{p'}$

$$\bar{\chi} \circ \gamma(g) = \bar{\chi}(\gamma(g)) = \sum_{\bar{\varphi} \in \mathrm{IBr}(\bar{B})} d_{\bar{\chi},\bar{\varphi}}\,\bar{\varphi}(\gamma(g)) = \sum_{\bar{\varphi} \in \mathrm{IBr}(\bar{B})} d_{\bar{\chi},\bar{\varphi}}\,\bar{\varphi} \circ \gamma(g).$$

Hence $\{\bar{\varphi} \circ \gamma \mid \bar{\varphi} \in \mathrm{IBr}(\bar{B})\} \subseteq \mathrm{IBr}(B)$.

Now assume that N is a normal p-subgroup of G. Then, by Corollary 3.6.3, every simple FG-module is the inflation of a simple $F\bar{G}$-module and hence $\mathrm{IBr}(G) = \{\bar{\varphi} \circ \gamma \mid \bar{\varphi} \in \mathrm{IBr}(\bar{G})\}$. If in addition $G = N\,\mathbf{C}_G(N)$ and H/N is a p'-subgroup of $\bar{G}$, then, by the Schur–Zassenhaus theorem, $H = N \times C$ for a p'-subgroup $C \leq H$. Therefore γ induces a bijection $G_{p'} \to \bar{G}_{p'}$. Hence, for $\varphi, \varphi' \in \mathrm{IBr}(G)$,

$$(\varphi, \varphi')_{G_{p'}} = \frac{1}{|G|} \sum_{g \in G_{p'}} \varphi(g)\varphi'(g^{-1}) = \frac{1}{|N|} \frac{1}{|\bar{G}|} \sum_{\bar{g} \in \bar{G}_{p'}} \bar{\varphi}(\bar{g})\bar{\varphi}'(\bar{g}^{-1}) = \frac{1}{|N|}(\bar{\varphi}, \bar{\varphi}')_{\bar{G}_{p'}}.$$

Denoting the Cartan matrices of G and $\bar{G}$ by $C(G)$ and $C(\bar{G})$, respectively, we obtain from Theorem 4.3.3, equation (4.18), that $C(G) = |N| \cdot C(\bar{G})$. Thus G and $\bar{G}$ have the same number of p-blocks, $\mathrm{Bl}_p(\bar{G}) \to \mathrm{Bl}_p(G)$, $\bar{B} \mapsto \mathrm{Inf}_\gamma(\bar{B})$ is bijective and $\mathrm{IBr}(\mathrm{Inf}_\gamma(\bar{B})) = \{\bar{\varphi} \circ \gamma \mid \bar{\varphi} \in \mathrm{IBr}(\bar{B})\}$. From Exercise 4.4.1 we conclude that $\mathrm{d}(\mathrm{Inf}_\gamma(\bar{B})) - \mathrm{d}(\bar{B}) = \nu_p(|G|) - \nu_p(|\bar{G}|) = \nu_p(|N|)$. $\qquad\square$

As a consequence we can describe the p-blocks $B \in \mathrm{Bl}_p(G)$ with defect groups D satisfying $G = D\,\mathbf{C}_G(D)$.

Theorem 4.6.12 *Let* $B \in \mathrm{Bl}_p(G)$ *with* $\mathrm{Def}(B) = \{D\}$ *and* $G = D\,\mathbf{C}_G(D)$. *Then there is a unique character* $\theta \in \mathrm{Irr}(B)$ *with* $D \leq \ker\theta$. *This character is usually called the* **canonical character** *of* B. *Furthermore* $\mathrm{IBr}(B) = \{\varphi\}$ *with* $\varphi = \theta|_{G_{p'}}$ *and* $\mathrm{Irr}(B) = \{\theta_\lambda \mid \lambda \in \mathrm{Irr}(D)\}$, *where* θ_λ *is given by*

$$\theta_\lambda(g) = \begin{cases} \lambda(g_p)\theta(g_{p'}) & \text{for } g \in G, \ g_p \in D, \\ 0 & \text{for } g \in G, \ g_p \notin D. \end{cases} \tag{4.24}$$

If in addition D *is cyclic, then the projective indecomposable* FG-*module belonging to* B *is "uniserial," that is, it has a unique composition series.*

Proof. Let $\gamma \colon G \to \bar{G} = G/D$ be the canonical homomorphism. By Lemma 4.6.11, $B = \mathrm{Inf}_\gamma(\bar{B})$ for a unique block $\bar{B} \in \mathrm{Bl}_p(\bar{G})$, and, since $\mathrm{d}(B) = \nu_p(|D|)$, it follows that $\bar{B}$ has defect zero. Theorem 4.4.14 implies that there is a unique character $\bar{\theta}$ with $\mathrm{Irr}(\bar{B}) = \{\bar{\theta}\}$ and $\mathrm{IBr}(\bar{B}) = \{\bar{\theta}|_{\bar{G}_{p'}}\}$. Hence $\theta := \bar{\theta} \circ \gamma$ is the only ordinary irreducible character of B having D in its kernel and $\varphi := \theta|_{G_{p'}}$ is the only irreducible Brauer character belonging to B.

For $\lambda \in \mathrm{Irr}(D)$ define $\theta_\lambda(g)$ by (4.24). Every elementary subgroup of G (see Definition 3.10.3) is contained in a subgroup of the form $P \times H$ with a p-subgroup $P \geq D$ of G and $H \leq G$ a p'-group. Then for $\mu \in \mathrm{Irr}(P)$ and $\psi \in \mathrm{Irr}(H)$

$$(\theta_\lambda|_{P \times H}, \mu \times \psi)_{P \times H} = \frac{1}{|P|} \sum_{x \in D} \lambda(x)\mu(x^{-1}) \cdot (\theta_H, \psi)_H$$

$$= \frac{1}{[P:D]} (\lambda, \mu_D)_D \cdot (\theta_H, \psi)_H \in \mathbb{Z}$$

because $[P:D] \mid (\theta_H, \psi)_H$ by Exercise 4.6.1. Theorem 3.10.9, together with Corollary 3.7.6, shows that θ_λ is a generalized character. From

$$(\theta_\lambda, \theta_\lambda)_G = \frac{1}{|G|} \sum_{\bar{g} \in \bar{G}} \sum_{d \in D} |\lambda(d)|^2 |\bar{\theta}(\bar{g})|^2 = \frac{1}{|\bar{G}|} \sum_{\bar{g} \in \bar{G}} |\bar{\theta}(\bar{g})|^2 = (\theta, \theta)_G = 1,$$

we conclude that $\theta_\lambda \in \mathrm{Irr}(G)$. Since $\theta_\lambda|_{G_{p'}} = \theta_{G_{p'}} = \varphi$, it follows that $\mathrm{Irr}(B) \supseteq \{\theta_\lambda \mid \lambda \in \mathrm{Irr}(D)\}$. We have equality because $C(B) = [|D|]$ by Lemma 4.6.11.

Let V be a projective indecomposable FG-module belonging to B. In order to prove that V is uniserial, it is sufficient to show that the length $r(V)$ of the radical series of V is not less than (and hence equal to) $\mathrm{l}(V) = |D|$, since $C(B) = [|D|]$. On the other hand, V_D is projective, hence free, and by Example 1.1.23 $r(V_D) = |D|$. Thus $r(V) \geq |D| = \mathrm{l}(V)$. $\qquad\square$

Exercises

Exercise 4.6.1 Let $\chi \in \mathrm{Irr}(G)$ with $\mathrm{d}_p(\chi) = d$. Show that $\chi(g)\frac{p^d}{|\mathbf{C}_G(g)|_p} \in R$ for every $g \in G$. Assume that $H \leq G$ with $p \nmid |H|$ and let $p^r = \min_{h \in H} |\mathbf{C}_G(h)|_p$. Conclude that $p^{r-d} \mid (\chi_H, \psi)_H$ for every $\psi \in \mathrm{Irr}(H)$ provided that $r \geq d$.

The following exercise shows that a defect class for a block might depend on the choice of the p-modular splitting system.

Exercise 4.6.2 Let G be the simple group $U_3(4)$ and let (K, R, F, η) be a standard 3-modular splitting system for G; see Definition 4.2.10.

(a) Let $g \in G$ have order 13. Show that there is a $B \in \mathrm{Bl}_3(G)$ such that g^G, $(g^{-1})^G$ are defect classes of B and $(g^5)^G$, $(g^{-5})^G$ are not defect classes of B.

(b) Using a compatible sequence containing $X^3 + X^2 - X + 1 \in \mathbb{F}_3[X]$ instead of the Conway polynomial $f_{3,3}$ show that there is a 3-modular splitting system for G such that $(g^5)^G$, $(g^{-5})^G$ are defect classes of B and not g^G, $(g^{-1})^G$.

The following exercise demonstrates that the number of defect classes with a given defect d may be less than the number of blocks of defect d.

Exercise 4.6.3 Let G be the simple group $U_3(3)$ and let $p = 2$. Show that there are exactly two 2-blocks B_1, B_2 of defect zero. Furthermore, show that there is exactly one G-conjugacy class, which is the unique defect class for B_1, and that this class is also the unique defect class for B_2.

Exercise 4.6.4 Let G be the sporadic simple group J_2 and let $p = 2$. Show that $\mathrm{Bl}_p(G) = \{B_0(G), B_1\}$, and that **3b, 5a, 5b** (in the notation of the ATLAS ([38]) or the GAP library of character tables) are defect classes for B_1. Conclude that a defect group $D \in \mathrm{Def}(B_1)$ is isomorphic to V_4 and that $N := \mathbf{N}_G(D) \cong A_5 \times A_4$ is a maximal subgroup of G (see [38]). Show that $\mathrm{Bl}_p(N) = \{B_0(N), b_1\}$ and $\mathrm{k}(b_1) = \mathrm{k}_0(b_1) = \mathrm{k}(B_1) = \mathrm{k}_0(B_1) = 4$, where $\mathrm{k}_i(B)$ denotes the number of ordinary irreducible characters of the block B having height i. Compute the decomposition matrix $D(b_1)$ and show that $D(b_1) \neq D(B_1)$ (for any ordering of $\mathrm{Irr}(B_1)$ and $\mathrm{IBr}(B_1)$) although $\mathrm{l}(b_1) = \mathrm{l}(B_1) = 3$. Furthermore, let $P \in \mathrm{Def}_p(B_0(G))$ and $N_0 := \mathbf{N}_G(P)$. Show that N_0 has only the principal 2-block b_0 and $(\mathrm{k}_i(b_0) \mid i \in \mathbb{N}_0) = (8, 2, 6, 3, 0, 0, \dots)$, compared with $(\mathrm{k}_i(B_0(G)) \mid i \in \mathbb{N}_0) = (8, 6, 2, 0, 4, 0, 0, \dots)$. Also verify that $\mathrm{l}(b_0) = 3$ and $\mathrm{l}(B_0(G)) = 7$.
Hint: N_0 may be computed using a permutation representation of G obtained in GAP by `PrimitiveGroup(100,1)`.

4.7   Brauer correspondence

In this section we assume that (K, R, F, η) is a p-modular splitting system for G and all its subgroups. Furthermore $S \in \{R, F\}$. We recall our convention that all SG-modules considered are free and finitely generated as S-modules. As before πR is the maximal ideal of R. We will now introduce a concept that allows us to relate under favorable circumstances blocks of a subgroup H to blocks of G. One of the highlights we look at is Brauer's first main theorem, which basically says that for a given p-subgroup P the blocks of $\mathbf{N}_G(P)$ and G with defect group P are in one-to-one correspondence.

Definition 4.7.1 For a ring S and a subgroup $H \leq G$ we define an S-linear map

$$s_H \colon \mathbf{Z}(SG) \to \mathbf{Z}(SH) \quad \text{by} \quad C^+ \mapsto (C \cap H)^+ \quad \text{for} \quad C \in \mathrm{cl}(G).$$

Observe that $C \cap H$ is a union of conjugacy classes of H, so that $(C \cap H)^+$ is indeed in $\mathbf{Z}(SH)$ for any conjugacy class $C \in \mathrm{cl}(G)$. As always, we follow the custom that an empty sum is zero.

Of course, s_H depends on the ring S, but that will be clear from the context. Note that $\eta \colon R \to F$ may be extended to ring homomorphisms $\mathbf{Z}(RG) \to \mathbf{Z}(FG)$ and $\mathbf{Z}(RH) \to \mathbf{Z}(FH)$ (see Exercise 4.4.4), which will also be denoted by η. It follows for instance that

$$s_H(\eta(z)) = \eta(s_H(z)) \qquad \text{for} \qquad z \in \mathbf{Z}(RG). \tag{4.25}$$

If b is a p-block of H then $\hat{\omega}_b \circ s_H \colon \mathbf{Z}(FG) \to F$ is an F-linear map. If this happens to be an F-algebra homomorphism then there is a p-block B of G with

$$\hat{\omega}_B = \hat{\omega}_b \circ s_H.$$

In this case we will say that "b^G **is defined**" and put $b^G := B$. The block b^G is then also called the block **induced** by b and we will write $\hat{\omega}_b^G := \hat{\omega}_{b^G}$.

We remark that in the literature one can find several different ways to define block induction, any two of which are equivalent on their common domain of definition. For a discussion, see [20].

Before turning to the question under what conditions b^G is defined, we show that the defect groups of b and b^G correspond nicely when b^G is defined.

Lemma 4.7.2 *Let b be a p-block of $H \leq G$. If b^G is defined then every defect group D_b of b is contained in some defect group of b^G.*

Proof. Let C be a defect class of b^G. Then

$$\hat{\omega}_b(s_H(C^+)) = \hat{\omega}_{b^G}(C^+) \neq 0.$$

In particular, $C \cap H \neq \emptyset$ and there is an $h \in C \cap H$ with $\hat{\omega}_b((h^H)^+) \neq 0$. The min–max theorem (Theorem 4.6.6) implies that $D_b \leq_H Q$ for $Q \in \mathrm{Syl}_p(\mathbf{C}_H(h))$. Hence Q is contained in $\mathbf{C}_G(h)$, and therefore in some $D \in \mathrm{Syl}_p(\mathbf{C}_G(h))$. Since $h \in C$ and C is a defect class of b^G, we have $D \in \mathrm{Def}(b^G)$. $\qquad\square$

We now come to the somewhat technical problem of how to decide when b^G is defined. In order to compute $\hat{\omega}_b \circ s_H$ the following lemma is quite useful. Here it is convenient to extend the definition of ω_χ to the case where χ is not necessarily irreducible.

Definition 4.7.3 If $\psi \in \mathrm{cf}(G, K)$ and $\psi(1) \neq 0$ we define a K-linear map

$$\omega_\psi \colon \mathbf{Z}(KG) \to K \,, \quad C^+ \mapsto \frac{|C|\psi(g)}{\psi(1)} \qquad \text{for} \qquad C = g^G \in \mathrm{cl}(G).$$

Lemma 4.7.4 *Let* $\psi = \sum_{\chi \in \mathrm{Irr}(G)} a_\chi \chi \in \mathrm{cf}(G, K)$, $\psi(1) \neq 0$ *and* $B \in \mathrm{Bl}_p(G)$.
Then

$$\psi(1)\, \omega_\psi = \sum_{\chi \in \mathrm{Irr}(G)} a_\chi\, \chi(1)\, \omega_\chi, \tag{4.26}$$

$$\omega_\psi(\epsilon_B C^+) = \frac{|C|\, \psi_B(g)}{\psi(1)} \qquad for \qquad g^G = C \in \mathrm{cl}(G). \tag{4.27}$$

Proof. Equation (4.26) follows immediately from the definition. For (4.27)
recall that ψ_B is the B-part of ψ (see Definition 4.4.4). Then for $C = g^G$

$$\psi(1)\, \omega_\psi(\epsilon_B C^+) = \sum_{\chi \in \mathrm{Irr}(G)} a_\chi \chi(1)\, \omega_\chi(\epsilon_B C^+) = \sum_{\chi \in \mathrm{Irr}(B)} a_\chi \chi(1)\, \omega_\chi(C^+)$$

$$= \sum_{\chi \in \mathrm{Irr}(B)} a_\chi |C| \chi(g) = |C| \psi_B(g).$$

$\square$

Lemma 4.7.5 *Let* $H \leq G$ *and* $\theta \in \mathrm{cf}(H, K)$ *with* $\theta(1) \neq 0$. *Then*

$$\omega_\theta \circ s_H = \omega_{\theta^G},$$

where $\theta^G \in \mathrm{cf}(G, K)$ *is the induced class function (see Definition 3.2.7). In
particular, if* $\theta \in \mathrm{Irr}(b)$ *for* $b \in \mathrm{Bl}_p(H)$ *and* $\theta^G \in \mathrm{Irr}(G)$, *then* b^G *is defined and*
$\theta^G \in \mathrm{Irr}(b^G)$.

Proof. Let $g^G \cap H = h_1^H \dot\cup \cdots \dot\cup h_r^H$ with $h_i \in H$. By Lemma 3.2.7 we have

$$\theta^G(g) = \sum_{i=1}^r \frac{|\mathbf{C}_G(g)|}{|\mathbf{C}_H(h_i)|} \theta(h_i).$$

On the other hand, let $C := g^G$, then

$$\omega_\theta \circ s_H(C^+) = \omega_\theta((C \cap H)^+) = \sum_{i=1}^r \frac{|H|}{|\mathbf{C}_H(h_i)|} \frac{\theta(h_i)}{\theta(1)}$$

$$= \frac{|H|}{|\mathbf{C}_G(g)|} \frac{1}{\theta(1)} \sum_{i=1}^r \frac{|\mathbf{C}_G(g)|}{|\mathbf{C}_H(h_i)|} \theta(h_i)$$

$$= \frac{|g^G| \theta^G(g)}{\theta^G(1)} = \omega_{\theta^G}(C^+).$$

If $\theta \in \mathrm{Irr}(b)$ then by (4.21), p. 324,

$$\hat\omega_b \circ s_H(C^+) = \eta(\omega_\theta(s_H(C^+))) = \eta(\omega_{\theta^G}(C^+)) = \hat\omega_B(C^+),$$

where the final equation holds only if θ^G is irreducible and belongs to the block
B.

$\square$

Clifford correspondence provides an example where Lemma 4.7.5 can be used, as seen in the following example.

Example 4.7.6 Assume that $N \trianglelefteq G$ and $\varphi \in \mathrm{Irr}(N)$ with $T := T_G(\varphi)$. By Corollary 3.6.8 we have a bijection $\mathrm{Irr}(\,T \mid \varphi\,) \to \mathrm{Irr}(\,G \mid \varphi\,)$, $\psi \mapsto \psi^G$. Let $\psi \in \mathrm{Irr}(\,T \mid \varphi\,)$ belong to $b \in \mathrm{Bl}_p(T)$. By Lemma 4.7.5 $B := b^G$ is defined and $\psi^G \in \mathrm{Irr}(B)$. ◆

If $H \leq G$ and the character tables of H and G are known together with the fusion map of the conjugacy classes of H to the conjugacy classes of G, then Lemma 4.7.5 can be used to compute $\hat{\omega}_b \circ s_H$ for all $b \in \mathrm{Bl}_p(H)$ and one can decide whether or not b^G is defined.

Theorem 4.7.7 *Let $H \leq G$ and $b \in \mathrm{Bl}_p(H)$. Then b^G is defined and $b^G = B \in \mathrm{Bl}_p(G)$ if and only if for some $\chi \in \mathrm{Irr}(B)$ and $\theta \in \mathrm{Irr}(b)$ and every $g \in G$ we have*

$$(\,\omega_{\theta^G}((g^G)^+) - \omega_\chi((g^G)^+)\,)^{\varphi(m)} \in pR, \qquad (4.28)$$

where m is the order of g and φ is the Eulerian function. In particular, this is independent of the choice of the p-modular splitting system (K, R, F, η).

Proof. By definition b^G is defined and $b^G = B \in \mathrm{Bl}_p(G)$ if and only if

$$a_g := \omega_{\theta^G}((g^G)^+) - \omega_\chi((g^G)^+) \in \ker\eta \qquad (4.29)$$

for all $g \in G$. Obviously (4.28) implies (4.29). Conversely, assume that (4.29) holds for all $g \in G$. From Corollary 2.3.3 and Exercise 4.7.2 it follows that $a_g \in R'_m$, the ring of algebraic integers in $\mathbb{Q}_m$, where m is the order of g. Hence $a_g \in \pi(R \cap \mathbb{Q}_m)$ for some prime element π. If $\gamma \in \mathrm{Gal}(\mathbb{Q}_m/\mathbb{Q})$ and $\gamma(\zeta_m) = \zeta_m^j$ then $\gamma(a_g) = a_{g^j} \in \pi(R \cap \mathbb{Q}_m)$. Let

$$f = \prod_{\gamma \in \mathrm{Gal}(\mathbb{Q}_m/\mathbb{Q})} (X - \gamma(a_g)) \in R'_m[X].$$

Since f is invariant under $\mathrm{Gal}(\mathbb{Q}_m/\mathbb{Q})$, it is in $\mathbb{Z}[X]$. Also all coefficients except for the leading coefficient are in $\pi R \cap \mathbb{Z} = p\mathbb{Z}$. Hence

$$0 = f(a_g) \equiv a_g^{\varphi(m)} \mod p.$$

$\square$

Example 4.7.8 We want to investigate for which 2-blocks b of subgroups of $G = \mathrm{M}_{11}$ or $G = \mathrm{M}_{12}$ the induced block b^G is defined. For this we first write a short GAP program for testing (4.28):

```
gap> testmodpi := function( t, y, p )
> # t character table, y classfunction, p a prime; the function tests
```

```
> # whether  for all i   IsIntegralCyclotomic( y[i]^m / p ) where
> # m = Phi( OrdersClassRepresentatives(t)[i] )
> local i, res ;
> res:= true;
> for i in [1..Length(y)] do
>    if IsInt( y[i] ) and not IsInt( y[i]/p ) then
>          res := false;
>    elif not IsInt( y[i] ) and  not
>      IsIntegralCyclotomic( y[i]^Phi( OrdersClassRepresentatives(t)[i] )/ p )
>       then res := false;
>    fi;
> od;
> return( res );
> end;
```

Now we take $G := \mathrm{M}_{11}$ and choose $p := 2$. The table of marks t of G is available in the GAP library, and for each conjugacy class of subgroups of G the generators of a representative group H are given, so we can compute (with the Dixon–Schneider algorithm) the character table cth of H. Also the fusion of the conjugacy classes of cth to those of the character table ct of the underlying group G of t is stored, so we can induce characters of cth to ct. Observe that the character table cth differs from the library table CharacterTable("M11") in the ordering of the conjugacy classes. For each p-block $B \in \mathrm{Bl}_p(G)$ we choose a character $\chi \in \mathrm{Irr}(B)$ and compute the central character ω_χ:

```
gap>   t := TableOfMarks( "M11" );;
gap>   ct := CharacterTable( UnderlyingGroup(t) );; p := 2;;
gap>   omegaBs := List( Irr(ct){[1,6,7]} , CentralCharacter );;
```

We could replace the last line of code by

```
gap>   pb := PrimeBlocks( ct, p );;
gap>   omegaBs := List( Irr(ct){List( [1..Length(pb.defect)],
>                        j -> Position(pb.block,j) )} ,CentralCharacter);;
```

Next we do the same for the character tables of the representatives of the conjugacy classes of proper subgroups of G and use Theorem 4.7.7, that is the program testmodpi, to find out whether or not the induced block is defined. We collect the results in the list inducedblocks:

```
gap>   inducedblocks := [];;
gap>   for i in [1..Length(OrdersTom(t))-1] do
>          h := RepresentativeTom( t, i );;
>          cth := CharacterTable(h); blh := PrimeBlocks( cth, p );
>          for k in [1..Length(blh.defect)] do
>            y := Irr(cth)[ Position( blh.block, k ) ];
>            y := InducedClassFunction( y, ct );
>            for z in omegaBs do
>                if testmodpi( ct, CentralCharacter(y) - z , p ) then
>                    Add( inducedblocks , [ i, k, Position( omegaBs , z) ] );
>                fi;
>            od;
```

```
>        od;
>        od;
gap>
```

We observe that for a proper subgroup $H \leq G$ the 2-block b^G is defined for $b \in \mathrm{Bl}_2(H)$ if and only if H is of even order and $b = B_0(H)$, in which case $b^G = B_0(G)$ in accordance with Theorem 4.7.15 below. On the other hand, using the above program for the Mathieu group $G := \mathrm{M}_{12}$ one finds that $B_0(H)^G$ is not defined for all subgroups of even order. For example, G contains two conjugacy classes $\mathcal{C}_1 = \langle (2,11)(3,5,12)(4,10,9,8,7,6) \rangle^G$ and $\mathcal{C}_2 = \langle (1,12,3,2,9,4)(5,8,10,6,7,11) \rangle^G$ of cyclic subgroups C_6 of order six. It is easily seen that $\mathrm{Bl}_2(\mathrm{C}_6) = \{ b_0 := B_0(\mathrm{C}_6), b_1, b_2 \}$. For $H \in \mathcal{C}_1$ one finds that only b_0^G is defined, while for $H \in \mathcal{C}_2$ only b_1^G and b_2^G are defined and $b_1^G = b_2^G = B_1$, where B_1 is the 2-block of defect two of G. ♦

The following lemma will be used in Theorem 4.7.10.

Lemma 4.7.9 *Let $H \leq G$, $b \in \mathrm{Bl}_p(H)$ and assume that $b^G = B$ is defined. Then $\epsilon_b s_H(\epsilon_B)$ is a unit in $\epsilon_b \mathbf{Z}(RH)$ and there is $z \in \epsilon_b \mathbf{Z}(RH)$ with*

$$z \epsilon_b \epsilon_B = \epsilon_b + c \qquad with \qquad c \in \langle (g^H)^+ \mid g \in G \setminus H \rangle_R. \tag{4.30}$$

In particular, $\epsilon_b \epsilon_B \neq 0$.

Proof. Put $u := \epsilon_B - s_H(\epsilon_B) \in \langle C^+ - (C \cap H)^+ \mid C \in \mathrm{cl}(G) \rangle_R$. Then

$$\epsilon_b \epsilon_B = \epsilon_b s_H(\epsilon_B) + \epsilon_b u, \qquad \epsilon_b u \in \langle (g^H)^+ \mid g \in G \setminus H \rangle_R.$$

By assumption, $\hat{\omega}_B = \hat{\omega}_b \circ s_H$, so

$$1 = \hat{\omega}_B(\hat{\epsilon}_B) = \hat{\omega}_b(\eta(s_H(\epsilon_B))) = \hat{\omega}_b(\eta(\epsilon_b s_H(\epsilon_B))).$$

Since $\epsilon_b \mathbf{Z}(RH)$ is a local ring, $\epsilon_b s_H(\epsilon_B)$ is a unit and there is $z \in \epsilon_b \mathbf{Z}(RH)$ with $z \epsilon_b s_H(\epsilon_B) = \epsilon_b$, and (4.30) follows with $c := z \epsilon_b u$. □

Theorem 4.7.10 (Conlon) *Let $H \leq G$, $b \in \mathrm{Bl}_p(H)$ and assume that $b^G = B$ is defined. If W is an indecomposable SH-module belonging to b, then*

$$W \mid (\dot{\epsilon}_B W^G)_H,$$

where $\dot{\epsilon}_B = \epsilon_B$ if $S = R$ and $\dot{\epsilon}_B = \hat{\epsilon}_B$ if $S = F$. In particular, there is an indecomposable SG-module V belonging to B such that $V \mid W^G$ and $W \mid V_H$.

Proof. Let $1 = g_1, \ldots, g_n$ be a transversal of H in G. We recall from the proof of Theorem 3.2.12 the SH-linear map $\tau \colon W^G \to W$, $\sum_i g_i \otimes w_i \mapsto w_1$. Using the notation of Lemma 4.7.9 we obtain an SH-linear map

$$\sigma \colon W \to \dot{\epsilon}_B W^G, \qquad w \mapsto z \, \dot{\epsilon}_b \dot{\epsilon}_B \otimes w.$$

Then for $w \in W$ we get, using (4.30),

$$\tau \circ \sigma(w) = \tau(z \, \dot{\epsilon}_b \, \dot{\epsilon}_B \otimes w) = \tau(\dot{\epsilon}_b \otimes w + c \otimes w) = \dot{\epsilon}_b w = w.$$

Thus $\tau \circ \sigma = \mathrm{id}_W$ and $W \mid (\epsilon_B W^G)_H$. $\qquad\qquad\square$

The following lemma is used in Corollary 4.7.12.

Lemma 4.7.11 *Let $H \le G$ and $\theta \in \mathrm{Irr}(b)$ for $b \in \mathrm{Bl}_p(H)$. Let $B \in \mathrm{Bl}_p(G)$. Then*

$$\frac{|g^G| \, (\theta^G)_B(g)}{\theta^G(1)} \in R \qquad for \qquad g \in G.$$

If b^G is defined and $\chi \in \mathrm{Irr}(B)$ then, for $g^G = C \in \mathrm{cl}(G)$,

$$\frac{|C| \, (\theta^G)_B(g)}{\theta^G(1)} \equiv \begin{cases} \omega_\chi(C^+) & \mod \pi & if \ b^G = B, \\ 0 & \mod \pi & if \ b^G \ne B. \end{cases}$$

Proof. By Lemma 4.7.5 (and Corollary 2.3.3) $\omega_{\theta^G}(C^+) = \omega_\theta((H \cap C)^+) \in R$. By (4.27) applied to $\psi := \theta^G$ the first assertion follows.

If b^G is defined and $\theta \in \mathrm{Irr}(b)$ we obtain, from (4.27) applied to $\psi := \theta^G$ and Lemma 4.7.5 as well as (4.25) and Exercise 4.4.4,

$$\eta\left(\frac{|C| \, (\theta^G)_B(g)}{\theta^G(1)}\right) = \eta(\omega_{\theta^G}(\epsilon_B C^+)) = \eta(\omega_\theta(s_H(\epsilon_B C^+))) = \hat{\omega}_b(\eta(s_H(\epsilon_B C^+)))$$

$$= \hat{\omega}_b \circ s_H(\eta(\epsilon_B C^+)) = \hat{\omega}_{b^G}(\hat{\epsilon}_B C^+) = \begin{cases} \hat{\omega}_B(C^+) & if \ b^G = B, \\ 0 & if \ b^G \ne B. \end{cases}$$

$\qquad\qquad\square$

Corollary 4.7.12 *Let $H \le G$ and $\theta \in \mathrm{Irr}(b)$ for $b \in \mathrm{Bl}_p(H)$. Assume that b^G is defined and that $B \in \mathrm{Bl}_p(G)$. Then*

$$\nu_p((\theta^G)_B(1)) \quad \begin{matrix} > \nu_p(\theta^G(1)) & if \ B \ne b^G, \\ = \nu_p(\theta^G(1)) & if \ B = b^G. \end{matrix}$$

In particular, there is some $\chi \in \mathrm{Irr}(b^G)$ with $(\chi, \theta^G)_G > 0$.

Proof. We apply Lemma 4.7.11 with $g = 1$. $\qquad\qquad\square$

Lemma 4.7.13 *Let $H \le G$ and $\theta \in \mathrm{Irr}(b)$ for $b \in \mathrm{Bl}_p(H)$. Assume that b^G is defined and let $\chi \in \mathrm{Irr}(G)$. Then*

$$\frac{\widetilde{(\chi|_H, \theta)_H}}{\theta(1)} \begin{cases} \equiv 0 & \mod \pi & if \ \chi \notin \mathrm{Irr}(b^G), \\ \not\equiv 0 & \mod \pi & if \ \chi \in \mathrm{Irr}(b^G) \ and \ \mathrm{ht}(\chi) = 0. \end{cases}$$

Proof. From Definition 4.2.18 it follows that

$$\widetilde{\chi|_H} = \frac{1}{[G:H]_p}\tilde{\chi}|_H.$$

Let $\chi \in \mathrm{Irr}(B)$ with $B \in \mathrm{Bl}_p(G)$. Since $\tilde{\chi} \in \langle\mathrm{Irr}(B)\rangle_{\mathbb{Z}}$ by Lemma 4.4.15(a), we have $(\tilde{\chi},\eta)_G = (\tilde{\chi},\eta_B)_G$ for all $\eta \in \mathrm{cf}(G,K)$. Then

$$\begin{aligned}
\frac{(\widetilde{\chi|_H},\theta)_H}{\theta(1)} &= \frac{1}{[G:H]_p}\frac{(\tilde{\chi}|_H,\theta)_H}{\theta(1)} = \frac{1}{[G:H]_p}\frac{(\tilde{\chi},\theta^G)_G}{\theta(1)}\\
&= \frac{[G:H]}{[G:H]_p}\frac{(\tilde{\chi},(\theta^G)_B)_G}{\theta^G(1)}\\
&= [G:H]_{p'}\frac{|G|_p}{|G|}\sum_{g^G\in\mathrm{cl}(G_{p'})}\frac{|g^G|\,(\theta^G)_B(g)}{\theta^G(1)}\chi(g^{-1})\\
&= \frac{1}{|H|_{p'}}\sum_{g^G\in\mathrm{cl}(G_{p'})}\frac{|g^G|\,(\theta^G)_B(g)}{\theta^G(1)}\chi(g^{-1}).
\end{aligned}$$

By Lemma 4.7.11 this is $0 \bmod \pi$ if $\chi \notin \mathrm{Irr}(b^G)$. If, on the other hand, $\chi \in \mathrm{Irr}(b^G)$ then we obtain from Lemma 4.7.11

$$\frac{(\widetilde{\chi|_H},\theta)_H}{\theta(1)} \equiv \frac{1}{|H|_{p'}}\sum_{g^G\in\mathrm{cl}(G_{p'})}\frac{|g^G|\,\chi(g)}{\chi(1)}\chi(g^{-1}) \equiv [G:H]_{p'}\frac{(\tilde{\chi},\chi)_G}{\chi(1)} \quad \bmod \pi.$$

Now, by Theorem 4.4.16 $\frac{(\tilde{\chi},\chi)_G}{\chi(1)} \not\equiv 0 \bmod \pi$ if and only if $\mathrm{ht}(\chi) = 0$. $\qquad\square$

The following observation by Okuyama is essential for the proof of Brauer's third main theorem.

Theorem 4.7.14 (Okuyama) *Let $H \leq G$ and let $B \in \mathrm{Bl}_p(G), b \in \mathrm{Bl}_p(H)$. Assume that $\chi \in \mathrm{Irr}(B)$ is such that $\chi|_H \in \mathrm{Irr}(b)$. If χ and $\chi|_H$ both have height zero, and if $b' \in \mathrm{Bl}_p(H)$ is such that b'^G is defined, we have*

$$b = b' \quad \text{if and only if} \quad b^G = B.$$

Proof. First assume that $b = b'$ and put $\psi := \chi|_H$. By assumption b^G is defined. Since $\mathrm{ht}(\psi) = 0$ we know from Theorem 4.4.16 that

$$\frac{(\widetilde{\chi|_H},\psi)_H}{\psi(1)} = \frac{(\tilde{\psi},\psi)_H}{\psi(1)} \not\equiv 0 \quad \bmod \pi.$$

By Lemma 4.7.13 $\chi \in \mathrm{Irr}(b^G)$, so $B = b^G$.

Now assume that $b^G = B$ and let $\theta \in \mathrm{Irr}(b')$. Using Lemma 4.7.13 we conclude that

$$(\widetilde{\chi|_H},\theta)_H = (\tilde{\psi},\theta)_H \neq 0.$$

By Lemma 4.4.15 we have $\theta \in \mathrm{Irr}(b)$ and hence $b = b'$. $\qquad\square$

Theorem 4.7.15 (Brauer's third main theorem) *Let $H \leq G$ and let $b_0 := B_0(H), B_0 := B_0(G)$ be the principal blocks of H, respectively G. If $b \in \mathrm{Bl}_p(H)$ is such that b^G is defined then*

$$b^G = B_0 \qquad \textit{if and only if} \qquad b = b_0.$$

Proof. This follows from Theorem 4.7.14, taking $\chi := \mathbf{1_G}$. $\qquad\qquad \square$

Up to now we have not seen any criterion which guarantees that b^G is defined for a given $b \in \mathrm{Bl}_p(H)$ when $H \leq G$. In fact, the question seems to be a bit mysterious, as we have seen in Example 4.7.8. Of course, if s_H is a homomorphism of algebras, then b^G is defined for any p-block b. There does exist an important case, in which s_H is, indeed, a homomorphism of F-algebras.

Theorem 4.7.16 *If P is a p-subgroup of G then the F-linear map*

$$\mathrm{Br}_P := s_{\mathbf{C}_G(P)} \colon \mathbf{Z}(FG) \to \mathbf{Z}(F\,\mathbf{C}_G(P)), \quad C^+ \mapsto (C \cap \mathbf{C}_G(P))^+$$

is a homomorphism of F-algebras, called the **Brauer homomorphism** *with respect to P.*

Proof. Let $C_1, C_2 \in \mathrm{cl}(G)$. We have

$$C_1^+ C_2^+ = \sum_{g \in G} \alpha_g g \quad \text{with} \quad \alpha_g = |\Omega|\, 1_F \quad \text{with} \quad \Omega := \{(x,y) \in C_1 \times C_2 \mid xy = g\}$$

and $\mathrm{Br}_P(C_1^+ C_2^+) = \sum_{g \in \mathbf{C}_G(P)} \alpha_g g$. On the other hand,

$$\mathrm{Br}_P(C_1^+)\, \mathrm{Br}_P(C_2^+) = \sum_{g \in \mathbf{C}_G(P)} \alpha_g' g \quad \text{with} \quad \alpha_g' = |\Omega'| 1_F \quad \text{with}$$

$$\Omega' := \{\, (x,y) \in (C_1 \cap \mathbf{C}_G(P)) \times (C_2 \cap \mathbf{C}_G(P)) \mid xy = g \,\}.$$

For $g \in \mathbf{C}_G(P)$ the p-group P acts by conjugation on Ω, and Ω' is the set of fixed points of Ω. Since the lengths of the orbits on $\Omega \setminus \Omega'$ are divisible by p, we have $|\Omega| \equiv |\Omega'| \mod p$ and hence $\alpha_g = \alpha_g'$ for $g \in P$. $\qquad \square$

Remark 4.7.17 Observe that any subgroup H with $\mathbf{C}_G(P) \leq H \leq \mathbf{N}_G(P)$ acts on $C \cap \mathbf{C}_G(P)$ by conjugation for any class $C \in \mathrm{cl}(G)$ and hence $(C \cap \mathbf{C}_G(P))^+ \in \mathbf{Z}(FH)$. This means that $\mathrm{im}(\mathrm{Br}_P) \subseteq \mathbf{Z}(FH)$, and Br_P may also be considered as a homomorphism from $\mathbf{Z}(FG)$ to $\mathbf{Z}(FH)$.

Corollary 4.7.18 *If P is a p-subgroup of G then*

$$\ker \mathrm{Br}_P = \langle C^+ \mid C \in \mathrm{cl}(G) \ \textit{with} \ P \nleq_G Q \ \textit{for} \ Q \in \mathrm{Def}_p(C) \,\rangle_F.$$

If B is a p-block of G and $D \in \mathrm{Def}(B)$ then

$$\mathrm{Br}_P(\hat{e}_B) \neq 0 \quad \textit{if and only if} \quad P \leq_G D.$$

Proof. (i) Let $Q \in \text{Def}_p(C)$ for some $C \in \text{cl}(G)$. Then $C \cap \mathbf{C}_G(P) \neq \emptyset$ if and only if $P \leq_G Q$. Thus $\text{Br}_P(C^+) = 0$ if and only if $P \not\leq_G Q$.

(ii) Let $\hat{\epsilon}_B = \sum_{C \in \text{cl}(G)} \alpha_C C^+$. Then by Theorem 4.6.6 (b) $\alpha_C = 0$ unless, for $Q \in \text{Def}_p(C)$, one has $Q \leq_G D$. So $\text{Br}_P(\hat{\epsilon}_B) \neq 0$ if and only if there is a class C with defect group Q with $P \leq_G Q$ (by (i)) and $Q \leq_G D$. $\qquad \square$

Theorem 4.7.19 *Let P be a p-subgroup of G and let $H \leq G$ satisfy*

$$P\,\mathbf{C}_G(P) \leq H \leq \mathbf{N}_G(P).$$

If b is a p-block of H then $B := b^G$ is defined and $\hat{\omega}_b^G = \hat{\omega}_b \circ \text{Br}_P$. Furthermore, if $B \in \text{Bl}_p(G)$, then $B = b^G$ for some block b of H if and only if P is contained in some defect group of B. In this case

$$\text{Br}_P(\hat{\epsilon}_B) = \sum_{b \in \text{Bl}_p(H),\, b^G = B} \hat{\epsilon}_b.$$

Proof. We will first show that $\hat{\omega}_b \circ s_H = \hat{\omega}_b \circ \text{Br}_P$, considering Br_P as a homomorphism from $\mathbf{Z}(FG)$ to $\mathbf{Z}(FH)$, as we may by assumption and Remark 4.7.17. For this we have to show that for any $C \in \text{cl}(G)$ we have

$$\alpha := \hat{\omega}_b((C \cap H)^+) - \hat{\omega}_b((C \cap \mathbf{C}_G(P))^+) = 0.$$

Since $\mathbf{C}_G(P) \trianglelefteq H$, we know that $U := (C \cap H) \setminus (C \cap \mathbf{C}_G(P))$ is a union of conjugacy classes of H and $\alpha = \hat{\omega}_b(U^+)$. But $P \trianglelefteq H$, so $P \leq \mathbf{O}_p(H)$ and $\mathbf{C}_H(\mathbf{O}_p(H)) \leq \mathbf{C}_G(P)$. Hence $U \cap \mathbf{C}_H(\mathbf{O}_p(H)) = \emptyset$ and by Theorem 4.6.10 $\hat{\omega}_b(U^+) = 0$. Thus we have shown that $\hat{\omega}_b \circ s_H = \hat{\omega}_b \circ \text{Br}_P$, and this is an algebra homomorphism by Theorem 4.7.16, so b^G is defined.

If $B \in \text{Bl}_p(G)$ and $B = b^G$ for some $b \in \text{Bl}_p(H)$ then by Lemma 4.7.2 any defect group D_b of b is contained in some $D \in \text{Def}(B)$. But by Theorem 4.6.10 $P \leq \mathbf{O}_p(H) \leq D_b$.

Conversely, if $B \in \text{Bl}_p(G)$ and $P \leq D$ for some $D \in \text{Def}(B)$, then by Corollary 4.7.18 $\text{Br}_P(\hat{\epsilon}_B) \neq 0$ and hence a non-trivial idempotent in $\mathbf{Z}(FH)$. Thus

$$\text{Br}_P(\hat{\epsilon}_B) = \hat{\epsilon}_{b_1} + \cdots + \hat{\epsilon}_{b_m}$$

for some block idempotents $\hat{\epsilon}_{b_1}, \ldots, \hat{\epsilon}_{b_m}$ of FH. Now, $B = b^G$ for $b \in \text{Bl}_p(H)$ if and only if $1 = \hat{\omega}_b^G(\hat{\epsilon}_B) = \hat{\omega}_b(\text{Br}_P(\hat{\epsilon}_B))$ and this holds if and only if $b \in \{b_1, \ldots, b_m\}$. $\qquad \square$

Lemma 4.7.20 *Let P be a normal p-subgroup of G and $B \in \text{Bl}_p(G)$. Then*

(a) $\hat{\epsilon}_B \in \langle\, C^+ \mid C \in \text{cl}(G),\ C \subseteq \mathbf{C}_G(P)\, \rangle_F$;

(b) *if $P \in \text{Def}(B)$ then $\hat{\epsilon}_B \in \langle\, C^+ \mid C \in \text{cl}(G),\ \text{Def}_p(C) = \{P\}\, \rangle_F$.*

Proof.   (a) We use Theorem 4.7.19 with $H := G$ and $b := B$. Since $B^G = B$ we have

$$\hat{\epsilon}_B = \mathrm{Br}_P(\hat{\epsilon}_B) \in \langle\, (C \cap \mathbf{C}_G(P))^+ \mid C \in \mathrm{cl}(G) \,\rangle_F,$$

and (a) follows, since $\mathbf{C}_G(P) \trianglelefteq G$.

(b) $g^G \subseteq \mathbf{C}_G(P)$ means that $P \leq \mathbf{C}_G(g)$, that is P is contained in any defect group of $C := g^G$. But if $P \in \mathrm{Def}(B)$ then by definition $\hat{\epsilon}_B \in \mathbf{Z}_P(FG)$, so (b) follows from (a).

$\square$

We will need the following group theoretical result.

Lemma 4.7.21 *If P is a p-subgroup of G, then the map $C \mapsto C \cap \mathbf{C}_G(P)$ defines a bijection*

$$f\colon \{C \in \mathrm{cl}(G) \mid P \in \mathrm{Def}_p(C) \,\} \to \{C \in \mathrm{cl}(\mathbf{N}_G(P)) \mid P \in \mathrm{Def}_p(C) \,\}.$$

Proof.   Let $C \in \mathrm{cl}(G)$ with $P \in \mathrm{Def}_p(C)$ and $x, y \in C \cap \mathbf{C}_G(P)$. Then there is $g \in G$ with $y = x^g$, so $P, P^g \in \mathrm{Syl}_p(\mathbf{C}_G(y))$. By the Sylow theorems there is $c \in \mathbf{C}_G(y)$ with $P^{gc} = P$. Hence $gc \in \mathbf{N}_G(P)$ and $x^{gc} = y$. So $C \cap \mathbf{C}_G(P)$ is a conjugacy class of $\mathbf{N}_G(P)$ and f is a well-defined (and injective) map. Now let $h \in \mathbf{N}_G(P)$ with $P \in \mathrm{Def}_p(h^{\mathbf{N}_G(P)})$. If $P \not\in \mathrm{Syl}_p(\mathbf{C}_G(h))$ then there is a $P_1 \leq \mathbf{C}_G(h)$ with $P \triangleleft P_1$. But this means $P_1 \leq \mathbf{N}_G(P)$, which contradicts $P \in \mathrm{Def}_p(h^{\mathbf{N}_G(P)})$. So $P \in \mathrm{Def}_p(h^G)$ and $h^{\mathbf{N}_G(P)} = h^G \cap \mathbf{C}_G(P) = f(h^G)$ by the first part of the proof. $\square$

Theorem 4.7.22 (Brauer's first main theorem) *If P is a p-subgroup of G, the map $b \mapsto b^G$ defines a bijection*

$$\{b \in \mathrm{Bl}_p(\mathbf{N}_G(P)) \mid P \in \mathrm{Def}(b)\} \to \{B \in \mathrm{Bl}_p(G) \mid P \in \mathrm{Def}(B)\}$$

called the **Brauer correspondence**. *Furthermore,* $\mathrm{Br}_p(\hat{\epsilon}_{b^G}) = \hat{\epsilon}_b$.

Proof.   Let $b \in \mathrm{Bl}_p(\mathbf{N}_G(P))$ with $P \in \mathrm{Def}(b)$. By Theorem 4.7.19 b^G is defined and $\hat{\omega}_b^G = \hat{\omega}_b \circ \mathrm{Br}_P$. We have to show that $P \in \mathrm{Def}(b^G)$. Let $C' \in \mathrm{cl}(\mathbf{N}_G(P))$ be a defect class for b. By Lemma 4.7.21 $C' = C \cap \mathbf{C}_G(P)$ for $C' \subseteq C \in \mathrm{cl}(G)$ and $P \in \mathrm{Def}_p(C)$. Hence

$$\hat{\omega}_{b^G}(C^+) = \hat{\omega}_b(\mathrm{Br}_P(C^+)) = \hat{\omega}_b((C')^+) \neq 0.$$

By the min–max theorem (Theorem 4.6.6) some $Q \in \mathrm{Def}(b^G)$ is contained in P. Since $\mathrm{Def}(b) = \{P\}$ we must have $Q = P$ by Lemma 4.7.2.

Next we show that the map $b \mapsto b^G$ is surjective. To this end let $B \in \mathrm{Bl}_p(G)$ have defect group P. By Theorem 4.7.19

$$\mathrm{Br}_P(\hat{\epsilon}_B) = \hat{\epsilon}_{b_1} + \cdots + \hat{\epsilon}_{b_m}, \tag{4.31}$$

with $\{b_1, \ldots, b_m\} = \{b \in \mathrm{Bl}_p(\mathbf{N}_G(P)) \mid b^G = B\}$ and $b_i \neq b_j$ for $i \neq j$. By Theorem 4.6.10 and Lemma 4.7.2 $\mathrm{Def}(b_j) = \{P\}$ for all $j = 1, \ldots, m$.

Finally, we show that the map $b \mapsto b^G$ is injective and hence $m = 1$ in (4.31). So let $b_1, b_2 \in \mathrm{Bl}_p(\mathbf{N}_G(P))$ have defect group P and satisfy $b_1^G = b_2^G$. From Theorem 4.7.19 we know that $\hat{\omega}_{b_1} \circ \mathrm{Br}_P = \hat{\omega}_{b_2} \circ \mathrm{Br}_P$, thus

$$\hat{\omega}_{b_1}((C \cap \mathbf{C}_G(P))^+) = \hat{\omega}_{b_2}((C \cap \mathbf{C}_G(P))^+) \qquad \text{for} \qquad C \in \mathrm{cl}(G).$$

By Lemma 4.7.21 $\hat{\omega}_{b_1}$ and $\hat{\omega}_{b_2}$ agree on $\{ (C')^+ \mid C' \in \mathrm{cl}(\mathbf{N}_G(P)), \ P \in \mathrm{Def}_p(C') \}$, and by Lemma 4.7.20 $\hat{\omega}_{b_1}(\hat{\epsilon}_{b_j}) = \hat{\omega}_{b_2}(\hat{\epsilon}_{b_j})$ for $j = 1, 2$; hence $b_1 = b_2$. $\qquad \square$

Corollary 4.7.23 *Let $H \leq G$, $b \in \mathrm{Bl}_p(H)$ with $D \in \mathrm{Def}(b)$ and assume that $\mathbf{C}_G(D) \leq H$. Then b^G is defined.*

Proof. Let $H_1 := \mathbf{N}_H(D)$. By Theorem 4.7.22 there is a unique $b_1 \in \mathrm{Bl}_p(H_1)$ with $D \in \mathrm{Def}(b_1)$ and $b = b_1^H$. Our assumption implies that $D\,\mathbf{C}_G(D) \leq H_1 \leq \mathbf{N}_G(D)$, so by Theorem 4.7.19 b_1^G is defined. The assertion now follows from Exercise 4.7.3. $\qquad \square$

Corollary 4.7.24 *If D is a defect group of a p-block of G, then $D = \mathbf{O}_p(\mathbf{N}_G(D))$.*

Proof. This follows from Theorem 4.7.22 because of Theorem 4.6.10. $\qquad \square$

There does not seem to be a group theoretical characterization of defect groups of p-blocks. The corollary just says that they are radical p-subgroup in the sense of the following definition.

Definition 4.7.25 A subgroup D of G is called a **radical p-subgroup** of G if $D = \mathbf{O}_p(\mathbf{N}_G(D))$.

Brauer's first main theorem compares the blocks of G with defect group P with the blocks of $N := \mathbf{N}_G(P)$ with defect group P. It is often useful to go one step further, namely to $P\,\mathbf{C}_G(P)$. Observe that $P\,\mathbf{C}_G(P) \trianglelefteq N$. Changing the perspective we put $N \trianglelefteq G$ and we will first study the relations of blocks of FG and FN.

Definition 4.7.26 Let $N \trianglelefteq G$ and $b \in \mathrm{Bl}_p(N)$. Then $B \in \mathrm{Bl}_p(G)$ is said to **cover** b if $\epsilon_B \epsilon_b \neq 0$, where $\epsilon_B \in \mathbf{Z}(RG)$ and $\epsilon_b \in \mathbf{Z}(RN)$ are the block idempotents of B and b, respectively. The set of p-blocks of G covering $b \in \mathrm{Bl}_p(N)$ will be denoted by $\mathrm{Bl}(G \mid b)$.

Remark 4.7.27 If $N \trianglelefteq G$ then G acts by conjugation on the block idempotents (and thereby on the blocks) of RN. If $\epsilon_b \in \mathbf{Z}(RN)$ is a block idempotent, then the orbit sum

$$\sigma_G(\epsilon_b) := \sum \{{}^g\epsilon_b \mid g \in G\} = \sum_{i=1}^t \epsilon_{b_i} = \sum_{i=1}^s \epsilon_{B_i} \in \mathbf{Z}(RN) \cap \mathbf{Z}(RG),$$

with mutually orthogonal block idempotents ϵ_{b_i} and ϵ_{B_i} of RN and RG, respectively. We put $T_G(b) := \{g \in G \mid \epsilon_b^g = \epsilon_b\}$, thus $t = [G : T_G(b)]$. Clearly $B \in \mathrm{Bl}(G \mid b)$ if and only if $\epsilon_B = \epsilon_{B_i}$ for some $i \in \{1, \dots, s\}$, and this is the case if and only if $\hat{\omega}_B(\sigma_G(\epsilon_b)) = 1$. Obviously every block $B \in \mathrm{Bl}_p(G)$ covers exactly one conjugacy class of blocks of N.

Lemma 4.7.28 *Assume* $N \trianglelefteq G$. *Let* $B \in \mathrm{Bl}_p(G)$ *and* $b \in \mathrm{Bl}_p(N)$.

(a) B *covers* b *if and only if for some (or every)* $\chi \in \mathrm{Irr}(B)$ *there is a* $\psi \in \mathrm{Irr}(b)$ *with* $(\psi, \chi_N)_N > 0$.

(b) B *covers* b *if and only if for some (or every)* $\varphi \in \mathrm{IBr}(B)$ *there is an irreducible constituent of* φ_N *in* $\mathrm{IBr}(b)$.

(c) $B \in \mathrm{Bl}(G \mid b)$ *if and only if* $\hat{\omega}_B((h^G)^+) = \hat{\omega}_b((h^G)^+)$ *for all* $h \in N$.

(d) *If* b^G *is defined then* b^G *covers* b.

(e) *If* $B \in \mathrm{Bl}(G \mid b)$ *and* $\mathbf{C}_G(D) \leq N$ *for* $D \in \mathrm{Def}(B)$ *then* B *is the only p-block of* G *covering* b *and* $B = b^G$.

(f) *If* G/N *is a p-group then* $|\mathrm{Bl}(G \mid b)| = 1$.

Proof.   (a) Let $\chi \in \mathrm{Irr}(B)$ and let ψ be an irreducible constituent of χ_N belonging to $b \in \mathrm{Bl}_p(N)$. By Exercise 3.6.6 $\omega_\chi|_Z = \omega_\psi|_Z$ for $Z = \mathbf{Z}(RN) \cap \mathbf{Z}(RG)$. In particular $\hat{\omega}_B(\sigma_G(\epsilon_b)) = 1$ and by Remark 4.7.27 B covers b and its conjugates.

(b) Let $\varphi \in \mathrm{IBr}(B)$ and let $\theta \in \mathrm{IBr}(N)$ be an irreducible constituent of φ_N. Choose $\chi \in \mathrm{Irr}(G)$ with $d_{\chi,\varphi} > 0$. Then $\chi \in \mathrm{Irr}(B)$ and χ_N has an irreducible constituent $\psi \in \mathrm{Irr}(N)$ with $d_{\psi,\theta} > 0$. Since ψ and θ belong to the same p-block of N, the result follows from (a).

(c) $(C^+ \mid C \in \mathrm{cl}(G), \ C \subseteq N)$ is a basis of $Z := \mathbf{Z}(FN) \cap \mathbf{Z}(FG)$. Thus (c) follows from (a) and Exercise 3.6.6.

(d) Since $N \trianglelefteq G$ we have $\hat{\omega}_{b^G}((h^G)^+) = \hat{\omega}_b(s_N((h^G)^+)) = \hat{\omega}_b((h^G)^+)$ for $h \in N$. So the result follows from (c).

(e) Assume $B \in \mathrm{Bl}(G|b)$ and $D \in \mathrm{Def}(B)$. Let $C \in \mathrm{cl}(G)$. If $\hat{\omega}_B(C^+) \neq 0$ then by Theorem 4.6.6 $D \leq_G Q$ for some $Q \in \mathrm{Def}_p(C)$ and consequently $\mathbf{C}_G(Q) \leq_G \mathbf{C}_G(D) \leq N$ by assumption. Hence $C \subseteq N$. It follows that $\hat{\omega}_B$ vanishes outside of N and so $\hat{\omega}_B = \hat{\omega}_b \circ s_N$, thus $B = b^G$.

(f) Let $B, B' \in \mathrm{Bl}(G \mid b)$. By part (c) we have $\hat{\omega}_B|_Z = \hat{\omega}_{B'}|_Z$. But by Osima's theorem (Theorem 4.4.7) $\hat{\epsilon}_B \in Z$, since $G_{p'} \subseteq N$. So $\hat{\omega}_{B'}(\hat{\epsilon}_B) = 1$ and $B' = B$.   □

Corollary 4.7.29 *Assume that* $\mathbf{C}_G(\mathbf{O}_p(G)) \leq \mathbf{O}_p(G)$. *Then* $|\mathrm{Bl}_p(G)| = 1$.

Proof.   Since $N := \mathbf{O}_p(G)$ has only one p-block, every $B \in \mathrm{Bl}_p(G)$ covers $B_0(N)$ by Remark 4.7.27. If $D \in \mathrm{Def}(B)$, then $N \leq D$ by Theorem 4.6.10, hence $\mathbf{C}_G(D) \leq \mathbf{C}_G(N) \leq N$ and $B = B_0(N)^G$ by Lemma 4.7.28(e).   □

Theorem 4.7.30 (Extended first main theorem) *If $B \in \mathrm{Bl}_p(G)$ has defect group P, then there is a unique $\mathbf{N}_G(P)$-orbit of p-blocks $b \in \mathrm{Bl}_p(P\,\mathbf{C}_G(P))$ such that $b^G = B$. All these p-blocks b have defect group P.*

Proof. By Theorem 4.7.19 there is a block $b \in \mathrm{Bl}_p(P\,\mathbf{C}_G(P))$ with $B = b^G$, and $b^{\mathbf{N}_G(P)}$ is defined also. Theorem 4.6.10 implies that $P \subseteq D_b \cap D$ for every $D_b \in \mathrm{Def}(b)$ and $D \in \mathrm{Def}(b^{\mathbf{N}_G(P)})$. By Exercise 4.7.3 $B = (b^{\mathbf{N}_G(P)})^G$ and by Lemma 4.7.2 D_b and D are contained in a G-conjugate of P, hence $P = D_b = D$.

If $b_1, b_2 \in \mathrm{Bl}_p(P\,\mathbf{C}_G(P))$ satisfy $(b_1)^G = (b_2)^G = B$ then $(b_1)^{\mathbf{N}_G(P)}$ and $(b_2)^{\mathbf{N}_G(P)}$ are blocks inducing B by Exercise 4.7.3, so by Brauer's first main theorem $B_1 := (b_1)^{\mathbf{N}_G(P)} = (b_2)^{\mathbf{N}_G(P)}$. Lemma 4.7.28(d) implies that B_1 covers b_1, b_2, hence by Remark 4.7.27 these blocks are conjugate in $\mathbf{N}_G(P)$. Conversely, if $b_1, b_2 \in \mathrm{Bl}_p(P\,\mathbf{C}_G(P))$ are conjugate in $\mathbf{N}_G(P)$ then it is obvious that $(b_1)^{\mathbf{N}_G(P)} = (b_2)^{\mathbf{N}_G(P)}$ and the result follows. $\square$

Corollary 4.7.31 *Let $B \in \mathrm{Bl}_p(G)$ and $P \in \mathrm{Def}(B)$. Then there is up to $\mathbf{N}_G(P)$-conjugacy a unique $b \in \mathrm{Bl}_p(P\,\mathbf{C}_G(P))$ and $\theta \in \mathrm{Irr}(b)$ with $b^G = B$ and $P \subseteq \ker\theta$. Often θ is also called the* **canonical character** *of B. Furthermore $T_{\mathbf{N}_G(P)}(\theta) = T_{\mathbf{N}_G(P)}(b) := \{g \in \mathbf{N}_G(P) \mid \epsilon_b^g = \epsilon_b\}$.*

Proof. Use Theorem 4.6.12. $\square$

We close this section by extending the Clifford correspondence to blocks.

Theorem 4.7.32 (Fong, Reynolds) *Let $N \trianglelefteq G$ and $b \in \mathrm{Bl}_p(N)$ with $T := \mathrm{T}_G(b)$.*

(a) *If $B \in \mathrm{Bl}(T \mid b)$ then B^G is defined and*

$$\mathrm{Irr}(B) \to \mathrm{Irr}(B^G), \quad \chi \mapsto \chi^G, \qquad \mathrm{IBr}(B) \to \mathrm{IBr}(B^G), \quad \varphi \mapsto \varphi^G$$

are bijections, and $d_{\chi\varphi} = d_{\chi^G\,\psi^G}$ for $\chi \in \mathrm{Irr}(B)$, $\varphi \in \mathrm{IBr}(B)$.

(b) $\mathrm{Bl}(T \mid b) \to \mathrm{Bl}(G \mid b)$, $B \mapsto B^G$ *is a bijection.*

(c) *If $B \in \mathrm{Bl}(T \mid b)$ and $D \in \mathrm{Def}(B)$ then $D \in \mathrm{Def}(B^G)$. Also $\mathrm{ht}_p(\chi) = \mathrm{ht}_p(\chi^G)$ for $\chi \in \mathrm{Irr}(B)$.*

Proof. If $\theta \in \mathrm{Irr}(b) \cup \mathrm{IBr}(b)$, then $\mathrm{T}_G(\theta) \leq T$ because $\theta^g \in \mathrm{Irr}(b^g) \cup \mathrm{IBr}(b^g)$ for every $g \in G$. Also, if θ^g belongs to b, then $g \in T$. Let $\theta_1, \ldots, \theta_r$ (respectively $\theta_1', \ldots, \theta_s'$) be representatives of the T-conjugacy classes in $\mathrm{Irr}(b)$ (or $\mathrm{IBr}(b)$, respectively). Then the θ_i (or θ_i') are pair-wise not conjugate in G, and by Clifford's theorem (Theorem 3.6.2) for $U \in \{T, G\}$ the sets $\mathrm{Irr}(U \mid \theta_i)$ (or $\mathrm{IBr}(U \mid \theta_i')$) are mutually disjoint. Therefore the Clifford correspondences

$$\mathrm{Irr}(T \mid \theta_i) \to \mathrm{Irr}(G \mid \theta_i), \qquad \mathrm{IBr}(T \mid \theta_i') \to \mathrm{IBr}(G \mid \theta_i')$$

(see 3.6.8 and 4.3.8) can be composed to bijections

$$X_T := \bigcup_{i=1}^{r} \mathrm{Irr}(T \mid \theta_i) \; \to \; X_G := \bigcup_{i=1}^{r} \mathrm{Irr}(G \mid \theta_i), \quad \chi \mapsto \chi^G,$$

$$Y_T := \bigcup_{i=1}^{r} \mathrm{IBr}(T \mid \theta_i') \; \to \; Y_G := \bigcup_{i=1}^{s} \mathrm{IBr}(G \mid \theta_i'), \quad \varphi \mapsto \varphi^G.$$

Let $B \in \mathrm{Bl}(T \mid b)$ and $\chi \in \mathrm{Irr}(B)$. By Lemma 4.7.28(a) $\chi \in \mathrm{Irr}(T \mid \theta)$ for some $\theta \in \mathrm{Irr}(b)$. Thus $\mathrm{Irr}(B) \subseteq X_T$ and by Lemma 4.7.5 B^G is defined. By restricting $X_T \to X_G$ we obtain an injective map $\mathrm{Irr}(B) \to \mathrm{Irr}(B^G)$, $\chi \mapsto \chi^G$. Similarly $\mathrm{IBr}(B) \subseteq Y_T$ by Lemma 4.7.28(b), and $\mathrm{IBr}(B) \to Y_G$, $\varphi \mapsto \varphi^G$ is injective. If $\varphi \in \mathrm{IBr}(B)$ then using Theorem 4.3.9

$$\Phi_\varphi = \sum_{\chi \in \mathrm{Irr}(B)} d_{\chi\varphi} \chi \qquad \text{hence} \qquad \Phi_{\varphi^G} = \Phi_\varphi{}^G = \sum_{\chi \in \mathrm{Irr}(B)} d_{\chi\varphi} \chi^G.$$

Thus $\varphi^G \in \mathrm{IBr}(B^G)$ and $d_{\chi,\varphi} = d_{\chi^G,\varphi^G}$. From Theorem 4.4.2 we conclude that $\{\chi^G \mid \chi \in \mathrm{Irr}(B)\} = \mathrm{Irr}(B^G)$ and $\{\varphi^G \mid \varphi \in \mathrm{IBr}(B)\} = \mathrm{IBr}(B^G)$.

(b) follows, since $X_T = \bigcup_{B \in \mathrm{Bl}(T|b)} \mathrm{Irr}(B)$ and $X_G = \bigcup_{B' \in \mathrm{Bl}(G|b)} \mathrm{Irr}(B')$.

(c) If $\chi \in \mathrm{Irr}(T)$ and $\chi^G \in \mathrm{Irr}(G)$ then clearly $\mathrm{d}_p(\chi) = \mathrm{d}_p(\chi^G)$. By Definition 4.4.12 $\mathrm{d}(B) = \mathrm{d}(B^G)$ for $B \in \mathrm{Bl}(T \mid b)$. So (c) follows from Lemma 4.7.2. $\qquad\square$

Example 4.7.33 Let $G := \mathrm{C}_2^4 \rtimes \mathrm{A}_5$ be the group considered in Example 3.8.15. We put $N := \mathrm{C}_2^4$ and have $\mathrm{Irr}(N) = \{\theta_1 := \mathbf{1}_N\} \dot\cup \{\theta_2^g \mid g \in G\} \dot\cup \{\theta_3^g \mid g \in G\}$ with $T_2 := \mathrm{T}_G(\theta_2) \cong N \rtimes \mathrm{A}_4$ and $T_3 := \mathrm{T}_G(\theta_3) \cong N \rtimes \mathrm{S}_3$.

We choose $p := 3$ and observe that there are three conjugacy classes of p-blocks in N with representatives b_i and $\mathrm{Irr}(b_i) = \{\theta_i\}$ for $i = 1, 2, 3$. Moreover, $\ker \theta_i \trianglelefteq T_i$ and

$$\tilde{T}_i := T_i / \ker \theta_i \cong \begin{cases} \mathrm{C}_2 \times \mathrm{A}_4 & \text{for } i = 2, \\ \mathrm{C}_2 \times \mathrm{S}_3 & \text{for } i = 3. \end{cases}$$

Let $\mathrm{Irr}(\mathrm{C}_2) = \{\mathbf{1}_{\mathrm{C}_2}, \lambda\}$ and let $\pi_i \colon T_i \to \tilde{T}_i$ be the natural projection ($i = 1, 2$). Then $\mathrm{Irr}(T_2 \mid \theta_2) = \{\mathrm{Inf}_{\pi_2}(\lambda \times \psi) \mid \psi \in \mathrm{Irr}(\mathrm{A}_4)\}$. From Lemma 4.4.17 and the knowledge of the p-blocks of A_4 we conclude that

$$\mathrm{Bl}(T_2 \mid b_2) = \{B_{20}, B_{21}\} \quad \text{with} \quad \mathrm{d}(B_{20}) = 0 \quad \text{and} \quad D_{B_{21}} = [1, 1, 1]^{\mathrm{T}}.$$

Similarly, since S_3 has just one p-block we get

$$\mathrm{Bl}(T_3 \mid b_3) = \{B_{31}\} \quad \text{with} \quad D_{B_{31}} = \begin{bmatrix} 1 & 0 \\ 0 & 1 \\ 1 & 1 \end{bmatrix}.$$

Using Exercise 4.7.6 for $\mathrm{Bl}(B \mid b_1)$ we see that in the notation of Example 3.8.15

$$\{\mathrm{Irr}(B) \mid B \in \mathrm{Bl}_p(G)\} = \{\{\chi_1, \chi_4, \chi_5\}, \{\chi_2\}, \{\chi_3\}, \{\chi_6, \chi_7, \chi_8\},$$
$$\{\chi_9\}, \{\chi_{10}, \chi_{11}, \chi_{12}\}\}$$

and

$$\{\Phi_\varphi \mid \varphi \in \mathrm{IBr}(G)\} = \{\chi_2, \ \chi_3, \ \chi_9, \ \chi_1 + \chi_5, \ \chi_4 + \chi_5, \ \chi_6 + \chi_7 + \chi_8,$$
$$\chi_{10} + \chi_{12}, \ \chi_{11} + \chi_{12}\}.$$

$\blacklozenge$

Exercises

Exercise 4.7.1 Let G be a group, let $H = \{1\}$, the trivial subgroup, and let b be the only p-block of H. Show that b^G is defined if and only if $G = \{1\}$.

Exercise 4.7.2 Let $H \leq G$ and $\theta \in \mathrm{Irr}(H)$. Show that $\omega_{\theta^G}(C^+)$ is an algebraic integer for any $C \in \mathrm{cl}(G)$.

Exercise 4.7.3 Let $H \leq U \leq G$ and $b \in \mathrm{Bl}_p(H)$.

(a) Suppose that b^U is defined. Show that b^G is defined if and only if $(b^U)^G$ is defined, and that in this case $b^G = (b^U)^G$.

(b) For $H := \langle (1,2)(3,4) \rangle \leq U := V_4 \leq G := A_4$ and $b := B_0(H)$ show that b^U is not defined, while b^G is defined.

Exercise 4.7.4 Let $N \trianglelefteq G$. Show that $B_0(G)$ covers only $B_0(N)$.

Exercise 4.7.5 Show that $|\mathrm{Bl}_2(S_n)| \leq |\mathrm{Bl}_2(A_n)|$ for any $n \in \mathbb{N}$.

Exercise 4.7.6 Let $N \trianglelefteq G$ with natural projection $\pi_N \colon G \to \bar{G} = G/N$. Assume that $p \nmid |N|$. Show that there is a bijection

$$\mathrm{Bl}(G \mid B_0(N)) \to \mathrm{Bl}_p(\bar{G}), \quad B \mapsto \bar{B} \qquad \text{with} \quad \mathrm{Irr}(B) = \{\mathrm{Inf}_{\pi_N}(\bar{\chi}) \mid \bar{\chi} \in \mathrm{Irr}(\bar{B})\}.$$

Exercise 4.7.7 Let $N_i \trianglelefteq G$ and $b_i \in \mathrm{Bl}_p(N_i)$ for $i = 1, 2$. Assume that $N_1 \leq N_2$ and that $B \in \mathrm{Bl}_p(G)$ covers b_1 and b_2. Show that b_2 covers b_1.

Exercise 4.7.8 Assume that $P \in \mathrm{Syl}_3(G)$ and $\mathbf{N}_G(P) \cong S_3 \times D_{10}$. Show that G has exactly four 3-blocks of defect one. Find the canonical characters of these blocks.

4.8 Vertices

In this section G will always be a finite group and S will denote a field of characteristic $p > 0$ or a complete discrete valuation ring with $\mathrm{char}(S/\mathrm{J}(S)) = p > 0$. As usual we assume that all SG-modules considered are free and finitely generated as S-modules. Later in this section we will assume that (K, R, F, η) is a p-modular splitting system for G. In analogy to Section 4.6, we will assign p-subgroups to any indecomposable SG-module called vertices and study their properties. Given a cyclic p-group P we will see that there are only finitely many FG-modules with vertex P (up to isomorphism), and we will analyze explicitly the situation of blocks with a normal defect group D of order p. In Section 4.12 the general case of a block with a defect group of order p will be described.

If V is an SG-module, $H \leq G$ and $G = \dot{\bigcup}_{i=1}^m g_i H$, then for any $g \in G$ there exist $\sigma \in S_m$ and $h_1, \ldots, h_m \in H$ such that

$$g g_i = g_{\sigma(i)} h_i \qquad \text{for all} \qquad h \in H.$$

Thus, if $v \in \mathrm{Inv}_H(V)$ (that is, $hv = v$ for all $h \in H$, see Definition 1.1.18) then $\sum_{i=1}^m g_i v \in \mathrm{Inv}_G(V)$. Hence the following definition makes sense.

Definition 4.8.1 Let H be a subgroup of G and let $G = \overset{\cdot}{\bigcup}_{i=1}^{m} g_i H$. For any SG-module V we define

$$\mathrm{Tr}_H^G \colon \mathrm{Inv}_H(V) \to \mathrm{Inv}_G(V), \quad v \mapsto \sum_{i=1}^{m} g_i v \qquad (v \in V).$$

The following remark is easily verified.

Remark 4.8.2 Let V be an SG-module and $H \leq G$. Then

(a) Tr_H^G is S-linear and independent of the choice of the transversal $\{g_1, \ldots, g_m\}$;

(b) if $H_1 \leq H_2 \leq G$ then $\mathrm{Tr}_{H_2}^G \circ \mathrm{Tr}_{H_1}^{H_2} = \mathrm{Tr}_{H_1}^G$.

Recall that for SG-modules V, W the S-module $\mathrm{Hom}_S(V, W)$ becomes an SG-module via $(g \cdot \varphi)(v) := g\varphi(g^{-1}v)$ for $g \in G$, $\varphi \in \mathrm{Hom}_S(V, W)$ and $v \in V$. Obviously $\mathrm{Inv}_H(\mathrm{Hom}_S(V, W)) = \mathrm{Hom}_{SH}(V, W)$ for $H \leq G$. So

$$\mathrm{Tr}_H^G \colon \mathrm{Hom}_{SH}(V, W) \to \mathrm{Hom}_{SG}(V, W).$$

Lemma 4.8.3 *Let V, W_1, W_2 be SG-modules and $H \leq G$.*

(a) *For $\theta \in \mathrm{End}_{SH} V$, $\psi \in \mathrm{Hom}_{SG}(W_1, V)$ and $\varphi \in \mathrm{Hom}_{SG}(V, W_2)$ one has*

$$\mathrm{Tr}_H^G(\theta \circ \psi) = \mathrm{Tr}_H^G(\theta) \circ \psi, \qquad \mathrm{Tr}_H^G(\varphi \circ \theta) = \varphi \circ \mathrm{Tr}_H^G(\theta).$$

Hence

$$\mathrm{Tr}_H^G \ (\mathrm{End}_{SH} V) \ \trianglelefteq \ \mathrm{End}_{SG} V.$$

(b) *If $\theta \in \mathrm{End}_{SG} V$, then $\mathrm{Tr}_H^G(\theta) = [G : H]\, \theta$.*

Proof. The proof is obvious.   □

Theorem 4.8.4 *Let $H \leq G$. The following conditions for an SG-module V are equivalent.*

(a) *V is H-projective, that is $V \mid W^G$ for some SH-module W (see Definition 3.2.15).*

(b) *$\mathrm{Tr}_H^G\ (\mathrm{End}_{SH} V) = \mathrm{End}_{SG} V$ or, equivalently, there exists $\theta \in \mathrm{End}_{SH} V$ with $\mathrm{Tr}_H^G(\theta) = \mathrm{id}_V$.*

(c)

$$V \mid (V_H)^G.$$

Proof. (a) $\Rightarrow$ (b) Let W be an SH-module with $V \mid W^G$. This means that we have SG-homomorphisms

$$\pi \colon W^G \to V \quad \text{and} \quad j \colon V \to W^G \quad \text{with} \quad \pi \circ j = \mathrm{id}_V.$$

We choose $g_i \in G$ with $G = \dot{\bigcup}_{i=1}^m g_i H$ and $g_1 = 1$ and define

$$\tau \colon W^G \to W^G, \quad \sum_{i=1}^m g_i \otimes w_i \mapsto 1 \otimes w_1 \qquad (w_i \in W).$$

It is easily checked that $\tau \in \mathrm{End}_{SH} W^G$ and $\mathrm{Tr}_H^G(\tau) = \mathrm{id}_{W^G}$. In fact,

$$\mathrm{Tr}_H^G(\tau)(\sum_{j=1}^m g_j \otimes w_j) = \sum_{i=1}^m g_i \tau(g_i^{-1}(\sum_{j=1}^m g_j \otimes w_j)) = \sum_{i=1}^m g_i(1 \otimes w_i).$$

Putting $\theta := \pi \circ \tau \circ j \in \mathrm{End}_{SH} V$ we get (using Lemma 4.8.3(a))

$$\mathrm{Tr}_H^G(\theta) = \pi \circ \mathrm{Tr}_H^G(\tau) \circ j = \pi \circ j = \mathrm{id}_V .$$

(b) $\Rightarrow$ (c) Let θ be in $\mathrm{End}_{SH} V$ with $\mathrm{Tr}_H^G(\theta) = \mathrm{id}_V$. We have an SG-homomorphism

$$\pi \colon (V_H)^G \to V, \quad \sum_{i=1}^m g_i \otimes v_i \mapsto \sum_{i=1}^m g_i v_i \qquad (v_i \in V),$$

where the $g_i \in G$ are as above. Also, we have an SH-homomorphism

$$\eta \colon V \to (V_H)^G, \quad v \mapsto 1 \otimes v \qquad \text{and} \qquad \pi \circ \eta = \mathrm{id}_V .$$

We now define $\tilde{\eta} := \mathrm{Tr}_H^G (\eta \circ \theta)$. Then $\tilde{\eta} \colon V \to (V_H)^G$ is an SG-homomorphism, and by Lemma 4.8.3

$$\pi \circ \tilde{\eta} = \mathrm{Tr}_H^G (\pi \circ \eta \circ \theta) = \mathrm{Tr}_H^G (\mathrm{id}_V \circ \theta) = \mathrm{id}_V .$$

Hence $V \mid (V_H)^G$.

(c) $\Rightarrow$ (a) This is trivial. $\qquad\qquad\qquad\qquad\qquad\qquad\qquad\qquad\qquad\qquad\square$

As a consequence we obtain a generalization of Maschke's theorem (Theorem 1.5.6).

Corollary 4.8.5 *If $[G : H]$ is a unit in S, then every SG-module is H-projective.*

Proof. Putting $\theta := [G : H]^{-1} \mathrm{id}_V$, one obtains $\mathrm{Tr}_H^G(\theta) = \mathrm{id}_V$. $\qquad\square$

Theorem 4.8.6 *Let V be an indecomposable SG-module. Then there is a unique conjugacy class $\mathrm{Vtx}(V)$ of subgroups of G, such that V is H-projective for $H \leq G$ if and only if H contains a subgroup $Q \in \mathrm{Vtx}(V)$. Any $Q \in \mathrm{Vtx}(V)$ is called a **vertex** of V. A vertex is always a p-group.*

Proof. Let $Q \leq G$ be such that V is Q-projective but not U-projective for any proper subgroup $U < Q$. Obviously, V is H-projective for any H containing Q or a conjugate of Q (see Exercise 3.2.2).

Now let H be an arbitrary subgroup of G such that V is H-projective. Then there is an SH-module W with $V \mid W^G$. By Mackey's theorem (Theorem 3.2.17)

$$V_Q \mid W^G{}_Q = \bigoplus_{g \in T} ((g \otimes W)_{gHg^{-1} \cap Q})^Q.$$

Here T is a set of representatives of the (Q, H)-double cosets, i.e. $G = \dot{\bigcup}_{g \in T} QgH$. Since V is Q-projective, it follows from Theorem 4.8.4 and the transitivity of induction (Lemma 3.2.14) that

$$V \mid V_Q{}^G \mid (W^G{}_Q)^G = \bigoplus_{g \in T} ((g \otimes W)_{gHg^{-1} \cap Q})^G.$$

Since V is indecomposable by assumption, there is a $g \in T$ such that

$$V \mid ((g \otimes W)_{gHg^{-1} \cap Q})^G,$$

that is, V is $gHg^{-1} \cap Q$-projective. Because of the minimality of Q we conclude that $gHg^{-1} \cap Q = Q$, hence $Q^g \leq H$. This shows at the same time the uniqueness of Q up to conjugacy. Finally, by Corollary 4.8.5 any SG-module is P-projective for $P \in \mathrm{Syl}_p(G)$. So by the first part of the theorem, vertices are p-groups. $\square$

Definition 4.8.7 Let V be an indecomposable SG-module. Any indecomposable SQ-module W for some $Q \in \mathrm{Vtx}(V)$ with $V \mid W^G$ is called a **source** of V.

Lemma 4.8.8 *Any two sources* W_1, W_2 *of an indecomposable SG-module V are conjugate in G, that is, there is a $g \in G$ such that $W_1 \cong g \otimes W_2$. If $Q \in \mathrm{Vtx}(V)$ and the SQ-module W is a source of V then*

$$V \mid W^G, \qquad W \mid V_Q, \qquad \mathrm{Vtx}(W) = \{Q\}.$$

Proof. We may assume that W_1, W_2 are SQ-modules for a fixed $Q \in \mathrm{Vtx}(V)$ (see Exercise 3.2.2). Since $V \mid V_Q{}^G$ by Theorem 4.8.4 and V is indecomposable, there is an indecomposable SQ-module W with $V \mid W^G$ and $W \mid V_Q$. Then

$$W \mid (W_i^G)_Q = \bigoplus_{g \in T} ((g \otimes W_i)_{gQg^{-1} \cap Q})^Q \qquad (i = 1, 2),$$

where $G = \dot{\bigcup}_{g \in T} QgQ$. Since W is indecomposable there is $g_i \in T$ with $W \mid ((g_i \otimes W_i)_{gQg^{-1} \cap Q})^Q$ for $i = 1, 2$. By the defining (minimality) property of a vertex we must have $g_i Q g_i^{-1} \cap Q = Q$, that is, $g_i \in \mathbf{N}_G(Q)$ and $W \mid g_i \otimes W_i$ for $i = 1, 2$. Since the modules W_i, and hence $g_i \otimes W_i$, are indecomposable, the result follows. $\square$

The following lemma, also called the two-out-of-three lemma, will be used in the proof of Green correspondence; see Section 4.9.

Lemma 4.8.9 *Let V be an indecomposable SG-module with $Q \in \mathrm{Vtx}(V)$ and let $Q \leq H \leq G$. Then for any two of the following conditions there is an indecomposable SH-module W fulfilling them:*

(a) $V \mid W^G$;   (b) $W \mid V_H$;   (c) $Q \in \mathrm{Vtx}(W)$.

Proof.   (a) and (b) Since V is H-projective, $V \mid (V_H)^G$, so there is some indecomposable summand W of V_H such that $V \mid W^G$.

(a) and (c) Assume that the SQ-module M is a source of V. Then $V \mid M^G = (M^H)^G$. Hence there is an indecomposable SH-module W with $V \mid W^G$ and $W \mid M^H$. Hence, if $Q_1 \in \mathrm{Vtx}(W)$, we have $Q \leq_G Q_1 \leq_H Q$, because $\{Q\} = \mathrm{Vtx}(M)$.

(b) and (c) Let M be as above. Then by Lemma 4.8.8 $M \mid V_Q = (V_H)_Q$, so there is an indecomposable SH-module W with $W \mid V_H$ and $M \mid W_Q$. If, as above, $Q_1 \in \mathrm{Vtx}(W)$, then we conclude from Theorem 4.8.10(a), applied to W and M and also to V and W, that $Q \leq_H Q_1 \leq_G Q$.   □

It will be an immediate consequence of Theorem 4.9.2 that in the situation of Lemma 4.8.9 it is actually possible to find an SH-module W fulfilling all three conditions.

Theorem 4.8.10 *Let V be an indecomposable SG-module with $Q \in \mathrm{Vtx}(V)$. Let $H \leq G$ and $V_H = U_1 \oplus \cdots \oplus U_m$ with indecomposable SH-modules U_i and $Q_i \in \mathrm{Vtx}(U_i)$ for $i = 1, \ldots, m$.*

(a) $Q_i \leq_G Q$ *for $i = 1, \ldots, m$.*

(b) *If V is H-projective, then $V \mid U_j{}^G$ for some $j \in \{1, \ldots, m\}$ and $Q_j =_G Q$. If $Q \leq H$, then $Q_j =_H Q$ for some j.*

(c) *If $Q_j =_G Q$, then V and U_j have a common source.*

Proof.   (a) Let W be an indecomposable SQ-module which is a source of V. We choose T with $G = \bigcup_{g \in T} HgQ$ and get

$$U_1 \oplus \cdots \oplus U_m = V_H \mid W^G{}_H = \bigoplus_{g \in T} ((g \otimes W)_{gQg^{-1} \cap H})^H. \qquad (4.32)$$

Hence every U_i is $g_i Q g_i^{-1} \cap H$-projective for some $g_i \in T$, and consequently $Q_i \leq_H g_i Q g_i^{-1} \cap H \leq_G Q$.

(b) By assumption we have $V \mid V_H{}^G$, hence $V \mid U_j^G$ for some j, which implies $Q \leq_G Q_j$. Using (a) we get $Q_j =_G Q$. If $Q \leq H$, then

$$V \mid V_Q{}^G = ((V_H)_Q)^G = (U_1 \oplus \cdots \oplus U_m)_Q{}^G.$$

Hence $V \mid W^G$ for some indecomposable summand W of some $(U_j)_Q$. We have $Q \in \mathrm{Vtx}(W)$, and on applying part (a) with U_j instead of V and W instead of U_i we get $Q \leq_H Q_j$; since $Q_j =_G Q$, the result follows.

(c) Let W be as in the proof of (a). From (4.32) we see that there is a $g \in T$ with

$$U_j \mid ((g \otimes W)_{gQg^{-1} \cap H})^H, \qquad gQg^{-1} \cap H \geq_H Q_j =_G Q.$$

Hence $gQg^{-1} \subseteq H$, and $g \otimes W$ is a common source for V and U_j.   □

Remark 4.8.11 For any group G let S_G denote the trivial SG-module.

(a) If P is a p-group and $Q \leq P$ then the permutation module $(S_Q)^P$ is indecomposable.

(b) $\mathrm{Vtx}(S_G) = \mathrm{Syl}_p(G)$.

(c) Every p-subgroup of G is a vertex of some indecomposable SG-module.

Proof.   (a) The trivial module SP is the only simple SP-module (Theorem 1.3.11). By Frobenius Nakayama reciprocity (Theorem 3.2.12) we have

$$\mathrm{Hom}_{SP}(S_Q{}^P, S_P) \cong \mathrm{Hom}_{SQ}(S_Q, S_Q) \cong S_Q.$$

Hence the head of $S_Q{}^P$ is simple and so $S_Q{}^P$ is indecomposable.

(b) Let $P \in \mathrm{Syl}_p(G)$. If $Q \in \mathrm{Vtx}(S_P)$ then $S_P \mid S_Q{}^P$. Since $S_Q{}^P$ is indecomposable by part (a), $Q = P$. From Theorem 4.8.10(a) applied to $V = S_G$ and $H = P$ it follows that any vertex of S_G is conjugate in G to P (and that S_P is a source of S_G).

(c) Let Q be a p-subgroup of G. By Mackey's theorem (Theorem 3.2.17) $S_Q \mid (S_Q{}^G)_Q$, so there is an indecomposable SG-module V with $V \mid S_Q{}^G$ and $S_Q \mid V_Q$. From Theorem 4.8.10(a) we see that Q is a vertex of V.   □

Theorem 4.8.12 *If F is a field and V is a non-projective indecomposable FG-module*

$$\mathrm{Vtx}(V) = \mathrm{Vtx}(\Omega(V)).$$

Proof. Since $V \cong_{FG} \Omega(\Omega^{-1}(V))$ by Theorem 1.6.27, it is enough to show that

$$V \ H\text{-projective} \quad \Longrightarrow \quad \Omega(V), \Omega^{-1}(V) \ H\text{-projective} \qquad (4.33)$$

for any $H \leq G$. Let W be an FH-module such that $V \mid W^G$. By Lemma 1.6.20 and Lemma 3.2.21(b) we have a short exact sequence

$$\{0\} \longrightarrow \Omega(W)^G \longrightarrow P(W)^G \longrightarrow W^G \longrightarrow \{0\}.$$

Since $V \mid W^G$ and (by Lemma 1.6.20) $P(V) \mid P(W)^G$, we conclude that $\Omega(V) \mid \Omega(W)^G$. Similarly $\Omega^{-1}(V) \mid \Omega^{-1}(W)^G$, and (4.33) follows.   □

In Example 1.1.23 we saw that a cyclic p-group P has exactly $|P|$ indecomposable FP-modules W_i $(1 \le i \le |P|)$ over any field F of characteristic p up to isomorphism. In fact we may choose the notation so that $\dim_F W_i = i$. On the other hand, Exercise 4.8.3 shows that $G := C_p \times C_p$ has infinitely many isomorphism classes of indecomposable FG-modules. The following result answers the question of which group algebras are "of finite representation type," that is, have only finitely many isomorphism classes of indecomposable modules.

Theorem 4.8.13 (D. Higman) *Let P be a p-subgroup of G and let F be a field of characteristic p.*

(a) *If P is cyclic then there are up to isomorphism at most $|G|$ indecomposable FG-modules with a vertex in P.*

(b) *If P is not cyclic then there are indecomposable FG-modules with arbitrary large dimension and a vertex in P.*

(c) *G has only finitely many isomorphism classes of indecomposable FG-modules if and only if a Sylow p-subgroup of G is cyclic.*

Proof. (a) If V is an indecomposable FG-module with vertex in P, then V is P-projective and $V \mid (V_P)^G$ by Theorem 4.8.4. Now let P be cyclic and $V_P \cong \bigoplus_{i=1}^n n_i W_i$ with W_i indecomposable with $\dim_F W_i = i$ and $n_i \in \mathbb{N}_0$ for $i = 1, \dots, n := |P|$ (see Example 1.1.23). Then $V \mid W_i^G$ for some $i \in \{1, \dots, n\}$. Since W_i^G has at most $[G : P]$ indecomposable direct summands by Exercise 4.8.2, there are at most $|P| \cdot [G : P] = |G|$ indecomposable FG-modules V up to isomorphism with a vertex in P.

(b) If P is not cyclic then P has $C_p \times C_p$ as a homomorphic image. By Exercise 4.8.3 and inflation, FP has an indecomposable FP-module W_n with $\dim_F W_n \ge n$ for all $n \in \mathbb{N}$. Since by Mackey's theorem $W_n \mid (W_n^G)_P$, there is an indecomposable FG-module V with $W_n \mid V_P$. Then $\dim_F V \ge n$, and by Theorem 4.8.10(a) the module V has a vertex in P.

(c) follows from (a) and (b) applied to $P \in \mathrm{Syl}_p(G)$. $\qquad\square$

For the rest of this section we will assume that (K, R, F, η) is a p-modular splitting system for G.

One can describe all the indecomposable FG-modules belonging to a p-block with cyclic defect group D. We will do this here for the case that D is a normal subgroup of order p.

Theorem 4.8.14 *Let $B \in \mathrm{Bl}_p(G)$ be a block of defect one having a normal defect group D. Then $B = b^G$ for some $b \in \mathrm{Bl}_p(\mathbf{C}_G(D))$ with $D \in \mathrm{Def}(b)$ and $|\mathrm{IBr}(B)| = e := [T_G(b) : \mathbf{C}_G(D)]$ (see Remark 4.7.27). Moreover, $e \mid p-1$. The projective covers $P(V_i)$ of the simple FG-modules V_i in B are all uniserial of length p. One may order the V_i in such a way that the composition factors of*

$P(V_i)$ *are as follows:*

$$V_i, V_{i+1}, \underbrace{\ldots, V_{i+e-1}, \ldots, V_i, V_{i+1}, \ldots}_{m}, V_{i+e-1}, V_i \qquad \text{with} \qquad m := \frac{p-1}{e},$$

where the indices should be read modulo e. Thus the Cartan matrix $C(B)$ is equal to $m \cdot \mathbf{J}_{e,e} + \mathbf{I}_e$, where $\mathbf{J}_{k,l} \in \mathbb{N}^{k \times l}$ denotes the all-1-matrix for $k, l \in \mathbb{N}$. Also $|\operatorname{Irr}(B)| = e + m$, and one can order $\operatorname{Irr}(B)$ and $\operatorname{IBr}(B)$ in such a way that the decomposition matrix of B has the form

$$D(B) = \left[\frac{\mathbf{I}_e}{\mathbf{J}_{m,e}} \right].$$

If U_{ij} denotes the factor module of $P(V_i)$ of composition length j, then

$$\{ U_{ij} \mid i = 1, \ldots, e, \ j = 1, \ldots, p \}$$

is the set of indecomposable FG-modules in B up to isomorphism.

Proof.   Observe that by Theorem 4.7.19 $B = b^G$ for some $b \in \operatorname{Bl}_p(\mathbf{C}_G(D))$ and $D \in \operatorname{Def}(b)$ because of Lemma 4.7.2 and Theorem 4.6.10. We abbreviate to $C := \mathbf{C}_G(D)$ and $T := T_G(b)$.

(a) We first consider T. By Theorem 4.6.12, $\operatorname{IBr}(b) = \{\varphi\}$ with $\varphi = \theta|_{C_{p'}}$, where θ is the canonical character of b. Lemma 4.7.28(e) shows that b^T is the only p-block of T covering b. Thus $\epsilon_b = \sigma_T(\epsilon_b) = \epsilon_{b^T}$, see Remark 4.7.27. Since G/C is cyclic of order dividing $|\operatorname{Aut}(D)| = p-1$ the canonical character θ has (by Theorem 3.6.13) e extensions to characters $\chi_1', \ldots, \chi_e' \in \operatorname{Irr}(T)$. Lemma 4.7.28 shows that $\{\chi_1', \ldots, \chi_e'\} \subseteq \operatorname{Irr}(b^T)$ and $\operatorname{IBr}(b^T) = \{\chi_1'|_{T_{p'}}, \ldots, \chi_e'|_{T_{p'}}\}$.

For $1 \leq i \leq e$ let W_i be a simple FT-module with Brauer character $\chi_i'|_{T_{p'}}$ and let $U_i := P(W_i)$. Then $(W_i)_C \cong W$, the simple FC-module in b. All the W_i are all C-projective by Corollary 4.8.5. Using Theorem 4.8.4(c) we see that $W^T \cong W_1 \oplus \cdots \oplus W_e$. Trivially, $\operatorname{Rad}(U_{iC}) \subseteq \operatorname{Rad}(U_i)$. On the other hand, $\operatorname{Rad}(U_{iC}) = \operatorname{J}(FC)U_i \leq_{FT} U_i$ (since $C \trianglelefteq T$) and

$$U_i/\operatorname{Rad}(U_{iC}) \mid (U_i/\operatorname{Rad}(U_{iC}))_C{}^T \cong W^T \oplus \cdots \oplus W^T$$

is semisimple, hence $\operatorname{Rad}(U_{iC}) = \operatorname{Rad}(U_i)$ and U_{iC} is the projective cover of W, which is uniserial with composition length p by Theorem 4.6.12. Hence U_i is also uniserial with composition length p.

Assume that the composition factors of U_1 are (up to isomorphisms)

$$W_1 = W_{i_1}, W_{i_2}, \ldots, W_{i_p} = W_1 \qquad \text{with} \qquad W_{i_j} \in \{W_1, \ldots, W_e\}.$$

Then $\operatorname{Rad}(U_1)$ is a homomorphic image of $P(W_{i_2})$, and thus the composition factors of $P(W_{i_2})$ are $W_{i_2}, \ldots, W_{i_p}, W_{i_2}$. By induction on j the composition factors of $P(W_{i_j})$ are $W_{i_j}, W_{i_{j+1}} \ldots, W_{i_{j+p}}$, where the subindices should be read modulo $p - 1$.

Let $f := \min\{\, j > 1 \mid W_{i_j} \cong W_1 \,\}$. Then $U_{i_1}, \ldots, U_{i_{f-1}}$ are pair-wise non-isomorphic and $U_{i_f} \cong U_1$. It follows that $W_{i_j} \cong W_{i_k}$ if and only if $j \equiv k$ mod f. Any composition factor of U_j for $1 \le j \le f$ is isomorphic to one of $\{W_{i_1}, \ldots, W_{i_f}\}$. Theorem 1.7.7 implies that $f = e$. We reorder the W_i, and thereby also the χ'_i, so that $W_j = W_{i_j}$ for $1 \le j \le e - 1$.

(b) Now we go up from T to G. By Theorem 4.7.19 $B := b^G$ is defined and by Exercise 4.7.3 $B = (b^T)^G$. Lemma 4.7.28(d) and (e) show that B is the only p-block of G covering b. Let $V_i := W_i^G$ for $i = 1, \ldots, e$. By Theorem 3.6.7 and Lemma 4.7.28(b) the V_i are representatives of the simple FG-modules belonging to B. Then V_i has Brauer character $\chi_i|_{G_{p'}}$ with $\chi_i := {\chi'_i}^G \in \mathrm{Irr}(G)$ (see Theorem 3.6.7). Let W, W' be FT-modules in b^T. Then $\hat{e}_b W = W$ and $\hat{e}_b W' = W'$ because of Remark 4.7.27. If $g \in G \setminus T$ then $\hat{e}_b \ne {}^g\hat{e}_b$. From ${}^g\hat{e}_b(g \otimes W') = g \otimes W'$ we conclude that $\hat{e}_b(g \otimes W') = 0$. Consequently

$$\mathrm{Hom}_{FT}(W, ((g \otimes W')_{{}^gT \cap T})^T) = \mathrm{Hom}_{FT}(\hat{e}_b W, \hat{e}_b((g \otimes W')_{{}^gT \cap T})^T) = \{0\}.$$

Lemma 3.2.21 implies

$$\mathrm{Hom}_{FT}(W, W') \cong_F \mathrm{Hom}_{FG}(W^G, W'^G).$$

Let

$$P(W_i) =: U_{i0} > U_{i1} > \cdots > U_{ip} = \{0\} \qquad (i = 1, \ldots, e)$$

be the composition series of $P(W_i)$. Then

$$\mathrm{Hom}_{FG}(U_{ij}^G, V_k) \cong_F \mathrm{Hom}_{FT}(U_{ij}, W_k) \cong_F \begin{cases} F & \text{if } i + j \equiv k \mod e, \\ \{0\} & \text{else.} \end{cases}$$

In particular, for $j < p$ each U_{ij}^G has a unique maximal submodule. Also $U_{i0}^G = P(W_i)^G$ is projective (see Corollary 3.2.16). Thus $U_{i0}^G \cong_{FG} P(V_i)$, the projective cover of V_i, has a unique composition series

$$P(V_i) \cong_{FG} U_{i0}^G > U_{i1}^G > \cdots > U_{ip}^G = \{0\} \qquad (i = 1, \ldots, e)$$

with composition factors $V_i, V_{i+1}, \ldots, V_{i+p-1}$, where the indices have to be taken modulo e.

By Lemma 4.7.28 $\chi \in \mathrm{Irr}(G)$ belongs to B if and only if χ_C has a constituent in $\mathrm{Irr}(b) = \{\theta_\lambda \mid \lambda \in \mathrm{Irr}(D)\}$. If $(\chi_C, \theta)_C \ne 0$ then $\chi \in \{\chi_1, \ldots, \chi_e\}$, and conversely. Let $\lambda, \lambda' \in \mathrm{Irr}(D) \setminus \{\mathbf{1}_D\}$. Then

$$T_G(\theta_\lambda) = C \qquad \text{and} \qquad \theta_\lambda{}^G \in \mathrm{Irr}(B).$$

Furthermore $\theta_\lambda{}^G = \theta_{\lambda'}{}^G$ if and only if θ_λ and $\theta_{\lambda'}$ are conjugate in G, hence in one orbit under T. Since the orbits of T on $\mathrm{Irr}(D) \setminus \{\mathbf{1}_D\}$ have length e, we obtain

$$|\{\theta_\lambda{}^G \mid \lambda \in \mathrm{Irr}(D) \setminus \{\mathbf{1}_D\}\}| = \frac{p-1}{e}.$$

Both $\theta_\lambda{}^G$ and $\chi_1 + \cdots + \chi_e = \theta^G$ vanish outside of C and agree on $C_{p'}$. Thus $\theta_\lambda{}^G|_{G_{p'}} = \chi_1|_{G_{p'}} + \cdots + \chi_e|_{G_{p'}}$. This shows that the decomposition matrix $D(B)$ is as indicated.

Finally we show that every indecomposable B-module is uniserial. Let V be a B-module and let $U \leq_{FG} V$ be a uniserial submodule of largest possible dimension. Let $W \leq_{FG} V$ be maximal with respect to the property that $W \cap U = \{0\}$. This implies that V/W has a simple socle L which, of course, belongs to B. By Theorem 1.6.27(e) V/W is isomorphic to a submodule of $P(L)$, hence uniserial. If $(V/W)/\operatorname{Rad}(V/W) \cong V_1$, say, then there is an epimorphism $\tilde\psi\colon P(V_1) \to V/W$ which can be lifted to a homomorphism $\psi\colon P(V_1) \to V$. Then $\psi(P(V_1))$ is a uniserial submodule of V of dimension $\dim V/W \geq \dim(U \oplus W)/W = \dim U$. From the choice of U we conclude that $V = U \oplus W$. So $V = U$ if V is indecomposable and then V is a homomorphic image of $P(V_1)$. Since isomorphic modules must have isomorphic socles and the same composition length, the result follows.   □

We will now show how a defect group of a block is related to the vertices of the indecomposable modules belonging to the block.

Theorem 4.8.15 *Assume that $S \in \{R, F\}$ and let V be an indecomposable SG-module belonging to a p-block B. Then V is D-projective for $D \in \operatorname{Def}(B)$. Thus any vertex of V is contained in a defect group of B.*

Proof. If $S = F$ put $\pi := 0$, otherwise let π be a generator for the unique maximal ideal of R, as before. Let $D \in \operatorname{Def}(B)$, and let $C_1, \ldots, C_m$ be the conjugacy classes of G which have defect groups $Q_i \in \operatorname{Def}_p(C_i)$ contained in D ($1 \leq i \leq m$). Let $\epsilon \in \mathbf{Z}(SG)$ be the block idempotent of B. By the min–max theorem (Theorem 4.6.6) we have $\epsilon = \sum_{i=1}^m a_i C_i^+ + \pi z$ with $a_i \in S$ and some $z \in \mathbf{Z}(SG)$. Since $\epsilon V = V$, and $\operatorname{End}_{SG} V$ is local, there must be an $i \in \{1, \ldots, m\}$ with $a_i \in S \setminus \pi S$ such that the left multiplication

$$\lambda_{C_i^+}\colon V \to V, \ v \mapsto C_i^+ v \ \in \operatorname{End}_{SG} V$$

is a unit, that is an automorphism. (Observe that $\pi\lambda_z\colon v \mapsto \pi z v$ is not surjective.) For each $g \in C_i$ the left multiplication $\lambda_g\colon V \to V$, $v \mapsto gv$ is in $\operatorname{End}_{S\mathbf{C}_G(g)} V$ and $\lambda_{C_i^+} = \operatorname{Tr}^G_{\mathbf{C}_G(g)}(\lambda_g)$. Then

$$\operatorname{id}_V = \operatorname{Tr}^G_{\mathbf{C}_G(g)}(\lambda_g) \circ (\lambda_{C_i^+})^{-1} = \operatorname{Tr}^G_{\mathbf{C}_G(g)}(\lambda_g \circ (\lambda_{C_i^+})^{-1}).$$

By Theorem 4.8.4 V is $\mathbf{C}_G(g)$-projective, and hence a vertex of V is contained in a defect group Q_i of C_i. Since $Q_i \leq D$, the result follows.   □

Corollary 4.8.20 will show that in Theorem 4.8.15 there is an indecomposable SG-module belonging to B having vertex D.

The following notation will be useful in Theorem 4.8.17 and Section 4.9.

Definition 4.8.16 Let $\mathcal{H}$ be a set of p-subgroups of G and let V be an SG-module; V is called $\mathcal{H}$-**projective** if and only if every indecomposable direct summand of V is H-projective for some $H \in \mathcal{H}$.

Theorem 4.8.17 (Nagao) *Let V be an indecomposable RG-lattice belonging to a p-block B of G with block idempotent $\epsilon_B \in \mathbf{Z}(RG)$. Assume that $P \leq G$ is a p-group and that $H \leq G$ satisfies $P\,\mathbf{C}_G(P) \leq H \leq \mathbf{N}_G(P)$ and let $\epsilon \in \mathbf{Z}(RH)$ be the uniquely defined central idempotent with $\hat{\epsilon} = \mathrm{Br}_P(\hat{\epsilon}_B)$ (see Theorem 4.7.19). Then*

$$V_H = \epsilon V_H \oplus W \qquad \text{with } W \text{ an } \mathcal{H}\text{-projective } RH\text{-module}$$

for $\mathcal{H} = \{\, Q \leq H \mid Q \text{ a } p\text{-subgroup with } P \not\leq Q \,\}$.

Proof. By assumption $V_H = \epsilon_B V_H = \epsilon V_H \oplus (1 - \epsilon)V_H$. We have to show that $(1 - \epsilon)V_H = (\epsilon_B - \epsilon)V_H$ is $\mathcal{H}$-projective. Since $\hat{\epsilon} = \mathrm{Br}_P(\hat{\epsilon}_B)$ we have

$$\hat{\epsilon}_B - \hat{\epsilon} \in \langle\, C^+ \mid C \in \mathrm{cl}(H),\ C \not\subseteq \mathbf{C}_G(P)\,\rangle_F.$$

Let $C \in \mathrm{cl}(H)$ satisfy $C \not\subseteq \mathbf{C}_G(P)$ and $Q \in \mathrm{Def}_p(C)$, say $Q \in \mathrm{Syl}_p(\mathbf{C}_H(h))$ with some $h \in C$. If $P \leq Q$ then $h \in \mathbf{C}_G(Q) \leq \mathbf{C}_G(P) \trianglelefteq H$ and, consequently, $C \subseteq \mathbf{C}_G(P)$, a contradiction. Hence $P \not\leq Q$. Let

$$\{C_1, \ldots, C_m\} = \{C \in \mathrm{cl}(H) \mid P \not\leq Q \text{ for } Q \in \mathrm{Def}_p(C)\,\}.$$

Then $\epsilon_B - \epsilon = \sum_{i=1}^m a_i C_i^+ + \pi z$ with $a_i \in R$ and $z \in \mathbf{C}_{RG}(H)$. For any indecomposable direct summand U of $(\epsilon_B - \epsilon)V_H$ we have $\lambda_{\epsilon_B - \epsilon} = \mathrm{id}_U$, where $\lambda_x \colon U \to U$, $u \mapsto xu$ is again the left multiplication with $x \in RH$. Since $\mathrm{End}_{RH}\, U$ is local, there must be an $i \in \{1, \ldots, m\}$ such that $\lambda_{C_i^+} \in \mathrm{End}_{RH}\, U$ is a unit. As in the proof of Theorem 4.8.15, we see that U is $\mathbf{C}_H(h)$-projective for $h \in C_i$ and hence Q-projective for $Q \in \mathrm{Def}_p(C_i)$. $\qquad\square$

In Section 4.3 we discussed properties of the characters of projective RG-lattices. In order to generalize some of these to characters of Q-projective RG-lattices for $Q \leq G$, we will use the following important theorem, which can also be viewed as a far reaching generalization of the observation made in Remark 4.8.11(a).

Theorem 4.8.18 (Green's indecomposability theorem) *Let $H \trianglelefteq G$ and let G/H be a p-group. Assume that F is algebraically closed and that W an indecomposable RH-module. Then W^G is also indecomposable.*

Proof. See [41], p. 466. $\qquad\square$

Theorem 4.8.19 *Let $Q \leq P \in \mathrm{Syl}_p(G)$ and let V be a Q-projective RG-lattice with character χ. Then*

(a) $[P : Q] \mid \chi(1)$;

(b) *if $g \in G$ with $g_p \nleq_G Q$ then $\chi(g) = 0$.*

Proof. We may assume without loss of generality that F is algebraically closed and that V is indecomposable.

(a) By Theorem 4.8.4 and Mackey's theorem (Theorem 3.2.17) we have

$$V_P \mid (V_Q{}^G)_P = \bigoplus_{g \in T} ((g \otimes V_Q)_{{}^gQ \cap P})^P = \bigoplus_{g \in T} \bigoplus_{i \in I_g} Y_{g,i}{}^P,$$

where T is defined by $G = \dot{\bigcup}_{g \in T} PgQ$ and

$$(g \otimes V_Q)_{{}^gQ \cap P} = \bigoplus_{i \in I_g} Y_{g,i} \qquad \text{with } Y_{g,i} \text{ indecomposable}.$$

By Theorem 4.8.18 the induced modules $Y_{g,i}{}^P$ are indecomposable, so V_P is a direct sum of some of these modules, which all have R-rank divisible by $[P : Q]$, hence $[P : Q] \mid \chi(1)$.

(b) Since V is Q-projective, $V \mid V_Q{}^G$, hence using Mackey's theorem (Theorem 3.2.17) we get

$$V_{\langle g \rangle} \mid (V_Q{}^G)_{\langle g \rangle} = \bigoplus_{t \in T} ((t \otimes V_Q)_{{}^tQ \cap \langle g \rangle})^{\langle g \rangle}, \tag{4.34}$$

where $G = \dot{\bigcup}_{t \in T} \langle g \rangle tQ$. By assumption $g_p \nleq_G Q$, so

$${}^tQ \cap \langle g \rangle \lneq \langle g_p \rangle \in \mathrm{Syl}_p(\langle g \rangle), \qquad \text{hence} \qquad {}^tQ \cap \langle g \rangle \leq \langle g_p{}^p \rangle \leq \langle g^p \rangle.$$

Consequently

$$((t \otimes V_Q)_{{}^tQ \cap \langle g \rangle})^{\langle g \rangle} = (((t \otimes V_Q)_{{}^tQ \cap \langle g \rangle})^{\langle g^p \rangle})^{\langle g \rangle}.$$

From (4.34) we see $V_{\langle g \rangle} \mid \bigoplus_{i=1}^n Y_i{}^{\langle g \rangle}$ for some indecomposable $R\langle g^p \rangle$-lattices Y_i. By Theorem 4.8.18 the modules $Y_i{}^{\langle g \rangle}$ are indecomposable and $V_{\langle g \rangle}$ is a direct sum of some of these. Consequently $\chi|_{\langle g \rangle}$ is a sum of some induced characters $\theta_i{}^{\langle g \rangle}$, where θ_i is the character of Y_i. But $\theta_i{}^{\langle g \rangle}(g) = 0$ for all i, since $g \notin \langle g^p \rangle$, hence $\chi(g) = 0$. $\qquad\square$

Corollary 4.8.20 *Let B be a p-block of G with defect group D. Then there is an indecomposable RG-lattice belonging to B with vertex D.*

Proof. For $\chi \in \mathrm{Irr}(B)$ let V_χ be an RG-lattice such that KV_χ affords χ. Then V_χ is indecomposable and belongs to B. By Theorem 4.8.15 V_χ has a vertex Q_χ contained in D. By Theorem 4.8.19(a)

$$[P : Q_\chi] = [P : D][D : Q_\chi] \mid \chi(1) = [P : D] p^{\mathrm{ht}_p(\chi)} q,$$

with $P \in \mathrm{Syl}_p(G)$ and $p \nmid q$. So $Q_\chi = D$ if χ has height zero. $\qquad\square$

Exercises

Exercise 4.8.1 Give a Proof of Lemma 4.8.3.

Exercise 4.8.2 Let $P = \langle g \rangle$ be a cyclic p-subgroup of G and let W be an indecomposable FP-module with $\dim_F W = n$. Show that W^G has at most $[G : P]$ indecomposable direct summands.
Hint: Let $\delta\colon P \to \mathrm{GL}(W)$ be the representation afforded by W. Show that the Jordan normal form of $\delta^G(g)$ has exactly $[G : P]$ Jordan blocks. On the other hand, show that $\delta^G(g)$ has at least r Jordan blocks if $W^G \cong V_1 \oplus \cdots \oplus V_r$ for some FG-modules V_i.

Exercise 4.8.3 Let F be an arbitrary field and let $N_n \in F^{n \times n}$ be the matrix with $(i, i-1)$-entry 1 (for $i = 2, \ldots, n$) and zeros elsewhere. Put

$$A_n := \begin{bmatrix} \mathbf{I}_n & \mathbf{0}_n \\ \mathbf{I}_n & \mathbf{I}_n \end{bmatrix} \quad \text{and} \quad B_n := \begin{bmatrix} \mathbf{I}_n & \mathbf{0}_n \\ N_n & \mathbf{I}_n \end{bmatrix}.$$

(a) Let $X \in F^{2n \times 2n}$ be a matrix which commutes with A_n and B_n. Show that X is of the form $X = \begin{bmatrix} Y & \mathbf{0}_n \\ Z & Y \end{bmatrix}$ with $Y, Z \in F^{n \times n}$ satisfying $Y N_n = N_n Y$.

(b) Show that $Y \in F^{n \times n}$ satisfies $Y N_n = N_n Y$ if and only if

$$Y = \begin{bmatrix} y_1 & 0 & 0 & \cdots & 0 \\ y_2 & y_1 & 0 & \cdots & 0 \\ y_3 & y_2 & y_1 & \ddots & \vdots \\ \vdots & \vdots & \ddots & \ddots & 0 \\ y_n & y_{n-1} & \cdots & y_2 & y_1 \end{bmatrix}.$$

(c) Now let char $F = p$ and $G \cong C_p \times C_p$ be an elementary abelian group with generators a, b. Show that $a \mapsto A_n$, $b \mapsto B_n$ defines an indecomposable representation $\delta_n\colon G \to F^{2n \times 2n}$.
Hint: Use (a) and (b) to compute the endomorphism ring of the representation module.

Exercise 4.8.4 Let $Q := \langle (1, 2, 3) \rangle \leq G := S_4$ and let W_i be an indecomposable $\mathbb{F}_3 Q$-module of dimension i for $i = 1, 2, 3$.

(a) Use GAP to write each W_i^G as a sum of indecomposable $\mathbb{F}_3 G$-modules.

(b) Show there are exactly eight indecomposable $\mathbb{F}_3 G$-modules up to isomorphism.

Exercise 4.8.5 Let P be a normal p-subgroup of G with canonical projection $\pi_P\colon G \to \bar{G} := G/P$.

(a) If $\bar{W}$ is a projective indecomposable $S\bar{G}$-module and $W := \mathrm{Inf}_{\pi_P}(\bar{W})$ then $\mathrm{Vtx}(W) = \{P\}$ and S_P is the source of W.

(b) Conversely, if W is an indecomposable SG-module with vertex P and trivial source S_P, then $W = \mathrm{Inf}_{\pi_P}(\bar{W})$ for a projective indecomposable $S\bar{G}$-module $\bar{W}$.

4.9   Green correspondence

As in Section 4.8 let G be a finite group and let S be a field of characteristic $p > 0$ or a complete discrete valuation ring with $\mathrm{char}(S/\,\mathrm{J}(S)) = p > 0$; (K, R, F, η) is assumed to be a p-modular splitting system for G. All SG-modules considered are assumed to be free and finitely generated as S-modules. In addition we fix the following notation for this section. Let P be a p-subgroup of G and let H satisfy

$$\mathbf{N}_G(P) \leq H \leq G.$$

We put

$$
\begin{aligned}
\mathcal{X} := \mathcal{X}(P, H, G) &:= \{P^g \cap P \mid g \in G \setminus H\}, \\
\mathcal{Y} := \mathcal{Y}(P, H, G) &:= \{P^g \cap H \mid g \in G \setminus H\}, \\
\mathcal{A} := \mathcal{A}(P, H, G) &:= \{Q \leq P \mid Q \nleq_G \mathcal{X}\}.
\end{aligned}
$$

Remark 4.9.1 Recall that $Q \nleq_G \mathcal{X}$ means that Q is not conjugate in G to any subgroup of a group in $\mathcal{X}$. Obviously $P \in \mathcal{A}$, because the groups in $\mathcal{X}$ are all proper subgroups of P. Also, a vertex of an indecomposable direct summand of an $\mathcal{X}$-projective SG-module cannot be in $\mathcal{A}$. Similarly, a vertex of an indecomposable direct summand of a $\mathcal{Y}$-projective SH-module, which is also P-projective, cannot be in $\mathcal{A}$.

We will denote the isomorphism class of an SG-module V by $[V]$ and denote the set of isomorphism classes of indecomposable SG-modules by $\mathcal{L}(SG)$, and similarly for SH-modules.

Theorem 4.9.2 (Green) *For every indecomposable SG-module V (SH-module W) with a vertex in $\mathcal{A}$ there is a unique indecomposable SH-module $\mathbf{f}(V)$ (respectively SG-module $\mathbf{f}^{-1}(W)$) with*

$$
\begin{aligned}
V_H &= \mathbf{f}(V) \oplus W' && \text{with} && W' && \mathcal{Y}\text{-projective,} && (4.35) \\
W^G &= \mathbf{f}^{-1}(W) \oplus V' && \text{with} && V' && \mathcal{X}\text{-projective.} && (4.36)
\end{aligned}
$$

The map $[V] \mapsto [\mathbf{f}(V)]$ *gives a bijection*

$$[\mathbf{f}] \colon \{[V] \in \mathcal{L}(SG) \mid \mathrm{Vtx}(V) \cap \mathcal{A} \neq \emptyset\} \to \{[W] \in \mathcal{L}(SH) \mid \mathrm{Vtx}(W) \cap \mathcal{A} \neq \emptyset\}$$

with inverse $[W] \mapsto [\mathbf{f}^{-1}(W)]$. *The modules $\mathbf{f}(V)$ and $\mathbf{f}^{-1}(W)$ are called the* **Green correspondents** *of V and W, respectively. Corresponding modules have a vertex in common and also a common source; $\mathbf{f}$ is called the* **Green correspondence** *w.r.t. (P, H, G).*

By the remarks preceding the theorem, $\mathbf{f}(V)$ and $\mathbf{f}^{-1}(W)$ are the only indecomposable direct summands of V_H and W^G, respectively, having a vertex in $\mathcal{A}$.

Before entering the proof, we need a few lemmas, where we always retain the above hypotheses.

Lemma 4.9.3 *For $Q \leq P$ the following conditions are equivalent:*

(a) $Q \leq_G \mathcal{X}$;　　　(b) $Q \leq_H \mathcal{X}$;　　　(c) $Q \leq_H \mathcal{Y}$.

Proof. (a) $\Rightarrow$ (b) $Q \leq_G \mathcal{X}$ means that there are $g \in G$, $x \in G \setminus H$ with $Q^g \leq P^x \cap P$. If g happens to be in H, then $Q \leq_H \mathcal{X}$. Otherwise $Q \leq P^{g^{-1}} \cap P \in \mathcal{X}$, because $Q \leq P$ by assumption.

(b) $\Rightarrow$ (c) This is obvious, since the groups of $\mathcal{X}$ are all subgroups of those in $\mathcal{Y}$.

(c) $\Rightarrow$ (a) Assume that $Q^h \leq P^g \cap H$ for $h \in H$, $g \in G \setminus H$. Then $Q \leq P^{gh^{-1}} \cap P \in \mathcal{X}$, because $gh^{-1} \notin H$. $\qquad\square$

Lemma 4.9.4 (a) *If W is a P-projective SH-module, then*

$$(W^G)_H \cong W \oplus Y \qquad \text{with} \qquad Y \text{ a } \mathcal{Y}\text{-projective } SH\text{-module.}$$

(b) *If V is a P-projective SG-module, then V is $\mathcal{X}$-projective if and only if V_H is $\mathcal{Y}$-projective.*

Proof. (a) Since W is P-projective, there is an SP-module M and an SH-module W_1 with $M^H \cong W \oplus W_1$. By Mackey's theorem (Theorem 3.2.17) we have

$$(W^G)_H \cong W \oplus Y \qquad \text{and} \qquad (W_1^G)_H \cong W_1 \oplus Y_1,$$

with Y, Y_1 being certain SH-modules. We then get

$$(M^G)_H \cong ((M^H)^G)_H \cong (W^G)_H \oplus (W_1^G)_H \cong M^H \oplus Y \oplus Y_1$$

and, on the other hand,

$$(M^G)_H \cong M^H \oplus \bigoplus_{g \in T \setminus H} ((g \otimes M)_{gPg^{-1} \cap H})^H,$$

where $G = \bigcup_{g \in T} PgH$. For $g \in T \setminus H$ we have $gPg^{-1} \cap H \in \mathcal{Y}$. Thus $Y \oplus Y_1$ and hence Y is $\mathcal{Y}$-projective.

(b) We may assume without loss of generality that V is indecomposable and has a vertex $Q \leq P$. We first assume that V is $\mathcal{X}$-projective, that is $Q \leq_G \mathcal{X}$. By Lemma 4.8.9 there is an indecomposable SH-module W with

$$V \mid W^G \qquad \text{and} \qquad Q \in \mathrm{Vtx}(W).$$

By Lemma 4.9.3 $Q \leq_H \mathcal{Y}$. Then $V_H \mid (W^G)_H = W \oplus Y$, where Y is $\mathcal{Y}$-projective by part (a). Hence V_H is $\mathcal{Y}$-projective.

Conversely, assume that V_H is $\mathcal{Y}$-projective. Again by Lemma 4.8.9 there is an indecomposable SH-module W with

$$W \mid V_H \qquad \text{and} \qquad Q \in \mathrm{Vtx}(W).$$

Therefore $Q \leq_H \mathcal{Y}$, and by Lemma 4.9.3 we also have $Q \leq_G \mathcal{X}$. Hence V is $\mathcal{X}$-projective. $\qquad\square$

Proof of Theorem 4.9.2. Let V be an indecomposable SG-module with vertex $Q \in \mathcal{A}$. By Lemma 4.8.9 there is an indecomposable SH-module W with

$$V \mid W^G \qquad \text{and} \qquad Q \in \text{Vtx}(W).$$

We have

$$V_H \mid (W^G)_H \cong W \oplus Y' \qquad \text{with} \qquad Y' \; \mathcal{Y}\text{-projective}$$

by Lemma 4.9.4. If $W \nmid V_H$ then V_H would be $\mathcal{Y}$-projective, so by Lemma 4.9.4(b) V would be $\mathcal{X}$-projective, contrary to our assumption. Hence

$$V_H \cong W \oplus W' \qquad \text{with} \qquad W' \mid Y' \; \mathcal{Y}\text{-projective}.$$

Observe that W is the only direct summand of V_H with vertex Q, in fact, the only one with vertex in $\mathcal{A}$ (by Remark 4.9.1). Thus we may put $\mathbf{f}(V) := W$. Then V and $\mathbf{f}(V)$ have the same vertex and the same source, because $V \mid \mathbf{f}(V)^G$ and (4.35) holds.

Let $W^G \cong V \oplus V'$, so

$$W \oplus Y' \cong (W^G)_H = V_H \oplus V'_H = W \oplus W' \oplus V'_H.$$

Since Y' is $\mathcal{Y}$-projective, so is V'_H. By Lemma 4.9.4(b) V' is $\mathcal{X}$-projective. Thus V is the only indecomposable direct summand of $\mathbf{f}(V)^G$ with a vertex in $\mathcal{A}$, which shows that $[\mathbf{f}]$ is injective. Writing again W instead of $\mathbf{f}(V)$ and putting $\mathbf{f}^{-1}(W) := V$ we obtain (4.36).

To show that $[\mathbf{f}]$ is also surjective, let W be an arbitrary indecomposable SH-module with vertex $Q \in \mathcal{A}$. Since $W \mid (W^G)_H$, there is an indecomposable direct summand V of W^G with $W \mid V_H$. Hence V is Q-projective and, on the other hand, by Theorem 4.8.10 $Q \leq_G Q' \in \text{Vtx}(V)$. Thus $Q \in \text{Vtx}(V)$. Since by the first part of the proof we know that $\mathbf{f}(V)$ is the only indecomposable direct summand of V_H with a vertex in $\mathcal{A}$, we conclude that $W \cong \mathbf{f}(V)$. $\qquad\square$

The following theorem shows that Brauer correspondence and Green correspondence fit together nicely.

Theorem 4.9.5 *Assume that (K, R, F, η) is a p-modular splitting system for G and $S \in \{R, F\}$. Let $N = \mathbf{N}_G(P)$ and let $\mathbf{f}$ be the Green correspondence w.r.t. (P, N, G). If V is an indecomposable SG-module belonging to $B \in \text{Bl}_p(G)$ with a vertex in $\mathcal{A}(P, N, G)$, then $\mathbf{f}(V)$ belongs to a block $b \in \text{Bl}_p(N)$ with $b^G = B$.*

Proof. Assume that $W := \mathbf{f}(V)$ belongs to $b \in \text{Bl}_p(N)$. If $D \in \text{Def}(b)$ then by Theorem 4.6.10 $P \leq D$, and hence $\mathbf{C}_G(D) \leq \mathbf{C}_G(P) \leq N$. By Corollary 4.7.23 $B' := b^G$ is defined. By Theorem 4.7.10 there is an indecomposable SG-module V' belonging to B' with $V' \mid W^G$ and $W \mid V'_H$. Hence V' and W have a common vertex, and by Theorem 4.9.2 $V' \cong \mathbf{f}^{-1}(W) \cong V$. So $B' = B$. $\qquad\square$

The Green correspondence is particularly simple and useful when P is a **TI-subgroup** in G, that is, if

$$P \cap P^g = \{1\} \qquad \text{for all} \qquad g \in G \setminus \mathbf{N}_G(P),$$

because in this case "$\mathcal{X}$-projective" simply means "projective."

Lemma 4.9.6 *Let P be a TI-p-subgroup of G and let $H = \mathbf{N}_G(P)$. Let $\mathbf{f}$ be the Green correspondence with respect to (P, H, G) and let V_1, V_2 be non-projective indecomposable SG-modules with vertices in P. If there is a non-split short exact sequence of SH-modules*

$$\{0\} \longrightarrow \mathbf{f}(V_1) \longrightarrow W \longrightarrow \mathbf{f}(V_2) \longrightarrow \{0\}, \tag{4.37}$$

then there is a non-split short exact sequence

$$\{0\} \longrightarrow V_1 \longrightarrow V \longrightarrow V_2 \longrightarrow \{0\} \tag{4.38}$$

of SG-modules.

Proof. Assume that there is a non-split short exact sequence (4.37) and put $W_i := \mathbf{f}(V_i)$ for $i = 1, 2$. We have $W_i^G \cong V_i \oplus X_i$ with projective SG-modules X_i for $i = 1, 2$. By Lemma 3.2.21(b) we may embed W_1^G into W^G for notational convenience, and we have $W^G/W_1^G \cong_{SG} W_2^G$. Let U be a submodule of W^G with $W^G/U \cong X_2$ and $U/W_1^G \cong V_2$. We claim that

$$\{0\} \longrightarrow W_1^G/X_1 \longrightarrow U/X_1 \longrightarrow V_2 \longrightarrow \{0\} \tag{4.39}$$

is non-split. Since $W_1^G/X_1 \cong V_1$ naturally, this proves the first assertion. To prove the claim, assume that W_1^G/X_1 has a complement U_1/X_1 in U/X_1. Since X_1 is projective, X_1 has a complement U_2 in U_1 by Theorem 1.6.27(d). The short exact sequence

$$\{0\} \longrightarrow U/U_2 \longrightarrow W^G/U_2 \longrightarrow X_2 \longrightarrow \{0\}$$

is split, since X_2 is projective. Hence there is a submodule U_3 such that U_3/U_2 is a complement to U/U_2 in W^G/U_2. So U_3 is a complement to W_1^G in W^G, and

$$\{0\} \longrightarrow W_1^G \longrightarrow W^G \longrightarrow W_2^G \longrightarrow \{0\}$$

is split, contrary to Lemma 3.2.21(c). $\qquad\qquad\square$

Theorem 4.9.7 *Let B be a p-block of G with defect group P, a TI-subgroup in G, and $H = \mathbf{N}_G(P)$. Then the Green correspondence $\mathbf{f}$ with respect to (P, H, G) induces a bijection between the isomorphism classes of non-projective indecomposble SG-modules belonging to B and those of SH belonging to b, the Brauer correspondent of B. If V is a non-projective indecomposable SG-module belonging to B and $W := \mathbf{f}(V)$, then*

$$V_H = W \oplus Y \qquad and \qquad W^G = V \oplus X \tag{4.40}$$

with a projective SG-module X and an SH-module Y which is a direct sum of projective modules and indecomposable modules not belonging to b. Furthermore, if V_1, V_2 are non-projective indecomposable SG-modules in B, then

$$\overline{\mathrm{Hom}}_{SG}(V_1, V_2) \cong_S \overline{\mathrm{Hom}}_{SH}(\mathbf{f}(V_1), \mathbf{f}(V_2)).$$

Proof. By assumption $\mathcal{X} := \{\{1\}\}$, $\mathcal{A} = \{Q \leq P \mid Q \neq \{1\}\}$. By Theorem 4.8.15 the vertex of any indecomposable SH-module W' in b is contained in P; since P is a TI-subgroup, W' cannot be $\mathcal{Y}$-projective unless it is projective. Put $W_i := \mathbf{f}(V_i)$ $(i = 1, 2)$, $W_1^G = V_1 \oplus X_1$ and $(V_2)_H = W_2 \oplus Y_2$ with X_1 projective and Y_2 $\mathcal{Y}$-projective as in (4.40). Then $\overline{\mathrm{Hom}}_{SG}(X_1, V_2) = \{0\}$ and $\overline{\mathrm{Hom}}_{SH}(W_1, Y_2) = \{0\}$, so

$$\begin{aligned}
\overline{\mathrm{Hom}}_{SG}(V_1, V_2) &\cong_S \overline{\mathrm{Hom}}_{SG}(V_1 \oplus X_1, V_2) \cong_S \overline{\mathrm{Hom}}_{SG}(W_1^G, V_2) \\
&\cong_S \overline{\mathrm{Hom}}_{SH}(W_1, (V_2)_H) \cong_S \overline{\mathrm{Hom}}_{SH}(W_1, W_2 \oplus Y_2) \\
&\cong_S \overline{\mathrm{Hom}}_{SH}(W_1, W_2)
\end{aligned}$$

using Corollary 3.2.22. $\square$

In Section 4.8 (Theorem 4.8.14) we investigated the simple and indecomposable FG-modules of a p-block b of defect one for the case where a defect group of b is normal in G. The Green correspondence allows us to lift some of the information to blocks of defect one without the assumption that a defect group is normal.

Theorem 4.9.8 *Let $B \in \mathrm{Bl}_p(G)$ be a block of defect one with $P \in \mathrm{Def}(B)$ and $H := \mathbf{N}_G(P)$. Let $b \in \mathrm{Bl}_p(H)$ be the unique block with $b^G = B$ and let $Y_1, \ldots, Y_e$ be the simple FH-modules belonging to b (up to isomorphism). Then there are exactly e simple FG-modules $V_1, \ldots, V_e$ in B which may be ordered such that*

$$\mathbf{f}(V_i)/\mathrm{Rad}(\mathbf{f}(V_i)) \cong_{FH} Y_i \qquad and \qquad \mathrm{Soc}(\mathbf{f}(V_i)) \cong_{FH} Y_{\pi(i)},$$

for some $\pi \in \mathrm{S}_e$, where $\mathbf{f}$ is the Green correspondence w.r.t. (P, H, G). In particular

$$\mathrm{l}(B) = \mathrm{l}(b). \tag{4.41}$$

Proof. Let U_{ij} be the factor module of $P(Y_i)$ of length j for $1 \leq i \leq e$, $1 \leq j \leq p$. By Theorem 4.8.14 these are all the indecomposable FH-modules belonging to b up to isomorphism. The proof of this theorem also shows that U_{ij} is a homomorphic image of U_{ik} if $j \leq k$ and by Lemma 1.6.30 $\overline{\mathrm{Hom}}_{FH}(U_{ik}, U_{ij}) \neq \{0\}$ for $j \leq k < p$. On the other hand, let V, V' be non-isomorphic simple FG-modules belonging to B with $\mathrm{l}(\mathbf{f}(V)) \geq \mathrm{l}(\mathbf{f}(V'))$. By Theorem 4.9.5, $\mathbf{f}(V), \mathbf{f}(V')$ belong to b, and using Theorem 4.9.7 we conclude that

$$\{0\} = \overline{\mathrm{Hom}}_{FG}(V, V') \cong_F \overline{\mathrm{Hom}}_{FH}(\mathbf{f}(V), \mathbf{f}(V')).$$

Thus $\mathbf{f}(V)/\mathrm{Rad}(\mathbf{f}(V)) \not\cong_{FH} \mathbf{f}(V')/\mathrm{Rad}(\mathbf{f}(V'))$. Similarly one sees that

$$\mathrm{Soc}(\mathbf{f}(V)) \not\cong_{FH} \mathrm{Soc}(\mathbf{f}(V')).$$

Now let $W_i := \mathbf{f}^{-1}(Y_i)$. By Theorem 4.9.5 W_i belongs to B. If V is a simple FG-module with $\mathrm{Hom}_{FG}(V, W_i) \neq \{0\}$ then, because of Lemma 1.6.30 and Theorem 4.9.7,

$$\{0\} \neq \overline{\mathrm{Hom}}_{FG}(V, W_i) \cong_F \overline{\mathrm{Hom}}_{FH}(\mathbf{f}(V), Y_i).$$

So every Y_i is a factor module of $\mathbf{f}(V)$ for some simple FG-module $V_i := V$ in B, which completes the proof of the theorem. $\qquad\square$

Remark 4.9.9 Equation (4.41) in Theorem 4.9.8 holds more generally if B is a block with cyclic defect group and b is its Brauer correspondent; see [1], Theorem 1, p. 123.

Example 4.9.10 Let $G := \mathrm{SL}_2(p)$ for a prime $p > 2$ and

$$P := \left\{ \begin{bmatrix} 1 & 0 \\ c & 1 \end{bmatrix} \mid c \in \mathbb{F}_p \right\}, \qquad H := \mathbf{N}_G(P) = \left\{ \begin{bmatrix} a & 0 \\ c & a^{-1} \end{bmatrix} \mid a \in \mathbb{F}_p^\times, c \in \mathbb{F}_p \right\}.$$

Observe that $P \in \mathrm{Syl}_p(G)$. For $i \in \{0, \ldots, p-1\}$ let Y_i be the one-dimensional $\mathbb{F}_p H$-module affording the representation

$$H \to \mathbb{F}_p, \qquad \begin{bmatrix} a & 0 \\ c & a^{-1} \end{bmatrix} \mapsto a^i.$$

Then $Y_0 = \mathbb{F}_p$, the trivial module, and $Y_i \otimes Y_j \cong Y_{i+j}$ if we take the indices modulo $p - 1$.

In Exercise 1.3.4 we saw that G has simple $\mathbb{F}_p G$-modules V_i of dimension i for $i = 1, \ldots, p$. In fact,

$$V_i = \langle X^{i-1}, X^{i-2}Y, \ldots Y^{i-1} \rangle_{\mathbb{F}_p} \subseteq \mathbb{F}_p[X, Y].$$

Recall that each $g := \begin{bmatrix} a & b \\ c & d \end{bmatrix} \in G$ acts as an algebra automorphism on $\mathbb{F}_p[X, Y]$ with

$$g \cdot X = aX + cY, \qquad g \cdot Y = bX + dY.$$

Let

$$g_s := \begin{bmatrix} a & 0 \\ 0 & a^{-1} \end{bmatrix} \in H \text{ and } g_u := \begin{bmatrix} 1 & 0 \\ c & 1 \end{bmatrix} \in P \text{ with } a, c \in \mathbb{F}_p^\times.$$

Let $\delta_i : G \to \mathrm{GL}_i(\mathbb{F}_p)$ be the matrix representation afforded by V_i with respect to the indicated basis. Then it is easily verified that

$$\delta_i(g_s) = \begin{bmatrix} a^{i-1} & 0 & \cdots & 0 \\ 0 & a^{i-3} & & 0 \\ \vdots & & \ddots & \\ 0 & 0 & & a^{-i+1} \end{bmatrix}, \quad \delta_i(g_u) = \begin{bmatrix} 1 & 0 & \cdots & 0 \\ d_2 & 1 & & 0 \\ * & d_3 & 1 & 0 \\ \vdots & & \ddots & \ddots \\ * & \cdots & * & d_i & 1 \end{bmatrix},$$

with $d_j := (i - j + 1)c \neq 0$ for $2 \leq j \leq i$. By Example 1.1.23 $(V_i)_P$ is indecomposable, and Theorem 4.8.14 shows that $V_{iH} := (V_i)_H$ is uniserial. Moreover, $\mathrm{Rad}(V_{iH}) = \langle X^{i-2}Y, \ldots, Y^{i-1} \rangle_{\mathbb{F}_p}$. Hence

$$(V_i)_H / \mathrm{Rad}(V_{iH}) \cong Y_{i-1} \qquad \text{and} \qquad \mathrm{Rad}(V_{iH}) / \mathrm{Rad}^2(V_{iH}) \cong_{FH} Y_{i-3},$$

where the indices should be read modulo $p - 1$. Since V_{iH} is a homomorphic image of $P(Y_{i-1})$, Theorem 4.8.14 shows that for any $i \in \{0, \ldots, p - 1\}$ the composition factors of $P(Y_i)$ are given by

$$Y_i, Y_{i-2}, \ldots, Y_{i-2(p-2)}, Y_{i-2(p-1)} = Y_i,$$

taking indices modulo $p - 1$.

Now Green correspondence comes into play, in order to lift the information we have about about $\mathbb{F}_p H$-modules to information about $\mathbb{F}_p G$-modules. For $i < p$ the simple $\mathbb{F}_p G$-module V_i is not projective, so $P \in \mathrm{Vtx}(V_i)$. Let $\mathbf{f}$ be the Green correspondence w.r.t. (P, H, G). Then $\mathbf{f}(V_i) = V_{iH}$.

Since $V_{iH} / \mathrm{Rad}(V_{iH}) \cong_{\mathbb{F}_p H} Y_{i-1}$, we conclude that V_{iH} is a homomorphic image of the uniserial projective cover $P(Y_{i-1})$, thus

$$V_{iH} \cong_{\mathbb{F}_p H} P(Y_{i-1}) / \mathrm{Rad}^i(P(Y_{i-1})).$$

$\mathrm{Rad}^i(P(Y_{i-1}))$ has head (top composition factor) Y_{p-i-2}, and hence is a homomorphic image of $P(Y_{p-i-2})$, and

$$\begin{aligned}(V_{p-i-1})_H &\cong_{\mathbb{F}_p H} P(Y_{p-i-2}) / \mathrm{Rad}^{p-i-1}(P(Y_{p-i-2})) \\ &\cong_{\mathbb{F}_p H} \mathrm{Rad}^i(P(Y_{i-1})) / \mathrm{Soc}(P(Y_{i-1})).\end{aligned}$$

Hence we have a non-split short exact sequence

$$\{0\} \longrightarrow (V_{p-i-1})_H \longrightarrow P(Y_{i-1}) / \mathrm{Soc}(P(Y_{i-1})) \longrightarrow (V_i)_H \longrightarrow \{0\}.$$

By Theorem 4.9.6 it follows that there is a non-split exact sequence of $\mathbb{F}_p G$-modules

$$\{0\} \longrightarrow V_{p-i-1} \longrightarrow V^{(i)} \longrightarrow V_i \longrightarrow \{0\}$$

for $1 \leq i \leq p - 1$. We obtain another non-split short exact sequence of $\mathbb{F}_p H$-modules by considering $W := P(Y_{i-1}) \oplus Y_{p-i}$. In fact, $\mathrm{Rad}^{i-1}(P(Y_{i-1})) \oplus Y_{p-i}$ contains $p - 1$ "diagonal" uniserial maximal submodules U such that

$$W / U \cong_{\mathbb{F}_p H} P(Y_{i-1}) / \mathrm{Rad}^i(P(Y_{i-1})) \cong V_{iH}.$$

The module U, having head Y_{p-i} and length $p - i + 1$, is isomorphic to

$$P(Y_{p-i}) / \mathrm{Rad}^{p-i+1}(P(Y_{p-i})) \cong_{\mathbb{F}_p H} (V_{p-i+1})_H.$$

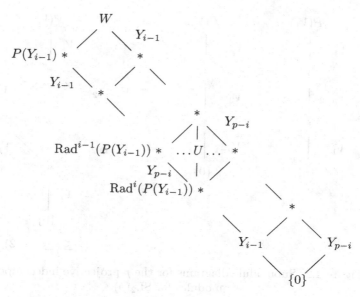

Figure 4.1. Non-split extension of $(V_{p-i+1})_H$ by V_{iH}.

We get a non-split exact sequence (see Figure 4.1):

$$\{0\} \longrightarrow (V_{p-i+1})_H \longrightarrow W \longrightarrow V_{iH} \longrightarrow \{0\}.$$

Using Theorem 4.9.6 again we obtain a non-split exact sequence of $\mathbb{F}_pG$-modules:

$$\{0\} \longrightarrow V_{p-i+1} \longrightarrow V_{\bullet}^{(i)} \longrightarrow V_i \longrightarrow \{0\} \qquad \text{for} \quad 1 \le i \le p-1.$$

Both $\mathbb{F}_pG$-modules $V^{(i)}$ and $V_{\bullet}^{(i)}$ must be homomorphic images of $P(V_i)$. We conclude that

$$\dim P(V_i) \ge 2p \qquad \text{for} \quad i = 2, \dots, p-2.$$

By Theorem 1.6.27 it is clear that $\dim V_{p-1} \ge 2p$. But by Theorem 1.6.24

$$p(p^2 - 1) = \dim \mathbb{F}_pG$$

$$= \sum_{i=1}^{p} \dim V_i \dim P(V_i) \ge p + (2 + \cdots + p - 1)2p + p^2 = p(p^2 - 1).$$

Hence $\dim P(V_1) = \dim V_p = p$ and $\dim P(V_i) = 2p$ for $i = 2, \dots, p-2$, and we have determined the complete submodule structure of all $P(V_i)$ (see Figure 4.2) and thereby of $\mathbb{F}_pG$:

♦

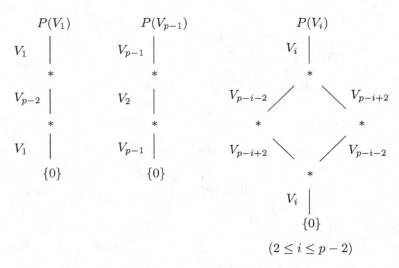

Figure 4.2. Submodule diagrams for the p-projective indecomposable modules for $\mathrm{SL}_2(p)$.

Exercises

Exercise 4.9.1 Show that an SG-module V is H-projective for some $H \leq G$ if and only if each short exact sequence of SG-homomorphisms

$$\{0\} \longrightarrow V_1 \xrightarrow{\alpha} V \xrightarrow{\beta} V_2 \longrightarrow \{0\},$$

for which the sequence of SH-homomorphisms

$$\{0\} \longrightarrow V_{1H} \xrightarrow{\alpha} V_H \xrightarrow{\beta} V_{2H} \longrightarrow \{0\}$$

splits, is itself split.

Exercise 4.9.2 Let $H \leq G$ with $p \nmid [G : H]$. Assume that V is an SG-module such that V_H is semisimple. Show that V itself must be semisimple.

4.10 Trivial source modules

As in the preceding sections, let $S \in \{R, F\}$, where (K, R, F, η) is a p-modular splitting system for the finite group G. We also assume that K and F are splitting fields for all subgroups of G. All modules considered are assumed to be free and finitely generated as S-modules.

Definition 4.10.1 An SG-module V is called a **trivial source module** if it is a direct sum of indecomposable SG-modules V_i having trivial sources S_{Q_i} (with $Q_i \in \mathrm{Vtx}(V_i)$).

The following lemma gives a different characterization of trivial source modules.

Lemma 4.10.2 *Let P be a p-subgroup of G and $N := \mathbf{N}_G(P)$. Then*

(a) *An SG-module V is a trivial source module if and only if it is a direct summand of a permutation module.*

(b) *If $P \trianglelefteq G$, then every indecomposable summand of the permutation module $(S_P)^G$ has vertex P.*

(c) *For each indecomposable trivial source SN-module W with $\mathrm{Vtx}(W) = \{P\}$ there exists an indecomposable SG-module V with $P \in \mathrm{Vtx}(V)$ and source S_P such that*

$$eV_N = W \oplus W',$$

with $P \notin \mathrm{Vtx}(W_i)$ for any indecomposable $W_i \mid W'$. If $P \leq H \leq N$ with $p \nmid [H : P]$, then

$$V_H = W_H \oplus W'_H,$$

where the indecomposable summands of W_H have vertex P, while those of W'_H have vertices strictly contained in P.

Proof. (a) It suffices to consider indecomposable modules. Let V be an indecomposable direct summand of $(S_H)^G$ for some $H \leq G$ and let $Q \in \mathrm{Vtx}(V)$ with $Q \leq H$. Then

$$V_Q \mid ((S_H)^G)_Q = \bigoplus_{g \in T} (S_{gHg^{-1} \cap Q})^Q \quad \text{if} \quad G = \dot{\bigcup}_{g \in T} QgH.$$

Thus all indecomposable direct summands of V_Q are trivial source modules, and by Theorem 4.8.10 (applied with $H := Q$) V is a trivial source module.

(b) Let $V \mid (S_P)^G$ be indecomposable and let $Q \in \mathrm{Vtx}(V)$ with $Q \leq P$. By Theorem 4.8.4 there is a $\theta \in \mathrm{End}_{SQ} V$ such that

$$\mathrm{Tr}_P^G \circ \mathrm{Tr}_Q^P(\theta) = \mathrm{Tr}_Q^G(\theta) = \mathrm{id}_V.$$

Since P acts trivially on V (Exercise 1.2.1), we have $\mathrm{Tr}_Q^P(\theta)(v) = [P : Q]\theta(v)$ for $v \in V$, thus $\mathrm{im}(\mathrm{Tr}_Q^G(\theta)) \subseteq [P : Q]V$, which is feasible only if $P = Q$.

(c) Let $V = \mathbf{f}^{-1}(W)$ be the Green correspondent of the given trivial source module W with respect to (G, N, P). Then the first assertion follows from Theorem 4.9.2, since $P \notin \mathcal{Y}$ in the notation of that theorem. Since $\mathrm{Syl}_p(H) = \{P\}$, a vertex of any direct summand of W'_H is contained in P (and in P^g for some $g \notin N$), and hence is strictly contained in P. If $W_0 \mid W_H$ is indecomposable, then

$$W_0 \mid (S_P)^N{}_H = \underbrace{(S_P)^H \oplus \cdots \oplus (S_P)^H}_{[N:H]}$$

by Mackey's theorem. So $\mathrm{Vtx}(W_0) = \{P\}$ by part (b). $\qquad\square$

A simple observation is given by the following corollary.

Corollary 4.10.3 *Let P be a p-subgroup of G. Then the following statements hold:*

(a) *The number of indecomposable trivial source FG-modules (up to isomorphism) with vertex P is equal to $|\mathrm{IBr}(\mathbf{N}_G(P)/P)|$.*

(b) *The total number of indecomposable trivial source FG-modules (up to isomorphism) is given by $\sum_Q |\mathrm{IBr}(\mathbf{N}_G(Q)/Q)|$, where Q runs through the p-subgroups of G up to G-conjugacy.*

Proof. Green correspondence, see Theorem 4.9.2, applied to $(G, \mathbf{N}_G(P), P)$ shows that there is a bijection between the indecomposable trivial source FG-modules having P as a vertex and the indecomposable trivial source $F\mathbf{N}_G(P)$-modules with vertex P. From Exercise 4.8.5 it follows that the indecomposable $F\mathbf{N}_G(P)$-modules with source F_P are the inflations to $\mathbf{N}_G(P)$ of the projective indecomposable $F\mathbf{N}_G(P)/P$-modules. Hence (a) and (b) follow immediately. $\qquad\square$

Definition 4.10.4 (a) Let $\mathcal{TS}_+(SG)$ be the set of isomorphism classes $[V]$ of trivial source SG-modules V. The abelian group with set of generators $\mathcal{TS}_+(SG)$ and defining relations

$$[V] + [W] - [V \oplus W] = 0 \qquad \text{for} \qquad [V], [W] \in \mathcal{TS}_+(SG)$$

will be denoted by $\mathcal{TS}(SG)$. It is a free group with $\mathbb{Z}$-basis consisting of the isomorphism classes of indecomposable trivial source modules. Since by Mackey's theorem the tensor product of trivial source modules is again a trivial source module, we can turn $\mathcal{TS}(SG)$ into a ring by defining a product

$$[V] \cdot [W] := [V \otimes_S W] \qquad \text{for} \qquad [V], [W] \in \mathcal{TS}_+(SG)$$

and $\mathbb{Z}$-linear extension.

(b) Let

$$\mathcal{C}_p(G) := \{H \leq G \mid H/\mathbf{O}_p(H) \text{ is a cyclic } p'\text{-group}\}.$$

Observe that any $H \in \mathcal{C}_p(G)$ is a split extension of $\mathbf{O}_p(H)$ by a cyclic p'-group by the Schur–Zassenhaus theorem.

(c) For $H \in \mathcal{C}_p(G)$ and $V \in \mathcal{TS}_+(FG)$ we write

$$V_H = (V_H)^{=P} \oplus (V_H)^{<P},$$

where each indecomposable direct summand of $(V_H)^{=P}$ has vertex $P := \mathbf{O}_p(H)$ and each indecomposable direct summand of $(V_H)^{<P}$ has vertex strictly contained in P. (Of course, it may happen that $(V_H)^{=P} = \{0\}$ or $(V_H)^{<P} = \{0\}$.) Since P acts trivially on $(V_H)^{=P}$ (see Exercise 1.2.1), $(V_H)^{=P}$ may be considered as an $F(H/P)$-module whose Brauer character we will denote by $\tilde{\varphi}_{V_H^{=P}}$. In fact, this is an ordinary character of H/P, since this is a p'-group. For $H \in \mathcal{C}_p(G)$ and $\tilde{c} \in H/\mathbf{O}_p(H)$ we define a $\mathbb{C}$-algebra homomorphism (see Corollary 3.2.16)

$$s_{H,\tilde{c}} : \mathbb{C} \otimes \mathcal{TS}(FG) \to \mathbb{C} \qquad \text{by} \qquad s_{H,\tilde{c}}([V]) := \tilde{\varphi}_{V_H^{=P}}(\tilde{c})$$

and linear extension; $s_{H,\tilde{c}}$ is called a **species**.

Remark 4.10.5 Let $H \in \mathcal{C}_p(G)$ with $P := \mathbf{O}_p(H)$ and $V := (F_U)^G$, the permutation module afforded by the G-set $\Omega := G/U$, for some $U \le G$. Then

$$V_H = \bigoplus_{g \in T} (F_{gUg^{-1} \cap H})^H, \qquad \text{where} \qquad G = \dot{\bigcup_{g \in T}} HgU.$$

A vertex of any indecomposable summand of $(F_{gUg^{-1} \cap H})^H$ is contained in $gUg^{-1} \cap P$. Hence

$$(V_H)^{=P} = \bigoplus_{g \in T_1} (F_{gUg^{-1} \cap H})^H \qquad \text{with} \qquad T_1 := \{g \in T \mid P \le gUg^{-1}\}.$$

Thus $(V_H)^{=P} = F \operatorname{Fix}_\Omega(P)$ is an FH-permutation module. Also, for $c \in H$ let $\tilde{c} := cP \in H/P$ and $H_c := \langle P, c \rangle \le H$. Then, writing $\Omega_P := \operatorname{Fix}_\Omega(P)$,

$$s_{H,\tilde{c}}([V]) = |\{\omega \in \Omega_P \mid c\omega = \omega\}| = |\{\omega \in \Omega \mid h\omega = \omega \text{ for all } h \in H_c\}|$$
$$= m_{G/U}(H_c), \tag{4.42}$$

where $m_{G/U}$ is the mark of the G-set G/U; see Definition 3.5.1. In particular, for $U = H_c$ we obtain from Lemma 3.5.3 $s_{H,\tilde{c}}([V]) = [\mathbf{N}_G(H_c) : H_c]$, so $s_{H,\tilde{c}}$ is certainly not the zero map.

Lemma 4.10.6 (a) *If $\hat{V}$ is a trivial source FG-module, then there exists a unique trivial source RG-lattice V such that $V/\pi V \cong \hat{V}$.*

(b) *V is an indecomposable trivial source RG-lattice if and only if $\hat{V} := V/\pi V$ is an indecomposable trivial source FG-module. Moreover, in this case* $\operatorname{vtx}(V) = \operatorname{vtx}(\hat{V})$.

Proof. (a) We assume without loss of generality that $\hat{V}$ is indecomposable. Let H be a subgroup of G such that $\hat{V}$ is isomorphic to a direct summand of the permutation FG-module $(F_H)^G$. Hence we can write $\hat{V} \cong (F_H)^G \hat{e}$ for an idempotent $\hat{e}$ in $E := \operatorname{End}_{FG}(F_H)^G$. Since

$$E \cong \operatorname{End}_{RG}(R_H)^G / \pi \operatorname{End}_{RG}(R_H)^G$$

by Exercise 4.1.4, we can lift $\hat{e}$ to an idempotent $e \in \operatorname{End}_{RG}(R_H)^G$, and $(R_H)^G e$ is an indecomposable RG-lattice with $(R_H)^G e / \pi (R_H)^G e \cong \hat{V}$. To show uniqueness assume that V_1 and V_2 are trivial source RG-lattices such that $\hat{V}_1 \cong \hat{V}_2 \cong \hat{V}$. Let $\hat{\varphi}$ be an FG-isomorphism from $\hat{V}_1$ onto $\hat{V}_2$. It follows again from Exercise 4.1.4 that there is an RG-homomorphism φ from V_1 to V_2 with $\widehat{(\varphi)} = \hat{\varphi}$. This implies that $\varphi(V_1) + \pi V_2 = V_2$, and we conclude that φ is surjective. As a homomorphism of RG-lattices of the same R-rank, φ has to be bijective.

(b) The first part of (b) follows easily from Exercise 4.1.4. For the second part, let P be a vertex of V. Since $V|(R_P)^G$, and hence $\hat{V}|(F_P)^G$, it follows that $\hat{V}$ is P-projective. On the other hand, if Q is a vertex of $\hat{V}$ then there is an FG-endomorphism $\hat{\varphi}$ of $\hat{V}$ such that $\operatorname{id}_{\hat{V}} = \operatorname{Tr}_Q^G(\hat{\varphi})$. Since $\hat{\varphi}$ can be lifted to an RG-endomorphism φ of V, we get that $\mathrm{id}_V \in \mathrm{Tr}_Q^G(\mathrm{End}_{RQ} V) + \pi \, \mathrm{End}_{RG} V$. This shows that

$$\mathrm{End}_{RG} V = \mathrm{Tr}_Q^G(\mathrm{End}_{RQ} V),$$

which means V is Q-projective. $\square$

An easy consequence of the lemma is the following.

Corollary 4.10.7 *Let* $\mathcal{T}\mathcal{R}_+(RG)$, *respectively* $\mathcal{T}\mathcal{F}_+(FG)$, *be the trivial source ring of trivial source RG-lattices, respectively trivial source FG-modules. Then the map* $\widehat{} \colon \mathcal{T}\mathcal{R}_+(RG) \to \mathcal{T}\mathcal{F}_+(FG)$ *defined by* $\widehat{[V]} := [V/\pi V]$ *for all trivial source RG-lattices V and extended $\mathbb{Z}$-linearly to* $\mathcal{T}\mathcal{S}_+(SG)$ *is an isomorphism of rings.*

Example 4.10.8 Let (K, R, F) be a 3-modular splitting system for $G := \mathrm{M}_{11}$. In the following we will describe the indecomposable trivial source RG-lattices and hence also the indecomposable trivial source FG-modules, since these lift uniquely to trivial source RG-lattices as stated in Lemma 4.10.6.

We refer to the table of marks of G which can be found in Example 3.5.24. We will denote representatives of the conjugacy classes of subgroups of G by $H_1, \ldots, H_{39}$, so that the i, j-entry in the table of marks is $|\mathrm{Fix}_{G/H_i}(H_j)|$. Up to conjugacy G has three 3-subgroups: $P_1 := H_1 = \{1\}$, $P_2 := H_3 \cong C_3$, $P_3 := H_{13} \cong C_3 \times C_3$. Indecomposable trivial source RG-modules with vertex $\{1\}$ are simply the projective indecomposable modules, whose ordinary characters can be obtained from the 3-decomposition matrix of G (see Example 4.4.18). In GAP their characters can be obtained as follows:

```
gap> ct := CharacterTable("M11");;  ctmod3 := ct mod 3;;
gap> projectives := Irr(ct) * DecompositionMatrix(ctmod3);;
```

The 3-projective indecomposable characters of M_{11} are as follows:

| $\lvert \mathbf{C}_G(g) \rvert$: | 7920 | 48 | 18 | 8 | 5 | 6 | 8 | 8 | 11 | 11 |
$G = \mathrm{M}_{11}$	1a	2a	3a	4a	5a	6a	8a	8b	11a	11b
Φ_1	99	3	.	-1	4	.	1	1	.	.
Φ_2	126	6	.	-2	1	.	.	.	β	$\bar{\beta}$
Φ_3	126	6	.	-2	1	.	.	.	$\bar{\beta}$	β
Φ_4	54	6	.	2	-1	.	.	.	-1	-1
Φ_5	81	-3	.	-1	1	.	α	$\bar{\alpha}$	γ	$\bar{\gamma}$
Φ_6	81	-3	.	-1	1	.	$\bar{\alpha}$	α	$\bar{\gamma}$	γ
Φ_7	99	3	.	-1	-1	.	1	1	.	.
Φ_8	45	-3	.	1	.	.	-1	-1	1	1

with $\alpha := 1 + \sqrt{-2}$, $\beta := \frac{-1+\sqrt{-11}}{2}$, $\gamma := \frac{-3+\sqrt{-11}}{2}$.

We want to investigate the Green correspondents of the other trivial source modules. We choose $P_2 := H_3$, the representative of the conjugacy class of subgroups of G of order three stored in the GAP table of marks of G (obtained via "RepresentativeTom(t,3)"). Then $N_2 := \mathbf{N}_G(P_2) = H_{26}$ is isomorphic to $S_3 \times S_3$. In GAP we obtain its character table as follows:

```
gap> t := TableOfMarks("M11");;
gap> ctn2 := CharacterTable( RepresentativeTom(t,26) ) ;;
```

$\lvert \mathbf{C}_{N_2}(g)\rvert:$	36	4	9	18	18	12	12	6	6
N_2	1a	2a	3a	3b	3c	2b	2c	6a	6b
ψ_1	1	1	1	1	1	1	1	1	1
ψ_2	1	-1	1	1	1	-1	1	1	-1
ψ_3	1	-1	1	1	1	1	-1	-1	1
ψ_4	1	1	1	1	1	-1	-1	-1	-1
ψ_5	2	.	-1	-1	2	2	.	.	-1
ψ_6	2	.	-1	-1	2	-2	.	.	1
ψ_7	2	.	-1	2	-1	.	-2	1	.
ψ_8	2	.	-1	2	-1	.	2	-1	.
ψ_9	4	.	1	-2	-2	.	.	.	.

Observe that N_2 has three conjugacy classes of subgroups of order three. We may assume that P_2 is generated by an element of 3b. (Working in GAP this choice is not arbitrary; in fact, by our definite choice of P_2, the generators of P_2 *are* in that class.) Then

$$\mathcal{X}(P_2, N_2, G) = \{\{1\}\},$$
$$\mathcal{Y} := \mathcal{Y}(P_2, N_2, G) = \{\{1\}\} \cup \{\langle h\rangle \mid h \in \mathbf{3a}\} \cup \{\langle h\rangle \mid h \in \mathbf{3c}\}.$$

By Green correspondence the isomorphism classes of indecomposable trivial source RG-modules with vertex P_2 correspond bijectively with the indecomposable trivial source RN_2-modules with vertex P_2. Since P_2 acts trivially on the latter modules, they can be considered as projective $R(N_2/P_2)$-modules. It is easy to see that there are four projective indecomposable $R(N_2/P_2)$-modules and that the inflations to N_2 have ordinary characters

$$\eta_1 := \psi_1 + \psi_8, \quad \eta_2 := \psi_2 + \psi_8,$$
$$\eta_3 := \psi_3 + \psi_7, \quad \eta_4 := \psi_4 + \psi_7.$$

This can also be obtained automatically by GAP: N_2/P_2 is solvable, so the Brauer characters and hence the decomposition matrices of this group can immediately be found from the ordinary character table (see Corollary 4.13.6 below). So we get the ordinary characters of projective indecomposable modules of $R(N_2/P_2)$ by

```
gap> ctf:=CharacterTable(RepresentativeTom(t,26)/RepresentativeTom(t,3));;
gap> ctfprojectives := Irr(ctf)* DecompositionMatrix( ctf mod 3 );
```

Let $S_1, \ldots, S_4$ be the irreducible (trivial source) RN_2-lattices with characters $\psi_1, \ldots, \psi_4$ and let $W_1, \ldots, W_4$ be the trivial source RN_2-lattices with character $\eta_1, \ldots, \eta_4$. Then $W_1 \cong (R_{\mathbf{C}_{N_2}(h)})^{N_2}$ is actually a permutation module, where h in the class 2b of N_2 and $W_i \cong S_i \otimes_R W_1$ for $i = 2, 3, 4$. (Observe that RN_2 has more trivial source modules with vertices, which are conjugate in G to P_2, for example $(R_{\mathbf{C}_{N_2}(h)})^{N_2}$ for $h \in \mathbf{2c}$.) We conclude that up to isomorphism the indecomposable trivial source modules for G with vertex P_2 are $\{\mathbf{f}^{-1}(W_i) \mid 1 \leq i \leq 4\}$, and we try to determine the (ordinary) characters. The Green correspondent $\mathbf{f}^{-1}(W_1)$ is easily determined as follows. Let $V_1 := R_{H_{37}}{}^G$ be the permutation module on the cosets of the maximal subgroup $H_{37} \cong L_2(11)$ of G. From the table of marks of G we easily find (see Exercise 3.5.3)

$$V_1|_{N_2} \cong R_{U_1}{}^{N_2} \oplus R_{U_2}{}^{N_2} \oplus R_{U_3}{}^{N_2},$$

where U_1, U_2 are centralizers of elements in **2b** and **2c**, respectively, and $U_3 \cong S_3$ contains elements of **3a**. By Corollary 4.8.5, $R_{U_2}{}^{N_2}$ and $R_{U_3}{}^{N_2}$ are $\mathcal{Y}$-projective, since $\mathrm{Syl}_3(U_2)$, $\mathrm{Syl}_3(U_3) \subseteq \mathcal{Y}$. Thus $V_1 \cong \mathbf{f}^{-1}(W_1)$. We may check that $\eta_1{}^G - 1_{H_{37}}{}^G = \Phi_2 + \Phi_3 + 2\Phi_4 + 2\Phi_7 + 2\Phi_8$ is indeed the character of a projective module, as it should be, because $\mathcal{X} = \{\{1\}\}$. But this alone, of course, would not prove that V_1 is the Green correspondent of W_1. In order to determine the other three indecomposable trivial source RG-modules with P_2 being a vertex, we proceed as follows.

For the maximal subgroup S_5 of G we take the sign RS_5-lattice, which is a trivial source lattice (see Exercise 4.10.2), and for the maximal subgroup $L_2(11)$ there are two trivial source lattices with ordinary character χ_4' and χ_5' (see Exercise 4.10.1). Inducing these lattices to G gives three trivial source lattices with ordinary characters $\chi_5 + \chi_{10}$, $\chi_3 + \chi_4 + \chi_9 + \chi_{10}$ and $\chi_2 + \chi_5 + \chi_8 + \chi_{10}$. The first trivial source lattice is indecomposable, since the degree of a proper summand would have to be divisible by three, and the second lattice splits as a direct sum of the projective indecomposable module with ordinary character χ_9 and an indecomposable trivial source lattice with ordinary character $\chi_3 + \chi_4 + \chi_{10}$. The vertex for both indecomposable trivial source lattices has to contain P_2, since three only divides once the rank of the lattices. We will show later, see Example 4.10.12, that the third trivial source lattice is actually indecomposable and hence P_2 is a vertex. We have therefore found the ordinary characters of all four indecomposable trivial source RG-lattices with P_2 as a vertex.

Finally, let $P_3 := H_{13} \in \mathrm{Syl}_3(G)$. Then $N_3 := \mathbf{N}_G(P_3) = H_{35}$, with character table given by

$\mid \mathbf{C}_{N_3(g)} \mid :$	144	4	8	16	18	12	8	8	6
N_3	1a	4a	4b	2a	3a	2b	8a	8b	6a
θ_1	1	1	1	1	1	1	1	1	1
θ_2	1	-1	1	1	1	-1	1	1	-1
θ_3	1	-1	1	1	1	1	-1	-1	1
θ_4	1	1	1	1	1	-1	-1	-1	-1
θ_5	2	.	-2	2	2	.	.	.	.
θ_6	2	.	.	-2	2	.	$-\alpha$	α	.
θ_7	2	.	.	-2	2	.	α	$-\alpha$	.
θ_8	8	.	.	.	-1	-2	.	.	1
θ_9	8	.	.	.	-1	2	.	.	-1

and $\alpha := \sqrt{-2}$. We quickly see that we have up to isomorphism seven trivial source modules W_i' with vertex P_3, having ordinary characters θ_i for $i = 1, \ldots, 7$ and moreover $\mathcal{X}(P_3, N_3, G) = \mathcal{Y}(P_3, N_3, G) = \{\{1\}\}$.

The ordinary characters of the induced trivial RG-lattices are given in Table 4.3, and the ordinary characters of their Green correspondents $V_1', \ldots, V_7'$ are given by Table 4.4. This follows readily for all but V_5'. The character

Table 4.3. Ordinary characters of the induced trivial RG-lattices

θ_1^G	θ_2^G	θ_3^G	θ_4^G	θ_5^G	θ_6^G	θ_7^G
$\chi_1 + \chi_2 + \chi_8$	$\chi_5 + \chi_8$	χ_{10}	$\chi_2 + \chi_9$	$\chi_5 + \chi_8 + \chi_{10}$	$\chi_3 + \chi_9 + \chi_{10}$	$\chi_4 + \chi_9 + \chi_{10}$

Table 4.4. Ordinary characters of the Green correspondents

V_1'	V_2'	V_3'	V_4'	V_5'	V_6'	V_7'
χ_1	$\chi_5 + \chi_8$	χ_{10}	χ_2	χ_5	$\chi_3 + \chi_{10}$	$\chi_4 + \chi_{10}$

of the trivial source RG-lattice V_5' can be obtained by inducing up the sign $R A_6 . 2_3$-lattice from $A_6 . 2_3$ (in ATLAS notation; see [38]). Since the character of the induced RG-lattice is χ_5, it is an indecomposable trivial source lattice with P_3 as a vertex, and it follows that it has to be isomorphic to V_5'. ◆

Lemma 4.10.9 Let $H \in \mathcal{C}_p(G)$ and $P := \mathbf{O}_p(H)$. Then $s_{H,\tilde{c}} = s_{H,\tilde{c}'}$ for $\tilde{c}, \tilde{c}' \in H/P$ if and only if $\tilde{c}, \tilde{c}'$ are conjugate in $\mathbf{N}_G(P)/P$.

Proof. We know that $\tilde{c}, \tilde{c}'$ are p-regular. By Theorem 4.3.3 they are conjugate in $\mathbf{N}_G(P)/P$ if and only if $\Phi(\tilde{c}) = \Phi(\tilde{c}')$ for all projective indecomposable characters Φ of $\mathbf{N}_G(P)/P$. Let W be a projective indecomposable $F(\mathbf{N}_G(P)/P)$-module with character Φ, inflated to an $F\mathbf{N}_G(P)$-module. Thus $\mathrm{Vtx}(W) = \{P\}$ and F_P is a source for W (see Exercise 4.8.5). By Lemma 4.10.2(c) there is an indecomposable FG-module V with $P \in \mathrm{Vtx}(V)$ and source F_P such that

$$V_N = W \oplus W',$$

with $P \notin \mathrm{Vtx}(W_i)$ for any indecomposable $W_i \mid W'$. By definition we have $s_{H,\tilde{c}}([W]) = \tilde{\varphi}_{W_H}(\tilde{c}) = \Phi|_H(\tilde{c})$. From this the result follows immediately. □

As we have observed, Green correspondence induces a bijection between the indecomposable trivial source FG-module V with vertex Q and the Green correspondents $\boldsymbol{f}(V)$ in $\mathbf{N}_G(Q)$, which can be considered as the projective indecomposable $F(\mathbf{N}_G(Q)/Q)$-modules. For a given p-subgroup Q, we will now define a set of species which is in bijection with the p-regular classes of $\mathbf{N}_G(Q)/Q$. This enables us to define the trivial source character table of G as a (square) matrix with complex entries whose columns are indexed by these species ranging over the conjugacy classes of p-subgroups and whose rows correspond to the indecomposable trivial source FG-modules. Moreover, this matrix is naturally partitioned by the conjugacy classes of p-subgroups of G.

Definition 4.10.10 Let $P_1, \ldots, P_r$ be representatives for the conjugacy classes of p-subgroups of G with normalizers $N_i := \mathbf{N}_G(P_i)$ for $i = 1, \ldots, r$. For each N_i choose representatives $\tilde{c}_{i,j}$, $j = 1, \ldots, l(i)$ for the p-regular conjugacy classes in N_i/P_i and define $H_{i,j} \leq N_i$ by $H_{i,j}/P_i = \langle \tilde{c}_{i,j} \rangle$. Let $V_{k,m}$, $m = 1, \ldots, t(k)$ be the indecomposable trivial source FG-modules with $P_k \in \mathrm{Vtx}(V_{k,m})$. We define

$$T_{i,k} := [s_{H_{i,j}, \tilde{c}_{i,j}}(V_{k,m})]_{1 \leq j \leq l(i), \, 1 \leq m \leq t(k)} \in \mathbb{C}^{l(i) \times t(k)}, \tag{4.43}$$

where $i, k = 1, \ldots, r$. Finally, we define the trivial source character table $T(G)$ as the (block) matrix $[T_{i,k}]_{1 \leq i,k \leq r}$.

The following observation gives a description of certain entries in $T(G)$.

Lemma 4.10.11 *Let G be a finite group and let V be an indecomposable trivial source FG-module with vertex P. Moreover, let $s_{H,\tilde{c}}$ be a species with $\mathbf{O}_p(H) = Q$ and $H/Q = \langle \tilde{c} \rangle$ and let $N := \mathbf{N}_G(Q)$. Then the following statements hold.*

(a) $V_N = (V_N)^{\supseteq Q} \oplus (V_N)^{\not\supseteq Q}$, *where $(V_N)^{\supseteq Q}$ denotes the direct sum of the indecomposable FN-summands of V_N whose vertices contain Q, and $(V_N)^{\not\supseteq Q}$ denotes the direct sum of the indecomposable FN-summands whose vertices do not contain Q. Moreover $s_{H,\tilde{c}}([V]) = \tilde{\varphi}_{(V_N)^{\supseteq Q}}(\tilde{c})$.*

(b) $s_{H,\tilde{c}}([V]) = 0$ *unless $Q \leq_G P$. In particular, $T_{i,k} = \mathbf{0}$ for $i < k$ in (4.43).*

(c) *If $Q \in \mathrm{Vtx}(V)$ we get $(V_N)^{\supseteq Q} = \mathbf{f}(V)$, the Green correspondent w.r.t. (Q, N, G), which is the inflation of a projective indecomposable $F(N/Q)$-module with character $\Phi_{\mathbf{f}(V)}$. Thus $s_{H,\tilde{c}} = \Phi_{\mathbf{f}(V)}(\tilde{c})$. In particular, $T_{i,i}$, $i = 1, \ldots, r$ is invertible and hence the rows of $T(G)$ are $\mathbb{C}$-linear independent.*

Proof. (a) Let U be an indecomposable summand of V_N with vertex Q_1. Then, $Q_1 \supseteq Q$ if and only if $Q_1^x \supseteq Q$ for any $x \in N$, since $Q \trianglelefteq N$. This proves the first part. Moreover, if $Q_1 \not\supseteq Q$, then the trivial source module U_Q does not have Q has a vertex and if $Q_1 \supseteq Q$ then all indecomposable summands of U_Q have vertex Q. Hence $(V_Q)^{=Q} = (V_N^{\supseteq Q})_Q$ and $s_{H,\tilde{c}}([V]) = \tilde{\varphi}_{(V_N)^{\supseteq Q}}(\tilde{c})$.

(b) If $Q \not\leq_G P$, then the vertices of any indecomposable summand of V_Q are properly contained in Q. Hence $(V_Q)^{=Q} = 0$.

(c) This follows directly from $V_N = \mathbf{f}(V) \oplus Y$, where Y is the sum of indecomposable FN-modules with vertices in $\mathcal{Y} := \{N^g \cap Q | g \in G \setminus N\}$ and $Q \notin \mathcal{Y}$, so $\mathbf{f}(V) = (V_N)^{\supseteq Q}$. The remaining statements follow from Exercise 4.8.5 and Theorem 4.3.3. $\square$

Example 4.10.12 Continuing the previous example, we now determine the full trivial source table for M_{11} and $p := 3$. There are three conjugacy classes of 3-subgroups, $P_1 := \{\{1\}\}$, P_2 and P_3. For N_1/P_1 we have eight 3-regular conjugacy classes denoted by $1a, 2a, 4a, 5a, 8a, 8b, 11a, 11b$. We abbreviate the corresponding species by $b_1, \ldots, b_8$. For N_2/P_2 we get four 3-regular conjugacy classes $1a, 2a, 2b, 2c$, and we denote the corresponding species denoted by $s_1, \ldots, s_4$. Finally N_3/P_3 has seven 3-regular conjugacy classes $1a, 4a, 4b, 2a, 2b, 8a, 8b$ with corresponding species denoted by $t_1, \ldots, t_7$.

As we have previously noted, $\mathcal{X}(P_3, N_3, G) = \{\{1\}\}$ and therefore the restriction of the indecomposable trivial source module V_i', $i = 1, \ldots, 7$ to N_3

is the direct sum $f(V_i') \oplus X_i$, where X_i is a projective FN_3-module. Using Lemma 4.10.11 it is easy to determine the diagonal blocks of trivial source character table $T(G)$ and also the values of $b_1, \ldots, b_8$ on all classes of indecomposable trivial source FG-modules. We are left to determine the values of the species $s_1, \ldots, s_4$ on the classes of the indecomposable FG-modules $V_1', \ldots, V_7'$. Since $P_3 \leq N_2 \leq N_3$, this reduces to the task of restricting the characters $\Phi_{f(V_i')}$ to N_2/P_3, which is easily done. The complete trivial source character table is given in Table 4.5.

Table 4.5. Trivial source character table of $M_{11}, p = 3$

b_1	b_2	b_3	b_4	b_5	b_6	b_7	b_8	s_1	s_2	s_3	s_4	t_1	t_2	t_3	t_4	t_5	t_6	t_7
99	3	−1	4	1	1	0	0	0	0	0	0	0	0	0	0	0	0	0
126	6	−2	1	0	0	β	$\bar\beta$	0	0	0	0	0	0	0	0	0	0	0
126	6	−2	1	0	0	$\bar\beta$	β	0	0	0	0	0	0	0	0	0	0	0
54	6	2	−1	0	0	−1	−1	0	0	0	0	0	0	0	0	0	0	0
81	−3	−1	1	α	$\bar\alpha$	γ	$\bar\gamma$	0	0	0	0	0	0	0	0	0	0	0
81	−3	−1	1	$\bar\alpha$	α	$\bar\gamma$	γ	0	0	0	0	0	0	0	0	0	0	0
99	3	−1	−1	1	1	0	0	0	0	0	0	0	0	0	0	0	0	0
45	−3	1	0	−1	−1	1	1	0	0	0	0	0	0	0	0	0	0	0
12	4	0	2	0	0	1	1	3	3	1	1	0	0	0	0	0	0	0
66	2	−2	1	0	0	0	0	3	3	−1	−1	0	0	0	0	0	0	0
75	−5	−1	0	1	1	−2	−2	3	−3	1	−1	0	0	0	0	0	0	0
120	8	0	0	0	0	−1	−1	3	−3	−1	1	0	0	0	0	0	0	0
1	1	1	1	1	1	1	1	1	1	1	1	1	1	1	1	1	1	1
10	2	2	0	0	0	−1	−1	1	−1	−1	1	1	1	−1	1	1	−1	−1
11	3	−1	1	−1	−1	0	0	2	0	0	2	2	2	0	−2	0	0	0
55	−1	−1	0	1	1	0	0	1	−1	−1	1	1	1	−1	1	−1	1	1
55	7	−1	0	−1	−1	0	0	1	1	1	1	1	1	1	1	−1	−1	−1
65	−3	−1	0	α	$\bar\alpha$	−1	−1	2	0	0	−2	2	−2	0	0	0	−δ	δ
65	−3	−1	0	$\bar\alpha$	α	−1	−1	2	0	0	−2	2	−2	0	0	0	δ	−δ

$\alpha := 1 + \sqrt{-2}$, $\beta := \frac{-1+\sqrt{-11}}{2}$, $\gamma := \frac{-3+\sqrt{-11}}{2}$ and $\delta := \sqrt{-2}$.

We leave it to the reader to determine the trivial source table if the trivial source module of degree 120 were not indecomposable and to show that such a table would not be consistent. ◆

As an immediate consequences from Lemma 4.10.11 we get the following corollaries.

Corollary 4.10.13 (Conlon) (a) *Let*

$$\mathrm{Sp}(G) := \{s_{H,\tilde c} \mid H \in \mathcal{C}_p(G), \ H/\mathbf{O}_p(H) = \langle \tilde c \rangle \}.$$

If $\xi, \eta \in \mathbb{C} \otimes TS(FG)$ are such that $s(\xi) = s(\eta)$ for all $s \in \mathrm{Sp}(G)$, then $\xi = \eta$.
(b) *If V, W are trivial source FG-modules, then $V \cong W$ if and only if $V_H \cong_{FH} W_H$ for all $H \in \mathcal{C}_p(G)$.*

Corollary 4.10.14 *Let Ω and Ω' be transitive G-sets. Then the permutation modules $F\Omega$ and $F\Omega'$ are isomorphic if and only if*

$$m_\Omega(H) = m_{\Omega'}(H) \qquad \text{for all} \qquad H \in \mathcal{C}_p(G),$$

where m_Ω and $m_{\Omega'}$ are the marks of the corresponding G-sets.

Proof. This follows from Corollary 4.10.13 in conjunction with Remark 4.10.5, in particular (4.42). □

Exercises

Exercise 4.10.1 For the group $G := L_2(11)$ and (K, R, F, η) a 3-modular splitting system, show that there is a trivial source RG-lattice with ordinary character χ'_4, respectively χ'_5.
Hint: Note that A_5 is a maximal subgroup of G. For $\eta_4 \in \mathrm{Irr}(A_5)$ with $\eta_4(1) = 4$ look at the character η_4^G.

Exercise 4.10.2 Let G be a cyclic group of order $n = p^a q$ with p a prime, $a \in \mathbb{N}$ and $(p, q) = 1$. Describe all indecomposable trivial source RG-lattices where (K, R, F, η) is a p-modular splitting system for G.

4.11   Generalized decomposition numbers

Throughout this section we assume that (K, R, F, η) is a p-modular splitting system for the finite group G. From the p-decomposition numbers of a group G and $\mathrm{IBr}(G)$, one may obtain $\{ \chi|_{G_{p'}} \mid \chi \in \mathrm{Irr}(G) \}$. In order to obtain the values of the ordinary irreducible characters of G on the p-singular elements, we have to introduce the "generalized decomposition numbers."
Definition 4.11.1 If $x \in G$ is a p-element then

$$\mathrm{sec}_p(x) := \{ \, g \in G \ \mid \ g_p \in x^G \, \}$$

is called the p-**section** of x. Let $\mathcal{S}_p(G)$ be a set of representatives of the G-conjugacy classes of p-elements of G.

Recall that g_p is the p-part of g (see Definition 4.2.1). Then $g = g_p g_{p'}$ with a p-regular $g_{p'} \in \mathbf{C}_G(g_p)$. Of course each p-section is a union of conjugacy classes, and $\mathrm{sec}_p(1) = G_{p'}$. We collect some obvious facts:

$$G = \bigcup\nolimits_{x \in \mathcal{S}_p(G)} \mathrm{sec}_p(x) \,, \qquad \mathbf{Z}(KG) = \bigoplus_{x \in \mathcal{S}_p(G)} \mathbf{Z}(KG)_x$$

with

$$\mathbf{Z}(KG)_x := \langle \, C^+ \mid C \in \mathrm{cl}(G), \ C \subseteq \mathrm{sec}_p(x) \, \rangle_K.$$

Lemma 4.11.2 *Let $x \in G$ be a p-element and let $y_1, \ldots, y_r$ be a set of representatives of the p'-classes of $\mathbf{C}_G(x)$. Then*

$$\mathrm{sec}_p(x) = (xy_1)^G \ \dot\cup \ \cdots \ \dot\cup \ (xy_r)^G.$$

For $\varphi \in \mathrm{IBr}(\mathbf{C}_G(x))$ we define

$$f_{x,\varphi} := \sum_{i=1}^{r} \varphi(y_i^{-1}) \left((xy_i)^G\right)^+ \in \mathbf{Z}(KG)_x. \tag{4.44}$$

Then $(\, f_{x,\varphi} \mid x \in \mathcal{S}_p(G), \ \varphi \in \mathrm{IBr}(\mathbf{C}_G(x)) \,)$ is a K-basis of $\mathbf{Z}(KG)$.

Proof. The first assertion is clear, since $(g^h)_p = (g_p)^h$ and $(g^h)_{p'} = (g_{p'})^h$ holds for $g, h \in G$. The second one follows because the matrix $[\varphi(y_i^{-1})]_{\varphi,i}$ is non-singular by Theorem 4.2.14. $\qquad\square$

Lemma 4.11.3 *Let $x \in G$ be a p-element and $H = \mathbf{C}_G(x)$. For $\chi \in \mathrm{Irr}(G)$ and $\varphi \in \mathrm{IBr}(H)$ there exist unique algebraic integers $d^x_{\chi\varphi}$ such that*

$$\chi(xy) = \sum_{\varphi \in \mathrm{IBr}(H)} d^x_{\chi\varphi}\, \varphi(y) \qquad for \qquad y \in H_{p'}.$$

The $d^x_{\chi\varphi}$ are called the **generalized decomposition numbers** *and $D^x :=$ $[d^x_{\chi\varphi}]_{\chi\in\mathrm{Irr}(G),\varphi\in\mathrm{IBr}(H)}$ is called the* **generalized decomposition matrix** *for x or the p-section $\mathrm{sec}_p(x)$. They can be written as*

$$d^x_{\chi\varphi} := \sum_{\theta\in\mathrm{Irr}(H)} \frac{(\chi|_H, \theta)_H\, \theta(x)}{\theta(1)}\, d_{\theta\varphi}. \tag{4.45}$$

Proof. Since $x \in \mathbf{Z}(H)$, any irreducible representation $\boldsymbol{\delta}\colon H \to \mathrm{GL}_n(K)$ must represent x by a scalar matrix $\boldsymbol{\delta}(x) = \zeta_m^j \mathbf{I}_n$ for some $j \in \{1, \ldots, m\}$, where m is the order of x. Then $\boldsymbol{\delta}(xy) = \zeta_m^j \boldsymbol{\delta}(y)$ for any $y \in H$. Consequently, if $\boldsymbol{\delta}$ has character $\theta \in \mathrm{Irr}(H)$, then

$$\theta(xy) = \zeta_m^j \theta(y) \qquad for \qquad y \in H \qquad and \qquad \zeta_m^j = \frac{\theta(x)}{\theta(1)}.$$

So

$$\chi(xy) = \sum_{\theta\in\mathrm{Irr}(H)} (\chi|_H, \theta)_H\, \theta(xy) = \sum_{\theta\in\mathrm{Irr}(H)} \frac{(\chi|_H, \theta)_H\, \theta(x)}{\theta(1)} \theta(y)$$

$$= \sum_{\varphi\in\mathrm{IBr}(H)} \Big(\sum_{\theta\in\mathrm{Irr}(H)} \frac{(\chi|_H, \theta)_H\, \theta(x)}{\theta(1)}\, d_{\theta\varphi} \Big)\, \varphi(y) = \sum_{\varphi\in\mathrm{IBr}(H)} d^x_{\chi\varphi}\varphi(y),$$

with $d^x_{\chi\varphi}$ as defined in (4.45). The uniqueness follows from Theorem 4.2.14. $\quad\square$

Observe that the D^1 is just the ordinary p-decomposition matrix. Hence the following lemma gives a generalization of the equation $C = D^{\mathrm{T}}D$ in Theorem 4.1.23, equation (4.6).

Lemma 4.11.4 *Let $x, x' \in \mathcal{S}_p(G)$. Then*

$$\overline{D^x}^{\mathrm{T}} D^x = \delta_{x,x'} C^{(x)} \qquad and \qquad \overline{D^x}^{\mathrm{T}} D^{x'} = \mathbf{O} \quad for \quad x \neq x',$$

where $C^{(x)}$ is the Cartan matrix of $F\mathbf{C}_G(x)$ and $\mathbf{O}$ is a zero matrix.

Proof. Let $\mathrm{sec}_p(x) = \dot{\bigcup}_{i=1}^r (xy_i)^G$ and $\mathrm{sec}_p(x') = \dot{\bigcup}_{i=1}^{r'} (x'z_i)^G$. We put

$$X^x := [\chi(xy_i)]_{\chi\in\mathrm{Irr}(G), 1\le i\le r}, \qquad X^{x'} := [\chi(x'z_i)]_{\chi\in\mathrm{Irr}(G), 1\le i\le r'}.$$

Then $X^x = D^x \Phi^x$ and $X^{x'} = D^{x'} \Phi^{x'}$, where Φ^x, $\Phi^{x'}$ are the Brauer character tables of $\mathbf{C}_G(x)$ and $\mathbf{C}_G(x')$, respectively. The ordinary orthogonality relations (Theorem 2.1.15) show that $\overline{X^x}^{\mathrm{T}} X^{x'} = \mathbf{O} \in \mathbb{Z}^{r \times r'}$ for $x \neq x'$ and

$$\overline{X^x}^{\mathrm{T}} X^x = \mathrm{diag}(|\,\mathbf{C}_G(xy_1)|, \ldots, |\,\mathbf{C}_G(xy_r)|).$$

But $\mathbf{C}_G(xy_i) = \mathbf{C}_{\mathbf{C}_G(x)}(y_i)$; hence, using Theorem 4.1.23, equation (4.6) for $\mathbf{C}_G(x)$,

$$\overline{\Phi^x}^{\mathrm{T}} \overline{D^x}^{\mathrm{T}} D^x \Phi^x = \overline{X^x}^{\mathrm{T}} X^x = \overline{\Phi^x}^{\mathrm{T}} D^{(x)\mathrm{T}} D^{(x)} \Phi^x = \overline{\Phi^x}^{\mathrm{T}} C^{(x)} \Phi^x,$$

where $D^{(x)}$ is the p-decomposition matrix of $\mathbf{C}_G(x)$. Since Φ^x is invertible, the result follows. $\qquad\square$

Theorem 4.11.5 *Let x be a p-element of G. Suppose V is an RG-lattice belonging to the p-block B with character χ_V. Then $\epsilon_b V$ is an $R\,\mathbf{C}_G(x)$-lattice for every p-block b of $\mathbf{C}_G(x)$ and*

$$\chi_V(xy) = \sum_{b \in \mathrm{Bl}_p(\mathbf{C}_G(x)), b^G = B} \chi_{\epsilon_b V}(xy) \qquad for \qquad y \in \mathbf{C}_G(x)_{p'}.$$

Proof. Let $Q := \langle x \rangle$ and $H := \mathbf{C}_G(x)$. By Theorem 4.7.19, $\mathrm{Br}_Q(\hat{e}_B) = \sum_{b \in \mathrm{Bl}_p(H), b^G = B} \hat{e}_b$. By Theorem 4.8.17, we have

$$V_H = \Big(\sum_{b \in \mathrm{Bl}_p(H), b^G = B} \epsilon_b \Big) V \oplus \bigoplus_{i=1}^m W_i$$

with indecomposable RH-lattices with vertices not containing x. By Theorem 4.8.19(b), it follows that $\chi_{W_i}(xy) = 0$ for all i. $\qquad\square$

Corollary 4.11.6 (Brauer's second main theorem) *Let B be a p-block of G, let $\chi \in \mathrm{Irr}(B)$ and let x be a p-element of G. If $\varphi \in \mathrm{IBr}(b)$ for $b \in \mathrm{Bl}_p(\mathbf{C}_G(x))$ and $b^G \neq B$, then $d^x_{\chi\varphi} = 0$. Furthermore,*

$$\chi(xy) = \sum_{\substack{b \in \mathrm{Bl}_p(\mathbf{C}_G(x)) \\ b^G = B}} \sum_{\varphi \in \mathrm{IBr}(b)} d^x_{\chi\varphi} \varphi(y) \qquad for \ all \qquad y \in (\mathbf{C}_G(x))_{p'}.$$

Proof. Let χ be afforded by the RG-lattice V. Furthermore, let $H := \mathbf{C}_G(x)$. By Theorem 4.11.5 we have

$$\chi(xy) = \sum_{b^G = B} \chi_{\epsilon_b V}(xy) = \sum_{\substack{b \in \mathrm{Bl}_p(H) \\ b^G = B}} \sum_{\theta \in \mathrm{Irr}(b)} (\chi|_H, \theta)_H \, \theta(xy)$$

$$= \sum_{\substack{b \in \mathrm{Bl}_p(H) \\ b^G = B}} \sum_{\theta \in \mathrm{Irr}(b)} \sum_{\varphi \in \mathrm{IBr}(b)} \frac{(\chi|_H, \theta)_H \theta(x)}{\theta(1)} d_{\theta\varphi} \, \varphi(y)$$

$$= \sum_{\substack{b \in \mathrm{Bl}_p(H) \\ b^G = B}} \sum_{\varphi \in \mathrm{IBr}(b)} d^x_{\chi\varphi} \, \varphi(y),$$

as in the proof of Lemma 4.11.3. □

Corollary 4.11.6 says that the generalized decomposition matrices D^x break up into (rectangular) submatrices D_B^x corresponding to the blocks $B \in \mathrm{Bl}_p(G)$ with rows indexed by $\mathrm{Irr}(B)$ and columns indexed by $\bigcup\{\mathrm{Irr}(b) \mid b \in \mathrm{Bl}_p(\mathbf{C}_G(x)), \ b^G = B\}$.

Example 4.11.7 Let $G := \mathrm{M}_{11}$ and $p := 2$. We have four p-sections in addition to $\mathrm{sec}_p(1) = G_{p'}$:

$$\mathrm{sec}_p(2a) = 2a \cup 6a, \quad \mathrm{sec}_p(4a) = 4a, \quad \mathrm{sec}_p(8a) = 8a, \quad \mathrm{sec}_p(8b) = 8b,$$

where we have replaced x by the name of its conjugacy class. From the ATLAS ([38]) we know that $H_1 := \mathbf{C}_G(2a) = Q_8 \rtimes S_3$, a split extension of a quaternion group Q_8 with S_3, and hence its 2-Brauer character table is given by

$$\begin{matrix} \psi_1 \\ \psi_2 \end{matrix} \begin{bmatrix} 1 & 1 \\ 2 & -1 \end{bmatrix}.$$

For the other classes we get $H_2 := \mathbf{C}_G(4a) = \mathbf{C}_G(8a) = \mathbf{C}_G(8b) \cong C_8$ having only the trivial Brauer character $\mathbf{1}$. We present all the generalized decomposition numbers in one matrix as follows:

section:	1a					2a		4a	8a	8b
	φ_1	φ_2	φ_3	φ_4	φ_5	ψ_1	ψ_2	$\mathbf{1}$	$\mathbf{1}$	$\mathbf{1}$
χ_1	1	0	0	0	0	1	0	1	1	1
χ_2	0	1	0	0	0	0	1	2	0	0
χ_3	0	1	0	0	0	0	-1	0	$\sqrt{-2}$	$-\sqrt{-2}$
χ_4	0	1	0	0	0	0	-1	0	$-\sqrt{-2}$	$\sqrt{-2}$
χ_5	1	1	0	0	0	1	1	-1	-1	-1
χ_8	0	0	1	0	0	2	1	0	0	0
χ_9	1	0	1	0	0	-1	-1	1	-1	-1
χ_{10}	1	1	1	0	0	-1	0	-1	1	1
χ_6	0	0	0	1	0	0	0	0	0	0
χ_7	0	0	0	0	1	0	0	0	0	0

◆

Lemma 4.11.8 Let x be a p-element of G. If a class function $\theta \in \mathrm{cf}(G, K)$ vanishes on the whole section $\mathrm{sec}_p(x)$, then the same holds for the B-part θ_B of θ for every $B \in \mathrm{Bl}_p(G)$.

Proof. Let $\theta = \sum_{\chi \in \mathrm{Irr}(G)} a_\chi \chi$ with $a_\chi \in K$. Then $\theta_B = \sum_{\chi \in \mathrm{Irr}(B)} a_\chi \chi$. Let $H := \mathbf{C}_G(x)$. By assumption we have for every $y \in H_{p'}$

$$0 = \theta(xy) = \sum_{\substack{B \in \mathrm{Bl}_p(G)}} \sum_{\chi \in \mathrm{Irr}(B)} a_\chi \sum_{\substack{b \in \mathrm{Bl}_p(H) \\ b^G = B}} \sum_{\varphi \in \mathrm{IBr}(b)} d_{\chi\varphi}^x \, \varphi(y)$$

$$= \sum_{b \in \mathrm{Bl}_p(H)} \sum_{\varphi \in \mathrm{IBr}(b)} \Big(\sum_{\chi \in \mathrm{Irr}(b^G)} a_\chi d_{\chi\varphi}^x \Big) \, \varphi(y).$$

Since $\mathrm{IBr}(H)$ is linearly independent, we conclude that $\sum_{\chi\in\mathrm{Irr}(b^G)} a_\chi d^x_{\chi\varphi} = 0$ for all $b \in \mathrm{Bl}_p(H)$. But then

$$\theta_B(xy) = \sum_{\substack{\chi\in\mathrm{Irr}(B)}} a_\chi \sum_{\substack{b\in\mathrm{Bl}_p(H)\\ b^G=B}} \sum_{\varphi\in\mathrm{IBr}(b)} d^x_{\chi\varphi}\,\varphi(y)$$

$$= \sum_{\substack{b\in\mathrm{Bl}_p(H)\\ b^G=B}} \sum_{\varphi\in\mathrm{IBr}(b)} \Big(\sum_{\chi\in\mathrm{Irr}(B)} a_\chi d^x_{\chi\varphi} \Big)\,\varphi(y) = 0.$$

$\square$

Theorem 4.11.9 (Block orthogonality) *If* $g, g' \in G$ *are in different p-sections then*

$$\sum_{\chi\in\mathrm{Irr}(B)} \chi(g)\overline{\chi(g')} = 0 \qquad \text{for any} \qquad B \in \mathrm{Bl}_p(G).$$

Proof. Put $\theta := \sum_{\chi\in\mathrm{Irr}(G)} \overline{\chi(g')}\chi$. Then θ vanishes on $\sec(g_p)$ by the ordinary orthogonality relations (Theorem 2.1.15). From Lemma 4.11.8 we conclude that θ_B also vanishes on $\sec(g_p)$. But $\theta_B(g) = \sum_{\chi\in\mathrm{Irr}(B)} \overline{\chi(g')}\chi(g)$. $\square$

The following important consequence of Brauer's second main theorem (and Theorem 4.11.9) shows that the basis of $\mathbf{Z}(KG)$ introduced in Lemma 4.11.2 is adapted to the block decomposition.

Theorem 4.11.10 *Let x be a p-element of G and $b \in \mathrm{Bl}_p(\mathbf{C}_G(x))$. Then for any $B \in \mathrm{Bl}_p(G)$ and $\varphi \in \mathrm{IBr}(b)$*

$$\epsilon_B \cdot f_{x,\varphi} = \begin{cases} 0 & \text{if} \quad B \neq b^G, \\ f_{x,\varphi} & \text{if} \quad B = b^G. \end{cases}$$

Proof. Recall that $f_{x,\varphi} \in \mathbf{Z}(KG)$ (see Lemma 4.11.2). We abbreviate $H := \mathbf{C}_G(x)$. Assume that $\sec_p(x) = (xy_1)^G \dot\cup \cdots \dot\cup (xy_r)^G$ with $y_i \in H_{p'}$. Then

$$\epsilon_B \cdot f_{x,\varphi} = \sum_{\chi\in\mathrm{Irr}(B)} \omega(f_{x,\varphi})\epsilon_\chi = \sum_{i=1}^r \varphi(y_i^{-1})|(xy_i)^G| \cdot \underbrace{\sum_{\chi\in\mathrm{Irr}(B)} \frac{\chi(xy_i)}{\chi(1)}\epsilon_\chi}_{=:a} \quad (4.46)$$

with (using Corollary 2.1.7)

$$a = \sum_{g\in G} \frac{1}{|G|} \sum_{\chi\in\mathrm{Irr}(B)} \chi(xy_i)\overline{\chi(g)}\,g = \frac{1}{|G|} \sum_{j=1}^r \sum_{\chi\in\mathrm{Irr}(B)} \chi(xy_i)\overline{\chi(xy_j)}\,((xy_j)^G)^+.$$

Here we have used Theorem 4.11.9. Next we invoke Brauer's second main theorem (Theorem 4.11.6) and subsequently Lemma 4.11.3, obtaining

$$
\begin{aligned}
a &= \frac{1}{|G|} \sum_{\substack{b,b' \in \mathrm{Bl}_p(H) \\ b^G = B}} \sum_{\varphi' \in \mathrm{IBr}(b)} \sum_{\psi \in \mathrm{IBr}(b')} \sum_{\chi \in \mathrm{Irr}(B)} d^x_{\chi,\varphi'} \overline{d^x_{\chi,\psi}} \varphi'(y_i) f_{x,\psi} \\
&= \frac{1}{|G|} \sum_{\substack{b,b' \in \mathrm{Bl}_p(H) \\ b^G = B}} \sum_{\varphi' \in \mathrm{IBr}(b)} \sum_{\psi \in \mathrm{IBr}(b')} c^{(x)}_{\psi,\varphi'} \varphi'(y_i) f_{x,\psi} \\
&= \frac{1}{|G|} \sum_{\substack{b \in \mathrm{Bl}_p(H) \\ b^G = B}} \sum_{\psi \in \mathrm{IBr}(b)} \Big(\sum_{\varphi' \in \mathrm{IBr}(b)} c^{(x)}_{\psi,\varphi'} \varphi'(y_i) \Big) f_{x,\psi} \\
&= \frac{1}{|G|} \sum_{\substack{b \in \mathrm{Bl}_p(H) \\ b^G = B}} \sum_{\psi \in \mathrm{IBr}(b)} \Phi_\psi(y_i) f_{x,\psi}.
\end{aligned}
$$

Plugging this into (4.46), we get

$$
\begin{aligned}
\epsilon_B \cdot f_{x,\varphi} &= \frac{1}{|H|} \sum_{\substack{b \in \mathrm{Bl}_p(H) \\ b^G = B}} \sum_{\psi \in \mathrm{IBr}(b)} \sum_{i=1}^{r} |y_i^H| \varphi(y_i^{-1}) \Phi_\psi(y_i) f_{x,\psi} \\
&= \sum_{\substack{b \in \mathrm{Bl}_p(H) \\ b^G = B}} \sum_{\psi \in \mathrm{IBr}(b)} \delta_{\varphi,\psi} f_{x,\psi}
\end{aligned}
$$

by Theorem 4.3.3. This completes the proof. $\qquad\square$

The following interesting consequence shows that $k(B) - l(B)$ (for the definition of $k(B)$ and $l(B)$ see Lemma 4.4.6) can be determined "locally," that is, by considering the centralizers of non-trivial p-elements.

Theorem 4.11.11 *Let $B \in \mathrm{Bl}_p(G)$. Then*

(a)
$$
k(B) - l(B) = \sum_{\substack{x \in S_p(G) \\ x \neq 1}} \sum_{\substack{b \in \mathrm{Bl}_p(\mathbf{C}_G(x)) \\ b^G = B}} l(b);
$$

(b) *if $D \in \mathrm{Def}(B)$ with $|D| = p$ and $B = b^G$ for some $b \in \mathrm{Bl}_p(\mathbf{C}_G(D))$ then*

$$
l(B) = e := [T_{\mathbf{N}_G(D)}(b) : \mathbf{C}_G(D)] \qquad and \qquad k(B) = e + \frac{p-1}{e}.
$$

Proof. (a) By Corollary 4.1.22, $k(B) = \dim_K \epsilon_B \mathbf{Z}(KG)$. On the other hand, by Lemma 4.11.2 and Theorem 4.11.10,

$$
\epsilon_B \mathbf{Z}(KG) = \sum_{x \in S_p(G)} \sum_{\varphi \in \mathrm{IBr}(\mathbf{C}_G(x))} \epsilon_B f_{x,\varphi} K = \sum_{x \in S_p(G)} \sum_{\substack{b \in \mathrm{Bl}_p(\mathbf{C}_G(x)) \\ b^G = B}} \sum_{\varphi \in \mathrm{IBr}(b)} f_{x,\varphi} K.
$$

Since the $f_{x,\varphi}$ are linearly independent (see Lemma 4.11.2), the result follows.

(b) For the case that $G = \mathbf{N}_G(D)$, see Theorem 4.8.14. Also, by Theorem 4.9.8, $l(B) = l(b^{\mathbf{N}_G(D)})$. So it suffices to prove $k(B) - l(B) = \frac{p-1}{e}$ in the general case.

By Corollary 4.7.31 there is a unique $\mathbf{N}_G(D)$-conjugacy class of p-blocks b of $\mathbf{C}_G(D)$ with $b^G = B$. Its length is $[\mathbf{N}_G(D) : T_{\mathbf{N}_G(D)}(b)]$. Theorem 4.6.12 shows that $l(b) = 1$ for each such b. By Theorem 4.7.19 for $x \in \mathcal{S}_p(G)$ there is a $b \in \mathrm{Bl}_p(\mathbf{C}_G(x))$ with $B = b^G$ if and only if $x^g \in D$ for some $g \in G$. In this case we may choose $x \in D$ and then $\mathbf{C}_G(x) = C_G(D)$ unless $x = 1$. The number of non-trivial p-sections intersecting D is $(p-1)/[\mathbf{N}_G(D) : \mathbf{C}_G(D)]$. Applying (a) we obtain

$$k(B) - l(B) = \frac{p-1}{[\mathbf{N}_G(D) : \mathbf{C}_G(D)]} \cdot [\mathbf{N}_G(D) : T_{\mathbf{N}_G(D)}(b)] = \frac{p-1}{e}.$$

$\square$

We would like to give a very simple application of Theorem 4.11.9. For this we need the following definition.

Definition 4.11.12 If $B \in \mathrm{Bl}_p(G)$ then the **kernel** of B is

$$\ker B := \bigcap_{\chi \in \mathrm{Irr}(B)} \ker \chi.$$

Obviously $\ker B = \{g \in G \mid g \, \epsilon_B = \epsilon_B\}$.

Theorem 4.11.13 *If B is a p-block of G and $\chi \in \mathrm{Irr}(B)$ then*

$$\ker B = \mathbf{O}_{p'}(\ker \chi).$$

In particular, for the principal block we get $\ker B_0(G) = \mathbf{O}_{p'}(G)$.

Proof. Let $\chi \in \mathrm{Irr}(B)$ and $H := \mathbf{O}_{p'}(\ker \chi)$. We first show that $H \subseteq \ker B$. Let

$$\epsilon_H := \frac{1}{|H|} \sum_{h \in H} h \in \mathbf{Z}(RG).$$

Observe that ϵ_H is in RG, since $p \nmid |H|$, and it is central because H is normal in G. Thus

$$\epsilon_H = \epsilon_1 + \cdots + \epsilon_r \qquad \text{with block idempotents} \quad \epsilon_i \in \mathbf{Z}(RG).$$

Since $H \subseteq \ker \chi$ we have $h \, \epsilon_\chi = \epsilon_\chi$ for all $h \in H$, hence $\epsilon_H \epsilon_\chi = \epsilon_\chi$. Consequently $\epsilon_B \in \{\epsilon_1, \ldots, \epsilon_r\}$ and

$$h \, \epsilon_B = h \, \epsilon_H \epsilon_B = \epsilon_H \epsilon_B = \epsilon_B \qquad \text{for all} \qquad h \in H.$$

Assume now that $\ker B \nsubseteq H$. Since $\ker B \trianglelefteq \ker \chi$, there must be a p-element $1 \neq x \in \ker B$. Theorem 4.11.9 (applied with $g := x, g' := 1$) gives

$$0 = \sum_{\xi \in \mathrm{Irr}(B)} \xi(x)\xi(1) = \sum_{\xi \in \mathrm{Irr}(B)} \xi(1)^2,$$

a contradiction. $\qquad\square$

Theorem 4.11.14 *Let u be a p-element of G and $v \in \mathbf{O}_{p'}(\mathbf{C}_G(u))$. Then*

$$\chi(uv) = \chi(u) \qquad \text{for every} \qquad \chi \in \mathrm{Irr}(B_0(G)).$$

Proof. By Brauer's second and third main theorem (Corollary 4.11.6 and Theorem 4.7.15) we have

$$\chi(uv) = \sum_{\varphi \in \mathrm{IBr}(b_0)} d_{\chi\varphi}^u \varphi(v),$$

where $b_0 = B_0(\mathbf{C}_G(u))$. But by Theorem 4.11.13 $\varphi(v) = \varphi(1)$ for $\varphi \in b_0$. From this the result follows. $\qquad\square$

Example 4.11.15 Let $G = \mathrm{M}_{22}$ and $p = 3$. There is just one conjugacy class of 3-elements, $u^G = \mathbf{3a}$, and one class, $x^G = \mathbf{6a}$, of elements of order six. Note that $\mathbf{C}_G(u)$ contains exactly three involutions because $[\mathbf{C}_G(x^2) : \mathbf{C}_G(x)] = 3$. Since there are no elements of order four in $\mathbf{C}_G(u)$, it follows that $\mathbf{C}_G(u)$ has a normal Sylow 2-subgroup (of order four). Theorem 4.11.14 implies that $\chi_i(\mathbf{3a}) = \chi_i(\mathbf{6a})$ for $i \in I := \{1, 5, 7, 10, 11, 12\}$ because $B_0(G) = \{\chi_i \mid i \in I\}$. Compare this with Exercise 4.11.1.

Since $G = \mathrm{M}_{22}$ has only one 2-block, we can also deduce from Theorem 4.11.14 that $\mathbf{O}_{2'}(\mathbf{C}_G(u)) = \{1\}$ for any 2-element $u \in G$. $\qquad\blacklozenge$

We are aiming at an interesting and important group theoretical application, which deals with finite groups containing a conjugacy class C of involutions such that every Sylow 2-subgroup of G contains exactly one element of C. Trivially (by the Sylow theorems) this condition is fulfilled whenever a Sylow 2-subgroup of G contains just one involution. The 2-groups with this property are all known.

Lemma 4.11.16 *If a 2-group P contains just one involution, then either P is cyclic or is a generalized quaternion group*

$$Q_n = \langle x, y \mid x^{2^n} = 1, y^2 = x^{2^{n-1}}, y^{-1}xy = x^{-1} \rangle.$$

If $P \in \mathrm{Syl}_2(G)$ is cyclic, then $G = \mathbf{O}_{2'}(G)P$.

Proof. See [88], Satz I.14.9 and Satz IV.2.8. The proof is purely group theoretical. Transfer is used for the last assertion. $\qquad\square$

So a finite simple group cannot have a cyclic Sylow 2-subgroup. It cannot have a generalized quaternion group as a Sylow subgroup either, as the following, much deeper, theorem shows.

Theorem 4.11.17 (Brauer–Suzuki) *Let $P \in \mathrm{Syl}_2(G)$ be a generalized quaternion group. Then $\mathbf{O}_{2'}(G) \neq \{1\}$ or $|\mathbf{Z}(G)| = 2$.*

Proof. For a proof see [126]. All known proofs require representation theory. The hardest case is when P is the quaternion group of order eight. Here block theory is used. □

The last two results imply the following: if $P \in \mathrm{Syl}_2(G)$ has a unique involution u, then

$$u \in \mathbf{Z}^\star(G) := \{g \in G \mid g\mathbf{O}_{2'}(G) \in \mathbf{Z}(G/\mathbf{O}_{2'}(G))\}.$$

The following theorem is a generalization of this.

Theorem 4.11.18 (Glauberman's $\mathbf{Z}^\star$-theorem) *Suppose that u is an involution in $P \in \mathrm{Syl}_2(G)$ with $u^G \cap P = \{u\}$. Then $u \in \mathbf{Z}^\star(G)$. In particular, G is not simple, unless $G \cong \mathrm{C}_2$.*

Before entering into the proof, we give a lemma.

Lemma 4.11.19 *Let u be an involution in $P \in \mathrm{Syl}_2(G)$. Then $u^G \cap P = \{u\}$ if and only if $h = u\,u^g$ has odd order for every $g \in G$.*

Proof. Observe that $D := \langle u, u^g \rangle$ is a dihedral group of order $2m$, if $h := u\,u^g$ has order m. If $m = 2k$ is even then $h^k \in \mathbf{Z}(D)$ and

$$x := u\,h^k = \underbrace{u^g u \dots u^g}_{k-1} u \underbrace{u^g u \dots u^g}_{k-1} \in u^G \cap P^y \qquad \text{for some} \qquad y \in G,$$

because $\langle u, x \rangle$ has order four. Moreover $u^G \cap P \neq \{u\}$. The converse is obvious. □

Proof of Theorem 4.11.18. We use induction on $|G|$ and assume that $u \in P \in \mathrm{Syl}_2(G)$ satisfies the hypothesis of the theorem.
(1) We may assume that $\mathbf{O}_{2'}(G) = \{1\}$ and that u is not contained in any proper normal subgroup of G. For, $\bar{u} := u\mathbf{O}_{2'}(G) \in \bar{G} := G/\mathbf{O}_{2'}(G)$ satisfies $\bar{u}^{\bar{G}} \cap \bar{P} = \{\bar{u}\}$ with $\bar{P} := P\mathbf{O}_{2'}(G)/\mathbf{O}_{2'}(G)$. If $\mathbf{O}_{2'}(G) \neq 1$ then, by induction, $\bar{u} \in \mathbf{Z}^\star(\bar{G}) = \mathbf{Z}(\bar{G})$, which means that $u \in \mathbf{Z}^\star(G)$. Likewise, if $u \in N \lhd G$, then by induction $u \in \mathbf{Z}^\star(N) = \mathbf{Z}(N)$, since $\mathbf{O}_{2'}(N) \leq \mathbf{O}_{2'}(G) = \{1\}$. From Lemma 4.11.19 we conclude that $u \in \mathbf{Z}(G)$, because for all $g \in G$ we have $uu^g \in \mathbf{Z}(N) \leq \mathbf{O}_2(G)$.

(2) If $t \in G \setminus u^G$ is an involution, then $y := ut$ has order $2m$ with m odd. If the order of y were odd, then $\langle u \rangle, \langle t \rangle \in \mathrm{Syl}_2(\langle u, t \rangle)$, so $t \in u^G$. So the order of y is even. But $y^2 = uu^t$ is of odd order by Lemma 4.11.19.

(3) If $\chi \in B_0(G)$ and $t \in P \setminus \{u\}$ is an involution, then

$$\chi(ut) = \chi(u't') \qquad \text{for all} \qquad u' \in u^G, \ t' \in t^G. \tag{4.47}$$

Since χ is a class function, we may assume that $u' = u$.

(a) If $t^g = t' \in \mathbf{C}_G(u)$, then $u, u^g \in \mathbf{C}_G(t^g)$, since $u \in \mathbf{Z}(P)$ and thus $u \in \mathbf{C}_G(t)$. By Lemma 4.11.19 there is an $h \in \langle u, u^g \rangle \leq \mathbf{C}_G(t^g)$ with $u^{gh} = u$. Then $(ut)^{gh} = ut^g = u't'$ and (4.47) holds for all $\chi \in \mathrm{Irr}(G)$.

(b) So assume $t' \notin \mathbf{C}_G(u)$ and put $y := ut'$ and $x := y^m$, where $2m$ is the order of y; see (2). Then x is a central involution in $\langle u, t' \rangle$. We put $H := \mathbf{C}_G(x) \geq \langle u, t' \rangle$. If $H = G$, then $u \langle x \rangle \in \mathbf{Z}^*(G/\langle x \rangle) = \mathbf{Z}(G/\langle x \rangle)$ and $\langle u, x \rangle \trianglelefteq G$, which gives a contradiction to (1). Thus $H < G$, and by induction $u \in \mathbf{Z}^*(H)$ and $y = ut' \equiv t'u \bmod \mathbf{O}_{2'}(H)$. Hence $y^2 \equiv ut't'u = 1 \bmod \mathbf{O}_{2'}(H)$ and $y = xh$ with $h \in \mathbf{O}_{2'}(G)$. By Theorem 4.11.14 we get

$$\chi(ut') = \chi(xh) = \chi(x) \qquad \text{for every} \qquad \chi \in \mathrm{Irr}(B_0(G)).$$

But $t'' := ux$ is an involution in $\langle u, t' \rangle$ which is not conjugate to u (by Lemma 4.11.19), hence conjugate to t'. Thus $x = ut''$ and, since $t'' \in \mathbf{C}_G(u)$, we can use part (a) to see that $\chi(x) = \chi(ut'') = \chi(ut)$ for all $\chi \in \mathrm{Irr}(G)$.

(4) If $t \in P \setminus \{u\}$ is an involution, then for any $\chi \in \mathrm{Irr}(B_0(G))$

$$\chi(t) = 0 \qquad \text{or} \qquad \chi(u) = \pm\chi(1).$$

Let $\omega := \omega_\chi$ be the central character corresponding to $\chi \in \mathrm{Irr}(B_0(G))$. Then

$$\omega((u^G)^+)\omega((t^G)^+) = \omega((u^G)^+(t^G)^+),$$

so

$$\chi(u)\chi(t) = \frac{\chi(1)^2}{|u^G||t^G|}\omega((u^G)^+(t^G)^+).$$

Note that $(u^G)^+(t^G)^+$ is a sum of $|u^G||t^G|$ elements of the form $u't'$ with $u' \in u^G$ and $t' \in t^G$. By (3) we have

$$\omega((u^G)^+(t^G)^+) = |u^G||t^G|\frac{\chi(ut)}{\chi(1)},$$

and consequently

$$\chi(u)\chi(t) = \chi(1)\chi(ut).$$

Note that ut is also an involution in $P \setminus \{u\}$. Replacing t by ut we get

$$\chi(u)\chi(ut) = \chi(1)\chi(t), \qquad \text{hence} \qquad \chi(u)^2\chi(t) = \chi(1)^2\chi(t),$$

from which the result follows.

(5) Final part of the proof. By Lemma 4.11.16 and Theorem 4.11.17, we may restrict ourselves to the case in which P contains an involution $t \neq u$. By (1)

we may also assume that $\chi(u) \neq \chi(1)$ for all non-trivial $\chi \in \mathrm{Irr}(G)$. Using block orthogonality (Theorem 4.11.9) twice we get

$$0 = \sum_{\chi \in B_0(G)} \chi(u)\chi(t) \qquad \text{and} \qquad 0 = \sum_{\chi \in B_0(G)} \chi(1)\chi(t).$$

Hence by (4) we have

$$0 = 2 + \sum_{\substack{\chi \in B_0(G) \\ \chi \neq 1_G}} (\chi(u) + \chi(1))\,\chi(t) = 2,$$

which is a contradiction.   □

Exercises

Exercise 4.11.1 Let $G := M_{11}$ and let u be a 3-element of G. From the character table of G (p. 177) one can see that $\mathbf{C}_G(u)$ contains an involution v. Verify that $\chi(uv) = \chi(u)$ does *not* hold for every $\chi \in \mathrm{Irr}(G)$ in the principal 3-block.

Exercise 4.11.2 Let $B \in \mathrm{Bl}_2(G)$ be a 2-block of defect one. (a) Show that

$$D_B = \begin{bmatrix} 1 \\ 1 \end{bmatrix} \qquad \text{and} \qquad D_B^x = \begin{bmatrix} 1 \\ -1 \end{bmatrix},$$

if $\langle x \rangle \in \mathrm{Def}(B)$.
Hint: Use Theorem 4.11.11.
    (b) Conclude that $\mathrm{Irr}(B) = \{\chi, \chi'\}$ and $\chi(g) = \chi'(g)$ for all $g \in G$ of odd order, whereas $\chi(g) = -\chi'(g)$ for all $g \in G$ of even order.
    (c) Assume that $\mathbf{C}_G(x) \cong C_2 \times A_5$. Show that $\mathrm{Irr}(B) = \{\chi, \chi'\}$ with $\{\chi(x), \chi'(x)\} = \{4, -4\}$.

Exercise 4.11.3 Let $G := J_1$, let x be an involution in G and let $g, h \in G$ be elements of order three and five respectively.

(a) Use the character table of G and the fact (proved in Example 4.12.10) that $\mathbf{C}_G(x) \cong C_2 \times A_5$ to verify that

$$D_{B_0(G)}^x = \begin{bmatrix} 1 & 0 & 0 \\ 1 & 1 & 1 \\ -1 & 0 & -1 \\ -1 & -1 & 0 \\ 1 & 1 & 1 \\ -1 & 0 & -1 \\ -1 & -1 & 0 \\ 1 & 0 & 0 \end{bmatrix}.$$

(b) Using the fact (proved in Example 4.12.10) that $\mathbf{C}_G(g) \cong \mathbf{C}_G(h) \cong C_3 \times D_{10}$, verify Theorem 4.11.14 for $p = 3$ and $p = 5$.

Exercise 4.11.4 (See ref. [133].) Let $B \in \mathrm{Bl}_p(G)$. Show that

(a) $\ker B = \{g \in G \mid g \cdot \hat{\epsilon}_B = \hat{\epsilon}_B\}$;

(b) $\ker B = \mathrm{Inv}_G(V)$ for any projective FG-module V belonging to B;

(c) $\ker B = \mathbf{O}_{p'}(\bigcap \{\mathrm{Inv}_G(V) \mid V \text{ simple } FG\text{-module belonging to } B\})$ if V is a simple FG-module belonging to B.

4.12   Brauer's theory of blocks of defect one

In this section we assume that (K, R, F, η) is a p-modular splitting system for the finite group G.

We are going to present one of the highlights of modular representation theory, Brauer's description of p-blocks of defect one, which was later extended by Thompson and Dade to blocks with cyclic defect groups. We will not give a proof of the theorem, but rather point to practical applications. In particular we will show the power of this method by computing the complete character table of a simple group of order 175560 purely by using a few group theoretic restrictions and the theory of blocks of defect one.

We collect the description of p-blocks of defect one in the following theorem. Note that the case $p = 2$ is not so interesting and has been dealt with in Exercise 4.11.2.

Theorem 4.12.1 (R. Brauer) *Let $p > 2$ and $B \in \mathrm{Bl}_p(G)$ be a p-block of defect one, $P \in \mathrm{Def}(B)$. We abbreviate $N := \mathbf{N}_G(P)$ and $C := \mathbf{C}_G(P)$. Let $b \in \mathrm{Bl}_p(C)$ such that $b^G = B$ with canonical character $\theta \in \mathrm{Irr}(b)$. Put $e := [T_N(\theta) : C]$ and $m := \frac{p-1}{e}$. Then we have the following.*

(a) $\mathrm{l}(B) = e$ *and* $\mathrm{k}(B) = e + m$.

(b) $\mathrm{Irr}(B) = \{\chi_1, \dots, \chi_e, \zeta_1, \dots, \zeta_m\}$ *with a family* $\{\zeta_1, \dots, \zeta_m\}$ *of p-conjugate characters (see Definition 2.2.12), called the **exceptional characters** of B, and $\chi_1, \dots, \chi_e$ being p-rational characters, called the **non-exceptional characters** of B. In particular $\zeta_i|_{G_{p'}} = \zeta_j|_{G_{p'}}$ for $i, j \in \{1, \dots, m\}$.*

(c) *All entries in the decomposition matrix $D(B)$ are zero or one. If Φ_j is a projective indecomposable character belonging to B $(1 \leq j \leq e)$ then $\Phi_j = \chi_{j_1} + \chi_{j_2}$ with $j_1 \neq j_2 \in \{0, \dots, e\}$, where $\chi_0 := \zeta_1 + \cdots + \zeta_m$.*

(d) *There are uniquely defined "signs" $\epsilon_i \in \{1, -1\}$ for $i = 0, \dots, e$ such that*

$$\chi_i(1)_{p'} \equiv \epsilon_i [G : N]_{p'} \frac{[N : C]}{e} \theta(1)_{p'} \quad \mathrm{mod}\ p \qquad (1 \leq i \leq e) \qquad (4.48)$$

and

$$\zeta_k(1)_{p'} \equiv \epsilon_0 [G : N]_{p'} [N : C] \theta(1)_{p'} \quad \mathrm{mod}\ p \qquad (1 \leq k \leq m). \qquad (4.49)$$

(e) *For $g \in G_{p'}$*

$$\epsilon_0 \zeta_k(g) = \sum_{i=1}^{e} \epsilon_i \chi_i(g) \qquad (1 \leq k \leq m).$$

(f) *Let RGe_j with $e_j^2 = e_j \in RG$ be a projective indecomposable RG-module belonging to B with character $\Phi_j = \chi_{j_1} + \chi_{j_2}$ as in (c). For $i \in \{0, \dots, e\}$ put $Y_{j,i} := KGe_j\epsilon_{\chi_i} \cap RGe_j$, where $\epsilon_{\chi_0} := \sum_{k=1}^{m} \epsilon_{\zeta_k}$. Then $KY_{j,i}$ has character χ_i for $i \in \{j_1, j_2\}$. Furthermore $\hat{Y}_{j,j_1}, \hat{Y}_{j,j_2}$ are uniserial submodules of $FG\hat{e}_j$ and*

$$\mathrm{Rad}(FG\hat{e}_j) = \hat{Y}_{j,j_1} + \hat{Y}_{j,j_2}, \qquad \hat{Y}_{j,j_1} \cap \hat{Y}_{j,j_2} = \mathrm{Soc}(FG\hat{e}_j).$$

Proof. A proof of parts (a) to (e) may be found in [69], sect. 11. For a proof of (f) see [56], sect. VII.8. Observe that part (a) of the theorem was already proved in Theorem 4.11.11. ☐

Corollary 4.12.2 *If in Theorem 4.12.1 $|G|_p = p$ then the congruences (4.48) and (4.49) may be replaced by*

$$\chi_i(1) \equiv \epsilon_i \frac{[N:C]}{e}\theta(1) \mod p \quad and \quad \zeta_k(1) \equiv \epsilon_0[N:C]\theta(1) \mod p.$$

If B is the principal p-block, then $\theta = 1_C$ and $e = [N:C]$. Thus the congruences (4.48) and (4.49) may be replaced by

$$\chi_i(1) \equiv \epsilon_i \mod p \quad and \quad \zeta_k(1) \equiv \epsilon_0[N:C] \mod p.$$

Proof. If $|G|_p = p$ then $P \in \mathrm{Syl}_p(G)$ and $[G:N] \equiv 1 \mod p$ by the Sylow theorems. Also p does not divide $\chi_i(1)$, $\theta(1)$ or $\zeta_j(1)$. ☐

Corollary 4.12.3 *Let $P \in \mathrm{Syl}_p(G)$ have order p. Then*

$$|\{\chi \in \mathrm{Irr}(G) \mid \chi(1) \equiv \pm i \mod p\}| = |\{\chi \in \mathrm{Irr}(\mathbf{N}_G(P)) \mid \chi(1) \equiv \pm i \mod p\}|$$

for $i = 1, \ldots, p-1$.

Proof. Every irreducible character χ of G or $N := \mathbf{N}_G(P)$ with $\chi(1) \not\equiv 0$ mod p belongs to a p-block with defect group P. By Brauer's first main theorem (Theorem 4.7.22) the blocks B of G with defect group P are in bijection to those of N with defect group P. So it suffices to compare the number of $\chi \in \mathrm{Irr}(B)$ with degree $\equiv \pm i \mod p$ to those of B' with the same property, where $B' \in \mathrm{Bl}_p(N)$ with $(B')^G = B$. If $b \in \mathrm{Bl}_p(\mathbf{C}_G(P))$ is a block with $b^G = B$ then b^N is also defined (by Theorem 4.7.19) and $b^N = B'$ (by Exercise 4.7.3 and Theorem 4.7.22). So the result follows on applying Theorem 4.12.1(d) to B and B'. ☐

Lemma 4.12.4 *Let p, q be odd primes with $p - 1 = 2q$ and $P \in \mathrm{Syl}_p(G)$ with $|\mathbf{C}_G(P)| \neq |\mathbf{N}_G(P)| = p(p-1)$. Then $\mathbf{N}_G(P)/P \cong C_{2q}$ or D_{2q}. Let $B_1, \ldots, B_r$ be the p-blocks of G with defect one and let $e^{(i)} := |\mathrm{IBr}(B_i)|$ for $i = 1, \ldots, r$. We also abbreviate $d_i = |\{\chi \in \mathrm{Irr}(G) \mid \chi(1) \equiv \pm i \mod p\}|$ for $i \in \{1, 2, q\}$. It is easily verified that there are just four possibilities for these parameters, as specified in Table 4.6.*

Corollary 4.12.5 *Let V_j be a simple FG-module with Brauer character $\varphi_j \in \mathrm{IBr}(B)$ and let $\Phi_j = \Phi_{\varphi_j} = \chi_i + \chi_k$ with $i \neq k \in \{0, \ldots, e\}$. Then $FG\hat{e}_j \cong P(V_j)$ has the submodule structure shown in Figure 4.3. Here $V_j, V_{j+1}, \ldots, V_{j+r-1}$ and $W_{j+1}, \ldots, W_{j+s-1}$ are simple FG-modules with $V_x \not\cong W_y$ for $x \in \{j+1, \ldots, j+r-1\}$ and $y \in \{j+1, \ldots, j+s-1\}$. Any uniserial submodule of $FG\hat{e}_j$ is contained in $\hat{Y}_{j,i}$ or $\hat{Y}_{j,k}$. If V_{j+l} has Brauer character $\varphi_{j'}$, then the composition factors of $\hat{Y}_{j',i}$ are $V_{j+l+1}, \ldots, V_{j+l+r}$, where the indices should be read modulo $j + r$.*

Table 4.6.

$[\mathbf{N}_G(P):\mathbf{C}_G(P)]$	$\mathbf{N}_G(P)/P$	r	$e^{(1)},\ldots,e^{(r)}$	d_1	d_2	d_q
$2\cdot q$	C_{2q}	1	$p-1$	p	.	.
q	C_{2q}	2	q,q	$2\cdot q$	.	4
2	C_{2q}	q	$2,\ldots,2$	$2\cdot q$	q^2	.
2	D_{2q}	$\frac{q+1}{2}$	$2,1,\ldots,1$	2	$q+\frac{(q-1)p}{2}$	.

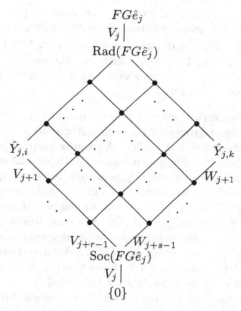

Figure 4.3. Submodule structure of $FG\hat{e}_j$.

Proof. The structure of $FG\hat{e}_j$ follows immediately from part (f) of Theorem 4.12.1. By part (c) of this theorem (and Brauer reciprocity, see Theorem 4.1.23(b)), χ_i and χ_k have only one modular constituent in common. Thus $\mathrm{Rad}(FG\hat{e}_j)/\mathrm{Soc}(FG\hat{e}_j)$ is the direct sum of two uniserial modules having no composition factor in common, and the poset of submodules is as indicated in the theorem (with exactly $r+s+2$ submodules). By Theorem 1.6.27 the uniserial module $\hat{Y}_{j,i}/\mathrm{Soc}(FG\hat{e}_j)$ embeds into $P(V_{j+r-1})$ and hence into $\hat{Y}_{j',i}$, if V_{j+r-1} has Brauer character $\varphi_{j'}$. Thus the composition factors of $\hat{Y}_{j',i}$ are $V_j, V_{j+1},\ldots,V_{j+r-1}$, that is, they are obtained from those of $\hat{Y}_{j,i}$ by a cyclic permutation. From this the last assertion follows. $\square$

Corollary 4.12.6 *If in Theorem* 4.12.1 χ_i *is real for some* $i \in \{0,\ldots,e\}$, *then* χ_i *has at most two real Brauer constituents.*

Proof. Let φ_j be a real Brauer constituent of χ_i and let $e_j \in RG$ be a primitive idempotent as in Theorem 4.12.1(f). By part (b), $\Phi_j = \chi_i + \chi_k$ for some $k \neq i \in \{0, \ldots, e\}$. Then $(FG\hat{e}_j)^\star \cong_{FG} FG\hat{e}_j$. By duality

$$(\hat{Y}_{j,i})^\star \cong (FG\hat{e}_j)^\star/(\hat{Y}_{j,i})^\circ \cong FG\hat{e}_j/\hat{Y}_{j,k}. \tag{4.50}$$

Let $V_{j+1}, \ldots, V_{j+r-1}, V_j$ be the composition factors of the uniserial FG-module $\hat{Y}_{j,i}$, then the composition factors of the uniserial FG-module $FG\hat{e}_j/\hat{Y}_{j,k}$ are $V_j, V_{j+1}, \ldots, V_{j+r-1}$. Hence, by (4.50),

$$V_j \cong (V_j)^\star, \; V_{j+1} \cong (V_{j+r-1})^\star, \; V_{j+2} \cong (V_{j+r-2})^\star, \ldots, V_{j+r-1} \cong (V_{j+1})^\star.$$

If $i \neq 0$ then $V_j, V_{j+1}, \ldots, V_{j+r-1}$ are pair-wise non-isomorphic and we see that $V_{j+r/2} \cong (V_{j+r/2})^\star$ if r is even, whereas V_j is the only selfdual composition factor of $\hat{Y}_{j,i}$ if r is odd. For $i = 0$ note that by Theorem 4.12.1(b) the modular irreducible constituents of χ_0 have multiplicity m. Hence $V_x \cong V_y$ if and only if $x \equiv y \mod s$, where $s = r/m$. As above we see that χ_0 has two real constituents if s is even and only one otherwise. $\qquad\square$

There is a nice geometrical way to present most of the information about p-blocks with cyclic defect groups. We recall that a **tree** is a finite, connected, undirected graph without loops or cycles. A **planar** graph is a graph together with an embedding into the Euclidean plane. Such an embedding induces a circular ordering of the edges emanating from each vertex by going around the vertex in a counter clockwise manner. The converse holds also, that is, such a a circular ordering of the edges determines a unique planar embedding.

By Theorem 4.12.1(c) every $\varphi_j \in \mathrm{IBr}(B)$ is a constituent of exactly two of the characters $\chi_0, \ldots, \chi_e$. Representing φ_j by an edge with end points χ_{j_1}, χ_{j_2} if $\Phi_j := \Phi_{\varphi_j} = \chi_{j_1} + \chi_{j_2}$, we obtain graph $\Gamma(B)$ with vertices $\chi_0, \ldots, \chi_e$ and edges the $\varphi_j \in \mathrm{IBr}(B)$. Theorem 1.7.7 shows that $\Gamma(B)$ is connected. Since $\Gamma(B)$ has e edges and $e + 1$ vertices, it must be a tree.

The planar embedding encodes the information contained in part (f) of Theorem 4.12.1. Assume that φ_j and $\varphi_{j'}$ are constituents of χ_i ($0 \leq i \leq e$). As shown in Corollary 4.12.5, the composition factors of $\hat{Y}_{j',i}$ are obtained from those of $\hat{Y}_{j,i}$ by a cyclic permutation. We thus obtain a unique cyclic ordering of the modular irreducible constituents of χ_i, i.e. of the edges of $\Gamma(B)$ emanating from χ_i.

Definition 4.12.7 Let B be a p-block of G as in Theorem 4.12.1. The **Brauer tree** $\Gamma(B)$ of B is a planar tree with vertices $\chi_0, \ldots, \chi_e$ ($\chi_0 = \zeta_1 + \cdots + \zeta_m$ is called the "exceptional vertex" and m its "multiplicity") and edges the irreducible Brauer characters belonging to B; χ_k is incident to $\varphi_i \in \mathrm{IBr}(B)$ if and only if φ_i is a constituent of $\chi_k|_{G_{p'}}$. The circular ordering of the edges emanating from χ_k is given by ordering the Brauer characters according to their occurrence as Brauer characters of the factors in the unique composition series of the uniserial module $\hat{Y}_{i,k}$, provided that χ_k is incident to φ_i.

Example 4.12.8 Let $B \in \mathrm{Bl}_p(G)$ be a block of defect one having a normal defect group. Then by Theorem 4.8.14 the Brauer tree $\Gamma(B)$ is a "star" with

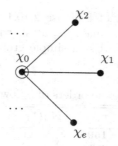

Figure 4.4. Brauer tree: a star.

the exceptional vertex in the center, and is as shown in Figure 4.4. Here for $i = 1, \ldots, e$ the vertex $\chi_i \in \mathrm{Irr}(B)$ satisfies $\chi_i|_{G'_p} = \varphi_i$, where φ_i is the Brauer character of the simple FG-module V_i in the notation of Theorem 4.8.14. We have omitted the labels for the edges, which is common practice.   ◆

Theorem 4.12.9 *Let B be a p-block $(p > 2)$ of defect one which is invariant under complex conjugation. Then the edges and vertices of $\Gamma(B)$ consisting of real characters and real Brauer characters, respectively, form a subgraph, which is an open polygon, called the **real stem** of $\Gamma(B)$.*

Proof. Complex conjugation induces an automorphism of order two of $\Gamma(B)$. Since there is a unique path in $\Gamma(B)$ joining two vertices, two real vertices must be joined by a path consisting of real edges and vertices. Moreover, the vertices that are incident with an edge cannot have the same degree, since $p > 2$. Thus vertices incident with a real edge are real. Hence the real vertices and edges of $\Gamma(B)$ form a connected subgraph, which is an open polygon by Corollary 4.12.6. Observe that, since e or $e + 1$ is odd, $\Gamma(B)$ always contains a real edge or a real vertex.   □

Example 4.12.10 We consider a simple group G of order $n := 2^3 \cdot 3 \cdot 5 \cdot 7 \cdot 11 \cdot 19$, which is the order of the sporadic simple Janko group J_1. In order to determine its character table we apply the theory of blocks of defect one and Brauer's main theorems on block theory. The main task is not to compute the character values, but to find the structure of the Sylow normalizers, often called the "local structure" of G. Block theory plays an essential role here and also in determining the centralizer orders of the elements of G. Once this is achieved, the computation of the characters presents no further difficulties.

We start by considering the possibilities for the number n_p of Sylow p-subgroups P_p and the orders of the normalizers of Sylow p-subgroups for $p = 19, 11, 7$ using purely group theoretical arguments, such as the Sylow theorems.

The simple GAP function syl of Exercise 2.10.1 may be used and gives the possible orders of the Sylow normalizers, as shown in Table 4.7. The correct values are printed in boldface. Using simple group theoretical arguments we will work through each in turn.

Table 4.7. Possible orders of Sylow normalizers

n_{19}	$|\mathbf{N}_G(P_{19})|$	n_{11}	$|\mathbf{N}_G(P_{11})|$	n_7	$|\mathbf{N}_G(P_7)|$
1540	$\mathbf{2 \cdot 3 \cdot 19}$	**1596**	$\mathbf{2 \cdot 5 \cdot 11}$	**4180**	$\mathbf{2 \cdot 3 \cdot 7}$
210	$2^2 \cdot 11 \cdot 19$	210	$2^2 \cdot 11 \cdot 19$	1254	$2^2 \cdot 5 \cdot 7$
77	$2^3 \cdot 3 \cdot 5 \cdot 19$	133	$2^3 \cdot 3 \cdot 5 \cdot 11$	330	$2^2 \cdot 7 \cdot 19$
20	$2 \cdot 3 \cdot 7 \cdot 11 \cdot 19$	56	$3 \cdot 5 \cdot 11 \cdot 19$	190	$2^2 \cdot 3 \cdot 7 \cdot 11$
				57	$2^3 \cdot 5 \cdot 7 \cdot 11$
				22	$2^2 \cdot 3 \cdot 5 \cdot 7 \cdot 19$

(a) $|\mathbf{N}_G(P_7)| = 7 \cdot 6$. $n_{19} = 20$ is impossible, because this would imply that G is isomorphic to a subgroup of the alternating group A_{20} and on the other hand would contain elements of order $11 \cdot 19$, which is absurd. Hence G has no elements of order $7 \cdot 19$. Looking at the possible orders of $\mathbf{N}_G(P_{11})$ one finds that there are no elements of order $7 \cdot 11$ either. Thus $|\mathbf{N}_G(P_7)|$ is coprime to $11 \cdot 19$ and hence equal to $7 \cdot 6$ or $7 \cdot 20$. But the latter implies that $P_5 \trianglelefteq \mathbf{N}_G(P_7)$ for $P_5 \in \mathrm{Syl}_5(\mathbf{N}_G(P_7))$, so $7 \cdot 20 \mid |\mathbf{N}_G(P_5)|$, and consequently $|\mathbf{N}_G(P_5)| = 7 \cdot 20 \cdot 19$, which is a contradiction since this would lead to elements of order $7 \cdot 19$ (in a 5-complement in $\mathbf{N}_G(P_5)$). Hence (a) follows.

(b) $|\mathbf{N}_G(P_{19})| = 19 \cdot 6$.   In the proof of (a) we have already excluded $n_{19} = 20$. We have to show that $n_{19} \notin \{77, 210\}$.

   (i) Assume that $n_{19} = 77$. Then $N := \mathbf{N}_G(P_{19}) = P_{19} \rtimes U$ with $|U| = 120$. It follows that $19 \mid |\mathbf{N}_G(P_5)|$. We already know from (a) that $7 \nmid |\mathbf{N}_G(P_5)|$ (observe that otherwise P_7 and P_5 must centralize each other). We use syl (see Exercise 2.10.1) to find the possible number of Sylow 5-subgroups of G:

```
gap> Filtered( syl( 175560, 5 ), x-> x mod 19 <> 0 and x mod 7 = 0 );
[ 21, 231 ]
```

Thus $3 \nmid |\mathbf{N}_G(P_5)|$, and so G has no elements of order 15. On the other hand, in a solvable group of order 120 a Sylow 2-complement (see [88], sect. VI.1) has order 15. So A_5 must be a composition factor of U, and consequently $U \in \{S_5, SL_2(5), A_5 \times C_2\}$ by [38], p. 2. Hence $[N : \mathbf{C}_G(P_{19})] = 2$, so $U \cong S_5$ or $U \cong A_5 \times C_2$, and there are elements of order $3 \cdot 19$. We consider the (five) possible orders of $\mathbf{N}_G(P_3)$:

```
gap> s := Filtered( syl(175560, 3), x-> x mod 19 <> 0 and x mod 5 = 0 );;
gap> for m in s do Print("19 * ",175560/(m*19),",    "); od;
19 * 6 ,    19 * 42 ,    19 * 132 ,    19 * 24 ,    19 * 168 ,
```

In each case $\mathbf{N}_G(P_3)$ would have just one Sylow 19-subgroup, thus $\mathbf{N}_G(P_3) \subseteq N$ and $\mathbf{N}_G(P_3) = \mathbf{N}_N(P_3)$. But $|\mathbf{N}_U(P_3)| = 12$, so $|\mathbf{N}_N(P_3)| = 19 \cdot 12$, which gives a contradiction.

(ii) Assume now that $n_{19} = 210$. Then $N := \mathbf{N}_G(P_{19}) = \mathbf{N}_G(P_{11})$ and $[N : \mathbf{C}_N(P_{19})] = [N : \mathbf{C}_N(P_{11})] = 2$. Obviously the commutator subgroup N' is cyclic of order $11 \cdot 19$, and so N has exactly four linear characters and $\chi(1) \mid 4$ for all $\chi \in \mathrm{Irr}(N)$ by Theorem 3.6.5. By Corollary 4.12.3, G must have exactly four irreducible characters of degree $\equiv \pm 1 \mod p$ for $p = 11, 19$. The principal 19-block B_0 has nine exceptional characters of degree b with $b \equiv \pm 2 \mod 19$, the trivial character and one further non-exceptional character of degree a with $a \equiv \pm 1 \mod 19$. Of course, a and b must be divisors of $|G|$ and $2 < a, b < \sqrt{|G|}$ (by Exercise 2.2.4), that is, a, b must be in the following list degs:

```
gap> degs := Filtered( [1..419], x ->  IsInt(175560/x) and x <>2 );;
gap> List([1,2] , i -> Filtered( degs, x -> x mod 19 in [i,19-i] ));
[ [ 1, 20, 56, 77, 132, 210 ], [ 21, 40, 55, 154, 264 ] ]
```

Thus $a \in \{20, 56, 77, 132, 210\}$ and $b \in \{21, 40, 55, 154, 264\}$. Theorem 4.12.1(e) implies $b = a \pm 1$, so $b \in \{21, 55\}$. Since G has only four irreducible characters of degree $\equiv \pm 1 \mod 11$, we conclude that $b \neq 21$. Lemma 4.12.4 applied with $p = 7$ implies that G has at most seven irreducible characters of degree $\equiv \pm 1 \mod 7$, hence $b \neq 55$. This contradiction proves (b).

(c) $\mathbf{C}_G(P_{19}) = P_{19}$ and G has three algebraically conjugate conjugacy classes 19a, 19b and 19c of elements of order 19. Also the principal 19-block B_0 is the only 19-block of defect one.

(i) Assume that $[\mathbf{N}_G(P_{19}) : \mathbf{C}_G(P_{19})] = 2$. Then G has elements of order $3 \cdot 19$. If there are also elements of order $3 \cdot 11$, that is, if $3 \mid \mid \mathbf{N}_G(P_{11})\mid$, it follows (using syl) that $\mid \mathbf{N}_G(P_3)\mid \in \{2 \cdot 3 \cdot 5 \cdot 11 \cdot 19 , \ 2^2 \cdot 3 \cdot 11 \cdot 19\}$, hence by the Sylow theorems $P_{19} \trianglelefteq \mathbf{N}_G(P_3)$, which gives a contradiction to (b). So $\mid \mathbf{N}_G(P_{11})\mid = 10 \cdot 11$, since the other possibility, $2^2 \cdot 11 \cdot 19$, would imply $11 \mid \mid \mathbf{N}_G(P_{19})\mid$. As in the proof of (b) part (ii), we see that the principal 19-block B_0 contains nine exceptional characters $\theta_1, \ldots, \theta_9$ of degree $b \in \{21, 55\}$ and a non-exceptional character of degree $a \in \{20, 56\}$. Again the case $b = 55$ can be ruled out by Lemma 4.12.4 with $p = 7$. Similarly, Lemma 4.12.4 applied with $p = 11$ and $q = 5$ shows that $(a, b) \neq (20, 21)$, because $20 \equiv -2 \mod 11$ implies that $[\mathbf{N}_G(P_{11}) : \mathbf{C}_G(P_{11})] = 2$, and $\theta_1, \ldots, \theta_9$ with $\theta_i(1) = 21 \equiv -1 \mod 11$ would have to be the only non-trivial non-exceptional characters in the five 11-blocks of defect one. Exactly one of these characters would have to be in the principal 11-block, which is a contradiction, since the θ_i are algebraically conjugate.

(ii) Now assume that $[\mathbf{N}_G(P_{19}) : \mathbf{C}_G(P_{19})] = 3$. The principal 19-block B_0 contains six exceptional characters of degree b with $7 \mid b$ or $b \equiv \pm 2 \mod 7$ by Lemma 4.12.4. The GAP program degblock of Exercise 4.12.2 shows that there is just one possibility for the degrees of the irreducible characters in B_0:

```
gap> d := List([1,3], i -> Filtered( degs, x -> x mod 19 in [i,19-i] ));;
gap> d[2]:= Filtered(d[2], x->6*x^2 < 175560 and x mod 7 in [0,2,5]);; d;
[ [ 1, 20, 56, 77, 132, 210 ], [ 35, 168 ] ]
```

```
gap> dd := degblock( 175560, d[1], d[2], 3, 19 );
[ [ [ -56, 1, 20 ], -35 ] ]
```

Thus the non-exceptional characters in B_0 have degree 1, 20 and 56 and the exceptional characters have degree 35, which contradicts Theorem 4.12.9. This proves (c).

(d) $|\mathbf{N}_G(P_{11})| = 11 \cdot 10$ or $[\mathbf{N}_G(P_{11}) : \mathbf{C}_G(P_{11})] = 2$ and $\mathbf{C}_G(P_{11}) \cong P_{11} \times A_5$. There is no $\chi \in \mathrm{Irr}(G)$ with $\chi(1) = 20$.

Since $19 \nmid |\mathbf{N}_G(P_{11})|$ we know $|\mathbf{N}_G(P_{11})| \in \{11 \cdot 10, \ 11 \cdot 120\}$. Assume that $|\mathbf{N}_G(P_{11})| = 11 \cdot 120$, that is, $N := \mathbf{N}_G(P_{11}) = P_{11} \rtimes U$ with U of order 120. Then U can have one or six Sylow 5-subgroups, so $|\mathbf{N}_U(P_5)| \in \{120, 20\}$ for $P_5 \in \mathrm{Syl}_5(U)$. We rule out 120, since $|\mathbf{N}_G(P_5)| \in \{10, 110, 60, 660\}$ as the following code shows:

```
gap> List(Filtered( syl(175560,5), x -> x mod (7*19) = 0 ), x-> 175560/x);
[ 10, 110, 60, 660 ]
```

Thus U has six Sylow 5-subgroups and as in (b) it follows that U is isomorphic to S_5, $SL_2(5)$ or $A_5 \times C_2$. Since U has a normal subgroup $U \cap \mathbf{C}_G(P_{11})$ of index 2, 5 or 10, we can exclude $SL_2(5)$. It follows that $\mathbf{C}_G(P_{11}) \cong P_{11} \times A_5$, and by Exercise 4.12.1 G has no irreducible character of degree 20.

If $|\mathbf{N}_G(P_{11})| = 110$ then by Lemma 4.12.4 applied with $p = 11, q = 5$ there would be zero or at least five irreducible characters of degree 20. On the other hand, any such character would be a non-exceptional character in the principal 19-block B_0. But by Theorem 4.12.1 B_0 has just five such characters $\neq 1_G$ and they cannot all have degree 20 because otherwise Theorem 4.12.1(e) would imply that the exceptional characters in B_0 would have degree $1 + 5 \cdot 20 = 101 \nmid |G|$.

(e) $\mathbf{C}_G(P_7) = P_7$ and G has just one conjugacy class 7a of elements of order seven. The principal block is the only 7-block of defect one.

Let **degs** be the set of positive divisors of $|G|$ congruent to 0, ± 1 or ± 6 modulo 19 which are less than $\sqrt{|G|}$ and different from 2 and 20. By (c) and (d) we know that $\chi(1) \in$ **degs** for $\chi \in \mathrm{Irr}(G)$. For $i = 1, 2, 3$ we consider $\mathtt{d[i]} := \{x \in \mathbf{degs} \mid x \equiv \pm i \mod 7\}$:

```
gap> degs := Filtered( degs, x-> x mod 19 in [0,1,6,13,18] and x <> 20);;
gap> d := List( [1,2,3], i-> Filtered(degs, x -> x mod 7 in [i,7-i]) );
[ [ 1, 6, 57, 76, 120, 132, 190, 209 ], [ 19, 44, 114, 152, 285, 380, 418 ],
  [ 38, 95, 165, 228 ] ]
gap> Print(degblock(175560,d[1],d[2],2,7),degblock(175560,d[1],d[3],3,7));
[ ][ ]
```

The degrees of the non-exceptional characters in the principal 7-block are in **d[1]**. If $[\mathbf{N}_G(P_7) : \mathbf{C}_G(P_7)] = 2$ or 3 then the degrees of the exceptional characters in the principal 7-block are in **d[2]** and **d[3]**, respectively. The GAP program **degblock** of Exercise 4.12.2 shows that these cases cannot occur, and (e) follows.

(f) The degrees of the irreducible characters in the principal 19-block B_0 are
$1, 56, 56, 56, 77, 210, 120, 120, 120$ or $1, 56, 56, 77, 77, 77, 120, 120, 120$.
$|\mathbf{N}_G(P_{11})| = 11 \cdot 10$, $\mathbf{C}_G(P_{11}) = P_{11}$, and G has just one class **11a** of elements
of order 11.

By (e) $\chi(1) \in \{x \in \mathbb{N} \mid x \equiv a \mod 7 \text{ for } a \in \{0, \pm 1\}\}$, so we easily see that
the non-exceptional characters of B_0 have degrees in $d_1 := \{1, 56, 77, 132, 210\}$
and the exceptional ones in $d_2 := \{6, 70, 120\}$. The first assertion follows by
applying again the GAP program `degblock` of Exercise 4.12.2:

```
gap> degs := Filtered(degs, x -> x mod 7 in [0,1,6]);;
gap> d := List( [1,6] , i-> Filtered(degs, x -> x mod 19 in [i,19-i]) );
[ [ 1, 56, 77, 132, 210 ], [ 6, 70, 120 ] ]
gap> degblock( 175560, d[1], d[2], 6, 19 );
[ [ [ -56, -56, -56, 1, 77, 210 ], 120 ],
  [ [ -56, -56, 1, 77, 77, 77 ], 120 ] ]
```

In both cases there are more than five irreducible characters of degree $\equiv \pm 1$
mod 11. Applying Exercise 4.12.1 we see that (d) implies $|\mathbf{N}_G(P_{11})| = 11 \cdot 10$.
Similarly as in (e), we see that $\mathbf{C}_G(P_{11}) = P_{11}$.

(g) G has 13 or 15 conjugacy classes and

$$(\chi(1) \mid \chi \in \mathrm{Irr}(G)) =$$
$$(1, 56, 56, 56, 76, 76, 77, 120, 120, 120, 133, 209, 210), \qquad (4.51)$$

or

$$(1, 56, 56, 76, 76, 77, 77, 77, 120, 120, 120, 133, 133, 133, 209). \quad (4.52)$$

By (f), eight (or six, respectively) of the characters in the principal 19-block
have 11-defect one. So the principal 11-block must also contain three (or five)
characters of 19-defect zero. These have degrees in $\{76, 133\}$ as the following
GAP code shows:

```
gap> Filtered( degs, x-> x mod 19 = 0 and x mod 11 in [1,10] );
[ 76, 133 ]
```

By Theorem 4.12.1 we have to find $x, y \in \mathbb{N}_0$ with

$$1 + 3 \cdot 56 + 210 - 3 \cdot 120 = \quad 19 \quad = \quad x \cdot 76 - y \cdot 133, \qquad x + y = 3$$

and

$$1 + 2 \cdot 56 - 3 \cdot 120 = -247 \quad = \quad x \cdot 76 - y \cdot 133, \qquad x + y = 5,$$

respectively. It is easily seen that the only solution is $(x, y) = (2, 1)$ in the first
case and $(x, y) = (2, 3)$ in the second case. We define

$$I := \{\chi \in \mathrm{Irr}(G) \mid d_{19}(\chi) = 1 \text{ or } d_{11}(\chi) = 1\}$$

and obtain in both cases

$$|G| - \sum_{\chi \in I} \chi(1)^2 = (11 \cdot 19)^2.$$

This means that there is a unique irreducible character of degree $11 \cdot 19 = 209$,
which is not of p-defect one for $p = 19$ or $p = 11$, and (g) follows.

(h) The degrees of the irreducible characters of G are given in (4.52). All $\chi \in \mathrm{Irr}(G)$ have Frobenius Schur indicator one and G has a single class **2a** of involutions with centralizer order 120.

From (c), (e) and (f) we know that P_q is self-centralizing for $q = 7, 11, 19$. Hence the centralizer of any p-element $\neq 1$ for $p \notin \{7, 11, 19\}$ has order dividing $2^3 \cdot 3 \cdot 5$. In particular there are at least $175560/(2^3 \cdot 3 \cdot 5) = 1463$ involutions. By Theorem 2.9.9, $\sum_{\chi \in \mathrm{Irr}(G)} \chi(1)$ is an upper bound for the number of elements of order ≤ 2 in G. For the sum of the degrees we find 1310 in (4.51) and 1464 in (4.52). Thus (h) follows.

We finally determine the normalizers of P_p for $p \in \{3, 5\}$. Using the proof of (h) we find for the possible orders of $\mathbf{N}_G(P_p)$:

```
gap> List( [3,5], p -> Filtered( List( syl(175560, p), x -> 175560/x ),
>                       y -> IsInt( 120/y ) ) );
[ [ 6, 60, 24 ], [ 10, 60 ] ]
```

From (h) we get $10 \mid |\mathbf{C}_G(P_5)|$, so $|\mathbf{N}_G(P_5)| = 60$. Consequently $5 \mid |\mathbf{C}_G(P_3)|$ and $|\mathbf{N}_G(P_3)| = 60$. It follows that

$$N := \mathbf{N}_G(P_5) \cong \mathbf{N}_G(P_3) \cong P_5 \rtimes U_5 \cong P_3 \rtimes U_3$$

with Sylow complements U_5 and U_3. From (4.52) we see that G has 12 irreducible characters of 5-defect one. By Theorem 4.12.1 a 5-block of defect one contains five (if e = 4 or 1) or four (if e = 2) irreducible characters. Hence G and, by Brauer's first main theorem, N has three 5-blocks of defect one, all with e = 2. It follows that $[N : \mathbf{C}_G(P_5)] = 2$. Since $|\mathrm{Irr}(N)| = 12$, it is clear that $\mathbf{C}_G(P_5)$ cannot be abelian, thus $\mathbf{C}_G(P_5) \cong P_5 \times S_3$. Similarly $\mathbf{C}_G(P_3) \cong P_3 \times D_{10}$, and we conclude that $N \cong S_3 \times D_{10}$.

Hence G has two conjugacy classes **5a, 5b** of elements of order five with centralizer order 30, two conjugacy classes **15a, 15b** of elements of order 15 with centralizer order 15, two conjugacy classes **10a, 10b** of elements of order ten with centralizer order 10, and a conjugacy class **3a** and **6a** with centralizer order 30 and 6, respectively. Observe that we were using the trivial fact that $\mathbf{C}_G(g) \subseteq \mathbf{C}_G(g^i)$ for any $i \in \mathbb{N}$ and any $g \in G$.

In total we have found the centralizer orders of 15 (that is, of all) conjugacy classes of G. Also the power maps are easily determined, since we know that the classes **5a, 5b** are algebraically conjugate as well as the classes of elements of order 19. Of course, we may assume that the cubes of elements in **15a** are in **5a** as well as the fifth powers of those of **10a**. As noted in Remark 3.2.27, the character table of J_1 is uniquely determined by this information, since one can just induce up the linear characters of the cyclic subgroups and use the LLL algorithm to extract all irreducible characters of G:

```
gap> t := CharacterTable("J1");;
gap> ind := InducedCyclic( t, "all" );; ll := LLL( t, ind );;
gap> List( ll.irreducibles, y -> y[1] );
[ 56, 56, 120, 120, 120, 76, 76 ]
```

```
gap> red := ReducedClassFunctions( t, [List([1..15],x->1)],ll.remainders);;
gap> List( red.irreducibles, y -> y[1] );
[ 77, 77, 77, 133, 133, 133, 209 ]
```

The LLL algorithm applied to all characters induced from the linear characters of cyclic subgroups yields seven irreducible characters. Reducing the remainders with the trivial character we obtain all the missing irreducible characters of G. We display the character table of G in Table 4.8.

Table 4.8. Character table of J_1

J_1	$\lvert G\rvert$	120	30	30	30	6	7	10	10	11	15	15	19	19	19
	1a	2a	3a	5a	b	6a	7a	10a	b	11a	15a	b	19a	b	c
χ_1	1	1	1	1	1	1	1	1	1	1	1	1	1	1	1
χ_2	56	.	2	α	α'	.	.	.	.	1	β	β'	-1	-1	-1
χ_3	56	.	2	α'	α	.	.	.	.	1	β'	β	-1	-1	-1
χ_4	76	4	1	1	1	1	-1	-1	-1	-1	1	1	.	.	.
χ_5	76	-4	1	1	1	-1	-1	1	1	-1	1	1	.	.	.
χ_6	77	5	-1	2	2	-1	.	.	.	.	-1	-1	1	1	1
χ_7	77	-3	2	β	β'	.	.	β	β'	.	β	β'	1	1	1
χ_8	77	-3	2	β'	β	.	.	β'	β	.	β'	β	1	1	1
χ_9	120	.	.	.	.	1	.	.	.	-1	.	.	γ	ϵ	δ
χ_{10}	120	.	.	.	.	1	.	.	.	-1	.	.	δ	γ	ϵ
χ_{11}	120	.	.	.	.	1	.	.	.	-1	.	.	ϵ	δ	γ
χ_{12}	133	5	1	-2	-2	-1	.	.	.	1	1	1	.	.	.
χ_{13}	133	-3	-2	$-\beta$	$-\beta'$	.	.	β	β'	1	$-\beta$	$-\beta'$	.	.	.
χ_{14}	133	-3	-2	$-\beta'$	$-\beta$	.	.	β'	β	1	$-\beta'$	$-\beta$	.	.	.
χ_{15}	209	1	-1	-1	-1	1	-1	1	1	.	-1	-1	.	.	.

$\alpha,\ \alpha' := 1 \mp \sqrt{5}$, $\beta,\ \beta' := \frac{1}{2}(-1 \pm \sqrt{5})$, $\gamma := \zeta_{19} + \zeta_{19}^{7} + \zeta_{19}^{8} + \zeta_{19}^{11} + \zeta_{19}^{12} + \zeta_{19}^{18}$,
$\delta := \eta_4(\gamma)$ and $\epsilon := \eta_2(\gamma)$, with $\eta_i \in \mathrm{Gal}(\mathbb{Q}(\zeta_{19})/\mathbb{Q})$ mapping ζ_{19} to ζ_{19}^{i}.

Of course, it would also be easy to complete the character table by hand just using the degrees and the information about the conjugacy classes we had proved above. For instance, χ_4, χ_5 (of degree 76) are of 2-defect one with defect group $\langle x\rangle$, where $x \in 2\mathrm{a}$. The centralizer orders and element orders show that $\mathbf{C}_G(x) \cong C_2 \times A_5$. In fact, since G has no elements of order 30 we conclude that $\mathbf{C}_G(x)$ is non-solvable, and since there are no elements of order four we know $\mathbf{C}_G(x) \not\cong 2.A_5$. From Exercise 4.11.2 we get $\{\chi_4(x), \chi_5(x)\} = \{4, -4\}$. The remaining values of χ_4, χ_5 are completely determined by the congruence method (see Remark 2.2.3). Observe that G has four 3-blocks of defect one, for which we may use Exercise 4.12.3. In particular, for the principal block $B_0 \in \mathrm{Bl}_3(G)$ with $\mathrm{Irr}(B_0) = \{\chi_1, \chi_4, \chi_6\}$, we see that χ_6 can be directly obtained from χ_1 and χ_4.

We continue by determining the Brauer tree of the principal 19-block $B_0(G)$. Since all characters are real it is an open polygon. The leaves correspond to irreducible characters which stay irreducible modulo $p = 19$, one of course being the trivial character and the other one necessarily being a character $\chi_a \in \mathrm{Irr}(G)$ of degree 77, because the non-trivial irreducible characters χ_i with $\epsilon_i = 1$ (in the notation of Theorem 4.12.1) all have degree 77. The reduction of constants of the neighbor of χ_a thus has a modular irreducible constituent of degree 77; this must be the exceptional vertex. So we know the degrees of the irreducible characters corresponding to the nodes of the Brauer tree:

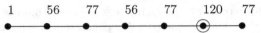

We will follow the custom of denoting the irreducible characters just by their degrees with an index $a, b, \ldots$ identifying them among characters of the same degree (in their order of appearance in the character table of G). We may assume that the neighbor of 1_a is 56_a, since we may switch the classes $5a, 5b$ and $15a, 15b$ if need be. In order to locate at least one of the characters of degree 77 in the Brauer tree, we compute the permutation character $1^G_{\mathbf{C}_G(z)}$ for an involution $z \in G$. It is a permutation character of degree $7 \cdot 11 \cdot 19$ and is uniquely determined by the character table of G:

```
gap> ct := CharacterTable("J1");; perm := PermChars(ct, 7*11*19);;
gap> b19 := Filtered([1..15], i -> Irr(ct)[i][15] <> 0);;
gap> MatScalarProducts( Irr(ct){b19}, perm );
[ [ 1, 1, 1, 2, 0, 0, 1, 1, 1 ] ]
```

The permutation character `perm[1]` is 19-projective, and since we have computed its scalar product with the characters in $\mathrm{Irr}(B_0)$ (which are precisely the irreducible characters that do not vanish on the class $19c$), this gives us the $B_0(G)$-part of `perm[1]`:

$$\texttt{perm[1]}_{B_0(G)} = (1_a + 56_a) + (56_b + 77_a) + (77_a + \sum_{x \in \{a,b,c\}} 120_x).$$

Thus the Brauer tree is as follows:

$$1_a \qquad 56_a \qquad 77_{b/c} \qquad 56_b \qquad 77_a \qquad 120 \qquad 77_{c/b}$$

In order to find the position of 77_b and 77_c in the Brauer tree we induce up the irreducible characters of $\mathbf{N}_G(P_{11})$ to G. We know the structure of $\mathbf{N}_G(P_{11})$, and the character table is easily determined. There are two possible class fusions of the character table `t11` of $\mathbf{N}_G(P_{11})$ into the character table `ct` of G, but both yield the same set of characters when inducing the irreducibles of `t11` to `ct`. We compute the scalar product of the first eight of these induced characters with the characters in $\mathrm{Irr}(B_0(G))$:

```
gap> gg := AllSmallGroups(110);;
gap> gg := Filtered(gg,g-> Size(Centralizer(g,SylowSubgroup(g,11)))=11);;
gap> t11 := CharacterTable(gg[1]);; pf:=PossibleClassFusions(t11,ct);
[ [ 1, 2, 4, 10, 8, 5, 9, 5, 9, 4, 8 ], [ 1, 2, 5, 10, 9, 4, 8, 4, 8, 5, 9 ] ]
gap> ind1 := InducedClassFunctionsByFusionMap(t11, ct, Irr(t11), pf[1]);;
gap> ind2 := InducedClassFunctionsByFusionMap(t11, ct, Irr(t11), pf[2]);;
gap> Set(ind1) = Set(ind2);
true
```

```
gap> MatScalarProducts(Irr(ct){b19},ind1{[1..8]});
[ [ 1, 1, 1, 2, 0, 0, 1, 1, 1 ], [ 0, 1, 1, 1, 1, 1, 1, 1, 1 ],
  [ 0, 0, 1, 0, 1, 1, 1, 1, 1 ], [ 0, 1, 0, 0, 1, 1, 1, 1, 1 ],
  [ 0, 1, 0, 0, 1, 1, 1, 1, 1 ], [ 0, 0, 1, 0, 1, 1, 1, 1, 1 ],
  [ 0, 0, 1, 1, 1, 0, 1, 1, 1 ], [ 0, 1, 0, 1, 0, 1, 1, 1, 1 ] ]
```

All these induced characters are 19-projective. Looking at the eighth induced character we see that the following character is 19-projective:

$$56_a + 77_c + 77_a + \sum_{x \in \{a,b,c\}} 120_x,$$

which finally shows that the Brauer tree is given by

$$\underset{1_a}{\bullet} \quad \underset{56_a}{\bullet} \quad \underset{77_c}{\bullet} \quad \underset{56_b}{\bullet} \quad \underset{77_a}{\bullet} \quad \underset{120}{\circledcirc} \quad \underset{77_b}{\bullet}$$

From this we immediately obtain $\mathrm{IBr}(B_0)$:

$$\varphi_1 := (1_a)|_{G_{p'}}, \qquad \varphi_5 := (56_a)|_{G_{p'}} - \varphi_1, \quad \varphi_2 := (77_c)|_{G_{p'}} - \varphi_5,$$
$$\varphi_3 := (56_b)|_{G_{p'}} - \varphi_2, \quad \varphi_4 := (77_a)|_{G_{p'}} - \varphi_3, \quad \varphi_8 := (77_b)|_{G_{p'}}.$$

Finally, we display the complete Brauer character table (Table 4.9), including also the defect-zero characters restricted to $G_{p'}$. As one can see from Table 4.9, we have several examples of non-rational Brauer characters for which an algebraic conjugate is not a Brauer character.

Table 4.9. 19-Brauer character table of J_1

$\mathrm{IBr}_{19}(J_1)$	1a	2a	3a	5a	5b	6a	7a	10a	10b	11a	15a	15b
φ_1	1	1	1	1	1	1	1	1	1	1	1	1
φ_2	22	-2	1	α	α'	1	1	$-\alpha$	$-\alpha'$	.	δ	δ'
φ_3	34	2	1	β	β'	-1	-1	α	α'	1	$-\beta'$	$-\beta$
φ_4	43	3	-2	$-\alpha$	$-\alpha'$	.	1	$-\alpha$	$-\alpha'$	-1	$-\alpha$	$-\alpha'$
φ_5	55	-1	1	γ	$-\gamma$	-1	-1	-1	-1	.	$-\beta'$	$-\beta$
φ_6	76	4	1	1	1	1	-1	-1	-1	-1	1	1
φ_7	76	-4	1	1	1	-1	-1	1	1	-1	1	1
φ_8	77	-3	2	α	α'	.	.	α	α'	.	α	α'
φ_9	133	5	1	-2	-2	-1	.	.	.	1	1	1
φ_{10}	133	-3	-2	$-\alpha$	$-\alpha'$	.	.	α	α'	1	$-\alpha$	$-\alpha'$
φ_{11}	133	-3	-2	$-\alpha'$	$-\alpha$	.	.	α'	α	1	$-\alpha'$	$-\alpha$
φ_{12}	209	1	-1	-1	-1	1	-1	1	1	.	-1	-1

$\alpha, \alpha' := \frac{-1 \pm \sqrt{5}}{2}, \quad \beta, \beta' := \frac{3 \pm \sqrt{5}}{2}, \quad \gamma := -\sqrt{5}, \quad \delta, \delta' := -1 \mp \sqrt{5}.$

♦

Example 4.12.11 We consider the 11-modular characters of the Mathieu group M_{11} of order $7920 = 2^4 \cdot 3^2 \cdot 5 \cdot 11$. The prime 11 divides the order of M_{11} exactly once, and, as we can read off easily from the ordinary character table of M_{11}, the structure of the normalizer of a Sylow 11-subgroup is a Frobenius group $11 \rtimes 5$ of order 55. This implies that the principal block B_0 of defect one is the only block not of defect zero. According to Brauer's theorem on blocks of defect one (Theorem 4.12.1), there are $e := (11 - 1)/2 = 5$ irreducible Brauer characters in B_0 and seven irreducible ordinary characters. The following table summarizes useful information about these characters:

χ_1	χ_2	χ_3	χ_4	χ_6	χ_7	χ_9
1_a	10_a	10_b	10_c	16_a	16_b	45_a

Note further that χ_6, χ_7 are the exceptional characters in the block and that all characters are real-valued on the 11-regular classes but the complex conjugate pair χ_3, χ_4. Hence the Brauer tree will have the following real stem with four vertices:

1_a $\qquad$ 45_a

Since the irreducible character χ_5 is not in B_0, it is of defect zero, and the restriction to the block of the tensor product $\chi_5 \otimes \chi_5 = \chi_1 + \chi_2$ is a projective character and obviously a projective indecomposable character. This shows that the vertex indexed by χ_2 is linked to the one for χ_1, and hence the real stem is forced to look as follows:

1_a $\quad$ 10_a $\quad$ 45_a $\quad$ $16_a, 16_b$

The vertices of the complex pair χ_3 and χ_4 with $\epsilon_3 = \epsilon_4 = -1$ (in the notation of Theorem 4.12.1) both have to be linked to a vertex with character χ_i on the real stem with $\epsilon_i = 1$. The only feasible choice for this vertex is the vertex corresponding to the character χ_9. Hence the Brauer tree is given as follows:

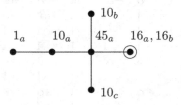

$\blacklozenge$

Exercises

Exercise 4.12.1 Let $p > 7$ be a prime and let $P \in \mathrm{Syl}_p(G)$ with $[\mathbf{N}_G(P) : \mathbf{C}_G(P)] = 2$ and $\mathbf{C}_G(P) \cong P \times A_5$. Show that G has no irreducible characters of degree $\equiv \pm 2$ mod p and exactly two irreducible characters of degree $\equiv \pm 1$ mod p.

Exercise 4.12.2 Let $P \in \mathrm{Syl}_p(G)$ be of order p and let $e := [\mathbf{N}_G(P) : \mathbf{C}_G(P)]$. Let B_0 be the principal block of G and $\epsilon_0, \dots, \epsilon_e, \in \{-1, 1\}$ be defined as in Theorem 4.12.1, that is, $\chi_i(1) \equiv \epsilon_i$ mod p for $i = 1, \dots, e$ and $\zeta(1) \equiv \epsilon_0 e$ mod p, where $\chi_1, \dots, \chi_e$ are the non-exceptional characters in B_0 and ζ is one of the exceptional characters in B_0. If $d_i := \epsilon_i \chi_i(1)$ for $i = 1, \dots, e$ and $d_0 := \epsilon_0 \zeta(1)$, then the d_i satisfy

$$\sum_{i=1}^{e} d_i = d_0 \qquad \text{and} \qquad \sum_{i=1}^{e} d_i^2 + \frac{p-1}{e} \cdot d_0^2 \le n := |G|. \tag{4.53}$$

Verify that the following program returns a list dd such that $[[d_1, \dots, d_e], d_0] \in$ dd provided that $\chi_i(1) \in$ degne for $i = 1, \dots, e$ and $\zeta(1) \in$ dege:

```
gap> degblock := function( n, degne, dege, e, p )
> local dd, d1, d2, de, tup, x;
> dd:= []; d1:= ShallowCopy( degne ); d2 := ShallowCopy( dege );
> for x in [1..Length(d1)] do
>    if d1[x] mod p <> 1 mod p then d1[x] := - d1[x]; fi;
> od;
> for x in [1..Length(d2)] do
>    if d2[x] mod p <> e  then d2[x]:= -d2[x]; fi;
> od;
> tup := UnorderedTuples(d1, e);;
> for x in tup do
>     if Sum(x) in d2 then Add( dd , [x, Sum(x)] ); fi;
> od;
> for x in dd do Sort( x[1] ); od; dd := Set(dd); Sort(dd);
> dd:=Filtered(dd, x -> Sum(List(x[1], a -> a^2)) + ((p-1)/e)*x[2]^2 < n);
> dd:=Filtered(dd, x -> Length(Positions(x[1],1)) = 1);
> return(dd);
> end;;
```

Of course, the program should only be used if e and $|$degne$|$ are small.

Exercise 4.12.3 Let $B \in \mathrm{Bl}_3(G)$ and $\mathbf{N}_G(P) \cong S_3 \times D_{10}$ for $P \in \mathrm{Def}(B)$.
(a) Show that

$$D_B = \begin{bmatrix} 1 & 0 \\ 0 & 1 \\ 1 & 1 \end{bmatrix}.$$

(b) Deduce that $\mathrm{Irr}(B) = \{\chi, \chi', \chi''\}$ with

$$\chi''(g) = \begin{cases} \chi(g) + \chi'(g) & \text{if } g \text{ is 3-regular,} \\ -\chi(g) = -\chi'(g) & \text{else.} \end{cases}$$

Exercise 4.12.4 Assume that the centralizer and element orders of $G := J_1$ are known as well as the degrees of the irreducible characters. Complete the character table of G just by using the congruence relations and block theory.

4.13   Brauer characters of p-solvable groups

In this section we consider a class of groups for which modular representation theory is easier and further developed. In fact, for p-solvable groups a number of fundamental problems in modular representation theory have been solved that in the general case still seem to be intractable. As usual we assume that G is a finite group, p is a prime and (K, R, F, η) is a p-modular splitting system for G.

Definition 4.13.1 G is called p-**solvable** if every chief factor of G is either a p-group or a p'-group.

Part (a) of the following result is often called the "Hall-Higman lemma."

Lemma 4.13.2 *For a group G let $\mathbf{O}_{p'p}(G) \leq G$ be defined by $\mathbf{O}_{p'p}(G)/\mathbf{O}_{p'}(G) = \mathbf{O}_p(G/\mathbf{O}_{p'}(G))$. Assume that G is p-solvable. Then*

(a) $\mathbf{C}_{G/\mathbf{O}_{p'}(G)}(\mathbf{O}_{p'p}(G)/\mathbf{O}_{p'}(G)) \leq \mathbf{O}_{p'p}(G)/\mathbf{O}_{p'}(G);$

(b) *if $g \in G \setminus \mathbf{O}_{p'p}(G)$ then $p \mid |g^G|$.*

Proof.   (a) We may assume that $\mathbf{O}_{p'}(G) = \{1\}$. Assume that $\mathbf{C}_G(\mathbf{O}_p(G)) \not\leq \mathbf{O}_p(G)$ and that $M \leq \mathbf{O}_p(G)\,\mathbf{C}_G(\mathbf{O}_p(G))$ such that $M/\mathbf{O}_p(G)$ is a chief factor of G. By assumption, $p \nmid [M : \mathbf{O}_p(G)]$, and by the Schur–Zassenhaus theorem (see [5], p. 70) $\mathbf{O}_p(G)$ has a complement C in M. Since $C \subseteq \mathbf{C}_G(\mathbf{O}_p(G))$, we have $M = C \times \mathbf{O}_p(G)$ and $C \leq \mathbf{O}_{p'}(G)$, a contradiction.
(b) If $p \nmid |g^G|$ then $\mathbf{C}_G(g)$ contains a Sylow p-subgroup P of G, hence $g \in \mathbf{C}_G(P)$. Since $\mathbf{O}_{p'p}(G) \leq \mathbf{O}_{p'}(G)P$,

$$g\mathbf{O}_{p'}(G) \in \mathbf{C}_{G/\mathbf{O}_{p'}(G)}(\mathbf{O}_{p'p}(G)/\mathbf{O}_{p'}(G)) \leq \mathbf{O}_{p'p}(G)/\mathbf{O}_{p'}(G),$$

where the last inclusion follows from part (a). Hence $g \in \mathbf{O}_{p'p}(G)$. $\square$

Clifford correspondence, which we have introduced for Brauer characters in Corollary 4.3.8, is useful in particular for the representation theory of p-solvable groups.

Lemma 4.13.3 *Let $N \lhd G$ with $p \nmid [G : N]$. Assume that $\psi \in \mathrm{Irr}(N)$ is such that $\psi' := \psi|_{N_{p'}} \in \mathrm{IBr}(N)$ and $T_G(\psi) = T_G(\psi')$. Then*

$$\tau \colon \mathrm{Irr}(\,G \mid \psi\,) \to \mathrm{IBr}(\,G \mid \psi'\,), \quad \chi \mapsto \chi|_{G_{p'}} \quad \text{is a bijection.}$$

Proof.   We use induction on $|G|$. Let $T := T_G(\psi) = T_G(\psi')$ and assume first that $T < G$. Then, using induction, we have a bijection

$$\mathrm{Irr}(T \mid \psi) \to \mathrm{IBr}(T \mid \psi'), \quad \theta \mapsto \theta|_{T_{p'}}.$$

Since $(\theta|_{T_{p'}})^G = (\theta^G)|_{G_{p'}}$, the result follows from the Clifford correspondences (Lemma 3.6.8 and Lemma 4.3.8) in this case.

We may assume that $T = G$, which means that ψ, ψ' are G-invariant. Let W be a simple FN-module with Brauer character ψ' and for $\varphi \in \mathrm{IBr}(G)$ let Y_φ be a simple FG-module with Brauer character φ. Using Corollary 3.2.18 we may write

$$\psi^G = \sum_{\chi \in \mathrm{Irr}(G|\psi)} a_\chi\, \chi \qquad \text{and} \qquad {\psi'}^G = \sum_{\varphi \in \mathrm{IBr}(G|\psi')} b_\varphi\, \varphi.$$

By Theorem 3.2.13, $\chi|_N = a_\chi \psi$. Since $p \nmid [G : N]$, we can deduce from Exercise 4.9.2 that W^G is also semisimple. Thus using Frobenius–Nakayama reciprocity (Theorem 3.2.12) we obtain for $\varphi \in \mathrm{IBr}(G \mid \psi')$

$$b_\varphi = \dim \mathrm{Hom}_{FG}(W^G, Y_\varphi) = \dim \mathrm{Hom}_{FN}(W, (Y_\varphi)_N)$$

and $(Y_\varphi)_N \cong \underbrace{W \oplus \cdots \oplus W}_{b_\varphi}$, because ψ' is G-invariant. Thus for $\chi \in \mathrm{Irr}(\, G \mid \psi\,)$

$$a_\chi \psi' = (\chi|_N)_{N_{p'}} = (\chi|_{G_{p'}})|_N = \sum_{\varphi \in \mathrm{IBr}(G)} d_{\chi\varphi} \varphi_N = \sum_{\varphi \in \mathrm{IBr}(G|\psi')} d_{\chi\varphi} b_\varphi \psi'.$$

Hence

$$a_\chi = \sum_{\varphi \in \mathrm{IBr}(G|\psi')} d_{\chi\varphi} b_\varphi \qquad \text{for} \qquad \chi \in \mathrm{Irr}(G \mid \psi)$$

and $d_{\chi\varphi} = 0$ for $\varphi \notin \mathrm{IBr}(G \mid \psi')$. Similarly

$$\psi^G|_{G_{p'}} = \sum_{\chi \in \mathrm{Irr}(G|\psi)} a_\chi \Big(\sum_{\varphi \in \mathrm{IBr}(G)} d_{\chi\varphi}\varphi \Big) = \sum_{\varphi \in \mathrm{IBr}(G)} \Big(\sum_{\chi \in \mathrm{Irr}(G|\psi)} d_{\chi\varphi} a_\chi \Big) \varphi.$$

Since $\psi^G|_{G_{p'}} = {\psi'}^G$ we get

$$b_\varphi = \sum_{\chi \in \mathrm{Irr}(G|\psi)} d_{\chi\varphi} a_\chi \qquad \text{for} \qquad \varphi \in \mathrm{IBr}(G \mid \psi').$$

Hence

$$a_\chi = \sum_{\varphi \in \mathrm{IBr}(G|\psi')} d_{\chi\varphi} \Big(\sum_{\xi \in \mathrm{Irr}(G|\psi)} d_{\xi\varphi} a_\xi \Big) = \sum_{\xi \in \mathrm{Irr}(G|\psi)} \Big(\sum_{\varphi \in \mathrm{IBr}(G|\psi')} d_{\chi\varphi} d_{\xi\varphi} \Big) a_\xi.$$

It follows that

$$\sum_{\varphi \in \mathrm{IBr}(G|\psi')} d_{\chi\varphi} d_{\xi\varphi} = \begin{cases} 1 & \text{for} \quad \xi = \chi \in \mathrm{Irr}(G \mid \psi), \\ 0 & \text{for} \quad \xi \neq \chi \in \mathrm{Irr}(G \mid \psi). \end{cases}$$

This means that $\chi|_{G_{p'}} \in \mathrm{IBr}(G \mid \psi')$ and also that $\chi|_{G_{p'}} \neq \xi|_{G_{p'}}$ for $\xi \neq \chi$ in $\mathrm{Irr}(G \mid \psi)$. So τ is injective.

Finally, if $\varphi \in \mathrm{IBr}(G \mid \psi')$ then

$$0 \neq b_\varphi = \sum_{\chi \in \mathrm{Irr}(G|\psi)} d_{\chi\varphi} a_\chi,$$

and hence there is a $\chi \in \mathrm{Irr}(G \mid \psi)$ with $d_{\chi\varphi} \neq 0$. By the above we have $\chi|_{G_{p'}} \in \mathrm{IBr}(G \mid \psi')$, so $\varphi = \chi|_{G_{p'}}$, which shows that τ is surjective as well. $\square$

Theorem 4.13.4 (Wolf) *Let G/N be p-solvable and let $\psi \in \operatorname{Irr}(N)$ be such that $\psi|_{N_{p'}} \in \operatorname{IBr}(N)$ and $T_G(\psi) = T_G(\psi|_{N_{p'}})$. Assume further that p does not divide $o(\psi)\psi(1)$, where $o(\psi)$ is the determinantal order of ψ, see Definition 3.6.18. Then for every $\varphi \in \operatorname{IBr}(G \mid \psi|_{N_{p'}})$ there is a $\chi \in \operatorname{Irr}(G \mid \psi)$ with $\chi|_{G_{p'}} = \varphi$.*

Proof. We use induction on $|G/N|$. If $T := T_G(\psi) = T_G(\psi|_{G_{p'}}) < G$, then let $\theta \in \operatorname{IBr}(T \mid \psi|_{G_{p'}})$ be the Clifford correspondent of φ. By induction there is a $\xi \in \operatorname{Irr}(T \mid \psi)$ with $\xi|_{G_{p'}} = \theta$. Then the Clifford correspondent ξ^G does the job. So we may assume that ψ is G-invariant.

Let M/N be a chief factor of G. If $p \nmid [M : N]$, then by Lemma 4.13.3 we have a bijection

$$\tau_M : \operatorname{Irr}(M \mid \psi) \to \operatorname{IBr}(M \mid \psi|_{N_{p'}}) \ , \ \xi \mapsto \xi|_{M_{p'}}.$$

So there is a $\xi \in \operatorname{Irr}(M \mid \psi)$ such that $\varphi \in \operatorname{IBr}(G \mid \xi|_{M_{p'}})$. Since τ_M is a bijection, we have $T_G(\xi) = T_G(\xi|_{M_{p'}})$. Since ψ is M-invariant, $\xi|_N = e\psi$ with e, the degree of an irreducible representation of M/N (see Theorem 3.7.5) dividing $[G : N]$. Thus $p \nmid \xi(1)$. Obviously $\ker(\det \psi) \leq \ker(\det \xi) \leq M$, hence $o(\xi) \mid o(\psi)[M : N]$ is coprime to p. Hence by induction there is a $\chi \in \operatorname{Irr}(G \mid \xi)$ with $\chi|_{G_{p'}} = \varphi$.

Now let M/N be a p-group. By Corollary 3.6.14 there is a unique $\theta \in \operatorname{IBr}(M)$ with $\theta|_N = \psi|_{N_{p'}}$. Also by Theorem 3.6.19 there is a unique $\xi \in \operatorname{Irr}(M)$ with $\xi|_N = \psi$ and $([M : N], o(\xi)) = 1$. The uniqueness properties ensure that both θ and ξ are G-invariant. Also $(\xi|_{M_{p'}})|_{N_{p'}} = (\xi|_N)|_{N_{p'}} = \psi|_{N_{p'}}$, hence $\xi|_{M_{p'}} = \theta$. Since $\varphi \in \operatorname{IBr}(G \mid \xi|_{M_{p'}})$, the inductive hypothesis implies the existence of $\chi \in \operatorname{Irr}(G \mid \xi)$ with $\chi|_{G_{p'}} = \varphi$. $\qquad\square$

Choosing $N = \{1\}$ in the theorem one obtains the following important result.

Theorem 4.13.5 (Fong–Swan theorem) *If G is p-solvable, then for every $\varphi \in \operatorname{IBr}(G)$ there is a $\chi \in \operatorname{Irr}(G)$ with $\varphi = \chi|_{G_{p'}}$.*

It follows that for p-solvable groups the irreducible Brauer characters can be found easily from the ordinary character table.

Corollary 4.13.6 *If G is p-solvable, then*

$$\operatorname{IBr}(G) = \{\chi|_{G_{p'}} \mid \chi \in \operatorname{Irr}(G) \text{ with } \chi|_{G_{p'}} \neq \chi_1|_{G_{p'}} + \chi_2|_{G_{p'}}, \chi_1, \chi_2 \in \operatorname{Char}_K(G)\}.$$

Theorem 4.13.7 (Fong) *Assume that G is p-solvable and $\theta \in \operatorname{Irr}(\mathbf{O}_{p'}(G))$ is G-invariant. Let $b \in \operatorname{Bl}_p(\mathbf{O}_{p'}(G))$ be the block with $\operatorname{Irr}(b) = \{\theta\}$. Then $\operatorname{Bl}(G \mid b) = \{B\}$ with $\operatorname{Irr}(B) = \operatorname{Irr}(G \mid \theta)$ and $\operatorname{IBr}(B) = \operatorname{IBr}(G \mid \theta)$. Furthermore $\operatorname{Def}(B) = \operatorname{Syl}_p(G)$.*

Proof. We may assume $p \mid |G|$. Let $b \in \operatorname{Bl}_p(\mathbf{O}_{p'}(G))$ with $\operatorname{Irr}(b) = \{\theta\}$. By Lemma 4.7.28 (a) and (b) it suffices to show that $\operatorname{Irr}(G \mid \theta) = \operatorname{Irr}(B)$ for some $B \in \operatorname{Bl}_p(G)$ with $\operatorname{Def}(B) = \operatorname{Syl}_p(G)$. We will prove this by induction on $|G|$.

Let N be a maximal normal subgroup of G containing $\mathbf{O}_{p'}(G)$. Assume $B \in \mathrm{Bl}(G \mid b)$ and let $b_1 \in \mathrm{Bl}_p(N)$ be a block covered by B. By Exercise 4.7.7 $b_1 \in \mathrm{Bl}(N \mid b)$. Since $\mathbf{O}_{p'}(G) = \mathbf{O}_{p'}(N)$, we may use induction and conclude that $\mathrm{Irr}(b_1) = \mathrm{Irr}(N \mid \theta)$ and $\mathrm{Def}(b_1) = \mathrm{Syl}_p(N)$. It suffices to show that $\mathrm{Bl}(G \mid b_1) = \{B\}$ and that $\mathrm{Def}(B) = \mathrm{Syl}_p(G)$.

If G/N is a p-group, then $\mathrm{Bl}(G \mid b_1) = \{B\}$ follows from Lemma 4.7.28(f). Also b is invariant in G, since b_1 is the unique p-block of N covering b. Thus $\hat{\epsilon}_B = \hat{\epsilon}_{b_1}$. Let h^G be a defect class for B and let $D \in \mathrm{Def}_p(h^G)$. By Theorem 4.4.7 $h^G \subseteq G_{p'} \subseteq N$. Then, since $[G : N] = p$,

$$h^G = h_1^N \dot{\cup} \cdots \dot{\cup} h_r^N \qquad \text{with} \qquad r = \begin{cases} 1 & \text{if } [\mathbf{C}_G(h) : \mathbf{C}_N(h)] = p, \\ p & \text{if } \mathbf{C}_G(h) = \mathbf{C}_N(h). \end{cases}$$

Note that $\hat{\omega}_{b_1}((h_i^N)^+) = \hat{\omega}_{b_1}((h^N)^+)$ for all i, since b is invariant in G. By Lemma 4.7.28(c) $\hat{\omega}_b((h^G)^+) = \hat{\omega}_B((h^G)^+) \neq 0$, so $r = p$ is impossible. By the min–max theorem $|N|_p \mid |\mathbf{C}_N(h)|$, hence $D \in \mathrm{Syl}_p(G)$.

Since G is p-solvable, we may thus assume that $p \nmid [G : N]$. Let $\psi \in \mathrm{Irr}(b_1)$ have height zero, thus $p \nmid \psi(1)$, because $\mathrm{Def}(b_1) = \mathrm{Syl}_p(N)$. Let $B \in \mathrm{Bl}(G \mid b_1)$. By Lemma 4.7.28(a) and Exercise 4.4.8 there is a $\chi \in \mathrm{Irr}(B)$ with $\chi \in \mathrm{Irr}(G \mid \psi)$. Then $p \nmid \chi(1)$ and hence $\mathrm{Def}(B) = \mathrm{Syl}_p(G)$. So it suffices to show that any two characters $\chi, \chi' \in \mathrm{Irr}(G \mid \psi)$ belong to the same p-block of G. For this we use the criterion of Theorem 4.4.8. For $C = g^G \subseteq N$ we have by Lemma 4.7.28(c)

$$\eta(\omega_\chi(C^+)) = \hat{\omega}_b(C^+) = \eta(\omega_{\chi'}(C^+)).$$

On the other hand, $\mathbf{O}_{p'p}(G) \leq N$, and Lemma 4.13.2(b) implies that for $g \in G \setminus N$ and $C := g^G$ we have $p \mid |C|$ and thus

$$\eta(\omega_\chi(C^+)) = \eta(|C| \frac{\chi(g)}{\chi(1)}) = 0 = \eta(|C| \frac{\chi'(g)}{\chi'(1)}) = \eta(\omega_{\chi'}(C^+)),$$

because $p \nmid \chi(1), \chi'(1)$. Hence $\eta \circ \omega_\chi = \eta \circ \omega_{\chi'}$ and χ, χ' belong to the same p-block B. $\qquad \square$

Corollary 4.13.8 *Let G be a p-solvable group, let $B \in \mathrm{Bl}_p(G)$ and assume that $\mathrm{k}(H/\mathbf{O}_{p'}(H)) \leq |H|_p$ holds for all $H \leq G$. Then*

$$\mathrm{k}(B) \leq |D| \qquad \text{for} \qquad D \in \mathrm{Def}(B). \tag{4.54}$$

Proof. We use induction on $|G|$ and assume that $B \in \mathrm{Bl}_p(G)$ and that B covers $b \in \mathrm{Bl}_p(\mathbf{O}_{p'}(G))$ with $\mathrm{Irr}(b) = \{\theta\}$.

If θ is G-invariant then by Fong's theorem (Theorem 4.13.7) $\mathrm{Def}(B) = \mathrm{Syl}_p(G)$ and $\mathrm{Irr}(B) = \mathrm{Irr}(G \mid \theta)$. From Theorem 3.7.32(b) we conclude that $\mathrm{k}(B) \leq \mathrm{k}(G/\mathbf{O}_{p'}(G))$ and (4.54) follows in this case.

So we may assume that $T := T_G(\theta) < G$. By the Fong–Reynolds theorem (Theorem 4.7.32) there is a unique block $B_1 \in \mathrm{Bl}_p(T \mid b)$ with $B_1^G = B$. Furthermore $\mathrm{k}(B_1) = \mathrm{k}(B)$ and if $D \in \mathrm{Def}(B_1)$ then $D \in \mathrm{Def}(B)$. By the induction hypothesis $\mathrm{k}(B_1) \leq |D|$, hence (4.54) follows. $\qquad \square$

The question whether or not (4.54) holds in general for all groups and all blocks is called Brauer's $\mathrm{k}(B)$-problem or $\mathrm{k}(B)$-conjecture; see Section 4.14.

Exercises

Exercise 4.13.1 Let G be p-solvable and let $\chi \in \mathrm{Irr}(G)$ such that $\chi(g) = 0$ for $g \in G \setminus G_{p'}$. Show that χ is a p-projective character (see Definition 4.3.1).

Exercise 4.13.2 Let G be p-solvable, $\varphi \in \mathrm{IBr}(G)$ and $\sigma \in \mathrm{Gal}(\mathbb{Q}_{|G|}/\mathbb{Q})$. Show that $\varphi^\sigma \in \mathrm{IBr}(G)$. Compare with Exercise 4.2.2.

Exercise 4.13.3 The following is the character table of $\mathrm{GL}_2(3)$, essentially in GAP format:

```
      2  4   4   3   1   1   2   3   3
      3  1   1   .   1   1   .   .   .

         1a  2a  4a  3a  6a  2b  8a  8b

X.1      1   1   1   1   1   1   1   1
X.2      1   1   1   1   1  -1  -1  -1
X.3      2   2   2  -1  -1   .   .   .
X.4      3   3  -1   .   .  -1   1   1
X.5      3   3  -1   .   .   1  -1  -1
X.6      2  -2   .  -1   1   .   A  -A
X.7      2  -2   .  -1   1   .  -A   A
X.8      4  -4   .   1  -1   .   .   .

A = -E(8)-E(8)^3  = -ER(-2) = -i2
```

Find the table of p-Brauer characters for $p = 2, 3$.

4.14   Some conjectures

One of the oldest and most famous conjectures in modular representation theory is the following (see prob. 20 in [13]).

Conjecture 4.14.1 (Brauer's $\mathrm{k}(B)$-conjecture) Let $B \in \mathrm{Bl}_p(G)$ with $D \in \mathrm{Def}(B)$. Then $\mathrm{k}(B) := |\mathrm{Irr}(B)| \le |D|$.

If $\mathrm{d}(B) = 0$ then by Theorem 4.4.14 $\mathrm{k}(B) = 1 = p^{\mathrm{d}(B)}$. For $\mathrm{d}(B) = 1$ we know from Theorem 4.11.11(b) that $\mathrm{k}(B) \le p$. More generally, the conjecture holds true for blocks with cyclic defect groups (see [56], theorem VII.2.12, p. 277). Furthermore, the conjecture is known to be true for blocks of defect two (see [14]). But the general case is still open.

The following purely group theoretical theorem is of interest in its own right, but its main importance is in connection with the $\mathrm{k}(B)$-conjecture.

Theorem 4.14.2 (k(GV)-theorem) *Let V be an elementary abelian p-subgroup and let G be a finite group with $p \nmid |G|$ acting faithfully and irreducibly on V. If GV denotes the corresponding semidirect product, then the number of conjugacy classes of GV is given by*

$$k(GV) \le |V|. \tag{4.55}$$

Note that by Corollary 3.6.6 the assertion (4.55) holds provided that G has a regular orbit on V, that is, if there is a $v \in V$ such that

$$\mathbf{C}_G(v) := \{g \in G \mid g\,v = v\} = \{1\}.$$

In fact, in this case one has k(GV) < $|V|$ unless G is abelian (see Corollary 3.6.6).

For more than 40 years Theorem 4.14.2 was called the k(GV)-**conjecture** or k(GV)-problem. The first major partial results on this problem were obtained by Knörr ([102]). He proved the conjecture for supersolvable groups. One of his fundamental results says that k(GV) $\le |V|$ holds provided that V contains a vector v such that $\mathbf{C}_G(v)$ is abelian. These methods were further refined by Gluck ([64]), Gow ([70]) and Robinson and Thompson ([152]), who showed that k(GV) $\le |V|$ holds if there is a $v \in V$ such that $V_{\mathbf{C}_G(v)}$ contains a faithful selfdual submodule. These authors were able to conclude that the k(GV)-conjecture holds whenever $p > 5^{30}$. This bound was lowered considerably by Robinson ([151]), before Gluck, Magaard and Riese ([65], [149]) and, using somewhat different methods, Köhler and Pahlings ([104]) proved that the conjecture holds whenever $p \notin \{3, 5, 7, 11, 13, 19, 31\}$. Later, Gluck and Magaard ([66]) were able to treat the case $p = 31$, and Riese and Schmid ([150]) proved the conjecture for all $p \ne 5$. Finally even this case could be settled by Gluck, Magaard, Riese and Schmid ([67]). For a complete proof of Theorem 4.14.2, see the monograph [155].

The relevance of the k(GV)-problem for the k(B)-conjecture was first noticed by Nagao ([124]).

Theorem 4.14.3 *If G is p-solvable and $B \in \mathrm{Bl}_p(G)$ with $D \in \mathrm{Def}(B)$ then* k(B) $\le |D|$.

Proof. Let G be a p-solvable group. We show by induction on $|G|$ that $k'(G) := k(G/\mathbf{O}_{p'}(G)) \le |G|_p$, the case $|G| = 1$ being trivial. The claim then follows from Corollary 4.13.8. If $\mathbf{O}_{p'}(G) \ne \{1\}$ then by induction

$$k'(G) = k'(G/\mathbf{O}_{p'}(G)) \le |G/\mathbf{O}_{p'}(G)|_p = |G|_p.$$

So we may assume that $\mathbf{O}_{p'}(G) = \{1\}$ and consequently $k'(G) = k(G)$. We define $\mathbf{O}_{pp'}(G) \trianglelefteq G$ by $\mathbf{O}_{pp'}(G)/\mathbf{O}_p(G) = \mathbf{O}_{p'}(G/\mathbf{O}_p(G))$. If $\mathbf{O}_{pp'}(G) \ne G$ then we may use induction and Exercise 2.1.10 and get

$$k(G) \le k(\mathbf{O}_{pp'}(G)) \cdot k(G/\mathbf{O}_{pp'}(G)) \le |\mathbf{O}_{pp'}(G)|_p |G/\mathbf{O}_{pp'}(G)|_p = |G|_p.$$

So we may assume that $G = \mathbf{O}_{pp'}(G)$. This means that $P := \mathbf{O}_p(G) \in \mathrm{Syl}_p(G)$, and by the Schur–Zassenhaus theorem P has a complement $H \cong G/P$ in G.

Let $V \leq P$ be a minimal normal subgroup of G and let $\mathbf{O}_{p'}(G/V) = Q/V$ with $V \leq Q \trianglelefteq G$. If $V \neq P$ then using induction and Exercise 2.1.10 again we get

$$\mathrm{k}(G) \leq \mathrm{k}(Q) \cdot \mathrm{k}(G/Q) = k'(Q) \cdot k'(G/Q) \leq |Q|_p \cdot |G/Q|_{p'} = |G|_p.$$

So we may assume that $V = P = \mathbf{O}_p(G)$, which means that the p'-group H acts faithfully (since $\mathbf{C}_G(V) = V$ by Lemma 4.13.2(a)) and irreducibly (since V is minimal) on the elementary abelian p-group V, and the assertion follows from Theorem 4.14.2. □

A long series of conjectures started with an observation by McKay. For any finite group G, let

$$m_p(G) := |\{\chi \in \mathrm{Irr}(G) \mid p \nmid \chi(1)\}| \qquad \text{for a prime } p.$$

Browsing through the ATLAS ([38]), or the character table library of GAP ([19]), one quickly finds that $m_2(G)$ is a power of two for many simple groups. We give evidence in Table 4.10 using the character tables of simple groups in the character table library of GAP. We leave aside the groups $\mathrm{L}_2(q)$ for odd q, because a glance at the generic character table on p. 197 shows that $m_2(\mathrm{L}_2(q)) = 4$ for odd q.

Of course, characters of odd degree belong to 2-blocks of maximal defect. In all the groups in Table 4.10, there is just one such 2-block (the principal block), with two exceptions, namely $\mathrm{U}_3(9)$ and $\mathrm{L}_3(11)$. Both of these groups have five 2-blocks of maximal defect, each containing four characters of odd degree. One may observe that in many of the cases $m_2(G) = [P : P']$ for $\mathbf{N}_G(P) = P \in \mathrm{Syl}_p(G)$. In these cases the above findings are "explained" by the following conjecture.

Conjecture 4.14.4 (McKay's conjecture) *For any finite group G*

$$m_p(G) = m_p(\mathbf{N}_G(P)) \qquad for \qquad P \in \mathrm{Syl}_p(G).$$

Recently an important step towards the proof of this conjecture has been made, see [91], by reducing it to a question about simple groups. The authors introduce the notion a group to be "good," which unfortunately is too complicated to be stated here, and show that if all simple groups involved in a group G are "good" then the McKay conjecture holds for G. In particular they give a proof that the McKay conjecture holds for groups with an abelian Sylow p-subgroup.

If $\chi \in \mathrm{Irr}(B)$ satisfies $p \nmid \chi(1)$, then χ belongs to a p-block B of maximal defect (that is, a block with defect group $P \in \mathrm{Syl}_p(G)$) and, of course, $\mathrm{ht}_p(\chi) = 0$. Hence

$$m_p(G) = \sum_{\substack{B \in \mathrm{Bl}_p(G) \\ \mathrm{Def}(B) = \mathrm{Syl}_p(G)}} \mathrm{k}_0(B).$$

Table 4.10. $m_2(G)$ for some simple groups G

G	$m_2(G)$
$A_5, A_6, A_7, L_3(3), M_{11}, U_3(5), L_3(7)$	4
$L_2(8), U_3(3), A_8, L_3(4), U_4(2), Sz(8), M_{12}, J_1, A_9, L_3(5), M_{22}, J_2,$ $A_{10}, U_4(3), G_2(3), S_4(5), L_4(3), M_{23}, {}^2F_4(2)', A_{11}, HS, J_3, U_3(11),$ $S_4(7), McL, G_2(5), L_4(5), R(27), U_4(5), Ru, L_5(3), ON, L_4(9), Ly$	8
$L_2(16), U_3(4), S_4(4), S_6(2), U_3(7), L_5(2), U_5(2), L_3(9), O_8^+(2),$ $O_8^-(2), {}^3D_4(2), A_{12}, M_{24}, G_2(4), A_{13}, He, O_7(3), U_6(2), Suz,$ $Co_3, O_8^+(3), A_{16}, Fi_{22}, A_{17}, O_7(5), HN, S_8(3), O_9(3), Th, Fi_{23}$	16
$U_3(9), L_3(11)$	20
$U_3(8)$	24
$L_2(32), Sz(32), S_6(3), L_6(2), A_{14}, S_8(2), A_{15}, O_8^-(3), O_{10}^+(2),$ $O_{10}^-(2), Co_2, S_6(5), F_4(2), Co_1, J_4, {}^2E_6(2), Fi_{24}'$	32
$L_2(64), L_3(8), L_4(4), U_4(4), S_4(8), S_6(4), L_7(2), S_{10}(2), E_6(2),$ ${}^2F_4(8), B, M$	64
$L_8(2), S_{12}(2)$	128

We follow the notation of the ATLAS ([38]).

Recall that $k_i(B) = |\{ \chi \in \mathrm{Irr}(B) \mid \mathrm{ht}_p(\chi) = i \}|$. Furthermore the Brauer correspondent of a block with maximal defect is again a block with maximal defect. Thus the following is a generalization of McKay's conjecture.

Conjecture 4.14.5 (Alperin–McKay conjecture) *If G is a finite group and B is a p-block of G with defect group $P \in \mathrm{Def}(B)$ and Brauer correspondent $b \in \mathrm{Bl}_p(\mathbf{N}_G(P))$, then $k_0(B) = k_0(b)$.*

Exercise 4.6.4 verifies the conjecture for $G := J_2$ and $p = 2$. It also shows that the conjecture would be false if one replaced $(k_0(B), k_0(b))$ by $(k(B), k(b))$ or by $(k_1(B), k_1(b))$.

An old related conjecture is the following (see prob. 23 in [13]):

Conjecture 4.14.6 (Brauer's height-zero conjecture) *The defect groups of a p-block $B \in \mathrm{Bl}_p(G)$ are abelian if and only if all $\chi \in \mathrm{Irr}(B)$ have height zero.*

This is obviously true for blocks of defect zero or one. One can also show that it holds for blocks with cyclic defect group. Moreover there is a theorem of Gluck and Wolf saying that Brauer's height-zero conjecture holds for p-solvable groups. For a proof, see [117].

There is also a conjecture concerning nonabelian defect groups as follows.

Conjecture 4.14.7 (Robinson's conjecture) *If* $B \in \mathrm{Bl}_p(G)$ *has a nonabelian defect group* D *then*

$$p^{\mathrm{ht}_p(\chi)} < [D : \mathbf{Z}(D)] \qquad \text{for all} \qquad \chi \in \mathrm{Irr}(B).$$

As an example consider $G = \mathrm{M}_{12}$. From the character table of G one can see that G has a non-abelian Sylow 3-subgroup P of order 3^3, so $[P : \mathbf{Z}(P)] = 3^2$. On the other hand the principal 3-block (the only one having a non-abelian defect group) has two characters of height one and none of larger height. Similarly for $p = 2$, the principal p-block B_0 and $P \in \mathrm{Syl}_p(G)$, one finds

$$\max\{p^{\mathrm{ht}_p(\chi)} \mid \chi \in \mathrm{Irr}(B_0)\} = 2^3 < 2^5 = [P : \mathbf{Z}(P)],$$

which is also in accordance with Robinson's conjecture.

The following conjecture is related to Brauer's height-zero conjecture and to the k(B)-conjecture.

Conjecture 4.14.8 (Olsson's conjecture) *If* $D \in \mathrm{Def}(B)$ *for* $B \in \mathrm{Bl}_p(G)$ *then*

$$\mathrm{k}_0(B) \leq [D : D'].$$

As a consequence of the proof of the k(GV)-conjecture, we mention the following result.

Theorem 4.14.9 *Let* $B \in \mathrm{Bl}_p(G)$ *and let* $D \in \mathrm{Def}(B)$. *Assume that* $b \in \mathrm{Bl}_p(\mathbf{N}_G(D))$ *is the Brauer correspondent of* B. *Then we can state the following.*

(a) *The Alperin–McKay conjecture implies Olsson's conjecture, that is,*

$$\mathrm{k}_0(B) = \mathrm{k}_0(b) \quad \Longrightarrow \quad \mathrm{k}_0(B) \leq [D : D'].$$

(b) *If* D *is abelian and* k(B) = k(b), *then* k(B) $\leq |D|$.

Proof. (a) was proved by Külshammer in [106].

For (b) see [155], theorem 12.3b.   □

It has been mentioned before that no group theoretical description of the number of p-blocks of a finite group G is known. The same is true for the number of p-blocks of defect zero, which is, of course, the number of irreducible defect-zero characters. For a fixed prime p and $d \in \mathbb{N}_0$ we define

$$\mathrm{k}^{(d)}(G) := |\{\chi \in \mathrm{Irr}(G) \mid \mathrm{d}_p(\chi) = d\}|.$$

Definition 4.14.10 A **p-weight** for G is a pair (P, χ), where P is a p-subgroup of G and $\chi \in \mathrm{Irr}(\mathbf{N}_G(P)/P)$ has p-defect zero. A p-weight (P, χ) **belongs to a p-block** $B \in \mathrm{Bl}_p(G)$ if $\mathrm{Inf}(\chi) \in \mathrm{Irr}(b)$ for $b \in \mathrm{Bl}_p(\mathbf{N}_G(P))$ with $b^G = B$.

Lemma 4.14.11 *Let* (P, χ) *be a p-weight of* G.

(a) *If* $\mathrm{Inf}(\chi) \in \mathrm{Irr}(b)$ *for* $b \in \mathrm{Bl}_p(\mathbf{N}_G(P))$ *then* b^G *is defined and* $P \leq D$ *for some* $D \in \mathrm{Def}(b^G)$.

(b) *P is a radical p-subgroup of G.*

Proof. (a) follows from Theorem 4.7.19.

    (b) By Theorem 4.6.10 $\mathbf{O}_p(\mathbf{N}_G(P))$ is contained in any defect group of b. On the other hand, $\nu_p(|P|) = \mathrm{d}(\mathrm{Inf}(\chi)) \leq \mathrm{d}(b)$. Hence $\mathrm{Def}(b) = \{P\}$ and $P = \mathbf{O}_p(\mathbf{N}_G(P))$.   □

    Observe also that G acts in a natural way on the set of p-weights of G and also on those belonging to a fixed p-block $B \in \mathrm{Bl}_p(G)$. The following conjecture was proposed by Alperin in [2].

Conjecture 4.14.12 (Alperin's weight conjecture)

(a) *The number of G-orbits of p-weights of G equals the number* $\mathrm{l}(G)$ *of p-regular conjugacy classes.*

(b) *If* $B \in \mathrm{Bl}_p(G)$, *then the number of G-orbits of p-weights of G belonging to B equals* $\mathrm{l}(B)$.

    Part (a) of the conjecture may be considered as expressing $\mathrm{k}^{(0)}(G)$ in terms of $\mathrm{l}(G)$ and the numbers of irreducible defect-zero characters of some smaller groups:

$$\mathrm{k}^{(0)}(G) = \mathrm{l}(G) - \sum_{P \backslash \{1\}} \mathrm{k}^{(0)}(\mathbf{N}_G(P)/P), \tag{4.56}$$

where the sum extends over a set of representatives of the conjugacy classes of non-trivial p-subgroups (or radical p-subgroups) of G.

Lemma 4.14.13 *Conjecture* 4.14.12(b) *holds if B has defect one.*

Proof. Let $B \in \mathrm{Bl}_p(G)$ be a block of defect one, let $D \in \mathrm{Def}(B)$ with $N := \mathbf{N}_G(D)$ and let $b \in \mathrm{Bl}_p(N)$ be the Brauer correspondent of B. By Theorem 4.9.8 we have $\mathrm{l}(B) = \mathrm{l}(b)$. If V is a simple FN-module belonging to b then, by Theorem 4.8.15, V has a vertex contained in D, hence V is D-projective. Since D is in the kernel of V, we have $V \mid (F_D)^N$. Thus V is the inflation of a projective $F(N/D)$-module $\bar{V}$ and yields a p-weight (D, χ), where $\chi \in \mathrm{Irr}(N/D)$ is the character of the projective $K(N/D)$-module belonging to $\bar{V}$.

    Conversely, if (D, χ) is a p-weight belonging to B, and $\bar{V}$ is a projective $F(N/D)$-module with Brauer character $\chi|_{(N/D)_{p'}}$, then $V := \mathrm{Inf}(\bar{V})$ is a simple FN-module belonging to b.   □

Remark 4.14.14 Using Remark 4.9.9 (instead of Theorem 4.9.8) in the proof of Lemma 4.14.13, one can see that Conjecture 4.14.12(b) holds more generally for p-blocks with cyclic defect groups.

    It is known that Alperin's weight conjecture holds for p-solvable groups (see [71]). It has also been verified for many other groups, including all finite groups of Lie type in defining characteristic p and all symmetric groups, for all primes.

Example 4.14.15 We check Alperin's weight conjecture for $G := M_{11}$ and $p = 2$. From the table of marks of G (see Example 3.5.24) we find the conjugacy classes of p-subgroups P and also $[\mathbf{N}_G(P) : P]$ (see Lemma 3.5.3(a)). If $[\mathbf{N}_G(P) : P]$ is a non-trivial power of p then P is obviously not a radical p-subgroup of G. Similarly $[\mathbf{N}_G(P) : P] = 24$ can be excluded for a radical p-subgroup P of G. Using Lemma 3.5.3(c) we see that there are four conjugacy classes of radical p-subgroups P, and we also get the isomorphism types of $\mathbf{N}_G(P)/P$ and thus the number of weights as shown in Table 4.11.

Table 4.11.

P	$\{1\}$	V_4	Q_8	QD_{16}	$1(G)$
$[\mathbf{N}_G(P): P]$	7920	6	6	1	
$\mathbf{N}_G(P)/P$	G	S_3	S_3	$\{1\}$	
Orbits of weights	2	1	1	1	5

Altogether we see that we have five G-orbits of 2-weights, and this is precisely the number of 2-regular conjugacy classes of G. All 2-weights (P, χ) belong to $B_0(G)$ (which is the only 2-block of non-zero defect) except for those with $P = \{1\}$. So Conjecture 4.14.12(b) holds also. ◆

The GAP library of tables of marks contains in many cases, for each conjugacy class of subgroups, generators for a representative of the class, so that concrete computations can be carried out for this group, for instance, the character table can be computed. In these cases one can automatically check Alperin's weight conjecture.

Example 4.14.16 The Higman–Sims group $G := $ HS has 589 conjugacy classes of subgroups, 250 conjugacy classes of 2-subgroups, but only nine classes of radical 2-subgroups. Seven of these contain subgroups P such that (P, χ) is a weight with some defect-zero character $\chi \in \mathrm{Irr}(\mathbf{N}_G(P)/P)$. There are two 2-blocks, the principal block B_0 and a block B_1 of defect two. Table 4.12 lists the number of weights (P, χ) belonging to these blocks with specified $|P|$:

Table 4.12.

$|P|$	4	16	64	128	256	256	512	$1(B)$
$[\mathbf{N}_G(P): P]$	60	720	168	6	6	6	1	
$\mathbf{N}_G(P)/P$	$C_{15} \rtimes C_4$	S_6	$L_2(7)$	S_3	S_3	S_3	$\{1\}$	
Weights in B_0	.	1	1	1	1	1	1	6
Weights in B_1	3	.	.	.	.	.	.	3

◆

While in Alperin's weight conjecture only defect-zero characters play a role, the Dade conjectures, to be introduced now, deal with characters of arbitrary p-defect. We need a few definitions first.

A p-chain $\mathcal{C}$ of G of length $|\mathcal{C}| := n \geq 0$ is any strictly increasing chain

$$\mathcal{C} : P_0 < P_1 < \cdots < P_n \qquad (4.57)$$

of p-subgroups of G. The group G acts on such chains by conjugation, and the stabilizer of $\mathcal{C}$ is defined as

$$\mathbf{N}_G(\mathcal{C}) := \bigcap_{i=0}^{n} \mathbf{N}_G(P_i).$$

The chain $\mathcal{C}$ is called "elementary" if all its members P_i are elementary abelian; $\mathcal{C}$ is called "normal" if all P_i are normal subgroups in P_n; and $\mathcal{C}$ is called a "radical p-chain" if

$$P_i = \mathbf{O}_p(\mathbf{N}_G(\mathcal{C}^{(i-1)})) \qquad \text{for} \qquad 1 \leq i \leq n, \qquad (4.58)$$

where $\mathcal{C}^{(i)}$ is the subchain

$$\mathcal{C}^{(i)} : P_0 < P_1 < \cdots < P_i.$$

The set of all elementary p-chains, normal p-chains and p-chains consisting of radical p-subgroups of G starting with $P_0 = \mathbf{O}_p(G)$ will be denoted by $\mathcal{E}(G)$, $\mathcal{N}(G)$ and $\mathcal{U}(G)$, respectively, the set of all radical p-chains of G beginning with $P_0 = \mathbf{O}_p(G)$ by $\mathcal{R}(G)$. We choose sets of representatives of the G-orbits on these sets of chains and denote them by $\mathcal{E}(G)/G$, $\mathcal{N}(G)/G$, $\mathcal{U}(G)/G$ and $\mathcal{R}(G)/G$, respectively.

Lemma 4.14.17 *With the notation introduced above we have*

(a) $\mathcal{E}(G) \cup \mathcal{R}(G) \subseteq \mathcal{N}(G)$;

(b) *if* $\mathcal{C} \in \mathcal{N}(G)$ *and* $b \in \mathrm{Bl}_p(\mathbf{N}_G(\mathcal{C}))$ *then* b^G *is defined.*

Proof.   (a) Obviously $\mathcal{E}(G) \subseteq \mathcal{N}(G)$. Let $\mathcal{C} \in \mathcal{R}(G)$ be as in (4.57). From (4.58) we conclude that

$$P_i \leq P_n \leq \mathbf{N}_G(\mathcal{C}^{(n-1)}) \qquad (0 \leq i < n).$$

(a) follows from this, since $P_i \trianglelefteq \mathbf{N}_G(\mathcal{C}^{(n-1)})$ for $i < n$.
   (b) Let $\mathcal{C} \in \mathcal{N}(G)$ be as in (4.57). Then

$$P_n \, \mathbf{C}_G(P_n) \leq \mathbf{N}_G(\mathcal{C}) \leq \mathbf{N}_G(P_n),$$

and (b) follows from Theorem 4.7.19. $\qquad\square$

Remark 4.14.18 Knörr and Robinson ([103], lemma 3.2) have shown that for an arbitrary p-chain $\mathcal{C}$ and $b \in \mathrm{Bl}_p(\mathbf{N}_G(\mathcal{C}))$ the block b^G is defined.

We put

$$k(\mathcal{C}, B, d) := \left| \bigcup_{\substack{b \in \mathrm{Bl}_p(\mathbf{N}_G(\mathcal{C})) \\ b^G = B}} \{\chi \in \mathrm{Irr}(b) \mid d_p(\chi) = d\} \right|.$$

Then the "ordinary Dade conjecture" can be stated as follows (see [46]).

Conjecture 4.14.19 (Ordinary Dade conjecture) *If* $\mathbf{O}_p(G) = \{1\}$ *and* $B \in \mathrm{Bl}_p(G)$ *with* $\mathrm{d}(B) > 0$ *then*

$$\sum_{\mathcal{C} \in \mathcal{R}(G)/G} (-1)^{|\mathcal{C}|} \, k(\mathcal{C}, B, d) = 0 \qquad \text{for any} \qquad d \in \mathbb{N}. \qquad (4.59)$$

Although all the conjectures of Dade are stated in terms of radical p-subgroups, it follows from results of Knörr and Robinson ([103], prop. 3.3) and Dade ([46], prop. 3.7) that one can replace $\mathcal{R}(G)$ in all of these by $\mathcal{E}(G)$ or $\mathcal{N}(G)$ or $\mathcal{U}(G)$. In fact it appears that for large groups $\mathcal{E}(G)/G$ is easier to compute than $\mathcal{R}(G)/G$ or $\mathcal{U}(G)/G$, although there are cases in which $\mathcal{R}(G)/G$ or $\mathcal{U}(G)/G$ is significantly smaller than $\mathcal{E}(G)/G$.

Example 4.14.20 We verify Conjecture 4.14.19 for $G := \mathrm{M}_{11}$ and $p := 2$. There are two p-blocks of defect zero and $B := B_0(G)$, and G has up to conjugacy just three elementary abelian 2-subgroups, $E_0 := \{1\}$, $E_1 \cong \mathrm{C}_2$ and $E_2 \cong \mathrm{V}_4$, with normalizers G, $\mathrm{GL}_2(3)$ and S_4, respectively (see Exercise 3.1.2). There is up to conjugacy just one elementary chain of length three, because the normalizer (S_4) of E_2 acts transitively on the involutions in E_2. Table 4.13 lists for each orbit $\mathcal{E}(G)/G$ a representative $\mathcal{C}$ (by the isomorphism types of its members), the isomorphism type of $\mathbf{N}_G(\mathcal{C})$, the parity of $|\mathcal{C}|$ and the number of irreducible characters of $\mathbf{N}_G(\mathcal{C})$ of defect d for $d = 0, \ldots, 4$. Except for $\mathcal{C} = \mathcal{C}_0$ the stabilizer $\mathbf{N}_G(\mathcal{C})$ has just one p-block, so that for $d = 0, \ldots, 4$ we have $k^{(d)}(\mathbf{N}_G(\mathcal{C})) = k(\mathcal{C}, B, d)$. Table 4.13 verifies Conjecture 4.14.19 and also shows that it would be false if blocks of defect zero were admitted.

Table 4.13.

$\mathcal{C}$		$(-1)^{|\mathcal{C}|}$	$\mathbf{N}_G(\mathcal{C})$	$[\, k^{(d)}(\mathbf{N}_G(\mathcal{C})) \mid d = 0 \ldots 4 \,]$
$\mathcal{C}_0:$	$\{1\}$	$+1$	G	$[\, 2,\ 0,\ 1,\ 3,\ 4 \,]$
$\mathcal{C}_1:$	$\{1\} < \mathrm{C}_2$	-1	$\mathrm{GL}_2(3)$	$[\, 0,\ 0,\ 1,\ 3,\ 4 \,]$
$\mathcal{C}_2:$	$\{1\} < \mathrm{V}_4$	-1	S_4	$[\, 0,\ 0,\ 1,\ 4,\ 0 \,]$
$\mathcal{C}_3:$	$\{1\} < \mathrm{C}_2 < \mathrm{V}_4$	$+1$	D_8	$[\, 0,\ 0,\ 1,\ 4,\ 0 \,]$
				$[\, 2,\ 0,\ 0,\ 0,\ 0 \,]$

On the other hand, we have already seen in Example 4.14.15 that G has four conjugacy classes of radical p-subgroups, (which are isomorphic to $\{1\}$, V_4, Q_8 and QD_{16}). We get the results shown in Table 4.14.

Table 4.14.

$\mathcal{C}$		$(-1)^{|\mathcal{C}|}$	$\mathbf{N}_G(\mathcal{C})$	$[\, \mathrm{k}^{(d)}(\mathbf{N}_G(\mathcal{C})) \mid d = 0 \ldots 4 \,]$
$\mathcal{C}_0:$	$\{1\}$	$+1$	G	$[\, 2,\ 0,\ 1,\ 3,\ 4 \,]$
$\mathcal{C}_1:$	$\{1\} < \mathrm{V}_4$	-1	S_4	$[\, 0,\ 0,\ 1,\ 4,\ 0 \,]$
$\mathcal{C}_2:$	$\{1\} < \mathrm{V}_4 < \mathrm{D}_8$	$+1$	D_8	$[\, 0,\ 0,\ 1,\ 4,\ 0 \,]$
$\mathcal{C}_3:$	$\{1\} < \mathrm{Q}_8$	-1	$\mathrm{GL}_2(3)$	$[\, 0,\ 0,\ 1,\ 3,\ 4 \,]$
$\mathcal{C}_4:$	$\{1\} < \mathrm{Q}_8 < \mathrm{QD}_{16}$	$+1$	QD_{16}	$[\, 0,\ 0,\ 0,\ 3,\ 4 \,]$
$\mathcal{C}_5:$	$\{1\} < \mathrm{QD}_{16}$	-1	QD_{16}	$[\, 0,\ 0,\ 0,\ 3,\ 4 \,]$
				$[\, 2,\ 0,\ 0,\ 0,\ 0 \,]$

Note also that QD_{16} occurs as a stabilizer of the chains $\mathcal{C}_4$ and $\mathcal{C}_5$. These chains have lengths of different parity, so their contributions to (4.59) cancel out, regardless of the character degrees of QD_{16}. ◆

Now assume that $\mathbf{Z}(G) = \{1\}$. Then G can be identified with the group $\mathrm{Inn}(G)$ of inner automorphisms of G, and $G \trianglelefteq A := \mathrm{Aut}(G)$ with $A/G \cong \mathrm{Out}(G)$. Note that A also acts on the chains $\mathcal{E}(G)$ or $\mathcal{N}(G)$ or $\mathcal{R}(G)$ or $\mathcal{U}(G)$, and we define for $\mathcal{C}$, as in (4.57), $\mathbf{N}_A(\mathcal{C}) := \bigcap_{i=0}^n \mathbf{N}_A(P_i)$. Obviously $\mathbf{N}_G(\mathcal{C}) \trianglelefteq \mathbf{N}_A(\mathcal{C})$, and we can embed $\mathbf{N}_A(\mathcal{C})/\mathbf{N}_G(\mathcal{C}) \subseteq \mathrm{Out}(G)$. For $\chi \in \mathrm{Irr}(\mathbf{N}_G(\mathcal{C}))$ let $T_{\mathbf{N}_A(\mathcal{C})}(\chi) \le \mathbf{N}_A(\mathcal{C})$ be the inertia subgroup of χ and $\bar{T}(\chi) := T_{\mathbf{N}_A(\mathcal{C})}(\chi)/\mathbf{N}_G(\mathcal{C})$ may be embedded in $\mathrm{Out}(G)$. For $B \in \mathrm{Bl}_p(G)$, $d \in \mathbb{N}$ and $H \le \mathrm{Out}(G)$ we define

$$\mathrm{k}(\mathcal{C}, B, d, H) := \Big| \bigcup_{\substack{b \in \mathrm{Bl}_p(\mathbf{N}_G(\mathcal{C})) \\ b^G = B}} \{\chi \in \mathrm{Irr}(b) \mid \mathrm{d}_p(\chi) = d,\ \bar{T}(\chi) = H\} \Big|.$$

We then arrive at the following refinement of Conjecture 4.14.19.

Conjecture 4.14.21 (Invariant Dade conjecture) *If* $\mathbf{Z}(G) = \mathbf{O}_p(G) = \{1\}$ *and* $B \in \mathrm{Bl}_p(G)$ *with* $\mathrm{d}(B) > 0$ *then*

$$\sum_{\mathcal{C} \in \mathcal{R}(G)/G} (-1)^{|\mathcal{C}|}\ \mathrm{k}(\mathcal{C}, B, d, H) = 0 \tag{4.60}$$

for any $d \in \mathbb{N}$ *and any subgroup* H *of* $\mathrm{Out}(G)$.

Example 4.14.22 We consider $p := 3$ and the sporadic simple McLaughlin group $G := \mathrm{McL}$ with $[\mathrm{Aut}(G) : G] = 2$. Apart from three blocks of defect zero there is only the principal 3-block $B := B_0(G)$ with $|\mathrm{Irr}(B)| = 21$. Here it is easier to work with chains consisting of radical 3-subgroups, because G has ten conjugacy classes of elementary abelian 3-subgroups, whereas there are just three non-trivial radical 3-subgroups R_i up to conjugacy. We use the table of marks of G in the GAP library and the notation of Definition 3.5.1 to obtain the content of Table 4.15.

Table 4.15. Non-trivial radical 3-subgroups of McL and their normalizers

Order	Name	$N_G(R_i)$	$|N_G(R_i)|$	$N_{\text{Aut}(G)}(R_i)$
3^4	$R_1 =_G H_{157}$	$M_1 =_G H_{367} \cong 3^4 : M_{10}$	58320	$3^4 : (M_{10} \times 2)$
3^5	$R_2 =_G H_{243}$	$M_2 =_G H_{368} \cong 3^{1+4}_+ : 2S_5$	58320	$3^{1+4}_+ : 4S_5$
3^6	$R_3 =_G H_{304}$	$M_3 =_G H_{352} \cong 3^4.3^2.Q_8$	5832	$3^4.3^2.Q_8.2$

The table of marks of G reveals that $R_3 \in \text{Syl}_3(G)$ contains just one subgroup conjugate to R_1 (in G) and also just one conjugate to R_2, whereas $R_1 \not\leq_G R_2$. So there are up to conjugacy only two chains of length two in $\mathcal{U}(G)$. Also M_1 and M_2 are maximal subgroups of G, and M_3 is a maximal subgroup of M_1 and of M_2. Thus we obtain representatives for $\mathcal{U}(G)/G$ and their normalizers in G and $\text{Aut}(G)$ as displayed in Table 4.16.

Table 4.16. Chains of radical 3-subgroups of McL and their normalizers

$\mathcal{C}$	$N_G(\mathcal{C})$	$N_{\text{Aut}(G)}(\mathcal{C})$	
$\mathcal{C}_0 = (\{1\})$	G	$\text{Aut}(G)$	
$\mathcal{C}_i = (\{1\} \leq R_i)$	M_i	$M_i.2$	$(i = 1, 2, 3)$
$\mathcal{C}_{1,3} = (\{1\} \leq R_1 \leq R_3)$	M_3	$M_3.2 \cong 3^4.3^2.Q_8$	
$\mathcal{C}_{2,3} = (\{1\} \leq R_2 \leq R_3)$	M_3	$M_3.2 \cong 3^4.3^2.Q_8$	

We see that the contributions of $\mathcal{C}_3$ and $\mathcal{C}_{1,3}$ to the sum in (4.60) cancel out, so these chains may be omitted in the following. The character tables of M_1, M_2 and $M_1.2$, $M_2.2$ (which are maximal subgroups of $\text{Aut}(G)$) are contained in the GAP library. The character tables of M_3 and $M_3.2$ are easily computed. It turns out that each of these groups has a single 3-block. We obtain the following results:

	$d \leq 2$	$d = 3$	$d = 4$	$d = 5$	$d = 6$
$k(\mathcal{C}_0, B, d, \{1\})$	.	2	.	2	4
$-k(\mathcal{C}_1, B, d, \{1\})$	.	.	.	-2	-4
$-k(\mathcal{C}_2, B, d, \{1\})$	.	-2	-2	-2	-4
$k(\mathcal{C}_{2,3}, B, d, \{1\})$	.	.	2	2	4

	$d \leq 2$	$d = 3$	$d = 4$	$d = 5$	$d = 6$
$k(\mathcal{C}_0, B, d, \text{Out}(G))$	.	1	3	1	8
$-k(\mathcal{C}_1, B, d, \text{Out}(G))$	.	.	-3	-1	-8
$-k(\mathcal{C}_2, B, d, \text{Out}(G))$	.	-1	-4	-1	-8
$k(\mathcal{C}_{1,3}, B, d, \text{Out}(G))$	.	.	4	1	8

As all the columns add up to zero, this verifies the invariant conjecture (and hence also the ordinary conjecture) for the prime $p = 3$. The invariant conjecture was proved for all primes dividing $|G|$ in [55].   ◆

Dade has formulated even stronger conjectures in the hope that for the verification of the strongest form it would be sufficient to consider only simple non-abelian groups. We just mention the following conjecture (see [47]) that takes account of projective representations (or characters of covering groups).

Conjecture 4.14.23 (Projective Dade conjecture) *Assume that* $\mathbf{O}_p(G) = \{1\}$ *and that* $\tilde{G}$ *is a factor group of a representation group of* G *with* $\tilde{G}/Z \cong G$ *for a cyclic subgroup* $Z \le \mathbf{Z}(\tilde{G})$ *and choose* $\lambda \in \mathrm{Irr}(Z)$. *For* $B \in \mathrm{Bl}_p(\tilde{G})$, *an integer* d *and a* p-*chain* $\mathcal{C}$ *of* G, *define*

$$k(\mathcal{C}, B, d, \lambda) := \Big| \bigcup_{\substack{b \in \mathrm{Bl}_p(\mathbf{N}_{\tilde{G}}(\mathcal{C})) \\ b^{\tilde{G}} = B}} \{\chi \in \mathrm{Irr}(b) \mid d_p(\chi) = d, \ \chi_Z = \chi(1) \cdot \lambda\} \Big|.$$

Then

$$\sum_{\mathcal{C} \in \mathcal{R}(G)/\tilde{G}} (-1)^{|\mathcal{C}|} \, k(\mathcal{C}, B, d, \lambda) = 0 \tag{4.61}$$

for any $d \in \mathbb{N}$ *provided that* $\mathrm{d}(B) > \nu_p(|Z|)$.

Observe that $\tilde{G}$ acts on p-chains of G. Also for $\lambda = 1_Z$ equation (4.61) coincides with (4.59) in Dade's ordinary conjecture.

Example 4.14.24 We consider again $G := \mathrm{McL}$ and $p = 3$. The Schur multiplier of G has order three, so we have a unique representation group $\tilde{G} \cong 3 \cdot G$. The principal block B_0 is the only 3-block of G with $\mathrm{d}(B) > 1$. Let $\lambda \in \mathrm{Irr}(Z)$ be faithful. The chains we need have already been determined in Example 4.14.22, but we need to find the preimages $\tilde{M}_i \le \tilde{G}$ of the stabilizers $M_i \le G$ for $i = 1, 2, 3$. It turns out that all $\tilde{M}_i$ are non-split extensions of Z and all have just one 3-block. The character tables of $\tilde{M}_i$ are contained in the GAP library for $i = 1, 2$ and in [55] for $i = 3$. We obtain

	$d \le 3$	$d = 4$	$d = 5$	$d = 6$	$d = 7$
$k(\mathcal{C}_0, B_0, d, \lambda)$	.	6	12	.	.
$-k(\mathcal{C}_1, B_0, d, \lambda)$	.	-6	-15	.	.
$-k(\mathcal{C}_2, B_0, d, \lambda)$	.	.	-12	.	.
$k(\mathcal{C}_{2,3}, B_0, d, \lambda)$	.	.	15	.	.

which verifies Conjecture 4.14.23 in this case.   ◆

Another important conjecture related to the previously mentioned conjectures was stated by Broué, see [22]. It relies on the following notion.

Definition 4.14.25 Let G and H be two groups, p a prime, (K, R, F, η) a p-modular splitting system for G and H and let $\mu \in \mathbb{Z}\,\mathrm{Irr}(G \times H)$ be a virtual character of $G \times H$. Then μ is called a **perfect character** if the following two conditions are satisfied:

(1) $\frac{\mu(g,h)}{|C_G(g)|} \in R$ and $\frac{\mu(g,h)}{|C_H(h)|} \in R$ for all $g \in G$ and all $h \in H$;

(2) if $\mu(g,h)$ is non-zero then g and h are both p-regular or both p-singular.

To a given perfect character we assign two K-linear maps. Define I_μ: $\mathrm{cf}(H,K) \to \mathrm{cf}(G,K)$ by

$$I_\mu(\alpha)(g) := \frac{1}{|H|} \sum_{h \in H} \mu(g, h^{-1}) \alpha(h)$$

and R_μ: $\mathrm{cf}(G,K) \to \mathrm{cf}(H,K)$ by

$$R_\mu(\beta)(h) := \frac{1}{|G|} \sum_{g \in G} \mu(g^{-1}, h) \beta(g)$$

for $\alpha \in \mathrm{cf}(H,K)$ and $\beta \in \mathrm{cf}(G,K)$. The map I_μ is called "generalized induction" and R_μ is called "generalized restriction." Moreover, it is easy to check that $(I_\mu(\alpha), \beta)_G = (\alpha, R_\mu(\beta))_H$. Condition (1) above shows that an R-valued class function is mapped to an R-valued class function by both I_μ and R_μ. Condition (2) above shows that functions that are zero on the p-singular, respectively p-regular, classes are mapped to functions that are zero on the p-singular classes, respectively p-regular classes. A motivation for this definition is the fact that if L is an (RG, RH)-bilattice such that the restriction to both RG and RH is projective, then the character μ_L is a perfect character; see [22].

Given a p-block of a group G and a p-block of a group H, a perfect isometry for the two blocks is now described as follows.

Definition 4.14.26 Let B be a p-block of a group G and let b be a p-block of a group H. Then an isometry I from $K \mathrm{Irr}(b)$ to $K \mathrm{Irr}(B)$ is called a **perfect isometry** if there is a perfect character $\mu \in \mathbb{Z}(\mathrm{Irr}(B) \times \mathrm{Irr}(b))$ with $I = I_\mu|_{K \mathrm{Irr}(b)}$. Note that in this case there is a bijection $J: \mathrm{Irr}(b) \to \mathrm{Irr}(B)$ and signs $\epsilon: \mathrm{Irr}(b) \to \{-1, 1\}$ such that $I_\mu(\zeta) = \epsilon(\zeta) J(\zeta)$ for all $\zeta \in \mathrm{Irr}(b)$ and $\mu = \sum_{\zeta \in \mathrm{Irr}(b)} \epsilon(\zeta) J(\zeta) \times \zeta$.

One of Broué's conjectures can now be stated as follows.

Conjecture 4.14.27 (Broué's conjecture) *Let G be a finite group, p a prime, B a p-block with abelian defect group D and b the p-block of $\mathbf{N}_G(D)$ with $b^G = B$ (the Brauer correspondent). Then B and b are perfectly isometric.*

It is shown in [22] that in this situation the Alperin weight conjecture, the Alperin–McKay conjecture and the ordinary Dade conjecture follow from Broué's conjecture.

Remark 4.14.28 Broué's conjecture is only the shadow of a more general module theoretic conjecture about a so-called equivalence of the bounded derived module categories of B and b; see [22].

In the following we are going to look at the sporadic simple group J_1 for $p = 2$. Before we do this, we discuss a necessary condition that an isometry has to satisfy and which is also easy to check.

Remark 4.14.29 Let I_μ be a perfect isometry as above and let $\zeta, \zeta' \in \mathrm{Irr}(b)$. From condition (2) for a perfect isometry we conclude (using Definition 4.2.18) that

$$(\zeta, \zeta')_{H_{p'}} = (\widehat{\zeta}, \widehat{\zeta'})_H = (\epsilon(\widehat{\zeta)J(\zeta)}, \epsilon(\widehat{\zeta')J(\zeta'})}_G = (J(\zeta), J(\zeta'))_{G_{p'}} \epsilon(\zeta)\epsilon(\zeta').$$

By Exercise 4.4.7 this equation can also be expressed as follows:

$$D(b)C(b)^{-1}D(b)^\mathrm{T} = MD(B)C(B)^{-1}D(B)^\mathrm{T}M^\mathrm{T},$$

where $D(b)$, respectively $D(B)$, is the decomposition matrix of b, respectively B; $C(b)$, respectively $C(B)$, is the Cartan matrix of b, respectively B; and finally M is the (monomial) matrix of I_μ with respect to the bases $\mathrm{Irr}(b)$ and $\mathrm{Irr}(B)$. Note that $M^\mathrm{T} = M^{-1}$.

Example 4.14.30 We study the case of the principal 2-block B_0 of the sporadic simple group $G := \mathrm{J}_1$. A Sylow 2-subgroup P is elementary abelian of order eight and its normalizer in G has structure $N := \mathbf{N}_G(P) = 2^3 : (7 : 3)$. The ordinary irreducible characters $\chi_1, \chi_6, \chi_7, \chi_8, \chi_{12}, \chi_{13}, \chi_{14}, \chi_{15}$ are the characters in the principal block B_0 of G, and all eight irreducible characters $\zeta_1, \ldots, \zeta_8$ of N are in the principal block b_0 of N. The following GAP code computes the inner product matrices on the 2-regular classes for these characters and then verifies that the isometry given by

ζ_1	ζ_2	ζ_3	ζ_4	ζ_5	ζ_6	ζ_7	ζ_8
χ_1	χ_6	χ_7	$-\chi_8$	$-\chi_{14}$	$-\chi_{15}$	$-\chi_{12}$	$-\chi_{13}$

satisfies the equality mentioned above. In order to simplify the display of the inner product matrices, both will be multiplied by $|P| = 8$ and T will be the matrix of the isometry.

We first write a short GAP program hat which computes $\widehat{\chi_i}$ for the ith irreducible character χ_i of a character table t:

```
gap> hat := function(t,i)
> local n,y,j ;   n:=Length(Irr(t));   y := List([1..n], x -> 0);
> for j in [1..n] do
> if not IsInt(OrdersClassRepresentatives(t)[j]/2) then y[j]:=Irr(t)[i][j];fi;
> od;
> return(y);end;;
gap> ct := CharacterTable("J1");;
gap> b1 := Positions( PrimeBlocks(ct,2).block, 1 );
[ 1, 6, 7, 8, 12, 13, 14, 15 ]
gap> hchi := List( b1, i -> hat(ct,i) );;
gap> res := 8*MatScalarProducts( ct, hchi, hchi );;
gap> Display(res);
[ [   5,   1,   1,   1,   1,   1,   1,  -3 ],
  [   1,   5,   1,   1,  -3,   1,   1,   1 ],
  [   1,   1,   5,   1,   1,  -3,   1,   1 ],
  [   1,   1,   1,   5,   1,   1,  -3,   1 ],
```

```
[    1,   -3,    1,    1,    5,    1,    1,    1 ],
[    1,    1,   -3,    1,    1,    5,    1,    1 ],
[    1,    1,    1,   -3,    1,    1,    5,    1 ],
[   -3,    1,    1,    1,    1,    1,    1,    5 ] ]
gap> ct1 := CharacterTable("2^3.7.3");;
gap> hxi := List( [1..8], i -> hat(ct1,i) );;
gap> resn:= 8*MatScalarProducts( ct1, hxi, hxi );;
gap> Display(resn);
[ [    5,    1,    1,   -1,   -1,    3,   -1,   -1 ],
  [    1,    5,    1,   -1,   -1,   -1,    3,   -1 ],
  [    1,    1,    5,   -1,   -1,   -1,   -1,    3 ],
  [   -1,   -1,   -1,    5,   -3,    1,    1,    1 ],
  [   -1,   -1,   -1,   -3,    5,    1,    1,    1 ],
  [    3,   -1,   -1,    1,    1,    5,    1,    1 ],
  [   -1,    3,   -1,    1,    1,    1,    5,    1 ],
  [   -1,   -1,    3,    1,    1,    1,    1,    5 ] ]
gap> T:=
> [ [ 1, 0, 0, 0, 0, 0, 0, 0 ], [ 0, 1, 0, 0, 0, 0, 0, 0 ],
>   [ 0, 0, 1, 0, 0, 0, 0, 0 ], [ 0, 0, 0, -1, 0, 0, 0, 0 ],
>   [ 0, 0, 0, 0, 0, 0, -1, 0 ], [ 0, 0, 0, 0, 0, 0, 0, -1 ],
>   [ 0, 0, 0, 0, -1, 0, 0, 0 ], [ 0, 0, 0, 0, 0, -1, 0, 0 ] ];;
gap> res = T*resn*TransposedMat(T);
true
```

We now construct the function μ according to Definition 4.14.26:

```
gap> J := [1,6,7,8,14,15,12,13];;    eps := [1,1,1,-1,-1,-1,-1,-1];;
gap> mu := function(i,j)     return( Sum( List( [1..8], k ->
>           eps[k] * Irr(ct)[ J[k] ][i] * Irr(ct1)[k][j])) ); end;;
gap> domain := Cartesian( [1..Length(Irr(ct))], [1..Length(Irr(ct1))] );;
```

Since $\frac{\mu(g,h)}{|C_G(g)|} \in R$ if $\frac{\mu(g,h)}{|C_G(g)|_p}$ is a cyclotomic integer, we compute the 2-parts of the centralizer orders and check that conditions (1) and (2) hold true:

```
gap> ctpord := List( SizesCentralizers(ct),
>                    x -> Product( Filtered(Factors(x), p -> p=2) ) );;
gap> ct1pord := List( SizesCentralizers(ct1),
>                    x -> Product( Filtered(Factors(x), p -> p=2) ) );;
gap> ForAll( domain ,
>          x -> IsIntegralCyclotomic(mu(x[1],x[2])/ctpord[x[1]]) and
>               IsIntegralCyclotomic(mu(x[1],x[2])/ct1pord[x[2]]) );
true
gap> nz := Filtered( domain, x -> mu(x[1],x[2]) <> 0);;
gap> ForAll( nz, x -> ( IsInt(OrdersClassRepresentatives(ct)[x[1]]/2)
>                and  IsInt(OrdersClassRepresentatives(ct1)[x[2]]/2) )
>         or        ( not IsInt(OrdersClassRepresentatives(ct)[x[1]]/2)
>               and not IsInt(OrdersClassRepresentatives(ct1)[x[2]]/2) ) );
true
```

◆

Exercise

Exercise 4.14.1 Write GAP programs to compute representatives for the conjugacy classes of elementary p-chains and p-chains consisting of radical p-subgroups of a group. Use these to verify the results of Examples 4.14.20, 4.14.22 and 4.14.24.

References

[1] J. L. Alperin, *Local Representation Theory*. Cambridge Studies in Advanced Mathematics, vol. 11 (Cambridge: Cambridge University Press, 1986).

[2] J. L. Alperin, Weights for finite groups. In *The Arcata Conference on Representations of Finite Groups (Arcata, Calif., 1986)*. volume 47 of Proc. Sympos. Pure Math., vol. 47 (Providence, RI: American Mathematical Society, 1987), pp. 369–379.

[3] J. L. Alperin and R. B. Bell, *Groups and Representations*, Graduate Texts in Mathematics, vol. 162 (New York: Springer-Verlag, 1995).

[4] F. W. Anderson and K. R. Fuller, *Rings and Categories of Modules*. Graduate Texts in Mathematics, vol. 13 (New York: Springer-Verlag, 1974).

[5] M. Aschbacher, *Finite Group Theory*. Cambridge Studies in Advanced Mathematics, vol. 10 (Cambridge: Cambridge University Press, 1986).

[6] E. Bannai and T. Ito, *Algebraic Combinatorics. I.* (Menlo Park, CA: The Benjamin/Cummings Publishing Co. Inc., 1984).

[7] U. Baum, Existenz und effiziente Konstruktion schneller Fouriertransformationen überauflösbarer Gruppen. Ph.D. thesis, University of Bonn (1991).

[8] U. Baum, Existence and efficient construction of fast Fourier transforms on supersolvable groups. *Comput. Complexity*, **1**(3):235–256 (1991).

[9] U. Baum and M. Clausen, Computing irreducible representations of supersolvable groups. *Math. Comp.*, **63**(207):351–359 (1994).

[10] D. J. Benson. *Representations and Cohomology. I.* Cambridge Studies in Advanced Mathematics, vol. 30 (Cambridge: Cambridge University Press, 1991).

[11] H. Besche, Die Berechnung von Charaktergraden und Charakteren endlicher auflösbarer Gruppen im Computeralgebrasystem GAP. Diplomarbeit, Lehrstuhl D für Mathematik, RWTH-Aachen (1992).

[12] O. Bonten, Über Kommutatoren in endlichen einfachen Gruppen. Ph.D. thesis, RWTH (1993).

[13] R. Brauer, Representations of finite groups. In *Lectures on Modern Mathematics, Vol. I*, ed. T. Saaty (New York: Wiley, 1963), pp. 133–175.

[14] R. Brauer and W. Feit, On the number of irreducible characters of finite groups in a given block. *Proc. Nat. Acad. Sci. U.S.A.*, **45**:361–365 (1959).

[15] T. Breuer and K. Lux, The multiplicity-free permutation characters of the sporadic simple groups and their automorphism groups. *Comm. Algebra*, **24**(7):2293–2316 (1996).

[16] T. Breuer, Subgroups of J_4 inducing the same permutation character. *Comm. Algebra*, **23**(9):3173–3176 (1995).

[17] T. Breuer. Computing possible class fusions from character tables. *Comm. Algebra*, **27**(6):2733–2748 (1999).

[18] T. Breuer, *Characters and Automorphism Groups of Compact Riemann Surfaces*. London Mathematical Society Lecture Note Series, vol. 280 (Cambridge: Cambridge University Press, 2000).

[19] T. Breuer, *The* GAP *Character Table Library, Version 1.1, 2004*. Available from http://www.math.rwth-aachen.de/~Thomas.Breuer/ctbllib/ (2004).

[20] T. Breuer and E. Horváth, On block induction. *J. Algebra*, **242**(1):213–224 (2001).

[21] T. Breuer and G. Pfeiffer, Finding possible permutation characters. *J. Symbolic Comput.*, **26**(3):343–354 (1998).

[22] M. Broué, Isométries parfaites, types de blocs, catégories dérivées. *Astérisque*, **181–182**:61–92 (1990).

[23] F. Buekenhout, The geometry of the finite simple groups. In *Buildings and the Geometry of Diagrams (Como, 1984)*, ed. L. Rosati. Lecture Notes in Mathematics, vol. 1181 (Berlin: Springer, 1986), pp. 1–78.

[24] W. Burnside, *Theory of Groups of Finite Order,* 2nd edn (Cambridge: Cambridge University Press, 1911), reprinted by Dover Publications Inc. (1955).

[25] R. W. Carter, *Finite Groups of Lie Type*. Pure and Applied Mathematics, vol. XXVIII (New York: John Wiley & Sons Inc., 1985).

[26] M. Clausen, A direct proof of Minkwitz's extension theorem. *Appl. Algebra Engrg. Comm. Comput.*, **8**(4):305–306 (1997).

[27] M. Clausen and U. Baum, *Fast Fourier Transforms* (Mannheim: Bibliographisches Institut, 1993).

[28] A. M. Cohen, H. Cuypers, and H. Sterk, eds., *Some Tapas of Computer Algebra.* Algorithms and Computation in Mathematics, vol. 4 (Berlin: Springer-Verlag, 1999).

[29] H. Cohen, *A Course in Computational Algebraic Number Theory.* Graduate Texts in Mathematics, vol. 138 (Berlin: Springer-Verlag, 1993).

[30] H. Cohn, R. Kleinberg, B. Szegedy and C. Umans, Group-theoretic algorithms for matrix multiplication. In *FOCS '05: Proceedings of the 46th Annual IEEE Symposium on Foundations of Computer Science* (Washington, D.C.: IEEE Computer Society, 2005), pp. 379–388.

[31] H. Cohn and C. Umans, A group-theoretic approach to fast matrix multiplication. In *Proceedings of the 44th Annual Symposium on Foundations of Computer Science* (Washington, D.C.: IEEE Computer Society, 2003), pp. 438–449.

[32] P. M. Cohn, *Algebra, Vol. 2*, 2nd edn (Chichester: John Wiley & Sons Ltd, 1989).

[33] M. J. Collins, *Representations and Characters of Finite Groups.* Cambridge Studies in Advanced Mathematics, vol. 22 (Cambridge: Cambridge University Press, 1990).

[34] M. D. E. Conder, Generators for alternating and symmetric groups. *J. London Math. Soc. (2)*, **22**(1):75–86 (1980).

[35] S. B. Conlon. Calculating characters of p-groups. *J. Symbolic Comput.*, **9**(5–6):535–550 (1990).

[36] S. B. Conlon, Computing modular and projective character degrees of soluble groups. *J. Symbolic Comput.*, **9**(5–6):551–570 (1990).

[37] J. H. Conway, Character calisthenics. In *Computational Group Theory (Durham, 1982)*, ed. M. D. Atkinson (London: Academic Press, 1984), pp. 249–266.

[38] J. H. Conway, R. T. Curtis, S. P. Norton, R. A. Parker and R. A. Wilson, *Atlas of Finite Groups* (Eynsham: Oxford University Press, 1985).

[39] H. S. M. Coxeter and W. O. J. Moser, *Generators and Relations for Discrete Groups*, 2nd edn (Berlin: Springer-Verlag, 1965).

[40] C. W. Curtis and I. Reiner, *Representation Theory of Finite Groups and Associative Algebras.* Pure and Applied Mathematics, vol. XI. (New York: Wiley-Interscience, 1962).

[41] C. W. Curtis and I. Reiner, *Methods of Representation Theory. Vol. I.* (New York: Wiley-Interscience, 1981).

[42] C. W. Curtis and I. Reiner, *Methods of Representation Theory. Vol. II.* Wiley Classics Library (New York: Wiley-Interscience, 1987).

[43] E. C. Dade, Answer to a question of R. Brauer. *J. Algebra*, **1**:1–4 (1964).

[44] E. C. Dade, Deux groupes finis distincts ayant la même algèbre de groupe sur tout corps. *Math. Z.*, **119**:345–348 (1971).

[45] E. C. Dade, Normal subgroups of M-groups need not be M-groups. *Math. Z.*, **133**:313–317 (1973).

[46] E. C. Dade, Counting characters in blocks. I. *Invent. Math.*, **109**(1):187–210 (1992).

[47] E. C. Dade, Counting characters in blocks. II. *J. Reine Angew. Math.*, **448**:97–190 (1994).

[48] J. D. Dixon, High speed computation of group characters. *Numer. Math.*, **10**:446–450 (1967).

[49] J. D. Dixon, Constructing representations of finite groups. In *Groups and Computation*, eds. L. Finkelstein and W. Kantor, DIMACS Ser. Discrete Math. Theoret. Comput. Sci., vol. 11 (Providence, RI: America Mathematical Society), pp.105–112.

[50] L. Dornhoff, *Group Representation Theory. Part A: Ordinary Representation Theory.* Pure and Applied Mathematics, vol. 7 (New York: Marcel Dekker Inc., 1971).

[51] L. Dornhoff, *Group Representation Theory. Part B: Modular Representation Theory.* Pure and Applied Mathematics, vol. 7 (New York: Marcel Dekker Inc., 1972).

[52] A. Dress, A characterisation of solvable groups. *Math. Z.*, **110**:213–217 (1969).

[53] B. Eick and J. Müller, On p-groups forming Brauer pairs. *J. Algebra*, **304**(1):286–303 (2006).

[54] E. W. Ellers and N. Gordeev, On the conjectures of J. Thompson and O. Ore. *Trans. Amer. Math. Soc.*, **350**(9):3657–3671 1998.

[55] G. Entz and H. Pahlings, The Dade conjecture for the McLaughlin group. In *Groups St Andrews 1997 in Bath, I*, eds. C.M. Campbell *et al.* London Mathematical Society Lecture Note Series, vol. 260 (Cambridge: Cambridge University Press, 1999), pp. 253–266.

[56] W. Feit, *The Representation Theory of Finite Groups*. North-Holland Mathematical Library, vol. 25 (Amsterdam: North-Holland Publishing Co., 1982).

[57] W. Feit, The computations of some Schur indices. *Israel J. Math.*, **46**(4):274–300 (1983).

[58] B. Fischer, Clifford-matrices. In *Representation Theory of Finite Groups and Finite-Dimensional Algebras (Bielefeld, 1991)*, Progr. Math., vol. 95 (Basel: Birkhäuser, 1991), pp. 1–16.

[59] P. Fong, On decomposition numbers of J_1 and $R(q)$. *Symposia Mathematica, Vol. XIII (Convegno di Gruppi e loro Rappresentazioni, INDAM, Rome, 1972)*, (London: Academic Press, 1974), pp. 415–422.

[60] J. S. Frame, Congruence relations between the traces of matrix powers. *Can. J. Math.*, **1**:303–304 (1949).

[61] G. Frobenius, Über die Charaktere der symmetrischen Gruppe. *Sitzungsberichte Akad. Wiss. Berlin*, pp. 516–534 (1900).

[62] The GAP Group. *GAP – Groups, Algorithms, and Programming, Version 4.4.10*, 2007. Available from: http://www.gap-system.org. (2007).

[63] M. Geck and G. Pfeiffer, *Characters of Finite Coxeter Groups and Iwahori-Hecke Algebras*. London Mathematical Society Monographs, New Series, vol. 21 (New York: The Clarendon Press, Oxford University Press, 2000).

[64] D. Gluck, On the $k(GV)$ problem. *J. Algebra*, **89**(1):46–55 (1984).

[65] D. Gluck and K. Magaard, The extraspecial case of the $k(GV)$ problem. *Trans. Amer. Math. Soc.*, **354**(1):287–333 (electronic) (2002).

[66] D. Gluck and K. Magaard, The $k(GV)$ conjecture for modules in characteristic 31. *J. Algebra*, **250**(1):252–270 (2002).

[67] D. Gluck, K. Magaard, U. Riese, and P. Schmid, The solution of the $k(GV)$-problem. *J. Algebra*, **279**(2):694–719, (2004).

[68] D. M. Goldschmidt. A group theoretic proof of the $p^a q^b$ theorem for odd primes. *Math. Z.*, **113**:373–375 (1970).

[69] D. M. Goldschmidt. *Lectures on Character Theory* (Wilmington, DE: Publish or Perish Inc., 1980).

[70] R. Gow, On the number of characters in a block and the $k(GV)$ problem for self-dual V. *J. London Math. Soc. (2)*, **48**(3):441–451 (1993).

[71] P. G. Gres', Fong's reduction, the correspondences of Brauer and Glauberman and Alperin's weight conjecture. In *Proceedings of the International Conference on Algebra, Part 1 (Novosibirsk, 1989)*, Contemp. Math., vol. 131 (Providence, RI: American Mathematical Society, 1992), pp. 137–158.

[72] L. C. Grove, *Algebra*. Pure and Applied Mathematics, vol. 10 (New York: Harcourt Brace Jovanovich Publishers 1983).

[73] L. C. Grove, *Groups and Characters*. Pure and Applied Mathematics (New York: Wiley-Interscience, 1997).

[74] R. M. Guralnick and J. Saxl, Primitive permutation characters. In *Groups, Combinatorics & Geometry (Durham, 1990)*, eds. M. W. Liebeck and J. Saxl. London Math. Soc. Lecture Note Ser., vol. 165 (Cambridge: Cambridge University Press, 1992).

[75] P. Hall, The eulerian function of a group. *Quart. J. Math. Oxford Ser.2*, **7**:134–151 (1936).

[76] P. Hall, The classification of prime-power groups. *J. Reine Angew. Math.*, **182**:130–141 (1940).

[77] L. S. Heath and N. A. Loehr, New algorithms for generating Conway polynomials over finite fields. *Proceedings of the Tenth Annual ACM-SIAM Symposium on Discrete Algorithms (Baltimore, MD, 1999)* (New York: ACM, 1999), pp. 429–437.

[78] N. S. Hekster, On the structure of n-isoclinism classes of groups. *J. Pure Appl. Algebra*, **40**(1):63–85 (1986).

[79] M. Hertweck, A counterexample to the isomorphism problem for integral group rings. *Ann. of Math. (2)*, **154**(1):115–138 (2001).

[80] D. G. Higman, Intersection matrices for finite permutation groups. *J. Algebra*, **6**:22–42 (1967).

[81] D. G. Higman, Coherent configurations. I. Ordinary representation theory. *Geometriae Dedicata*, **4**(1):1–32 (1975).

[82] G. Hiss, The modular atlas homepage. Available from: `http://www.math.rwth-aachen.de/homes/MOC` (2007).

[83] G. Hiss, C. Jansen, K. Lux and R. A. Parker, *Computational Modular Character Theory*. Available from: `http://www.math.rwth-aachen.de/homes/MOC/CoMoChaT/` (1997).

[84] D. F. Holt, B. Eick and E. A. O'Brien, *Handbook of Computational Group Theory*. Discrete Mathematics and its Applications (Boca Raton, FL: Chapman & Hall/CRC, 2005).

[85] D. F. Holt and W. Plesken, *Perfect Groups.* Oxford Mathematical Monographs (New York: The Clarendon Press, Oxford University Press, 1989).

[86] D. F. Holt and S. Rees, Testing modules for irreducibility. *J. Austral. Math. Soc. Ser. A*, **57**(1):1–16 (1994).

[87] A. Hulpke, Zur Berechnung von Charaktertafeln. Diplomarbeit, Lehrstuhl D für Mathematik, RWTH-Aachen (1993).

[88] B. Huppert, *Endliche Gruppen. I.* Die Grundlehren der Mathematischen Wissenschaften, Band 134 (Berlin: Springer-Verlag, 1967).

[89] B. Huppert and N. Blackburn, *Finite Groups. III*, Fundamental Principles of Mathematical Sciences, vol. 243 (Berlin: Springer-Verlag, 1982).

[90] I. M. Isaacs, Commutators and the commutator subgroup. *Amer. Math. Monthly*, **84**(9):720–722 (1977).

[91] I. M. Isaacs, G. Malle and G. Navarro, A reduction theorem for the McKay conjecture. *Invent. Math.*, **170**(1):33–101 (2007).

[92] I. M. Isaacs. *Character Theory of Finite Groups.* Pure and Applied Mathematics, no. 69 (New York: Academic Press, Harcourt Brace Jovanovich Publishers, 1976).

[93] G. Ivanyos and K. Lux, Treating the exceptional cases of the MeatAxe. *Experiment. Math.*, **9**(3):373–381 (2000).

[94] N. Jacobson, *Basic algebra. II* (San Francisco, CA: W. H. Freeman and Co., 1980).

[95] G. D. James, The modular characters of the Mathieu groups. *J. Algebra*, **27**:57–111 (1973).

[96] G. James and A. Kerber. *The Representation Theory of the Symmetric Group*, Encyclopedia of Mathematics and its Applications, vol. 16 (Reading, MA: Addison-Wesley Publishing Co., 1981).

[97] G. James and M. Liebeck, *Representations and Characters of Groups.* Cambridge Mathematical Textbooks (Cambridge: Cambridge University Press, 1993).

[98] C. Jansen, K. Lux, R. Parker and R. Wilson, *An Atlas of Brauer Characters.* London Mathematical Society Monographs, New Series, vol. 11 (New York: The Clarendon Press, Oxford University Press, 1995).

[99] G. J. Janusz, Primitive idempotents in group algebras. *Proc. Amer. Math. Soc.*, **17**:520–523 (1966).

[100] G. A. Jones, Characters and surfaces: a survey. In *The Atlas of Finite Groups: Ten Years On (Birmingham, 1995)*. London Mathematical Society Lecture Note Series, vol. 249 (Cambridge: Cambridge University Press, 1998), pp. 90–118.

[101] P. B. Kleidman and R. A. Wilson, The maximal subgroups of J_4. *Proc. London Math. Soc. (3)*, **56**(3):484–510 (1988).

[102] R. Knörr, On the number of characters in a p-block of a p-solvable group. *Illinois J. Math.*, **28**(2):181–210 (1984).

[103] R. Knörr and G. R. Robinson, Some remarks on a conjecture of Alperin. *J. London Math. Soc. (2)*, **39**(1):48–60 (1989).

[104] C. Köhler and H. Pahlings, Regular orbits and the $k(GV)$-problem. In *Groups and Computation, III (Columbus, OH, 1999)*, Ohio State Univ. Math. Res. Inst. Publ., vol. 8 (Berlin: de Gruyter, 2001), pp. 209–228.

[105] E. A. Komissartschik and S. V. Tsaranov. Intersections of maximal subgroups in simple groups of order less than 10^6. *Comm. Algebra*, **14**(9):1623–1678 (1986).

[106] B. Külshammer, A remark on conjectures in modular representation theory. *Arch. Math. (Basel)*, **49**(5):396–399 (1987).

[107] H. Kurzweil and B. Stellmacher, *Theorie der endlichen Gruppen.* (Berlin: Springer-Verlag, 1998).

[108] T. Y. Lam, *A First Course in Noncommutative Rings*, 2nd edn. Graduate Texts in Mathematics, vol. 131 (New York: Springer-Verlag, 2001).

[109] P. Landrock, *Finite Group Algebras and their Modules.* London Mathematical Society Lecture Note Series, vol. 84 (Cambridge: Cambridge University Press, 1983).

[110] S. Lang, *Algebra*, 3rd edn. Graduate Texts in Mathematics, vol. 211 (New York: Springer-Verlag, 2002).

[111] A. K. Lenstra, H. W. Lenstra, Jr. and L. Lovász, Factoring polynomials with rational coefficients. *Math. Ann.*, **261**(4):515–534 (1982).

[112] W. J. LeVeque, *Fundamentals of Number Theory*, (Reading, MA: Addison-Wesley Publishing Co., 1977).

[113] F. Lübeck, Conway polynomials for finite fields. Available from: `http://www.math.rwth-aachen.de/~Frank.Luebeck/data/ConwayPol` (2001).

[114] K. Lux, Algorithmic methods in modular representation theory. Habilitation thesis, RWTH-Aachen (1997).

[115] S. MacLane and G. Birkhoff, *Algebra*, 2nd edn (New York: Macmillan Inc., 1979).

[116] G. Malle and B. H. Matzat, *Inverse Galois Theory*. Springer Monographs in Mathematics (Berlin: Springer-Verlag, 1999).

[117] O. Manz and T. R. Wolf, *Representations of Solvable Groups*, London Mathematical Society Lecture Note Series, vol. 185 (Cambridge: Cambridge University Press, 1993).

[118] H. Matsuyama, Solvability of groups of order $2^a p^b$. *Osaka J. Math.*, **10**:375–378 (1973).

[119] B. H. Matzat, *Konstruktive Galoistheorie*. Lecture Notes in Mathematics vol. 1284 (Berlin: Springer-Verlag, 1987).

[120] J. McKay, Cartan matrices, finite groups of quaternions, and Kleinian singularities. *Proc. Amer. Math. Soc.*, **81**(1):153–154 (1981).

[121] T. Minkwitz, Extensions of irreducible representations. *Appl. Algebra Engrg. Comm. Comput.*, **7**(5):391–399 (1996).

[122] J. Müller, M. Neunhöffer, F. Röhr and R. Wilson, Completing the Brauer trees for the sporadic simple Lyons group. *LMS J. Comput. Math.*, **5**:18–33 (electronic) (2002).

[123] H. Nagao, On the groups with the same table of characters as symmetric groups. *J. Inst. Polytech. Osaka City Univ. Ser. A.*, **8**:1–8 (1957).

[124] H. Nagao, On a conjecture of Brauer for p-solvable groups. *J. Math. Osaka City Univ.*, **13**:35–38 (1962).

[125] H. Nagao and Y. Tsushima, *Representations of Finite Groups* (Boston, MA: Academic Press Inc., 1989). Translated from the Japanese.

[126] G. Navarro, *Characters and Blocks of Finite Groups*. London Mathematical Society Lecture Note Series, vol. 250 (Cambridge: Cambridge University Press, 1998).

[127] J. Neubüser, H. Pahlings and W. Plesken, CAS; design and use of a system for the handling of characters of finite groups. In *Computational Group Theory (Durham, 1982)*, ed. M. D. Atkinson (London: Academic Press, 1984), pp. 195–247.

[128] J. Neukirch, *Algebraic Number Theory*. Fundamental Principles of Mathematical Sciences, vol. 322 (Berlin: Springer-Verlag, 1999). Translated from the 1992 German original.

[129] E. A. O'Brien, The groups of order 256. *J. Algebra*, **143**(1):219–235 (1991).

[130] O. Ore, Some remarks on commutators. *Proc. Amer. Math. Soc.*, **2**:307–314 (1951).

[131] P. Orlik and L. Solomon, Arrangements defined by unitary reflection groups. *Math. Ann.*, **261**(3):339–357 (1982).

[132] H. Pahlings, On the character tables of finite groups generated by 3-transpositions. *Comm. Algebra*, **2**:117–131 (1974).

[133] H. Pahlings, Groups with faithful blocks. *Proc. Amer. Math. Soc.*, **51**:37–40 (1975).

[134] H. Pahlings, Characterization of groups by their character tables. I, II. *Comm. Algebra*, 4(2):111–153, 155–178 (1976).

[135] H. Pahlings, Some sporadic groups as Galois groups. *Rend. Sem. Mat. Univ. Padova*, **79**:97–107 (1988).

[136] H. Pahlings, Character polynomials and the Möbius function. *Arch. Math. (Basel)*, **65**(2):111–118 (1995).

[137] H. Pahlings, The character table of $2_+^{1+22}.Co_2$. *J. Algebra*, **315**(1):301–325 (2007).

[138] H. Pahlings, cliffordmatrices.g. Available from: http://www.gap-system.org/Packages/Contrib/contrib.html (2007).

[139] H. Pahlings and W. Plesken, Group actions on Cartesian powers with applications to representation theory. *J. Reine Angew. Math.*, **380**:178–195 (1987).

[140] H. Pahlings, Normal p-complements and irreducible characters. *Math. Z.*, **154**(3):243–246 (1977).

[141] H. Pahlings, On the Möbius function of a finite group. *Arch. Math. (Basel)*, **60**(1):7–14 (1993).

[142] R. A. Parker, The computer calculation of modular characters (the meataxe). In *Computational Group Theory (Durham, 1982)*, ed. M. D. Atkinson (London: Academic Press, 1984), pp. 267–274.

[143] R. A. Parker, An integral meataxe. In *The Atlas of Finite Groups: Ten Years On (Birmingham, 1995)*, London Mathematical Society Lecture Note Series., vol. 249 (Cambridge: Cambridge University Press, 1998).

[144] A. E. Parks, A group-theoretic characterization of M-groups. *Proc. Amer. Math. Soc.*, **94**(2):209–212 (1985).

[145] G. Pfeiffer, Character tables of Weyl groups in GAP. *Bayreuth. Math. Schr.*, **47**:165–222 (1994).

[146] G. Pfeiffer, The subgroups of M_{24}, or how to compute the table of marks of a finite group. *Experiment. Math.*, **6**(3):247–270 (1997).

[147] W. Plesken, Solving $XX^{\mathrm{tr}} = A$ over the integers. *Linear Algebra Appl.*, **226/228**:331–344 (1995).

[148] W. Plesken and D. Robertz, Representations, commutative algebra, and Hurwitz groups. *J. Algebra*, **300**(1):223–247 (2006).

[149] U. Riese, The quasisimple case of the $k(GV)$-conjecture. *J. Algebra*, **235**(1):45–65 (2001).

[150] U. Riese and P. Schmid, Real vectors for linear group and the $k(GV)$-problem, *J. Algebra*, **267**(2): 725–755 (2003).

[151] G. R. Robinson, Further reductions for the $k(GV)$-problem. *J. Algebra*, **195**(1):141–150 (1997).

[152] G. R. Robinson and J. G. Thompson, On Brauer's $k(B)$-problem. *J. Algebra*, **184**(3):1143–1160 (1996).

[153] S. Roman, *Advanced Linear Algebra*, 2nd edn. Graduate Texts in Mathematics, vol. 135 (New York: Springer, 2005).

[154] A. I. Saksonov, An answer to a question of R. Brauer. *Vesci Akad. Navuk BSSR Ser. Fiz.-Mat. Navuk*, **1967**(1):129–130 (1967).

[155] P. Schmid, *The Solution of the $k(GV)$ Problem.* ICP Advanced Texts in Mathematics, vol. 4 (London: Imperial College Press, 2007).

[156] G. J. A. Schneider, Dixon's character table algorithm revisited. *J. Symbolic Comput.*, **9**(5-6):601–606 (1990).

[157] I. Schur, Untersuchungen über die Darstellungen der endlichen Gruppen durch gebrochen lineare Substitutionen. *J.Reine Angew.Math.*, **130**:85–137 (1907).

[158] L. L. Scott Integral equivalence of permutation representations. In *Group Theory (Granville, OH, 1992)*, eds. S. Sehgal and R. Soloman (River Edge, NJ: World Scientific, 1993), pp. 262–274.

[159] S. K. Sehgal and A. E. Zalesskiĭ, Multiplicities of irreducible components of restrictions of complex representations of finite groups to certain subgroups. *Comm. Algebra*, **21**(1):37–51 (1993).

[160] J.-P. Serre, *Topics in Galois Theory.* Research Notes in Mathematics, vol. 1 (Boston, MA: Jones and Bartlett Publishers, 1992).

[161] A. Shalev, Commutators, words, conjugacy classes and character methods. *Turkish J. Math.*, **31**(suppl.):131–148 (2007).

[162] E. Skrzipczyk, Charaktertafeln von p-Gruppen. Diplomarbeit, Lehrstuhl D für Mathematik, RWTH-Aachen, Aachen (1992).

[163] J. G. Thompson, Some finite groups which appear as $\mathrm{Gal}\,L/K$, where $K \subseteq \mathbf{Q}(\mu_n)$. In *Group Theory, Beijing 1984*. Lecture Notes in Mathematics, vol. 1185 (Berlin: Springer, 1986) pp. 210–230.

[164] A. Turull, Schur indices of perfect groups. *Proc. Amer. Math. Soc.*, **130**(2):367–370 (electronic) (2002).

[165] W. R. Unger, Computing the character table of a finite group. *J. Symbolic Comput.*, **41**(8):847–862 (2006).

[166] B. L. van der Waerden, *Algebra. Vol. II.* Fundamental Principles of Mathematical Sciences, vol. 34 (Berlin: Springer-Verlag, 1959).

[167] H. Völklein, *Groups as Galois Groups.* Cambridge Studies in Advanced Mathematics, vol. 53 (Cambridge: Cambridge University Press, 1996).

[168] L. C. Washington, *Introduction to Cyclotomic Fields*, 2nd edn. Graduate Texts in Mathematics, vol. 83 (New York: Springer-Verlag, 1997).

[169] W. Willems, Metrische G-Moduln über Körpern der Charakteristik 2. *Math. Z.*, **157**(2):131–139 (1977).

[170] R. A. Wilson, An atlas of group representations. Available from: `http://brauer.maths.qmul.ac.uk/Atlas/` (1996).

[171] R. A. Wilson, Standard generators for sporadic simple groups. *J. Algebra*, **184**(2):505–515 (1996).

[172] R. A. Wilson, The Monster is a Hurwitz group. *J. Group Theory*, **4**(4):367–374 (2001).

[173] P.-H. Zieschang, *Theory of Association Schemes.* Springer Monographs in Mathematics, (Berlin: Springer-Verlag, 2005).

Notation index

Subject index

Printed in the United States
By Bookmasters

Printed in the United States
By Bookmasters